GEOCHEMISTRY OF HYDROTHERMAL ORE DEPOSITS

GEOCHEMISTRY OF HYDROTHERMAL ORE DEPOSITS

Second Edition

Edited by
HUBERT LLOYD BARNES

The Pennsylvania State University
Ore Deposits Research Section
University Park, Pennsylvania

A Wiley-Interscience Publication
JOHN WILEY & SONS

New York / Chichester / Brisbane / Toronto

Copyright © 1979 by John Wiley & Sons, Inc.

All rights reserved. Published simultaneously in Canada.

Reproduction or translation of any part of this work beyond that permitted by Sections 107 or 108 of the 1976 United States Copyright Act without the permission of the copyright owner is unlawful. Requests for permission or further information should be addressed to the Permissions Department, John Wiley & Sons, Inc.

Library of Congress Cataloging in Publication Data

Barnes, Hubert Lloyd.
    Geochemistry of hydrothermal ore deposits.

    "A Wiley-Interscience publication."
    Includes bibliographies and indexes.
    1. Hydrothermal deposits. 2. Geochemistry.
I. Title.

QE390.B37 1969   553'.19   79-354
ISBN 0-471-05056-3

Printed in the United States of America

10 9 8 7 6 5 4 3 2

CONTRIBUTORS

Hubert L. Barnes, Professor of Geochemistry and Director, Ore Deposits Research Section, The Pennsylvania State University, University Park, Pennsylvania

Paul B. Barton, Jr., Geologist, U.S. Geological Survey, Reston, Virginia

Patrick R. L. Browne, Senior Lecturer, Department of Geology, University of Auckland, Auckland, New Zealand

C. Wayne Burnham, Professor of Geochemistry and Head, Department of Geosciences, The Pennsylvania State University, University Park, Pennsylvania

Donald M. Burt, Assistant Professor of Mineralogy and Ore Deposits, Department of Geology, Arizona State University, Tempe, Arizona

Lawrence M. Cathles, III, Associate Professor of Geosciences, Department of Geosciences, The Pennsylvania State University, University Park, Pennsylvania

Bruce R. Doe, Chief, Branch of Isotope Geology, U.S. Geological Survey, Denver, Colorado

A. James Ellis, Director, Chemistry Division, Department of Scientific and Industrial Research, Petone, New Zealand.

Jeffrey S. Hanor, Professor of Geology, Department of Geology, Louisiana State University, Baton Rouge, Louisiana

Harold C. Helgeson, Professor of Geochemistry, Department of Geology and Geophysics, University of California, Berkeley, California

Heinrich D. Holland, Professor of Geochemistry, Department of Geological Sciences, Harvard University, Cambridge, Massachusetts

Sergey D. Malinin, Senior Research Associate, Vernadsky Institute of Geochemistry and Analytical Chemistry, Moscow, U.S.S.R.

Denis L. Norton, Associate Professor of Geosciences, Department of Geosciences, University of Arizona, Tucson, Arizona

Hiroshi Ohmoto, Professor of Geochemistry, Department of Geosciences, The Pennsylvania State University, University Park, Pennysylvania

Edwin Roedder, Geologist, U.S. Geological Survey, Reston, Virginia

Arthur W. Rose, Professor of Geochemistry and Director, Mineral Conservation Section, The Pennsylvania State University, University Park, Pennsylvania

Robert O. Rye, Geologist, U.S. Geological Survey, Denver, Colorado

Terry M. Seward, Geochemist, Chemistry Division, Department of Scientific and Industrial Research, Petone, New Zealand

Brian J. Skinner, Eugene Higgins Professor of Geology and Geophysics, Yale University, New Haven, Connecticut

Hugh P. Taylor, Jr., Professor of Geology, Division of Geological and Planetary Sciences, California Institute of Technology, Pasadena, California

Byron G. Weissberg, Geochemist, Chemistry Division, Department of Scientific and Industrial Research, Petone, New Zealand

Robert E. Zartman, Geologist, U.S. Geological Survey, Denver, Colorado

PREFACE

The purpose of this edition remains the explanation and critical evaluation of various types of evidence on ore-forming processes. The discussions were invited from authorities in each of the subdisciplines now making major contributions to concepts of the genesis of hydrothermal ore deposits. "Hydrothermal" here is used simply to mean hot, aqueous fluids. Excluded from this review are purely igneous processes, such as the accumulation of chromite by fractional crystallization within a magma chamber, and also purely sedimentary processes, such as the precipitation of manganese in nodules on the sea floor. Largely because of insufficient investigation, absent also are thorough discussions of processes of ore deposition by replacement and of the effects on ores of both mechanical stress and thermal metamorphism.

The sequence of chapters in this edition also follows, in general, that of events during the evolution of an ideal ore solution, from an early emphasis on sources of the solutions, to their reactions with host rocks, then the products of precipitation, and finally the nature of the residual fluid.

In other respects, this second edition is dominantly a new book. New authors have written more than half of the chapters, including added discussions of mass-transfer phenomena and of diagenetic sources for hydrothermal fluids. The publisher is also different as a consequence of an amicable agreement with the publisher of the first edition, Holt, Rinehart, and Winston, Inc., whose corporate goals now lie elsewhere.

Areas of disagreement between the authors will soon become evident to the serious reader. Because these reflect the current state of the science, no attempt has been made here either to resolve or disguise these healthy conflicts. Instead, their effects, ideally, should be to stimulate and guide further research into the more important, controversial areas.

To assist the reader, a standard notation has been adopted for every chapter except that on Mass Transfer Among Minerals And Hydrothermal Solutions, Chapter 11. The unique requirements of this chapter for a more complex notation have been satisfied dominantly by the addition of other symbols. These are defined within this chapter where they are first used. The general notation, given following the Table of Contents, is a compromise between current usage in geochemistry and that recommended by the International Union of Pure and Applied Chemistry. Values given here in calories and bars or atmospheres remain directly comparable with data in the present literature although the arguments reviewed by M. L. McGlashan (1973, *Ann. Rev. Phys. Chem.*, **24,** 51-76) on principles of notation and units are persuasive that some changes are desirable.

HUBERT LLOYD BARNES

UNIVERSITY PARK, PENNSYLVANIA
FEBRUARY 1979

CONTENTS

Notation   xv

1 THE MANY ORIGINS OF HYDROTHERMAL MINERAL DEPOSITS   1

Historical roots   3
The scene today   6
A synopsis   19
References   21

2 PLUMBOTECTONICS THE PHANEROZOIC   22

Background   24
Dynamic model of lead isotope evolution   34
Lead isotopes and geologic environments   43
Acknowledgments   65
References   66

3 MAGMAS AND HYDROTHERMAL FLUIDS   71

Volatiles in magmas   72
Thermodynamic relations in magmas   78
The generation of calc-alkaline magmas   91

Shallow emplacement and solidification of hydrous magma 104
The magmatic aqueous phase 116
Subsolidus hydrolysis equilibria 125
Acknowledgments 133
References 133

4 THE SEDIMENTARY GENESIS OF HYDROTHERMAL FLUIDS 137

Diagenesis of sediments 138
Properties of water in aqueous solutions in sedimentary basins 139
Properties of some ore-forming fluids 159
Flow of subsurface brines 163
Conclusions 167
Acknowledgments 168
References 168

5 HYDROTHERMAL ALTERATION 173

Theory of metasomatic processes in hydrothermal alteration 176
Chemistry of alteration processes 183
Stability of alteration minerals 193
Classification of alteration 202
Alteration in hot springs and epithermal ore deposits 205
Alteration around veins in granitic rocks 207
Alteration at porphyry copper deposits 210
Alteration in skarn deposits 218
Alteration in greisen deposits 226
Acknowledgments 227
References 227

6 OXYGEN AND HYDROGEN ISOTOPE RELATIONSHIPS IN HYDROTHERMAL MINERAL DEPOSITS 236

Isotopic notation and standards 237
Equilibrium isotope fractionations 238
Isotopic variations in natural waters 241
Interactions between surface-derived waters and igneous intrusions 248
Amounts of meteoric water involved in hydrothermal systems 250

"Epithermal" ore deposits in volcanic terranes	253
Porphyry copper and molybdenum deposits	256
Supergene alteration of porphyry copper deposits	261
Mississippi Valley-type lead-zinc-fluoride deposits	262
Volcanogenic massive sulfide deposits	263
Metamorphic ore deposits	263
Other hydrothermal ore deposits	265
Conclusions	271
References	272

7 SULFIDE MINERAL STABILITIES — 278

Limitations of chemical and physical conditions	279
Selection of significant natural systems	279
Methods of examining and studying ore minerals	288
Methods of studying phase relations	291
Presentation of data	344
Specific sulfide systems	354
Acknowledgments	390
References	390

8 SOLUBILITIES OF ORE MINERALS — 404

Complex-forming ligands	406
Concentrations in ore solutions	411
Solubility data	415
Ore deposition	434
Acknowledgments	454
References	454

9 THE SOLUBILITY AND OCCURRENCE OF NON-ORE MINERALS — 461

The system $NaCl$-H_2O	461
The solubility of quartz and other SiO_2 polymorphs	465
The solubility of fluorite	471
The solubility of alkaline-earth carbonate minerals	474
The solubility of calcium, strontium, and barium sulfates	490

Summary	500
References	501

10 ISOTOPES OF SULFUR AND CARBON — 509

Analytical techniques	511
Sulfur isotope geothermometry	513
Sulfur isotopic variation in hydrothermal systems	521
Sulfur isotopic data from hydrothermal deposits	544
Carbon isotope geothermometry	550
Carbon isotopic variation in hydrothermal systems	553
Acknowledgments	561
References	561

11 MASS TRANSFER AMONG MINERALS AND HYDROTHERMAL SOLUTIONS — 568

Chemical reactions and mass transfer	570
Chemical affinity and reaction progress	575
Partial and local equilibrium	583
Conservation of mass and charge	587
Calculation of mass transfer	589
Discussion	600
Concluding remarks	605
Acknowledgments	606
References	606

12 THERMAL ASPECTS OF ORE DEPOSITION — 611

Heat transfer processes	612
Thermal energy sources	619
Thermal aspects of alteration and sulfide deposition	624
Acknowledgments	629
References	629

CONTENTS xiii

13 EXPLORED GEOTHERMAL SYSTEMS **632**

Occurrence of geothermal systems 633
Characteristics of systems 635
Fluid compositions 639
Hydrothermal alteration 668
Summary 675
References 678

14 FLUID INCLUSIONS AS SAMPLES OF ORE FLUIDS **684**

Mechanism of trapping 688
Nature of the fluid trapped 694
Leakage into or out of inclusions 697
Changes since trapping 698
Composition of fluid inclusions 704
Temperature, pressure, and density of ore fluids 718
Acknowledgments 730
References 731

15 ORE METALS IN ACTIVE GEOTHERMAL SYSTEMS **738**

Metal-rich geothermal systems 739
Mercury in hot springs 757
Broadlands geothermal system—a case study 759
Conclusions 774
Acknowledgment 774
References 774

INDEX 781

NOTATION

å	Ion size parameter of Debye-Hückel equation	H	Enthalpy
a	Chemical activity	J	Flux
(aq)	Aqueous ionic or molecular species	k	Permeability
		k	Rate constant
A	Constant of Debye-Hückel equation	K	Kelvin temperature; thermal conductivity
B	Constant of Debye-Hückel equation	K	Equilibrium constant
		(l)	Liquid molecular species
cal	Calorie (1.0 cal = 4.184 J)	L	Liquid
c	Concentration	ln	Logarithm to the base, e
C	Heat capacity	log	Logarithm to the base, 10
D	Diffusion coefficient	m	Molality (moles/kg of solvent)
e	Naperian constant (2.7183)	M	Mass
e^-	Electron equivalents	n	Number of moles
$\mathcal{E}$	Galvanic cell electromotive force	N	Number of molecules; normality (equivalents/liter of solution)
Eh	Oxidation potential	N_0	Avogadro's number (6.023 × 10^{23} mole^{-1})
f	Fugacity (in atmospheres or bars)	P	Pressure
$\mathcal{F}$	Faraday constant (23.0623 kcal/volt equivalent)	pK	Negative logarithm of the equilibrium constant
g	Acceleration due to gravity (980.665 cm/sec^2)	pH	Negative logarithm of the chemical activity of the hydrogen ion
(g)	Ionic or molecular species in the gaseous phase	Q	Heat content; heat flow
G	Gibbs free energy	r	Mean radius of ion or particle

R Gas constant (1.987 cal/deg mole; 8.314 J/deg·mole; 0.08206 liter·atm/deg/·mole)
(s)   Solid molecular species
S   Solid
S Entropy
t Time
T Temperature
v Velocity
V   Vapor
V Volume
X Mole fraction
z_i Valence of ionic species, i
z Compressibility factor of a gas ($z = PV/nRT$)
α Isotopic fractionation factor: the isotopic ratio in one phase divided by the same ratio in a reference phase; coefficient of thermal expansion
γ Activity coefficient
δ Deviation, in permil, of an isotopic ratio from a standard given by

$$\delta = 1000 \left\{ \frac{\text{sample ratio}}{\text{standard ratio}} - 1 \right\}$$

Δ Incremental change
ϵ Dielectric constant
η Viscosity
κ Thermal diffusivity
λ Decay constant representing the fraction of radioactive atoms that disintegrate per unit time
μ Chemical potential; ionic strength; micro-units ($\times 10^{-6}$)
π 3.14159
ρ Density
ϵ Kinematic viscosity, equals η/ρ

ξ Progress variable
σ Porosity
Σ Summation

SUPERSCRIPTS

°   Standard state
⁻   Molal quantity
′   Effective value
l, v, s, m   Phases: liquid, vapor, solid, or melt
^{34}S Isotopic number of the element, S for example
S^{2-} Valence of an aqueous species, the sulfide ion for example

SUBSCRIPTS

A, B …   Phase
c   Critical state value
f   Value for the reaction of formation from the elements
i, j, k …   Components; species in solution
P   Constant pressure
S   Constant entropy
t   Total
T   Constant temperature
V   Constant volume
W   Property of water or an aqueous solution
ϵ Element in a phase or in a species in solution
ν Stoichiometric reaction coefficient

GEOCHEMISTRY OF HYDROTHERMAL ORE DEPOSITS

1

The Many Origins of Hydrothermal Mineral Deposits

BRIAN J. SKINNER
Yale University

A *mineral deposit* is any natural, but locally restricted, concentration of minerals in the Earth's crust. The definition draws neither distinctions between kinds of minerals concentrated, nor distinctions between how the concentrations happen; indeed almost every geologic process can, under suitable conditions, produce a mineral concentration of some kind. If the minerals concentrated contain substances of material significance to people and if the concentrations are rich enough to warrant mining, a mineral deposit is accorded the special appellation *ore deposit*. That, however, is an economic and social definition, and although it is a perfectly valid definition it is not one by which we wish to be restricted in a discussion of mineral deposits. The same processes that produce rich deposits also produce lean deposits, and we will draw our evidence from both the rich and the lean.

Mineral deposits form in many different ways, some well understood, some poorly understood. Among the well-understood ways by which minerals become concentrated are the precipitation of salts from lakes and seawater, the separation of minerals by gravitational fractionation in cooling magmas, the formation of placers by separation in flowing water of minerals of different density, the formation of laterites by leaching during weathering, and deposition of caliches by evaporation of upward-moving capillary waters. However, none of these well-understood deposit-forming processes are discussed in this volume. Instead, the book concentrates on *hydrothermal mineral deposits,* a special family of deposits that share a complex and only partly understood heritage. Both the geologic setting of hydrothermal mineral deposits and the minerals concentrated within them range widely; yet there are common connecting threads. Each member of the family formed when a hot, aqueous solution, commonly called a *hydrothermal solution,* flowed through a defined channel in the crust, or over a restricted portion of the surface of the crust, and precipitated a localized mass of minerals from its dissolved load. Just how the solutions form, flow, and react, and what they deposit, are the topics addressed in subsequent chapters. Most of the discussion is very chemical in presentation, and sometimes it may seem to be far-removed from the central theme. In large part, this happens because different problems have attracted different kinds of specialists. This opening chapter provides a framework within which the reader, possibly unfamiliar with the specialties, can relate the many topics discussed, and see how they, in fact, do all address some aspect of the larger question. This chapter also attempts to identify some of the problems that have yet to receive adequate study.

Hydrothermal mineral deposits are small features by comparison with most geologic features—even the largest is less than a cubic mile in volume. At first glance they may seem to be geologic accidents, one-of-a-kind, small distortions caused in some way by local aberrations of larger, Earth-scale processes. But the accidents cannot be entirely accidental or random, because deposits can be classified into obvious families and individual family members occur with much higher frequency in some portions of the crust than others. Furthermore, despite an enormous variability in the geologic habitats where deposits are found, and in the number of minerals known in nature, there is a chemical consistency between hydrothermal deposits that is best expressed by the *limited range of minerals,* mostly sulfides, found concentrated within them. The variability of habitat may cause us to despair of finding a common thread for hydrothermal mineral deposits, but the chemical consistency expressed by the minerals tells us that there must be a commonality. The commonality also implies that relatively few chemical processes are important in the formation of hydrothermal mineral deposits. Therein lies the encouragement we need to decipher the origins of deposits, and this book is a record of the achievements of some of those so encouraged.

When a new hydrothermal mineral deposit is examined, we are seemingly presented with evidence sufficient to answer all questions. But long study has shown that we are presented at best with partial evidence, and commonly with evidence biased in such a way that conjecture continues to surround our understanding of the formative processes of even the best studied deposits.

The innumerable qualitative and quantitative questions raised about hydrothermal mineral deposits can be encompassed within four general categories: (1) Where did the water originate? (2) What was the nature of the hydrothermal aquifer and what was the driving force? (3) Where and how did the solution accumulate its dissolved constituents? (4) What caused the solution to deposit its minerals? We will probably never be able to provide unambiguous, incontrovertible answers to all the questions, at least not with all the detail we might desire. But some of the questions are yielding answers, slowly but surely, through the application of geochemistry; in sum, the advances of our understanding of the *details* of ore-forming processes have been spectacular over the last 25 years. The advances have been a long time coming, because the broad outlines of how hydrothermal deposits must form have been known for centuries. It is instructive, therefore, to briefly review the heritage of ideas on which our modern work rests.

HISTORICAL ROOTS

Early speculations regarding the origins of hydrothermal mineral deposits are lost in the murky gloom of unwritten history. The first person to draw conclusions from recorded field observations, and therefore the first to study mineral deposits in a scientific way, was Georg Bauer (1494-1555), known best by his pen name, Agricola. Many of Agricola's deductions now seem ludicrous, but his observations on the flow of underground water, on the decomposition of rocks by rainwater, and on the deposition of mineral matter from waters encountered in mines led him to a perceptive conclusion. He decided that the ores he saw being mined must have formed when subterranean openings were infilled by deposition from meteoric waters that descended from the surface and circulated deep in the crust. The circulating waters became heated and selectively leached constituents from rocks through which they passed; eventually they reached a site where precipitation occurred. Agricola's was the first reasoned conclusion that metalliferous ores could form from hydrothermal solutions of meteoric origin, and his statement is the first clear exposition of what later came to be called *formation by lateral secretion.*

A completely different conclusion for the source of mineral deposits, including those we now recognize as hydrothermal deposits, was offered by Réné Descartes (1596-1650), who proposed that the Earth is a small, partly cooled star with a cold, rocky crust but a hot interior. Descartes postulated that fractures and openings in the crust were formed by adjustments during cooling; the mineral matter filling the fractures condensed or precipitated from vapors released during cooling and crystallization of the molten interior. Descartes apparently did not speculate about the composition of the vapors, but his suggestion is the first statement, albeit a crude one, of what later came to be called the formation of ores by *magmatic hydrothermal solutions.*

Thus, more than 300 years ago, the seeds of a controversy had been planted; water from above versus water from below. In subsequent centuries, many students of ore deposits adopted stances that served to widen the controversy, but two of them, Hutton (1726-1797) and Werner (1750-1817) clarified part of the picture by introducing concepts for the formation of certain kinds of mineral deposits not formed by hydrothermal solutions. Hutton made the extreme proposal that no metalliferous ores were deposited from solution, but that all were formed by injection and crystallization of dry, metal- and sulfur-rich magmas. Clearly he overstated the case, but his proposal was correct for a class of deposits of which he was unaware—the copper, nickel, and platinum deposits formed by the separation and segregation of immiscible sulfide magmas from mafic, silicate magmas. Werner, the classic Neptunist, attributed formation of all metalliferous deposits to infilling of fractures and other openings by mineral-depositing waters flowing directly down from above and believed the ores we now classify as epigenetic were merely an extension of those we recognize as syngenetic. Most of the mineral matter precipitated was, Werner believed, already present in surface waters and did not need to be extracted from rocks as required by the lateral secretion theory. Like Hutton's, Werner's position was extreme, but his ideas live on in our recognition that some stratiform mineral deposits are chemical, sedimentary rocks; some geologists even now hold to the idea that one class of stratiform sulfide deposits may have formed by deposition from seawater.

As the nineteenth century dawned, the seemingly simple outlines of origin had been completed. The mineral matter in hydrothermal deposits is either locally derived from country rock surrounding the deposits, or it is transported from distant sources by solution. If from distant sources, the transporting solutions must have either descended from the surface or ascended from depth. If they ascended, the solutions are either primary, released from a cooling magma, or secondary, released from enclosing sedimentary or metamorphic rock. Around those straight-forward possibilities there followed a century and a half of dogmatic debate, much of it

HISTORICAL ROOTS

acrimonious and some of it based more on hunches and rhetoric than on reliable observation and measurement.

The theory of magmatic hydrothermal ores was the easiest case to defend. The close spatial relationship between many mineral deposits and igneous rocks is obvious. Additionally, during the first half of the nineteenth century, G. P. Scrope and Henry Sorby demonstrated that water is a constituent of magmas, and it was quickly realized that magmatic water must be released during volcanism or subsurface crystallization of intrusives. It was only a small step to propose that the escaping waters carry with them, in solution, the material needed to form ore deposits. Discussion soon centered on the form of the escaping fluids—were they liquids like those seen in hot springs or were they supercritical fluids in that region of temperature and pressure where meniscus-forming liquids do not exist? We now know that fluids do not undergo sudden, drastic changes in chemical properties as they pass from a liquid to a supercritical fluid and that there is no reason to separate so-called pneumatolytic deposits as a class on the basis of the physical properties of the depositing solution, but even as recently as 20 years ago, the topic was debated heatedly by some geologists. In the middle and later years of the nineteenth century, many of the classic magmatic-hydrothermal relationships were developed; among the principal spokesmen were T. Scheerer, Élie de Beaumont, and L. de Launay in France, Bernhard Von Cotta in Germany, and F. Pošepný writing both in Europe and in the United States. The lateral secretion theory carried within it grounds for considerable dissent—surface waters can undergo either deep or shallow circulation and can collect their dissolved load either close to or distant from the site of deposition. Among the leading spokesmen who debated the possibilities were Gustav Bischof and F. Sandberger in Germany, A. Daubrée in France, T. S. Hunt in Canada, and S. F. Emmons in the United States. Little accord was reached.

One prescient student of the nineteenth century, J. A. Phillips, studied the tin deposits worked in the ancient mines of Cornwall, England. Phillips realized that a continuum might exist between a purely magmatic solution and a purely meteoric one. Rising waters would mix with and be indistinguishable from descending waters. Phillips further realized that analyses of nearby rocks did nothing to solve the question of sources of metals. All of the chemical elements found in the Cornish tin veins were present in both the adjacent igneous and metamorphic rocks. Because the veins demonstrably traversed both kinds of rocks, Phillips prudently concluded that, in the absence of a way to identify sources, each had probably contributed but the relative contributions could not be determined. The compromise reached by Phillips presaged the stances of the most influential thinkers in the early years of the twentieth century, among whom C. R. Van Hise,

J. F. Kemp, and Waldemar Lindgren were the most prominent. Each of the three men recognized that meteoric and magmatic waters could both play a role, that metals could come from many sources, that solutions could travel long distances, and that unambiguous resolution of the problems was not possible with the tools at their disposal. Thus, conditioned by evidence drawn from the deposits they had individually studied, each tended to give more or less emphasis to the importance of differing sources of water and of the dissolved constituents. Indeed, their conditioning, and hence their emphases, changed throughout their working years as a consequence of growing observational experience.

The flexible and reasonable approach of Lindgren and his contemporaries remains with us today. Although such factors as the eloquence of a given teacher, or limited experience, conditioned geologists to emphasize one water source over another, it is probably correct to say that by the middle of the twentieth century, all students of hydrothermal mineral deposits accepted a multiplicity of sources for both waters and dissolved loads. This accord was a triumph of reason built on countless hours of work and observation. But definitive proof was still lacking. The key to unlock the dilemma was finally provided through stable and radiogenic isotopes. The successes of isotopic geochemistry over the last 15 years have been truly spectacular, and they have shown that ambience was indeed the correct stance.

THE SCENE TODAY

With our historic overview, let us now return to the four major questions concerning hydrothermal solutions, and briefly review our present understanding of the problems.

Sources of Hydrothermal Fluids

When we find a mineral deposit, the hydrothermal solution from which it precipitated has normally long departed. But vestiges of the solution can sometimes still be found trapped as tiny inclusions in the precipitated minerals. Studies of the fluids in these microscopic bottles inside transparent crystals such as quartz, sphalerite, calcite, and fluorite have revealed more about the nature of hydrothermal solutions than we have learned from any other source. From analyses of fluid inclusions we know that hydrothermal solutions vary widely in composition and concentration, but that all are brines of some sort. This information in turn tells us that deposit-forming fluids are chemically similar to fluids encountered in present-day geothermal systems. To amplify our understanding of ore-forming fluids, therefore, we can turn to studies of modern geothermal systems and the hot solutions coursing through them. There may seem to be a conceptual problem in this approach; why study a solution that is not presently forming a mineral deposit? The answer lies in the low frequency with which solutions encounter suitable combinations of flow patterns together with the chemical and physical changes effective in causing local precipitation. Solutions that can do the trick, given the right set of circumstances, are apparently common. Although few hydrothermal solutions seem to encounter the right combination of circumstances to form mineral deposits, the fluids themselves are widespread throughout the crust.

If hydrothermal solutions are common, it should be possible to classify them. As shown by White (1974) this is true. From innumerable observations of water in mines, tunnels, drill holes, and hot springs we can conclude that the source of subsurface hydrothermal waters can be separated into four types. The four are (1) surface water (including rainwater, lake and river water, seawater and groundwater), (2) connate and deeply penetrating groundwater, (3) metamorphic water, and (4) magmatic water. Connate water may once have been surface water, but long burial in sediments, and reaction with its enclosing minerals, eventually gives it a mark and character of its own. If it were possible to study an entire hydrothermal system, we could identify the source or sources of water in it by direct observation. But we have never been able to examine an entire system; they are always too large, too inaccessible, and too easily disrupted by the study itself. We must turn, therefore, to secondary evidence and, here, isotopic chemistry becomes invaluable. The two chemical elements in water, hydrogen and oxygen, both have more than one stable isotope. Whenever water is involved in a change of state, for example evaporation or a chemical reaction such as an interaction between hydrothermal solution and enclosing wall rocks, the isotopes are subject to exchanges and the H/D and $^{16}O/^{18}O$ ratios in the water are changed. By exchange and fractionation, each of the four types of water can develop a characteristic isotopic composition.

Measurement of the relative abundances of the isotopes of hydrogen and oxygen in hydrothermal solutions, both in fluid inclusions and in modern hydrothermal systems, therefore, can provide invaluable information on the sources of the water. But two kinds of problems introduce uncertainties into our interpretation of isotopic data. The first arises from rock-water reactions. The reactions can be complete or partial. For example, a deeply buried connate water in a pile of compacting sediments, or a metamorphic water driven out by metamorphic dehydration, may once have been surface water. Yet the connate and metamorphic waters will have new isotopic compositions acquired by subsurface reactions. If only partial isotopic fractionation has occurred, the isotopic composition will be intermediate between that of surface water and a completely equilibrated metamorphic water. The second problem arises from the tendency of water from several sources to mix and to produce a solution with an isotopic composition intermediate between the composition of the end-member components. An isotopic continuum can result and, as D. E. White (1974) has succinctly pointed out, it is still often impossible to prove that a *single* kind of water produced a given ore deposit. Many deposits are, indeed, demonstrably formed from waters of at least two parentages. Despite the many uncertainties still surrounding isotopic identification of water sources, some very important observations can and have been made. One observation of tremendous significance is that similar deposits can be formed from quite different kinds of water. Demonstration of multiple parentage for the water in hydrothermal solutions implicitly leads us to draw an interesting conclusion; the actual source of the water, interesting though it may be, is apparently not the *controlling* factor in the formation of hydrothermal mineral deposits.

Hydrothermal Channelways

Within a mineral deposit, the solution channelways are usually obvious because precipitated minerals and altered wall-rocks remain as mute evidence. The direction in which the solutions flowed, especially in flat-lying deposits, is usually less obvious, but in many cases can be inferred from mineral zoning or similar evidence. Before a hydrothermal solution reaches the site of deposition, however, it may have travelled many tens of kilometers and reacted with hundreds of cubic kilometers of rock. With a defunct hydrothermal system, we are faced with an almost impossible task in trying to identify the tortuous passageways beyond the mine area. Consequently, we are forced to examine modern hydrothermal systems, whether they are forming mineral deposits or not, and to reason by analogy where similar aquifers may have formed ores in the past.

Many modern hydrothermal systems are reasonably well understood. The passage of fluids through porous strata in compacting sedimentary basins, for example, is known from petroleum exploration and production; that such fluids can, under suitable conditions, apparently produce Mississippi Valley-type deposits has recently been demonstrated by Carpenter et al. (1974). Similarly, flow of fluids through porous layers in piles of volcanic rocks is increasingly well understood as a result of hot-spring studies and development of geothermal fields for power plants. Such studies have had an important impact on our thinking and will probably play an even greater role in the future.

Despite encouraging advances we have yet to reconstruct the paleohydrology of a single ore-forming system, and this imposes a severe limitation to our overall understanding of hydrothermal ore deposits. A tantalizing example of how important paleohydrology can be is seen in the work of White (1971) and others, who have been able to reconstruct some of the paleohydrology for the Copper Harbor Conglomerate aquifer that brought in the solutions responsible for the White Pine copper deposits in Michigan. Reconstruction of paleohydrologic systems is undoubtedly one of the most fruitful areas for continuing research. Unfortunately it is also one of the most difficult, but the rewards are clearly worth the effort, because the prize will be the ability to pinpoint the location of buried ore deposits.

When we tackle the question of the force that drives a hydrothermal fluid through its channelways, we again face ambiguity. Modern hydrothermal systems flow for six reasons:

1. Flow occurs when there is a difference in hydrostatic head between source and outlet of an aquifer. An example of deposits probably formed by fluids flowing under a hydrostatic head can be found in the Tennessee zinc deposits.
2. Flow occurs when the lithostatic pressure of a compacting rock pile reduces porosity and extrudes contained fluids upward. The Michigan copper deposits were probably formed by waters driven out from the vast pile of compacting Keeweenawan lavas, and some Mississippi Valley-type deposits seem to owe their origin, at least in part, to waters driven out from compacting sedimentary basins.
3. Another driving force, possibly also important in the origin of Mississippi Valley-type deposits, is osmotic pumping. Osmotic pressures developed across natural membranes, such as shales, can produce the necessary force to drive waters from sedimentary basins.
4. Flow occurs when density differences are induced by local heat sources such as intrusive igneous rocks, causing hot, low-density water to rise. Local heat sources may even cause convecting hydrothermal systems; many kinds of deposits, among them epithermal Ag-Au veins, porphyry coppers, and Cyprus-type massive sulfide bodies, seem to be associated with convecting systems.
5. Flow occurs when very saline, and therefore very dense, fluids sink and displace less dense fluids upward. So far, we have been unable to uniquely identify a deposit formed from fluids driven in this fashion.
6. Upward flow occurs when dissolved fluids are released from cooling

magmas. This is the classic, magmatic hydrothermal solution and examples of deposits formed from such solutions can be seen in the early, high-temperature mineralizations at Butte, Bingham, Climax, and many other deposits.

Many additional questions can be asked about hydrothermal systems, but as with the questions of channelways and driving mechanisms, we can do little more than speculate and reason by analogy with modern systems. Two questions are of paramount importance because they have very practical consequences. The first concerns the lifetimes of ore-forming systems. A short-lived system will have insufficient time to form large ore deposits. However, we are still in doubt as to how long is long enough, because the length of time needed to form hydrothermal deposits is very poorly understood. The question is one to which much more attention should be paid.

Abundant evidence is available from radiometric dates to show that for mineral deposits associated with an event such as volcanism or intrusion, the magmatism and mineralization are essentially coincident within limits of measurement (Livingston et al., 1968). However, we are limited by the precision of dating techniques, because none is sufficiently precise to date the duration of mineralization. Nevertheless, it must commonly be less than a million years, because some porphyry coppers in the Pacific are little more than a million years old. Some indirect evidence is available. Modern hot springs driven by local igneous heat sources have lifetimes of a few thousand to approximately one million years. Presumably paleohydrothermal systems associated with igneous rocks had similar lifetimes. Roedder (1960), reasoning from compositions of fluid inclusions, concluded that times of a few thousand years suffice to form magmatic-hydrothermal ores. Finally, there is suggestive evidence from ores such as those of the Kuroko-type, the New Brunswick-type, and the Cyprus-type, each of which apparently precipitated from hydrothermal fluids debouched on the sea floor, that formation times could not have been more than a few thousand years. This conclusion is reached because the ores contain little detritus, even though rocks above and below are normal detrital sediments.

The second vital question concerning hydrothermal systems is the depth at which ore deposits can form. The best way information can be gathered on this point is to study the pressure-temperature stability relations of the minerals deposited by the fluids and of the fluid inclusions in the minerals. Unfortunately, many mineral deposits have been metamorphosed after formation and the evidence is destroyed. Where later metamorphism has not obscured the evidence, however, mineral assemblages indicate that pressures in hydrothermal systems rarely exceed approximately 1000 bars. If we make the unlikely assumption that the pressure in even the deepest system is due to a hydrostatic head rather than the lithostatic head, the maximum depth at which the ores could have formed is only approximately 10 km. The depth of most sites of deposition is probably much less. It is increasingly evident, for example, that in some ores deposition has occurred from boiling solutions, which could only happen within about 1 km of the surface. The problem of depth needs much closer study, but, from present evidence, we are led to conclude that most hydrothermal ores form within the upper 3 km of the crust, and that deeper regions can only contain ores buried subsequent to formation or localized at depths down to 10 km by a special situation such as reaction between the solution and a favorable horizon. The depth question has a very practical aspect, because it suggests that the deeper we search in the crust, the fewer hydrothermal ores we are likely to find.

Source of Dissolved Constituents

I mentioned earlier that all hydrothermal solutions are brines. Despite the fact that the salinities of hydrothermal solutions vary widely, from as little as 3% to as much as 50% dissolved solids by weight, there is surprisingly good agreement about the general character of ore-forming fluids. With respect to the major elements in solution, we can place considerable confidence in the agreement because concurrent data come from many different sources. Most data come from analyses of fluid inclusions. They confirm what is known from modern hydrothermal systems such as the Salton Sea geothermal brines (Muffler and White, 1969) and the brines discovered beneath the Cheleken Peninsula on the Caspian Sea (Lebedev and Nikitina, 1968), as well as information from sources such as hot springs and oilfield brines.

A hydrothermal ore-fluid is an aqueous solution containing, as major components, sodium, potassium, calcium, and chlorine. Other elements must also be present; for example magnesium, bromine, sulfur (as either sulfate or sulfide or both), strontium, and sometimes iron, zinc, carbon (as HCO_3 and CO_2), and nitrogen (as NH_4) can be present in amounts greater than 1000 ppm (Table 1-1). The source of the major constituents in solution can rarely, if ever, be identified with certainty. In a few cases, a special property such as a unique bromine/iodine ratio will indicate that a specific evaporite bed has supplied some of the constituents. For most solutions, however, we can do little more than conclude that the solution obtained its constituents from a cooling magma or it dissolved salts from the rocks through which it passed. Additionally, we conclude from both field and experimental studies (White, 1965) and from $^{87}Sr/^{86}Sr$ ratios (Hedge, 1974) that anion exchange reactions between solution and min-

TABLE 1-1 Compositions of some modern and ancient hydrothermal solutions

Concentrations in ppm. 1 = Salton Sea geothermal brine (Muffler and White, 1969); 2 = Cheleken geothermal brine (Lebedev and Nikitina, 1968); 3 = oil field brine, Gaddis Farms D-1 well, Lower Rodessa reservoir, central Mississippi, 11,000 ft (Carpenter et al., 1974); 4 = fluid inclusion in fluorite, Cave-in-Rock District, Ill. (Roedder et al., 1963); 5 = fluid inclusion in sphalerite, OH vein, Creede Colo., recalculated by Skinner and Barton (1973); 6 = fluid inclusions, core zone at Bingham Canyon (Roedder, 1971).

	Modern solutions			Ancient solutions		
Element	1	2	3	4	5	6
Cl	155,000	157,000	158,200	87,000	46,500	295,000
Na	50,400	76,140	59,500	40,400	19,700	152,000
Ca	28,000	19,708	36,400	8,600	7,500	4,400
K	17,500	409	538	3,500	3,700	67,000
Sr	400	636	1,110	—	—	—
Ba	235	—[a]	61	—	—	—
Li	215	7.9	—	—	—	—
Rb	135	1.0	—	—	—	—
Cs	14	0	—	—	—	—
Mg	54	3,080	1,730	5,600	570	—
B	390	—	—	<100	185	—
Br	120	526.5	870	—	—	—
I	18	31.7	—	—	—	—
F	15	—	—	—	—	—
NH_4	409	—	39	—	—	—
HCO_3^-	>150	31.9	—	—	—	—
H_2S	16[b]	0	—	—	—	—
SO_4^{2-}	5	309	310	1,200	1,600	11,000
Fe	2,290	14.0	298	—	—	8,000
Mn	1,400	46.5	—	450	690	—
Zn	540	3.0	300	10,900	1,330	—
Pb	102	9.2	80	—	—	—
Cu	8	1.4	—	9,100	140	—

[a] Not determined.
[b] Sulfide present; all S reported as H_2S.

erals such as feldspars, micas, and clays must proceed continually so that a solution can change in composition as it moves. A hydrothermal fluid is therefore an evolving or changing entity. Influenced principally by temperature and the rocks through which it passes, but also by mixing with other solutions and by effects such as boiling and membrane filtration (Hanshaw, 1972), a solution can change both its composition and salinity.

Ore metals, such as copper, lead, zinc, tin, molybdenum, and silver, are rarely present as major constituents in solution. Rather they seem most commonly to be present at levels up to a few tens or hundreds of parts per million. Like the major elements in solution, the source of ore metals can rarely be specified with certainty. But one metal, lead, does have a variable isotopic composition, and this property has been used to supply some extraordinarily valuable information about sources. The isotopes ^{235}U, ^{238}U, and ^{232}Th are all radioactive and each decays to a different stable isotope of lead. The isotopic composition of lead from different rocks, and even from specific minerals in the same rock, is therefore sensitive not only to the initial lead-isotopic ratios but also to differences in thorium and uranium and to the length of time the radioactive elements have had to change the lead-isotopic ratios. The distinctive isotope ratios can be used as characteristic fingerprints. The results are somewhat surprising. Doe and Delavaux (1972) showed that the source of lead in the Old Lead Belt in southeast Missouri is probably the feldspars in the La Motte Sandstone, the main aquifer for hydrothermal fluids in the district. The lead was apparently transferred from solid solution in the feldspars, to solution in the hydrothermal fluid, by rock-water reactions. Similarly, Zartman (1974) showed that many of the Mesozoic and Cenozoic ores in the western United States derived their lead in part or wholly from metamorphic basement rocks beneath the ores or from adjacent sedimentary rocks. Stacey et al. (1968) discovered that lead in the high-temperature copper ores of Utah was derived from associated igneous rocks presumably by release of a magmatic fluid, while Doe and Stacey (1974) showed the lead in the lower temperature ores at Creede, Colorado, was not derived from adjacent igneous rocks but came instead from the underlying Precambrian country rocks.

We are therefore presented, once again, with evidence that hydrothermal fluids can change their character as they move. Clearly they are capable of removing lead from rocks they pass through, but apparently, too, a fluid released by a crystallizing magma can also carry "magmatic" lead with it. Although we cannot prove the point from isotopic studies with other metals, we are led to conclude from the lead evidence that hydrothermal fluids must collect all metals in the same way—in part from a magma, if the solution originated from a magma, as well as from all the rocks they pass through in the hydrothermal channelways. Of the two mechanisms, magma release and channelway reactions, lead isotopes suggest that the latter is the more important of the two sources.

Elements such as lead, zinc, molybdenum, tin, and silver, the geochemically scarce elements with crustal abundances below 0.01%, rarely form separate phases in ordinary rocks. Instead, the metals are present in solid solution in common silicate minerals; most of the lead, for example, is carried in potassium feldspars, whereas zinc, copper, and tin are present in clays, micas, pyroxenes, and amphiboles. There are innumerable reactions that could occur between a traversing solution and the minerals with which it is in contact, but the specific reactions that release the metals from solid solution in their host minerals remain in doubt. Transfer may require recrystallization of the host minerals and the production of entirely new minerals, or it might happen by simple ion-exchange reactions. Possibly several kinds of reaction proceed together; but again we face a question that requires more definitive research.

It is apparent, then, that any rock can serve as a source of geochemically scarce metals provided a hydrothermal solution undergoes the reactions that will extract them. There have been many attempts to discover, by careful chemical analysis, whether a given rock body is depleted in one or more scarce metals and is therefore the source for metals in an adjacent ore body. None has been successful. The most successful study was that of Mackin (1968) who proved that iron (a geochemically abundant element) in the iron ores of the Iron Springs District, Utah, just equaled the mass of the iron leached from adjacent wallrocks. For less abundant elements, however, analytic precision becomes a limiting factor. For example, if 20% of the lead in a rock is scavenged by a traversing solution, the lead content of an average crustal rock will drop from 10 to 8 ppm. No one has yet attained sufficient precision in both sampling and analyzing a rock mass to detect the change.

A major question remains to be asked; must a rock mass be abnormally enriched in a given scarce metal before it can serve as a source for a scavenging hydrothermal solution? The answer, apparently, is sometimes yes, sometimes no. For elements with crustal abundances in the range 0.001–0.01%, the answer is no. The detrital sediments that are the source of the metals in the Salton Sea geothermal brines, for example, and the La Motte Sandstone, are ordinary rocks, not enriched to a significant extent. The important factor, apparently, is whether or not the right silicate minerals are present and whether, as they react with the hydrothermal solution, they release their contained trace elements. Thus, a potassium-feldspar-rich arkose or rhyolite (or even a rhyolitic obsidian) will yield a lead-rich solution, whereas a pyroxene- and olivine-rich basalt will yield a copper-rich solution, but the rocks need not be unusually rich in lead or copper. The kind of source-rock is obviously important, however, because it determines the reactions that occur, and thus can strongly influence the composition of the solution and eventually the kind of ore deposit formed.

Potassium-feldspar- and mica-rich rocks will tend to produce lead-zinc deposits, intermediate igneous rocks containing pyroxenes and amphiboles may yield copper-zinc deposits, whereas mafic rocks could produce copper deposits.

The geochemically very scarce elements, such as tin, silver, and mercury with crustal abundances below 0.001%, seem to present a different case. Their abundance in most rocks is so low that most hydrothermal solutions do not become sufficiently enriched to lead to deposition of tin, mercury, or silver ore bodies. For elements such as these, a concentration prior to solution extraction does seem to be needed. Some of the granitic rocks that are sources for the tin in the Far East tin deposits, for example, appear to be abnormally rich in tin. The question of prior enrichment remains in doubt, however, because the geochemical data presently available are too sparse and too crude. The question requires much closer study. If primary enrichment is proved to be necessary, a further question, the reason for the primary enrichment, also remains to be answered.

Transport by, and Precipitation from, Hydrothermal Solutions

It should be apparent from the preceding discussion that hydrothermal solutions must be common and widespread in the crust. Indeed, because of the interaction between the hydrosphere and the lithosphere, the formation of hydrothermal solutions must be the rule rather than the exception. Why, then, are hydrothermal ore deposits so scarce and so small? The answer lies, apparently, with the way metals are transported and precipitated. Many of the chapters in this volume bear eloquent testimony to the spectacular advances made over the last 25 years in our knowledge of how a solution transports and deposits its dissolved load. Indeed, later chapters discuss the topics in such detail that only the briefest discussion is warranted here.

Most of the common ore minerals are sulfides, and they share an important property; they are extremely insoluble in water. The problem of insolubility has been known for a long time, and many attempts have been made to find a way of transporting significant amounts of metals and sulfide sulfur in the same solution without causing precipitation. More than a century ago, the role of transport was assigned to "mineralizers." Commonly they were thought of as fugitive constituents that could greatly increase the solubility of the recalcitrant ore minerals but were no longer present for examination and therefore open to extreme speculation. With few exceptions, such as the role of fluorine in transporting tin, specific mineralizers could not be pinpointed. The mystery was solved about 20 years ago, when it was shown that formation of complex ions in solution could lead to the required solubility. The complex ions, formed between metals and ligands in solution, serve as shields, or carriers for the metals, consequently inhibiting precipitation of sulfide minerals.

Two kinds of complex ions seem to be important in hydrothermal solutions. The first involves the very species that causes the solubility problem —reduced sulfur species. If a solution is very sulfur-rich, species such as $Zn(HS)_3^-$ and $HgS(H_2S)_2$ can form and carry large amounts of metals in solution. The problem with sulfide complexes is that, in order for the complexes to form in the first case, the concentration of reduced sulfur atoms in solution (as distinct from oxidized sulfate atoms) must greatly exceed that of the metal species. Many fluid inclusion analyses, and the compositions of fluids in modern hydrothermal systems, indicate that the metal/sulfide ratio is a difficult restriction. The second class of complex ions involves chlorine. Aqueous species such as $ZnCl_2$ and $CuCl_3^{2-}$ will form in chloride-rich solutions under the right conditions, and they are quite capable of transporting metals in the presence of low concentrations of dissolved sulfide ions. With chloride complexes, the total number of metal atoms in solution must greatly exceed the reduced sulfur atoms, and, in this sense, fluid inclusions and modern systems support chloride complexes. One problem with chloride complexes is that when precipitation occurs, it can only continue until the sulfide ions are consumed. Unless more reduced sulfur is added to the solution, most of the dissolved metals will not precipitate and form an ore deposit.

Controversy still surrounds the relative importance of these two families of complexes. Certainly other ligands, such as fluoride, hydroxide, sulfate, carbonate, amine, and organic complexes sometimes play a role, but from present evidence the role of chloride complexes seems more important than other complexes. The evidence comes from modern hydrothermal fluids such as the Salton Sea geothermal brines, from fluid inclusion compositions, and from experimental studies on the stability conditions of various complexes. We have already seen, however, that evidence from hydrothermal systems can be scanty, and this means it might also be highly biased. Data are accumulating rapidly, however, and the remaining questions of ore mineral transport will probably be answered in the near future.

The question of cause of precipitation still remains to be considered. Fortunately, we know a great deal more about precipitation than we do about the sources of mineral constituents. Yet even with the question of precipitation we are faced with uncertainty. At least four different, and quite independent, changes in a hydrothermal solution can lead to precipitation, and in any given solution several different effects may operate at once. Because each of the four effects discussed below will cause the same minerals to precipitate, we can rarely identify the dominant cause of precipitation.

1. Temperature changes cause precipitation in three ways: (a) by affecting the solubility products of sulfide and oxide minerals; (b) less obviously, by affecting the formation and stability of the complex ions transporting the metals; (c) by influencing the hydrolysis constants of ligands, such as Cl^-, involved in formation of complexes. The effects of temperature on solubility are therefore complicated, but it is now possible, through the work of Helgeson (1969) and others, to estimate (although often with considerable uncertainties) the sum of the different effects and therefore to predict the change in solubility of a given mineral that will accompany a rise or fall in temperature. In most cases, but not all, a decline in temperature will cause a reduction in solubility and can, provided a solution reaches saturation, lead to precipitation of ore minerals.

Although falling temperatures can lead to precipitation, a temperature drop of 20°C or more is probably needed to remove a lot of material from solution. To form an ore deposit, the temperature drop must occur in a localized area; this leads to a difficulty—how to consume or lose sufficient heat to cause a significant temperature drop. Many possible ways have been suggested, including endothermic reactions between solutions and minerals, but with only three exceptions, all effects seem to be too small to account for the temperature drop.

The first, and probably the most important, effect causing a temperature drop is the mixing of a hot, rising solution with a cold near-surface water mass. The effect must be common not only on land, but also in the sea, because mixing must always occur, to some extent, when a hydrothermal solution is debouched onto the sea floor. The second cooling effect is due to adiabatic decompression, an effect commonly called throttling. A rapid, local temperature drop will occur when the pressure on a fluid approaching the surface changes over a short distance from lithostatic to hydrostatic. The third, and probably least effective, means of dropping temperatures is by heat loss to wallrocks. The mechanism is not effective because the heat loss will happen over a long distance, leading to sparsely distributed mineralization.

2. Pressure changes can cause solubility changes, but the effects seem to be unimportant if there is no separation of a separate vapor phase. Large pressure changes, of the order of 1000 bars, are needed to cause significant precipitation, and there are few circumstances where such large changes happen over a localized region. Two possible circumstances where pressure effects might be important are in the formation of gold-quartz veins during metamorphism and the formation of pegmatites.

While discussing pressure effects, it is appropriate to point out the importance of a pressure-controlled isothermal effect—boiling—in causing precipitation. Boiling has two important effects; it directly increases solution concentration, but, more importantly it is a mechanism whereby volatile constituents can be removed from solution, leaving the residue more alkaline and less capable of metal transport.

3. Chemical reactions between moving solution and the rocks lining the channelways must be a major cause of precipitation. The common field observation that certain ores are preferentially associated with a specific wallrock is evidence of the importance of such reactions.

The reactions of certain types of wallrocks have been extensively studied, but others, such as those involved in skarn deposits and greisens, still require a great deal more work. Nevertheless, it is apparent that three different kinds of reactions are important. First, there are the reactions that cause hydrogen ions in solution to be exchanged with the wallrocks. The chemistry of hydrothermal solutions indicates that most are weak acids, so extraction of hydrogen ions from solution (or, as it is commonly stated, hydrogen ion metasomatism of the wallrocks) tends to reduce the stability of chloride complexes and lead to sulfide mineral precipitation. The most common kinds of hydrogen ion-consuming reactions are (a) solution of carbonates and (b) hydrolysis of feldspars and mafic minerals to form micas and clays.

The second important class of wallrock reaction involves the addition of a component from wallrock to the solution. An obvious example is the addition of a reduced sulfur species, such as H_2S or sulfur from pyrite in a black shale. The addition of reduced sulfur causes the immediate precipitation of sulfide minerals.

The third class of wallrock reaction involves a change in the oxidation state of a solution, thereby changing the valence of certain metals such as copper, uranium, and vanadium or the stability of certain complex ions. Two important processes by which oxidation state can be influenced are through reduction due to serpentinization and the addition of carbonaceous matter. The effects of carbonaceous compounds are particularly important because they can not only influence the state of metals in solution, but can also cause the reduction of $(SO_4)^{2-}$ in solution to H_2S. This, of course, has the same effect as adding reduced sulfur from an external source. In certain cases it is apparent that a reducing agent, such as methane (CH_4), can even be carried for long distances in the same hydrothermal solution as $(SO_4)^{2-}$ despite extreme chemical incompatibility. The rate of the $(SO_4)^{2-}$ reduction by CH_4 is very slow and thus, a single solution can carry metastably within it the cause of sulfide mineral precipitation. In some cases, the rate of sulfate reduction may be greatly speeded up by the action of bacteria. However, bacteria can only work effectively at temperatures below about 80°C, so their involvement can only be important in low-temperature deposits. One example where this may be going on today is in the precipitation of sulfide minerals from the dense hydrothermal brines erupting onto the floor of the Red Sea (Degens and Ross, 1969).

4. The chemical changes due to solution mixing that takes place when a hydrothermal fluid mixes with a solution of different composition are probably common causes of ore deposition. The effects are, however, exceedingly difficult to demonstrate, and even where solution mixing can be reasonably assumed, it is usually impossible to separate thermal from chemical effects. Aside from thermal effects, mixing can add a precipitating component, such as H_2S, to a solution, or it can change acidity and concentration in such a way as to reduce the complex ion stability and lead to precipitation.

A SYNOPSIS

Some of you who read this book may be familiar with one or more mineral deposits and no doubt you will ask the question, "How does this or that deposit fit into what is being discussed here?" Hydrothermal solutions, as we have seen, can evolve in several ways, and their dissolved loads can come from many sources. The solutions are forever changing, reacting with wallrocks, replacing earlier formed minerals and, sometimes, forming ore deposits. The kind of deposit depends on what is in solution, and where and how precipitation occurs. The different environments of ore deposition are identified on Figure 1.1 and examples of well-known deposits are shown for each environment. Perhaps you will not agree with their designations—they reflect, after all, the prejudices of the author as to the relative importance of many different effects. Pick other examples, if you wish, or rearrange the choices if you evaluate the evidence differently. But keep Figure 1.1 in mind as you proceed through this volume, because it will help you place a lot of complex, and seemingly unrelated factors into a framework of understanding.

You are now ready to proceed to the technical chapters. Each discusses one or more of the problems mentioned in this first chapter, but it will soon be apparent to you that the experts' knowledge is still fragmental, opinions still biased, and that disagreements are common. This makes the study of mineral deposits an exciting endeavor. If any doubts remain about opportunities for fruitful work in the future, this book will soon dispel them; there is an abundant need for new workers with new approaches and new ideas.

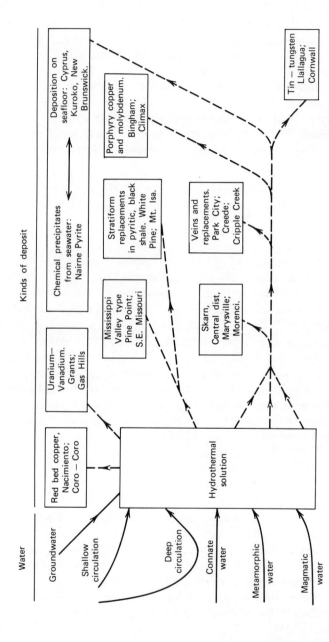

Fig. 1.1 Schematic diagram relating types of mineral deposits and sources of hydrothermal fluids. Adapted from a similar diagram by Skinner and Barton (1973).

REFERENCES

Carpenter, A. B., M. L. Trout, and E. E. Pickett (1974) Preliminary report on the origin and chemical evolution of lead- and zinc-rich oil field brines in Central Mississippi: *Econ. Geol.,* **69,** 1191-1206.

Degens, E. T. and D. A. Ross, editors (1969) *Hot Brines and Recent Heavy Metal Deposits in the Red Sea:* New York: Springer-Verlag.

Doe, B. R. and M. H. Delavaux (1972) Source of lead in southeast Missouri galena ores: *Econ. Geol.,* **67,** 409-425.

_____ and J. S. Stacey (1974) The application of lead isotopes to the problems of ore genesis and ore prospect evaluation: A review: *Econ. Geol.,* **69,** 757-776.

Hanshaw, B. B. (1972) Clay membrane phenomena: *The Encyclopedia of Geochemistry and Environmental Sciences,* ed. R. W. Fairbridge, New York: Van Nostrand Reinhold, 172-176.

Hedge, C. E. (1974) Strontium isotopes in economic geology: *Econ. Geol.,* **69,** 823-825.

Helgeson, H. C. (1969) Thermodynamics of hydrothermal systems at elevated temperatures and pressures: *Am. J. Sci.,* **267,** 729-804.

Lebedev, L. M. and I. B. Nikitina (1968) Chemical properties and ore content of hydrothermal solutions at Cheleken: *Dokl. Akad. Nauk, SSSR,* Engl. transl. *Earth Sci. Sect.,* **183,** 180-182.

Livingston, D. E., R. L. Mauger, and P. E. Damon (1968) Geochronology of the emplacement, enrichment and preservation of Arizona porphyry copper deposits: *Econ. Geol.,* **68,** 30-36.

Mackin, J. H. (1968) Iron ore deposits of the Iron Springs District, Southwestern Utah: *Ore Deposits of the United States, 1933-1967, Vol. II,* ed. J. D. Ridge, New York: Am. Inst. Mining Metall. Petrol. Eng., 992-1019.

Muffler, L. J. P. and D. E. White (1969) Active metamorphism of Upper Cenozoic sediments in the Salton Sea geothermal field and the Salton Trough, southeastern California: *Bull. Geol. Soc. Am.,* **80,** 157-182.

Roedder, E. (1960) Fluid inclusions as samples of ore-forming fluids: *Int. Geol. Cong., XXI Session, Norden;* Part XVI, 218-229.

_____ (1971) Fluid inclusion studies on the porphyry-type ore deposits at Bingham, Utah, Butte, Montana and Climax, Colorado: *Econ. Geol.,* **66,** 98-120.

_____, B. Ingram, and W. E. Hall (1963) Studies of fluid inclusions. III. Extraction and quantitative analysis of inclusions in milligram range: *Econ. Geol.,* **58,** 353-374.

Skinner, B. J. and P. B. Barton, Jr. (1973) Genesis of mineral deposits: *Annual Review of Earth and Planetary Science Vol. 1,* Palo Alto, Calif., Annual Reviews, 183-211.

Stacey, J. S., R. E. Zartman, and I. T. Nkomo (1968) A lead isotope study of galenas and selected feldspars from mining districts in Utah: *Econ. Geol.,* **63,** 796-814.

White, D. E. (1965) Saline waters of sedimentary rocks: *Am. Assoc. Petrol. Geol. Mem.,* **4,** 342-366.

_____ (1974) Diverse origins of hydrothermal ore fluids: *Econ. Geol.,* **69,** 954-973.

White, W. S. (1971) A paleohydrologic model for mineralization of the White Pine copper deposits, northern Michigan: *Econ. Geol.,* **66,** 1-13.

Zartman, R. E. (1974) Lead isotopic provinces in the cordillera of the western United States and their geologic significance: *Econ. Geol.,* **69,** 792-805.

2

Plumbotectonics, The Phanerozoic

BRUCE R. DOE and ROBERT E. ZARTMAN
U.S. Geological Survey

In recent years lead isotopes have found increasing use in studies of the provenances of igneous rocks and ore deposits (see, for example, Cooper and Richards, 1966; Doe, 1967, 1968; Tatsumoto, 1969; Oversby and Gast, 1970; Oversby and Ewart, 1972; Zartman, 1974). These applications have generally revealed that if a time interval as geologically short as the Mesozoic-Cenozoic is examined, then distinct patterns of lead isotopic abundances are seen to be associated with specific environments. That is, superimposed upon the isotopic evolution of lead, which results from the radioactive decay of uranium and thorium, are effects related to fundamental chemical differences between source regions.

Although distinctions between major environments do persist, fortunately much local isotopic variation is lost at the scale of igneous and ore-forming processes. Consequently, observed patterns fit remarkably well with a general model of an evolving mantle, upper continental crust, and lower continental crust. In this paper we place emphasis on Phanerozoic—especially on Mesozoic and Cenozoic—deposits, for which we have (1) by far the greatest number of analyses on the largest variety of materials, (2) the best geographic and geologic coverage, and (3) the closest control over mineralization age. With only a few notable exceptions, the necessary documentation does not exist for Precambrian deposits, which must await further study before being compared meaningfully to the model in any detail.

Although our primary concern here is with concentrations of lead in considerable excess of the clarke, both rock and ore lead usually acquire their isotopic composition from rather ordinary source materials. An ore deposit is often a final, localized occurrence of certain economically important minerals produced by petrogenetic processes affecting much larger volumes of surrounding rock. Accordingly, any interpretation of its origin also must take into account the total geologic environment that contributed to the mineralization.

Our approach stresses in a more comprehensive manner the role of dynamic geologic processes of the mantle and crust in determining lead isotopic composition, albeit certain aspects of such a treatment have appeared elsewhere (Patterson and Tatsumoto, 1964; Wasserburg, 1966; Gast, 1967, Armstrong, 1968; Russell, 1972; Armstrong and Hein, 1973; Cumming and Richards, 1975). The advent of plate tectonics has particularly stimulated interest in mechanisms that provide for continuing communication between the diverse chemical systems of the mantle and crust. Although not dependent upon any particular plate tectonic model in detail, we do use the concept of a long-term, chemically evolving Earth in our interpretation.

The theoretical treatment for evolution of lead isotopes by radioactive decay has been amply documented, and the reader wishing to review these mathematical formulations is referred to the works of Russell and Farquhar (1960), Kanasewich (1968), Doe (1970), Doe and Stacey (1974), Stacey and Kramers (1975), and Cumming and Richards (1975). Some of the important parameters and equations, which are pertinent to the ensuing discussion, are given in Table 2-1.

Much of the isotopic resolution presented in this work depends on the capability to accurately determine subtle distinctions in composition. This capability may be largely attributed to the many laboratory innovations of the past decade that have greatly improved analytic precision and increased productivity. The rock and ore data used in this synthesis are stored in the Lead Isotope Data Bank (LIDB) of the U.S. Geological Survey (Doe, 1976). Previously unpublished analyses from our laboratory with analytic uncertainties not exceeding 0.1%/ratio are given in Table 2-2. A summary plot of all lead isotopic ratios for Phanerozoic rocks and ore deposits, defining fields of broad geologic environments, is shown in Figure 2.1. The relatively unique positioning of these fields—a feature whose importance

TABLE 2-1 Some parameters and equations used in lead isotope studies

Nuclide	Decay constant	Symbol	Reference
(a) Decay constants			
^{238}U	0.155125×10^{-9} yr^{-1}	λ	[Jaffey et al., 1971]
^{235}U	0.98485×10^{-9} yr^{-1}	λ'	[Jaffey et al., 1971]
^{232}Th	0.049475×10^{-9} yr^{-1}	λ''	[LeRoux and Glendenin, 1963]
Present-day ratio ^{238}U/^{235}U = 137.88			[Atomic Energy Commission, 1962]

(b) Primordial lead isotopic ratios [Tatsumoto et al., 1973.]

^{206}Pb/^{204}Pb = 9.307; ^{207}Pb/^{204}Pb = 10.294; ^{208}Pb/^{204}Pb = 29.476

(c) Closed-system evolution equations for lead isotopes (between arbitrary times t_1 and t_2[a])

$$\left(\frac{^{206}\text{Pb}}{^{204}\text{Pb}}\right)_{t_2} = \left(\frac{^{206}\text{Pb}}{^{204}\text{Pb}}\right)_{t_1} + \left(\frac{^{238}\text{U}}{^{204}\text{Pb}}\right)_{t_1 \to t_2} (e^{\lambda t_1} - e^{\lambda t_2})$$

$$\left(\frac{^{207}\text{Pb}}{^{204}\text{Pb}}\right)_{t_2} = \left(\frac{^{207}\text{Pb}}{^{204}\text{Pb}}\right)_{t_1} + \left(\frac{^{235}\text{U}}{^{204}\text{Pb}}\right)_{t_1 \to t_2} (e^{\lambda' t_1} - e^{\lambda' t_2})$$

$$\left(\frac{^{208}\text{Pb}}{^{204}\text{Pb}}\right)_{t_2} = \left(\frac{^{208}\text{Pb}}{^{204}\text{Pb}}\right)_{t_1} + \left(\frac{^{232}\text{Th}}{^{204}\text{Pb}}\right)_{t_1 \to t_2} (e^{\lambda'' t_1} - e^{\lambda'' t_2})$$

[a] Subscripts t_1 and t_2 refer to the starting and ending times of the system, respectively.
Note: Throughout this paper we follow the established conventions of counting geologic time backward from the present and of evaluating abundances of uranium and thorium isotopes in terms of their equivalent present-day value; i.e., normalized for radioactive decay.

has not been fully recognized in the past—provides the impetus for this exercise in plumbotectonics.

BACKGROUND

Early mathematical treatments of the isotopic compositions of rock and ore lead suggested, in many cases, existence of a nearly constant uranium/lead and thorium/lead reservoir throughout geologic time. Attention often focused on making the best fit of the data to rather simple closed-system

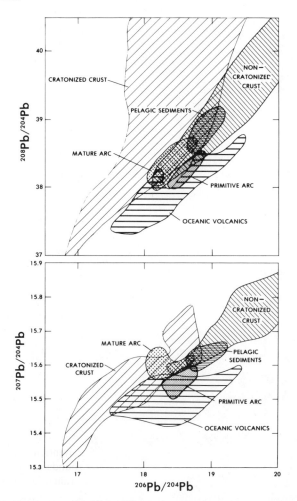

FIG. 2.1 Summary plots for $^{207}Pb/^{204}Pb$ and $^{208}Pb/^{204}Pb$ versus $^{206}Pb/^{204}Pb$. These environments are illustrated in greater detail in other figures: oceanic volcanics, emphasizing spreading centers (Figure 2.4); pelagic sediments (Figure 2.4); upper continental crust (Fig. 2.5, 2.6, 2.8, and 2.10;) lower continental crust (Figures 2.7 and 2.8); primitive arcs (Figure 2.9); mature arcs, and submarine exhalative deposits (Figure 2.9). References to data are given in the captions of the detailed figures and the Lead Isotope Data Bank.

models, thereby distinguishing between *ordinary* and *anomalous* occurrences depending on their adherence to such models. Unfortunately, for many years the relationship between these primarily mathematical models and the actual geologic processes of igneous rock and ore genesis was poorly understood. Emphasis has recently shifted toward an effort to view isotopic composition as a consequence of environment and the geochemical

TABLE 2-2 Locations, lead isotopic compositions, and geologic relationships of newly analyzed galenas

All data by the triple-filament thermal ionization technique, Maryse H. Delevaux—analyst, unless otherwise noted. Analytic uncertainties are less than 0.1% of each ratio.

District	Mine or prospect	Coordinates (Lat., Long.)	$\dfrac{^{206}Pb}{^{204}Pb}$	$\dfrac{^{207}Pb}{^{204}Pb}$	$\dfrac{^{208}Pb}{^{204}Pb}$	Geologic relationships
Austria						
Bleiberg-Kreuth	Antoni Shaft	46°25′N, 13°45′E	18.378	15.672	38.557	Pb-Zn ore in Triassic Wettersteinkalk and Triassic dolomite of the Carnian Stage.
	West Antoni Shaft	46°25′N, 13°45′E	18.371	15.661	38.517	
Canada						
Central mineral belt	Buchans mine	48°50′N, 56°50′W	17.844	15.507	37.670	Zn-Pb bearing coarse barite ore in altered tuff belonging either to the Silurian Springdale Group or Roberts Arm Group of Middle Ordovician(?) to Early Silurian(?) age.
Cyprus						
Troodos massif	Skouriotissa mine	30°05′N, 32°53′E	18.476	15.571	38.405	Massive sulfide deposit of Late Cretaceous age (massive sulfide sample concentrations by isotope dilution: U = 0.24, Th ≤ 0.01, Pb = 176 ppm)
Egypt						
Along Red Sea Coast: East-Central Egypt	Bir Ranga mine	24°22′N, 35°13′E	18.595	15.589	38.405	Replacement of middle Miocene gypsum bed.
	Um Gheig mine	25°40′N, 34°33′E	19.155	15.626	38.699	Large strata bound ore deposit underlying a thick Miocene gypsum bed.
	Taleit Eid mine	25°40′N, 34°20′E	20.755	15.694	41.005	Quartz vein in slightly metamorphosed Precambrian-Z schist.

Indonesia

Belitung	Kelapa Kampit area, Diamond drill hole DDH22	2°56′S, 108°03′E	18.609	15.714	38.948	Cassiterite-sulfide mineralization possibly related to Late Triassic granitic intrusion.
	Selumar mine	2°56′S, 108°09′E	18.497	15.711	38.878	Lode tin deposit in sediments possibly related to Late Triassic granitic intrusion.
Java:						
East Priangan	Gunung Sawal mine	~8°S, 108°W	18.608	15.719	38.961	Lead-zinc bearing vein in volcanic rocks of Tertiary age. Lately worked deposit. Approximately 210 km ESE of South Bantem.
Cirotan	South Banten mine	~7°S, 106°E	18.790	15.695	39.188	Cu-Au-Ag bearing veins in Tertiary andesitic rocks.
Buitenzorg	Cikondang	~7°S, 107°E	18.597	15.618	38.729	Au-Ag bearing veins in Neogene andesite. Veins also contain Zn, Pb, and some Cu. Approximately 75 km ESE of South Bantem.
Sulawesi: Sassak	Pariward vein	3°01′S, 119°38′W	18.199	15.590	39.019	Porphyry copper deposit
Sumatra: Palembang	Sungei Tuboh prospect	~3°S, 102°E	18.360	15.582	38.411	Pb-Zn-Cu in skarn adjacent to hornblende porphyry dikes and diorite. One of the larger Pb-Zn showings. Located approximately 100 km NE of Simau gold mine.
Lebong	Simau gold mine	~3°S, 102°E	18.419	15.580	38.462	Au-Ag bearing veins in volcanic rocks older than young Tertiary. One of outstanding Ag-Au producers of S.E. Asia.
West Coast	Muara Sipongi mine	~0.5′N, 100°E	18.663	15.634	38.850	Veins in Tertiary volcanic rocks. Tertiary(?) contact metasomatic deposit in calc-silicate host rock. Approximately 225 km NE of Simau gold mine. Small Au-Ag-Cu producer.

TABLE 2-2: Continued

District	Mine or Prospect	Coordinates (Lat., Long.)	$\frac{206\text{Pb}}{204\text{Pb}}$	$\frac{207\text{Pb}}{204\text{Pb}}$	$\frac{208\text{Pb}}{204\text{Pb}}$	Geologic Relationships
		Japan				
Hokkaido: Toya-ko	Toya #2 mine	42°39.4'N, 140°55.9'E	18.455	15.582	38.552	Miocene Kuroko (black) ore.
Honshu: Hokuroku, Akita Prefecture	Kosaka mine, Uchinotai Western deposit	40°21.1'N, 140°45.6'E	18.463	15.589	38.623	Bedded massive Cu-Zn-Pb ore in Miocene silicic green tuffs of the tholeiitic zone (Kuroko-type deposit).
Yamagata Prefecture	Yoshino mine, Nirasawa no. 2 deposit	38°09.7'N, 140°10.3'E	18.475	15.607	38.633	Small Kuroko-type deposit
Ottate deposit, Iwate Prefecture	Taro mine	39°45.9'N, 141°55.8'E	18.712	15.621	38.669	Kuroko-type deposit of Cretaceous age close to serpentinite body.
Shimane Prefecture	Wanibuchi mine	35°25.2'N, 132°43.0'E	18.245	15.569	38.417	Small Kuroko-type gypsum deposit.
Tsushima Island (near Kyushu): Nagasaki Prefecture	Taishu mine	34°13.3'N, 129°13.2'N	18.470	15.646	38.966	One of major Pb-Zn producers in Japan, Cretaceous Pb-Zn vein.
		Mexico				
Zacatecas	Cueva Santa vein, Fresnillo mine	23°10.3'N, 102°53.5'W	18.840	15.646	38.778	Post monzonite vein possibly of Tertiary age.
	Lower LaFontana manto, Fresnillo mine	do.	18.832	15.641	38.771	Manto deposit cut by a monzonite stock. Deposit is of Early Cretaceous (?) age.
		New Zealand				
Broadlands	Broadland drillhole Br16	38°32.5'S, 176°19.4'E	18.834	15.625	38.713	Galena from between 289 and 332 m. depth in geothermal well. Temperature in well now is 160°C.

	Broadland drillhole Br7	38°32.3′S, 176°19.1′E	18.966	15.640	38.836	Galena from between 784 and 786 m. depth in geothermal well. Temperature in well now is 272°C.
		Peru				
Central Andes	Casapulca mine	~11°36′S, 76°10′W	18.822	15.649	38.871	Ag-Pb-Zn-Cu vein system in Tertiary red beds and volcanic rocks.
South-central Andes	Caudalosa mine	~13°12′S, 75°14′W	18.628	15.618	38.629	Pb-Ag-Zn-Barite vein deposit in Tertiary andesitic rocks.
North-central Andes	Pasto Bueno mine[a]	8°9.5′S, 77°51′W	18.827	15.671	38.882	W-base metal vein deposit spanning Miocene quartz monzonite stock and adjacent Jurassic-Cretaceous shale and quartzite.
		United States				
Alaska: Fairbanks	Open cut, Pedro Dome Busty Bell mine, Pedro Dome	65°2′N, 147°30′W do.	19.118 19.126	15.688 15.693	39.146 39.177	Deposits associated with the granite of the Mesozoic Pedro dome emplaced in schists and gneisses of Precambrian and early Paleozoic age.
	Steamboat Creek mine	do.	19.132	15.685	39.160	
Arizona: Bisbee	Cole mine	31°25′N, 109°54′W	17.136	15.464	37.837	Mesozoic replacement deposit in Cambrian Abrigo Limestone along fault approximately 3000 ft. south of main intrusive stock.
California: Shasta, West East	Mammoth mine Afterthought mine, #2 level	40°45.9′N, 122°27.5′W 40°44.2′N, 122°4.1′W	17.897 17.893	15.462 15.454	37.493 37.453	Massive sulfide Cu-Zn-(Pb) ores in Devonian (West) and Triassic (East) volcanic rocks and tuffs.
French Gulch	Halycon mine	40°43′N, 122°40′W	18.622	15.585	38.328	Fissure Au-quartz vein near contact of Paleozoic Bragdon Fm. and Copley Greenstone. Possibly related to the Shasta Balley batholith of Jurassic age.
Kernville	Prospect near Big Blue mine	35°44′N, 118°27′W	19.479	15.755	39.197	Fissure vein in Mesozoic alaskite and granodiorite along northeast trending shear zone.

TABLE 2-2: *Continued*

District	Mine or Prospect	Coordinates (Lat., Long.)	$\frac{^{206}\text{Pb}}{^{204}\text{Pb}}$	$\frac{^{207}\text{Pb}}{^{204}\text{Pb}}$	$\frac{^{208}\text{Pb}}{^{204}\text{Pb}}$	Geologic Relationships
Colorado:						
Central City	Eureka mine	39°48.3′N, 105°31.9′W	18.144	15.548	38.272	Laramide Au-Ag vein deposit.
Idaho Springs	Franklin-Silver mine	39°45.4′N, 105°30.5′W	18.174	15.542	38.260	Laramide Au-Ag vein deposit.
Georgia:						
Dahlonega	Battle Branch mine	34°28.5′N, 84°2.5′W	18.293	15.646	38.176	Gold-galena-chlorite mineralization generally parallel to schistocity of mica schists and associated with quartz lenses. Richest mineralization is at cross-fractures, however. Mine has been intermittently worked since 1831.
Maine:						
Hancock County	Black Hawk mine (Blue Hill mine)	44°23′N, 68°37′W	18.079	15.616	37.958	Bedded Zn-Cu-(Pb) replacement body in Cambrian(?) and Ordovician(?) Ellsworth Schist at contact of metamorphosed border zone with Devonian granitic Sedgwick pluton. Currently operating mine.
Acton, York County	Silver Wave mine	43°28′N, 70°55′W	18.458	15.640	38.346	Fracture-filling quartz veins in Lower Devonian(?) Ridgemere Formation; Devonian(?) biotite granite stock nearby.
Massachusetts:						
Newbury, Essex County	Unnamed prospect	42°46.6′N, 70°53.9′W	18.515	15.653	38.330	Comb quartz veins occupying open faults at contact of granodiorite dikes with upper Precambrian-middle Paleozoic(?) gneissic country rock.
Mississippi:						
Pisgah Field	Not known	~32.5°N, 90°W	18.885	15.642	38.834	Lead scale deposited from saline brine on oil well pipe apparently derived from the Dorcheat Member of the Schuler Formation of the Upper Jurassic Cotton Valley Group.

Location	Mine	Coordinates				Description
New Hampshire: Coos County	Shelburne mine	44°25.2′N, 71°07.8′W	18.507	15.624	38.454	Fracture-filling brecciated vein in middle Paleozoic schistose and granitic rock.
Madison Township	Madison mine (Silver Lake mine)	43°51.2′N, 71°09.2′W	18.620	15.628	38.542	Fracture-filling brecciated vein in Devonian Littleton Formation. Mineralization may be associated with the Triassic or Jurassic Conway Granite stock nearby. Largest Pb-Zn mine in New Hampshire (currently inoperative).
New Mexico: Sierra Nacimiento	Nacimiento mine	35°59.5′N, 106°53.9′W	20.035	15.754	38.874	Chalcocite ore replacement of carbonaceous fossil log in the Aqua Zarca Sandstone Member of the Triassic Chinle Formation. (Sample concentrations by isotope dilution: $U = 4.2$, $Th = 3.0$, $Pb = 720$ ppm).
North Carolina: Vance County	Hamme tungsten deposit	36°30.5′N, 78°28.5′W	18.337	15.600	38.071	Tungsten bearing quartz veins in albite granodiorite near the contact with early Paleozoic(?) schist.
Oklahoma: Pitcher field, Tri-State	Blue Goose mine	36°55′N, 94°50′W	21.942 21.901	15.920 15.922	41.076 41.072	Two analyses from opposite sides of a 5 cm. galena crystal from the main ore horizon near the Miami trough in Mississippian limestone.
Oregon: Quartzville	Accident claim	44°35.5′N, 122°17.2′W	18.842	15.566	38.502	Vein cutting basalt, andesite, breccias and tuff of the Pliocene and Miocene Sardine Fm. possibly related to numerous plutons of silicic and intermediate compositions.
Virginia: Mineral	Armineous mine	38°2′N, 77°54′W	18.205	15.643	38.121	Bedded massive sulfide deposit in the felsic volcanic rocks of the Cambro-Ordovician Chopawamsic Formation.

[a] Mean of five analyses by the silica gel method, S. S. Sun—analyst. Data from unpublished study by G. P. Landis and S. S. Sun and cited by written permission (Landis, written communication, December 5, 1974).

cycles of lead, uranium, and thorium. Many of the discoveries serving to influence thought on lead isotopes in a dynamic Earth are already recorded in the literature and require only brief mention here. Reference to Figure 2.1 will be helpful in relating isotopic fields to environment, as several of the fundamental suppositions that underlie our model are summarized in the following paragraphs.

The isotopic characteristics of oceanic volcanic rocks and related ultramafic xenoliths almost certainly have ruled out the modern mantle as the direct source of lead found in continental igneous rocks and ore deposits (Armstrong, 1968; Richards, 1971; Zartman and Tera, 1973; Doe and Stacey, 1974). The best access to mantle lead probably occurs at oceanic rises and oceanic intraplate "hotspots", where voluminous outpourings of tholeiitic basalt are found. The lead isotopic field of tholeiitic basalt is measurably deficient in *both* ^{206}Pb and ^{207}Pb as compared to much of the continental crust; this deficiency suggests the mantle has been somewhat isolated from the continental crust for a long period of geologic time. Additionally, complicated lead isotopic patterns are found in some oceanic island volcanic rocks, but this lead may also be explained most readily as having evolved in the mantle from a tholeiitic precursor (Oversby, 1974; Sun and Hanson, 1975).

Controversy continues over the nature and timing of the primary chemical differentiation of the Earth and subsequent maturation of the continental crust. For our purposes we need make only very general assumptions in this regard, which seem to be quite consistent with geologic observation. Because the main concern in this paper is with the Phanerozoic portion of the time scale, our presentation will not depend critically upon precise knowledge of more primitive conditions. Without entering the debate over the age of the Earth or the nature of the earliest planetary differentiation, we shall simply use the empirical data of Stacey and Kramers (1975) to broadly constrain our curve for orogene lead (see later discussion). More pertinent here is the accumulating evidence from the oldest sialic rocks that sometime around 4.0 b.y. ago an accretional stage of continental growth began. Further evolution of the continental crust involved more or less evenly spaced orogenies that mixed together some fraction of the mantle and portions of pre-existing continental crust to form new segments of crust. Just how regular continental growth has been is not well known, although for simplicity we shall assume it has been constant in magnitude over time. Our model tolerates moderate deviation from this assumption and still maintains the *relative* position of lead isotope fields for different environments.

The isotopic composition of most of the lead in the continental crust could not result from a closed-system source operating over all of geologic time (Patterson and Tatsumoto, 1964; Doe, 1967; Russell, 1972; Zartman, 1974; Stacey and Kramers, 1975). Such lead might be better explained as a product of sources experiencing a somewhat variable but gradually increasing ^{238}U/^{204}Pb ratio over time. In addition the upper crust on the average has a high ^{238}U/^{204}Pb ratio, which produces *in situ* a rather radiogenic lead with the passage of time (Doe, 1970; Rosholt et al., 1973). Especially where further enhanced by selective leaching, this situation can give rise to the very radiogenic type of lead found in the Mississippi Valley ore deposits (Heyl et al., 1966).

The lower portion of established continental crust is frequently composed of pyroxene granulite or other highly metamorphosed rocks that, whether studied in surface outcrop or as xenoliths in volcanic rocks, have quite low uranium contents (0.1-1 ppm) and ^{238}U/^{204}Pb ratios (0.5-5) (Moorbath et al., 1969; Manton and Tatsumoto, 1971; Gray and Oversby, 1972; Heier, 1973). Thorium and lead contents (1-10 ppm and 2-20 ppm, respectively) do not appear to be so drastically low compared to upper crustal values; the ^{232}Th/^{204}Pb ratio is often highly variable, but averages slightly less than its upper crustal value. Apparently, this preferential depletion of uranium arises during major orogenic episodes associated with formation of new segments of continental crust. The accumulation of large quantities of lead relatively unsupplemented by subsequent radioactive decay must have a significant impact on future uranium-lead and thorium-lead systematics in the lower continental crust. Conversely, the complementary excess of uranium would have the opposite effect in the upper continental crust (Doe et al., 1968; Moorbath and Welke, 1968; Zartman and Wasserburg, 1969; Zartman, 1974). These differences between the upper and lower continental crust will be shown to have important consequences for successive stages of crustal production.

Oceanic crust is too short-lived to be given reservoir status in our model, but its role in contributing to orogene development undoubtedly has considerable geologic importance. Hereafter unless otherwise stated, the term crust will be meant to imply only continental crust. Likewise, we shall not separately distinguish that region of the mantle beneath continents lying between the Moho and the seismic low-velocity zone, which may be attached to the overlying continental crust. This rock may have chemical characteristics different from the mantle of oceanic areas, but the overall geochemical cycle of lead is judged not to depend critically on including it.

From the preceding discussion, we draw the following conclusions about the geochemical cycle of lead. (1) A minimum of three broad environments, having appreciable amounts of lead, uranium, and thorium and sufficient longevity, are required to describe major isotopic characteristics. These include the mantle, the upper crust, and the lower crust. (2) The mantle displays a distinctively retarded lead isotopic evolution as compared to the crust, requiring some degree of isolation between these two systems after an early gross differentiation of the Earth. (3) A complementary partitioning of lead, uranium, thorium—particularly reflected in the ^{238}U/^{204}Pb ratio—is established between the upper and lower crust soon after a given segment of new crust is formed. To the extent that recycling of the continents by erosion into younger orogenes operates more effectively on the upper crust, a bias giving uranium greater "visibility" than lead is introduced that, in effect, accelerates isotopic evolution in the orogene. Although such partitioning of these elements has been known for some time, we now propose that this feature is quantitatively sufficient to explain observed deviation from closed-system models in continental crust. (4) The diverse chemistry of the mantle, upper crust, and lower crust cause isotopic heterogeneities to arise during periods of isolation of one reservoir from another. On the other hand, dynamic processes allowing communication between reservoirs will tend to reduce these differences. Observed isotopic characteristics can be understood in terms of an early differentiation of mantle and protocrust followed by recurring continental accretion of orogenes in which portions of mantle and pre-existing crust are brought together.

DYNAMIC MODEL OF LEAD ISOTOPE EVOLUTION

Our treatment bears similarity, especially in terminology, to that of Armstrong (1968) and Armstrong and Hein (1973), who first applied the concepts of plate tectonics to explain lead isotope evolution. The basic difference lies in our emphasis on the complementary nature of the upper and lower crust, rather than of the crust and mantle, to achieve a heterogeneous uranium, thorium, and lead distribution. Also, to keep our treatment as simple as possible, we have not attempted to simulate the variability of natural occurrences within reservoirs, but rather have constructed only average lead isotope evolution curves.

We will proceed by considering the transfer of matter—first in terms of total mass, and then of uranium, thorium, and lead contents—among three terrestrial reservoirs designated *mantle, upper (continental) crust,* and *lower (continental) crust.* In the specific model to be derived, 11 evenly spaced *orogenies* occur between 4.0 b.y. ago and the present (at 4.0 b.y., 3.6 b.y., ..., present), which convert portions of pre-existing reservoirs into new upper and lower crust and returned mantle. Following Armstrong (1968) we choose to consider only the upper mantle to 500 km depth, and this is regarded as a single, homogeneous reservoir. The upper and lower crust produced during each orogeny become distinct crustal segments capable of participating independently in subsequent stages of model development.

Initially, all matter resides in the mantle, from which the first segment of crust is created 4.0 b.y. ago. Thereafter, both the mantle and older crust contribute in the formation of new crust during orogenies. As a first approximation to accretion rates, the same amounts of upper and lower crust are produced during all orogenies. The amount of crust remaining after each subsequent orogeny, however, will be only that fraction escaping destruction in the formation of new crust. Our intent is to arrive at a final apportionment of continental crust, including a distribution of ages for the segments, that gives a simple approximation of present Earth conditions.

Except for a minimum requirement needed to form new crust, the extent to which the mantle contributes to an orogeny is not rigidly fixed by our model. However, the assumptions of an early, rapid differentiation of Earth and of a later, largely independent evolution of lead isotopes for the mantle restrict a substantial involvement of the mantle to the first several orogenies. Furthermore, to the extent that partial melting may appreciably enrich a magma in certain elements over their intrinsic mantle concentration, the amount of mantle involvement relates to the volume of extracted source material rather than the volume of resultant volcanic rock. Thus, the contribution of the mantle to the orogene, expressed as a mass fraction of the mantle to 500 km depth, is somewhat arbitrarily assigned values of $\frac{1}{2}$ at 4.0 b.y., $\frac{1}{4}$ at 3.6 b.y., $\frac{1}{8}$ at 3.2 b.y., $\frac{1}{16}$ at 2.8 b.y., and $\frac{1}{32}$ thereafter. Such values do provide that any major partitioning of uranium, thorium, and lead between the mantle and the crust takes place early, with later interchange only slightly influencing the mantle. This restriction is not to deny that the mantle has gone on to develop a complex internal heterogeneity, nor that it continues to be an important ingredient in later orogenes.

To simulate actual geologic conditions associated with an orogeny, two functions are defined for the contribution from the older crust. One function represents erosional processes, which reduce each prior segment of upper crust in turn exponentially down to some base level. [During the jth orogeny we may define the remaining height, h, of each upper crustal segment formed during an earlier orogeny, i, relative to its original height, h_0, as $h = h_0 e^{-k(j-i)}$, where k is the erosional decay constant, and the orogenies are numbered from 1 to 11, consecutively, with decreasing age.] The other function represents the amount of total crust consumed by overlapping of tectonic provinces and by continental foundering, and is related in like fashion to the surface area of older crust. [During the jth orogeny we may define the remaining surface area, a, of each crustal segment formed during an earlier orogeny, i, relative to its original surface area,

a_0, as $a = a_0 e^{-m(j-i)}$, where m is the areal decay constant, and the orogenies are numbered as above.] Good agreement between model and Earth is achieved if (1) the height of older upper crustal segments is reduced by $3/10$ (i.e., $k = -\ln 0.7$) and (2) the area of older crust is reduced by $1/10$ (i.e., $m = -\ln 0.9$, of their preceding values) during every subsequent orogeny.

The pertinent features of the model are delineated in Table 2-3 and are also shown in Figure 2.3, which gives the final configuration of the reservoirs and identifies some geologic environments relevant to later discussion. The model can now be used to establish the distribution of any chemical element among its reservoirs as a function of geologic time. A computer program of the model has been written for the lead, uranium, and thorium concentrations and for the lead isotopic composition existing at the time of each orogeny. For our purposes an orogeny can be considered as an instantaneous event so that radioactive decay takes place only in the reservoirs. During an orogeny lead, uranium, and thorium are (1) extracted from each reservoir in proportion to the fraction of that reservoir entering the orogeny; (2) mixed chemically and isotopically; and (3) redistributed into newly formed upper and lower crust and the mantle according to partitioning coefficients. By selecting appropriate values for the abundances of lead, uranium, and thorium and for the partitioning coefficients of the reservoirs, geochemical and isotopic parameters of the model could be closely matched with best estimates of their corresponding terrestrial values. In practice this objective was accomplished by successive iterations through the model calculations to optimally adjust the variables.

The lead evolution diagram generated by the model is given in Figure 2.3; modern lead isotopic fields are keyed to the equivalent geologic environments shown in Figure 2.2. We believe that this model offers some important insights into the general features of lead isotope evolution and, in particular, encompasses those dynamic processes that give rise to broad isotopic distinctions among provenances. Of primary concern in Figure 2.3 are: the curve for the mantle (a); the curve for the orogene (b) representing a balance of input from mantle, upper crust, and lower crust sources; the curve reflecting contribution from upper crust to the orogene (c); and the curve reflecting contribution from lower crust to the orogene (d). The latter three sources can be strictly evaluated only at the time of an orogeny, but the curves would approximate a continuous version of the model. Also exemplified in Figure 2.3 are hypothetical two-stage occurrences in which the second stage displays accelerated evolution in 1.6-b.y.-old upper crust (c'), and retarded evolution in 1.6 b.y.-old lower crust (d').

The relative distribution of lead, uranium, and thorium between upper and lower crust of a maturing continent is the aspect that most determines

TABLE 2-3 Parameters of the lead isotope evolution model

I. Starting conditions (all material in undifferentiated upper mantle to 500 km depth (800×10^{24} g) at 4.0 b.y. ago)
   A. Lead isotopic composition: $^{206}Pb/^{204}Pb = 10.36$; $^{207}Pb/^{204}Pb = 12.12$; $^{208}Pb/^{204}Pb = 30.55$.
   B. Element abundances: $^{204}Pb - 38.0 \times 10^{15}$ moles; $^{238}U = 349 \times 10^{15}$ moles; $^{232}Th - 1335 \times 10^{15}$ moles.

II. Eleven orogenies evenly spaced in time; i.e., at 4.0, 3.6, 3.2, ..., 0.0 b.y. During each orogeny:
   A. Of the material entering the orogene, (1) the mantle contributes ½ of itself, at 4.0 b.y., ¼ at 3.6 b.y., ⅛ at 3.2 b.y., ¹⁄₁₆ at 2.8 b.y., and ¹⁄₃₂ thereafter; (2) the upper crust contributes ³⁄₁₀ of remaining amounts of each older segment (erosional function); and (3) the total crust contributes ¹⁄₁₀ of remaining amounts of each older segment [areal function]. Components (2) and (3) do not pertain to first orogeny.
   B. The contents of the orogene are chemically and isotopically homogenized.
   C. 2.6×10^{24} g of new upper crust and 2.6×10^{24} g of new lower crust are created, and remaining material returns from the orogene to the mantle. Lead, uranium, and thorium distribute themselves according to the following partitioning ratios.

Element	Mantle	Upper Crust	Lower Crust
Lead	0.007	0.770	0.223
Uranium	0.006	0.870	0.124
Thorium	0.005	0.800	0.195

III. Ending conditions (prevailing today at time of final orogeny)
   A. Lead isotopic composition:

Environment	$\frac{^{206}Pb}{^{204}Pb}$	$\frac{^{207}Pb}{^{204}Pb}$	$\frac{^{208}Pb}{^{204}Pb}$	$\frac{^{238}U}{^{204}Pb}$	$\frac{^{232}Th}{^{238}U}$
Mantle	18.10	15.42	37.70	8.92	3.57
Orogene	18.86	15.62	38.83	10.87	3.64
Upper crust contributed to orogene	19.33	15.73	39.08	12.24	3.42
Lower crust contributed to orogene	17.27	15.29	38.57	5.89	5.98

B. Element abundances and concentrations:

Reservoir	Mass ($\times 10^{24}$ g)	Abundances ($\times 10^{15}$ moles)			Average concentrations (ppm)		
		^{204}Pb	^{238}U	^{232}Th	Pb	U	Th
Mantle	774	20.1	179	639	0.40	0.055	0.19
Upper crust	8	10.0	123	417	19.9	3.65	12.1
Lower crust	18	7.9	47	279	6.5	0.62	3.6

the various lead isotopic patterns among broad geologic environments. On the other hand and in contrast to most earlier dynamic models, we maintain an almost identical ^{238}U/^{204}Pb ratio for the mantle as for the *total* crust. As the crust evolves, the ^{238}U/^{204}Pb ratio and the ^{232}Th/^{204}Pb ratio (to a lesser extent) are fractionated, so that they yield high values in the upper crust and low values in the lower crust. Compounding the situation is a considerably more efficient storage of lead in the lower crust relative to upper crust as processes of extraction act unevenly during subsequent

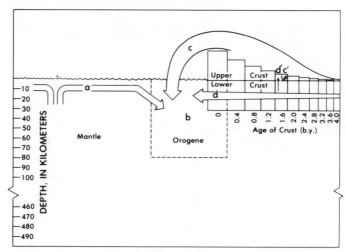

Fig. 2.2 Final configuration of terrestrial reservoirs showing apportionment of mantle, upper continental crust, and lower continental crust. The arrows represent the transport of material into the orogene during the last orogenic cycle, which, in turn, produced the stippled block of new (0.0-b.y.-old) continental crust. Specific environments discussed in text are a, oceanic mantle (and crust); b, orogene; c, upper crust contributed to orogene; d, lower crust contributed to orogene; c', remobilized 1.6-b.y.-old upper crust; and d', remobilized 1.6-b.y.-old lower crust.

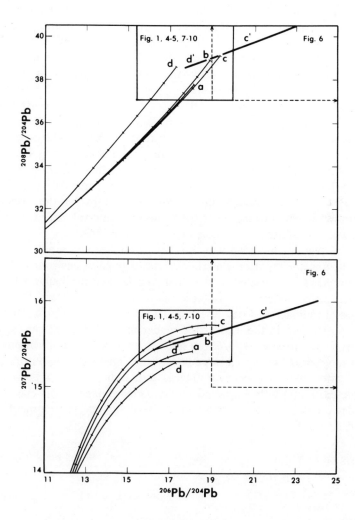

Fig. 2.3 Lead isotope evolution curves generated by our model for the mantle (a), orogene (b), upper crust contributed to orogene (c), and lower crust contributed to orogene (d). The end points labeled by letters represent the modern isotopic fields for these environments and may be compared with Figure 2.2. Tick marks along each curve indicate progressively older time in 0.4 b.y. increments. Also shown are the isotopic fields for lead from remobilized 1.6-b.y.-old upper crust (c′) and remobilized 1.6-b.y.-old lower crust (d′).

orogenies. The upper crust is acted upon by both the erosional and areal functions, whereas the lower crust only experiences the areal function. Consequently, the $^{238}U/^{204}Pb$ and $^{232}Th/^{204}Pb$ values of the upper crust increase, and isotopic evolution accelerates, as uranium and thorium are preferentially recycled relative to lead. Because the available amount of these elements that can be transferred to the upper crust is not limitless, however, the model does not allow accelerated isotopic evolution to occur indefinitely. As an equilibrium state of crustal creation and destruction is approached, more and more of the complementary lower crust with low $^{238}U/^{204}Pb$ and $^{232}Th/^{204}Pb$ ratios is reclaimed into the orogenic budget. The mantle also adds a somewhat less radiogenic lead to the orogene than does the upper crust; particularly obvious is the effect of this mantle lead on the $^{207}Pb/^{204}Pb$ ratio. The combined result of these braking factors becomes evident in our model over the past billion years, during which time the $^{207}Pb/^{204}Pb$ ratio has become virtually constant for the upper crust contributed to the orogene (over the same time interval closed-system evolution would produce a noticeable increase in this ratio).

This section is concluded with a comparison between predicted and observed lead isotopic patterns as they relate to geologic environments. Hereafter emphasis is restricted to the Phanerozoic, and particular attention is directed to the modern lead isotopic fields (see Table 2-3 and Figure 2.3). Because of radioactive decay, the time factor becomes significant for Paleozoic deposits, and in these cases the reader may refer to the corresponding 0.4 b.y. fields. In later reports, we hope to explore the ramifications of the model more broadly over all of geologic time. The following outline gives a classification of geologic environments and indicates the type(s) of lead isotopic pattern(s) to be expected for each environment. This classification becomes the basis of our discussion of the individual environments in the second part of the paper.

Geologic Environment	Model Field (Figure 2.3)
Oceanic	
Oceanic rises	a
Intraplate	a, b
Continental	
Basin-platform and cratonized upper crust	c, c'
Rejuvenated craton and lower crust	d', c'
Noncratonized upper crust	c
The orogene (island and continental arcs)	
Submarine	a, b
Nonsubmarine	a, b, c, d(?)
Transition	a, b, c, d(?)

Oceanic. The lead isotopic field of tholeiite and of mantle-derived ultramafic xenoliths closely approximates field a (compare Figures 2.1 and 2.3), and we accept this isotopic composition as being representative of a large proportion of the mantle. A more complex situation does exist for lead from some oceanic island basalts, especially of the more differentiated varieties, which apparently derives from more peculiar source material. Since we believe the source material for these island basalts probably does not represent a significant perturbation, either in volume or duration, to the overall mantle lead budget, the problem of its special origin needs no further discussion here.

Mantle-type lead has been recognized in several kinds of mineral deposits, especially where contributions from other sources have been minimal. Specifically, it has been identified in massive sulfide deposits intercalated with tholeiites, in metalliferous sediments from near-oceanic spreading centers, in certain intraplate oceanic deposits, and perhaps in primitive island arc occurrences with a high proportion of volcanic component. By contrast, only in unusual cases does this lead penetrate established continental crust without substantial modification in isotopic composition.

Continental. Modern isotopic analyses of rocks and ores clearly document a dispersion in $^{207}Pb/^{204}Pb$ ratios for a given $^{206}Pb/^{204}Pb$ ratio, and the highest values are quite consistently associated with lead of continental affinity, especially in sediments derived from older upper crustal rocks. These source terranes are characterized by high average values of $^{238}U/^{204}Pb$ and contribute more efficiently than the complementary lower crust to the orogenic recycling process. Lead characterized by an exaggerated isotopic evolution (i.e., field c of Figure 2.3) is most prevalent where magma is generated within thick sections of miogeosynclinal rocks, and subsequently intrudes and mineralizes these same rocks. This feature is sometimes further enhanced by selective extraction of radiogenic lead through lateral secretion or weathering.

Also given in our model are the special environments found in older crust, which are responsible for producing the two-stage lead isotopic patterns observed in igneous rocks and ore deposits of rejuvenated cratons and certain basin-platform occurrences. Within the lower crust of stable Precambrian terranes is preserved the most retarded rock lead for a given geologic age—a lead that may approach the original isotopic composition generally existing at the time the terrane was formed. These low $^{238}U/^{204}Pb$ ratio rocks, of course, have been used themselves to gain important insight into the isotopic evolution of lead throughout geologic time. If later incorporated into younger igneous rock or ore, such lead would be recognized by its anomalously nonradiogenic isotopic composition. In contrast, the high ^{238}U/^{204}Pb ratio rocks of the upper crust would have experienced an accelerated evolution of their lead isotopic composition. Depending on the circumstance, mobilization processes may operate either (1) in association with the introduction of nonradiogenic lower crustal lead, or (2) as distinct, separate mineralization events without any deeper supply of material. Hypothetical situations giving rise to two-stage lead deposits are exemplified in Figures 2.2 and 2.3 for the case of a 1.6-b.y.-old terrane presently undergoing mineralization. Field d' represents a typical isotopic array to be derived from the lower crust by magmatic mobilization; related epigenetic processes in the upper crust commonly will extend this field to join with field c'. Field c' may also be generated solely by total or partial extraction of lead from entirely within the upper crust, as perhaps is best demonstrated in deposits of the Mississippi Valley.

The Orogene. A large volume of silicic igneous rocks and ore deposits undoubtedly originates in the orogenic environment. Here an efficient homogenizing process of sedimentation, volcanism, plutonism, metamorphism, and rapid erosional turnover operates, which tends to erase much of the isotopic diversity accruing in the mantle, upper crust, and lower crust. Consequently, an average growth curve is generated that suggests a rather uniform uranium/lead and thorium/lead source, belying the complexities of its individual constituents (modern lead isotopic composition for the orogene occupies field b of Figure 2.3). In detail, however, this growth curve apparently does deviate to a small degree from what would be predicted by an exact closed-system evolution. According to our model, this deviation, which requires a slight increase in effective ^{238}U/^{204}Pb over geologic time toward the present, is explained by the transitory buildup of the lower crust. Storage of lead relatively unsupported by radioactive decay of uranium in the lower crust creates a correspondingly higher uranium/lead value in the upper crust, which is then recycled through the orogene.

The most obvious manifestation of the orogenic environment is in geologically active island and continental arcs. Into these natural mixers come great quantities of tholeiitic basalt and pelagic sediments of the oceanic crust, clastic and chemical detritus of adjacent continental crust, and the lower crustal and mantle material that constitutes the overlying wedge above subduction zones. The lead isotopic composition of such an arc ought to reflect the proportions of its contributing components; thus, lead from a primitive island arc surrounded by open ocean may plot in the lower part of the field, and lead from a continental arc receiving abundant input from cratonic sources may plot in the upper part of the field.

LEAD ISOTOPES AND GEOLOGIC ENVIRONMENTS

Oceanic Environments

RISES

A number of important massive sulfide ore deposits are of the submarine volcanic emanative type associated with mafic volcanism that apparently originated at the crests of oceanic rises or other spreading centers—for example, Cyprus; Maden, Turkey; Hixbar, Philippines; and Island Mountain, California (Hutchinson, 1973). The lead isotopic nature of these ore deposits, frequently called Cyprus-type after the best studied example, is not yet well characterized, probably as a consequence of their lead-poor nature.

The best studied deposits, from a lead-isotope standpoint, are the potentially economic metalliferous sediments forming at ridge crests, such as those of the East Pacific Rise (Dymond et al., 1973; Table 2-2) and the Red Sea (Cooper and Richards, 1969a; Delevaux and Doe, 1974). The metalliferous sediments from the East Pacific Rise and some samples from the Red Sea (Figure 2.4) have lead isotopic compositions more similar to those of ocean-ridge tholeiites, which are inferred to represent oceanic mantle lead (area a of Figure 2.3), than to common pelagic sediments. There is uncertainty whether the lead in these metalliferous sediments is derived largely through leaching of oceanic volcanics or is a direct product of the mantle such as a magmatic differentiate. The pattern for the $^{208}Pb/^{204}Pb$ plot is particularly distinctive for those samples with mantle values for $^{207}Pb/^{204}Pb$ (Figure 2.4); it overlaps only with plots representing primitive island arcs and ores of the igneous rocks of the southwestern United States. Some Red Sea metalliferous sediments have lead isotopic compositions similar to pelagic sediments. Cooper and Richards (1969a) suggested that one of the brine pools overflowed to bring hot acid brines in contact with calcareous pelagic sediments. An alternative hypothesis might be that all the brines are of a basin emanative origin. Whether the brine had basaltic or pelagic sediment lead in it could then simply reflect varying source materials in the basin.

Presented on Figure 2.4 is, to the best of our knowledge, the first lead isotopic analysis of a Cyprus-type ore deposit. The data barely lie in the oceanic mantle field for uranogenic lead but do lie well within the oceanic mantle field for thorogenic lead. The lead isotopic data therefore support genesis of such deposits directly or indirectly from the oceanic asthenosphere.

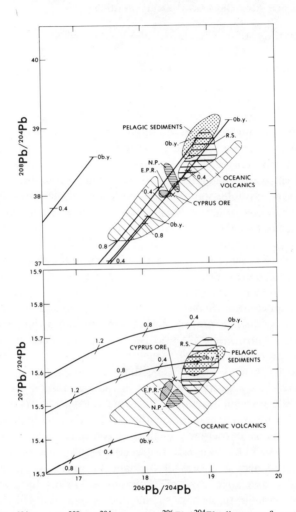

Fig. 2.4 $^{207}Pb/^{204}Pb$ and $^{208}Pb/^{204}Pb$ versus $^{206}Pb/^{204}Pb$ diagrams for various types of Cenozoic ocean ridge and oceanic intraplate materials: oceanic volcanic rocks are from the eastern Pacific Ocean, Hawaii, Mid-Atlantic Ridge, Canary Islands, Red Sea, and Indian Ocean; pelagic sediments are from the North and South Pacific Ocean, South Atlantic Ocean, Indian Ocean and Caribbean Sea; metalliferous sediments are from the East Pacific Rise (E.P.R.), Nazca plate (N.P.) and Red Sea (R.S.). Lines are as given on Figure 2.3 in billions of years. Only data obtained by the silica gel "emitter technique" or "double spiking technique" are included to maintain a uniform quality. References to data: Church and Tatsumoto (1975), Cooper and Richards (1969a), Delevaux and Doe (1974), Doe and Stacey (1974), Dymond and others (1973), Reynolds and Dasch (1971), Unruh and Tatsumoto (1975), Zartman and Tera (1973), and this paper.

INTRAPLATE

Although the intraplate environment involves mixtures of asthenospheric and continental leads, it illustrates the uniqueness of the asthenosphere within the oceanic environment and may conveniently be discussed here. We know of no fossil oceanic intraplate ore deposits, although such material may have furnished low-grade source material for some ores through metamorphic exhalations in the orogene environment. Pelagic sediments, for example, contain approximately 70 ppm lead, a value perhaps triple the average value in continental crustal rocks and roughly a factor of 100 greater than ocean ridge tholeiites. Manganese nodules also are a low-grade ore at the surface of the ocean floor and can contain up to 2% each of copper and nickel (Horn and Delach, 1972).

Very recently, data have become available on the oceanic intraplate metalliferous sediments of the Nazca plate in the South Pacific (Leg 34, sites 319 and 321 of the Deep Sea Drilling Project, Unruh and Tatsumoto, 1976). Leads in these sediments, which contain as much as 30% metals, are of the oceanic mantle type (area a of Figure 2.3), and Unruh and Tatsumoto suggest that the metals originated at the Galapagos Rise. Likewise, lead in an evaporite sequence (Leg 23, site 228) in the Red Sea that contains sphalerite in shaly partings and underlies a pelagic sediment sequence also has oceanic mantle characteristics (Delevaux and Doe, 1974).

In general, modern isotopic data (Reynolds and Dasch, 1971; Church, 1973) show pelagic sediments, including manganese nodules, to be very uniform in lead isotopic composition, especially in localities remote from continents. Values of $^{207}Pb/^{204}Pb$ are approximately 0.75% greater than for the oceanic volcanics (Figure 2.4 and area b of Figure 2.3). A large component of lead must be derived from the continents to account for this difference. The constancy of the mixture suggests that pelagic sediments indicate the best value for lead in the average orogene.

These potential ore deposits are isotopically well characterized, but their application to the understanding of the formation of fossil ore deposits is speculative.

Continental Environments

EPICONTINENTAL BASINS AND PLATFORMS

Considered here are ore deposits enclosed in sedimentary rocks with continental affinities, in which the probability of a close genetic association with igneous activity seems remote. The environments discussed include intracratonic basins and possibly marginal seas. The major deposits have strong stratiform and stratabound characteristics. For the most part, they are thought to have formed syngenetically or perhaps shortly after the surrounding sediments, through action of early diagenesis or postdepositional infiltration probably as a result of basin emanations. If mineralization closely follows sediment deposition, lead isotope measurements cannot easily distinguish between syngenetic and diagenetic mineralizations. Therefore, the phrase, "syngenetic/diagenetic" will be used to refer to them collectively. Epigenetic will be used to denote mineralization long after sediment deposition, generally hundreds of millions of years later. The lead isotope characteristics of deposits in the basinal and platform environment are shown in Figures 2.5 and 2.6. Deposits will be discussed in groupings of silicate and carbonate host rocks; however, some general comments are in order first.

Leads in these deposits never seem to have model ages older than the age of the enclosing sediments. Some may have model ages in excellent agreement with the age of the host rocks (Figure 2.5 and area b in Figure 2.3), but also other, highly radiogenic leads called J-type leads may represent negative or future model ages ($^{206}Pb/^{204}Pb$ > ~18.8, $^{208}Pb/^{204}Pb$ > ~38.8) as shown in Figure 2.6 and areas c–c' in Figure 2.3. Where data plots overlap with the orogene evolution curve on the figures, the data tend to have $^{207}Pb/^{204}Pb$ values equal to or greater than those for pelagic sediments. On the plot involving thorogenic lead ($^{208}Pb/^{204}Pb$), deposits in the J-lead range tend to plot below an extension of the orogene evolution curve. This behavior appears to be a characteristic of upper crustal leads in cratonic regions and is believed to result from enrichment of uranium (relative to thorium) in the upper crust (see c', Figure 2.3) with complementary loss of uranium from the lower crust.

Sandstone. Few lead isotope data are available for base metal ore deposits in multiply cycled sandstone far from the source regions; however, some data exist on lead contained in heavy metal-rich brines. Lead scale was deposited from brine derived from a Jurassic arkose in Mississippi; its lead isotopic composition approximates that of a modern orogene (Figure 2.5 and area b in Figure 2.3). The lead in brines underlying the Salton Sea of California is derived from Pliocene sediments (Doe et al., 1966) and is a bit more radiogenic (area c in Figure 2.3) than the Mississippi lead. Although igneous activity is prevalent in the Salton Sea area, this lead is included here because of its deduced origin from sediments. The inference is that lead derived by leaching of multiply reworked sediments far from the source areas, shortly after deposition, can be expected, like pelagic sediments, to approximate evolution in the orogene.

The first lead isotopic data on a "red bed" copper deposit in single-cycle

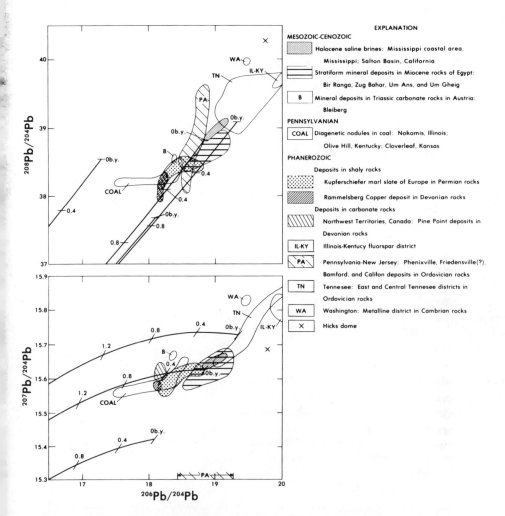

Fig. 2.5 ^{207}Pb/^{204}Pb and ^{208}Pb/^{204}Pb versus ^{206}Pb/^{204}Pb for mineral deposits and selected samples in the sedimentary environment. The only data derived by the lead tetramethyl or triple-filament techniques, thought to be within 0.1% of the absolute ratios, concern the samples from the Mississippi Coastal area, Bleiberg-Kreuth, and the stratiform deposits—such as Pine Point of Canada, Kupferschiefer and Rammelsberg of Europe and also the analyses from Egyptian deposits. Fields involving large or extreme variations in ^{207}Pb/^{204}Pb should be verified by modern analytic techniques before they should be accepted. Data on the deposits of Pennsylvania are not of sufficient quality so that any use may be made of the ^{207}PB/^{204}Pb values although the ^{206}Pb/^{204}Pb and ^{208}Pb/^{204}Pb values are adequate. Therefore note that the data from Pennsylvania are plotted on the ^{206}Pb/^{204}Pb axis in the plot involving ^{207}Pb/^{204}Pb. Lines are as stated in Figure 2.3 caption. References to data: Cumming and Robertson (1969), Doe, Hedge, and White (1966), Wedepohl, Delevaux and Doe (1978), Heyl and others (1966), Russell and Farquhar (1960), and this paper.

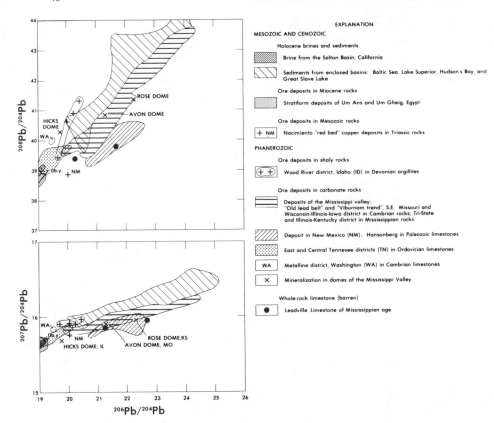

Fig. 2.6 $^{207}Pb/^{204}Pb$ and $^{208}Pb/^{204}Pb$ versus $^{206}Pb/^{204}Pb$ for Phanerozoic radiogenic ore deposits and selected Holocene brines and sediments from enclosed basins (note overlap with Figure 2.5). The only data derived by the lead tetramethyl or triple-filament techniques, thought to be within 0.1% of the absolute ratios, are those for the samples from the stratiform deposits of Egypt and the samples from the Nacimiento ore, the Wood River district, S.E. Missouri district and Tri-State district. The analysis from the Metalline district is not of modern quality and the large value of $^{207}Pb/^{204}Pb$ should be confirmed. Lines are as stated in Figure 2.3 caption. References to data: Chow (1965), Doe, Hedge, and White (1966), Doe and Delevaux (1972), Engel and Patterson (1957), Hall, Rye, and Doe (1978), Hart and Tilton (1966), Heyl and others (1966), Russell and Farquhar (1960), Slawson and Austin (1962), and this paper.

sediments are presented here. The lead, in a replaced log from the Nacimiento, New Mexico, deposit (Figure 2.6), is highly radiogenic and is included in the plot for sediments derived exclusively from much older terranes (area c' in Figure 2.3). The Holocene sediments in the epicontinental basins (Figure 2.6), for example, are generally derived from Archean terranes. To give an example closer to New Mexico, consider the composite of the 1.8-b.y.-old Idaho Springs Formation discussed by Lipman et al.

(1978). The whole-rock ratios are 19.4 for $^{206}Pb/^{204}Pb$, 15.7 for $^{207}Pb/^{204}Pb$, and 38.6 for $^{208}Pb/^{204}Pb$—rather radiogenic. Lead recovered in a hot 6N-HCl leach, however, yields exceedingly radiogenic ratios—28.4 for $^{206}Pb/^{204}Pb$, 16.7 for $^{207}Pb/^{204}Pb$, 49.9 for $^{208}Pb/^{204}Pb$—reflecting the easily mobilized nature of much of the uranium and its daughter lead in crystalline rocks. The lead isotope data for the Nacimiento deposit then are exactly what is expected in the leaching of lead from Precambrian detritus, and they give excellent support for a lateral secretion origin for the deposit, as has been proposed by Woodward and others (1974).

Shales. Shales, generally carbonaceous, are a prime environment for the formation of base metal deposits, such as the Permian Kupferschiefer of Europe and Devonian Rammelsberg of Germany. Rammelsberg is considered by some to be in the island-arc, submarine volcanic emanative category; however, the tuffs are distal and sparse, and the inclusion here seems more reasonable than inclusion in a submarine volcanic association. Kupferschiefer (Permian), Rammelsberg (Devonian), and diagenetic galena in coal (Pennsylvanian) have lead isotopic compositions in general agreement with their mineralization ages (area b in Figure 2.3). Commonly then, leads in shales formed in epeirogenic seas will approximate leads found in the orogene; however, shales developed in intracratonic basins in much older terranes—such as present-day sediments in Lake Superior, Hudson Bay, Great Slave Lake, and the Baltic Sea (Figure 2.6)—may have highly radiogenic lead. A notable district that illustrates this point, although it is not of a lateral secretion origin, is the moderate sized Wood River district in upper Paleozoic argillites adjacent to the Mesozoic Idaho batholith (Hall and Czmanski, 1972; Hall, Rye and Doe, 1978). The ore of this district, which has been interpreted to be of a magmatothermal origin (where igneous activity appears to be the source of heat in the ore forming process, but may or may not be the source of the metals in the ore fluids), is highly radiogenic like lead in the intracratonic basins or lead leached from single-cycle sediments. Heavy sulfur isotopic compositions suggest that ore in the Wood River district also more probably had an epicontinental than igneous origin. It is doubtful whether the nearby igneous rocks contributed significantly to the lead in the ore deposits either through magmatic fluids or through postmagmatic leaching processes associated with circulating cells of meteoric water. Most of the data on recent sediments in intracratonic basins concern basins surrounded by Archean rocks. The fact that the data on the ores of the Wood River district plot with these data on sediments suggests the lead in the ores was derived from radiogenic source material (perhaps detritus) of Archean age. The lead acquisition stage was therefore outside the stocks. Archean basement that might furnish appropriate source material is known in the region.

Carbonate host rocks. Carbonate rocks constitute an unusually good environment for deposition of base metal ore deposits, especially lead and zinc, and more than one mode of ore formation is to be expected. Those ores genetically related to igneous activity are discussed in the section on rejuvenated craton. For the remaining deposits, the lead isotope patterns are similar to those found for mineral deposits in host rocks composed predominantly of silicates. Sufficient data are available for deposits in the carbonate environment to show that there is a continuum in $^{206}Pb/^{204}Pb$ and in $^{208}Pb/^{204}Pb$ values, which runs (1) from deposits with leads that approximate orogene or single-stage model leads (Pine Point, Bleiberg), such as those represented by area b in Figure 2.3, (2) through those that have leads only slightly radiogenic for their ages (Phenixville, Um Gheig), which plot in area c of Figure 2.3, (3) to those that are highly radiogenic for their ages, such as the deposits of the Mississippi Valley, represented by area c' in Figure 2.3.

Beales and Jackson (1968) favored an early diagenetic, lateral-secretion origin for the Pine Point deposits. The lead isotopes are compatible with such an origin, in which the epicontinental basin incorporated multiply reworked sediments far from their source, thereby approximating an orogene.

The deposits of the Mississippi Valley are characterized by highly radiogenic leads with $^{206}Pb/^{204}Pb > 19$, and $^{208}Pb/^{204}Pb > 39.5$ (East and Central Tennessee, Illinois-Kentucky, Wisconsin-Illinois-Iowa, Southeast Missouri, and Tri-State districts). Of this group, only the Illinois-Kentucky fluorspar district is suspected of having any significant igneous involvement in its formation, and it does contain some of the least radiogenic leads of the group. Various workers have preferred the age of ore deposition for ores of the Mississippi Valley to be close to the age of formation of the host rocks (for East and Central Tennessee, see Wedow, 1971; for Southeast Missouri, see Beales et al., 1974). Most lead isotope investigators, however, have traditionally interpreted the mineralization to be young. The reason for this is that the regression of $^{207}Pb/^{204}Pb$ versus $^{206}Pb/^{204}Pb$ (approximately 0.09) is what is expected from lead derived from the local basement as the mineralization age approaches the present day (Heyl et al., 1966). Doe and Delevaux (1972), however, found that even though the basement rocks they analyzed had lost a significant amount of lead, there is a systematic difference between the $^{207}Pb/^{204}Pb$ values of basement and ore leads. They interpreted this difference to mean that the basement had contributed only subordinate amounts of lead, at most, to the ores. They did find that lead isotope trends in present-day lead were nearly duplicated in ore from the Upper Cambrian Lamotte Sandstone and in one of three samples of Upper Cambrian Bonneterre Dolomite, host rock of the ore. They concluded that lead isotope trends in present-day lead were nearly duplicated in ore from the Upper Cambrian Sandstone fairly recently, which permitted 1.4-b.y.-old source rocks known to underlie the area to contribute lead to the shelf sediments.

In an early-diagenetic interpretation, mineralization would have occurred about 500 m.y. ago. Such a hypothesis changes interpretation of the lead isotope data considerably. A syngenetic origin is still excluded because lead in the carbonate host rocks is barely radiogenic enough today to contribute lead to the ores. Considering the amount of radiogenic lead generated from uranium since 500 m.y. ago, lead in the carbonates at that time would have been sufficiently radiogenic to account for few, if any, of the ore samples. With such an old mineralization age, the age of the source material is calculated to be approximately 1.05 b.y. This age of source material further excludes leaching from the known basement, which is older than this. The relationship of the Lamotte Sandstone to ore becomes less clear, but Ojakangis (1963) thought the sandstone was derived from clastics of Keweenawan rocks (a billion years old) to the north. The regressions in this case, ignoring the Cambrian sedimentation event, should be approximately 0.07, rather than the 0.08–0.09 observed. Complicated models could be derived, however, whereby the sedimentation event did not disturb the original regression, so that local sources are not necessarily excluded. The sources also could well be the host rocks of basinal brines, as was suggested by Beales et al. (1974). There is one last bit of information that might support a billion-year age for source material and a 500 m.y. age for mineralization; that is, the values of $^{207}Pb/^{204}Pb$ trends of the ore are greater than those of the 1.4-b.y.-old basement. The systematics of lead isotopes are such that the observed difference would be expected if the ore were derived from billion-year-old source material rather than 1.4-b.y.-old source material. If basins are involved, the lead isotope data suggest the basins involved are more likely to the east rather than the west, because little billion-year-old basement is known west of Missouri to provide possible sources of appropriate detritus to the basins, whereas the Grenville Province to the east and the clastics of the Keweenawan to the north might provide abundant detritus of the appropriate age. The lead isotopic composition of the samples from the Tri-State district is compatible with a mineralization age approaching the depositional age of the Mississippian host rocks (approximately 320 m.y.), utilizing the same billion-year-old source material as did the Southeast Missouri deposits. Unfortunately, the difference between the 320-m.y. and 500-m.y. mineralization ages would cause only a difference of 0.2% in $^{207}Pb/^{204}Pb$ at an equivalent $^{206}Pb/^{204}Pb$, which is only twice analytic uncertainty. Data from the Illinois-Kentucky district have a steeper regression, reflecting a significantly different derivation. There, the basins involved may be different and the possibility of complications from igneous activity exists. For the

Southeast Missouri and Tri-State deposits, a valid independent determination of the age of mineralization could have a profound effect on the interpretation of the lead isotopes.

Ore deposits with lead isotope characteristics similar to deposits of the Mississippi Valley are not common elsewhere (see Cannon and Pierce, 1967). The best documented examples are the ores of the Wood River district (discussed on p. 00), the Sierra Nacimiento district (see p. 00), the deposits in the Hansonberg district of New Mexico, and the stratiform Um Gheig and associated deposits, the youngest examples, in Miocene sediments of Egypt (Figure 2.5). The mineralization at Hansonberg may be related to uplift of a lineament (Slawson and Austin, 1962) that caused brines to flow. Um Gheig, because it must be younger than early Miocene (<25 m.y.), belongs to the syngenetic/diagenetic category by our age definition. The Um Gheig leads are isotopically similar only to the least radiogenic leads, however, in the major lead-bearing deposits of the Mississippi Valley, and the lead in the Egyptian deposits would appear to be derived from highly evolved sediments approaching conditions in the orogene. The isotopic composition of the lead in the Egyptian deposits is, in fact, similar to that in present-day basinal brines in highly evolved sediments elsewhere in the world.

REJUVENATED CRATON

The isotopic composition of lead in magmatothermal ore deposits within the rejuvenated cratonal environment varies broadly. In individual districts, even where galena is a major phase of the deposits, values of $^{206}Pb/^{204}Pb$ typically range from 17.0 to 19.1, and noneconomic mineralizations can vary even more. The range for economic deposits, however, tends to be more restricted at the radiogenic end. Among the epigenetic deposits of the Western United States, for example, all Mesozoic and Cenozoic base metal ores have $^{206}Pb/^{204}Pb$ values of 19.1 or less (Figure 2.7), except for samples from the moderately economic Wood River district, Idaho, and the epicontinental basinal-platform deposits of the Hansonburg district, New Mexico.

A surprising number of large Mesozoic and Cenozoic deposits of this category in the Rocky Mountain and Colorado Plateau Regions ("main spectrum" deposits in Figure 2.7 herein; also area I of Zartman, 1974 and Figure 2.8 herein; area d' in Figure 2.3) contain leads that are nonradiogenic for their age and have primary isochron model lead ages averaging approximately 500 m.y. These include not only representatives of the porphyry copper type (Bisbee, Arizona, $^{206}Pb/^{204}Pb$ ~ 17.1; Bingham Canyon, Utah, $^{206}Pb/^{204}Pb$ ~ 17.7; and Butte, Montana, $^{206}Pb/^{204}Pb$ ~

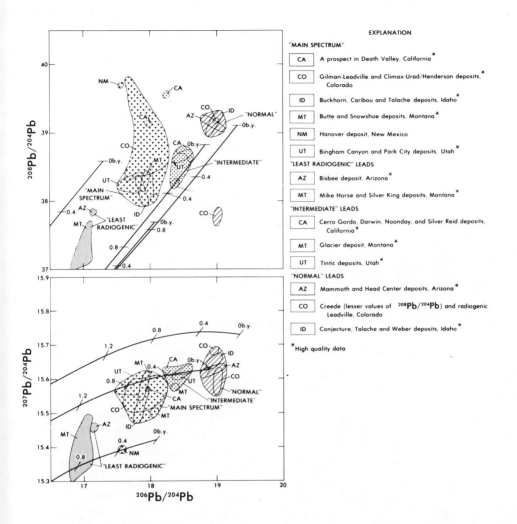

Fig. 2.7 ^{207}Pb/^{204}Pb and ^{208}Pb/^{204}Pb versus ^{206}Pb/^{204}Pb for young ore deposits in a rejuvenated craton (Rocky Mountain type). All deposits are Mesozoic or Cenozoic. Samples marked (*) were analyzed by the lead tetramethyl or triple-filament techniques and their values are thought to be within 0.1% of the absolute ratios; other data are believed to be within 1% of the absolute ratios. Some data from Gilman, Colorado have excessively great values of ^{207}Pb/^{204}Pb and are out of the diagram; these data are not of modern quality and little attention should be paid to the ^{207}Pb/^{204}Pb values. All data are normalized to absolute values where possible, but much of the data not marked (*) had to be used as published; however, for the purposes of the present discussion the data are satisfactory except for the ^{207}Pb/^{204}Pb values. Lines are as stated in Figure 2.3 caption. References to data: Delevaux, Pierce, and Antweiler (1966), Doe and others (1968), Doe and Stacey (1974), Russell and Farquhar (1960), Radabaugh, Merchant, and Brown (1968), Stacey, Zartman, and Nkomo (1968), Zartman (1974), Zartman and Stacey (1971), and this paper.

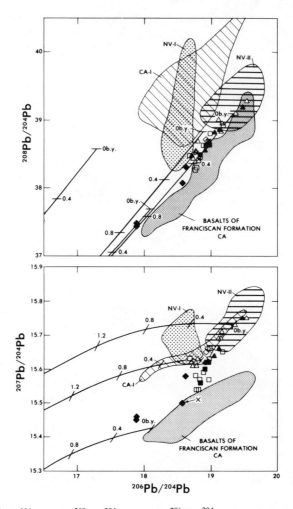

Fig. 2.8 $^{207}Pb/^{204}Pb$ and $^{208}Pb/^{204}Pb$ versus $^{206}Pb/^{204}Pb$ for Mesozoic and Cenozoic igneous rocks and ores of California (CA), Oregon, and Nevada (NV) that are in Area III of Zartman (1974) including data on basalts from the Franciscan Formation of California. Also included are fields for data from Area II of Nevada (NV-II) in addition to fields for Area I of California (CA-I) and Nevada (NV-I) taken from Zartman (1974). All analyses are of the highest quality and obtained by the silica-gel or triple filament methods. Lines are as stated in the Figure 2.3 caption. The symbol ($\times$) refers to present day data before correction to initial ratios. *Cenozoic and Mesozoic:* ◊ Cenozoic volcanic rocks of Mount Shasta. ▫ Other Cenozoic igneous rocks of California and Oregon. △ Mesozoic igneous rocks of the Sierra Nevada batholith and the Salinian block. ■ Ores of Oregon: Accident claim, Gaylord tunnel, and Musick prospect. ▲ Ores of California and Nevada: Bodie, Kernville and Zaca mines of California; Dominion, Leadville and Union mines of Nevada. *Pre-Cenozoic:* ◆ Igneous rocks and ores of the Shasta area. References to data: Church and Tilton (1973), Doe and Delevaux (1973), Sinha and Davis (1971), Zartman (1974), and this paper.

18.0) and "Moly ores" (Climax and Urah/Henderson, Colorado), but also those in carbonate rocks closely associated with igneous activity (Leadville/Gilman, Colorado, $^{206}Pb/^{204}Pb \sim 17.9$; Hanover, New Mexico, $^{206}Pb/^{204}Pb \sim 17.6$; Park City, Utah, $^{206}Pb/^{204}Pb \sim 17.9$). Production from many of these "main spectrum" deposits has been in the hundreds or thousands of millions of dollars, so the category includes the "nation savers."

To elaborate on the deposits in Colorado, the main ore at Gilman, Colorado (Radabaugh et al., 1968) is in Mississippian and Devonian carbonate rocks with subordinate production from the Cambrian Sawatch Quartzite, although some veins extend down into the underlying Precambrian basement. The lead isotopic composition of the carbonate rocks ($^{206}Pb/^{204}Pb \sim 21.22$) is similar to that of the Paleozoic carbonate rocks in Missouri, Nevada, Wyoming, and elsewhere (see Figure 2.6), and is much more radiogenic than the ore ($^{206}Pb/^{204}Pb \sim 17.9$). Igneous rocks are yet to be analyzed from the area, but the isotopic composition of the ore is typical of Mesozoic and Cenozoic igneous rocks elsewhere in Colorado. The Gilman sill of Laramide age, which overlies and predates the ore, is altered but not mineralized. Similarly the major production at Leadville, Colorado (Tweto, 1968), has been from carbonate units above the Sawatch Quartzite, commonly on the underside of Cenozoic sills older than the ore, although some veins extend down into the Precambrian strata. Only a few of the intrusives are mineralized themselves. The isotopic composition of lead at Leadville is similar to that at Gilman. Even though these ores, with a combined production in excess of 750 million dollars, are largely in carbonate rocks, the lead isotopic data indicate these deposits clearly belong to the magmatothermal classification rather than to the basinal-platform category. The intrusions responsible for the mineralization underlie the area at some unknown depth.

Much or all of the lead in these deposits, regardless of the kind of host rock involved, has been derived either directly from the intrusive rocks (perhaps through alteration by circulating heated ore fluids if not by magmatic differentiation) or from the Precambrian basement which the circulating heated ore fluids traversed. The lead isotopes, for the most part, appear to reflect a deeper ultimate source for the lead than was inferred for the deposits in epicontinental environments. This source, presumably the lower crust (see d', Figure 2.3), has isotopic characteristics similar to rocks depleted in uranium as a result of granulite facies metamorphism, unlike the upper crustal uranium-enriched sources for deposits in epicontinental environments.

Although few of the deposits in this grouping have $^{206}Pb/^{204}Pb$ values in excess of 19.1, not all have leads that are grossly nonradiogenic for their age. Those deposits with "normal" lead isotopic ratios tend to be smaller than the "main-spectrum" deposits, rarely if ever producing more than 150 million dollars worth of ore. The major ore deposits of the Tintic district, Utah, are in carbonate rocks and have values of $^{206}Pb/^{204}Pb \sim 18.3$-18.8. The major ores at Creede, Colorado are in Cenozoic volcanic rocks of the San Juan Mountains and are among the most radiogenic ($^{206}Pb/^{204}Pb \sim 19.1$) of the genre, more radiogenic in fact than any Mesozoic or Cenozoic igneous rock in the Rocky Mountain provinces, which have values of $^{206}Pb/^{204}Pb \le 18.8$. As has been pointed out by Doe and Stacey (1974), and specifically for Utah by Stacey et al. (1968), however, numerous prospects and marginal ore deposits in the magmatic hydrothermal districts have values of $^{206}Pb/^{204}Pb > 19.1$. The best of these—the Wood River district of Idaho with $^{206}Pb/^{204}Pb = 19.5$-20.4 (Figure 2.6)—has a total recorded production of only 38 million dollars distributed over many mines (Hall et al., 1978). The circulating systems in the magmatic hydrothermal regime appear to be either too small and/or too hot to form major base metal ore deposits with isotopic compositions like those of some low-temperature, lateral-secretion ores. A small system would not gather sufficient quantities of radiogenic lead, whereas an especially hot system might react with feldspars, which will contain nonradiogenic lead.

$^{208}Pb/^{204}Pb$ values are more variable than $^{206}Pb/^{204}Pb$ values. There is a tendency for $^{208}Pb/^{204}Pb$ to be radiogenic and to plot above the orogene evolution curve, which is interpreted to represent the depletion of uranium in the lower crust. Only ores from Creede, Colorado, plot distinctly below the evolution curve for thorogenic lead and might be confused with oceanic deposits. This association between ores and volcanic rocks in Southeast Colorado and oceanic rocks presumably means that the lower crust and continental lithosphere were never raised to granulite facies or higher rank metamorphism (Lipman et al., 1978). Plotting above the curve seems to be a characteristic cratonic feature that apparently evolves through depletion of uranium in the lower crust and corresponding enrichment in the upper crust during granulite facies and higher rank metamorphism.

We are not aware of any economic base metal deposits of the magmatic hydrothermal type with values of $^{208}Pb/^{204}Pb > 40$ except for the marginal Wood River district, although some prospects have such values (for examples in Utah, see Stacey et al., 1968; for Colorado, see Delevaux et al., 1966; for Montana, see Mudge et al., 1968).

NONCRATONIZED UPPER CRUST

A third continental environment is represented by the upper bounding curve and the area c of Figure 2.3. This case is equivalent to area II of Zartman (1974). As a prime example, he used the ores associated with

Mesozoic and Cenozoic igneous rocks in much of the Paleozoic miogeosynclinal terrane of Nevada (Figure 2.8). Base metal deposits in this environment tend to be moderate in size; however, lead from the major gold ore at Carlin, Nevada is represented. The lead isotopic compositions are rather uniform, with $^{208}Pb/^{204}Pb$ values that plot along or under the evolution curve and have a tendency toward great values of $^{207}Pb/^{204}Pb$, perhaps up to 15.8. The igneous rocks have these characteristics either because they were derived from upper crust consisting predominantly of the erosional component from the continents, or because the magmas were contaminated by such crust. Certain igneous rocks and ores in this group are not clearly part of the miogeosynclinal assemblage, however, and they may have risen through thick assemblages of Precambrian basement. Examples are the Sierra Nevada batholith of California north of the Garlock fault, and igneous activity on the east side of the batholith in the Mono Craters area of California. This hypothetical Precambrian source might develop the appropriate lead isotopic composition if it were never cratonized—never metamorphosed above upper amphibolite facies—or if it is composed of rocks more silicic than andesite or quartz diorite. Because of this uncertainty concerning the source material primarily involved in this region, we will not distinguish between Paleozoic miogeocynclinal and Precambrian sources; rather, we will group both together as noncratonized upper crust. The category is important because it was largely the recognition of these great $^{207}Pb/^{204}Pb$ values by Zartman (1974) that prompted the development of the model presented in this paper.

The patterns in the plot involving $^{208}Pb/^{204}Pb$ distinguish the noncratonized upper crust from the cratonized crust almost as well as for the plot involving solely uranogenic lead ($^{207}Pb/^{204}Pb$).

Island and Continental Arc Environments—The Orogene

SUBMARINE VOLCANIC "EXHALATIVE"

Submarine volcanic "exhalative" massive sulfide deposits in the island arc grouping (Figure 2.9) have played a major role in lead isotope geochemistry because they formed the basis for the development of the hypothesis concerning single-stage evolution of lead isotopes by Stanton and Russell (1959). In treating these deposits, we will accept deposits so classified by Stanton and Russell (1959) and Hutchinson (1973). Deposits in the island arc regime are frequently called Kuroko-type after the youngest occurrence, and unlike the Cyprus-type deposits, they are associated with silicic volcanism and may have considerable lead accompanying zinc in deposits on the flanks of submarine volcanoes (Kuroko, Japan and Buchans,

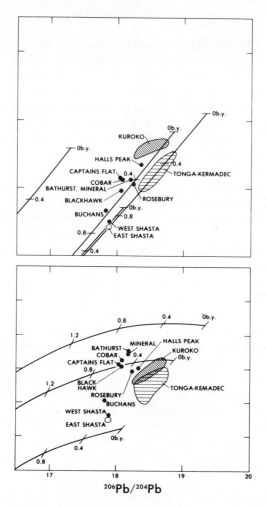

Fig. 2.9 $^{207}Pb/^{204}Pb$ and $^{208}Pb/^{204}Pb$ versus $^{206}Pb/^{204}Pb$ for Phanerozoic "submarine volcanic exhalative" massive sulfide deposits in the island-arc environment, compared with volcanic rocks of the primitive Tongo-Kermadec arc for comparison. Massive sulfide deposits are: the Miocene Kuroko deposits from Japan (including the Cretaceous Taro mine of similar nature); the Triassic East Shasta ores, California; the Upper Permian(?) Halls Peak deposit, New South Wales, Australia; the Devonian West Shasta deposits, California; the Silurian Cobar and Captains Falt deposits, New South Wales, Australia; the Ordovician deposits at Bathurst, New Brunswick, Canada; the Ordovician to Silurian(?) deposit at Buchans, Newfoundland, Canada; the Cambro-Ordovician deposits near Mineral, Virginia, and tentatively those at Blackhawk, Maine; and the Cambrian Rosebery deposit, Tasmania, Australia. As before, data on rocks were produced by the precise and accurate "emitter technique" or "double-spike technique" only. Ore data were derived by the precise and accurate thermal emission and lead tetramethyl "looping" technique (ratios thought to be within 0.1 percent of absolute). Explanation of lines are as in Figure 2.3 caption. References to data: Oversby and Ewart (1972), Russell and Farquhar (1960), Stacey, Delevaux, and Ulrych (1969), and this paper.

Canada), especially those of Phanerozoic age (Sangster, 1972; Lambert and Sato, 1974). Stanton and Russell showed that the evolution of lead isotopes and Kuroko-type deposits has been sufficiently systematic throughout geologic time to permit approximate ages of ore formation to be calculated through so-called model lead ages. A "best model lead-age single-stage" evolution model has recently been published (Doe and Stacey, 1974) that permits "formation ages" to be calculated with an uncertainty of approximately 150 m.y. This is as good a single-stage model as can be constructed for uranogenic lead. An unsatisfactory feature of the model is that it requires an older "age of formation" of the Earth for the ^{232}Th/^{204}Pb system (4.57 b.y.) than for the ^{238}U/^{204}Pb system (4.43 b.y.), a relationship that can be explained by a younger differentiation. The two-stage model of Stacey and Kramers (1975) offers some improvement of model ages by suggesting that a change in ^{238}U/^{204}Pb and ^{232}Th/^{204}Pb occurred approximately 3.7 b.y. ago. The 3.7-b.y. event is probably related to the time of formation of the first permanent crust and, in fact, 3.7 b.y. would fit better in our own model than 4.0 b.y. Unfortunately, the 400 m.y. intervals in our approximation make this difficult.

The new data on the Devonian and Triassic deposits of the Shasta mining district of California and middle Paleozoic deposit at Buchans in Newfoundland are compatible with an oceanic mantle evolution, because they plot near the lower-bound asthenosphere evolution curve (area a of Figure 2.3) thought to approximate the mantle. The conclusion is in agreement with the generally ensimatic nature of the geology. The published data on ores from Bathurst, New Brunswick, Canada, and Cobar, New South Wales, Australia, and the new data on Mineral, Virginia, and Blackhawk, Maine, are above the orogene evolution curve in ^{207}Pb/^{204}Pb, which may reflect an unusually great involvement of erosional product from the continents (area c of Figure 2.3). This conclusion is reasonable in view of the ensialic nature of the environments. Inasmuch as Bathurst and Cobar are the two largest known Phanerozoic deposits in the class, great ^{207}Pb/^{204}Pb values in submarine volcanic exhalative deposits in orogenes may eventually be found to have some economic significance. That is, the largest deposits may be in the ensialic parts of the orogene. No major deposits have yet been found in the ensialic submarine volcanics of the Appalachian provinces of Virginia or Maine; however, these deposits tend to cluster, and the great days of the Virginia and Maine mining camps may well be in the future rather than in the past. Other deposits are intermediate in ^{207}Pb/^{204}Pb between mantle and pelagic sediments (area b of Figure 2.3). It is tempting to suggest that these leads were derived from a mixture of oceanic mantle material and pelagic sediments, with a distinct bias toward the sediments, except for the Shasta and Buchans deposits which do appear to have predominantly mantle-derived lead.

Originally these deposits were believed to have lead derived from the oceanic mantle (Stanton and Russell, 1959; Ostic et al., 1967); however, data from most of these deposits do not fit an oceanic mantle trend (Richards, 1971; Russell, 1972) and are more compatible with the sort of lead isotope evolution that resulted in modern-day pelagic sediments (Armstrong, 1968; Brown, 1965; Figure 2.2 of this report). Even Richards' (1968) candidates for mantle-derived leads (Rosebury and Hall's Peak, Australia) fit our intermediate category better than an oceanic mantle category. Oversby (1974) developed a "least radiogenic oceanic mantle" evolution model much like our "asthenospheric" lead isotope evolution and with uranium-lead characteristics much like the first stage of the Stacey and Kramers (1975) model. However, she could not explain lead isotope evolution in Phanerozoic deposits of the island-arc submarine volcanic exhalative category, which is now explained by our observation that lead isotope evolution is rarely the same in orogenes as it is at spreading centers. In primitive island arcs (primitive orogenic belts), such as Tonga-Kermadec (Oversby and Ewart, 1972), remote from continents, contamination of magma with pelagic sediments or brines that have leached such sediments appears to be subordinate or absent. Samples from the island of Eua in the Tonga Islands are isotopically similar to pelagic sediments and might be "totally" contaminated with pelagic sediment lead; however, there is some question whether Eua belongs to the arc. In mature island arcs such as Japan, the picture is more complex. Data from the Kuroko-type ores follow the steep regression equivalent to a source age of approximately 3.5 b.y. for uranogenic lead, as noted by Tatsumoto (1969) for Pliocene and Quaternary volcanics, and more recently by Sato (1975) for the Kuroko ores; however, in the plot involving $^{208}Pb/^{204}Pb$, the data do not intersect oceanic mantle values. Because no rocks of such great antiquity are known in or near Japan, the great regression is probably an artifact of mixing, as suggested by Tatsumoto (1969). However, if one component of the mix is pelagic sediments, the other may be continental material rather than oceanic mantle. Thus the volcanic rocks and ores in the mature island-arc regime appear to be contaminated or derived largely from sources other than the oceanic mantle. There is no known conflict between this conclusion and the hypothesis of Sangster (1972) that the ore metals are derived through alteration of the volcanic rocks or volcanoclastic material associated with the ore. We therefore separate the island-arc submarine volcanic exhalative category into two subdivisions:

1. Primitive arc (Tonga-Kermadec): East and West Shasta and perhaps Buchans are examples.
2. Mature arc (Japan): Kuroko deposits are examples.

Irregularities in lead isotope evolution as observed in the orogene are not surprising, because individual orogenes are of limited scope and different mixtures are to be expected; for example, the Bathurst deposit, in Middle Ordovician rocks and the Maine and Virginia deposits in Cambro-Ordovician rocks are more radiogenic in their lead isotopes than are the Silurian deposits of Australia. More impressive is the great efficiency of the mixing in individual orogenes spread over the world.

NONSUBMARINE

Deposits in this category are abundant in Mesozoic and Cenozoic strata. Among the most notable examples are the huge Andean porphyry copper belt (for which analyses are unfortunately lacking), a few large zinc-lead deposits such as the Tertiary Casapulca deposit of Peru, the tin deposits on the island of Belitung, Indonesia, and the Taishu mine which is the oldest operating mine in Japan. Other deposits in this group tend to be small and less well known.

Nonsubmarine ores from island arcs and continental margins (Figure 2.10) usually display lead isotope evolution similar to that of the submarine volcanic exhalative massive sulfide ores. The primary difference is that, unlike the submarine volcanic exhalative deposits, the epigenetic mineral deposits have representatives with isotopic values that plot above the orogene evolution curves for both uranogenic and thorogenic lead plots; these trends are characteristic of ore from continental Precambrian environments (areas c and d of Figure 2.3). These trends need not reflect the Precambrian basement but may indicate variable quantities of detritus of continental Precambrian material entering the orogene. Examples of deposits showing these trends follow; some of these may be distinguished on both lead isotope plots. The great ^{208}Pb/^{204}Pb values of the Taishu deposit in Japan and the Sassak, Sulawesi, porphyry copper deposit in Indonesia, when compared to their ^{206}Pb/^{204}Pb values, indicate continental crust rather than oceanic mantle sources for these deposits. The ^{207}Pb/^{204}Pb values of the tin deposits of Belitung, Indonesia (which are on the continental shelf), the Gunung Sawal Pb-Zn vein, and the South Banten gold deposit of Java also indicate a "continental character" for those localities. The results are rather a surprise, perhaps, in regard to the Gunung Sawal area of Java, and in regard to Sassak, Sulawasi. The nonsubmarine deposits in mature island arcs such as Japan and Indonesia are seen to be much more similar to those deposits well onto continental margins rather than to those deposits formed in the submarine environment. Regressions on the uranogenic lead plot are steep, equivalent to source material ages older than any rocks known in the area, somewhat like that observed for the Kuroko

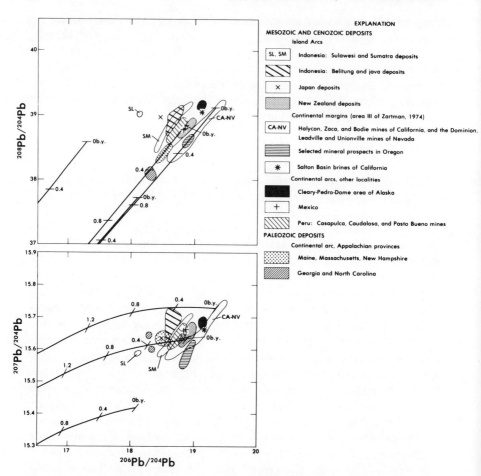

Fig. 2.10 $^{207}Pb/^{204}Pb$ and $^{208}Pb/^{204}Pb$ versus $^{206}Pb/^{204}Pb$ diagrams for non-submarine (epigenetic) mineral deposits of island and continental arcs. All analyses are of comparable quality and were performed by the thermal emission method except those for the Salton Basin brines. The plotted value for these brines represents the average of 6 analyses by the old surface emission technique. All data are normalized to absolute ratios where necessary. Lines are as stated in Figure 2.3 caption. References to data: Cooper and Richards (1969b), Doe and Delevaux (1973), Zartman (1974), and this paper.

ores but shifting to slightly greater values of $^{207}Pb/^{204}Pb$. The Sassak deposit, in fact, has lead indistinguishable from that in the rejuvenated craton category. Of the epigenetic deposits, only the marginally economic deposits of Oregon might be confused with oceanic deposits, and only some of the deposits on Sumatra might be confused with Kuroko-type deposits on the basis of lead isotopes. In general, lead isotopes successfully show mature island arcs to be more related to continents through the orogene than to the ocean basins. Although a more exhaustive study is desirable on metalliferous sediments of the Nazca plate, existing lead isotope data (shown on Figure 2.4) do not immediately confirm that they could have supplied a major portion of lead to Andean ore deposits (those of Peru on Figure 2.10) through subduction. As in mature island arcs, a more intimate continental association is suggested by the lead isotope data.

Model lead ages on many nonsubmarine deposits (those analyzed from Peru, for example) are every bit as good as those for submarine volcanic exhalative deposits (± 0.15 b.y.). In California, Oregon, Washington, Alaska, and the relevant part of Nevada (area III of Zartman, 1974), however, the primary isochron ages of the deposits tend to be negative or in the future by approximately 0.2 b.y. on the average with an uncertainty of approximately ± 0.2 b.y.; for $^{208}Pb/^{204}Pb$ the average is just about 0.0 m.y. or the present day with an uncertainty of about 250 m.y. skewed to ages that are too young.

A number of Paleozoic epigenetic deposits from the Appalachian provinces are included on Figure 2.10. These deposits are all more radiogenic than the Cambro-Ordovician massive sulfides of the Appalachian provinces (Bathurst, New Brunswick; Blackhawk, Maine; and Mineral, Virginia). This isotopic difference is roughly compatible with the known difference in age between the massive sulfide and epigenetic deposits. The Paleozoic and early Mesozoic epigenetic deposits of the Appalachian provinces therefore conform to the findings on younger orogenes, and there is no isotopic evidence for involvement of cratonized Precambrian material. Unfortunately, the geologic relations are generally not well enough known, because of metamorphic and structural complexities, to warrant discussion in any greater detail at this time.

PRIMITIVE ISLAND ARC-CONTINENTAL MARGIN TRANSITION

The Klamath Mountains furnish an excellent locality in which to study what happens when a primitive island-arc environment is incorporated into a continental margin. The oldest known rocks are Ordovician in age and possibly are composed of oceanic floor and island-arc materials (Hamilton, 1969). In the island-arc sequences of Devonian to Late Jurassic age, submarine volcanism and plutonism resulted in the formation of the massive sulfide deposits of West and East Shasta, California. There is no suggestion in either lead or strontium isotopes (this paper; Doe and Delevuax, 1973; Kistler and Peterman, 1973) that any component other than oceanic mantle or mafic crustal lead was involved until sometime in the Jurassic, when lead isotope values began to fluctuate between oceanic mantle and pelagic sediment values during the magmatic and ore-forming process. The possible contribution of mafic crust is mentioned in view of the conclusion of Arth and Hanson (1972), based on trace element studies, that the Jurassic trondjhemite of the Shasta area was derived from a basaltic eclogite. Nonsubmarine igneous activity in this area began in the Middle Jurassic and is represented by a sample from the Halycon Mine (Table 2-2), a mineral deposit possibly related to the Shasta Bally batholith of this age. The lead isotopic compositions of the sample from this mine and the younger volcanic rocks (Figure 2.8) are quite different from the older ore and igneous rock samples, except for one Upper Jurassic trondjhemite, which has lead similar in isotopic composition to lead in pelagic sediments. The significance of this similarity is not subject to unique interpretation, but the cycling of lead isotopic composition between the compositions of oceanic mantle and pelagic sediment is also commonly observed to the north, in the Cascade Mountains of Oregon (Church and Tilton, 1973).

The strict oceanic mantle character of lead isotopes in the older submarine volcanic rocks and ores (the primitive island arc stage) of the Shasta area, with little pelagic sediment involvement, is in accord with the observations of Oversby and Ewart (1972) in their study of the primitive Tonga-Kermadec arc complex. By the time the Shasta area emerged from the ocean, something drastic had happened, as if lead from pelagic sediments were being mixed with some of the magmas and ore bodies. Subduction of unconsolidated pelagic sediments is thought to be impossible, but Hamilton (1969) has suggested that the Paleozoic and Triassic submarine section could have been subducted (a possibility that is subject to testing by isotopic means). The probability that such rocks were well lithified, a process that apparently takes 100-200 m.y., may have permitted their subduction. Whether an island arc has essentially oceanic mantle or pelagic sediment leads in its igneous rocks and ores may well depend on the kind of lithified material being subducted.

Other mechanisms provide possible explanations for the extreme isotopic mixing, and we mention them without elaboration in view of the uncertain nature of plate tectonics. One possibility would be downwarping of the Paleozoic and Triassic section to the appropriate horizon where melting occurs around Middle Jurassic time, either as a result of a subduction process, or by other means. Another possibility is that the Shasta area overlaid an oceanic mantle until sometime in the Jurassic when it drifted over a continental mantle, where it has remained. Few basalts in the oceans have lead like that found in the sample from the Halycon mine and in the volcanic rocks of Mt. Shasta (the volcanic rocks of Gough Island on the Mid-Atlantic Ridge are probably an exception), so oceanic mantle alternatives seem unlikely to produce the kind of lead found in pelagic sediments.

The Cretaceous plutons in the northern part of the Sierra Nevada batholith, including a trondjhemite, have lead isotopic compositions much like those found in the continental margin domain in the Klamath Mountains area. Cycling of lead isotopic compositions down to oceanic mantle values has not yet been found in the early plutons of the Sierra Nevada batholith, but its discovery might yet be anticipated. This prospect is supported by data on the Franciscan Formation. Erosion of early phases of the batholith is thought to have produced the sediments of the Franciscan Formation, and the isotopic data of Sinha and Davis (1971) on these sediments spans most of the spectrum from oceanic mantle values to the most radiogenic values. The plutons and ores with values of $^{207}Pb/^{204}Pb$ greater than those shown by pelagic sediments apparently have been influenced by a thick section of Precambrian rocks with isotopic characteristics similar to the "erosional component." A particularly attractive feature of the model presented herein is that it nicely explains the steep regressions found in data on ores and igneous rocks formed in the orogene through mixing of variable proportions of asthenospheric, upper crustal and lower crustal lead.

ACKNOWLEDGMENTS

We wish to thank the following people for their interest, for discussions, and for contributing samples (locations of these samples are given in parentheses): Jacob E. Gair (North Carolina and Virginia), Wayne E. Hall (East and West Shasta, California, and the Wood River district, Idaho), Allen Heyl (Pitcher district, Oklahoma), Frank Lesure (Georgia), George Phair (Colorado), Bruce L. Reed (Belitung, Indonesia), Robert O. Rye (Peru), Andrew F. Shide (Massachusetts), Harold K. Stager (Kernville deposit and Halycon mine, California), and Harry A. Tourtelot (Cyprus), all of the U.S. Geological Survey; Jeffrey Abbot (Bisbee, Arizona) of Phelps Dodge Corporation; Cyrus Field (Oregon) of Oregon State University; Robert B. Forbes (Alaska) of the University of Alaska, P. R. L. Browne (New Zealand) of the New Zealand Geological Survey; Mel Jones (Belitung, Indonesia), University of Adelaide; Arthur Kinkel (Maine and New Hampshire), consulting geologist, St. Petersburg, Florida; Joseph Kowalik, University of Minnesota (Buchans, Newfoundland); K. Sato and T. Tatsumi (Kuroko ores, Japan) of the University of Tokyo; A. Sasaki and T. Sato (Taishu Mine) of the Geological Survey of Japan; Totong Suhanda (Indonesia) of the Direcktorat Geologi, Indonesia; Hans Wedepohl (Bleiberg-Krueth) of der Universität Gottingen, Germany; Lee A. Woodward (Nacimiento, New Mexico) of the University of New Mexico. We particularly have benefited from discussions with Peter W. Lipman of the U.S. Geological Survey, who pointed out that the Paleozoic and Triassic setting of the Shasta mining districts was similar to Tonga-Kermadec; Ronald W. Kistler of the U.S. Geological Survey who so patiently discussed the general highly complex tectonics of California; Warren B. Hamilton of the U.S. Geological Survey for many valuable discussions on Indonesia; and Richard Hutchinson and Bruce Bouley of the University of Western Ontario, for discussion of massive sulfide deposits in Maine.

REFERENCES

Armstrong, R. L. (1968) A model for the evolution of strontium and lead isotopes in a dynamic earth: *Rev. Geophys.*, **6**, 175-1991.

―― and S. M. Hein (1973) Computer simulation of Pb and Sr isotope evolution of the Earth's crust and upper mantle: *Geochim. Cosmochim. Acta*, **37**, 1-18.

Arth, J. G. and G. N. Hanson (1972) Quartz diorites derived by partial melting of eclogite or amphibolite at mantle depth: *Contrib. Mineral. Petrol.*, **37**, 161-174.

Atomic Energy Commission (1962) *Atomic Energy Commission Advisory Committee on Standards*: Report No. 11, July 18.

Beales, J. C., J. C. Carracedo, and D. W. Strangeway (1974) Paleomagnetism and the origin of Mississippi Valley-type ore deposits: *Can. J. Earth Sci.*, **11**, 211-223.

―― and S. A. Jackson (1968) Pine Point—A stratigraphical approach: *Bull. Can. Inst. Min. Metall.*, **61**, 867-878.

Brown, John S. (1965) Oceanic lead isotopes and ore genesis: *Econ. Geol.*, **60**, 47-68.

Cannon, R. S. and A. P. Pierce (1967) Isotopic varieties of lead in stratiform deposits; Genesis of Stratiform Lead-Zinc-Barite-Fluorite Deposits (Mississippi Valley Type Deposits)—A Symposium, New York, 1966: *Econ. Geol., Monographs*, **3**, 427-433.

Chow, T. J. (1965) Radiogenic leads of the Canadian and Baltic Shield regions; Symposium on Marine Geochemistry, 1964: *Rhode Island Univ. Narragansett Marine Lab. Occasional Pub. 3—1965*, 169-184.

Church, S. E. (1973) Limits of sediment involvement in the genesis of orogenic volcanic rocks: *Contr., Mineral. Petrol.*, **39**, 17-32.

―― and M. Tatsumoto (1975) Pb isotope relations in oceanic basalts from the Juan de Fuca-Gordo Rise area, N.E. Pacific Ocean: *Contrib. Mineral Petrol.*, **53**, 253-280.

―― and G. R. Tilton (1973) Lead and Strontium isotopic studies in the Cascade Mountains: Bearing on andesite genesis: *Bull. Geol. Soc. Am.*, **84**, 431-454.

Cooper, J. A. and J. R. Richards (1966) Lead isotopes and volcanic magma: *Earth Planet. Sci. Lett.*, **1**, 259-269.

―― and J. R. Richards (1969a) Lead isotope measurements on sediments from Atlantis II and Discovery Deep areas. *Hot Brines and Recent Heavy Metal Deposits in the Red Sea*, New York: Springer-Verlag, p. 499-511.

―― and J. R. Richards (1969b) Lead isotope measurements on volcanics and associated galenas from the Coromandel-Te Aroha region, New Zealand: *Geochem.*, **3**, 1-14.

Cumming, G. L. and J. R. Richards (1975) Ore lead isotope ratios in a continuously changing Earth: *Earth Planet. Sci. Lett.*, **28**, 155-171.

REFERENCES

_____ and D. K. Robertson (1969) Isotopic composition of lead from the Pine Point deposit: *Econ. Geol.*, **64**, 731-732.

Delevaux, M. H. and B. R. Doe (1974) Preliminary report on uranium, thorium, and lead contents and lead isotopic composition in sediment samples from the Red Sea: *Initial Reports of the Deep Sea Drilling Project* (Chap. 31) Washington: U.S. Gov't. Printing Office, **XXIII**, p. 943-946.

_____, A. P. Pierce, and J. C. Antweiler (1966) New isotopic measurements of Colorado ore leads: Geological Survey Research 1966: *U.S. Geol. Survey, Prof. Paper 550-C*, C178-C186.

Doe, B. R. (1967) The bearing of lead isotopes on the source of granitic magma: *J. Petrol.*, **8**, 51-83.

_____ (1968) Lead and strontium isotopic studies of Cenozoic volcanic rocks in the Rocky Mountain region—A summary: *Colo. Sch. Mines Quart.*, **63**, 149-174.

_____ (1970) Lead Isotopes: *Minerals, Rocks, and Inorganic Materials.* Heidelberg: Springer-Verlag, **3**, 137 p.

_____ (1976) Lead isotope data bank: 2,624 samples and analyses cited: *U.S. Geol. Survey Open File Report No. 76-201.*

_____ and M. H. Delevaux (1972) Source of lead in southeast Missouri galena ores: *Econ. Geol.*, **67**, 409-425.

_____, _____ (1973) Variations in lead isotopic compositions in Mesozoic granitic rocks of California: A preliminary investigation: *Bull. Geol. Soc. Am.*, **84**, 3513-3526.

_____, C. E. Hedge, and D. E. White (1966) Preliminary investigation of the source of lead and strontium in deep geothermal brines underlying the Salton Sea geothermal area: *Econ. Geol.*, **61**, 462-483.

_____ and J. S. Stacey (1974) The application of lead isotopes to the problems of ore genesis and ore prospect evaluation: A review: *Econ. Geol.*, **69**, 757-776.

_____, R. I. Tilling, C. E. Hedge, and M. R. Klepper (1968) Lead and strontium isotope studies of the Boulder batholith, southwestern Montana: *Econ. Geol.*, **63**, 884-906.

Dymond, J., J. B. Corliss, G. R. Heath, C. W. Field, E. J. Dasch, and H. H. Veeh (1973) Origin of metalliferous sediments from the Pacific Ocean: *Bull. Geol. Soc. Am.*, **84**, 3355-3372.

Engel, A. E. J., and C. C. Patterson (1957) Isotopic composition of lead in Leadville limestone, hydrothermal dolomite, and associated ore: *Bull. Geol. Soc. Am.*, **68**, 1723.

Gast, P. W. (1967) Isotopic geochemistry of volcanic rocks: *A Treatise on Rocks of Basaltic Composition,* New York: Interscience Publishers, p. 325-358.

Gray, C. M. and V. M. Oversby (1972) The behaviour of lead isotopes during granulite facies metamorphism: *Geochim. Cosmochim. Acta,* **36**, 939-952.

Hall, W. E., and G. K. Czamanske (1972) Mineralogy and trace element content of the Wood River Lead-Zinc deposits, Blaine County, Idaho: *Econ. Geol.*, **67**, 350-361.

_____, R. O. Rye, and B. R. Doe (1978) Stable isotope and fluid inclusion studies of the Wood River district, south central Idaho: *J. Res. U.S. Geol. Surv.*, **6**, 579-592.

Hamilton, W. (1969) Mesozoic California and the underflow of Pacific mantle: *Bull. Geol. Soc. Am.*, **80**, 2409-2430.

Hart, S. R., and G. R. Tilton (1966) The isotope geochemistry of strontium and lead in Lake Superior sediments and water; The Earth Beneath the Continents: *Am. Geophys. Union, Geophys. Mon. Ser.*, **10**, 127-137.

Heier, K. S. (1973) Geochemistry of granulite facies rocks and problems of their origin: *Phil. Trans. R. Soc. London A,* **273,** 429–442.

Heyl, A. V., M. H. Delevaux, R. E. Zartman, and M. R. Brock (1966) Isotopic study of galenas from the Upper Mississippi Valley Mineral Districts: *Econ. Geol.,* **61,** 933–961.

Horn, D. R., B. Horn, and M. N. Delach (1972) Worldwide distribution and metal content of deep-sea manganese deposits; Manganese Nodule Deposits in the Pacific: Symp./Workshop Proc., Honolulu, Hawaii, October 16–17, 46–60.

Hutchinson, R. W. (1973) Volcanogenic sulfide deposits and their metallogenic significance: *Econ. Geol.,* **68,** 1223–1246.

Jaffey, A. H., K. F. Flynn, L. E. Glendenin, W. C. Bentley, and A. M. Essling (1971) Precision measurement of half-lives and specific activities of ^{235}U and ^{238}U: *Phys. Rev. C,* **4,** 1889.

Kanasewich, E. R. (1968) The interpretation of lead isotopes and their geological significance: *Radiometric Dating for Geologists,* New York: Interscience, p. 147–224.

Kistler, R. W. and Z. E. Peterman (1973) Variations in Sr, Rb, K, Na, and in initial Sr^{87}/Sr^{86} in Mesozoic granitic rocks and intruded wall rocks in California: *Bull. Geol. Soc. Am.,* **84,** 3489–3512.

Lambert, I. B. and T. Sato (1974) The Kuroko and associated ore deposits of Japan: A review of their features and metallogenesis: *Econ. Geol.,* **69,** 1215–1236.

LeRoux, L. J. and L. L. Glendenin (1963) Half-life of thorium-232: *Proc. Natl. Conf. Nuclear Energy, Pretoria,* April.

Lipman, P. W., B. R. Doe, C. E. Hedge, T. A. Steven (1978) Petrologic evolution of the San Juan volcanic field, southwestern Colorado; Lead and strontium isotope evidence: *Bull. Geol. Soc. Am.* **89,** 59–82.

Manton, W. I., and M. Tatsumoto (1971) Some Pb and Sr isotopic measurements on eclogites from the Roberts Victor mine, South Africa; *Earth Planet. Sci. Lett.,* **10,** 217–226.

Moorbath, S., and H. Welke (1968) Isotopic evidence for the continental affinity of the Rockall Bank, North Atlantic: *Earth Planet. Sci. Lett.,* **5,** 211–216.

———, ———, and N. H. Gale (1969) The significance of lead isotope studies in ancient, high grade metamorphic basement complexes, as exemplified by the Lewisian rocks of northwest Scotland: *Earth Planet. Sci. Lett.,* **6,** 245–256.

Mudge, M. R., R. L. Erickson, and D. Kleinkopf (1968) Reconnaissance geology, geophysics and geochemistry of the southeastern part of the Lewis and Clark Range, Mont.: *U.S. Geol. Survey Bull. 1252,* E1–E35.

Ojakangis, R. W. (1963) Petrology and sedimentation of the Upper Lamotte Sandstone in Missouri: *Bull. Geol. Soc. Am.,* **65,** 201–221.

Ostic, R. G., R. D. Russell, and R. L. Stantan (1967) Additional measurements of the isotopic composition of lead from stratiform deposits: *Can. J. Earth Sci.,* **4,** 245–269.

Oversby, V. M. (1974) New look at the lead isotope growth curve: *Nature,* **248,** 132–133.

——— and A. Ewart (1972) Lead isotopic compositions of Tonga-Kermadec volcanics and their petrogenetic significance: *Contrib. Mineral. Petrol.,* **37,** 181–210.

——— and P. W. Gast (1970) Isotopic composition of lead in oceanic islands: *J. Geophys. Res.,* **75,** 2097–2114.

Patterson, C. C. and M. Tatsumoto (1964) The significance of lead isotopes in detrital feldspar with respect to chemical differentiation within the Earth's mantle: *Geochim. Cosmochim. Acta,* **28,** 1–22.

REFERENCES

Radabaugh, R. E., J. S. Merchant, and J. M. Brown (1968) 30. Geology and ore deposits of the Gilman (Red Cliff, Battle Mountain) District, Eagle County, Colorado, Chapter 30, Ore Deposits of the United States, 1933-1967; *Amer. Inst. Min., Metall., Petrol. Eng.*, 641-664.

Reynolds, P. H. and E. J. Dasch (1971) Lead isotopes in marine manganese nodules and the ore-lead growth curve: *J. Geophys. Res.*, **76**, 5124-5129.

Richards, J. R. (1971) Major lead ore bodies—Mantle origin? *Econ. Geol.*, **66**, 425-434.

Richards, J. R. (1968) "Primary" leads: *Nature*, **219**, 258-259.

Rosholt, J. N., R. E. Zartman, and I. T. Nkomo (1973) Lead isotope systematics and uranium depletion in the Granite Mountains, Wyoming: *Bull. Geol. Soc. Am.*, **84**, 989-1002.

Russell, R. D. (1972) Evolutionary model for lead isotopes in conformable ores and in ocean volcanics: *Revs. Geophys. and Space Physics*, **10**, 529-549.

―――― and R. M. Farquhar (1960) *Lead Isotopes in Geology*, New York: Interscience.

Sangster, D. F. (1972) Precambrian volcanogenic massive sulfide deposits in Canada: A review: *Geol. Surv. Can.*, Paper 72—22.

Sato, K. (1975) Unilateral isotopic variation of Miocene ore leads from Japan: *Econ. Geol.*, **70**, 800-805.

Sinha, A. K. and T. E. Davis (1971) Geochemistry of Franciscan volcanic and sedimentary rocks from California: *Carnegie Institute of Wash. Year Book*, **69**, 1969-1970, 394-400.

Slawson, W. F. and C. F. Austin (1962) A lead isotope study defines a geological structure: *Econ. Geol.*, **57**, 21-29.

Stacey, J. S., M. E. Delevaux, and T. K. Ulrych (1969) Some triple-filament lead isotope ratio measurements and an absolute growth curve for single-stage leads: *Earth Planet. Sci. Lett.*, **6**, 15-25.

――――, R. E. Zartman, and I. T. Nkomo (1968) A lead isotope study of galenas and selected feldspars from mining districts in Utah: *Econ. Geol.*, **63**, 796-814.

―――― and J. D. Kramers (1975) Approximation of terrestrial lead isotope evolution by a two-stage model: *Earth Planet. Sci. Lett.*, **26**, 207-221.

Stanton, R. L. and R. D Russell (1959) Anomalous leads and the emplacement of lead sulfide ores: *Econ. Geol.*, **54**, 588-607.

Sun, S. S. and G. N. Hanson (1975) Evolution of the mantle: Geochemical evidence from alkali basalt: *Geology*, **3**, 297-302.

Tatsumoto, M. (1969) Lead isotopes in volcanic rocks and possible ocean-floor thrusting beneath island arcs: *Earth Planet. Sci. Lett.*, **6**, 369-376.

――――, T. J. Knight, and C. J. Allegre (1973) Time differences in the formation of meteorites as determined from the ratio of lead-207 to lead-206: *Science*, **180**, 1279-1283.

Tweto, O. (1968) Geologic setting and interrelationships of mineral deposits in the mountain province of Colorado and south-central Wyoming, Chapter 27, Ore Deposits in the United States, 1933-1967: *Am. Inst. Min., Metall., Petrol. Eng.*, 551-588.

Unruh, D. M. and M. Tatsumoto (1976) Lead isotopic composition and uranium, thorium, and lead concentrations in sediments and basalts from the Nazca plate; a preliminary report. *Initial Reports of the Deep Sea Drilling Project*, Washington, D.C.: U.S. Gov't. Printing Office, **XXXIV**, 341-347.

Wasserburg, G. J. (1966) Geochronology and isotopic data bearing on development of the continental crust. *Adv. Earth Sci., Contrib. Internat. Conf.*, Cambridge, Mass., 1964, 431-459.

Wedepohl, K. H., M. H. Delevaux, and B. R. Doe (1978) The potential source of lead in the Permian Kupferschiefer bed of Europe and some selected Paleozoic mineral deposits in the Federal Republic of Germany: *Contrib. Mineral. Petrol.*, **65,** 273-281.

Wedow, H., Jr. (Chr.) (1971) A paleoaquifer and its relation to economic mineral deposits: The lower Ordovician Kingsport Formation and Mascot Dolomite: *Econ. Geol.*, **66,** 695-810.

Woodward, L. A., W. H. Kaufman, O. L. Schumacher, and L. W. Talbott (1974) Stratabound copper deposits in Triassic sandstone of Sierra Nacimiento, New Mexico: *Econ. Geol.*, **69,** 108-120.

Zartman, R. E. (1974) Lead isotopic provinces in the Cordillera of the Western United States and their geologic significance: *Econ. Geol.*, **69,** 792-805.

―――― and J. S. Stacey (1971) Lead isotopes and mineralization ages in Belt Supergroup rocks, Northwestern Montana and Northern Idaho: *Econ. Geol.*, **66,** 849-860.

―――― and F. Tera (1973) Lead concentration and isotopic composition in five peridotite inclusions of probably mantle origin: *Earth Planet. Sci. Lett.*, **20,** 54-66.

―――― and G. J. Wasserburg (1969) The isotopic composition of lead in potassium feldspars from some 1.0 b.y. North American igneous rocks: *Geochim. Cosmochim. Acta,* **33,** 901-942.

3

Magmas and Hydrothermal Fluids

C. WAYNE BURNHAM
The Pennsylvania State University

Numerous features of many ore deposits recur with such frequency on a worldwide basis as to suggest a consanguinity of the ores, the fluids that transported them, and the shallow-seated bodies of intrusive igneous rocks with which they are intimately associated. The case for consanguinity is very strong for the porphyry-type copper and molybdenum deposits that generally (1) are localized in the highly fractured upper parts of felsic porphyry stocks or superjacent wallrocks, (2) are centered on and exhibit a zonal relationship to the intrusive bodies, and (3) provide evidence of contemporaneity between hydrothermal mineralization-alteration and magmatic intrusion. The case is further strengthened by evidence presented later in this chapter that the extensive fracture systems, in which much of the ore is concentrated, were produced mainly by the action of aqueous fluids during, and immediately after, their separation from crystallizing magmas. However, it is weakened, in some minds at least, by isotopic evidence that indicates that local meteoric waters may have been involved in late-stage hydrothermal alteration and mineralization of some porphyry copper deposits. The extent of their involvement is highly variable, even within a given system, but their overall impact appears to be limited largely to local modification of features, including the distribution of metals, that were produced previously by processes operating during the solidification of hydrous magmas. These processes, as numerous and complexly interrelated as they are, by no means suffice to produce a porphyry copper deposit. Others must operate during magma generation and emplacement, for example, if the magmas undergoing crystallization at shallow levels are to have the requisite physical and chemical properties, such as heat and H_2O contents.

In the majority of processes of interest here, whether they operate in a hydrothermal system, in a solidifying magma, or in an amphibolite undergoing partial melting at great depths, volatiles such as H_2O play a prominent role. For this reason, the topic of volatiles in magmas, with especial emphasis on H_2O, will receive early attention in this chapter. The relationships established in these early discussions, notably the thermodynamic properties of hydrous magmas, will then form the basis for ensuing discussions of magma generation, emplacement, and crystallization. Throughout these discussions, the magma compositions that will receive greatest attention are those that form the calc-alkaline suite of igneous rocks broadly defined, ranging from diorite (andesite) to granite (rhyolite). Characteristically, the world's major porphyry copper-molybdenum deposits are intimately associated, both in space and in time, with the intermediate members of this suite, the quartz diorites, granodiorites, and quartz monzonites. This is fortunate, in the context of the present chapter, because these felsic melts are better understood than those of more mafic composition in terms of their thermodynamic properties and the role of H_2O.

VOLATILES IN MAGMAS

The volatile constituents of felsic magmas that are of primary concern here include H_2O, H_2S, CO_2, HCl, HF, and H_2. H_2O is by far the most abundant, as evidenced by the abundance of hydrous minerals (amphiboles and micas) in felsic igneous rocks and by analyses of volcanic gases and fluid inclusions. Owing to this relatively greater abundance of H_2O, the role of volatiles in magmas is largely the role of H_2O, although the other volatiles play a correspondingly more important role at the stage when an aqueous phase (hydrothermal fluid) separates from a crystallizing magma. The initial H_2O contents of the magmas with which extensive hydrothermal activity is associated generally range from about 2.5 to 6.5 wt. %, with a median probably close to 3.0 wt. %. The lower limit is set by the almost ubiquitous occurrence of either hornblende or biotite (sometimes both) as a phenocryst phase and the upper limit is the saturation value at about 2.1 kb, which is a pressure equivalent to a lithostatic load of approximately 8 km and, for reasons discussed below, is probably close to the upper pressure limit of extensive hydrothermal activity. Despite its small mass fraction, however, H_2O has an enormous effect on the physical and chemical properties of these magmas. In the system $NaAlSi_3O_8-SiO_2-H_2O$ at 2.0 kb pressure, for example, addition of only 6.4 wt.% H_2O is sufficient to lower the melting temperature of albite more than 300°C, whereas the same weight percent SiO_2 lowers it only approximately 20°C. This phenomenal difference is a direct consequence of the manner in which H_2O dissolves in aluminosilicate melts, as it is molal quantities, not relative masses, that govern their physical chemistry.

Reaction of H_2O with Aluminosilicate Melts

Magmas of intermediate composition near their liquidi are aluminosilicate melts composed mainly of $(Al,Si)O_4$ tetrahedra in which most of the oxygen atoms, hereinafter referred to as O^{2-}, are shared (form bridges) between neighboring tetrahedra. Approximately 90% of the O^{2-} in dioritic melts are thus shared and virtually all of them are shared in granitic melts. This linking of neighboring tetrahedra through the oxygen bridges produces highly polymerized three-dimensional networks that closely resemble the framework structures of the feldspars and silica minerals (Burnham, 1975a). This high degree of polymerization of granitic melts accounts for their high viscosities, which are in the order of 10^8-10^{10} poises at 1000°C and atmospheric pressure. The addition of only 6.4 wt. % H_2O, however, lowers the viscosity by approximately 10^5 poises (Burnham, 1967, Figure 2.5), an enormous effect that can be accounted for only if H_2O interacts with the melt in a way that breaks oxygen bridges and thereby depolymerizes it. Oxygen bridges may be broken by replacing a single bridging O^{2-} with two OH^- according to the hydrolysis reaction scheme:

$$H_2O(v) + O^{2-}(m) \rightleftharpoons 2\,OH^-(m)$$

where (v) and (m) refer to the aqueous fluid and melt phases, respectively.

Direct evidence that OH^- is formed when H_2O dissolves in silicate melts comes from infrared spectral studies of quenched glasses (Orlova, 1964); indirect evidence comes from the thermodynamic properties of H_2O in aluminosilicate melts (Burnham and Davis 1971, 1974). The thermodynamic relations also indicate that the solution reaction with these melts which contain trivalent cations in tetrahedral coordination, such as Al^{3+} in feldspathic melts, involves dissociation of the H_2O. The solution reaction also involves exchange of H^+ from H_2O with the nontetrahedrally coordinated cations (exchangeable cations) in the melt, which must be present to balance the charge on the tetrahedral groups that contain trivalent cations (Burnham, 1975a). In the system $NaAlSi_3O_8—H_2O$, for example, this solution reaction can be written:

$$H_2O(v) + O^{2-}(m) + Na^+(m) \rightleftharpoons OH^-(m) + ONa^-(m) + H^+(m)$$

where it is understood that the $H^+(m)$ is associated with one of the four $O^{2-}(OH^-)$ coordinating Al^{3+} in $AlO_3(OH)^{4-}$ groups.

The involvement of ions other than O^{2-} in the solution reaction of H_2O with aluminosilicate melts places constraints on the mass that constitutes a mole of anhydrous melt. Having selected one gram formula weight of H_2O (18.02 g) as a mole of water substance, the unit of silicate melt with which the first mole of H_2O interacts is the mass that is equivalent to one mole of $NaAlSi_3O_8$, in terms of exchangeable cations. Thus, a mole of feldspar melt is one gram formula weight ($NaAlSi_3O_8$, 262.2 g; $KAlSi_3O_8$, 278.3 g; $CaAl_2Si_2O_8$, 278.2 g), but a mole of compositionally more complex igneous-rock melt is not as simply defined. It can be computed from a chemical analysis of the rock, however, and placed on an $NaAlSi_3O_8$-equivalent basis by observing the constraints imposed by the stoichiometry of $NaAlSi_3O_8$.

With these constraints, Burnham (1979) calculated the equimolal masses of Columbia River basalt, Mt. Hood andesite, and Harding (New Mexico) pegmatite for the purpose of comparing the experimentally determined weight percent H_2O solubilities of Burnham and Jahns (1962) and Hamilton et al. (1964). The results of this comparison at 1100°C are presented in Figure 3.1a, from which it will be observed that the equimolal solubilities of H_2O in melts of a wide compositional range are virtually identical to those in $NaAlSi_3O_8$ melt (solid line) despite the fact that weight percent solubilities are markedly different (Figure 3.1b)

The identical solubilities of H_2O in rock melts of widely different compositions, when a mole of anhydrous rock melt is defined in accordance with the solution model, provide confirmation of the model's general validity, as the model was developed primarily from thermodynamic relations in the simple system $NaAlSi_3O_8—H_2O$. They also provide access to the whole area of the thermodynamic relations in felsic magmas, but

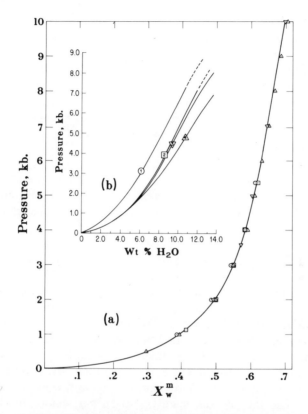

Fig. 3.1 The solubility of H_2O in silicate melts: (a) Equimolal solubility of H_2O in basalt (squares), andesite (circles), Li-pegmatite (upright triangles), and albite (inverted triangles) melts at 1100°C, (b) weight percent solubility of H_2O in (1) basalt (1100°C), (2) andesite (1100°C), (3) albite (700°-870°C), and (4) Li-pegmatite (660°-720°C) melts. See text for sources of data and Burnham (1979) for method of computing equimolal masses of igneous rock melts.

before turning to this topic, it is instructive to apply the concepts of the model to the solution of volatiles other than H_2O, especially in H_2O-bearing magmas.

Solution of Other Volatiles in Aluminosilicate Melts

The volatiles other than H_2O that have important roles in ore-forming processes may be divided into two groups on the basis of whether or not the solution mechanism is similar to that for H_2O. The acid volatiles (H_2S, HCl, and HF) are similar to H_2O in that they undergo hydrolysis with the O^{2-} of the melt while maintaining overall electrostatic neutrality. Volatiles such as CO_2, SO_2, and H_2, on the other hand, cannot react with the bridging O^{2-} and preserve electrostatic neutrality. Consequently, the solution of these volatiles in feldspathic melts is largely molecular, except in the special case of H_2 in Fe^{3+}-bearing melts where reduction can promote hydrolysis according to the reaction:

$$\tfrac{1}{2} H_2 + O^{2-} + Fe^{3+} \rightleftharpoons OH^- + Fe^{2+}$$

Unpublished experimental solubility measurements of sulfur in H_2O-saturated granitic melts made in the writer's laboratory and by Althaus and Skinner (Skinner, personal communication 1974) indicate that H_2S is only moderately less soluble than H_2O. This is as predicted from the model, because the reaction,

$$H_2S(v) + O^{2-}(m) + Na^+(m) \rightleftharpoons SH^-(m) + ONa^-(m) + H^+(m)$$

is analogous to the H_2O solution reaction discussed previously. The apparent lower solubility is understandable from volumetric and bond energy considerations; SH^- is larger than OH^- and the Si—SH bond is almost certainly weaker (higher free energy) than the Si—OH bond.

The solution reactions of the other acid volatiles, HCl and HF, are analogous to those of H_2O and H_2S, although there are notable differences in their relative solubilities. A comparison of the preliminary sulfur data with those of Kilinc and Burnham (1972) on partitioning of chloride between aqueous solutions and granitic melts suggests that the solubilities of HCl and H_2S are comparable, again in accordance with model predictions based on both volumetric and bonding considerations. Moreover, the partition coefficient of chloride ($K_P{}^{Cl} = m_{Cl}{}^v/m_{Cl}{}^m$) increases essentially linearly with H_2O content, as does $K_P{}^S$, in response to a mass action or common ion (OH^-) effect. However, at pressures above six kilobars, where the H_2O content is greater than about 65 mole %, $K_P{}^{Cl}$ decreases markedly with increasing H_2O content. This maximum in $K_P{}^{Cl}$ at approximately 6 kb (Kilinc and Burnham, 1972, Figure 1) perhaps indicates the onset of significant "molecular" solubility of NaCl and KCl, promoted by the depolymerizing effects of the large amounts of dissolved H_2O.

The effects of HF, in addition to H_2O, in lowering liquidus temperatures in the system $NaAlSi_3O_8$—H_2O—HF (Wyllie and Tuttle, 1961) imply that HF is even more soluble than H_2O in silicate melts. Indeed, by application of the H_2O solution model and the corresponding Raoult's law analogue (discussed in the next section) to the experimental data of Wyllie and Tuttle, the calculated saturation mole fraction of HF ($X_{HF}{}^m$) is approximately 1.25 times that of H_2O ($X_w{}^m$). This result is qualitatively predictable from the smaller size of F^-, relative to OH^-, and the greater stability of the SiF_4 complex. Thus, owing to somewhat greater solubility of HF, increasing pressure would not be expected to have as great an affect on $K_P{}^F = m_F{}^v/m_F{}^m$ as it does on $K_P{}^{Cl}$, and this is consistent with the findings of Burnham (1967, p. 40).

The solubility of CO_2 in aluminosilicate melts is very low in comparison with that of H_2O and the other acid volatiles. In $NaAlSi_3O_8$ melts at 20 kb CO_2 pressure, for example, Eggler (1973) obtained a saturation CO_2 content of less than 1.0 wt. % (<5.7 mole %) in the absence of H_2O, whereas the solubility of H_2O at the same pressure is greater than 25 wt. % (>83 mole %). These findings are readily explained by the present model; CO_2 is incapable of depolymerizing the three-dimensional network by hydrolyzing O^{2-}, all of which are bridging in $NaAlSi_3O_8$ melts. Moreover, the network of Si—O—Si and Si—O—Al bridges cannot expand readily to accommodate the large CO_2 molecule. Thus model considerations lead to the prediction that CO_2 solubility should be inversely correlated with the molal proportions of bridging to nonbridging O^{2-}, and this is precisely what has been observed (Eggler et al., 1974). At 20 kb, the solubility of CO_2 in $CaMgSi_2O_6$ melt, in which only half the O^{2-} form Si—O—Si bridges, is about four times that in $NaAlSi_3O_8$ melt at the same CO_2 pressure, and there is evidence (Mysen, 1975) that the nonbridging O^{2-} form $CO_3{}^{2-}$ with CO_2.

The effects of H_2O on the partitioning of CO_2 between melt and aqueous phase also stands in strong contrast to its effects on partitioning of the more acid volatiles. Holloway and Lewis (1974) have shown that the CO_2 content of $NaAlSi_3O_8$ melts, in equilibrium with an H_2O—CO_2 mixed-fluid phase at 5 kb total pressure, increases with the mole ratio of $H_2O:CO_2$ in the fluid phase up to about 0.35 and then decreases at higher mole ratios. These results are readily explained by the model and the fact that the solubility of any substance in a given solvent is a function of the activity or fugacity of that substance in the system. Thus, the addition of H_2O to the system $NaAlSi_3O_8$—CO_2 breaks Si—O—Si bridges and thereby increases the expansibility of the liquid to more readily accommodate large CO_2 molecules, some of which may form $CO_3{}^{2-}$ and $HCO_3{}^-$ with the nonbridging ONa^- and OH^-. However, as the CO_2-rich fluid phase becomes more dilute with addition of H_2O, the fugacity (and therefore "solubility") of CO_2 decreases, the two opposing effects approximately cancelling each other at an $H_2O:CO_2$ mole ratio in the fluid phase of approximately 0.35 at 5.0 kb. Qualitatively similar results were obtained by Mysen (1975) at higher pressures.

The experimental results of Haughton et al. (1974) indicate that the mechanism of solution for SO_2 in aluminosilicate melts is basically similar to that for CO_2, although it is complicated at low values of f_{O_2} by redox reactions of the type:

$$SO_2(v) + O^{2-}(m) \rightleftharpoons S^{2-}(m) + \tfrac{3}{2} O_2(v).$$

Thus, the solubility increases with the proportion of nonbridging O^{2-}, especially those produced by the addition of FeO to the melt. In the presence of H_2O, however, SO_2 hydrolyzes to more soluble H_2S, the proportions of which increase directly with f_w and inversely with f_{O_2}.

Very few experimental data are available on the solubility of H_2 in silicate melts and none, to the writer's knowledge, that are applicable to melts of interest here. Reasoning from the solution model, however, leads to the conclusion that molecular H_2 should be considerably more soluble than CO_2 in aluminosilicate melts owing to its much smaller size. In the presence of Fe^{3+}, as mentioned previously, the solubility of H_2 should be appreciably enhanced by coupled reduction and hydrolysis reactions. The presence of H_2O in the melt also should increase the solubility of H_2, just as it does for CO_2, but the effects should be proportionately larger, again owing to the much smaller size of the H_2 molecule.

THERMODYNAMIC RELATIONS IN MAGMAS

Thermodynamic Properties of H_2O in Magmas

The fact that the molal solubility of H_2O in igneous-rock melts of a wide range in composition is the same as that in $NaAlSi_3O_8$ melt at constant pressure and temperature (Figure 3.1a), provided the mole of rock melt is defined in accordance with the $NaAlSi_3O_8$—H_2O solution model, makes possible a quantitative application of the thermodynamic relations in this simple system to compositionally complex magmas. Along the solubility (or saturation) curves of Figure 3.1, H_2O-saturated silicate melt is in equilibrium with silicate-saturated H_2O fluid which, owing to the very low solubilities of silicates, can be regarded as essentially pure H_2O for the present purposes. The fundamental criterion for this, or any other equilibrium, is that the chemical potential of each component, in this case H_2O (μ_w), be the same in all phases. Therefore, if the effects of the very small amount of silicates dissolved in the H_2O are neglected, $\mu_w = G_w^0$, where G_w^0 is the molal Gibbs free energy of pure H_2O (Burnham et al., 1969). Furthermore, inasmuch as the fugacity of H_2O (f_w) is defined, at constant temperature, by the relation $d\mu_w \equiv RT\, d\ln f_w$, where R is the gas constant and T is in Kelvins, the condition of equilibrium also can be expressed by $f_w = f_w^0$, where f_w^0 is the fugacity of pure H_2O. Division of f_w by f_w^0 at pressure and temperature defines a_w, the activity of H_2O in the system; therefore, $f_w/f_w^0 = a_w \approx 1$ at equilibrium saturation.

The activity of H_2O in $NaAlSi_3O_8$ melts is given, for $X_w^m \leq 0.5$, by the relation (Burnham, 1975b)

$$(a_w)_{X_w^m \leq 0.5} = k (X_w^m)^2 \qquad (3.1)$$

and, for $X_w^m > 0.5$, by

$$(a_w)_{X_w^m > 0.5} = 0.25\, k \exp\left[\left(6.52 - \frac{2667}{T}\right)\left(X_w^m - 0.5\right)\right] \qquad (3.2)$$

where k is a constant that is dependent only on temperature and pressure, as shown in Figure 3.2. From earlier discussions on the reaction of H_2O with silicate melts, it is apparent that k is the analogue of a Henry's law activity constant for a dissociated solute. Thus, at a given pressure on the saturation curve of Figure 3.1a, $a_w = 1$ and X_w^m is the same for all silicate compositions represented; therefore, k is independent of aluminosilicate composition.

The importance of this conclusion is difficult to overemphasize. With values of k obtained from Figure 3.2, which was constructed from the data of Burnham and Davis (1974, Figure 13) and Burnham et al. (1969), the solubilities of H_2O in magmas of various compositions can be readily calculated from equation (3.1) or equation (3.2). Also, because the other thermodynamic functions for H_2O in $NaAlSi_3O_8$ melts, such as partial molal volumes, entropies, and enthalpies, are derivable from equation (3.1) or equation (3.2) and the corresponding properties of pure H_2O, these quantities can be rigorously evaluated for hydrous magmas of diverse compositions, provided the mole of anhydrous melt is defined in accordance with the solution model.

Thermodynamic Properties of the Silicate Components in Magmas

The thermodynamic properties of the $NaAlSi_3O_8$ component in $NaAlSi_3O_8$ —H_2O solutions have been determined by Burnham and Davis (1974) through the use of published thermochemical data on $NaAlSi_3O_8$ glass (Robie and Waldbaum, 1968) and application of the Gibbs-Duhem relation to equations (3.1) and (3.2). Thus, from equation (3.1) for $X_w^m \leq 0.5$

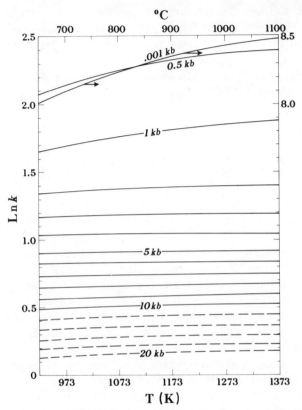

Fig. 3.2 Ln k in eqs. 3.1 and 3.2 as a function of temperature and pressure. The dashed isobars above 10 kb were obtained by extrapolation (Burnham, 1979). Data from Burnham and Davis (1974) and Burnham et al. (1969).

$$a_{ab}^{hm} = (X_{ab}^{hm})^2 = (1 - X_w^m)^2 \qquad (3.3)$$

and from equation (3.2) for $X_w^m > 0.5$

$$\ln a_{ab}^{hm} = \left(6.52 - \frac{2667}{T}\right)\left[\ln\left(1 - X_w^m\right) + X_w^m\right] - \frac{515}{T} - 0.127 \qquad (3.4)$$

where the subscript ab refers to the NaAlSi$_3$O$_8$ component (hereinafter referred to as the ab component) and the superscript hm refers to the hydrous melt. Equation (3.3) expresses the analogue of Raoult's law for the solvent in which the solute (H$_2$O) dissociates into two species.

The Raoult's law analogue behavior of the ab component implies that any two NaAlSi$_3$O$_8$—H$_2$O melts, in which $X_w^m \le 0.5$, will mix with neither heat nor volume effects and in this sense may be regarded as ideal solutions. Furthermore, inasmuch as equations (3.1) and (3.2) are equally applicable to all aluminosilicate components of the igneous-rock melts represented in Figure 3.1, it would be intuitively expected that these aluminosilicate components also mix ideally with each other. That this is indeed the case, provided the components are properly chosen, has been shown by Burnham et al. (1978); therefore, in these and other rock melts (magmas) whose compositions are such that an aluminosilicate mineral (chiefly feldspar) is a stable liquidus phase, $a_a^{am} = X_a^{am}$, where the subscript a refers to the specific aluminosilicate component and the superscript am refers to the anhydrous melt. It also can be shown for melts of these compositions that the nonaluminosilicate components, such as di ($Ca_{1.33}Mg_{1.33} \cdot Si_{2.67}O_8$), fo ($Mg_4Si_2O_8$), and qz ($Si_4O_8$) obey Henry's law ($a_i^{am} = k_iX_i^{am}$), provided they are chosen on the basis of 8.0 moles of O^{2-} per mole of component. Some petrologic implications of these mixing relations will be discussed below; others are discussed by Burnham (1979).

Thermodynamics of Melting

The fundamental criterion for heterogeneous equilibrium, as mentioned above, is that the chemical potential of a given component (μ_i) must be the same in all phases; hence, for equilibrium between a crystalline phase (s) and a melt (m) containing component i, $\mu_i^m = \mu_i^s$. For either phase m or s at equilibrium

$$\mu_i = \mu_i^0 + RT \ln a_i \tag{3.5}$$

where μ_i^0 is the chemical potential of component i in its standard state at pressure (P), temperature (T), and unit activity ($a_i = 1$). Inasmuch as the pressure dependence of crystal-melt equilibrium also is of interest here, however, it is convenient to choose standard states at T, $P = 1$ bar, and $a_i = 1$ in both pure phases (m) and (s) of component i. With this choice of standard states, the equation of equilibrium at P and T becomes

$$\mu_i^m - \mu_i^s = 0 = \frac{\Delta G_{mi}^0 + P\Delta V_{mi}}{RT} + \ln\left(\frac{a_i^m}{a_i^s}\right) \tag{3.6}$$

where $\Delta G_{mi}^0 = \mu_i^{0m} - \mu_i^{0s}$ is the Gibbs free energy of fusion at T (K) and one bar, and $P\Delta V_{mi}$ is a first approximation to the pressure integral of $\overline{V}_i^m - \overline{V}_i^s$ (the difference in partial molal volumes of component i in melt and solid phases, respectively). The potential error introduced by this approximation is generally less than uncertainties on ΔG_{mi}^0 in the pressure range of present interest.

Substitution of equations (3.3) and (3.4), and the relation $a_a^{am} = X_a^{am}$ into equation (3.6) yields, for $X_w^m \leq 0.5$ in aluminosilicate melts

$$0 = \frac{\Delta G_{ma}^0 + P\Delta V_{ma}}{RT} + \ln\left(\frac{X_a^{am}(1 - X_w^m)^2}{a_a^s}\right) \quad (3.7)$$

and, for $X_w^m > 0.5$ in these same melts

$$0 = \frac{\Delta G_{ma}^0 + P\Delta V_{ma}}{RT} + \ln\left(\frac{X_a^{am}}{a_a^s}\right) + \left(6.52 - \frac{2667}{T}\right)$$

$$\cdot \left[\ln\left(1 - X_w^m\right) + X_w^m\right] - \frac{515}{T} - 0.127 \quad (3.8)$$

These equations, together with previously published data on ΔG_{mab}^0 and ΔV_{mab} for albite, were used by Burnham (1979) to calculate crystal-melt equilibrium temperatures in the system $NaAlSi_3O_8$—H_2O, as a function of P, X_w^m, and a_w, to pressures of 20 kb. Up to approximately 16 kb, the results of these calculations were within 20°C of the experimentally determined equilibrium temperatures—a clear confirmation of the validity of equations (3.7) and (3.8), as well as of the combined accuracy of the thermodynamic data involved. At higher pressures along the H_2O-saturated solidus [analogous to the $S(a_w \approx 1.0)$ curves in Figures 3.3 and 3.4], however, experimentally determined equilibrium temperatures become progressively lower than the calculated values as the albite $\rightleftharpoons$ jadeite + quartz reaction boundary at 17 kb, 910K (637°C) is approached (Boettcher and Wyllie, 1969). This departure to lower crystal-melt equilibrium temperatures with increasing pressure is interpreted to reflect conversion in the melt of part of the tetrahedrally coordinated Al (Al^{IV}) in the ab component to octahedral coordination (Al^{VI}) in jadeite-like ($Na_{1.33}Al_{1.33}^{VI}Si_{2.67}O_8$) structural units. Thus, as Al^{IV} is converted to Al^{VI} ($Al^{IV} \rightarrow Al^{VI}$ shift) in this now pseudobinary system, $a_{ab}^{am} = X_{ab}^{am}$ becomes increasingly less than one and this results in depression of equilibrium temperatures [equations (3.7) and (3.8)]. This $Al^{IV} \rightarrow Al^{VI}$ shift with pressure, especially in the presence of H_2O, has important consequences for the generation of calc-alkaline magmas, hence the phenomenon will receive further attention below.

Having established the reliability of the thermodynamic properties of the system $NaAlSi_3O_8$—H_2O, equation (3.6) for ab component may be com-

THERMODYNAMIC RELATIONS IN MAGMAS

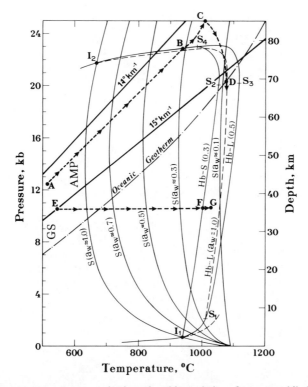

Fig. 3.3 Pressure-temperature projection of melting relations for an amphibolite of olivine tholeiite composition. GS and AMP represent approximate conditions of the upper greenschist and amphibolite facies of metamorphism, respectively. S ($a_w \approx 1.0$) to S ($a_w \approx 0.1$) are the beginning of melting (solidus) curves for activities of H_2O in the range 1.0–0.1, respectively. Hb-S (0.3) is the beginning of melting of basaltic amphibolite in the absence of a pore fluid, which occurs at a constant value of $X_w^m \approx 0.3$. Hb-L (0.5) represents the maximum stability of hornblende which occurs at $X_w^m \approx 0.5$ at pressures above S_1, and Hb-L ($a_w \approx 1.0$) is the maximum stability of hornblende at H_2O saturation. I_1 and I_2 are pseudo-invariant points, and S_1-S_4 are pseudo-singular points. Paths A-B-C-D and E-F-G, as well as sources of data, are discussed in the text.

bined with an equivalent expression for the an ($CaAl_2Si_2O_8$) component, in the form

$$\ln\left(\frac{a_{ab}^m a_{an}^s}{a_{an}^m a_{ab}^s}\right) = \frac{\Delta G_{man}^0 - \Delta G_{mab}^0 + P(\Delta V_{man} - \Delta V_{mab})}{RT} \qquad (3.9)$$

and the experimental data of Bowen (1913) to calculate internally consistent values for $\Delta G_{man}^0/RT$ at atmospheric pressure (Burnham, 1979, Figures 16-7

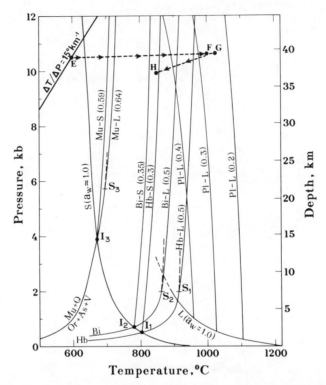

Fig. 3.4 Pressure-temperature projection of melting relations for average hornblende-biotite granodiorite composition. Mu + Q ⇌ Or + As + V is the upper stability of the assemblage muscovite plus quartz; S ($a_w \approx 1.0$) and L ($a_w \approx 1.0$) are the H_2O-saturated solidus and liquidus (plagioclase), respectively. Mu-S (0.59), Bi-S (0.35), and Hb-S (0.3) are the fluid-absent solidi for assemblages containing muscovite, biotite and hornblende, respectively. Mu-L (0.64), Bi-L (0.5), Hb-L (0.5), and Pl-L (0.X) are approximate maximum thermal stabilities (liquidi) of muscovite, biotite, hornblende, and plagioclase, respectively; numbers in parentheses are values of X_w^m under the indicated equilibrium conditions. I_1-I_3 are pseudo-invariant points and S_1-S_3 are pseudo-singular points as in Figure 3.3. See text for sources of data and discussion of path E-F-G-H.

and 16-8). The results of these calculations confirm Bowen's conclusion that all solutions in the system $NaAlSi_3O_8$—$CaAl_2Si_2O_8$, both melt and plagioclase, are essentially ideal at solidus temperatures and atmospheric pressure. Similar results were obtained using the experimental data of Yoder et al. (1957) and Yoder (1965) at P_w = 5.0 kb for $T \geq 1370K$ and anorthite-rich compositions. At lower temperatures, a_{ab} remains equal to X_{ab} in both melt and solid solutions, but $a_{an}^s \approx 1.25\ X_{an}^s$ ($\gamma_{an}^s = a_{an}^s/X_{an}^s \approx 1.25$). Furthermore, H_2O-saturated melts of intermediate composition are in equilibrium with plagioclase at temperatures as much as 40°C below those calculated from equation (3.8) using data for the an component. This phenomenon is interpreted again to reflect an $Al^{IV} \rightarrow Al^{VI}$ shift; in this case, however, the shift involves Al only from the an component to form zoisite-like $(Ca_{1.23}Al^{VI}_{1.85}Si_{1.85}O_{7.38}OH_{0.62})$ structural units in the melt. It is important to note that this $Al^{IV} \rightarrow Al^{VI}$ shift occurs only in melts of intermediate plagioclase composition; there is no evidence for such a shift in melts of either end-member composition at $P_w = 5$ kb. At 10 kb, on the other hand, anorthite-rich plagioclase melts incongruently to corundum $(Al^{VI}_{5.33}O_8)$ and Ca-silicate enriched melt, both in the presence (Yoder, 1965) and in the absence (Lindsley, 1967) of H_2O. Furthermore, at $P_w \geq 7$ kb, plagioclase of intermediate composition melts incongruently to zoisite and ab-enriched melts (Furst, 1978).

The phenomenon of an $Al^{IV} \rightarrow Al^{VI}$ shift in aluminosilicate melts is not restricted to high pressures. It can be demonstrated to occur even at atmospheric pressure in aluminosilicate melts that contain octahedrally coordinated cation sites. Thus, application of equation (3.6) for the ab component to the experimental data of Kushiro (1973) on the diopside solid solution-plagioclase-melt equilibrium in the system $NaAlSi_3O_8$—$CaAl_2Si_2O_8$—$Ca_{1.33}Mg_{1.33}Si_{2.67}O_8$ yields calculated Raoult's-law values of $a_{ab}^{am} = X_{ab}^{am}$ that are within ± 0.01 of the experimental values. However, application of equation (3.6) for the an component yields values of a_{an}^{am} that are consistently lower than the experimental values of X_{an}^{am}. In fact, $(X_{an}^{am} - a_{an}^{am})/X_{di}^{am} = R_{di}^{an} = 0.31 \pm 0.02$ for the melts analyzed by Kushiro. In accordance with the present quasi-crystalline model of silicate melts, this discrepancy between X_{an}^{am} and a_{an}^{am} is attributed to the $Al^{IV} \rightarrow Al^{VI}$ shift, which results in the formation of "Ca-Tschermak's" component $(CaAl_2Si_2O_8 \rightleftharpoons 0.75\ Ca_{1.33}Al_{2.67}Si_{1.33}O_8 + 0.25\ Si_4O_8)$ that also occurs in the coexisting diopside solid solution and in which half of the Al atoms are octahedrally coordinated. This interpretation is strongly supported by the fact that the ratio of octahedral, to octahedral plus tetrahedral sites (potential $Al^{VI}/Al^{VI} + Al^{IV}$) in a 1:1 mole mixture of di and ab + an components $(2\ Ca_{1.33}Mg_{1.33}Si_{2.67}O_8/NaAlSi_3O_8 + CaAl_2Si_2O_8)$ is 0.32, just 0.01 greater than the above average value of R_{di}^{an}.

It will be observed that, because $a_{ab}^{am} = X_{ab}^{am}$, the Al in $NaAlSi_3O_8$ does not participate in the $Al^{IV} \rightarrow Al^{VI}$ shift to form cts. In silica-undersaturated systems, however, this is no longer the case. In the system $NaAlSi_3O_8$—$CaAl_2Si_2O_8$—$Mg_4Si_2O_8$(fo), for example, both ab and an participate in reactions with fo to form other components (in addition to cts) that contain octahedral sites, such as $Mg_2Al_4O_8$(sp), $Ca_{1.33}Mg_{1.33} \cdot Si_{2.67}O_8$(di), $Mg_{2.67}Si_{2.67}O_8$(en), and $Mg_{0.89}Al_{1.78}Si_{2.22}O_8$ (cd). However, after taking cognizance of these reactions (proper identification of melt components), it may be shown that the equality between $R_{fo}^{ab} + R_{fo}^{an}$ and potential $Al^{VI}/(Al^{VI} + Al^{IV})$ sites holds without exception in every system for which the required experimental data are available. Moreover, the equality between $R_{i,...}^{ab} + R_{i,...}^{an}$, where i represents fo, di, en, or a properly weighted combination of the three, and potential $Al^{VI}/(Al^{VI} + Al^{IV})$ sites also holds for systems that contain two or more of these non-aluminosilicate components. In $Fe_4Si_2O_8$(fa)-bearing systems, on the other hand, it does not hold; but R_{fa}^{an} is found to be consistently approximately $0.38\ R_{fo}^{an}$. This lower occupancy of octahedral sites by Al (inhibited $Al^{IV} \rightarrow Al^{VI}$ shift) is interpreted to reflect a much greater octahedral-site preference of Fe^{2+} at atmospheric pressure.

The fact that $R_{i,...}^{ab}$ and $R_{i,...}^{an}$ can be quantitatively evaluated from composition data, properly recast in terms of mole fractions of melt components, means that a_{ab}^{am} and a_{an}^{am} also can be evaluated in igneous-rock melts, but the R_i^a factors have been determined only at atmospheric pressure. It is well known from experimental investigations of numerous aluminosilicate systems, including igneous-rock melts, that increasing pressure tends to expand the liquidus fields of Al^{VI}-bearing phases, such as spinel, aluminous pyroxenes, garnets, and corundum. In keeping with the quasi-crystalline melt model, this expansion implies that the activities of Al^{VI}-bearing components in the melt increase relative to those of Al^{IV}-bearing components (ab and an), which in turn implies that $R_{i,...}^{ab}$ and $R_{i,...}^{an}$ increase with pressure. As $R_{i,...}^{ab}$ and $R_{i,...}^{an}$ increase, of course, a_{ab}^{am} and a_{an}^{am} must decrease in a melt of fixed bulk composition. Precisely how much a_a^{am} is lowered for a given increment in pressure in fo-bearing systems is not known, as the experimental data required for evaluation are not available at present. In silica-saturated, di-bearing systems, however, the data of Yoder (1965) indicate that R_{di}^{an} increases at an average rate of approximately $0.077\ kb^{-1}$ (Burnham, 1979).

Inasmuch as $a_{ab}^{am} = X_{ab}^{am}$ and, hence, $R_{di}^{ab} = 0$ in these same silica-saturated, di-bearing melts, increasing pressure has a pronounced differential effect on a_{ab}^{am} and a_{an}^{am} that is proportional to X_{di}^{am}. As a consequence, $X_{ab}^{am}/X_{an}^{am} = a_{ab}^{am}/a_{an}^{am}$ in equation (3.9) increases with pressure and, if X_{di}^{am} is more than sufficient to compensate for the $P(\Delta V_{man} - \Delta V_{mab})/RT$ term, the temperature-pressure slope (dT/dP) of the plagioclase liquidus becomes negative. Concomitant with this decrease in temperature, the first-formed plagioclase naturally becomes more albitic.

Negative temperature-pressure slopes for plagioclase liquidi are not restricted to those melts that contain di or other pyroxene-like components. As discussed previously, addition of H_2O to a melt of intermediate plagioclase composition apparently results in formation of zoisite-like structural units, which contain $AlO_4(OH)_2^{7-}$ octahedra. The formation of zo component from part of the an component and dissolved H_2O lowers a_{an}^m

below the value obtained from equations (3.3) and (3.4), and the Raoult's law relation, $a_{an}^{am} = X_{an}^{am}$. Calculations using the experimental results of Yoder et al. (1957) and Whitney (1975a) indicate that the extent to which a_{an}^{am} is lowered in plagioclase-bearing melts by this mechanism, alone, may be approximated by the relation

$$a_{an}^{am} = X_{an}^{am} - 0.077\, P\, X_{OH^-}^m \left[\frac{X_{an}^{am}}{X_{an}^{am} + X_{ab}^{am}} \left(1 - \frac{X_{an}^{am}}{X_{an}^{am} + X_{ab}^{am}} \right) \right]$$

where $X_{OH^-}^m = 2 X_w^m / (1 + X_w^m)$ and P is in kilobars. Thus, even in melts that lack silicate components that contain octahedrally coordinated sites, such as those investigated by Whitney (1975a), the presence of dissolved H_2O (as OH^-) in amounts greater than about 10 mole % (~ 0.7 wt. % H_2O) is sufficient to cause plagioclase liquidi to take on negative temperature-pressure slopes. Furthermore, the value of dT/dP becomes increasingly negative with increasing X_w^m at a given high pressure.

The combined effects of pressure, H_2O, and silicate components that contain octahedral sites on plagioclase liquidus temperatures are clearly illustrated by the experimental results of Eggler and Burnham (1973) on the Mt. Hood andesite. At 5 kb and $X_w^m = 0.43$ (4.7 wt. % H_2O), the experimentally determined liquidus temperature is 1040°C, whereas at 10 kb and the same H_2O content it is 1000°C. These temperatures are well within experimental error of those calculated from the composition and thermodynamic data of Burnham (1979), using equations (3.6) and (3.9), the above equation for the effects of H_2O, and the pressure-dependent R_i^{an} factors for the pyroxene-like components. Such precise correspondence between experimental and calculated liquidus temperatures in melts as compositionally complex as those of the Mt. Hood andesite corroborates the validity of all of the thermodynamic relations employed in the calculations.

The experimental results on Mt. Hood andesite composition (Eggler and Burnham, 1973, Figure 2) also reveal that, as the slope of the plagioclase liquidus decreases with pressure, dT/dP for the clinopyroxene liquidus increases. An obvious explanation for this phenomenon is that the activity of the clinopyroxene solid solution is increased by the increased activity of cts component, which in turn is brought about by an increased $Al^{IV} \rightarrow Al^{VI}$ shift with pressure.

Certain implications of these plagioclase melting relationships for the generation and emplacement of H_2O-bearing (but undersaturated) magmas will be discussed in subsequent sections, but one implication of a more general nature should be noted here. The igneous plagioclase thermometer of Kudo and Weill (1970), as well as that of Mathez (1973), involves an assumption that $a_{ab}{}^m$ and $a_{an}{}^m$ are independent of melt composition, but it has just been shown that $a_{an}{}^m$ is both composition and pressure dependent. Therefore, this thermometer can yield reliable temperature estimates only in those rocks that have equilibrated at very low pressures and then only if allowance is made for effects of the $Al^{IV} \rightarrow Al^{VI}$ shift on $a_a{}^m$.

Before turning to the subject of the generation of magmas, the melting behavior of other common rock-forming minerals should be briefly examined. Among the common anhydrous silicates of interest here, perhaps the most complicated melting relations, except those of plagioclase, are those of the alkali feldspars, mainly because they are coupled through a common component, $NaAlSi_3O_8$, to plagioclase which generally begins to crystallize at higher temperatures in melts of present interest. In magmas (liquid plus crystals) of given bulk composition, liquidus temperatures of alkali-feldspar solid solutions increase with pressure, from only a few degrees to as much as $12°C/kb^{-1}$, depending upon the composition of the melt and the nature of the coexisting crystalline phases. Liquidus temperatures of the pyroxenes also increase with pressure, as noted above for the Mt. Hood andesite composition, and may displace plagioclase as the primary liquidus phase at high pressure.

The melting relations of quartz are of special interest because of the very large positive effect of pressure on melting temperatures. This large dT/dP for the liquidus arises from the very low entropy of melting of silica minerals relative to other framework minerals that have comparable volume changes on melting, such as the feldspars. For this same reason, however, dilution of the melt by H_2O, with its attendant relatively large partial molal entropy of mixing terms, reduces the Clapeyron slope. The net result in melts of interest here is that quartz liquidus temperatures increase about $20°C$ to $30°C$ kb^{-1}, the latter value being approached in relatively silicic melts with low H_2O contents. Thus, an isothermal or adiabatic reduction in pressure on a granitic magma initially in equilibrium with quartz may result in resorption of quartz phenocrysts, a commonly observed feature in porphyry-copper porphyries.

Melting relations of the other major oxide, magnetite, are complex, owing mainly to the effects of oxygen fugacity on the valence states of iron. The results of Eggler and Burnham (1973) indicate that the liquidus temperature of magnetite decreases with increasing pressure in H_2O-bearing andesitic magmas of given bulk composition and under the oxygen fugacity of the quartz-fayalite-magnetite (QFM) buffer. The decrease is very pronounced, because increasing pressure greatly increases the hydrogen fugacity without appreciably increasing the "buffered" oxygen fugacity; this causes the reaction $\frac{1}{2} H_2 + O^{2-} + Fe^{3+} \rightleftharpoons OH^- + Fe^{2+}$ to proceed to the right, which lowers the activity of Fe^{3+} in the melt and, in turn, the equilibrium temperature of magnetite.

The melting relations of the hydrous minerals, hornblende and biotite, doubtless are the most complicated of the common rock-forming minerals. They may be regarded as complex limited solid solutions of several components that melt (or dissolve in silicate melts) incongruently, and their thermal stabilities are sensitive to the activities of certain essential components in the melt, notably H_2O. In terms of the components in silicate melts as presently defined, the average hornblende of diorites and quartz diorites (Leake, 1968), for example, is composed principally of $NaAlSi_3O_8$ (ab), $CaAl_2Si_2O_8$(an), $Ca_{1.33}Al_{2.67}Si_{1.33}O_8$(cts), $Ca_{1.33}Mg_{1.33}Si_{2.67}O_8$(di), $Mg_2Fe_2Si_2O_8$(ol), $Mg_{2.67}Si_{2.67}O_8$(en), Fe_6O_8(mt), $Fe_{2.67}Ti_{2.67}O_8$(il), and H_2O(w). Of these nine components, at least five (ab, an, di, w, and ol or mt) are regarded as essential to the formation of hornblende in dioritic melts and the conditions of crystal-melt equilibrium require that the chemical potentials of these essential components be the same in both phases. However, the fact that hornblende is a limited solid solution means that the chemical potential or activity of each of these essential components must exceed a certain minimum value, at a given pressure and temperature, for hornblende to be stable.

This activity effect is well illustrated in the system andesite (diorite)—H_2O—CO_2 (Eggler and Burnham, 1973, Figure 4). At 5.0 kb total pressure, the maximum thermal stability of hornblende (940°C) occurs where the melt contains approximately 4.6 wt. % H_2O ($X_w^m = 0.43$) which is equivalent to an activity of H_2O (a_w) of 0.45 (Burnham, 1975b). At this point, hornblende is in equilibrium with plagioclase, orthopyroxene, clinopyroxene, ilmenite, magnetite, and a granitic melt. If a_w is now lowered by somehow reducing X_w^m, the hornblende liquidus temperature decreases abruptly, because $a_w = 0.45$ apparently is the minimum value for hornblende stability at 940°C and 5.0 kb. The experimental data indicate that lowering X_w^m to 0.38, which lowers a_w to about 0.36, lowers the equilibrium temperature of hornblende to about 880°C, at which point the diorite solidus is intersected. On the other hand, if a_w is increased beyond 0.45 by addition of H_2O to the melt, the liquidus temperature again decreases, because increasing X_w^m decreases (by dilution) the mole fractions and, as discussed earlier, the activities of the other hornblende components in the melt. This decrease in liquidus temperature is gradual, however, and occurs only after the activity of at least one of the other hornblende-essential components falls below the critical value, as by disappearance of the corresponding crystalline phase into the melt.

The experimental results of Holloway (1973) on the system pargasite—H_2O—CO_2 in the temperature range 1000°–1100°C indicates that the maximum in the stability of pure pargasite occurs at $X_w^m = 0.48$, irrespective of pressure in the range 2.0–8.0 kb. This is 0.05 higher than that obtained in the system andesite—H_2O—CO_2 at 940°C and 5.0 kb and yields a value of a_w, which is approximately 0.12 higher. The reason for this discrepancy appears to be that magnetite (Fe_6O_8), an essential component of dioritic hornblende, disappears into the melt of the andesite system as X_w^m is increased beyond 0.43 (Eggler and Burnham, 1973, Figure 4), thereby causing the liquidus temperature of hornblende to be relatively lower at higher values of X_w^m. Had magnetite remained stable to higher values of X_w^m, hornblende probably would have persisted to a thermal maximum at about 950°C and $X_w^m \approx 0.48$.

This phenomenon of a thermal maximum in the stability of a hydrous mineral as a function of X_w^m (or a_w) also was observed in the system phlogopite—H_2O by Yoder and Kushiro (1969), and is to be expected in all cases where a hydrous-mineral solid solution melts incongruently. In the case of phlogopite, calculations based on the experimental data of Yoder and Kushiro indicate that the thermal maximum occurs at $X_w^m \approx 0.5$, which, considering the approximations involved, is essentially the same as for pargasite. Furthermore, a reduction in X_w^m below 0.5 also abruptly lowers the liquidus temperature of phlogopite, and an increase in X_w^m also lowers the temperatures more gradually, as in the hornblende system.

The effect of pressure on the liquidus temperature of hornblende (or biotite) in a magma of fixed bulk composition, including H_2O, generally is small and positive (Eggler and Burnham, 1973, Figure 2), although the slope eventually becomes negative at high pressures as a result of changing coordination of Al in the melt and other factors. The small pressure effect again is due mainly to the large partial molal entropies of mixing in the melt of some of the hornblende components whose mole fractions are very small.

Emphasis has been placed here on the melting relations of hornblende for several reasons, one of the more important of which is that they place lower limits on the H_2O content of magmas that are associated with porphyry copper-molybdenum deposits, as hornblende is a common phenocryst mineral in these porphyries. According to the experimental results of Eggler and Burnham (1973, Figure 2), hornblende is stable up to approximately 925°C at 1.2 kb in H_2O-saturated (4.7 wt. %, $X_w^m = 0.43$) melt of the Mt. Hood andesite. If X_w^m is now lowered at this pressure to 0.30, the stability temperature of hornblende decreases abruptly, as discussed above, to approximately 850°C. However, at 1.2 kb and 850°C, $X_w^m = 0.3$ yields a value of $a_w = 0.5$, and at $a_w = 0.5$ the solidus temperature of andesite is about 870°C. Therefore, the minimum value of X_w^m at which hornblende can coexist stably with melt in this system is approximately 0.3, which is roughly equivalent to 3.0 wt. % H_2O. This minimum H_2O content is almost independent of magma composition and, from the work of Yoder and Kushiro (1969), also appears to hold for biotite.

It should be emphasized that 3 wt. % H_2O is a minimum value for the crystallization of hornblende when only a few percent of interstitial melt is present. For hornblende to form euhedral or even subhedral crystals requires a considerably greater percentage of melt, which can be produced only by depressing the solidus through increasing X_w^m. Thus, raising X_w^m to 0.4 ($\sim$4.3 wt. % H_2O) at 1.2 kb pressure lowers the solidus of the Mt. Hood andesite to approximately 820°C and raises the hornblende stability to approximately 920°C, thereby yielding 35–50% melt at the first appearance of hornblende. Under these latter conditions, all essential components of hornblende are present as solid phases, most of them in anhydrous solid solutions. The crystallization of hornblende, therefore, results from reaction between these solid phases and the hydrous melt. Furthermore, as the hornblende-forming reaction proceeds with falling temperatures, X_w^m is constrained to remain above 0.3.

THE GENERATION OF CALC-ALKALINE MAGMAS

Modes of Generation

Much has been written in recent years about the origin of calc-alkaline rocks and the old controversy over whether andesitic magmas are generated by differentiation from more mafic basaltic magmas or by partial fusion (anatexis) has received new impetus from the precepts of plate tectonics. The characteristic spatial association in the circum-Pacific region, as well as other parts of the world, of andesitic to dacitic volcanoes with underlying Benioff zones lends strong support to the view that calc-alkaline magmas are generated by partial fusion of oceanic crustal material as it descends in subduction zones defined by deep-focus earthquakes of the Benioff zones. This does not preclude the possibility that fractional crystallization-differentiation has played an important role, however, as many of the major andesitic volcanoes in the island arcs of the southwestern Pacific are built on a base of tholeiitic basalts that are less primitive than the basalts of the ocean basins (Jakeš and White, 1969). It may be, therefore, that the initial magmas generated in the upper parts of a subducting plate and overlying lithosphere were more basaltic and that the andesitic-dacitic magmas were produced from them by differentiation. It also may be that the initial magmas were andesitic and the earlier phases of intrusion (and extrusion) became more mafic by passage through, and reaction with, the immediately overlying periodititic lithosphere that probably was as hot as, or hotter than the magma at its source (Minear and Toksöz, 1970). In any case, it is useful to examine some of the processes that can be inferred to occur near the upper margins of a subducting plate, as the andesitic-dacitic terranes overlying active or inferred fossil subduction zones contain a major portion of the world's copper resources in porphyry-type deposits.

Generation of Magmas in Subduction Zones

Models for the consumption of oceanic crust and lithosphere in subduction zones that underlie arc-trench systems vary widely in detail, but there is general agreement that oceanic lithosphere, carrying with it basaltic crust and minor intercalated sediments, descends in the vicinity of oceanic trenches at angles ranging from 20° to 50° (Allen et al., 1975) or more and at rates up to 10 cm yr^{-1}. In the models of Minear and Toksöz (1970) and Oxburgh and Turcotte (1970), frictional (shear strain) heating along the slip zone raises temperatures near the upper surface of the descending plate to approximately 1000°C at depths as shallow as 36 km. However, the results of Turcotte and Schubert (1973), using the average widths of arc-trench gaps and dips of the Benioff zones as indicators of the depth of magma generation, suggest that the earlier depth estimates may be low by a factor of three. This latter estimate, on the other hand, does not take into consideration the possibility that seismic activity in Benioff zones may be concentrated in the cooler inner parts of the descending plate, yielding dips of the Benioff zones that are steeper than the dips of the slip zones. In any case, a temperature of 1000°C can be well within the melting range of basaltic rocks, depending mainly upon the amount and mode of occurrence of H_2O.

Prior to descent of oceanic crust, H_2O occurs as water in interstitial pores of the sediments and in fractures and vesicles of the basalts, as well as hydroxyl in clays of the sediments and chloritic minerals of the basalts. Presumably, most of the water-rich sediments are "scraped off" the descending plate and accumulate as a thick, narrow wedge that forms the inner wall of the trench, hence the average total H_2O content of the uppermost few kilometers of the plate at greater depths probably does not normally exceed two or three percent by weight. As the plate descends, it must undergo metamorphism, especially in the H_2O-rich upper part. The initial metamorphic reactions that occur in the basaltic rocks as temperature and pressure increase are predominantly hydration reactions that yield greenschist facies mineral assemblages. These reactions can consume far more H_2O (up to 13 wt. %) than is generally available and they generate heat (up to 60 cal gm^{-1} of rock hydrated). Consequently, essentially all of the

THE GENERATION OF CALC-ALKALINE MAGMAS

initial pore water becomes bound in hydrous minerals at this stage, except for a very small amount in a concentrated brine that may be trapped in fluid inclusions or interstitially. Even this fluid largely disappears at higher temperatures with the formation of Cl-bearing scapolite. Thus, prior to descent of the incompletely hydrated portion of the plate into the higher-grade environment of the amphibolite facies, essentially all of the volatile constituents, including CO_2, are bound in crystalline phases and the pore volume approaches zero.

Upon further descent of the plate and heating to temperatures of 550–600°C, the upper part passes into the amphibolite facies of metamorphism. The more highly hydrated mineral assemblages of the greenschist facies react at this stage to produce larger amounts of less hydrated amphibole, which generally contains 2.0–3.0 wt. % H_2O. Unless the bulk H_2O content of the rocks exceeds this amount, however, significant quantities of free H_2O are not released from the system, because basalts can be converted almost entirely into compositionally equivalent amphibole (Yoder and Tilley, 1962). In the event that bulk H_2O contents exceed approximately 2 wt. %, on the other hand, the excess is likely to escape into the overlying anhydrous lithosphere where it would be immediately absorbed in serpentinization and other hydration reactions. Therefore, an H_2O-rich pore fluid generally will not be present in these amphibolitized rocks at pressures of 10–15 kb.

The minor amounts of intercalated sediments that might be carried down with the descending plate may contain carbonates and the possibility remains that a CO_2-rich fluid phase could be produced by decarbonation. H_2O-bearing siliceous carbonates may react to produce amphibole and CO_2 if the thermal gradient along the slip zone is more than twice the recent estimates of Turcotte and Schubert (1973). Any CO_2 that might be produced in this fashion, however, is likely to be absorbed by reaction with plagioclase to form carbonate scapolites, which are stable under these conditions (Goldsmith, 1976). Thus, it seems highly probable that, as an amphibolitized descending plate approaches the onset of melting, a separate fluid phase generally will not be present, except for a very minor amount trapped in intracrystalline inclusions. In the following paragraphs, attention will therefore be focused on melting relationships under these conditions.

The melting relations of an amphibolite of olivine tholeiite composition plus H_2O are shown in pressure-temperature projection in Figure 3.3, together with an average geothermal gradient in the slip zone ($\Delta T/\Delta D$) of 11°C km^{-1} of depth (Turcotte and Schubert, 1973). The beginning of melting (solidus) for $a_w \approx 1.0$ and the hornblende liquidus (Hb-L) curves are based on the experimental results of Allen et al. (1972), Boyd (1959),

Eggler and Burnham (1973), Hamilton et al. (1964), Helz (1973), Hill and Boettcher (1970), Holloway (1973), Holloway and Burnham (1972), and Yoder and Tilley (1962). The $a_w < 1.0$ and "fluid-absent" (Hb-S) solidus curves have been calculated from the thermodynamic relations discussed previously and the $a_w \approx 1.0$ solidus relations. The reactions involving hornblende are not univariant, as implied, owing to changes in hornblende composition with pressure, temperature, and the nature of the coexisting phases; the curves have been drawn near the high-temperature boundary of the reaction zone. The thermal minima in the isoactivity solidus curves are caused, in part, by pressure-induced phase changes in the subsolidus phases, from feldspar-rich assemblages at lower pressures to denser aluminous pyroxenes at higher pressures. However, the solidi for $a_w \leq 0.5$ would pass through thermal minima in this pressure range even without solid-phase changes, as the small negative ΔV of solution for H_2O at these high pressures, when weighted by X_w^m, is more than compensated by the positive ΔV of the melting reactions. Also, the abrupt change in slope of the hornblende melting reactions between approximately 20 and 22 kb is due to the incoming of much denser pyrope-rich garnet. Thus, at higher pressures and temperatures above the $a_w \approx 1.0$ solidus, the stable assemblage consists mainly of garnet, olivine, aluminous pyroxenes, and hydrous melt.

An amphibolitized olivine tholeiite of essentially zero porosity (no pore fluid present) can descend along a pressure-temperature path such as A-B in Figure 3.3 to a depth of approximately 78 km and a temperature of 940°C without melting. At point B on this path, which is near the average gradient estimated by Turcotte and Schubert (1973), hornblende reaches its maximum-pressure stability limit and reacts to produce garnet peridotite plus hydrous melt. For melt to form here, $a_w \geq 0.3$, and this requires that $X_w^m \geq 0.5$ (≥ 6.4 wt. % H_2O). Therefore, the maximum amount of melt that can be produced at this point from an amphibolite that initially contained 1.5 wt. % H_2O is approximately 20%. For an amphibolite of lower total H_2O content, the maximum amount of melt produced is correspondingly less. It is also less if the average geothermal gradient is lower, as higher activities of H_2O (hence higher values of X_w^m) are required to produce melt at lower temperatures. The actual amount of melt produced at point B depends upon bulk composition, but it is likely to be close to the maximum for the prevailing H_2O content of the amphibolite.

According to the experimental results of Allen et al. (1972), the slope of the reaction curve at point B ($dP/dT = \Delta H_r / T \Delta V_r$) is positive. This implies that ΔH_r is negative (exothermic), as the reason for the thermal maxima in the hornblende stability curves (Figure 3.3) is that ΔV_r changes from positive to negative with the incoming of dense pyrope-rich garnet. However, the small slope of the reaction curve suggests that the amount of heat produced by this incongruent melting reaction also is small, and this is indicated by the deflection of path B-C toward slightly higher temperatures in Figure 3.3.

A noteworthy feature of path A-B-C, or any other path with an average gradient ($\Delta T/\Delta D$) between 8.5 and 13.5°C km^{-1}, is that melting occurs abruptly at a nearly constant pressure of 22–23 kb (75–80 km depth). The maximum amount of melt produced from a fully hydrated amphibolite in this narrow pressure (depth) interval ranges from approximately 5% at 700°C to 40% at 1050°C and the corresponding melt compositions range from trondhjemitic to dioritic. Continued descent of the partially melted plate along a path such as B-C, and beyond, results in further melting, although the heating rate probably declines due to loss of shear strength. In any event, the amount of melt that can be produced at point B and beyond is such that it may be expected to separate from the plate and ascend, perhaps diapirically, into the overlying lithosphere. If separation occurs near point B, the melt will contain at least 6.4 wt. % H_2O; if at point C, it will contain at least 5.4%. In either case, the production of copious melt in a very narrow depth interval provides a rational explanation for the remarkable alignment of volcanic centers at a uniform distance from a given oceanic trench, a feature that was emphasized by Turcotte and Schubert (1973).

Attention has been focused thus far on the partial melting of an amphibolite with essentially zero porosity, but the picture is not altered appreciably even if the porosity prior to the onset of melting was an exceptionally high 3%. The fluid in these interstitial pores almost certainly would not be pure H_2O, but a mixture predominantly of CO_2 and H_2O in proportions that would be determined by pressure, temperature, and the metamorphic mineral assemblage. If it is assumed, for example, that H_2O and CO_2 are present in approximately equal molal proportions, melting of the amphibolite would begin along path A-B at approximately 800°C. Upon first appearance of melt, H_2O partitions into it preferentially to CO_2, owing to its higher solubility. This lowers a_w and arrests the melting process; therefore, only a vanishingly small amount of melt forms at this temperature. Continued heating is required to produce significant melting; indeed, only approximately 5% melt is produced on the low-pressure side of point B. This is insufficient to dissolve all the CO_2 originally in the pore fluid, but upon incongruent melting of hornblende all of the fluid phase will dissolve in the melt and the contribution of original pore fluid to the total amount of melt produced is minor. Had the original pore fluid contained a larger or smaller proportion of H_2O, the results would have differed only in the amounts of melt produced along the path A-B and in the temperature at which melting would begin.

In the event the geothermal gradient in undisturbed lithosphere overlying the subduction zone is typical of oceanic regions (Ringwood et al., 1964), as shown in Figure 3.3, melt that separates at a point such as C will be heated as it ascends. By virtue of its high H_2O content, this melt is in disequilibrium with the low-melting fraction of the hotter overlying lithosphere; the greater the H_2O content and temperature difference, the greater the disequilibrium. Consequently, reactive assimilation, which operates to restore equilibrium, will tend to proceed in the direction of the steepest thermal gradient. This direction is approximately normal to the dip of the plate, but buoyancy forces tend to cause the less dense melt to rise vertically. Hence, the resultant path of ascent may be curved convex in the direction of plate descent, approaching the vertical when thermal and chemical equilibrium are established between magma and lithosphere, as at point D in Figure 3.3.

The process of assimilation involves primarily the dissolution of the low-melting fraction (perhaps 10% or less) of the lithosphere and recrystallization of the disaggregated refractory residuum, principally pyroxenes, olivine, and garnet. As this crystalline residuum sinks, it transfers heat to the melt, promotes mixing, and displaces the melt upward. By the time this syntectic melt reaches equilibrium at point D, for example, it will have assimilated lithospheric material equivalent in mass to approximately 80% of its initial mass at point C, contain approximately 3.0 wt. % H_2O, and be dioritic (andesitic) in composition. It will, therefore, have nearly doubled in total mass and become more mantle-like in certain trace element and isotopic characteristics. In the event the rate of equilibration is more rapid than is indicated by the path C-D, owing to the efficiency of mixing and heat-transfer processes, the actual path might intersect the geotherm at a higher temperature (and greater depth). As a consequence, more assimilation will occur and the melt will contain correspondingly less H_2O. In any event, an expected consequence of the subduction of oceanic crustal material is the generation of hydrous calc-alkaline magmas over a relatively narrow depth interval that have the requisite properties to give rise to explosive island-arc volcanism (H_2O) and the associated copper porphyries.

Generation of Magmas under Continents

The distribution of porphyry copper-molybdenum deposits and associated quartz monzonitic intrusives in the North American cordillera suggests a spatial association with Paleozoic geosynclines, but there is only weak evidence for the existence of subcontinental subduction zones in most of this region during late Mesozoic and early Cenozoic time to which these intrusives might be related. The existence of these intrusives, their associated volcanics, and the widespread volcanism that has extended from the late Mesozoic almost to the present day are nonetheless clear evidence that calc-alkaline magmas have been and perhaps are being generated under this region on a large scale. Furthermore, the geophysical evidence indicates that there are large areas of high heat flow, moderately thin crust, and anomalously light upper mantle in this region, features that are suggestive of abnormally high geothermal gradients. Whatever the ultimate cause of these features, the widespread occurrence of basaltic volcanoes suggests that their immediate cause is the invasion of the continental crust by mafic magmas generated in the underlying upper mantle.

These mafic magmas apparently are not the immediate parents of the copper-molybdenum porphyry magmas because they are generally too poor in potassium to yield the porphyries by differentiation alone, despite the evidence (B. W. Chappell, personal communication, 1974) that the porphyry magmas appear to have been derived from "igneous-type" source rocks (including metavolcanics and metagreywackes). Therefore, it is postulated that the porphyry magmas were produced mainly by the interaction of mantle-derived magmas with lowermost crustal rocks which, according to Boettcher (1971), Mehnert (1975), and others, consist mainly of mafic amphibolites, their dehydrated residua (granulites), and intercalated gneisses of more felsic compositions (which increase in abundance upward). The principal role of the mantle-derived magmas is, therefore, to provide large quantities of heat for partial melting and assimilation of lower crustal rocks.

Emplacement of an H_2O-poor mafic magma at a temperature of 1200°C into lower crustal amphibolites at a depth of approximately 40 km and a temperature of 600°C (average geothermal gradient of 15° km^{-1}) contains enough total heat, including latent heat of crystallization, to bring equal masses of magma and amphibolite to an equilibrium temperature of approximately 1025°C. From Figure 3.3 (point G), it can be seen that this temperature is slightly above the beginning of melting of mafic amphibolite in the absence of an aqueous pore fluid (point F); consequently, melting may be expected. The amount of anatectic melt produced is dependent mainly on the H_2O content of the amphibolite; for a moderately hydrated amphibolite (~1.5 wt. % H_2O), as much as 50% melt can be formed between F and G. Therefore, under the postulated optimum conditions, the emplacement and crystallization of a mantle-derived mafic magma into lower crustal amphibolites is capable of generating in them anatectic melts equal in mass to nearly one-half the mass of intruding magma. Moreover, these melts contain approximately 3.0 wt. % H_2O, are quartz dioritic in composition and may or may not be sufficiently potassic to yield the copper-molybdenum porphyries by direct crystallization, depending upon the bulk composition of the amphibolite.

This example obviously is not geologically realistic, as mantle-derived magma is not likely to be emplaced in the amphibolites and there crystallize without interaction. Instead, as magma rises into the amphibolites it reactively assimilates and partially melts them at the contacts, the heat required being derived from cooling and crystallization of olivine, pyroxene, and calcic plagioclase from the magma. These crystals, together with the olivine- and pyroxene-rich residuum from the amphibolite, sink near the walls of the intrusive and displace hotter magma upward in the interior, where it mixes with slightly cooler, but more H_2O-rich, melt produced from the amphibolite. By this mechanism, magma bores its way upward, and, by the time a given mass of initial magma has reacted with an equivalent mass of amphibolite, as much as half that mass of dioritic melt containing approximately 3.0 wt. % H_2O may remain. If the amphibolite initially contains less H_2O, correspondingly less melt will remain, but it will have the same H_2O content. Also, the melt will continue to reactively assimilate any hornblende with which it comes into contact until a temperature corresponding to point F is reached.

The emplacement of the same mantle-derived mafic magma into a felsic gneiss complex that consists of muscovite-, biotite-, and amphibole-bearing assemblages will result in extensive assimilation, as well as partial melting (anatexis) of the muscovite-bearing rocks some distance from the intrusive contacts. The pertinent phase relationships in a granodioritic gneiss are shown in Figure 3.4; they are based on the thermodynamic relations discussed previously and the results of experiments reported by Blencoe (1974), Burnham (1967), Piwinskii (1968), and Whitney (1975a). Thus it can be seen that heating a muscovite-bearing gneiss along a path such as E-F-G in Figure 3.4, in the absence of an aqueous pore fluid, results in partial melting at a temperature of approximately 710°C. The melt produced here is very H_2O-rich ($X_w^m \approx 0.59$), hence, only approximately 0.5% melt of granitic composition is produced for each percent of muscovite in the original rock. Should this melt be produced in sufficient quantities to separate from its source rock (and its kyanite reaction product), it could rise to depths of 10 or 15 km before solidifying, provided it did not cool significantly in the process. Cooling below the muscovite solidus, once the melt has been separated from kyanite, results in recrystallization of only minor amounts of muscovite, and melt therefore persists to the H_2O-saturated solidus ($a_w \approx 1.0$ in Figure 3.4) where it may undergo pegmatitic crystallization. This is probably the process by which pegmatitic segregations are formed in migmatite terrains. It is of small consequence in the formation of the copper-molybdenum porphyries, but it illustrates the important fact that H_2O-rich magmas are readily produced at moderately low temperatures from rocks containing hydrous minerals, in the absence of an aqueous pore fluid. It also illustrates the fact that such H_2O-rich magmas cannot ascend to very shallow depths before crystallizing completely.

Continued heating of this gneiss complex to approximately 820°C approximately doubles the amount of melt formed. At this temperature, biotite-bearing assemblages begin to melt and the amount of melt is further increased by about 1.1% for each percent biotite in the original rock. Therefore, a granodioritic gneiss that contains 20% total micas will yield approximately 22% granitic melt in which $X_w^m = 0.35$ (~3.6 wt. %). Should this melt separate from its source rocks and rise without loss of heat (adiabatically) it could reach a depth as shallow as approximately 2.0 km before solidifying. Moreover, it would become saturated with H_2O at a depth near 3.0 km and would ultimately crystallize as a biotite-free granite that may contain fayalite or cordierite, depending upon the degree of Al saturation and f_{O_2}. On the other hand, should the magma, once separated from the solid residuum of incongruent melting of biotite, undergo cooling below the biotite solidus for $X_w^m = 0.35$, only a small percent of biotite would crystallize and melt would again persist to the H_2O-saturated solidus ($a_w \approx 1.0$) for the prevailing pressure. This, again, might well be the dominant process by which plutonic two-mica granites and some pegmatites are formed, as well as possibly some of the molybdenum porphyries, but the magmas are generally too poor in calcium to be the parents of most copper porphyries.

Heating along path E-F (Figure 3.4) of hornblende-bearing granodioritic gneisses, which also generally contain biotite but not muscovite, to a temperature of about 860°C, results in the incongruent melting of hornblende. This process yields 0.6-0.7% melt, in which $X_w^m = 0.3$, for each percent hornblende in the rock and it occurs approximately 140°C below the corresponding hornblende solidus of the mafic amphibolite, a phenomenon that results partly from the different composition of the hornblende, but mainly from the much lower activities of hornblende-forming components in granodioritic melts, as compared with gabbroic melts. Continued heating of this system to 1000°C (point F), the approximate temperature of melting of the mafic amphibolite, causes additional melting of the granodioritic gneiss and consequent reduction in the H_2O content of the melt. The equilibrium H_2O content of the melt at point F is approximately 1.2 wt. % ($X_w^m = 0.15$) and the amount of melt produced in this temperature interval is approximately 2.5 times the amount present at 860°C, provided the total H_2O content of the original gneiss did not exceed 1.0 wt. %. In the event the initial H_2O content exceeded this amount, which is equivalent to approximately 25% biotite or 50% hornblende, somewhat more melt with a higher H_2O content would be present at point F. However, if the original granodioritic gneiss is assumed to have the H_2O content of Nockolds' (1954) average biotite-hornblende granodiorite (0.69 wt. %), approximately 58% of the rock will have melted to yield a granitic melt with 1.2 wt. % H_2O.

Thus far consideration has been given to the production of anatectic melts from felsic gneisses without chemical interaction between these gneisses and mantle-derived mafic magmas. Unlike the mafic amphibolites, these mantle-derived magmas are in gross chemical disequilibrium with nearly all the phases of the gneisses, except perhaps the pyroxenes and spinels formed by incongruent melting of hornblende and biotite. They will therefore melt and assimilate these rocks until equilibrium is established near the hornblende solidus ($X_w^m = 0.3$). At this point, which is near 860°C on path E-F, the melt is granitic in composition and has lost most of the chemical characteristics of the mafic magma, except for certain "incompatible" elements that partition strongly toward the melt. Moreover, the amount of syntectic melt that remains here, per unit mass of original mafic magma, is about twice the amount of quartz-dioritic melt that can be present at 1000°C in equilibrium with a mafic amphibolite of the same total H_2O content.

In the event that a syntectic quartz-dioritic melt produced by assimilation of mafic amphibolite at point F in Figure 3.4 is separated from its source rock and emplaced at slightly shallower depths in felsic gneisses, it also will react with them, but the conditions are now different from those considered previously. The melt, in which $X_w^m = 0.3$, is saturated with hornblende and a plagioclase of $Ab_{35}An_{65}$ composition; hence any reduction in temperature, as along path F-H in Figure 3.4, causes these minerals to crystallize. Crystallization, in turn, increases the H_2O content of the melt, approximately 0.03% for each percent of plagioclase and less for hornblende. However, dilution by assimilation of phases with which the melt is undersaturated, especially the alkali feldspars and quartz, lowers the activity of hornblende "components" in the melt and therefore markedly lowers its stability temperature, as can be seen by comparing the "Hb-L" curves in Figures 3.3 and 3.4. Therefore, only a very small amount of hornblende crystallizes along path F-H until a temperature near 900°C is reached. At this temperature, $X_w^m \approx 0.33$ and the melt is in equilibrium with hornblende in the gneiss, but it still will react with the biotite. Consequently, a further drop in temperature to approximately 850°C is required to bring the melt, now granitic in composition with about 3.75 wt. % H_2O ($X_w^m = 0.36$), into equilibrium with the gneissic wallrocks.

In summary, essentially anhydrous mafic magmas rising from the upper mantle into typical lower crustal rocks of continental regions have the capacity, in reaching equilibrium with these rocks, to generate relatively large quantities of H_2O-bearing melts by anatexis and assimilation. The melts generated under conditions of equilibrium with hydrous minerals range in composition from granitic to dioritic, depending upon the nature of the source rocks, and their H_2O contents range from approximately 3.0-9.0 wt. %. Melts produced at temperatures above those for equilibrium with hydrous minerals are more mafic and have correspondingly lower H_2O contents than their equilibrated counterparts. Whether or not they are in equilibrium with hydrous minerals, the more quartz dioritic of these melts are compositionally appropriate to yield the copper-molybdenum porphyries upon crystallization, with little or no differentiation. Moreover, they can be readily generated in lower crustal rocks in the absence of an aqueous pore fluid and have the requisite heat contents to carry them into the near-surface environment without solidifying.

In the event that lower crustal metamorphic rocks contain a mixed H_2O-CO_2 pore fluid, melting will begin at lower temperatures than the various solidi in Figure 3.4, except that for $a_w \approx 1.0$, and more melt is produced at a given temperature and pressure than in the fluid-absent cases. As in the case of partial melting in a subduction zone, a geologically reasonable amount of pore fluid has relatively slight impact on the melts produced by assimilation. However, owing to the very low solubilities of CO_2 in felsic melts under continental-crust pressures, there is a possibility that some melts produced from metasedimentary rocks may be nearly saturated with CO_2 (Holloway, 1976). This could be a significant factor in the formation of a fluid phase during emplacement of magma at shallow levels and will be discussed further below. First, however, it would be useful to examine the behavior of sulfur and metal sulfides in magma generation processes.

Behavior of Metal Sulfides in Magma Generation

The worldwide association of porphyry copper-molybdenum mineralization with calc-alkaline volcanism, which in turn appears to be commonly associated with convergent plate boundaries, led Sillitoe (1972) to postulate that the porphyry copper magmas have been produced by partial melting of abnormally metalliferous oceanic sediments and basalts in subduction zones. Although such a model is generally consistent with that developed here, abnormally metalliferous source rocks for the porphyry magmas do not appear to be required. Mafic rocks, whether they are metamorphosed oceanic basalts or lower crustal amphibolites, normally contain sufficient copper to ultimately yield a porphyry copper deposit, provided extensive hydrothermal processes operate in the porphyry system at shallow crustal levels to concentrate the metal. Petrographic observations indicate that unaltered plutonic hornblende-biotite granodiorites and quartz diorites, which contain 100 ppm or more copper also generally contain chalcopyrite, commonly poikilitically enclosed with pyrrhotite in early crystallized minerals. These features suggest that the parent magmas became saturated with a cupriferous pyrrhotite solid solution at a relatively early stage of crystallization and that a major factor in determining the sulfur content of a magma is the solubility of metal sulfides in the melt at its source.

Haughton et al. (1974) have shown experimentally that the solubility of sulfur in anhydrous basaltic melts at constant temperature and fugacities of O_2 and S_2 is directly dependent on the FeO content of the melt. For a Hawaiian basalt melt at 1200°C, the f_{O_2} of the QFM buffer, and f_{S_2} of approximately 1 bar, they calculated a solubility of 0.1 wt. % sulfur, which is in good agreement with that measured in quenched glasses by Moore and Fabbi (1971). They also claim, in agreement with earlier workers, that the solubility decreases by a factor of about five for each 100°C decrease in temperature between 1200 and 1000°C; this would imply a solubility of only 0.004 wt. % (~40 ppm sulfur) at 1000°C.

On the other hand, unpublished results of experiments conducted in the writer's laboratory on Fe-poor granitic melts at 850°C, in equilibrium with pyrrhotite, the QFM buffer, and a solution initially 1.0 m in NaHS, indicated sulfur contents as high as 0.17 wt. % at 2.0 kb H_2O pressure. Moreover, sulfur contents were found to decrease systematically with H_2O pressure through 6.0 kb. These results are consistent with those of Althaus and Skinner (B. J. Skinner, personal communication, 1974) who also observed similar sulfur solubilities and a systematic decrease in sulfur content with H_2O pressure above 1.0 kb.

These apparently discrepant results are readily reconciled by application of the solution model discussed earlier. At very low values of f_{H_2} and under the f_{O_2} conditions of the QFM buffer or lower, as in the experiments of Haughton et al. (1974), the predominant sulfur-bearing gaseous species is SO_2. Like CO_2, SO_2 is only very slightly soluble in aluminosilicate melts because it does not react with the bridging O^{2-}. In the presence of H_2O under the same f_{O_2} conditions, however, the ratio f_{H_2S}/f_{SO_2} increases with increasing f_w, because f_{H_2} increases in proportion to f_w at constant f_{O_2}, and H_2S is relatively soluble in the melt as SH^-. Thus, increasing f_w from some low value also causes f_{H_2S}, hence $X_{SH^-}^m$, to increase to a point where either (1) essentially all aqueous sulfur species are converted to H_2S and f_{H_2S} is unbuffered or (2) f_{H_2S} is buffered by an appropriate equilibrium phase assemblage. The experiments cited above are of the first type, where essentially all of the initial iron content of the melt (0.3 wt. % iron) was fixed in pyrrhotite and f_{H_2S} was therefore unbuffered. As a consequence, $X_{SH^-}^m$ increased with f_w to an estimated maximum of 0.012 (0.2 wt. % sulfur) at $f_w = 0.7$ kb and then decreased as equilibrium in the reaction

$$H_2O(v) + SH^-(m) \rightleftharpoons H_2S(v) + OH^-(m)$$

was displaced to the right. On the other hand, in more iron-rich, H_2O-undersaturated melts, such as those formed by partial melting of amphibolite, equilibria such as

$$2Fe_4Si_2O_8(m) + 8SH^-(m) \rightleftharpoons 8FeS(s) + 8OH^-(m) + Si_4O_8(m)$$

serve to buffer f_{H_2S}, hence $X_{SH^-}^m$, at levels determined by pressure, temperature, and the equilibrium phase assemblage. Therefore, instead of $X_{SH^-}^m$ passing through an isothermal, isobaric maximum with increasing f_w, as it did in the iron-poor, H_2O-saturated experimental melts, it may be expected to increase at $f_w > 0.7$ kb ($X_w^m > 0.35$), although at a rate much less than that of $X_{OH^-}^m$.

It is apparent from the equilibrium reaction immediately above that $a_{SH^-}^m$ varies inversely with the fourth root of a_{fa}^m (activity of the $Fe_4Si_2O_8$ component). This means that a 10-fold increase in X_{fa}^m causes $X_{SH^-}^m$ to be reduced only by approximately 50% because the activity coefficients of both fa and SH^- are insensitive to their respective small mole fractions in aluminosilicate melts. Therefore, a sulfur content of approximately 0.2 wt. % in an iron-poor, H_2O-saturated melt at 0.7 kb, 850°C, and $X_w^m = 0.35$ probably is close to the saturation value in a more iron-rich melt of the same H_2O content, but at temperatures in the range 1000–1100°C and pressures in the range 10–20 kb where f_w (at $X_w^m = 0.35$) ranges ~4–17 kb.

It will be noticed that this H_2O content ($X_w^m = 0.35$) is not greatly different from that of melts produced by fluid-absent melting of mafic amphibolites at pressures up to approximately 20 kb (Figures 3.3 and 3.4). Hence, inasmuch as a mafic amphibolite that contains as much as 1.5 wt. % H_2O can yield only about 50% melt at a temperature corresponding to points D or G in Figure 3.3, an initial sulfur content of 0.1 wt. % may be sufficient to yield a "sulfur-saturated" melt. If the initial H_2O content is less than 1.5 wt. %, less melt is produced and a correspondingly lower sulfur content of the source rock suffices to yield a "sulfur-saturated" magma. Thus, 20% melting of the metamorphosed equivalent of an average oceanic basalt (400 ppm sulfur, 87 ppm copper, and 0.6 wt. % H_2O) can yield a "sulfur-saturated" quartz dioritic melt that may contain as much as 400 ppm copper. This copper content is two or more times greater than that generally found in unaltered porphyries associated with the porphyry copper deposits, hence it is apparently unnecessary to call upon abnormally metalliferous source rocks for porphyry-copper magmas.

SHALLOW EMPLACEMENT AND SOLIDIFICATION OF HYDROUS MAGMAS

Emplacement of Hydrous Magmas

Emphasis in the foregoing discussions has been placed on the generation of H_2O-bearing, but undersaturated, calc-alkaline magmas, and it has been shown that such magmas may be the normal products of partial melting of amphibolites, even in the absence of pore fluids. Those magmas that are generated by the high-pressure breakdown of hornblende in subduction zones, as at point B in Figure 3.3, are incapable of direct ascent without being heated, because a decrease in pressure takes them back into the hornblende stability field, where they would completely solidify. Therefore, further heating and consequent melting is necessary to lower a_w below that required to stabilize hornblende and to provide the magmas with the capability of ascending to the surface. Once the pressure-temperature path of ascent, such as path C-D in Figure 3.3, crosses the ambient lithospheric geotherm, the thermal regime changes from one of heating to one of cooling and thereafter the marginal parts of the magma bodies undergo partial crystallization if equilibrium is attained. The extent of crystallization in a given magma body obviously depends upon the rate of cooling (ascent); the rate must exceed about $2°C\ km^{-1}$ ascent in order for pyroxene to crystallize. However, as this magma ascends, plagioclase will crystallize even without cooling, because reduction in pressure results in an increase in $a_{an}{}^m$ through reversal of the $Al^{IV} \rightarrow Al^{VI}$ shift discussed earlier. As a consequence, a body of magma ascending through oceanic lithosphere from a depth of 75 km theoretically could lose heat to its wallrocks at the rate of $0.6\ cal \cdot gm^{-1}\ km^{-1}$ of ascent (equivalent to cooling $2.2°C\ km^{-1}$) and still have more than half its initial mass arrive at the surface with the same temperature as at its source. Moreover, owing to loss of plagioclase, the melt has become slightly more mafic and H_2O-rich. Also, cooling of such a magma in a shallow chamber, at constant pressure, causes precipitation mainly of ferromagnesian minerals, because increasing a_w does not depress their liquidi as much as it does the liquidus of plagioclase.

In contrast, the intrusion of a syntectic H_2O-bearing dioritic magma into quartzo-feldspathic rocks of the continental crust results in direct solution of quartz and potassium feldspar, because the melt is undersaturated with respect to these phases. The heat required for this process, which is endothermic, is obtained by slight cooling and crystallization of plagioclase and pyroxene, minerals with which the magma is saturated. However, because the heats of solution of quartz and potassium feldspar are much smaller than the heats of crystallization of plagioclase and pyroxene, per unit mass, crystallization of 1% of these latter minerals can yield enough heat to dissolve approximately 2% quartz or 1.5% potassium feldspar. Consequently, calc-alkaline magmas emplaced at shallow depths in continental crustal rocks tend to be widely variable in composition, especially in silica content. Moreover, the more contaminated of these magmas generally should contain less H_2O (owing to dilution) than their less contaminated counterparts. This might explain why the early "rhyolite" intrusives (76% SiO_2, 5.1% K_2O) at El Salvador, Chile (Gustafson and Hunt, 1975) lack associated hydrothermal activity, whereas extensive hydrothermal activity accompanied the later granodiorite porphyries (62% SiO_2, 3.6% K_2O).

A possible exception to the above generalization that assimilation of crustal rocks results in a diminution in H_2O content is when a magma containing less than about 3.0 wt. % H_2O intrudes muscovite-rich rocks at pressures greater than about 4.0 kb. In this case, the melt may gain a small amount of H_2O. However, it should be noticed that because the degree of crystallinity increases from the interior of an intruding magma toward the cooler wallrocks through which it passes, the H_2O content of the melt also increases in the same direction. This buildup in H_2O content of the residual melt is proportional to the degree of crystallinity until the H_2O content reaches approximately 3.0 wt. %, whereupon hornblende or biotite becomes stable. Thereafter the H_2O content of the melt remains approximately constant until either one of the already crystallized minerals involved in the hornblende- or biotite-forming reaction, or the melt, disappears. For a melt of Nockold's (1954) average biotite-hornblende granodiorite composition that initially contained less than about 0.7 wt. % H_2O, the melt, which constituted less than 25% of the magma just prior to the appearance of hornblende, would disappear first at a temperature of 780 °C or above, depending upon total pressure (Figure 3.4). For a more H_2O-rich initial melt of the same silicate composition, melt would persist to lower temperatures (toward the margins of the intrusion) and eventually become saturated with H_2O.

In the above discussion regarding the distribution of H_2O in an intruding magma, the homogenizing effects of diffusion and turbulent flow of the magma have been neglected. The neglect of diffusion is justified, because it has been shown experimentally (see Carmichael et al., 1974, p. 140 for summary) that the diffusivity of H_2O in felsic magmas is too small to effect a significant redistribution of H_2O within a geologically reasonable length of time. Mixing induced by turbulent flow of magma tends to reduce thermal gradients in the flowing interior and steepen them near the walls. This, in turn, has an inverse effect on gradients in X_w^m, but does not affect the overall distribution of total H_2O content.

Differentiation, on the other hand, whether caused by gravitational settling of crystals or by differential viscous drag effects on crystals during intrusion ("filter pressing") may result in marked redistribution of H_2O in a magma body. For example, in a magma initially containing 3.0 wt. % H_2O, the settling of 10 vol. % plagioclase from level A to level B, which displaces an equal volume of melt upward from level B to level A, produces a magma at level A with 3.27 wt. % total H_2O and one at level B with only 2.7 wt. % total H_2O. Thus, differentiation can cause marked concentration of H_2O in the upper parts of a magma body, whether during or after emplacement. Also, of course, once a magma becomes saturated with a volatile component, further reduction in pressure causes bubbles to form and these bubbles will rise at a relatively rapid rate (Burnham, 1967).

Although the marginal parts of ascending magma bodies must lose large amounts of heat, both by conduction into, and assimilative reactions with the wallrocks through which they pass, the interior parts may not be significantly affected. Inasmuch as such magma bodies of dioritic composition were generally in equilibrium with calcic plagioclase in their source regions and dT/dP for the plagioclase liquidus is negative (Figure 3.4), reduction in pressure during ascent causes plagioclase to crystallize. This process is exothermic, owing to the latent heat of crystallization, hence interior magma temperatures might be expected to increase. The rate of increase is approximately 2.0–2.5°C for every percent of plagioclase crystallized, but this is partially offset by the cooling effects of adiabatic (isentropic) expansion, which amounts to about $1.5°C/kb^{-1}$ of decompression. For the granodioritic magma in Figure 3.4, from which approximately 10–15% plagioclase would crystallize in ascending isothermally from a 42- to a 4-km depth (from 11 to 1 kb), it is apparent that the latent heat is more than sufficient to offset decompression effects. Thus, it is reasonable to expect that bodies of calc-alkaline magma may reach shallow crustal levels with interior temperatures at least as high as prevailed in the source regions. Furthermore, if gravitational settling of plagioclase occurs during and after ascent, the upper parts of the magma body becomes enriched in H_2O. Crystal settling also may lead to formation of cumulate plagioclase bodies (anorthosites) at deeper levels in magma chambers of large vertical extent.

In considering the process of volatile separation in an ascending magma, attention also must be given to volatiles other than H_2O. Should a dioritic melt, which contains 3.0 wt. % H_2O and 0.6 wt. % CO_2 at its source, ascend to depths near 18.5 km (~5 kb), it will become saturated with CO_2, but not with H_2O. Further ascent to a depth of 15 km will cause approximately 17% of the CO_2 (0.1% of the total mass) to separate as bubbles. The generation of this fluid phase, in which H_2O is soluble, will result in transfer of some H_2O from the melt to the CO_2-rich bubbles. Under these conditions, the equilibrium H_2O content of the bubbles is about 10 wt. %; consequently only about 0.01 wt. % H_2O is transferred. At lower pressures, however, the dessicating effect of CO_2 is much larger; for example, continued ascent of the magma to a depth of 3.7 km will result in transfer of approximately 0.14 wt. % H_2O to the fluid phase which constitutes, by weight, approximately 0.55% of the total and contains about 35% H_2O. Thus, although this dessicating effect of CO_2 is relatively small per unit mass of magma, the upward migration of the fluid-phase bubbles over large vertical distances (near 15 km in this hypothetical example) could result in the concentration of significant amounts of H_2O-bearing fluids in the upper parts of magma chambers. Just how important this process is in the formation of porphyry copper-molybdenum systems is difficult to assess; the evidence from fluid inclusions, which homogenize at near-magmatic temperatures (Roedder, 1971) and generally contain less than approximately 12 wt. % CO_2, strongly suggests that the first fluid phase to separate from porphyry magmas is H_2O-rich, not CO_2-rich as are fluids from more basaltic magmas (Carmichael et al., 1974, p. 300). Further discussion of the distribution of components between coexisting silicate and fluid phases will be postponed until some of the processes associated with the solidification of H_2O-bearing magmas at shallow depths have been examined.

Solidification of Hydrous Magmas

Since publication of the first edition of this book in 1967, numerous descriptions of porphyry copper-molybdenum deposits have been published in which emphasis was placed first on similarities to the "typical" porphyry system (Lowell and Guilbert, 1970) and then on the differences. The differences were regarded as of minor genetic significance by Gustafson and Hunt (1975) who appropriately characterized them as "variations on a theme," a theme that recurs with such frequency as to imply a commonness of genetic process. The basic element of this theme is a stock, generally composite in nature, of granodioritic porphyry that forms a near-vertical digital protrusion from the top of a larger igneous body (Gustafson and Hunt, 1975, Figure 28A). The shapes of these stocks, coupled with the shallowness of their emplacement and other geologic features, such as breccia pipes, suggest that many of them solidified in or near volcanic conduits during the waning stages of eruptive activity. Thus, the development of a porphyry copper-molybdenum hydrothermal system is visualized as one of the end products of the solidification process, during the period when its main locus of operation was receding toward an underlying magma chamber. Attention in this section consequently will be focused on the solidification of a hydrous magma in this shallow environment, placing special emphasis on those processes that are important precursors to the generation of a magmatic-hydrothermal system.

For purposes of the following discussions, the initial state of the magmatic system, as depicted in Figure 3.5a, will be taken at that stage when a stock-like body of granodioritic magma, initially containing 3.0 wt. % H_2O, has been emplaced in a subvolcanic environment and already has undergone cooling to below solidus temperatures everywhere outside the line labelled "S_1." It is assumed that prior to this stage the system has been open to intrusion and, perhaps, extrusion of magma, as well as to escape of any volatiles generated during solidification of the magma outside the solidus boundary, but at this stage the magma system becomes closed, except to conductive loss of heat to the wallrocks. It is further assumed that the maximum temperature in the interior of the stock at this stage is 1025°C and that the 1000°C isotherm extends up to a depth of 2.5 km. Under these conditions, the 90% melt contour is roughly coincident with the 1000°C isotherm in Figure 3.5a, regardless of depth. Temperatures along the solidus boundary are not constant, however, but range from 800°C at 2.0 km to 700°C at 8.0 km depth (Figure 3.4).

Upward and outward from the 1000°C isotherm, the percentage of crystals obviously must increase, as must the H_2O content of the residual melt. Upward the melt is saturated with H_2O (~3.3 wt. %) and between this depth and that of the solidus boundary, the system consists of a pyroxene-bearing crystalline assemblage, decreasing amounts of residual melt of increasingly granitic composition, and an aqueous fluid phase. Neither hornblende nor biotite is stable in this interval, although hornblende may appear just at the solidus. Outward from the 1000°C isotherm at greater depths, however, hornblende appears in the temperature interval between 800 and 900°C, and is followed by biotite at temperatures between approximately 780 and 850°C. The actual temperatures in these intervals at which hornblende and biotite appear, as shown in Figure 3.4, depend upon pressure (depth) and the H_2O content of the melt (degree of crystallinity), but hornblende precedes biotite in this bulk composition. Hornblende forms by reaction between hydrous melt and already crystallized plagioclase, pyroxene, and magnetite-ilmenite, whereas biotite may form by reaction between the residual melt and hornblende. Owing to the low silica contents of both hornblende and biotite, their formation leads to pronounced enrichment of the residual melt in silica and resultant early crystallization of quartz. Their formation also removes some H_2O from the melt, but for a granodioritic magma with an initial H_2O content of 3.0 wt. %, generally less than 20% of it can be removed in this fashion.

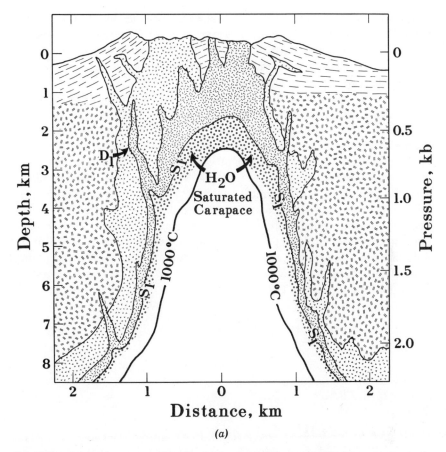

Fig. 3.5a Schematic cross-section through a hypothetical granodiorite porphyry stock and associated dike (D_1). S_1 represents the H_2O-saturated solidus (Figure 3.4) at this arbitrarily chosen initial stage in the development of a porphyry copper system and the circle pattern represents the zone of H_2O-saturated magma (H_2O-saturated carapace).

Consequently, the interstitial melt becomes saturated with H_2O as the solidus in Figure 3.5a is approached. At this early stage in the development of a porphyry system, therefore, the largely molten H_2O-undersaturated interior of the stock is completely encased in a crystalline rind of granodiorite, which has an inner liner of H_2O-saturated interstitial melt (see Whitney, 1975b).

This H_2O-saturated liner, or carapace, which is thickest at shallow levels, and thins markedly with depth as indicated in Figure 3.5a, plays a critical role in the future development of the porphyry copper-molybdenum system. It serves as a barrier to the migration of volatiles, either outward to the wallrocks or inward from the wallrocks, whether by diffusion or by fluid flow. Equally important, however, this H_2O-saturated carapace, especially in its upper parts, is the site of large-scale generation of an aqueous fluid phase by second boiling as the system cools and evolves toward the stage represented in Figure 3.5b.

The process of second boiling (resurgent boiling) is a natural consequence of cooling a melt that is saturated with respect to H_2O and one or more crystalline phases. The overall reaction, H_2O-saturated melt → crystals + "vapor," takes place with the evolution of heat which is essentially the heat of crystallization, because the heat of vaporization of H_2O from the melt is negligible (Burnham and Davis, 1974). More important, it takes place with the production of mechanical energy $(P\Delta V_r)$, because the volume per unit mass of crystals plus "vapor" is greater than the volume of an equal mass of H_2O-saturated melt. The total change in volume (ΔV_r) accompanying this reaction depends primarily upon total pressure (P_t), which in this case may be greater than lithostatic load pressure (P_l). From the data of Burnham and Davis (1971; 1974), ΔV_r may be closely approximated for a total pressure less than 2.0 kb by

$$\Delta V_r = (1 - 2.3 \times 10^{-4} P_t) \left(\frac{RTX_w^m}{P_t} \right)$$
$$- \Delta V_m (1 - X_w^m) \text{ cal} \cdot \text{bar}^{-1} \text{ mole}^{-1} \quad (3.10)$$

and for $P_t \geq 2.0$ kb by

$$\Delta V_r = 0.54 \left(\frac{RTX_w^m}{P_t} \right) - \Delta V_m (1 - X_w^m) \text{ cal} \cdot \text{bar}^{-1} \text{ mole}^{-1} \quad (3.11)$$

where T is in Kelvins, P_t is in bars, ΔV_m is the weighted average ΔV of melting of the crystalline phases (~ 0.21 cal bar^{-1} for dioritic compositions and ~ 0.22 cal bar^{-1} for silicic granites), and one mole of mixture is defined in accordance with the H_2O solution model discussed earlier. By substituting saturation values of X_w^m (and corresponding values of P_t) from Figure 3.1a into equations (3.10) and (3.11), it can be seen that ΔV_r decreases regularly with pressure. For example, at $P_t = 0.5$ kb the maximum ΔV_r is about 60% and at $P_t = 1.0$ kb it is 30%. However, at the shallow depths corresponding to these pressures (assuming $P_t = P_l$ in Figure 3.5a), the wallrocks and outer crystalline rind of the stock, owing to their rigidity, cannot generally accommodate this increase in volume by plastic deformation. Consequently, the internal pressure in the carapace must increase as cooling and crystallization proceed.

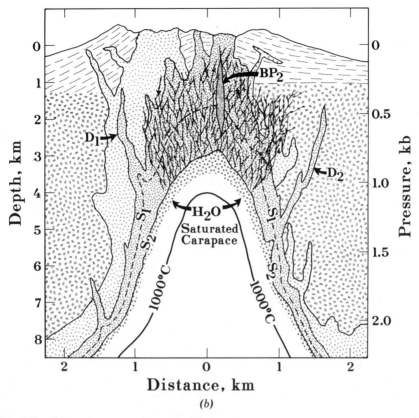

Fig. 3.5b Schematic cross-section as in Figure 3.5a, except at a later (second) stage of solidification. BP_2 and D_2 schematically represent a breccia pipe and dike that formed as a result of wallrock failure between stages 1 and 2. Chaotic line pattern represents extensive fracture system that also developed during this period of activity and retreat of the H_2O-saturated carapace.

The theoretical maximum internal overpressure ($\Delta P_{in} = P_t - P_l$) that can be generated simply by the magma cooling from the onset of second boiling to the H_2O-saturated solidus [$S(a_w \approx 1.0)$ in Figure 3.4], under conditions of constant total volume ($\Delta V_r = 0$), is closely approximated by the relation

$$\Delta P_{in} = \frac{0.54\,RT}{\Delta V_m}\left(\frac{X_w^m}{1 - X_w^m}\right) - P_l \text{ bars} \qquad (3.12)$$

Now, from Figure 3.1a it is apparent that, as P_t (ΔP_{in} at constant P_l) increases, so does X_w^m at saturation; hence, as $X_w^m \to 1$ in equation (3.12),

$\Delta P_{in} \to \infty$. Of course, this situation is never fully realized in natural systems for several reasons, but it illustrates the fact that extremely high internal overpressures can be generated in crystallizing hydrous magmas. Perhaps the highest internal overpressures in natural systems are reached in isolated small pockets of H_2O-saturated magma undergoing pegmatitic crystallization; under these conditions, values of $\Delta P_{in} \geq 5$ kb are not inconceivable.

Under the conditions depicted in Figure 3.5a, one of the principal factors that limits ΔP_{in} is the compressibility of the entire magma body that is contiguous with the H_2O-saturated carapace. As ΔP_{in} increases, the overpressure is transmitted throughout the contiguous magma system, which, in response, undergoes small but finite compression. Hence, $\Delta V_r > 0$ and ΔP_{in} is much less than the maximum value. A more important factor, however, is the tensile strength of the roof and wallrocks at shallow depths. Most igneous rocks deform quasielastically at low confining pressures and their tensile strengths, which are only slightly temperature dependent in the range 25–700°C, rarely exceed 400 bars (Ryan, 1979) under optimum conditions (homogeneous specimens free of macrocracks). Thus, a ΔP_{in} of this magnitude in a relatively large body of magma undergoing second boiling, as in the H_2O-saturated carapace of Figure 3.5a, generally is sufficient to induce brittle failure of the enclosing rocks (Burnham, 1972; Koide and Bhattacharji, 1975).

The dips of the ensuing fractures, which are concentrated in and above the apical parts of the stock as depicted in Figure 3.5b tend to be steep, because expansion of the system occurs in the direction of least principal stress and this direction, owing to the quasielastic behavior of the enclosing rocks at shallow depths, lies essentially in the horizontal plane. Its orientation within this plane, however, depends upon the regional stress field (tectonic setting). In the Arizona porphyry copper province, for example, the least principal stress appears to have been oriented NNW-SSE during emplacement of most of the porphyry stocks (Rehrig and Heidrick, 1973).

An additional overpressure that arises from differences in density between magma and wallrocks is superimposed on ΔP_{in} under quiescent conditions. The magnitude of this overpressure, commonly referred to as "telluric pressure," is dependent upon the height of the magma column above the depth of isostatic compensation; that is, the depth at which the wallrocks yield by plastic deformation under lithostatic pressure. It increases approximately 30–40 bars/km^{-1} height above this compensation level and therefore is greatest at the top of the magma column. It is this "telluric" overpressure that has been the principal driving force for intrusion up to the stage depicted in Figure 3.5a and that also has prevented any reduction in magma pressure that otherwise would have resulted from reduction in volume by crystallization [$-\Delta V_m$ in equations (3.10) and (3.11)] prior to the onset of second boiling.

The maximum mechanical energy ($P_t \Delta V_r$) released in the reaction H_2O-saturated melt → crystals + "vapor" is enormous, as can be seen by multiplying equation (3.11) through by P_t, although it is only approximately 1% of the total thermal energy content of the magma. As P_t increases along the H_2O-saturated solidus in Figure 3.4, $P_t \Delta V_r$ passes through a maximum of about 620 cal mole^{-1} of H_2O-saturated melt at $P_t = 0.7$ kb. This is equivalent to nearly 3×10^{23} ergs km^{-3} and considerably more than the estimated average kinetic energy released, per cubic kilometer of material erupted, in explosive volcanic eruptions (Burnham, 1972). Thus, if allowance is made for a small but finite tensile strength of the volcanic edifice, it is evident that the maximum mechanical energy is released at depths of only 1 or 2 km. Moreover, about 75% of this amount could be generated at depths of less than 0.5 km, whereas at 8.0 km depth only approximately half the maximum amount (~ 300 cal mole^{-1}) could be generated. At the depths considered in Figures 3.5a-c, then, the mechanical energy thus released from the H_2O-saturated carapace presumably is expended mainly in fracturing a much larger volume of wallrocks.

In the earlier stages of fracturing, the carapace of H_2O-saturated interstitial melt may not be breached, but stretched laterally instead. This action places superjacent roofrocks to which the carapace is strongly coupled under even greater tensile stress and results in localization of numerous smaller fractures in this region. The attendant expansion also results in a reduction of fluid pressure in the carapace, which, in turn, causes more of the H_2O-saturated interstitial melt to crystallize and evolve even more aqueous fluid phase. Moreover, the fluid phase evolved from outer parts of the carapace penetrates this myriad of fractures and extends them outward and upward by hydraulic action (hydrofracting). This action, in itself, results in further lowering of fluid pressure in each fracture to below lithostatic pressure, except near the top of the fracture. Thus, owing to the drop in fluid pressure attending development of fractures and consequent crystallization ("pressure quench"), the H_2O-saturated carapace retreats to progressively deeper levels in the stock.

Most of the fluid-filled fractures produced by the above mechanism are too narrow to be intruded by highly viscous magma ($\geq 10^6$ poises; Burnham, 1967), but some of the larger fractures opened in the lateral "stretching" of the system may breach the H_2O-saturated carapace. If the breach occurs in the thickened upper part of the carapace, where large volumes of aqueous fluid phase have accumulated, breccia dikes and pipes are likely to result (BP_2 in Figure 3.5b). If the breach occurs on the thinner flanks of the carapace, on the other hand, more normal dike intrusion of magma containing plagioclase and hornblende may result (D_2 in Figure 3.5b). The mechanics of breccia pipe formation are visualized much as described by Norton and Cathles (1973), except that tensile stresses in the crystalline rind of the roof are here regarded as due primarily to internal overpressure in the carapace, instead of to contraction on cooling. In dike emplacement, the driving force for intrusion is again internal overpressure of the magma, but emplacement is essentially noncatastrophic (except initially), owing to high magma viscosities. Of course, as pressure decreases and heat is lost to the wallrocks, the dike magma "boils off" its initial H_2O content and eventually freezes. In response to this overall enlargement of the intrusive, more magma rises into the stock system until internal pressures are restored to near-previous values.

At this stage, as depicted in Figure 3.5b, the magma system has been restored to much the same state as existed prior to fracturing, except the H_2O-saturated carapace now lies at a greater depth in the stock and the wallrocks have been weakened by the earlier fracturing. Also, the myriad of narrow fractures outside the solidus boundary, in which the fugacity of H_2O is appreciably less than it is in the H_2O-saturated interstitial melt of the carapace, promotes an accelerated rate of seepage and diffusive loss of H_2O and heat to the fracture system until the fractures become healed by the precipitation of mineral material, chiefly quartz. Beyond this stage of reclosing the system, further cooling of the magma leads to reactivation of the same processes that operated before. The end result is a chimney-like fracture system, as indicated in Figure 3.5c, that serves to channel ore-bearing fluids and heat from the underlying magma system to higher levels in the stock. It should be noticed, however, that as the H_2O-saturated carapace retreats to greater depths, hence greater P_1, the amount of expansion and extent of fracturing diminishes. An H_2O-saturated carapace that originally contained 3.0 wt. % H_2O theoretically could expand approximately 30% upon complete crystallization at a depth of 3.0 km, but not more than approximately 5% at a depth of 8.0 km. Furthermore, much of this latter expansion may be taken up in plastic deformation of the wallrocks at this depth and persist in the solidified granodiorite as miarolitic cavities. Porphyry copper-molybdenum fracture systems are thus restricted to the uppermost few kilometers of the earth's crust, where expansion of the H_2O-saturated carapace is large and the wallrocks yield by brittle fracture.

Whether or not the downward retreat of the H_2O-saturated carapace is distinctly episodic or semicontinuous depends upon numerous factors that probably vary widely from system to system. In some, notably the Climax and Urad-Henderson, Colorado, systems (Wallace, 1974) the

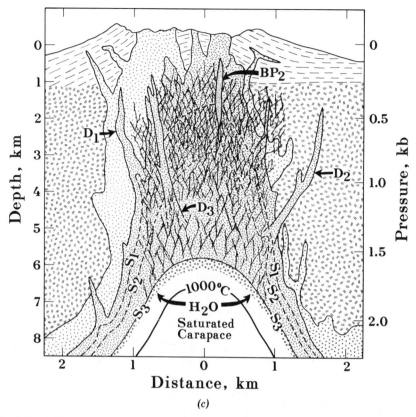

Fig. 3.5c Schematic cross-section as in Figure 3.5a,b, except at a stage of waning magmatic activity in the development of a porphyry copper-molybdenum system.

retreat appears to have been distinctly episodic, whereas in others (San Manuel-Kalamazoo, Arizona, and Bingham, Utah) the retreat seems to have been almost continuous. At El Salvador, Chile (Gustafson and Hunt, 1975), the processes appear to have been distinctly episodic, but telescoped in space as a result of large-scale intrusion of porphyry magma following earlier episodes of fracturing and hydrothermal activity. Despite such complications, which Gustafson and Hunt appropriately regarded as "variations on a theme," there is clear evidence for the operation in these systems of processes, critical to the formation of porphyry copper-molybdenum deposits, that are natural consequences of the solidification of hydrous magmas in a shallow crustal environment. The orchestration of these processes into the theme in a way such as to produce an ore deposit, however, represents a remarkable coincidence of numerous factors, some of which are imposed by the external environment.

THE MAGMATIC AQUEOUS PHASE

The Aqueous Phase in Equilibrium with Melt

As the H_2O-saturated carapace and its enveloping fracture system retreats downward, the internal fluid pressure required to initiate second boiling increases, owing to the strong pressure dependence of the solubility of H_2O (Figure 3.1). In a magma of approximately uniform bulk composition, this increased fluid pressure results from increased concentration of H_2O and other volatile constituents in the residual melt as crystallization of plagioclase and ferromagnesian minerals advances. Concomitant with this increase in concentration, the activities of the volatiles chlorine, sulfur, and fluorine, as well as the metals of interest here, also increase. Inasmuch as the partition coefficients of these substances are not strongly pressure dependent, the aqueous phase formed in early stages of second boiling becomes more enriched in these volatile constituents with depth. In later stages of second boiling near the H_2O-saturated solidus, however, the aqueous phase may become enriched or depleted in a particular volatile element, relative to its initial concentration, depending upon pressure, the magnitude of the partition coefficient, and whether or not the system is open or closed to the escape of the aqueous phase. Thus, the composition of the magmatic aqueous phase is complexly dependent upon the state of the system and varies in space at a given time, as well as in time at a given place.

The volatile elements of especial interest here are chlorine and sulfur: chlorine because of its important role in wallrock alteration and the transport of metals, and sulfur because of its obvious role in the precipitation of metal sulfides. Fluorine is not generally an abundant constituent of the magmatic aqueous phase in most porphyry copper systems, as indicated by the common occurrence of andalusite instead of topaz. The principal reason for its low abundance in these calcium-bearing granodioritic systems is the high thermal stability (low solubility) of fluorite which maintains f_F at a very low level, even at magmatic temperatures, and the ready substitution of F^- for OH^- in hydrous minerals. In some calcium-poor porphyry molybdenum and tin systems, on the other hand, fluorine may become a more abundant constituent of the magmatic aqueous phase, especially toward the end stages of crystallization. Even in these cases, however, its importance in the transport of metals probably is greatly diminished by the high stability of the SiF_4 complex.

Chlorine in the Magmatic Aqueous Phase

In contrast to fluorine, chlorine forms neither stable chloride minerals nor complexes with silicon or aluminum at magmatic temperatures in silicic systems, although very small amounts of chlorine may be taken up in apatite and biotite at high values of f_{Cl}. As a consequence of this and because NaCl, KCl, HCl, CaCl$_2$, and (FeCl$_2$ + FeCl$_3$) are stable complexes in aqueous solution at magmatic temperatures and HCl is the only one of these complexes that is appreciably soluble in the melt phase at moderate pressures, the partition coefficient of chlorine is very large in favor of the aqueous phase. Thus, as HCl partitions into the aqueous phase during boiling, hydrolysis reactions such as

$$2HCl(v) + 2NaAlSi_3O_8(m)$$
$$\rightleftharpoons 2NaCl(v) + Al_2SiO_5(m) + 5SiO_2(m) + H_2O(v)$$

reduce m^v_{HCl}, hence a^v_{HCl}, to only a small fraction of $m^v_{\Sigma Cl}$, and equilibrium is therefore not re-established until $m^v_{\Sigma Cl}/m^m_{\Sigma Cl} = K_P{}^{Cl}$ reaches 30 or 40 (Kilinc and Burnham, 1972).

This large value of $K_P{}^{Cl}$ in favor of the aqueous phase results in large and highly variable concentrations of chlorine in the magmatic aqueous phase, depending upon conditions at the time of separation from the magma. For example, if the chlorine content of the original granodioritic magma discussed in the last section, which contained 3.0 wt. % H$_2$O, is a conservative 0.1 wt. % (see Carmichael et al., 1974, p. 316), the first-formed fluid phase in the H$_2$O-saturated carapace at a total pressure of 0.6 kb (~2 km depth, $X_w{}^m = 0.32$, 90% melt) would contain approximately 4.6 wt. % chlorine. At $P_t = 2.1$ kb (~8.0 km) and $X_w{}^m = 0.51$, however, where boiling does not begin until the magma is about 53% crystalline, the first-formed aqueous phase contains about 7.5 wt. % chlorine, or about 12 wt. % NaCl equivalent. Now, if the systems in these two examples remain closed as crystallization and second boiling proceed, the chlorine content of the aqueous phase will decrease by a factor of as much as two, as the solidus is approached. If the system is open, the chlorine content of the evolving aqueous phase in the higher pressure example also will decrease, but by a factor of 20 or more, whereas in the lower pressure example it will actually increase somewhat (Kilinc and Burnham, 1972).

The strong partitioning of chlorine toward the aqueous phase lowers $a_w{}^v$ and causes more H$_2$O to separate from the melt to re-establish equilibrium, which in turn results in further crystallization of the melt. Although this effect is not large, even for an aqueous phase containing 18 wt. % chlorine ($X_{Cl}{}^v \approx 0.1$), it does accelerate the growth of the aqueous phase, initially. Thus, instead of an infinitesimal amount of fluid phase being formed at the onset of second boiling, on the order of 1 vol. % or more may be formed, especially if other volatiles such as CO$_2$ and H$_2$S are taken into consideration. Moreover, experimental results obtained in connection with the volumetric measurements of Burnham and Davis (1971) clearly indicate that aqueous phase separation occurs in a matter of minutes, even at very small degrees of H_2O supersaturation.

The large partition coefficient for chlorine in favor of the aqueous phase, coupled with the fact that the critical plait-point curve in the system $NaCl-H_2O$ intersects the solidus in Figure 3.4 at about 1.4 kb and 730°C, led Holland (1972) to postulate that a very NaCl-rich aqueous phase can separate from granitic magmas at higher temperatures and lower pressures. From a consideration of the phase relations in the system $NaCl-H_2O$ (Sourirajan and Kennedy, 1962), the partition coefficient for chlorine (Kilinc and Burnham, 1972), and the equilibrium fluid phase compositions (Kilinc, 1969), it appears that two fluid phases might coexist with silicate melt if the chlorine content of this melt exceeds approximately 0.5 wt. %, but only if it is assumed that the aqueous-phase relations of the other chlorides, mainly potassium, iron, and hydrogen, are the same as those in the system $NaCl-H_2O$. However, it is highly unlikely that the chlorine content of a felsic melt reaches 0.5 wt. %, except in H_2O-poor magmas where second boiling occurs only after 75-85% of the magma has crystallized. Moreover, the observation (Roedder, 1971, and others) that highly saline fluid inclusions in porphyries always homogenize at temperatures below 725°C clearly indicates that only a single fluid phase separated from these magmas, as the highest homogenization temperature (725°C) is below even the H_2O-saturated solidus (Figure 3.4) at pressures below 1.4 kb, and addition of alkali and iron chlorides to the aqueous phase raises solidus temperatures. It is therefore concluded that the highly saline fluid inclusions found at Bingham, Utah (Roedder, 1971) and in other porphyry bodies represent the high density—and high salinity— condensate from an initially homogeneous fluid phase that separated from the melt, in accordance with the model discussed previously, at temperatures generally in the range 800-950°C.

A comparison of experimental data on compositions of aqueous chloride solutions with those of chloride-free aqueous solutions in equilibrium with felsic rocks and their melts (Burnham, 1967; Kilinc, 1969; Holland, 1972) reveals that the molal sum of Na, K, (2.7-3.0) Fe, H, 2Ca, 2Mg, and 2Mn is equal to the total molality of chlorine. Of these, magnesium and manganese together account for less than 4.0% of the total chlorine in solution and, at the lower pressures and high temperatures of interest here, calcium accounts for less than another 4.0%. Consequently, the compositional features of the magmatic aqueous phase, excluding silica contents, are dominated by the chlorides of sodium, potassium, iron, and hydrogen as associated (essentially un-ionized) complexes at magmatic temperatures. The relative proportions of these four chlorides in the magmatic aqueous

THE MAGMATIC AQUEOUS PHASE

phase under a given set of pressure-temperature conditions, however, are complex functions of the various hydrolysis equilibria, the nature of the equilibrium phase assemblage, and the bulk composition of the system. In what follows, an attempt will be made to qualitatively assess the importance of these factors in the compositional evolution of the magmatic aqueous phase as it separates from a high-level granodioritic magma, as discussed in previous sections and depicted in Figures 3.5a–c.

The initial partitioning of chlorine into the aqueous phase, in accordance with the solution model presented earlier, is determined by the reaction:

$$Cl^-(m) + OH^-(m) \rightarrow HCl(v) + O^{2-}(m)$$

Most of the HCl produced in this reaction undergoes hydrolysis with the sodium-, potassium-, iron-, and calcium-bearing components of the melt and/or crystalline phases. In terms of the components of silicate melts defined earlier, these hydrolysis equilibria may be written in the following manner, on the assumption that the chlorides exist in the aqueous phase as monomeric complexes

$$2HCl(v) + \underset{ab}{2NaAlSi_3O_8(m)}$$

$$\rightleftharpoons 2NaCl(v) + \underset{as}{5/8\,Al_{3.2}Si_{1.6}O_8(m)} + \underset{qz}{5/4\,Si_4O_8(m)} + H_2O(v) \quad (3.13)$$

$$2HCl(v) + \underset{or}{2KAlSi_3O_8(m)}$$

$$\rightleftharpoons 2KCl(v) + \underset{as}{5/8\,Al_{3.2}Si_{1.6}O_8(m)} + \underset{qz}{5/4\,Si_4O_8(m)} + H_2O(v) \quad (3.14)$$

$$2HCl(v) + \underset{an}{CaAl_2Si_2O_8(m)}$$

$$\rightleftharpoons CaCl_2(v) + \underset{as}{5/8\,Al_{3.2}Si_{1.6}O_8(m)} + \underset{qz}{1/4\,Si_4O_8(m)} + H_2O(v) \quad (3.15)$$

$$4HCl(v) + \underset{di}{3/4\,Ca_{1.33}Mg_{1.33}Si_{2.67}O_8(m)}$$

$$\rightleftharpoons CaCl_2(v) + MgCl_2(v) + \underset{qz}{1/2\,Si_4O_8(m)} + 2H_2O(v) \quad (3.16)$$

$$8HCl(v) + \underset{mt}{1/2\,Fe_6O_8(m)} \rightleftharpoons FeCl_2(v) + 2FeCl_3(v) + 4H_2O(v) \quad (3.17)$$

and

$$6HCl(v) + \tfrac{3}{8}Fe_{5.33}O_8(m) \rightleftharpoons 2FeCl_3(v) + 3H_2O(v) \qquad (3.18)$$
$$\text{hm}$$

The main purpose in writing reactions (3.13)–(3.15) in this fashion is to emphasize the fact that hydrolysis of the "feldspar" components of the melt produces aluminum silicate (as) component which is virtually insoluble in chloride solutions (Burnham, 1967; Kilinc, 1969). In nearly all calc-alkaline magmas, the atomic ratio (Na + K + 2Ca − 2Mg)/Al ≤ 1.0; consequently, the preferential extraction of the alkalies into the chlorine-rich aqueous phase insures that the alkalies remaining in the melt exist mainly as the "feldspar" components, ab and or. This factor is very important in determining the distribution of potassium and sodium in a melt-vapor system and, hence, the composition of the magmatic aqueous phase.

The equilibrium constants for reactions (3.13) and (3.14), after substituting the expression $a_a{}^m = X_a{}^{am}(1 - X_w{}^m)^2$ derived earlier [equation (3.7)], may be written

$$K_{13} = \frac{(X_{NaCl}^v)(X_{as}{}^{am})^{5/16}(X_{qz}{}^{am})^{5/8} f_w{}^{1/2} \gamma_{NaCl}^v}{(X_{HCl}^v)(X_{ab}{}^{am})(1 - X_w{}^m)^{1/8} \gamma_{HCl}^v} \qquad (3.13a)$$

and

$$K_{14} = \frac{(X_{KCl}^v)(X_{as}{}^{am})^{5/16}(X_{qz}{}^{am})^{5/8} f_w{}^{1/2} \gamma_{KCl}^v}{(X_{HCl}^v)(X_{or}{}^{am})(1 - X_w{}^m)^{1/8} \gamma_{HCl}^v} \qquad (3.14a)$$

Division of equation (3.14a) by equation (3.13a), with the further assumption that $\gamma_{KCl}^v/\gamma_{NaCl}^v$ is constant (Holland, 1972; Thompson and Waldbaum, 1968), yields the distribution constant

$$K_D^{K\text{-}Na} = \frac{(X_{KCl}^v)(X_{ab}{}^{am})}{(X_{NaCl}^v)(X_{or}{}^{am})} \qquad (3.19)$$

for the exchange reaction $NaCl(v) + KAlSi_3O_8(m) \rightleftharpoons KCl(v) + NaAlSi_3O_8(m)$, which should be close to unity, provided $\gamma_{KCl}^v/\gamma_{NaCl}^v$ also is close to unity. The experimental data of Kilinc (1969) at 750°C and 2.3 kb, where the melt and aqueous phases coexisted with an aluminum silicate, give $K_D^{K\text{-}Na} = 1.01$!

These results might appear to be at variance with those of Holland (1972) that yield a $K_D^{K\text{-}Na}$ of 0.75 ± 0.06. However, the atomic ratio (Na + K + 2Ca − 2Mg)/Al in the melts of his experiments was greater than unity and this resulted in formation of other alkali-bearing components in the melt, in addition to ab and or. Hence, $K_D^{K\text{-}Na}$ in this case is governed by the nature and relative "solubilities" of these other alkali-bearing components (bulk composition effects) and not by equilibria (3.13) and (3.14).

The experimentally confirmed deduction that $K_D^{K\text{-}Na} \approx 1.0$ at equilibrium between a calc-alkaline melt and an aqueous phase implies, from equation (3.19), that $X_{KCl}^v/(X_{KCl}^v + X_{NaCl}^v) \approx X_{or}^{am}/(X_{or}^{am} + X_{ab}^{am})$ which, for a typical quartz-dioritic to granodioritic melt (Nockolds, 1954), lies in the range 0.20–0.35. As discussed previously, cooling of such melts results in crystallization mainly of plagioclase, which reduces X_{ab}^{am} relative to X_{or}^{am}; consequently $X_{KCl}^v/(X_{KCl}^v + X_{NaCl}^v)$ in a coexisting magmatic aqueous phase increases to satisfy the chlorine stoichiometry. A further and more pronounced increase in this ratio occurs when the magma is cooled into the stability field of hornblende (Figure 3.4), because the coexistence of hornblende, plagioclase, and melt of a given bulk composition "buffers" a_{NaCl}^v, hence X_{NaCl}^v, but not X_{KCl}^v. Therefore, X_{KCl}^v again increases to satisfy the chloride stoichiometry of the aqueous phase; values of $X_{KCl}^v/(X_{KCl}^v + X_{NaCl}^v)$ as high as 0.75 were obtained by Kilinc (1969) in a hornblende-bearing system where $X_{or}^{am}/(X_{or}^{am} + X_{ab}^{am})$ was only 0.23.

Upon further cooling of the magma into the stability field of biotite, a_{KCl}^v also becomes buffered at a value which, according to Kilinc's experiments, yields a value of X_{KCl}^v approximately equal to X_{NaCl}^v. Therefore, at equilibrium with hornblende, biotite, plagioclase, and a melt of granitic composition, $X_{KCl}^v/(X_{KCl}^v + X_{NaCl}^v) \approx 0.5$, even though $X_{or}^{am}/(X_{or}^{am} + X_{ab}^{am})$ may have a different value. Moreover, additional cooling of the magma into the stability field of alkali feldspar does not greatly affect $X_{KCl}^v/(X_{KCl}^v + X_{NaCl}^v)$, because the composition of the alkali feldspar that crystallizes is determined mainly by equilibrium relations between the melt and other crystalline phases.

In summary, $X_{KCl}^v/(X_{KCl}^v + X_{NaCl}^v)$, a highly important factor in metasomatic and alteration processes, may change markedly during the evolution of a magmatic aqueous phase from cooling magma. At near-liquidus temperatures, this ratio may be as low as 0.2 and, if brought into contact with already crystallized porphyry at slightly subsolidus temperatures, the fluids are capable of producing sodium metasomatism (albitization). As cooling of the magma proceeds and hornblende becomes stable, this ratio increases dramatically to values as high as 0.75 and the fluids thus gain the potential for extensive potassium metasomatism. This behavior provides a rational explanation for the extensive potassium feldspar-biotite alteration (potassium metasomatism) associated with some potassium-poor hornblende diorite porphyries, such as at Panguna, Bougainville.

Just as $X^v_{KCl}/(X^v_{KCl} + X^v_{NaCl})$ varies with bulk composition and the nature of the coexisting phase assemblage, so does $(X^v_{KCl} + X^v_{NaCl})/X^v_{\Sigma Cl}$. The principal cause for variations in this latter ratio is variations in the proportions of total chlorine that is complexed with iron, the third most abundant cationic constituent of the magmatic aqueous phase (excluding silicon) at low pressures. In fact, the experimental results of Kilinc (1969) indicate that, in the presence of either magnetite or hematite at two kilobars and magmatic temperatures, approximately as much chlorine is complexed with iron as with either sodium or potassium. These same results also indicate that $K_P^{Fe} = (m_{Fe}{}^v/m_{Fe}{}^m)(m^v_{\Sigma Cl})^3$ (hematite) or $(m_{Fe}{}^v/m_{Fe}{}^m) \cdot (m^v_{\Sigma Cl})^{2.67}$ (magnetite) is generally greater than $K_P^{Na} = (m_{Na}{}^v/m_{Na}{}^m)m^v_{\Sigma Cl}$, reflecting the relatively high stability of the iron chloride complexes. Moreover, K_P^{Fe} increases as either temperature or pressure (f_w) decreases; the effect of temperature is due to the decreasing solubility of the iron oxides in the melt with decreasing temperature, whereas the effect of pressure is one of increasing f_w, as is evident from equations (3.17) and (3.18).

It is also evident from equations (3.17) and (3.18) that, in the presence of magnetite or hematite at constant temperature, melt composition, f_w, and f_{O_2}, the ratio of the activities of the iron chlorides to that of HCl in the aqueous phase is buffered. Changing any one of these variables, however, effects a corresponding change in this ratio, either through changing the activities of the iron chlorides, or the activity of HCl, or both, as all of these hydrolysis equilibria (3.13-3.18) are coupled through HCl. The data (Kilinc, 1969) suggest that in a granodioritic system at approximately 2 kb and temperatures above the first appearance of a hydrous mineral, but where plagioclase and magnetite coexist with the melt, $X^v_{\Sigma Fe}/(X^v_{HCl})^{2.67}$ is between 33 and 100. Lowering the temperature into the stability field of hornblende or biotite, on the other hand, buffers a^v_{HCl} at lower values and causes $X^v_{\Sigma Fe}/(X^v_{HCl})^{2.67}$ to increase by a factor of 15 or more.

Similarly, X^v_{KCl}/X^v_{HCl}, the parameter that most influences subsolidus hydrolysis equilibria in wallrock alteration, may be as low as 2.0 at high temperatures, but increases to values between 10 and 15 at temperatures within the stability fields of hornblende and biotite. Another important parameter in subsolidus hydrothermal alteration, $X^v_{CaCl_2}/(X^v_{HCl})^2$, also increases from approximately 2.5 to 4.0 upon entry into the stability field of hornblende. Multiplication of these ratios by the corresponding values for $(X^v_{HCl}/X^v_{\Sigma Cl})^2$ yields a value for $X^v_{CaCl_2}/(X^v_{\Sigma Cl})^2$ of 0.02 which, judging from the general agreement with Holland's (1972) experimental results, may be taken as typical for the magmatic aqueous phase in equilibrium with a plagioclase-bearing granodioritic magma, whether or not pyroxene or hornblende is present.

Thus, despite the complexities occasioned by the coupling of equilibria

(3.13)-(3.18), bulk composition effects, and the nature of the equilibrium phase assemblage, it is possible to place rather narrow constraints on the composition of the magmatic aqueous phase in equilibrium with a granodioritic magma under specified conditions of temperature, pressure, and chlorine content of the melt. For example, if the chlorine content of the melt in equilibrium with plagioclase and magnetite at 900°C and 1.0 kb (Figure 3.4) is 0.1 wt. %, the coexisting aqueous phase will be approximately 1.0 m in ΣCl, with the major cationic constituents (excluding silicon) distributed as follows: 0.40 m sodium, 0.23 m potassium, 0.1 m ΣFe, 0.09 m hydrogen, and 0.02 m calcium. At somewhat lower temperatures in the stability field of hornblende, but at the same pressure and chlorine content of the melt, the distribution changes to 0.17 m sodium, 0.47 m potassium, 0.11 m ΣFe, 0.03 m hydrogen, and 0.02 m calcium. At even lower temperatures, in the stability field of biotite, the distribution changes again to 0.32 m sodium, 0.32 m potassium, 0.11 m ΣFe, 0.03 m hydrogen, and 0.02 m calcium.

Evidence presented by Holland (1972) clearly indicates that the more chalcophile elements, such as zinc, have a much stronger preference for aqueous chloride solutions than for melts, relative to the more lithophile elements; that is, $K_p{}^m/(m_{\Sigma Cl}^v)^n > 1.0$, where n is the valence of the metal cation (m). This suggests that copper, either cuprous or cupric, also is preferentially concentrated in the aqueous phase. Unfortunately, no experimental data are available for copper, but if $K_p{}^{Cu}/m_{\Sigma Cl}^v$ or $K_p{}^{Cu}/(m_{\Sigma Cl}^v)^2$ is comparable to $K_p{}^{Zn}/(m_{\Sigma Cl}^v)^2 \approx 2.0$, most of the copper can be extracted from a crystallizing porphyry magma as the H_2O-saturated carapace retreats downward (Figure 3.5a-c).

Sulfur in the Magmatic Aqueous Phase

The behavior of sulfur during second-boiling in the H_2O-saturated carapace resembles that of chlorine in one respect, but differs fundamentally in most others. As discussed previously, the bulk of the sulfur dissolved in a hydrous magma occurs as SH^- and, upon separation of an aqueous phase, partitions in accordance with the equilibrium

$$SH^-(m) + OH^-(m) \rightleftharpoons H_2S(v) + O^{2-}(m).$$

On the basis of experimental data and the solution mechanisms discussed previously, the equilibrium constant for this reaction is thought to be comparable to that for HCl and about twice that for H_2O at the same temperature and pressure, but here the similarities cease. Under the f_{O_2} conditions typical of granodioritic magmas (Ohmoto, Chapter 10 of this book; Carmichael et al., 1974, p. 330), approximately 50–90% of the sulfur exsolved from the magma may form SO_2 through a reaction such as

$$SH^-(m) + 5OH^-(m) \rightleftharpoons 3O^{2-}(m) + SO_2(v) + 3H_2(v)$$

depending upon pressure (f_w). Moreover, equilibria in the H_2S analogues of reactions (3.13)–(3.16) lie far to the left, especially in the presence of HCl and, of course, the H_2S analogue of reaction (3.17) precipitates pyrrhotite at magmatic temperatures.

The very strong inverse dependence of $f^v_{SO_2}/f^v_{H_2S}$ on f_w implies also a strong inverse dependence of $K_p^S = m_{\Sigma S}^v/m_{\Sigma S}^m$ on f_w, because SO_2 is virtually insoluble in aluminosilicate melts at low to moderate pressures. In the deeper parts of the porphyry system in Figure 3.5 ($P_t = 2.0$ kb), where magnetite and hornblende or biotite coexist with a melt ("buffered" f_{O_2}), $f^v_{SO_2}/f^v_{H_2S} < 1.0$ and $K_p^S \approx 4.0$. In the presence of the same phase assemblage at 0.5 kb, however, $f^v_{SO_2}/f^v_{H_2S} \approx 10$ and $K_p^S \approx 40$. Consequently, the total sulfur content of the aqueous phase that separates from a melt that contains 0.17 wt. % S (almost saturated with pyrrhotite) may be expected to range from approximately 0.2 m ΣS in the deeper parts of a porphyry system to as much as 2.0 m ΣS at shallow levels.

The formation of a magmatic aqueous phase in the H_2O-saturated carapace, regardless of the depth at which it occurs, results in a lowering of f_{H_2S} in the system, both by passage of H_2S from the melt into the newly created and expanding aqueous phase and the formation of $SO_2(v)$, to reach the above partition equilibrium. The marked decrease in f_{H_2S} causes dissolution of already crystallized sulfides, a process that becomes even more pronounced as the magmatic aqueous phase separates from the H_2O-saturated carapace and permeates the superjacent fracture system in which fluid pressures (f_w) are considerably lower, but in which temperatures may be only moderately lower. Thus, as the H_2O-saturated carapace retreats downward, it and the immediately overlying rocks are scrubbed of their sulfur, which is transported generally upward to lower temperature parts of the system mainly as SO_2. As temperatures decrease to well below the H_2O-saturated solidus, however, the SO_2 hydrolyzes to produce H_2S and H_2SO_4 in a 1:3 ratio (Holland, 1967); the resulting increase in f_{H_2S} causes precipitation of sulfides from the metal-chloride complexes, and the corresponding increase in $f_{H_2SO_4}$ causes precipitation of anhydrite, a common and occasionally very abundant mineral in the higher-temperature alteration assemblages of porphyry copper deposits (Gustafson and Hunt, 1975). These subsolidus reactions will be examined more closely in the next section, but first additional comments are in order regarding the diffusivity of H_2 in magmas and its effects on f_{SO_2}/f_{H_2S}.

The diffusivity of H_2O and O_2 in silicate melts has been found by experiment (see Carmichael et al., 1974, p. 137-141 for summary) to be too low to effect a significant redistribution of H_2O in the H_2O-saturated carapace of the model porphyry system. This is not true for H_2, however, whose diffusivity is more than three orders of magnitude greater than that of H_2O. Therefore, the magmatic aqueous phase that coexists with magnetite, hornblende, plagioclase, and melt of a given composition in the H_2O-saturated carapace may lose significant amounts of H_2 to the superjacent wallrocks. The immediate effect of this H_2 loss is an increase in f_{O_2} to maintain the $H_2O \rightleftharpoons H_2 + \frac{1}{2}O_2$ equilibrium, but f_{O_2} is buffered to a large extent by the coexisting magnetite and hornblende of specific composition. Consequently, either the hornblende changes composition (increase in ferric/ferrous ratio) or, more likely, reacts with the melt and other phases to produce phlogopitic biotite, magnetite and plagioclase, as apparently happened at Bingham, Utah (Lanier et al., 1975). The net effect of the H_2 loss, therefore, is an increase in the equilibrium f_{O_2} which locally may reach the stability field of hematite (see Figure 10.5). Concomitant with this increase in f_{O_2}, f_{SO_2}/f_{H_2S} increases, perhaps to 1000 or more, and this results in essentially complete extraction of sulfur from the melt and solid phases. Moreover, f_{SO_2}/f_{H_2S} will remain approximately constant upon isenthalpic expansion of the aqueous phase into the superjacent fracture system, thus maintaining the metal and sulfur transport capacity of a chlorine-bearing hydrothermal fluid until the thermal gradient exceeds the isenthalpic adiabat. Under appreciably steeper gradients, the hydrolysis of SO_2 mentioned above becomes increasingly important and the resulting increase in f_{H_2S} may cause precipitation of metal sulfides.

SUBSOLIDUS HYDROLYSIS EQUILIBRIA

The composition of the magmatic aqueous phase generated in the H_2O-saturated carapace of a porphyry system (Figure 3.5) was shown in the last section to be complexly dependent upon pressure, temperature, f_{O_2}, bulk composition, and the nature of the phase assemblage. As this aqueous phase permeates the superjacent fracture system in the already crystallized porphyry and approaches equilibrium with it, bulk composition ceases to be a factor, except insofar as it affects the nature of the igneous mineral assemblage and the compositions of individual mineral solid solutions. With this exception, then, the principal factors that control the equilibrium bulk composition of the aqueous phase at moderately subsolidus temperatures (550-750°C) and a given fluid pressure are the initial composition of the aqueous phase at its magmatic source, temperature, and the nature of the coexisting mineral assemblage, which, again except for unusually iron-poor assemblages, also controls f_{O_2} and the f_{SO_2}/f_{H_2S} ratio. Moreover, the difference between the aqueous phase composition in equilibrium with a particular subsolidus assemblage, at a given temperature and fluid pressure, and its initial composition is an indicator of the progress of the various exchange and hydrolysis reactions that dominate subsolidus equilibria. The extent of reaction (the relative masses of alteration products), however, is dependent not only on this difference, but also on the relative masses of fluid and wallrock reacted.

Perhaps the most important compositional parameter of the magmatic aqueous phase, in terms of metasomatism and wallrock alteration, is m^v_{KCl}/m^v_{HCl}, which is approximately proportional to f_{KCl}/f_{HCl} in the temperature range of interest here and to $a_{K^+}^v/a_{H^+}^v$ at low temperatures. Values for this ratio in the hypersolidus region, which were obtained mainly from the experimental results of Kilinc (1969) and Burnham (1967) and discussed in the last section, are presented in Figure 3.6 (points A, D, F, and H) and taken as initial values in the following discussion of subsolidus reactions.

Phase relations in the hypersolidus region of Figure 3.6 are from Figure 3.4, whereas those in the subsolidus region are based mainly on the data of Shade (1974) and Hemley and Jones (1964). Also, it will be noticed that the phase relations in Figure 3.6 are for a pressure of 1.0 kb; raising or lowering the pressure by 0.5 kb causes mainly vertical (temperature) shifts in the phase boundaries, but very little lateral $[\log(m^v_{KCl}/m^v_{HCl})]$ shift in isobarically invariant equilibria, such as point I (Shade, 1974, Figure 3). At pressures below about 0.5 kb, however, neither hornblende nor biotite is stable above the solidus (Figure 3.4); consequently, m^v_{KCl}/m^v_{HCl} in the magmatic aqueous phase can be even lower than indicated by point A in Figure 3.6.

It is apparent from path A-B-C in Figure 3.6 that should a magmatic aqueous phase, initially at a temperature above the stability field of hornblende and a pressure of 1.0 kb, escape from the H_2O-saturated carapace and permeate cooler wallrocks without extensive reaction, it will pass into the stability field of aluminum silicate (andalusite) below the solidus. At some temperature above approximately 600°C, such as point B, this solution will therefore react with the wallrocks mainly in accordance with the subsolidus equivalents of reaction schemes (3.13)–(3.15) to reach equilibrium on the andalusite (as)-potassium feldspar (or) boundary. On the other hand, at lower temperatures, such as point C, equilibrium is established in potassium-feldspar-bearing wallrocks only when the solution composition has shifted to point E on the muscovite-potassium feldspar boundary. Locally, as in the selvages of fractures, potassium feldspar may be consumed early, in which case the equilibrium m^v_{KCl}/m^v_{HCl} ratio shifts back to the andalusite-muscovite boundary. Petrographic evidence in

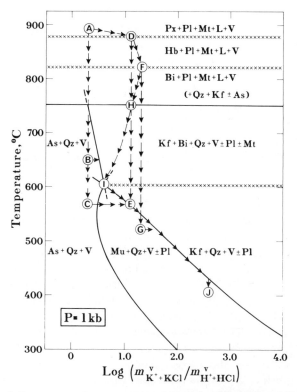

Fig. 3.6 Log ($m^v_{K^+ + KCl}/m^v_{H^+ + HCl}$) of aqueous chloride solutions in equilibrium or reaction relationship (arrows) with hypersolidus granodioritic magma (above horizontal solid line through point H) and with the indicated subsolidus mineral assemblages at 1.0 kb. Point I is an isobaric invariant point for equilibrium reaction 3.23; other labeled points are as described in the text. As = aluminum silicate (chiefly andalusite), Bi = biotite, Hb = hornblende, Kf = K-rich alkali feldspar, L = melt, Mt = magnetite, Mu = muscovite, Pl = plagioclase, Qz = quartz, and V = aqueous fluid phase.

altered porphyries of selective alteration in the more anorthitic zones of plagioclase phenocrysts and chemical evidence of calcium-depletion, coupled with potassium-enrichment, indicate that the subsolidus analogue of reaction 3.15, which increases m^v_{KCl}/m^v_{HCl} by consumption of HCl, is a major factor in establishing equilibrium, especially in the case of solutions initially at point B. In the latter case (C → E) where muscovite is formed, a reaction such as

$$CaAl_2Si_2O_8(s) + NaAlSi_3O_8(s) + 2HCl(v) + KCl(v)$$
$$\text{an} \qquad\qquad \text{ab}$$
$$\rightleftharpoons KAl_3Si_3O_{10}(OH)_2(s) + 2SiO_2(s) + CaCl_2(v) + NaCl(v) \qquad (3.20)$$
$$\text{ms} \qquad\qquad \text{qz}$$

which consumes HCl and KCl in a 2:1 ratio, also contributes to the increase in m^v_{KCl}/m^v_{HCl}, despite the fact that m^v_{KCl} actually decreases.

In contrast, an aqueous solution originating in the stability field of hornblende, as at point D in Figure 3.6, will not pass into the stability field of andalusite upon cooling to point E without reaction. Consequently equilibrium may be established at point E, on the muscovite-potassium feldspar boundary, principally by reaction (3.17) or a combination of it and the reaction

$$CaAl_2Si_2O_8(s) + 2KCl(v) + 4SiO_2(s) \rightleftharpoons 2KAlSi_3O_8(s) + CaCl_2(v) \quad (3.21)$$
$$\text{an} \qquad\qquad\qquad\qquad \text{qz} \qquad\qquad \text{or}$$

Similarly, a magmatic aqueous phase originating at point F, which is in the stability fields of both hornblende and biotite and has about the highest m^v_{KCl}/m^v_{HCl} ratio obtainable under hypersolidus conditions in a granodioritic system, also will pass directly into the stability field of muscovite and reach equilibrium on the muscovite-potassium feldspar boundary principally through reaction (3.20). Conceivably, solutions originating at A, D, or F may be cooled without appreciable reaction to temperatures in the range 350–400°C where pyrophyllite, instead of andalusite, is stable. Under these conditions, which will be discussed more fully elsewhere in this volume (Rose and Burt, Chapter 5), local equilibrium in the fracture selvage may be established either on the pyrophyllite-muscovite or muscovite-potassium feldspar boundary, depending upon the availability of feldspars.

In the event a magmatic aqueous phase originating at either A, D, or F reaches equilibrium with granodiorite porphyry at solidus temperatures and a pressure in excess of 0.5 kb, m^v_{KCl}/m^v_{HCl} will be between 10 and 15 (point H), having arrived at this value via point F by hypersolidus conversion of hornblende to biotite as discussed previously. Passage of this solution into cooler granodiorite porphyry wallrocks at a rate sufficiently slow to maintain a close approach to local equilibrium results in further lowering of m^v_{KCl}/m^v_{HCl} largely by reaction (3.21) but also in part by the subsolidus conversion of hornblende to biotite, neither of which affects m^v_{HCl}. Also, a reduction in anorthite content of the plagioclase solid solution by reaction (3.21) causes a corresponding increase in the activity of albite (a_{ab}^{pl}), which in turn decreases m^v_{KCl}/m^v_{HCl} through the exchange reaction

$$NaAlSi_3O_8(s) + KCl(v) \rightleftharpoons KAlSi_3O_8(s) + NaCl(v) \quad (3.22)$$
$$\text{ab} \qquad\qquad\qquad\quad \text{or}$$

Moreover, equilibrium in this reaction is shifted even further to the right upon cooling an igneous alkali feldspar-plagioclase assemblage below solidus temperatures because $a_{ab}{}^{af}/a_{or}{}^{af}$ increases with decreasing temperature at constant composition (Carmichael et al., 1974, Figure 4-30). The net result of all these reactions, therefore, is to constrain an aqueous chloride solution of magmatic origin, which equilibrates with granodioritic wallrocks or magma above point I, Figure 3.6, to intersect the muscovite-potassium feldspar boundary near its high-temperature limit (at I or between I and E).

Those solutions, which previously equilibrated along the andalusite-potassium feldspar boundary, intersect the muscovite-potassium feldspar boundary at I, an isobaric invariant point, through the equilibrium reaction

$$4KAlSi_3O_8(s) + Al_2SiO_5(s) + H_2O(v) + 2HCl(v)$$
or as
$$\rightleftharpoons 2KAl_3Si_3O_{10}(OH)_2(s) + 7SiO_2(s) + 2KCl(v) \quad (3.23)$$
$$\text{ms} \qquad \text{qz}$$

whereas those cooling solutions that approach the muscovite-potassium feldspar (+ quartz) boundary from the potassium-feldspar field intersect between I and E, where they equilibrate through the isobaric univariant reaction

$$3KAlSi_3O_8(s) + 2HCl(v)$$
or
$$\rightleftharpoons KAl_3Si_3O_{10}(OH)_2(s) + 6SiO_2(s) + 2KCl(v) \quad (3.24)$$
$$\text{ms} \qquad \text{qz}$$

With further cooling in the presence of feldspars under quasiequilibrium conditions, reactions (3.20) and (3.24) assure that $m_{\Sigma K}{}^v/m_{\Sigma H}{}^v$ will increase along the muscovite-potassium feldspar boundary of Figure 3.6.

It will be noticed that reactions (3.21), (3.22), and those that result in replacement of ferromagnesian minerals with biotite fix potassium in solid phases, thereby producing the potassium metasomatic effects that are such a prominent feature of the inner potassium feldspar-biotite (potassic) alteration zones in porphyry copper deposits the world over (Lowell and Guilbert, 1970). It also will be noticed that reactions (3.23) and especially (3.24) result in the fixing of hydrogen in muscovite ("sericite"), thereby producing the lower-temperature hydrogen metasomatism (Hemley and Jones, 1964) that is another prominent feature of those porphyries where a phyllic alteration zone is strongly developed. However, the fact that m_{KCl}^v/m_{HCl}^v increases in these reactions does not necessarily mean that the potassium content of the solution increases with decreasing temperature.

Reactions such as (3.20), (3.21), and (3.22) serve to buffer the potassium content at constant ΣCl, provided plagioclase or sodium-bearing alkali feldspar is present.

At pressures below about 0.5 kb, where neither hornblende nor biotite is stable above the solidus, aqueous fluids in which m^v_{KCl}/m^v_{HCl} is approximately that of point A predominate. Also, under these conditions the solidus is raised to temperatures above 800 °C and the isobaric invariant point I (Figure 3.6) is lowered to temperatures well below 600 °C; consequently, the temperature range over which aqueous fluids may equilibrate along the andalusite-potassium feldspar boundary is greatly expanded. Such equilibration involves mainly the subsolidus analogues of reactions (3.13)–(3.15), which produce aluminum silicates (andalusite at temperatures below approximately 750 °C). Thus, extensive aluminum silicate alteration, including lower temperature pyrophyllitization, may reflect a low-pressure source of the magmatic aqueous phase.

An alternative explanation for aluminum-silicate alteration, as well as an explanation for other important features of altered porphyries, such as anhydrite formation, involves the behavior of sulfur-bearing species in the magmatic aqueous phase on cooling. It was shown in the last section, in agreement with the calculations by Ohmoto (Chapter 10), that the total sulfur content of an aqueous phase in equilibrium with a granodioritic magma, under typical f_{O_2} conditions, ranges from approximately 0.2 m at 2.0 kb to 2.0 m or more at 0.5 kb fluid pressure. It also was shown that higher than normal values of f_{O_2}, caused by diffusive loss of H_2, could result in much higher total sulfur contents. Concomitant with the increase in total sulfur, f_{SO_2}/f_{H_2S} also increases to values of 10 or more, but upon cooling to subsolidus temperatures the SO_2 undergoes hydrolysis according to the scheme

$$4SO_2(v) + 4H_2O(v) \rightarrow H_2S(v) + 3H_2SO_4(v) \tag{3.25}$$

The resulting increase in $f_{H_2SO_4}$, coupled with a concomitant increase in f_{CaCl_2} through alteration reactions (3.21) and (3.20), causes precipitation of anhydrite by

$$CaCl_2(v) + H_2SO_4(v) \rightarrow \underset{ah}{CaSO_4(s)} + 2HCl(v) \tag{3.26}$$

thus explaining the observation by Camus (1975) that much of the anhydrite at El Teniente, Chile was deposited in open spaces. Progress of reaction

(3.25), which is promoted by precipitation of anhydrite in (3.26), also increases f_{H_2S} and, in turn, results in precipitation of metal sulfides through homogeneous reactions between H_2S and metal chlorides (chiefly iron chlorides) in the aqueous phase, as well as through sulfidization of iron bearing oxides and silicates in the wallrocks.

All of the reactions that precipitate metal sulfides from aqueous chloride soltions, as well as reaction (3.26), produce HCl in amounts that can far exceed the original HCl content of the magmatic aqueous phase. Consequently, m^v_{KCl}/m^v_{HCl} may follow a subsolidus path on cooling from point H in Figure 3.6 that is more sharply curved toward lower values, intersecting the andalusite-potassium feldspar boundary above isobaric invariant point I. Moreover, continued precipitation of sulfides, with or without anhydrite, after the solutions enter the stability field of muscovite produces "hydrogen metasomatism" (phyllic alteration) the extent of which may completely overshadow that produced by the original HCl content of the magmatic aqueous phase. Thus, although meteoric water may become involved in the later stages of phyllic alteration in some porphyry systems, as isotopic evidence suggests, its presence is not required for extensive phyllic alteration. Indeed, owing to the generally positive pressure ($P_{fluid} > P_{hydrostatic}$) in a porphyry fracture system being supplied by fluids from an H_2O-saturated magma carapace, it is probable that meteoric water gains major access to the system only after the supply from the magmatic source has been greatly diminished, as by waning magmatic activity or by choking of the fluid conduits from the H_2O-saturated carapace.

Very little has been said thus far about the sodium analogue of the system represented in Figure 3.6 mainly because the sodium-rich mica, paragonite, is extremely rare in phyllically altered porphyries, even in sodium-rich dioritic rocks. In accordance with the present model, this rarity of paragonite is a result of the fact that its maximum thermal stability in the presence of quartz (equivalent to isobaric invariant point I in Figure 3.6) is 50–100°C lower than that of muscovite + quartz. Thus, as fluids from a magmatic source are cooled along paths such as D-E, F-G, or H-I (Figure 3.6), they enter the stability field of muscovite + quartz with an $m^v_{\Sigma K}/(m^v_{\Sigma K} + m^v_{\Sigma Na}) \approx X^v_{KCl}/(X^v_{KCl} + X^v_{NaCl})$ (ionization constants for KCl, NaCl, and HCl are negligibly small at temperatures above 500°C), in the range 0.2–0.4. The isobaric univariant equilibrium between muscovite (+ quartz) and potassium feldspar of given composition buffers $m_{\Sigma K}{}^v/m_{\Sigma H}{}^v$ at a unique value for a given pressure and temperature (Figure 3.6), but $m^v_{\Sigma Na}/m_{\Sigma H}{}^v$ is not similarly buffered. Equilibrium is therefore established between muscovite, potassium-rich alkali feldspar, and sodium-rich plagioclase at an $m^v_{\Sigma Na}/m_{\Sigma H}{}^v$ that is well within the albite stability field, where it remains as temperature falls along the muscovite-potassium feldspar boundary (Hemley and Jones, 1964, Figure 3). Thus, despite the fact that $m_{\Sigma K}{}^v/m_{\Sigma H}{}^v$ increases with falling temperature along the muscovite-potassium feldspar boundary, $m^v_{\Sigma Na}/m_{\Sigma H}{}^v$ also increases by reactions such as (3.20) and $m_{\Sigma K}{}^v/(m_{\Sigma K}{}^v + m^v_{\Sigma Na})$ therefore decreases, to approximately 0.17 at 500°C, 1.0 kb, and to much lower values at low temperatures. Consequently, only if potassium feldspar is completely consumed and the solutions thereafter descend into the muscovite field, as at point J in Figure 3.6, will paragonite become stable. Further cooling beyond point J, in the absence of potassium feldspar, results in formation of paragonite, in accordance with the sodium analogue of equilibrium reaction (3.24), or of montmorillonite at temperatures below about 330°C (Hemley and Jones, 1964, Figure 2).

Reheating of these sodium-rich solutions in contact with the assemblage muscovite-quartz-sodic plagioclase, as by circulation in a convecting hydrothermal system, results in reconversion of the micas to alkali-rich feldspars. On the other hand, reheating of these same solutions in contact with a feldspar-free rock, such as a pelitic metasediment, may ultimately result in the formation of paragonite. This may be an important process in the formation of paragonite-bearing schists in many regionally metamorphosed terranes, but the virtual absence of paragonite from hydrothermally altered porphyries and the relatively minor abundance of sodium-montmorillonite indicate that it is not an important alteration process. Hence, phyllic alteration in porphyry systems is here regarded as having been produced mainly by interaction with acid volatiles from a higher temperature source. Accordingly, the isotopic imprint of meteoric water found in many phyllically altered porphyries probably was impressed by ion exchange, perhaps at temperatures as high as 500°C in some cases.

Finally, arguments were presented earlier against the separation of volatiles from the majority of porphyry magmas as two immiscible phases, one an H_2O-rich fluid and the other a chlorine-rich brine. On the other hand, evidence from fluid inclusion studies (Roedder, 1971) also was cited, which clearly indicates that in some places, notably Bingham, Utah, subsolidus separation (condensation) has occurred at temperatures as high as 700°C and at pressures approaching 1 kb. Condensation under these conditions can be brought about most readily by a modest (~100 bar) decrease in pressure (Sourirajan and Kennedy, 1962), as would attend escape of the magmatic aqueous phase from the H_2O-saturated carapace into the superjacent fracture system. As long as this alkali- and metal chloride-rich condensate coexists with the volatile-enriched (HCl, H_2S, SO_2, and CO_2) aqueous phase, solid-phase equilibria are not grossly affected. However, owing to differences in density, the more acid, sulfur-rich aqueous phase tends to be concentrated in the upper parts of a fracture system where it may enhance localization of phyllic alteration and sulfide deposition.

ACKNOWLEDGMENTS

This presentation has benefited greatly from the critical comments of Professors H. L. Barnes, D. M. Kerrick, and H. Ohmoto.

REFERENCES

Allen, J. C., A. L. Boettcher, and G. Marland (1975) Amphiboles in andesite and basalt: I. Stability as a function of $P\text{-}T\text{-}f_{O_2}$: *Am. Mineral.,* **60,** 1069-1085.

———, P. J. Modreski, C. Haygood, and A. L. Boettcher (1972) The role of water in the mantle of the earth: The stability of amphiboles and micas: *24th Intl. Geol. Congr. Sec.,* **2,** 231-240.

Blencoe, J. G. (1974) An experimental study of muscovite-paragonite stability relations: Unpublished PhD thesis, Stanford University.

Boettcher, A. L. (1971) The nature of the crust of the Earth, with special emphasis on the role of plagioclase: *Geophys. Mono. Series,* **14,** 261-277.

——— and P. J. Wyllie (1969) Phase relationships in the system $NaAlSiO_4\text{-}SiO_2\text{-}H_2O$ to 35 kilobars pressure: *Am. J. Sci.,* **267,** 875-909.

Bowen, N. L. (1913) The melting phenomena of the plagioclase feldspars: *Am. J. Sci.,* **35,** 577-599.

Boyd, F. R. (1959) Hydrothermal investigations of amphiboles: in *Researches in Geochemistry,* P. Abelson, ed., New York: Wiley, p. 377-396.

Burnham, C. Wayne (1967) Hydrothermal fluids at the magmatic stage: in *Geochemistry of Hydrothermal Ore Deposits,* H. L. Barnes, ed., New York: Holt, Rinehart and Winston, p. 34-76.

——— (1972) The energy of explosive volcanic eruptions: *Earth Mineral Sci.* (The Pennsylvania State University), **41,** 69-70.

——— (1975a) Water and magmas; a mixing model: *Geochim. Cosmochim. Acta,* **39,** 1077-1084.

——— (1975b) Thermodynamics of melting in experimental silicate-volatile systems: *Fortschr. Miner.,* **52,** 101-118.

——— (1979) The importance of volatile constituents: in *The Evolution of the Igneous Rocks: Fiftieth Anniversary Perspectives.* H. S. Yoder, Jr., ed., Princeton: Princeton Univ. Press (in press).

———, L. S. Darken, and A. C. Lasaga (1978) Water and magmas: Application of the Gibbs-Duhem equation; a response: *Geochim. Cosmochim. Acta,* **42,** 277-280.

——— and N. F. Davis (1971) The role of H_2O in silicate melts: I. P-V-T relations in the system $NaAlSi_3O_8\text{-}H_2O$ to 10 kilobars and 1000°C: *Am. J. Sci.,* **270,** 54-79.

——— and ——— (1974) The role of H_2O in silicate melts: II. Thermodynamic and phase relations in the system $NaAlSi_3O_8\text{-}H_2O$ to 10 kilobars, 700° to 1100°C: *Am. J. Sci.,* **274,** 902-940.

———, J. R. Holloway, and N. F. Davis (1969) Thermodynamic properties of water to 1000°C and 10,000 bars: *Geol. Soc. Am. Spec. Pap.* 132.

——— and R. H. Jahns (1962) A method for determining the solubility of water in silicate melts: *Am. J. Sci.,* **260,** 721-745.

Camus, Francisco (1975) Geology of the El Teniente orebody with emphasis on wall-rock alteration: *Econ. Geol.,* **70,** 1341-1372.

Carmichael, I. S. E., F. J. Turner, and J. Verhoogen (1974) *Igneous Petrology*: New York: McGraw-Hill.

Eggler, D. H. (1973) Role of CO_2 in melting processes in the mantle: *Carnegie Inst. Washington Year Book,* **72,** 457-467.

―― and C. W. Burnham (1973) Crystallization and fractionation trends in the system andesite-H_2O-CO_2-O_2 at pressures to 10 kb: *Bull. Geol. Soc. Am.,* **84,** 2517-2532.

Eggler, D. H., B. O. Mysen, and M. G. Seitz (1974) The solubility of CO_2 in silicate liquids and crystals: *Carnegie Institution Washington Yearbook,* **73,** 226-228.

Furst, G. A. (1978) The melting of plagioclase in the system Na_2O-CaO-Al_2O_3-SiO_2-H_2O at high pressure and temperature: Unpublished PhD thesis, The Pennsylvania State University, University Park, Pa.

Goldsmith, J. R. (1976) Scapolites, granulites, and volatiles in the lower crust: *Bull. Geol. Soc. Am.,* **87,** 161-168.

Gustafson, L. B. and J. P. Hunt (1975) The porphyry copper deposit at El Salvador, Chile: *Econ. Geol.,* **70,** 857-912.

Hamilton, D. L., C. W. Burnham, and E. F. Osborn (1964) The solubility of water and effects of oxygen fugacity and water content on crystallization in mafic magmas: *J. Petrology,* **5,** 21-39.

Haughton, D. R., P. L. Roeder, and B. J. Skinner (1974) Solubility of sulfur in mafic magmas: *Econ. Geol.,* **69,** 451-467.

Helz, R. T. (1973) Phase relations of basalts in their melting range at $P_{H_2O} = 5$ kb as a function of oxygen fugacity: Part I. Mafic phases: *J. Petrol.,* **14,** 249-302.

Hemley, J. J. and W. R. Jones (1964) Chemical aspects of hydrothermal alteration with emphasis on hydrogen metasomatism: *Econ. Geol.,* **59,** 538-569.

Hill, R. E. T. and A. L. Boettcher (1970) Water in the earth's mantle: Melting curves of basalt-water and basalt-water-carbon dioxide: *Science,* **167,** 980-981.

Holland, H. D. (1967) Gangue minerals in hydrothermal deposits: in *Geochemistry of Hydrothermal Ore Deposits,* H. L. Barnes, ed., New York: Holt, Rinehart, and Winston, 382-436.

―― (1972) Granites, solutions, and base metal deposits: *Econ. Geol.,* **67,** 281-301.

Holloway, J. R. (1973) The system pargasite-H_2O-CO_2: A model for melting of a hydrous mineral with a mixed-volatile fluid—I. Experimental results to 8 kbar: *Geochim. Cosmochim. Acta.,* **37,** 651-666.

―― (1976) Fluids in the evolution of granitic magmas: Consequences of finite CO_2 solubility: *Bull. Geol. Soc. Am.,* **87,** 1513-1518.

―― and C. W. Burnham (1972) Melting relations of basalt with equilibrium-water pressure less than total pressure: *J. Petrol.,* **13,** 1-29.

―― and C. F. Lewis (1974) CO_2 solubility in hydrous albite liquid at 5 kbar (abstract): *EOS, Trans. Am. Geophys. Union,* **55,** 483.

Jakeš, P. and A. J. R. White (1969) Structure of the Melanesian arcs and correlation with distribution of magma types: *Tectonophysics,* **8,** 223-236.

Kilinc, I. A. (1969) Experimental metamorphism and anatexis of shales and graywackes: Unpublished PhD thesis, The Pennsylvania State University, University Park, Pa.

―― and C. W. Burnham (1972) Partitioning of chloride between a silicate melt and coexisting aqueous phase from 2 to 8 kilobars: *Econ. Geol.,* **67,** 231-235.

Koide, H. and S. Bhattacharji (1975) Formation of fractures around magmatic intrusions and their role in ore localization: *Econ. Geol.,* **70,** 781-799.

REFERENCES

Kudo, A. M. and D. F. Weill (1970) An igneous plagioclase thermometer: *Contr. Mineral. Petrol.*, **25**, 52-65.

Kushiro, I. (1973) The system diopside-anorthite-albite: Determination of compositions of coexisting phases: *Carnegie Institution Wash. Yearbook,* **72**, 502-507.

Lanier, G., R. B. Folsom, and S. Cone (1975) Alteration of equigranular quartz monzonite, Bingham district, Utah: in *Guide Book to the Bingham Mining District,* R. E. Bray and J. C. Wilson, ed., Bingham Canyon, Utah: Soc. Econ. Geol., 73-97.

Leake, B. E. (1968) A catalog of analyzed calciferous and subcalciferous amphiboles together with their nomenclature and associated minerals: *Geol. Soc. Amer.,* Spec. Pap. 98.

Lindsley, D. H. (1967) Melting relations of plagioclase at high pressures: *Carnegie Institution Wash. Yearbook,* **65**, 204.

Lowell, J. D. and J. M. Guilbert (1970) Lateral and vertical alteration-mineralization zoning in porphyry ore deposits: *Econ. Geol.,* **65**, 373-408.

Mathez, E. A. (1973) Refinement of the Kudo-Weill plagioclase thermometer and its application to basaltic rocks: *Contrib. Mineral. Petrol.,* **41**, 61-72.

Mehnert, K. R. (1975) The Ivrea zone, a model of the deep crust: *N. Jb. Miner. Abh.,* **125**, 156-199.

Minear, J. W. and M. N. Toksöz (1970) Thermal regime of a downgoing slab and the new global tectonics: *J. Geophysics. Res.,* **75**, 1397-1420.

Moore, J. G. and B. P. Fabbi (1971) An estimate of the juvenile sulfur content of basalt: *Contrib. Mineral. Petrol.,* **33**, 118-127.

Mysen, B. O. (1975) Solubility of volatiles in silicate melts at high pressure and temperature: The role of carbon dioxide and water in feldspar, pyroxene, and feldspathoid melts: *Carnegie Institution Wash. Yearbook,* **74**, 454-468.

Nockolds, S. R. (1954) Average chemical compositions of some igneous rocks: *Bull. Geol. Soc. Am.,* **65**, 1007-1032.

Norton, D. L. and L. M. Cathles (1973) Breccia pipes-products of exsolved vapor from magmas: *Econ. Geol.,* **68**, 540-546.

Orlova, G. P. (1964) The solubility of water in albite melts: *Int. Geol. Rev.,* **6**, p. 254-258.

Oxburgh, E. R. and D. L. Turcotte (1970) Thermal structure of island arcs: *Bull. Geol. Soc. Am.,* **81**, 1665-1688.

Piwinskii, A. J. (1968) Experimental studies of igneous rock series, central Sierra Nevada batholith, California: *J. Geol.,* **76**, 548-570.

Rehrig, W. A. and T. L. Heidrick (1973) Regional fracturing in Laramide stocks of Arizona and its relationship to porphyry copper mineralization: *Econ. Geol.,* **67**, 198-212.

Ringwood, A. E., I. D. MacGregor, and F. R. Boyd (1964) Petrologic composition of the upper mantle: *Carnegie Institution Wash. Yearbook,* **63**, 174-152.

Robie, R. A. and D. R. Waldbaum (1968) Thermodynamic properties of minerals and related substances at 298.15°K (25.0°C) and one atmosphere (1.013 bars) pressure and at higher temperatures: *Bull. U. S. Geol. Surv.,* 1259.

Roedder, E. (1971) Fluid inclusion studies on the porphyry-type ore deposits at Bingham, Utah, Butte, Montana, and Climax, Colorado: *Econ. Geol.,* **66**, 98-120.

Ryan, M. P. (1979) High temperature mechanical properties of basalt: Unpublished PhD thesis, The Pennsylvania State University, University Park, Pa.

Shade, J. W. (1974) Hydrolysis reactions in the SiO_2-excess portion of the system K_2O-Al_2O_3-SiO_2-H_2O in chloride fluids at magmatic conditions: *Econ. Geol.,* **69**, 218-228.

Sillitoe, R. H. (1972) A plate tectonic model for the origin of porphyry copper deposits: *Econ. Geol.*, **76**, 184-197.

Sourirajan, S. and G. C. Kennedy (1962) The system H_2O-NaCl at elevated temperatures and pressures: *Am. J. Sci.*, **260**, 115-141.

Thompson, J. B. and D. R. Waldbaum (1968) Mixing properties of sanidine crystalline solutions: I. Calculations based on ion-exchange data: *Am. Mineral.*, **53**, 1965-1999.

Turcotte, D. L. and G. Schubert (1973) Frictional heating on the descending lithosphere: *J. Geophys. Res.*, **78**, 5876-5886.

Wallace, S. R. (1974) The Henderson ore body-elements of discovery, reflections: *Am. Inst. Min. Engr. Trans.*, **256**, 216-227.

Whitney, J. A. (1975a) The effects of pressure, temprature and X_{H_2O} on phase assemblage in four synthetic rock compositions: *J. Geol.*, **83**, 1-27.

―――― (1975b) Vapor generation in a quartz monzonite magma: A synthetic model with application to porphyry copper deposits: *Econ. Geol.*, **70**, 346-358.

Wyllie, P. J. and O. F. Tuttle (1961) Experimental investigations of silicate systems containing two volatile components: Part II. The effects of NH_3 and HF, in addition to H_2O on the melting temperatures of albite and granite: *Am. J. Sci.*, **259**, 128-143.

Yoder, H. S., Jr. (1965) Diopside-anorthite-water at five and ten kilobars and its bearing on explosive volcanism: *Carnegie Institution Wash. Yearbook*, **64**, 82-89.

―――― and I. Kushiro (1969) Melting of a hydrous phase: phlogopite: *Am. J. Sci., 267-A*, Schairer vol., 558-582.

――――, D. B. Stewart, and J. R. Smith (1957) Ternary feldspars: *Carnegie Institution Wash. Yearbook*, **55**, 206-214.

―――― and C. E. Tilley (1962) Origin of basalt magmas: An experimental study of natural and synthetic rock systems: *J. Petrol.*, **3**, 342-532.

4

The Sedimentary Genesis of Hydrothermal Fluids

JEFFREY S. HANOR
Louisiana State University

In the previous chapter, Burnham discussed the generation of hydrothermal fluids associated with igneous activity. In this chapter, we discuss an alternative, but less widely recognized, process for generating hydrothermal fluids, that is, sediment diagenesis.

Approximately 20% of the total volume of the unmetamorphosed sediments of the earth's crust consists of pore water. Most of this mass of fluid has evolved during the normal course of basin development and diagenesis. These waters have been heated during progressive burial and are kept warm in large part by the simple conductive flow of heat from the interior of the earth. Hydrothermal formation waters formed in this way have long been considered potential ore-forming solutions, particularly in the genesis of Mississippi Valley-type lead-zinc-barite-fluorite deposits (Lindgren, 1933). With the development of fluid inclusion techniques (Chapter 14) and the application of isotope geochemistry to problems of the origin of ore deposits (Chapters 2, 6, and 10), a convincing body of evidence has been developed that identifies subsurface sedimentary waters as an important component of some ore-forming fluids (White, 1974). Many important questions remain, however. For example, are normal processes of sediment diagenesis and migration of subsurface formation waters sufficient to generate ore-forming fluids, or is it necessary to invoke additional sources of metals and heat?

Largely because of interest in the occurrence of hydrocarbons in young sedimentary basins, such as the Northern Gulf of Mexico, an increasing body of information is being obtained on processes of sediment diagenesis during the active depositional phase of basin development. Many of these processes appear to be critical to our understanding of the migration and chemical evolution of formation waters. In this chapter we discuss controls on the physical and chemical properties of sedimentary waters and evaluate whether diagenetic processes alone are sufficient to generate potential ore-forming fluids. Finally, we will discuss how hydrothermal fluids formed in sedimentary basins might have migrated into sites favorable for ore deposition.

DIAGENESIS OF SEDIMENTS

A wide range of chemical and physical changes occur in sediments during and after their burial, changes which alter solid mineral phases and interstitial fluids. Many of these changes are encompassed in the term *diagenesis*. The thermal limits of diagenesis range from a few degrees below 0°C, when saline waters freeze, to approximately 250–300°C, temperatures observed in deeply buried sediments of the Gulf Coast. The upper thermal limit represents an apparent kinetic threshold for equilibration of argillaceous sediments. Below these temperatures reaction rates are generally slow. Above these temperatures a succession of characteristic mineral assemblages forms rapidly in response to increases in temperature and pressure. The details of these metamorphic processes are beyond the scope of this discussion.

Because most sedimentary basins are dynamic, open systems, where both energy (heat) and material (water and dissolved substances) are exchanged with the external environment and between adjacent layers of sediment, significant thermal, compositional, and fluid potential gradients can be maintained for indefinite periods of time. In addition, progressive burial exposes a given volume of sediment to ever-increasing temperatures and pressures. Thus, much of sediment diagenesis, including the migration and chemical evolution of sedimentary fluids, involves a complex series of events, controlled to a significant degree by nonequilibrium, irreversible processes. Some of these processes occur continuously over extended periods of time, but the diagenetic history of a sedimentary mass can be punctuated by episodic events, such as the loss of fluid, dissolved material, and heat that may occur during tectonism.

PROPERTIES OF AQUEOUS SOLUTIONS IN SEDIMENTARY BASINS

A review of the important physical and chemical properties of aqueous solutions in sedimentary basins is essential to an understanding of the potential role of these fluids as ore-forming solutions. Controls on the distribution of ore-forming components will be discussed in the latter part of this section.

Sources of Water

Water present in the pores of sediments includes water trapped at the time of deposition, water produced or exchanged during diagenetic chemical reactions, and water introduced from outside a basin by regional fluid flow. Metamorphic dehydration reactions occurring at high temperatures may provide an additional source of water at great depth (Fyfe, 1973).

During normal diagenesis, dehydration of clay minerals, gypsum, and organic matter can provide unbound water. One of the dominant diagenetic reactions occurring in the argillaceous sediments of the Gulf Coast is the dehydration of montmorillonite or mixed-layer montmorillonite-illite clays by the coupled removal of interlayer water and the conversion of montmorillonite to illite (Perry and Hower, 1972). Sources of potassium required for the conversion of montmorillonite to illite may include subsurface water, feldspars (Weaver and Beck, 1971), or detrital micas (Perry and Hower, 1972). Because barium and lead can both substitute for potassium in feldspars and micas, diagenetic destruction of those mineral phases during the formation of illite provides a potential source of these elements for formation waters.

The coupled dehydration and alteration of montmorillonite occurs rapidly when sediments reach certain threshold temperatures during progressive burial. One phase of water expulsion appears to occur at a temperature of approximately 90°C, and a second phase of dehydration occurs at a temperature in the vicinity of 120°C (Johns and Shimoyama, 1972). The migration of fluids caused by dehydration and the generation of differences in fluid pressure may provide a means of flushing hydrocarbons (Powers, 1967; Burst, 1969) and metals from argillaceous source rocks.

Presumably, the diagenetic dehydration and destruction of montmorillonite has also been significant in older sedimentary basins. This is suggested by the fact that, in contrast to the montmorillonite-rich clays of the Gulf Coast and other Mesozoic-Cenozoic basins, illite and chlorite dominate the argillaceous sediments of Paleozoic and Precambrian basins. Weaver and Beck (1971) suggest that temperatures in the vicinity of 200°C and the upward migration of potassium, magnesium, and iron from depth were necessary for the diagenetic production of the typical clay mineral suites observed in older shales. In many regions, such as the Illinois basin, this would have required temperatures significantly in excess of those presently observed. It is possible, however, that the observed differences in present mineralogy may also reflect in part differences in the composition and mineralogy of the original sediments (Millot, 1970).

Reduction in Porosity and Fluid Volume with Burial

The reduction of porosity during burial and diagenesis is an important process by which fluids are expelled from a sedimentary basin. Several factors affecting this process also change because many physical and some chemical properties of sediments are functions in part of porosity.

The reduction in porosity of argillaceous sediments as a function of burial depth is shown in Figure 4.1, based on data in the compilation by Perrier and Quiblier (1974). Initial porosities at the time of deposition range from 0.6 to 0.9, and most of the reduction in porosity occurs within the upper 300 m of burial. At depths of 3000 m, porosity ranges from 0.2 to nearly 0.0. The compaction behavior of an argillaceous sediment depends on a large number of variables, including rate of deposition, clay mineralogy, particle-size distribution and lithologic sequence. For example, clays interbedded with porous sands that are connected hydrologically with the

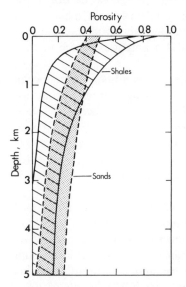

Fig. 4.1 Range in porosity of shales (ruled area) and sands (stippled area) as a function of depth of burial. Based on data of Perrier and Quiblier (1974).

surface may dewater more readily with burial than clays deposited in a monolithic sequence. Stratigraphic sequences are known in the Gulf Coast in which there is a reversal in shale porosity with depth. These zones of porosity reversal are also characterized by a marked increase in fluid pressure. Magara (1975) suggests that these zones represent sediments that have not followed the usual compaction-depth sequence because their low permeability has prevented the escape of interstitial water at normal rates with burial. The dehydration of montmorillonite may be an additional factor in maintaining both high porosities and fluid pressures.

The initial porosities of quartzose sands range from approximately 0.4 to 0.5 and are thus significantly lower than the initial porosities of finer-grained sediments. The reduction of porosity of sands with depth is less than that of clays, however, and at 3000 m, average porosities of sandstones range from 0.3 to 0.1 (Perrier and Quiblier, 1974). Porosity reduction in sands results both from mechanical compaction and the filling of void space by cements (Pettijohn et al., 1973).

The variation in porosity of carbonate sediments with depth of burial is more complex than in shales and sandstones (Choquette and Pray, 1970). Extensive filling of void space in some carbonate sediments can occur soon after deposition. Further cementation, recrystallization, and the development of stylolites are effective in reducing porosity during burial, but porosity reduction is not uniform, and significant variations can exist even within a single lithologic unit. Initial porosities of many carbonate sediments range 0.4–0.7. The porosities of ancient limestones, however, are as low as 0.05 (Choquette and Pray, 1970).

It is obvious that a significant volume of water is released during normal sediment compaction and diagenesis. The average shale, based on the data in Figure 4.1, can yield approximately 3.5×10^3 l water during compaction for every cubic meter of solid deposited. Seventy-five percent of this water is normally expelled during shallow burial from 0 to 100 m. The average sandstone can release 0.7×10^3 l water/m^3 of solid. Burial to a depth of 3000 m is required before 75% of this water is expelled. The calculated average rate of water release for shales and sands as a function of burial depth is shown in Figure 4.2. Superimposed on the compaction curve for shales are two idealized peaks representing water release due to dehydration of pure montmorillonite. These peaks are based on dehydration curves proposed by Johns and Shimoyama (1972), and Burst's (1969) estimate of the volume of water produced by dehydration. The actual position and magnitude of these peaks would depend on several factors, including the geothermal gradient and the relative abundance of montmorillonite in the shale.

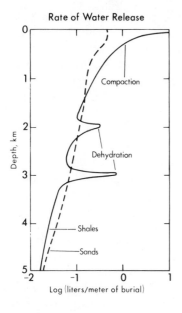

Fig. 4.2 Average rate of water release per cubic meter of solid during burial compaction of shales and sands. Calculated from data in Figure 4.1. Dehydration peaks for pure montmorillonite based on data of Johns and Shimoyama (1972).

Physical Properties

TEMPERATURE

Temperature-depth relations of fluids in sedimentary basins are known in most detail in areas of hydrocarbon production, where deep drilling has been extensive. Simple measurement of bottom-hole temperatures, however, does not always provide a reliable indication of *in-situ* sediment temperature because during drilling, the circulation of drilling mud acts as a heat sink. During production, adiabatic expansion of fluids in the well also can cause significant cooling (Jam et al., 1969). Some reported temperatures for sedimentary basins, including those discussed below, may be less than true, *in-situ* temperatures.

Figure 4.3 shows minimum temperatures as a function of depth for several sedimentary basins, as determined from the data of Graf et al. (1966), Hitchon and Friedman (1969), and Jones and Wallace (1974). In each region, with the exception of the Texas Gulf Coast profile, the variation in temperature with depth is approximately linear, and each geothermal gradient is thus nearly constant. The intercepts and slopes of the gradients, however, vary from basin to basin. The temperature intercept at zero depth is defined by the mean annual temperature and varies in the profiles illustrated from 9°C for the Alberta Basin to over 30°C for the Texas Gulf Coast to the south. Although diurnal and annual variations in surface tem-

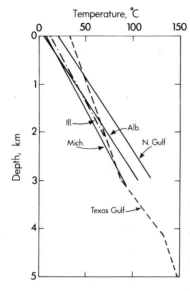

Fig. 4.3 Observed variation in temperature with depth in the Illinois, Michigan, and Northern (Mississippi) Gulf basins (Graf et al., 1966); the Alberta Basin (Hitchon and Friedman, 1969); and a portion of the Texas Gulf Coast (Jones and Wallace, 1974).

perature do occur, these short-term oscillations are damped out at shallow depths (Carslaw and Jaeger, 1959). However, long-term variations in climatic conditions occurring during continental glaciations or induced by the slow migration of continental masses into different latitudinal zones could result in a temperature difference of as much as 20°C or more at any given depth within a sedimentary basin.

The average geothermal gradient in older sedimentary basins is 15–40°C/km (Levorsen, 1967). If heat transfer is strictly by steady-state conduction, then the geothermal gradient, dT/dz, is related to the heat flow, Q, and the thermal conductivity of the sediments, K, by

$$Q = K(dT/dz) \tag{4.1}$$

The thermal conductivity of sediments varies by nearly an order of magnitude and is a function of lithology, porosity, temperature, and pressure (Clark, 1966). For a given lithology or rock type, the most significant variable is porosity, with the more porous sediments of higher fluid content being less conductive. A freshly deposited mud can have a thermal conductivity of less than 1.7×10^{-3} cal cm^{-1} sec^{-1}, while a shale of low porosity will have a conductivity of 4.5×10^{-3} or more. Such variations can have a profound effect on the geothermal gradient. For example, in the Cenozoic sediments of the Gulf Coast, gradients as high as 100°C/km exist at depth in stratigraphic zones characterized by unusually high porosities. Figure 4.3 shows the increase in geothermal gradient associated with such a zone in an area of the Texas Gulf Coast (Jones and Wallace, 1974). A temperature of 273°C has been observed at a depth of 5859 m in sediments in this region (Jones, 1970).

Heat transfer by processes other than conduction may be significant during some phases of sedimentation and basin development. Even in a tectonically dormant and apparently thermally uniform region of the crust, such as the Illinois basin, deep circulation of water, here driven by differences in regional hydrostatic head, moves heat up and out of the basin (Cartwright, 1970). In the Gulf Coast, heat is lost from depth by the diapiric intrusion of salt and by the migration of deep, hot formation waters from zones of high fluid pressure up along regional growth faults (Jones and Wallace, 1974). As fluids are expelled from deep sediments and sediment porosities in thermally insulating zones decrease, geothermal gradients should also decrease and the basin will cool off.

It is probable that significant variations in heat flow regime have occurred in older sediments, such as those deposited in the Alberta, Michigan, and Illinois basins. Deep waters in these basins could have been significantly warmer in the past, for example, simply if the rapid deposition of thick sequences of clays and sands produced an overlying blanket of initially porous, thermally-insulating sediment.

PRESSURE

The pressure exerted by a subsurface fluid against the solid mineral grains of the sediment is known variously as the fluid pressure, reservoir pressure, or formation pressure. If the fluid has hydrologic continuity upward to the air-water interface through interconnected pores, then the fluid pressure is equivalent to the hydrostatic pressure and will be a single function of depth and the density of the overlying water column ($\rho \approx 1.0$). If, however, a body of water at depth is completely enclosed by impermeable rock, then the fluid pressure will be equal to the pressure exerted by the overlying water-saturated rock column ($\rho \approx 2.3$), the lithostatic or geostatic pressure (Figure 4.4).

Stratigraphic intervals in which the measured fluid pressure is significantly in excess of calculated hydrostatic pressure are known variously as "abnormally pressured" or "geopressured" zones. Such zones occur in a number of Cenozoic sedimentary basins and are being intensively studied because of their association with the migration and accumulation of hydrocarbons (Vockroth, 1974). The stippled region in Figure 4.4 shows the range of observed fluid pressures as a function of depth in Gulf Coast sediments of Texas and Louisiana (Jones, 1968). In different regions of the Gulf Coast significant deviations from hydrostatic pressures begin to occur

PROPERTIES OF AQUEOUS SOLUTIONS IN SEDIMENTARY BASINS

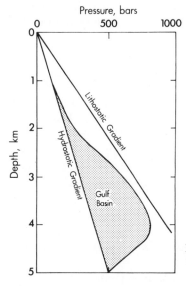

Fig. 4.4 Observed fluid pressures in the Gulf Coast Basin (stippled region). Data from Jones (1968).

at depths ranging from 1.4 to 4.5 km and are associated with abnormally high sediment porosities and increases in the geothermal gradient (Figure 4.3).

The following are some of the mechanisms that have been invoked to account for high fluid pressures in sediments:

1. Rapid deposition or mechanical compression of sediments having a low permeability (Bredehoft and Hanshaw, 1968; Hubbert and Rubey, 1959).
2. Increase in temperature through progressive burial of a confined body of water (Barker, 1972).
3. Dewatering of montmorillonite or other hydrous phases during burial diagenesis in sediments having a low permeability (Hanshaw and Bredehoft, 1968).
4. Osmotic diffusion of water through semipermeable strata into bodies of more saline fluids.

The first two mechanisms appear to be primary factors controlling the distribution of high fluid pressures in the Gulf (Bradley, 1975; Magara, 1975). Other factors may be of more local importance. The low permeability of rapidly deposited shales impedes the expulsion of water from shales and intercalated sand bodies. The interstitial fluids support part of the weight of the overlying rock column, and high fluid pressures are the result. A rupture or slow failure of the essentially impermeable seal that is required to maintain any zone of high fluid pressure will permit the migration of fluids to areas of lower pressure or fluid potential [equation (4.4)].

DENSITY AND VISCOSITY

The density of subsurface waters is a function of composition, temperature, and, more significantly than is often thought, pressure. Increases in salinity and pressure with depth cause densities to increase; however, T also increases with depth, and this has the opposite effect on density. On the basis of PVT data for NaCl solutions (Rowe and Chou, 1970), the maximum densities of brines per given depth in the Alberta, Illinois, Michigan, and northern Gulf basins have been estimated as shown by Figure 4.5. Densities increase, but density gradients decrease with depth, and significant heating at depth could produce lower densities and induce gravitational instability (Hanor, 1973). Densities for high calcium-chloride waters of the Michigan basin are probably in excess of those shown in Figure 4.5, but the same general density-depth relation holds.

The viscosity of aqueous solutions in sedimentary environments is a function largely of salinity and temperature. Figure 4.6 shows estimated values for viscosity of waters in the Michigan and northern Gulf of Mexico basins, based on tabulated viscosity values for pure water (Clark, 1966) and NaCl solutions (International Critical Tables). Comparatively, temperature is the dominant variable, and viscosities should be lower by as much as a factor of ½ to ¼ at depth. In highly saline brines, such as characterize the Mich-

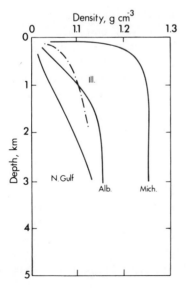

Fig. 4.5 Calculated densities of brines in the Illinois, Michigan, Northern Gulf and Alberta basins, corrected for T and P.

PROPERTIES OF AQUEOUS SOLUTIONS IN SEDIMENTARY BASINS 147

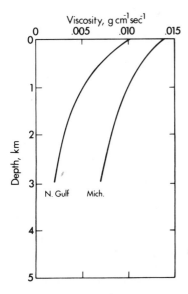

Fig. 4.6 Calculated viscosities of brines in the Northern Gulf and Michigan basins, corrected for T and P.

igan basin, viscosities are higher than for fresher waters at comparable depths and temperatures.

Chemical Properties

SALINITY

The average total dissolved solids content of formation waters in most Mesozoic and Paleozoic sedimentary basins increases with depth (Figure 4.7). Typically the rate of salinity increase diminishes with depth, and maximum observed salinities level off to a value that is characteristic for each basin (Dickey, 1969). In the Illinois basin, for example, this value is approximately 200‰. In the Michigan basin, however, which is characterized by the presence of evaporites, salinities of deep waters exceed 400‰. In older sedimentary basins, the areal variations in salinity of waters within a given formation tend to reflect basin geometry. Isocons, contours of equal salinity, often parallel regional basement depth contours, with the most saline waters occupying the deepest parts of the basin.

Significant reversals in salinity with depth have been found in Cenozoic sands and shales of the Gulf Coast (Schmidt, 1973). In the Manchester field of Louisiana, salinities range 120–200‰ to a depth of approximately 3 km, and then drop to seawater values of 35‰ and less in sediments of abnormally high fluid pressures (Schmidt, 1973; Osmaston, 1975). The amount of water produced by the dehydration of montmorillonite at depth

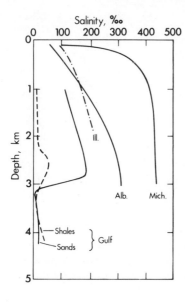

Fig. 4.7 Maximum observed salinities at various depths in the Illinois, Michigan, and Alberta basins. The two profiles on the left show the variation in salinity in shales and sands in the Manchester field, Louisiana Gulf Coast (Schmidt, 1973; Osmaston, 1975).

is probably insufficient to have caused this salinity reversal, and Osmaston (1975) suggests that the waters in the high-pressure zones simply were never as concentrated as the water lying immediately above. The problem is far from being resolved, however, for in other regions of high fluid pressure in the Gulf Coast, salinities are equivalent to or greater than salinities of overlying waters in zones characterized by hydrostatic pressures (Dickey et al., 1972).

MAJOR DISSOLVED SPECIES

Every sedimentary basin is characterized by a unique suite of water compositions, as shown clearly in DeSitter's (1947) evaluation of the diagenesis of sedimentary waters. However, some useful generalizations regarding the variation in water composition can still be made.

Relative to chloride content, most subsurface saline waters in sandstones and carbonates show an enrichment in calcium, strontium, barium, and fluoride, and a depletion in sodium, potassium, magnesium, and sulfate compared to modern seawater. The behavior of bicarbonate is more complex. Although bicarbonate concentrations are often low in deep brines, some waters of intermediate depth and salinity are dominated by bicarbonate. Most high-sulfate formation waters are confined to relatively shallow depths (White et al., 1963). An example of many of these general compositional trends is illustrated in Figure 4.8, which shows the apparent enrichment or depletion of major dissolved species during chemical evolution of brines of the Illinois basin, based on data from Graf et al., (1966). Because chloride increases with depth, we have an approximate depth sequence, observed in many other basins, starting with HCO_3-(SO_4) shallow waters and progressing from Na-Cl waters to Na-Ca-Cl waters with increasing depth.

Owing to difficulties in sample extraction, much less is known about the composition of subsurface waters in argillaceous sediments than in the more permeable sandstones and carbonates. It would appear from the studies of Weaver and Beck (1971) and Schmidt (1973) that, at least in Gulf Coast sediments, both the salinities (Figure 4.7) and the relative compositions of sand and shale waters are significantly different. In zones characterized by hydrostatic fluid pressures, salinities in shales are a factor of ¼ to ¹⁄₁₀ lower than that in adjacent sands. In some zones of high fluid pressure salinities of both are uniformly low (Figure 4.7). Weaver and Beck found that sodium and potassium in shale waters were more concentrated than in seawater, calcium and magnesium were less. The relative abundance of anions in shale interstitial waters is $HCO_3 > SO_4 > Cl$ and $SO_4 > HCO_3 > Cl$, in contrast to waters in sands, which are dominated by chloride.

ACIDITY AND OXIDATION STATE

The stability of many ore and gangue minerals in aqueous solutions can be conveniently related to the acidity and oxidation state of the system (Chapters 2 and 9). In this context, measurements of pH, Eh, pE, or P_{O_2} on formation waters are important. With the increasing departure of solutions from equilibrium surface conditions, however, the direct measurement of these parameters becomes increasingly difficult.

Most pH measurements of subsurface waters are made at well-head or laboratory temperatures, which are usually often less than true sediment temperatures. This pH must be corrected for changes in dissociation constants with temperature to calculate true pH values (Chapter 9). The pH of pure water, for example, will decrease on simple heating from 7.0 at 25°C to 5.8 at 150°C. The pH values discussed here have *not* been corrected to conditions at depth.

Published pH values for saline formation waters range from approximately 5 to 9. Analyses in the compilation of White et al., (1963) show a relation between bulk brine compositions and average pH values: high SO_4-HCO_3 waters, 8.0; sodium-chloride waters, 7.1; and sodium-calcium-chloride waters, 6.7. The standard deviation for each is 0.5 pH units. As discussed previously, this compositional sequence roughly corresponds to increasing depth, and the variation in pH thus suggests the progressive release of H^+ during diagenesis.

Very little information is available on the oxidation state of formation waters. In theory, the Eh could be calculated from data on pH, temperature, and relative concentrations of members of dissolved redox pairs, such as methane-carbon dioxide, nitrite-nitrate, or sulfide-sulfate. Some subsurface waters, for example, contain both abundant dissolved sulfide and sulfate (White et al., 1963). Although this should fix the oxidation potential of these waters, sulfate is inert to inorganic reduction below approximately 250°C and equilibrium could not be expected. Satisfactory evidence could be found if several pairs of redox reactions gave similar results, but rarely will that occur at low temperatures (Stumm and Morgan, 1970).

Controls on Major Element Geochemistry

The processes that control the composition of formation waters are as of yet imperfectly understood. Any satisfactory model must explain both the general increase in salinity of waters with depth (Figure 4.7) and the variations in relative proportions of dissolved constituents (Figure 4.8).

INTERACTION WITH EVAPORITES

Some saline formation waters, including metal-rich brines in Central Mississippi (Carpenter et al., 1974), have undoubtedly been derived in part from the solution of evaporites at depth or from the infiltration of hypersaline waters originally formed in evaporative surface environments (Rittenhouse, 1967). Modification of these brine compositions may take place during subsequent diagenesis. The formation of authigenic potassium feldspar (Carpenter et al., 1974), for example, would be effective in reducing the potassium content of evaporite brines.

Interaction with evaporites is not a universal solution to the problem of the origin of subsurface brines. Saline waters are known from evaporite-free basins and the proportions of major dissolved species in some brines cannot be explained simply on the basis of the evaporation and subsequent alteration of marine waters (White, 1965).

MEMBRANE FILTRATION

In 1947, DeSitter suggested that the increase in salinity with depth in sedimentary basins could be explained if fine-grained, argillaceous sediments behaved as semipermeable membranes, which allow water molecules to escape upward during burial and compaction, but which retard or prevent the migration of dissolved species. This mechanism has become known as membrane-, salt-, or hyperfiltration.

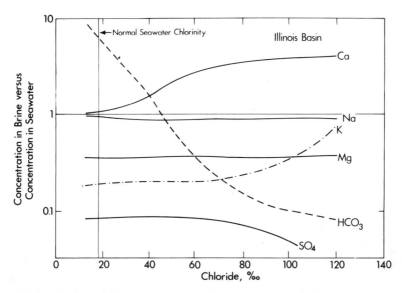

Fig. 4.8 Variation in major element concentration of waters in the Illinois basin as a function of dissolved chloride content. Values are normalized with respect to sea water of the same chloride content.

The theoretical basis for predicting that shales should behave as semipermeable membranes begins with the observation that most clay-mineral particles have a net negative surface charge. This surface charge is in part due to the substitution of aluminum for silicon and magnesium for aluminum in internal lattice sites (Grim, 1968). In aqueous solution, clay surfaces thus repel anions. The zone around a clay particle in which the concentrations of anions has been significantly lowered due to repulsion is known as the "diffuse" or "Gouy" layer (Berner, 1971). If an argillaceous sediment has been sufficiently compacted so that Gouy layers of adjacent clay mineral particles overlap, then the concentration of anions are effectively reduced in the interstitial pore waters of the sediment. Such a sediment can then act as a semipermeable membrane by discriminating against the flow of anions. Neutral molecules, such as H_2O, H_2CO_3, and H_2S can pass through the membrane, but anions are retarded. Because electrical neutrality must be maintained in the solutions external to the membrane, the retardation of anions will also result in complementary retardation of cations (Hanshaw and Coplen, 1973).

The thickness of the Gouy layer and, hence, the degree of compaction that is necessary to form a semipermeable membrane is a function of temperature, the charge density on individual clay particles, and the concentration and types of electrolytes in the pore solutions. Berner (1971) estimates that for waters of normal seawater salinity, modern marine sediments will exhibit membrane properties if compaction has been sufficient to reduce porosities to approximately 0.3 or less. From Figure 4.1, this could correspond to a burial depth of anywhere from 300 to 2700 m. An increase in the salinity of pore waters or a decrease in charge density on the clays requires a greater degree of shale compaction to produce membrane effects.

Some charged species may pass more readily than others through a clay membrane. Factors that control mobility of ions include selective adsorption or repulsion by the membrane and the relative velocities of dissolved species during flow of fluid. For example, cations more strongly attracted to exchange sites on clays should be more strongly retarded. Ions of large hydration radii should show greater hydraulic drag and slower flow rates. In systems in which flow rates are slow and hydraulic drag is a secondary effect, theory favors the passage of small, monovalent cations, Li > Na > K and the retardation of large, divalent cations, Mg < Ca < Sr < Ba. Elements present largely as neutral aqueous complexes should pass through a membrane more readily than elements existing primarily as dissociated ions.

Experimental work, summarized by Hanshaw and Coplen (1973) and Kharaka and Berry (1973), has confirmed the membrane properties of compacted clays. Kharaka and Berry found that the retardation of monovalent and of divalent cations in clay membranes followed the general sequences described above. At pressure gradients characteristic of subsurface conditions, calcium should be strongly retarded with respect to sodium. Chloride and sulfate are more strongly retarded than bicarbonate.

Flow of water across a sediment membrane will occur in response to differences in the fugacity of water. The direction of flow will be from the high- to the low-fugacity side of the membrane, and the rate of flow will depend both on the magnitude of the fugacity gradient and the ease with which water molecules can pass through the membrane, that is, the membrane conductivity. Fugacity differences can result from differences in hydraulic head, Δh [equation (4.4)]; total dissolved solids, Δc; electrical potential, Δe; and temperature, ΔT. Liquid movement induced by these mechanisms is known as hydraulic flow, osmosis, electroosmosis, and thermoosmosis, respectively. Olsen (1972) has studied the movement of aqueous solutions through compacted kaolinite under hydraulic, osmotic, and electric gradients. On the basis of Olsen's work, net fluid flow, J, across a kaolinite membrane can be described by:

$$J = (-K_h \cdot \Delta h + K_c \cdot \Delta \log c - K_e \cdot \Delta e)/L \tag{4.2}$$

where K_h is the hydraulic conductivity ($K_h = \xi\rho g/\eta$), and K_c and K_e are osmotic and electroosmotic conductivities respectively. L is the thickness of the clay. With the increasing compaction of the clay the ratios (K_c/K_h) and (K_e/K_h) both increase, and the relative effects of osmotic and electro-osmotic gradients become increasingly important.

Shales could act to concentrate salt in deep waters in the manner suggested by DeSitter if sufficient hydraulic pressure gradients exist to drive water upward against osmotic gradients through compacted shale membranes. As flow of water continues and membrane filtration of charged species takes place, deeper waters become progressively saltier and selectively enriched in those cations and anions which are discriminated against by the membranes. By this mechanism deeper waters should become not only more saline, but enriched, for example, in calcium and potassium relative to sodium, compositional trends that exist in some basins (Figure 4.8). It must be kept in mind that other processes, such as ion-exchange on disseminated, nonmembrane clays could also give rise to a fractionation of dissolved species. Ion-exchange alone, however, would not be important in increasing the total salinity of subsurface waters.

Bredehoft et al., (1963) have suggested that sufficient hydraulic gradients for membrane-filtration could be generated in the centers of tilted or depressed basins where deeper aquifers have recharge areas that outcrop at higher elevations than the recharge areas of more shallow aquifers confined to the center of the basin. The difference in head caused by differences in elevation of recharge areas could, under ideal circumstances, permit waters to enter a basin along its uplifted margins, flow laterally and down through deep aquifers and then flow up and out through a sequence of overlying shale membranes in the center of the basin. Significant vertical hydraulic gradients are also produced in regions of rapid sedimentation, such as the Gulf Coast (Figure 4.4). Some of these areas of high fluid pressure, however, are apparently sealed by shales of very low permeability, and it is conceivable that the rate of flow across some of these shales is too slow to result in significant filtration. Rates of flow in the Gulf have not yet been determined.

Observed variations in the δD and $\delta^{18}O$ content of North American formation waters have been interpreted to indicate that meteoric surface waters are the dominant source of H_2O in deep, saline brines (Clayton et al., 1966; Hitchon and Friedman, 1969). This interpretation is of critical importance because it implies that the continued deep circulation of fresh surface waters has taken place with a flushing out of marine H_2O and the retention of dissolved salts in deep basin formations. Such a process is thus consistent with the mechanism of membrane filtration as visualized by Bredehoft et al. (1963). We still have much to learn about the effects of dehydration, exchange, and fractionation on the isotopic composition of formation waters (Taylor, 1974), and if these processes were more clearly understood, a more quantitative assessment could be made of the contribution of meteoric H_2O to deep brines.

Membrane filtration should become progressively less efficient with depth as waters become saltier and the charge density on clays is reduced by diagenesis. Olsen's work shows also that greater hydraulic gradients would be required at high clay compactions to overcome osmotic forces, which would drive water downward toward the saltier side of the membrane. The observed leveling off of salinity values with depth in some basins (Figure 4.7) may reflect the progressive inefficiency of shale membranes with deep burial. The membrane-filtration model is consistent with many features of formation water chemistry. One must agree with Manheim (1970), however, that further data on hydraulic gradients and flow rates are necessary on a case-by-case basis to demonstrate the probable role of filtration in specific basins.

OTHER MECHANISMS

By the Soret effect, dissolved salts will diffuse down a thermal gradient and become concentrated in the colder part of a solution. In a sedimentary basin, this would result in the formation of the most saline and dense brines at the shallowest depths, a hydrodynamically unstable situation. In the presence of an ion-exchange medium, such as disseminated clay, the effects of Soret migration will be reversed and ions will diffuse toward the warmer part of the solution (Mangelsdorf et al., 1970). These authors estimate that a 5% enrichment in chloride per 100 m depth could occur under steady-state conditions. Dandurand et al. (1972) describe a mechanism for concentrating salts and ore-forming components by Soret migration. Their process requires a horizontal thermal gradient, however, and is not generally applicable to sedimentary environments.

Ore-Forming Components in Sedimentary Waters

In 1835, Forchhammer suggested that mineral veins could be formed by circulating waters that had derived their dissolved materials from the minute quantities of metals present in all rocks (cited in Adams, 1954). Subsequent mass-balance calculations (e.g., Krauskopf, 1967; Billings et al., 1969) have shown that leaching a small fraction of metals present in common rocks, including sediments, may provide sufficient material for generating ore deposits. Roedder (1960) has calculated that in order to form a base metal ore deposit within a geologically reasonable time of 10^5 years at geologically reasonable fluid flow rates, a minimum of 1 ppm of metal would have to be precipitated out of solution. For a sedimentary brine to qualify as a potential ore-forming solution, it would probably have to contain a minimum 1-10 ppm of the element or metal in question. We will look briefly at possible controls on the distribution of six elements characteristic of Mississippi Valley-type deposits: sulfur, lead, zinc, barium, strontium, and fluorine.

SULFUR

Lead and zinc in Mississippi Valley-type ore deposits occur predominantly as the sulfides galena and sphalerite; barium and strontium as sulfates in the barite-celestite solid solution series. The availability and oxidation state of sulfur are obviously important factors in determining whether or not a sedimentary brine could be a potential ore-forming solution.

The dissolved sulfate content of subsurface brines is highly variable but usually substantially lower than the 2700 ppm level found in seawater (White et al., 1963; Figure 4.8). The bacterial reduction of dissolved sulfate to sulfide is a dominant process in the early diagenesis of organic-rich marine sediments (Berner, 1971). The precipitation of iron sulfides and the upward diffusion of H_2S is effective in lowering the total dissolved sulfur concentration in interstitial sediment waters.

The predominant dissolved sulfur species in most subsurface waters is sulfate. However, in surface spring waters which are similar in composition to sodium chloride and sodium-calcium chloride subsurface brines, dissolved sulfide species often dominate (White et al., 1963, Tables 15 and 16). H_2S is also a common and often abundant constituent of oil-field gases. H_2S-rich gases characteristically occur in carbonate sequences, but may have been generated originally in organic-rich basinal shales (Hitchon, 1968).

The occurrence of reduced sulfur species in subsurface fluids is of importance in our discussion for two reasons. First, as has been stated, sulfide is necessary for the precipitation of galena and sphalerite. Second, in sulfide-rich, mildly-alkaline brines, bisulfide becomes a progressively important complexing agent for metals, a factor that would enhance both the extraction and transport of lead and zinc (Barnes and Czamanske, 1967). Unfortunately, less is known about the quantitative chemistry and distribution of sulfide-rich waters than many other water types, owing in part to difficulties in sampling and analysis. For example, with some exceptions (Carpenter et al., 1974), there is a general lack of information on the concentration of metals and dissolved sulfide in brines associated with H_2S-rich natural gas accumulations (Hitchon, 1968). Such data would contribute greatly to an assessment of the role of sulfide-rich sedimentary waters in ore-formation.

LEAD AND ZINC

Some of our most detailed information on the occurrences of lead and zinc in sedimentary formation waters comes from two areas in which the precipitation of these metals in wells has been sufficient to interfere with the production of fluids. In the Cheleken Peninsula of the Caspian Sea, USSR, brines in a 2800-m-thick sequence of Neogene redbeds contain 0.2–5.4 ppm Zn and 2–77 ppm Pb (Lebedev, 1972). Three different types of formation waters have been found in the Cheleken sequence. The deepest, with temperatures in excess of 105°C are Na—Cl—HCO_3 waters, containing 35 to 40 g/l T.D.S. The waters precipitate calcite and barite in well pipes during production. The high lead and zinc concentrations are found in overlying sodium chloride waters, which have bottom-hole temperatures of 80°C. These waters are considerably more saline (150–290 g/l) and precipitate native lead during production. The shallowest waters in the redbed sequence are saline (250 g/l) sodium-calcium chloride brines which contain abundant H_2S. The lead and zinc content of these waters has not been published, but pyrite is precipitated during their production.

At the Cheleken well field, high-metal and high-H_2S brines are mixed in a surface tank, and abundant sphalerite, galena, and pyrite are precipitated out. Growth rates of these sulfide precipitates can be rapid: a 2.5-mm-thick crust of sphalerite was observed to form within a period of 3 months (Lebedev, 1972).

Metal-rich oil field brines containing up to 370 mg/l Zn and 92 mg/l Pb occur over a large area of central Mississippi (Carpenter et al., 1974). These sodium-calcium chloride waters contain 160–340 g/l T.D.S. and range in temperature from approximately 100 to 140°C. The host rocks are Cretaceous carbonates occurring at a depth of 2700–4400 m. Various of the waters precipitate barite, metallic lead, and galena as well scale. Some of the brines are high in H_2S and contain a fine-grained black precipitate at the well-head.

Both the Cheleken and Central Mississippi regions are characterized by apparently normal geothermal gradients. In neither area is there any evidence that igneous activity has in any way been responsible for generating the high metal content of the brines.

Lead and zinc-rich brines are also known in the Salton Sea geothermal area (White, 1968) and in the Red Sea deeps (Degens and Ross, 1969). These are both regions of active crustal rifting and high heat flow. Although it appears likely that the metals, at least in the Salton Sea area, have been derived from sedimentary sources, extraction processes have probably been greatly enhanced by the high flux of heat through the systems.

In a study of the metal content of brines in older sediments, Billings et al. (1969) report that the zinc content of formation waters in the Alberta Basin ranges 0.01–290 mg/l with a weighted mean of 0.30. The lead content is not known, and the relation between zinc content and the concentration of the various major dissolved species is not discussed. In other basins, the dissolved metal content may be considerably less. Slentz (cited by Wedepohl, 1972) gives an average zinc content of only 0.01 ppm for miscellaneous oil-field waters.

It appears that under certain circumstances, ordinary sediment processes are sufficient to generate metal-rich fluids. Not enough is known about the overall distribution of lead and zinc in formation waters, however, to be able to identify with certainty the processes that can lead to their concentration. Wedepohl's (1972) discussion of the abundance of lead and zinc in sediments shows quite clearly that the average concentrations of these metals in common sedimentary rock types do not deviate greatly from average crustal values. The identification of possible sources of lead and zinc, then, may be less one of finding sediments unusually enriched in these metals than one of finding diagenetic environments and water compositions which favor the release of metals into aqueous solutions.

One potential source of metals is the organic fraction of sediments. Dissolved organic complexes are effective transport agents of metals in freshwater environments (Gardner, 1974). During the fluvial introduction of these complexes into the marine environment, however, many of the organic-metal complexes will flocculate out and be deposited with the fine-grained fraction of the fluvial suspended load. During burial and diagenesis of organic material (Johns and Shimoyama, 1972) various metals could preferentially partition into (1) a mobile organic phase dominated by hydrocarbons, (2) a nonmobile refractory organic phase, (3) an inorganic solid phase, or (4) an aqueous brine. Porphyrin compounds, a constituent of crude oils, preferentially complex nickel and vanadium; this may inhibit the incorporation of these particular elements into subsurface waters. Black, organic-rich shales, on the other hand, are enriched in a larger number of heavy metals, including zinc and lead (Vine and Tourtelot, 1970). These may represent sediments where metals have been trapped in a nonmobile organic or sulfide phase.

Other potential sources of lead and zinc in sediments include silicate, carbonate, and other inorganic phases. Lead, for example, substitutes for potassium in feldspar and micas and could be released during burial diagenesis or alteration of these phases. The isotopic composition of lead in

Mississippi Valley-type ores is consistent with the lead having been derived from Paleozoic sandstones and/or shallow Precambrian basement rocks (Heyl et al., 1974; Chapter 2).

The metal-rich waters of central Mississippi, the Salton Sea, and the Red Sea appear to have derived in part by interaction with evaporites (Carpenter et al., 1974; White, 1974). The Cheleken redbeds contain anhydrite (Lebedev, 1972) and the brines there may also have been influenced by evaporites. This apparent association is an important area for further study. The evidence to date suggests that high temperatures ($>80°C$) and high salinities ($>200‰$) or high dissolved sulfide concentrations favor the partitioning of metals into sedimentary fluids. Such conditions permit the formation of stable aqueous-metal complexes and appear to enhance the diagenetic destruction of the minerals or organic compounds which contain the metals.

BARIUM AND STRONTIUM

In contrast to seawater, which contains only 1×10^{-2} to 5×10^{-2} ppm dissolved barium (Puchelt, 1972), many subsurface brines contain from 1 to nearly 6000 ppm dissolved barium (White et al., 1963; Puchelt, 1967) and are clearly potential barite-forming solutions. The precipitation of $BaSO_4$-scale from formation waters has been recorded from most of the oil-producing areas of the United States (Weintritt and Cowan, 1967). Although much barite scale is formed during water-flooding operations (Gates and Caraway, 1965), precipitation in some deep wells (Weintritt and Cowan, 1967) has probably occurred simply as a result of adiabatic cooling during pumping of production waters up from depth. This phenomenon indicates that in some natural waters enough barium and sulfate can be carried simultaneously in solution to form, on cooling, significant quantities of barite.

Barium concentration is probably limited initially by saturation with respect to $BaSO_4$. For example, high sulfate concentrations in seawater appear to be a limiting factor in marine barium concentrations (Hanor, 1969). During early diagenesis, bacterial reduction removes dissolved sulfate, and pore waters have a greater capacity for carrying barium in solution. Barium values of 100 ppm can be achieved in formation waters having salinities no greater than seawater values of $35‰$. The sources of dissolved barium in formation waters, however, have not been positively identified. Although barium is preferentially adsorbed on clay minerals, studies by Hanor and Chan (1977) show that adsorbed barium is quantitatively removed by cation exchange during fluvial introduction of clays into the marine environment. More likely sources include interlayer exchange positions in clays and interior lattice sites in potassium silicates. Destruction of potassium micas and potassium feldspars during diagenesis would be effective in releasing barium. In addition, direct bacterial reduction and dissolution of barite is possible (Puchelt, 1967).

Most natural barite contains from 1 to 25 mole % $SrSO_4$ in solid solution (Hanor, 1968). Oil-field scale shows a comparable range in composition (Weintritt and Cowan, 1967). Subsurface brines have a molar Sr/Ba ratio of 10^{-2} to 10^2 (Hanor, 1966), which is probably sufficient to account for much of the observed range in solid composition. The differences in the Sr/Ba ratio of sedimentary waters may reflect the independent control of strontium by diagenesis of calcium carbonates and sulfates (Butler, 1973) and of barium by the diagenesis of potassium minerals.

FLUORINE

The dissolved fluoride content of common types of oil-field brines ranges from approximately 0.1 to 10 ppm (White et al., 1963). As a general rule, the highest average fluoride concentrations and highest F/Cl ratios are found in relatively dilute, HCO_3-SO_4 waters rather than in the sodium chloride and sodium-calcium chloride brines. Each brine type, however, contains waters which on the basis of fluoride content, may be potential fluorite-forming solutions. Possible diagenetic sources for fluorine include fluorite and OH-bearing minerals such as apatite and mica (Koritnig, 1972). The solubility of fluorite increases with increasing temperature and electrolyte concentration (Strübel, 1965, Chapter 9), and this mineral could dissolve during burial diagenesis. The high concentration of calcium in some deep brines, however, may limit how much dissolved fluoride these waters can contain. Less saline or calcium-rich waters may be more likely transport agents for dissolved fluoride.

PROPERTIES OF SOME ORE-FORMING FLUIDS

Having reviewed some important physical and chemical properties of sedimentary waters, it is now appropriate to discuss briefly properties of similar fluids that are known to have been involved in the formation of ore deposits. Most of our information regarding the nature of these ore-forming fluids comes from physical and chemical measurements made on fluid inclusions in ore and gangue minerals (Chapter 14).

Isotopic Composition of H_2O

The measured δD and $\delta^{18}O$ values of fluid inclusions from the Illinois, Kentucky, Upper Mississippi Valley, and Tri-State districts (Hall and

Friedman, 1963, 1969; Hall, 1973, cited by White, 1974) fall within the range of isotopic compositions of waters from the Michigan, Illinois, and Gulf Coast basins (Clayton et al., 1966; Hitchon and Friedman, 1969). Although waters derived from sources other than subsurface sedimentary brines could have been involved in the formation of Mississippi Valley-type deposits, it is not necessary to invoke other sources to account for the isotopic composition of the ore-forming fluids.

Chemical Composition

Chemical analyses indicate that the fluids precipitating sulfides and fluorite in Mississippi Valley-type districts were largely sodium-calcium chloride, low-sulfate brines having salinities of 100–300‰ (Hall and Friedman, 1963; Roedder et al., 1963; Roedder, 1971). Later-stage carbonates and barites were deposited from waters having salinities less than 100‰. The bulk composition of Mississippi Valley-type fluid inclusions is very similar to that of sedimentary brines (White, 1974) although some significant differences do occur. For example, the average K/Na ratio of subsurface brines is lower than in the fluids responsible for the deposition of Mississippi Valley-type deposits, as noted by White (1965), Sawkins (1968), and others. Because the Na/Cl ratio of waters in many basins is fairly constant (Figure 4.8), variations in K/Cl ratios parallel variations in K/Na. Figure 4.9 shows a plot of potassium versus chloride for waters from sandstones and carbonates in a number of sedimentary basins. We see that although each basin has a unique trend of values, the apparent enrichment of potassium with increasing chloride content, perhaps due to membrane-filtration or interaction with evaporites, is a general characteristic. Although the *average* K/Na or K/Cl ratio of subsurface waters may be lower than fluid inclusion ratios, their value in the more saline, deeper, and hotter waters of some basins may be comparable. The measured K/Cl ratios of fluid inclusions in late-stage carbonates and barite are too high to be explained in terms of a deep basin sandstone or carbonate water. It should be noted, however, that pore waters in deeply buried shales have much higher K/Cl ratios and are less saline than waters in adjacent sands (Weaver and Beck, 1971), possibly as the result of membrane effects. Whether shale waters ever have compositions strictly comparable to those determined for carbonate and barite inclusions is not yet known.

There is a regional zoning of fluorite and strontium-bearing minerals in Mississippi Valley-type deposits of eastern North America (Hanor, 1966) that parallels both the structural grain of the Precambrian basement and regional facies variations in lower Paleozoic sediments. It is possible that the regional mineral zoning is related to regional variations in the

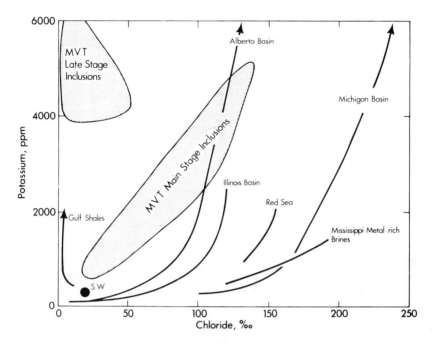

Fig. 4.9 Relation between potassium and chloride concentrations in Mississippi Valley-type fluid inclusions and various sedimentary brines. S.W. marks the position of seawater. Data from references cited in Figure 4.3, and Hall and Friedman (1963), Degens and Ross (1969), Weaver and Beck (1971), and Carpenter et al. (1974).

chemistry of the ore-forming solutions. These variations in water composition probably reflect selective distribution of source rocks or facies types that yield particular elements during diagenetic alteration and not portions of the crust unusually enriched or depleted in these elements.

Temperature-Salinity Relations

Figure 4.10 shows a plot of the range of homogenization temperatures versus the range of freezing temperatures measured for minerals from various Mississippi Valley-type ore districts (Sawkins, 1968; Roedder, 1971; Leach, 1973). These data give a general idea of the range of temperatures and salinities of the fluids that produced Mississippi Valley-type districts in comparison with the temperatures and salinities of sedimentary waters (Figures 4.3 and 4.7). The following limitations apply to Figure 4.10: (1) homogenization temperatures represent a minimum temperature

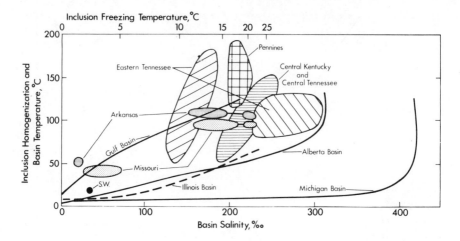

Fig. 4.10 Temperature-salinity relations in ore-forming fluids from Mississippi Valley-type districts (patterned areas), as deduced from fluid inclusion data, and temperature-salinity relations for waters in the Illinois, Michigan, Alberta, and Mississippi Gulf basins. S.W. marks the position of surface seawater. References cited in text.

of deposition and have not been corrected here for pressure effects. (2) The salinities indicated in Figure 4.10 are based on the freezing points of NaCl solutions; if the inclusions contain significant dissolved $CaCl_2$, then the total salinity may be higher than indicated. (3) It is not always technically possible to make homogenization and freezing temperature measurements on the same inclusions. If such data were available, however, the ranges for each district might be more narrowly defined. (4) Of necessity, the inclusion data is restricted to nonopaque mineral phases. Thus, galena and other opaque minerals are not represented. (5) Inclusions in late-stage minerals, such as calcite and barite commonly do not have a visible vapor phase in their inclusions and, thus, a minimum temperature of formation cannot be determined, but the maximum temperature of formation of these minerals would be approximately 50°C.

Despite these restrictions, important observations result. First, the fluids involved in the formation of each mineral district are characterized by a distinctive range of temperatures and salinities. There is, therefore, no unique fluid, at least in terms of temperature and bulk composition, which has been responsible for the formation of all these districts. The uniformity of temperatures and salinities within some districts supports derivation of the fluid from regional sources, rather than mixing of many different water types. Second, the present temperature-salinity trends of subsurface waters in the Illinois, Alberta, and Michigan basins plot consistently below those for the Mississippi Valley deposits. If waters in these Paleozoic-Mesozoic basins were responsible for the formation of such deposits, Figure 4.10 shows that they would have to have been approximately 50–100°C warmer than the waters of comparable salinity presently in the basins. As we will discuss further, temperatures during the active depositional phase of these basins may have been significantly warmer in the past. In the Northern Gulf Coast, for example, the temperature and salinity of waters plot within the average range of inclusion values. From the known high subsurface temperatures in the Texas Gulf Coast (Jones and Wallace, 1974) it is quite possible that the entire range of inclusion temperatures and salinities on Figure 4.10 lies within the range of sedimentary waters of the Gulf basin.

Sites of Ore-Deposition

Many Mississippi Valley-type ore districts in central North America are emplaced on, or around the margins of, broad regional uplifts, such as the Ozark, Jessamine, and Nashville domes. Barite deposits, in particular, show a nearly perfect correspondence with areas of regional uplift (Hanor, 1966). The sedimentary brines most like the fluids that deposited these Mississippi Valley-type mineral suites are presently found as much as 250 km laterally and 3 km deeper than the sites of ore deposition. Available evidence indicates that in individual districts much of the mineralization occurred within comparatively short periods of time: as little as 2.5×10^5 y for major ore deposits in the Wisconsin lead-zinc district (Lavery and Barnes, 1971); a few hundred thousand years for the stratiform barite deposits of Arkansas (Hanor and Baria, 1977). The restricted extent of clay-mineral wallrock alteration in Mississippi Valley-type deposits (Heyl, 1969) is likewise consistent with limited periods of flow of hydrothermal fluids.

If sedimentary waters are to invoked as ore-forming fluids, then a mechanism has to be found that will cause the rapid flushing of deep waters out through the margins of a basin. We will examine some possible mechanisms in the next section.

FLOW OF SUBSURFACE BRINES

Theory

A subsurface brine will flow in response to differences in the total fluid potential, Φ. For a homogeneous fluid in a chemically inert and homogeneous porous medium,

$$\Phi = gz + P/\rho \tag{4.3}$$

where g is the gravitational constant, z is the elevation of the pressure measurement relative to an arbitrary datum (for example, mean sea-level), P is static fluid pressure, and ρ is the density of the fluid (Hubbert, 1940). The hydraulic head or elevation of the potentiometric surface, h, at any point is given by

$$h = \Phi/g = z + (P/\rho g) \qquad (4.4)$$

In principle, fluid flow will occur down a potentiometric gradient in a direction perpendicular to equal elevations of the hydraulic head. For nonturbulent flow, the rate of water movement can be described by Darcy's Law:

$$J = \xi \cdot \frac{\rho}{\eta} \cdot \left(\frac{d\Phi}{dz}\right) \qquad (4.5)$$

where J is the fluid flux per unit area and unit time and ξ is the permeability of the sediment. From Figures 4.5 and 4.6 we can see that ρ/η, the reciprocal of kinematic viscosity, should increase by about a factor of two to five toward the base of a sedimentary basin. Thus a deep brine could flow approximately two to five times faster in response to the same difference in head than a shallow subsurface water in a sediment of comparable permeability.

The above concepts are useful in evaluating fluid flow in a variety of ground water systems. However, they do not appear to be adequate in predicting movement of some deep brines. Bond (1972) has discussed the limitations of using hydraulic potentiometric fields to evaluate groundwater flow in an analysis of fluid flow in deep formations of the Illinois basin. Equation (4.5) is predicated on the assumptions of a uniform fluid and a homogeneous porous medium. If, however, water densities vary, and the aquifer is not physically uniform, either because of structural or textural heterogeneities, then these assumptions are no longer valid. Because fluid densities are variable in basin-wide hydrologic systems and because real aquifers tend to be locally and regionally heterogeneous, hydraulic potentiometric surfaces and gradients may not be reliable indicators of the regional direction and magnitude of flow.

In addition to these physical factors, we have seen from Olsen's (1972) work that total fluid potential across a semipermeable membrane is dependent not only on differences in hydraulic head, but also on differences in salinity, electrical potential, and temperature [equation (4.2)]. Within deep basins where there are significant differences in salinity across clay beds and shales (e.g., Figure 4.7), we thus have possible local mechanisms for inducing fluid flow, flow that could be in a direction opposite to the hydraulic gradient, and flow that in principle could move water either into or out of a basin.

Although present understanding does not permit quantitative analysis of the time and rate of movement of deep basin brines, it is possible to assess qualitatively geologic factors that may be important in stimulating flow of hydrothermal solutions from basins.

Flow during Compaction

Most of the water trapped during slow deposition of sediments is flushed out by simple compaction (Figure 4.2) within the upper kilometer of burial, where maximum salinities range from 70,000 to 170,000 ppm and temperatures are less than 50°C. Although such waters could conceivably contain enough dissolved barium to qualify as barite-forming solutions, their salinities and temperatures are significantly lower than (1) sedimentary waters known to contain high concentrations of metals, and (2) inclusion waters trapped at the time of deposition of main-stage mineralization in Mississippi Valley-type deposits. Simple compaction of more deeply buried sediments should result in displacement of shallower waters out of the basin, not in surface discharge of hot, saline brines.

Flow during Tectonic Deformation

Faulting of a sedimentary basin could, under some circumstances, provide a permeable conduit for the discharge of fluids from depth. If fluid pressure at depth exceeds hydrostatic pressure, then deep waters will tend to migrate upward. Where permeability is high, and the hydrostatic gradient is great, then discharge could be rapid [equation (4.5)]. Jones and Wallace (1974) have described evidence for the present upward movement of hot brines along growth faults in the northern Gulf Basin. The brines are migrating from zones of high porosity and fluid pressure up into sediments characterized by normal hydrostatic pressures (Figure 4.4). The rates of flow have not yet been determined.

Release of fluids during faulting may have been significant in the formation of the Illinois-Kentucky district (Dozy, 1970), where the 38th parallel lineament (Heyl, 1972), a high-angle regional fault zone, transects the deepest part of the Illinois Basin. Minor faulting in the Ouachita geosyncline in the Mississippian may have released barium-rich brines that mixed with seawater to form the stratiform barite deposits of Central Arkansas (Hanor and Baria, 1977).

With a greater degree of tectonic deformation, other forms of fluid release are possible. Uplift and deformation of the western margin of the Alberta Basin, for example, conceivably could have driven deep brines eastward toward sites of mineral deposition (Billings et al., 1969). Leach (1973) has suggested that the mineral deposits of the Ozark region in Missouri and Arkansas were derived from fluids expelled during Pennsylvanian uplift and metamorphism of the Ouachita syncline to the south.

Flow during Evolution in Thermal Regime

Most recent sedimentary basins are in a state of thermal evolution. As we have seen in the Gulf Basin, rapid deposition of clays and sands has produced thermally insulating stratigraphic sequences characterized by high porosities, high fluid pressures, and high geothermal gradients. As hot brines are expelled upward, the sediments compact, porosities are reduced, and geothermal gradients decrease. The older portions of the basin are therefore cooling off.

There is no reason to believe that the thermal evolution of Mesozoic and Paleozoic basins has not been similarly complex. In the Mid-Continent region of North America, much of the sediments deposited during the early Paleozoic are predominately platform carbonates, sands, and shales. In contrast, during the late Paleozoic, deposition in many areas was characterized by the rapid emplacement of relatively thick sequences of deltaic and marine shales and sands. As the older Paleozoic sediments were buried in subsiding basins by this thermally insulating mass of material, geothermal gradients and bottom temperatures should have increased substantially. Hot brines would have migrated from regions of high fluid pressures in the centers of the basins laterally through more permeable, deeper aquifers toward areas of hydrostatic pressure along the basin margins. Highly permeable facies (Billings et al., 1969) or basement structures such as regional faults would have been effective in localizing the escape of deep waters. The flow of deep brines could be enhanced by the generation of free water at depth during the dehydration of montmorillonite (Dozy, 1970).

Significant variations in the heat-flow and hydrodynamic regimes of a basin can thus occur without the necessity of invoking variations in heat flux from the basement. The latter possibility, however, should not be neglected (Dunham, 1970). Although the interior craton of the North American continent is often referred to as the "Central Stable Region", this area underwent complex development of sedimentary basins and structural highs throughout the Paleozoic; a history punctuated by the emplacement of scattered igneous intrusive bodies (Heyl, 1969). Whether or not systematic variations in regional heat flow regime accompanied the tectonic evolution of this portion of the earth's crust is not known. Differential regional metamorphism of coal beds in the Illinois basin has been described and may be related to Mesozoic intrusive activity in the southern part of the basin (Damberger, 1971). Any additional source of heat such as salt diapirs, igneous intrusives, exothermic diagenetic processes, or increases in heat flow from the basement provides a potential means for inducing fluid flow. As noted previously, because density gradients for brines in many basins approach zero with depth, an increase in the geothermal gradient could cause a reversal in density gradient and produce gravitational instability (Hanor, 1973). In theory, this should cause upward migration of hot, saline, but less dense waters. Heat applied to the bottom or margins of a basin should be more effective in driving out deep brines than heat applied under the structurally elevated sites of ore deposition. A heat source in these latter areas would probably develop a localized convective cell of meteoric water, which would not involve circulation of deep basin brines.

If underlying crystalline basement rocks are sufficiently permeable, there is no reason why waters could not flow through them, possibly extracting dissolved material. The continuation of mineral veins in Ordovician limestones downward into Precambrian granite in the S.E. Ontario district (Guillet, 1963) is direct evidence that at least in this region hydrothermal fluids have migrated through basement rocks.

CONCLUSIONS

Hydrothermal sedimentary brines develop as a natural consequence of burial diagenesis. Some of these waters derive sufficient metals during burial to be classified as potential ore-forming solutions. The addition of dissolved sulfur from other sources, however, may be necessary in some instances for the precipitation of galena, sphalerite, or barite. Further study of the geochemistry of sulfide-rich waters, in particular, is of critical importance in our understanding of sedimentary ore-forming solutions.

The expulsion of fluids from deeply buried sediments into potential sites of ore-deposition appears to be a normal aspect of basin development (Jackson and Beales, 1967) and may occur by a variety of mechanisms. Basins developed along the margins of crustal plates are susceptible to episodes of crustal deformation, such as faulting or differential uplift, which may be sufficient to induce flow. Mechanisms for flushing out deep brines in undeformed or in intracratonic basins, such as the Illinois basin are less obvious. Waters which have produced some Mississippi Valley-type deposits may have been heated and flushed out during a late stage of basin development characterized by the rapid deposition of thick sequences of terrigenous clastics. The increase in thermal gradient and fluid pressure resulting from this deposition may have been sufficient to increase temperatures, reduce viscosities, and cause rapid migration of fluids. Whether or not a second generation of ore-forming solutions could form after the dewatering and cooling off of such a basin is an open question.

Much of the past evaluation of the role of sedimentary waters as in ore-forming fluids has been made on the basis of the properties of waters presently residing in ancient sediments. Of equal or greater importance, however, may be processes that occur during the active, depositional phase of basin development. We are learning from the Gulf Coast that profound changes in thermal regime, fluid potential, and chemical properties are all normal aspects of burial diagenesis. As research on this and other recent sedimentary basins continues, we will be able to evaluate on a more quantitative basis the factors that control water chemistry and the time and rates at which subsurface fluids migrate.

ACKNOWLEDGMENTS

I am indebted to H. L. Barnes for his critical review of this manuscript. The writing of this paper was supported by the Louisiana Water Resources Research Institute, Project A-031-LA, from funds granted by the Office of Water Research and Technology, U.S. Department of the Interior, under P.L. 88-379.

REFERENCES

Adams, F. D. (1954) *The Birth and Development of the Geological Sciences*: New York: Dover Publications.

Barnes, H. L. and G. K. Czamanske (1967) Solubilities and transport of ore-minerals: in *Geochemistry of Hydrothermal Ore Deposits*, H. L. Barnes, ed., New York: Holt, Rinehart and Winston, p. 334-381.

Barker, C. (1972) Aquathermal pressuring-role of temperature in development of abnormal-pressure zones: *Am. Assoc. Petrol. Geol. Bull.*, **56**, 2068-2071.

Berner, R. A. (1971) *Principles of Chemical Sedimentology*: New York: McGraw-Hill.

Billings, G. K., S. E. Kesler, and S. A. Jackson (1969) Relation of zinc-rich formation waters, Northern Alberta to the Pine Point ore deposit: *Econ. Geol.*, **64**, 385-391.

Bond, D. C. (1972) Hydrodynamics in deep aquifers of the Illinois basins: *Ill. Geol. Surv. Circ.*, *470*.

Bradley, J. S. (1975) Abnormal formation pressure: *Am. Assoc. Petrol. Geol. Bull.*, **59**, 957-973.

REFERENCES

Bredehoft, J. D., C. R. Blyth, W. A. White, and G. B. Maxey (1963) Possible mechanism for concentration of brines in subsurface formations: *Am. Assoc. Petrol. Geol. Bull.*, **47**, 257-269.

―― and B. B. Hanshaw (1968) On the maintenance of anomalous fluid pressures: I. Thick sedimentary sequences: *Geol. Soc. Am. Bull.*, **79**, 1097-1106.

Burst, J. F. (1969) Diagenesis of Gulf Coast clayey sediments and its possible relation to petroleum migration: *Am. Assoc. Petrol. Geol. Bull.*, **53**, 73-93.

Butler, G. P. (1973) Strontium geochemistry of modern and ancient calcium sulfate minerals: in *The Persian Gulf*, B. H. Porser, ed., New York: Springer-Verlag, p. 423-452.

Carpenter, A. B., M. L. Trout, and E. E. Pickett (1974) Preliminary report on the origin and chemical evolution of lead- and zinc-rich oil field brines in central Mississippi: *Econ. Geol.*, **69**, 1191-1206.

Carslaw, H. S. and J. C. Jaeger (1959) *Conduction of Heat in Solids*, Oxford: Clarendon.

Cartwright, K. (1970) Groundwater discharge in the Illinois Basin as suggested by temperature anomalies: *Water Resources Res.*, **6**, 912-918.

Choquette, P. W. and L. C. Pray (1970) Geological nomenclature and classification of porosity in sedimentary Carbonates: *Am. Assoc. Petrol. Geol. Bull.*, **54**, 207-250.

Clark, S. P., Jr., ed. (1966) Handbook of physical constants: *Geol. Soc. Am. Memoirs*, **97**.

Clayton, R. N., I. Friedman, D. L. Graf, T. K. Mayeda, W. F. Meents, and N. F. Shimp (1966) The origin of saline formation waters, I. Isotopic composition: *J. Geophys. Res.*, **71**, 3869-3882.

Damberger, H. H. (1971) Coalification pattern of the Illinois basin: *Econ. Geol.*, **66**, 488-494.

Dandurand, J. L., J. P. Fortuné, R. Perámi, J. Schott and F. Pollon (1972) On the importance of mechanical action and thermal gradient in the formation of metal-bearing deposits: *Mineral. Depos.*, **7**, 339-350.

Degens, E. T. and D. A. Ross, eds. (1969) *Hot Brines and Recent Heavy Metal Deposits in the Red Sea*: New York: Springer-Verlag.

DeSitter, L. U. (1947) Diagenesis of oil-field brines: *Am. Assoc. Petrol. Geol. Bull.*, **31**, 2030-2040.

Dickey, P. A. (1969) Increasing concentration of subsurface brines with depth: *Chem. Geol.*, **4**, 361-370.

―― A. G. Collins, and I. Fajardo, M. (1972) Chemical composition of deep formation waters in Southwestern Louisiana: *Am. Assoc. Petrol. Geol. Bull.*, **56**, 1530-1533.

Dozy, J. J. (1970) A geological model for the genesis of the lead-zinc ores of the Mississippi Valley, USA: *Trans. Inst. Min. Metall.*, **79**, B163-B170.

Dunham, K. C. (1970) Mineralization by deep formation waters: A review: *Trans. Inst. Min. Metall.*, **79**, B137-B136.

Fyfe, W. A. (1973) Dehydration reactions: *Am. Assoc. Petrol. Geol.*, **57**, 190-197.

Gardner, L. R. (1974) Organic versus inorganic trace metal complexes in sulfidic marine waters—Some speculative calculations based on available stability constants: *Geochim. Cosmochim. Acta*, **38**, 1297-1302.

Gates, G. L. and W. H. Caraway (1965) Oil well scale formation in water-flood operations using ocean brines, Wilmington, Calif.: *U.S. Bur. Mines, Rept. Inv.* 6658.

Graf, D. L., W. F. Meents, I. Friedman, and N. F. Shimp (1966) The origin of saline formation waters, III: Calcium chloride waters: *Ill. Geol. Surv. Circ. 397*.

Grim, R. E. (1968) *Clay Mineralogy*, (2nd ed.), New York: McGraw-Hill.

Guillet, G. R. (1963) Barite in Ontario: *Ontario Dept. Mines, Ind. Min. Rept.*, **10**.

Hall, W. E. and I. Friedman (1963) Composition of fluid inclusions, cave-in-rock fluorite district, Illinois, and Upper Mississippi Valley zinc-lead district: *Econ. Geol.,* **58,** 886-911.

____ (1969) Oxygen and carbon isotopic composition of ore and host rock of selected Mississippi Valley deposits: *U.S. Geol. Surv. Prof. Paper, 650-C,* C140-C148.

Hanor, J. S. (1966) The origin of barite, Ph.D. Thesis, Harvard University.

____ (1968) Frequency distribution of compositions in the barite-celestite series: *Am. Min.,* **53,** 1215-1222.

____ (1969) Barite saturation in seawater: *Geochim. Cosmochim. Acta,* **33,** 894-898.

____ (1973) The role of *in-situ* densities in the migration of subsurface brines: *Geol. Soc. Am. Abst.,* **5,** 651-652.

____ and L. R. Baria (1977) Controls on the distribution of barite deposits in Arkansas: in *Symposium on the Geology of the Ouachita Mountains,* **2,** G. C. Stone, ed., Arkansas Geol. Comm., 42-49.

____ and L. H. Chan (1977) Non-conservative behavior of barium during mixing of Mississippi River and Gulf of Mexico waters: *Earth Plan. Sci. Let.,* **37,** 242-250.

Hanshaw, B. B. and J. D. Bredehoft (1968) On the maintenance of anomalous fluid pressures, II. Source layer at depth: *Geol. Soc. Am. Bull.,* **79,** 1107-1120.

____ and T. B. Coplen (1973) Ultrafiltration by a compacted clay membrane-II. Sodium in exclusion at various ionic strengths: *Geochim. Cosmochim. Acta,* **37,** 2311-2328.

Heyl, A. V. (1969) Some aspects of genesis of zinc-lead-barite-fluorite deposits in the Mississippi Valley, U.S.A.: *Inst. Min. Metall. Trans.,* **78,** B148-B160.

____ (1972) The 39th parallel lineament and its relationship to ore deposits: *Econ. Geol.,* **67,** 879-894.

____, G. P. Landis, and R. E. Zartman (1974) Isotopic evidence for the origin of Mississippi Valley-type mineral deposits: A review: *Econ. Geol.,* **69,** 992-1006.

Hitchon, B. (1968) Geochemistry of natural gas in western Canada: *Am. Assoc. Petrol. Geol. Mem.,* **9,** 1995-2025.

____ and I. Friedman (1969) Geochemistry and origin of formation waters in the western Canada sedimentary basin—I. Stable isotopes of hydrogen and oxygen: *Geochim. Cosmochim. Acta.,* **33,** 1321-1349.

Hubbert, M. K. (1940) The theory of ground-water motion: *J. Geol.,* **48,** 785-944.

____ and W. W. Rubey (1959) Role of fluid pressure in mechanics of overthrust faulting, I. Mechanics of fluid-filled porous solids and its application to overthrust faulting: *Geol. Soc. Am. Bull.,* **70,** 115-166.

Jackson, S. A. and F. W. Beales (1967) An aspect of sedimentary basin evolution: the concentration of Mississippi Valley type ores during the late stages of diagenesis: *Can. Assoc. Petrol. Geol.,* **15,** 483-422.

Jam, P., P. A. Dickey, and E. Tryggvason (1969) Subsurface temperatures in south Louisiana: *Am. Assoc. Petrol. Geol. Bull.,* **53,** 2141-2149.

Johns, W. D., and A. Shimoyama (1972) Clay minerals and petroleum-forming reactions during burial and diagenesis: *Am. Assoc. Petrol. Geol. Bull.,* **56,** 2160-2167.

Jones, P. H. (1968) Geochemical hydrodynamics—a possible key to the hydrology of certain aquifer systems in the northern part of the Gulf of Mexico Basin: *Report 23rd International Geol. Congress,* **17,** 113-125.

____ (1970) Geothermal resources of the northern Gulf of Mexico Basin: in *U.N. Symposium on Development and Utilization of Geothermal Resources, Pisa,* **2,** 14-26.

REFERENCES

_____ and R. H. Wallace, Jr. (1974) Hydrogeologic aspects of structural deformation in the northern Gulf of Mexico Basin: *J. Res. U.S. Geol. Surv.*, **2**, 511-517.

Kharaka, Y. K. and F. A. F. Berry (1973) Simultaneous flow of water and solutes through geological membranes—I. Experimental investigation: *Geochim. Cosmochim. Acta*, **37**, 2577-2604.

Koritnig, S. (1972) Fluorine: in *Handbook of Geochemistry*, K. H. Wedepohl, ed., New York: Springer-Verlag.

Krauskopf, K. B. (1967) Source rocks for metal bearing fluids: in *Geochemistry of Hydrothermal Ore Deposits*, H. L. Barnes, ed., New York: Holt, Rinehart, and Winston, p. 1-33.

Lavery, N. G. and H. L. Barnes (1971) Zinc dispersion in the Wisconsin zinc-lead district: *Econ. Geol.*, **66**, 226-242.

Leach, D. (1973) A study of the barite-lead-zinc deposits of central Missouri and related mineral deposits in the Ozark Region: Unpublished Ph.D. thesis, University of Missouri.

Lebedev, L. M. (1972) Minerals of contemporary hydrotherms of Cheleken: *Geochem. Int.*, **9**, 485-504.

Levorsen, A. I. (1967) *Geology of Petroleum*, 2nd ed., San Francisco: Freeman.

Lindgren, W. (1933) *Mineral Deposits*, 4th ed., New York: McGraw-Hill.

Magara, K. (1975) Reevaluation of montmorillonite dehydration as a cause of abnormal pressure and hydrocarbon migration: *Am. Assoc. Petrol. Geol. Bull.*, **59**, 292-302.

Mangelsdorf, P. C., F. T. Manheim, and J. M. T. M. Gieskes (1970) Role of gravity, temperature gradients, and ion-exchange media in formation of fossil brines: *Am. Assoc. Petrol. Geol. Bull.*, **54**, 617-626.

Manheim, F. T. (1970) Critique of membrane-filtration concepts as applied to origin of subsurface brines (abst.): *Am. Assoc. Petrol. Geol. Bull.*, **54**, 858.

Millot, G. (1970) *Geology of Clays*, New York: Springer-Verlag.

Olsen, H. W. (1972) Liquid movement through kaolinite under hydraulic, electric and osmotic gradients: *Am. Assoc. Petrol. Geol. Bull.*, **56**, 2022-2028.

Osmaston, M. F. (1975) Interstitial water composition and geochemistry of deep Gulf Coast shales and sandstones: Discussion: *Am. Assoc. Petrol. Geol. Bull.*, **59**, 715-726.

Perrier, R. and J. Quiblier (1974) Thickness changes in sedimentary layers during compaction history; Methods for quantitative evaluation: *Am. Assoc. Petrol. Geol. Bull.*, **58**, 507-520.

Perry, E. A., and J. Hower (1972) Late-stage dehydration in deeply buried pelitic sediments: *Am. Assoc. Petrol. Geol. Bull.*, **56**, 2013-2021.

Pettijohn, F. J., Potter, P. E., and Siever, R. (1973) *Sand and Sandstone*, New York: Springer-Verlag.

Powers, M. C. (1967) Fluid-release mechanisms in compacting marine rocks and their importance in oil exploration: *Am. Assoc. Petrol. Geol. Bull.*, **51**, 1240-1254.

Puchelt, H. (1967) *Zur Geochemie de Bariums im exogenen Zyklus*, Berlin: Springer-Verlag.

_____ (1972) Barium: in *Handbook of Geochemistry*, K. H. Wedepohl, ed., Berlin: Springer-Verlag.

Rittenhouse, G. (1967) Bromine in oil field waters and its use in determining possibilities of origin of these waters: *Am. Assoc. Petrol. Geol. Bull.*, **51**, 2430-2440.

Roedder, E. (1960) Fluid inclusions as samples of the ore-forming fluids: *Int. Geol. Cong., 21st, Copenhagen, Part XVI*, 218-229.

———, B. Ingram, and W. E. Hall (1963) Studies of fluid inclusions III: Extraction and quantitative analysis of inclusions in the milligram range: *Econ. Geol.,* **58,** 353-374.

——— (1971) Fluid inclusion evidence of the environment of formation of mineral deposits in the southern Appalachian Valley: *Econ. Geol.,* **66,** 777-791.

Rowe, A. M., Jr. and J. C. S. Chou (1970) Pressure-volume-temperature-concentration relation of aqueous NaCl solutions: *J. Chem. Eng. Data,* **15,** 61-66.

Sawkins, F. J. (1968) The significance of Na/K and Cl/SO$_4$ ratios in fluid inclusions and subsurface waters, with respect to the genesis of Mississippi Valley type ore deposits: *Econ. Geol.,* **63,** 935-942.

Schmidt, G. W. (1973) Interstitial water composition and geochemistry of deep Gulf Coast shales and sands: *Am. Assoc. Petrol. Geol. Bull.,* **57,** 321-337.

Strübel, G. (1965) Quantitative Untersuchungen über die hydrothermales Löslichkeit von Flussspat (CaF$_2$): *N. Jh. Mineral.,* **3,** 83-95.

Stumm, W. and J. J. Morgan (1970) *Aquatic Chemistry.* New York: Wiley-Interscience.

Taylor, H. P., Jr. (1974) The application of oxygen and hydrogen isotope studies to problems of hydrothermal alteration and ore deposition: *Econ. Geol.,* **69,** 843-883.

Vine, J. D. and E. B. Tourtelot (1970) Geochemistry of black shale deposits—A summary report: *Econ. Geol.,* **65,** 253-272.

Vockroth, G. B. (1974) Abnormal subsurface pressure: *Am. Assoc. Petrol. Geol. Reprint Series,* **11.**

Weaver, C. E. and K. C. Beck (1971) Clay water diagenesis during burial: How mud becomes gneiss: *Geol. Soc. Am. Special Pap., 134.*

Wedepohl, K. (1972) Zinc: in *Handbook of Geochemistry,* K. H. Wedepohl, ed., New York: Springer-Verlag.

Weintritt, D. J. and J. C. Cowan (1967) Unique characteristics of barium sulfate scale deposition: *J. Petroleum Tech.,* **1967,** 1381-1394.

White, D. E., J. D. Hem, and G. A. Waring (1963) Chemical composition of subsurface waters: *U.S. Geol. Surv., Prof. Pap. 440-F.*

——— (1965) Saline waters of sedimentary rocks: *Am. Assoc. Petrol. Geol. Mem.,* **4,** 342-366.

——— (1968) Environments of generation of some base-metal ore deposits: *Econ. Geol.,* **63,** 301-335.

——— (1974) Diverse origins of hydrothermal ore fluids: *Econ. Geol.,* **69,** 954-973.

5

Hydrothermal Alteration

ARTHUR W. ROSE
The Pennsylvania State University

DONALD M. BURT
Arizona State University

In many hydrothermal ore deposits, changes in mineralogy and texture of wall rocks enclosing ore are far more extensive and obvious than ore itself. Students of ore deposits long ago recognized the usefulness of this hydrothermal alteration both as a guide to ore and as an indicator of the character of solutions associated with ore deposition. This chapter discusses processes of hydrothermal alteration, chemical theory, experimental results useful in interpreting alteration, and classical and recent field studies of altered rocks.

One essential feature of hydrothermal alteration is conversion of an initial mineral assemblage to a new set of minerals more stable under the hydrothermal conditions of temperature, pressure, and most importantly, fluid composition. Changes in chemical composition of the altered rock as it reacts with the solution are common, as are regular zonal patterns reflecting changes in the composition of the fluid with time or extent of reaction with rock. In some cases, as in alteration of carbonate rocks, alteration results in essentially monomineralic zones, but in others, such as at Butte, new minerals selectively replace certain minerals of the parent and leave others relatively untouched, or completely different alteration products replace different minerals. The original texture of the rock may be only slightly modified or completely obliterated in this process. Several periods of alteration, apparently developed at different times, are also common.

Alteration envelopes around veins at Butte, Montana, furnish a classic example of the effect of hydrothermal alteration (Sales and Meyer, 1948, 1950). Around each vein of the "Main Stage," the enclosing quartz monzonite exhibits a zoned series of alteration products extending from a few centimeters to perhaps 10 m away from the vein (Figure 5.1). From the unaltered rock inward toward the vein, successive zones involve alteration of plagioclase to montmorillonite, kaolinite, sericite, and (locally) pyrophyllite or dickite. Concurrent changes occur in other minerals. The argillic and sericitic zones at Butte show large depletions of calcium,

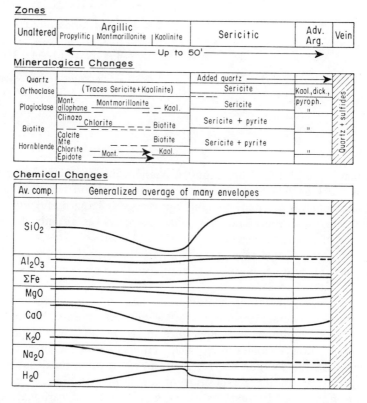

Fig. 5.1 Diagrammatic summary of alteration around Main-Stage veins at Butte, Montana (after Meyer et al., 1968).

sodium, and magnesium and additions of hydrogen (as structural OH in the clays and micas). Silica is depleted from the argillic zone and added in the sericitic and vein zones. Small additions of CO_2 and sulfur occur in all zones, and there is little change in potassium and aluminum. At depth, the above Main-Stage veins and alteration cut several types of older veins containing various combinations of quartz, molybdenite, chalcopyrite, magnetite, and orthoclase with either no alteration envelopes or alteration to alkali feldspar, quartz, sericite, biotite, anhydrite, andalusite, corundum, and sulfides (Brimhall, 1977; Roberts, 1973).

Silicate alteration of more or less pure limestone adjacent to faults at the Linchburg mine, Magdalena district, New Mexico, provides a second example of effects of hydrothermal processes (Titley, 1961, 1963). In this case, the original wallrock consisted of calcium carbonate, so that gains and losses of material are obvious. An inner zone, in and near major faults, consists mainly of barren quartz. Next to this intensely silicified zone is a relatively thin zone of coarsely crystalline andraditic garnet, also without commercial mineralization. A wide transition zone of andradite, replacing hedenbergitic clinopyroxene with abundant sphalerite and somewhat less galena, gives way outward to a zone of coarse manganoan hedenbergite, commonly showing alteration to cummingtonite or chlorite and containing abundant galena and less sphalerite. Its outer edge is marked by hematitic and chloritic alteration of marble. The zoned alteration varies from 20 m to several centimeters in width, although in places, the inner or outer zones may be missing.

Skarn zones at the Linchburg mine are believed to have formed as part of a continuous process (Titley, 1961, 1963), with the early garnet-clinopyroxene alteration having been affected by later, lower temperature amphibole-chlorite-hematite alteration as conditions gradually changed. The intimate relation of skarn with ore and the lack of ore in limestone beyond the skarn are typical of this type of deposit.

Hydrothermal alteration forms a readily visible empirical guide to hydrothermal activity and thus attracts the exploration geologist because of the possibility of associated ore. However, wallrock alteration may be used as a detailed guide to ore only if its spatial and temporal relations to ore deposition are understood. In most hydrothermal ore deposits, at least some hydrothermal alteration is contemporaneous with ore deposition, because hydrothermal fluids normally do not deposit ore when they are chemically in equilibrium with wallrock. The greater extent of the alteration makes it useful as a guide to ore. Alteration that preceded the ore may be useful in exploration if the ore-bearing solutions utilized the same passageways as the altering solutions (Lovering, 1949). Closer relationships to ore may exist if the early alteration increased the porosity and permeability of the rock so as to channel later ore fluids ("ground preparation"), or if the mineral assemblages of the early alteration are capable of chemically promoting ore deposition (Burt, 1972a). However, in a great many cases of ore that follows alteration, the ore is controlled structurally by later fracturing.

If several types of alteration formed sequentially, with cross-cutting relations, as at Tintic, Utah (Lovering, 1949), then the alteration type most closely associated with ore presumably formed nearly contemporaneously. If the several alteration types formed simultaneously in zones, then the zone containing the ore may be contemporaneous with ore, as at Butte (Sales and Meyer, 1948, 1950) or may be earlier than ore and was, for chemical or textural reasons, an especially favorable site of ore deposition (Burt, 1972a). The latter interpretation might be favored if the zone that contains the ore is in many places barren of ore (presumably due to lack of access for later ore-bearing solutions).

Altered rocks have also become increasingly useful for interpretation of chemical and physical conditions of ore deposition. The assemblage of hydrothermal minerals forms a valuable supplement to the actual ore minerals in determining age relations among several generations of ore and gangue minerals and in estimating physical and chemical conditions under which ores were deposited. Such studies are aided by the wider range of elements and minerals present in the alteration as compared to the ore itself.

THEORY OF METASOMATIC PROCESSES IN HYDROTHERMAL ALTERATION

Hydrothermal alteration is a type of metamorphism involving recrystallization of a parent rock to new minerals more stable under changed conditions. Therefore, most tools and techniques of metamorphic petrology can be applied to hydrothermally altered rocks if certain distinctive features are taken into account. The major distinctive feature of hydrothermal alteration is the importance of the hydrothermal fluid in transferring chemical constituents and heat. In regional and contact metamorphism, fluids rich in H_2O and/or CO_2 are generated by breakdown of minerals during prograde dehydration and decarbonation, but these fluids are generally considered to be small in amount relative to the solid phases of the rock and flow slowly through pores and fractures. As a result, the fluids are in approximate chemical and thermal equilibrium with surrounding solid phases, and metasomatic effects are usually subordinate.

Hydrothermal alteration, in contrast, typically involves relatively large amounts of fluid traversing rocks of considerable permeability in the form of fissures or connected pore space. As a result, flow rates are probably higher than for fluids in other types of metamorphism, as are thermal gradients, and the hydrothermal fluid thus tends to be considerably out of equilibrium with adjacent wallrocks. Unusual sources of fluids, such as crystallizing magmas or unusual combinations of rock types along the flow path, may also result in hydrothermal alteration.

Fluid inclusion studies (Roedder, Chapter 14) and observations indicating alternating periods of deposition and leaching of minerals in ore deposits suggest that fluids involved in wallrock alteration fluctuate in their chemical properties over short periods, probably due to changes in the plumbing of the fissure systems they traverse. These fluctuations are another feature that sets alteration apart from typical metamorphism.

Metasomatic effects, plus observations of relics and other evidence for two or more stages of differing alteration mineralogy even within single hand specimens and thin sections, have caused some workers to consider alteration as a "disequilibrium" process. As a consequence, the applicability of the phase rule and other thermodynamic principles has been questioned. However, Korzhinskii (1959) and Thompson (1959) have shown that the assumption of local or mosaic equilibrium allows treatment of many metasomatic processes using modified versions of traditional thermodynamic relationships. The essence of local equilibrium is that if a small enough volume of rock is taken, it contains no incompatible phases and is substantially in thermodynamic equilibrium. Local equilibrium may be assumed if reaction rates between solids and fluid are rapid relative to migration of material by flow or diffusion (Thompson, 1959; Fisher and Elliott, 1974). One criterion for a system in local equilibrium is that no incompatible phases are in contact. Some hydrothermally altered rocks satisfy this criterion. For instance, Burt (1974) shows that many skarns can be understood on the basis of local equilibrium. In contrast, many argillized aluminosilicate rocks contain incompatible phases apparently in contact, so that Hemley and Jones (1964) suggest that equilibrium exists only between the solution and new phases being formed rather than between all phases in the rock. Of course, natural as well as experimental mineral assemblages may represent metastable equilibria, but even metastable equilibria can be informative if they are recognized as such (Barton and Skinner, Chapter 7).

Korzhinskii (1936) and Thompson (1955, 1959, 1970) have shown that in metasomatic situations, the Gibbs phase rule and the mineralogical phase rule can be modified to read $P = C - M$, where P is the maximum number of phases present in an assemblage at an arbitrary choice of temperature, pressure, and fluid composition, C is the total number of components in the system, and M is the number of "mobile" components for which the concentration or chemical potential is controlled externally to the system rather than by an initial bulk composition (Vidale and Hewitt, 1973). In a great many hydrothermal systems, many of the components are mobile, and as a result the number of phases present in the altered rock is small, in some cases unity. Even if metasomatism is minor, the number of phases tends to decrease from zone to zone as the vein or other source of fluids is approached.

Mass Transport and Zoning

The transport of the materials involved in wallrock alteration can occur either by infiltration (mass movement due to fluid flow through the rock) or by diffusion (transport by diffusion of chemical species through stagnant pore fluids), or by a combination of both processes. The theory behind these two limiting models for metasomatic zoning has been developed by Korzhinskii (1970) and elucidated by Helgeson (1968, 1970a, 1970b), Hofmann (1972, 1973), Fisher (1973), Frantz and Mao (1974, 1975), and Brady (1975). A geologically reasonable combination of both processes near a fissure is discussed by Fletcher and Hofmann (1974).

Infiltration was probably the dominant means of mass transport if the geologic evidence shows that material moved large distances, predominantly in one direction, and if the mineral solid solutions in each alteration zone tend to have a constant composition that changes abruptly between zones (Hofmann, 1972). On the other hand, diffusion was probably the major process if material moved in two directions (was exchanged between adjacent contrasting rock types); or if the composition of mineral solid solutions changes regularly in each zone without abrupt breaks between zones (with exceptions described by Hofmann, 1972).

In any actual ore depositing environment, of course, diffusion and infiltration occur simultaneously. Infiltration may dominate in and near a fracture through which fluid is flowing, and diffusion through stagnant pore fluids may dominate a short distance away. This feature tends to make the "correct" choice of models for a given geologic situation ambiguous. Satisfactory models for zoned skarns have been attained by Burt (1974) by assuming that the zoning patterns will look predominantly like diffusion, with infiltration near veins and fractures merely tending to "spread the zones out" (Fletcher and Hofmann, 1974, p. 251).

Models for the simultaneous development of alteration zones generally are based on assumptions of local equilibrium and graphic or analytic representations of the gradients in concentration, activity, or chemical potential in the fluid and solids. Models for pure infiltration and pure diffusion have been formulated.

Figure 5.2, after Frantz and Weisbrod (1974), indicates the calculated results of infiltration by a fluid that is not in equilibrium with the wallrock. The mineral assemblages and composition of solution after an arbitrary length of time are shown on Figure 5.2, assuming local equilibrium, no solubility of aluminum, a temperature of 500°C, and a pressure of 1000 bars.

The two fronts, defined by abrupt changes in KCl and HCl content of the fluid and by changes in the solid assemblages, migrate through the rock in the direction of fluid flow. Near the fronts, small domains of local equilibrium are separated by an infinitesimal zone of reaction, the dimensions of which are related in a real case to the rate of solid-fluid reaction, as compared to the rate of transport in the fluid. The proportions of minerals and the amount of pore space in the initial rock, as well as the composition of the fluid, affect the final proportions of minerals and the nature of zoning (Frantz and Weisbrod, 1974).

In an analogous manner, reaction fronts are formed if transport is by diffusion, except that fluid composition varies continuously between fronts, and transport can take place in both directions. Rates of advance of the fronts depend on rate of diffusion, as controlled by the applicable concentration gradients, diffusion coefficients, and pore geometry. Figure 5.3 illustrates the characteristics of such a system in which several phases exhibiting limited solid solution are contacted with a solution rich in one end member of the system. Note that the concentration gradient in the solution is nearly linear between fronts because the new mineral grows at a slow rate relative to the diffusion rate (termed a quasi-steady state). Fletcher and Hofmann (1974) illustrate more complicated cases involving both infiltration and diffusion, in which patterns of pure infiltration are rounded

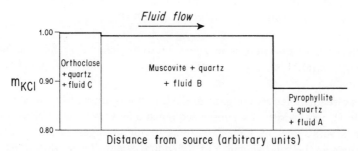

Fig. 5.2 Concentration of KCl in fluid infiltrating to the right through a porous pyrophyllite-quartz rock. The initial fluid entering at the left is in equilibrium with quartz and orthoclase at 500°C, 1 kb. The initial rock contains 28% porosity; no movement of Al is allowed; and the initial solution contains $0.998\ m$ KCl, $0.002\ m$ HCl, and $0.04\ m$ SiO_2 (after Frantz and Weisbrod, 1974). Note that the alteration product, prophyllite is metastable at 500°C, 1 kb.

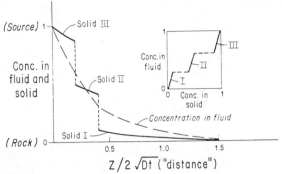

Fig. 5.3 Calculated results of diffusion of a single component into a semi-infinite volume lacking the component, which forms three solid solution series with the solid occupying the volume. The compositional relations between the solid and the fluid are shown in the inset. Z = distance, D = modified diffusion coefficient, t = time (after Fletcher and Hofmann, 1974).

off and distorted if appreciable diffusion is allowed. Also, some zones that exist for pure diffusion may not appear in the infiltration case.

If diffusion is assumed to be the dominant means of mass transport, saturation with various alteration phases may conveniently be shown either on chemical potential diagrams (Korzhinskii, 1959) or on diagrams showing the result of extended reaction between "inexhaustible" masses of various limiting compositions under a given set of externally imposed conditions (Thompson, 1959). The two types of graphic representation are theoretically and topologically nearly equivalent, as indicated by the examples in Figure 5.4, and both may be used to explain or predict zoning sequences in nature.

Comparison of diffusion and infiltration models with natural alteration sequences leads to the following generalizations:

1. Given sufficient time, the zones tend to become monomineralic if a large number of components are "mobile" and other conditions are constant.
2. As already noted with regard to the Linchburg mine, New Mexico (Titley, 1961), total width of a zoning sequence may vary from a few centimeters to several tens of meters without major changes in the sequence of zones or the relative width of individual zones. These variations in total width are presumably due to lateral changes in physical parameters at the reaction fronts (e.g., the porosity of the rock and the amount of fluid available for mineral solution and redeposition).

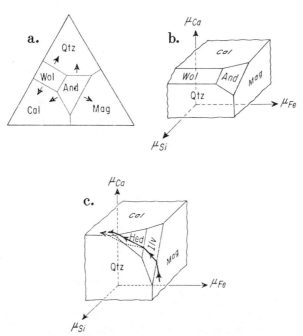

Fig. 5.4 (a) Geometry of phases formed as a result of extended reactions between inexhaustible prisms of quartz, calcite, and magnetite (after Burt, 1974). (b) Saturation surface in μ_{Si}-μ_{Fe}-μ_{Ca}-space for system corresponding to Fig. 5.4a. (c) Saturation surface in μ_{Ca}-μ_{Fe}-μ_{Si}-space for the zoning sequences at Temperino, Italy (externally controlled μ_{O_2}, μ_{CO_2}, and μ_{H_2O}). Solid line, normal sequence; dotted line, local sequence. Qtz = quartz, Wol = wollastonite, And = andradite, Cal = calcite, Mag = magnetite, Hed = hedenbergite, Ilv = ilvaite.

3. Alternatively, individual zones may vary in width laterally and may be missing entirely in places. Such changes presumably reflect lateral changes in chemical parameters (e.g., composition of wallrocks or fluid).
4. Reversals of the zoning sequence (i.e., of the order of appearance of zones from the vein outward) are also possible laterally. Such reversals reflect a change in the topology of the equilibrium diagram; that is, a change in compatibilities of minerals due to changing chemical parameters, as represented by reactions such as, for skarns (Burt, 1972a):

$$2CaCO_3 + 4CaFeSi_2O_6 + O_2 \rightleftharpoons 2Ca_3Fe_2Si_3O_{12} + 2SiO_2 + 2CO_2$$
$$\text{calcite} \quad \text{hedenbergite} \quad\quad\quad \text{andradite} \quad\; \text{quartz}$$

(5.1)

Thus at Campiglia Marittima, Tuscany, Italy (Bartholome and Dimanche, 1967) and at the Sasagatani mine, Shimane Prefecture, Japan (Burt, 1972b) a thin (1-5 mm) zone of andradite (with or without quartz) occurs between massive hedenbergite skarn and marble. This zone presumably developed retrogressively as reaction (5.1) went to the right.

5. "Uphill" diffusion (i.e., against concentration or chemical potential gradients) may occur in multicomponent natural systems, so that a quartz-rich zone, for example, can occur several zones away from the presumed source of silica (Cooper, 1974; Burt, 1974).

6. The concentration of an element or component in a given metasomatic zone cannot be used to infer its relative chemical potential (fugacity; activity) in pore fluid that precipitated it, without more knowledge of other chemical potentials involved than is generally available. For example, presence of minerals of greatly differing oxidation state (such as, in a zoned skarn, andraditic garnet and hedenbergitic clinopyroxene) in different parts of a zoning sequence does not necessarily indicate a gradient in chemical potential (or fugacity) of oxygen across the zoning sequence during its development (Burt, 1974).

7. Absence of a mineral from a given zoning sequence does not necessarily indicate that the mineral was unstable with respect to P, T, or volatile fugacities. It may indicate a failure of the chemical potential gradient of the solution to intersect the saturation surface of that mineral in chemical potential space (see Figure 5.4). It may also indicate supersaturation of the mineral.

8. Absence of a given element or component (e.g., NaCl) from the minerals of a wallrock alteration sequence does not necessarily indicate that this element or component was not abundant in the fluid phase.

In most situations involving diffusion in wallrock adjacent to a vein, diffusion takes place in a nearly isothermal environment (Lovering, 1950). This conclusion is reached because the mathematical treatment of diffusion and heat conduction is the same, and the rate of thermal conduction is much greater than the rate of chemical diffusion. In the halo around a vein, both processes operate due to gradients between the potential (chemical or thermal) in flowing fluid in the vein and the potential in an initially homogeneous wallrock. Heat effects of chemical reactions will tend to be damped out by the relatively rapid thermal conduction. Although minor temperature gradients undoubtedly exist in the altering wallrock, they probably amount to only a few degrees in the zone of significant transport by diffusion.

CHEMISTRY OF ALTERATION PROCESSES

Exchange Processes and Exchange Operators

Wallrock alteration and concomitant ore deposition are a process of irreversible chemical exchange between hydrous solutions and adjacent wallrock. Certain components are selectively leached from wallrock and are added to the fluid, and other components (including ore metals) are selectively taken up by wall rock (or form a coating on the wallrock) and are removed from the hydrothermal fluid. The net result depends on the physical conditions at the wall rock-fluid reaction interface, on the composition of wallrock and fluid, and on the relative amounts of fluid and wallrock involved in the exchange process. Only altered wallrock is visible as a result of this process, because the altered fluid is removed from the system (except possibly for fluid inclusions in precipitated minerals).

Some types of hydrothermal alteration involve only a one-way transfer of constituents from the fluid to the rock or *vice versa*. For example, hydration/dehydration, carbonation/decarbonation, oxidation/reduction, and sulfidation/reduction involve a gain or loss, respectively, of H_2O, CO_2, O_2, or S_2. The components H_2O, CO_2, O_2, and S_2 are the "operators" that perform the change, and the tendency toward the change can be expressed in terms of their pressure, fugacity, concentration, activity, or chemical potential. Precipitation or dissolution of a mineral is similarly a one-way process, and solubility is the controlling parameter.

Note also that the concentration of "operators" such as O_2 or S_2 in the reacting system may be negligible. The "operators" are a form of chemical shorthand for the species actually present, and are used because they are convenient as components in balancing reactions.

Many other reactions involve an exchange of components in two directions between rocks and fluids. Two simple types of exchange processes (or reactions) are distinguished: oxidation-reduction, in which one or more elements change nominal oxidation states, and acid-base. The two types of reactions can occur in combination.

If equilibria are written in terms of ions, the operator (component) that performs oxidation-reduction is conventionally chosen as the electron e^-. Two typical oxidation-reduction reactions involving the electron are

$$Cu^+ + e^- \rightleftharpoons Cu \qquad (5.2)$$

$$Fe^{2+} + 2e^- \rightleftharpoons Fe \qquad (5.3)$$

The second reaction can be subtracted from two times the first to arrive at the oxidation-reduction exchange reaction by which copper metal can commercially be recovered from mine leach solutions using iron scrap, namely

$$2Cu^+ + Fe \rightleftharpoons Fe^{2+} + 2Cu \qquad (5.4)$$

If the iron scrap (reducing agent) is considered as the wallrock, and the leach solution (oxidizing agent) is considered as the hydrothermal fluid, the process is a very simplified example of an exchange involving oxidation and reduction that results in ore deposition (of copper metal).

When electrically neutral components are chosen, the operator that performs the oxidation is conventionally chosen as oxygen O_2 or analogously for sulfides, as sulfur S_2 (vapor). Reactions analogous to (5.1) and (5.2) then are

$$Cu_2S \rightleftharpoons 2Cu + \tfrac{1}{2}S_2 \qquad (5.5)$$

$$FeS \rightleftharpoons Fe + \tfrac{1}{2}S_2 \qquad (5.6)$$

Subtracting (5.6) from (5.5) yields

$$Cu_2S + Fe \rightleftharpoons FeS + 2Cu \qquad (5.7)$$

Reaction (5.7) is by most definitions an oxidation-reduction reaction, although it explicitly involves neither oxygen nor electrons.

The second group of reactions, acid-base reactions, can be chosen to include all reactions except oxidation-reduction reactions. Acid-base reactions involve sharing of electron pairs (or of electron cloud density) in the formation of chemical bonds. The electron density acceptor is the acid, and the electron density donor is the base, in the same way the electron acceptor (Cu^+) was the oxidizing agent, and the electron donor (Fe) was the reducing agent in reaction (5.3).

Included among acid-base reactions are all ion exchange reactions. Examples include Mg^{2+} for Ca^{2+} (cation exchange), H^+ for cations (a special type of cation exchange generally called hydrolysis or protolysis) or F^- for $(OH)^-$ (anion exchange). These reactions can be written in terms of a conventional ionic formulation, with the exchanged species on opposite sides, or else, as here, in terms of "exchange operators" (Burt, 1974), such as, for hydrolysis, $H_n^+ C_{-1}^{n+}$, where C^{n+} is a cation of valence n (e.g., HK_{-1}, H_2Ca_{-1}, etc.). These exchange operators are not ordinary chemical species, but they are perfectly usable as components. Their main advantage

CHEMISTRY OF ALTERATION PROCESSES

compared with other choices of components is that they are intrinsically acidic or basic. They also lend themselves readily to a graphic representation and are in principle no more imaginary then the oxidation-reduction operators e^-, O_2, and S_2 discussed above.

If it is not obvious that a cation exchange reaction such as Mg^{2+} for Ca^{2+}, also expressible by $MgCa_{-1}$ (pronounced "Mg, Ca minus one," written with the ionic charges omitted and with the more acidic cation first) is an acid-base reaction, consider, for example, the compound $Ca(OH)_2$, portlandite. In this mineral, Ca^{2+} is six-coordinated to hexagonal close-packed $(OH)^-$. If Mg^{2+} is substituted into this structure, the isostructural compound $Mg(OH)_2$, brucite, results. Because the smaller Mg^{2+} has a greater attraction for the electrons on the anion, brucite is less basic than portlandite. (The electrons "spend more time" with Mg^{2+} than they did with Ca^{2+}, decreasing the effective negative charge on $(OH)^-$.)

In the reaction

$$Mg^{2+} + \underset{\text{Portlandite}}{Ca(OH)_2} \rightleftharpoons \underset{\text{Brucite}}{Mg(OH)_2} + Ca^{2+} \tag{5.8}$$

Mg^{2+} is the acid, because it accepts electron cloud density, and $Ca(OH)_2$ is the base, because it gives it up. In the more compact exchange operator notation, the reaction may be expressed as

$$Ca(OH)_2 + MgCa_{-1} \rightleftharpoons Mg(OH)_2 \tag{5.9}$$

In this reaction, $MgCa_{-1}$ is an acidic cation-exchange operator, because it changes a more basic compound into a less basic one. Its "reciprocal" operator $CaMg_{-1}$ must be basic, because it has the opposite effect. Note that $Mg(OH)_2$ will be stable in more acid solutions than $Ca(OH)_2$, and that the ion-exchange reaction can be driven to the left or right by changing the acidity of the solution at a constant ratio of the concentrations of Mg^{2+} and Ca^{2+} (Korzhinskii, 1956).

This example illustrates that the acidic strengths of ions can be related to their charge to radius ratio (sometimes termed for cations "ionic potential"; related terms for anions are "softness" or "polarizability"). The proton H^+ is considered as the most acidic cation, because of its extremely small radius, whereas the electron e^- for the same reason may be considered as the most basic anion. A universal "order of acidity" of other cations or anions does not exist, because "acidity" and "basicity" are relative terms. The order of acidity of cations can be quantitatively measured relative to a given anion exchange operation, and *vice versa*. For more details, consult Van der Werf (1961) or a recent inorganic chemistry text such as Huheey (1972).

Oxidizing agents cannot always be distinguished from acids, nor reducing agents from bases, except by their function in a given reaction. Increasing oxidation of a compound corresponds to increasing acidity, and thus H_2SO_4 is more acidic than H_2S, and $FeCl_3$ is more acidic than $FeCl_2$.

Some ion-exchange operators can correspond to real phases or species. For example, CO_2 can be written as the acidic anion-exchange operator $(CO_3^{2-})(O_{-1}^{2-})$ in the neutralization reaction $CaO + CO_2 = CaCO_3$; and it is therefore not surprising that wallrock alteration reactions involving fluorite and F_2O_{-1} in greisens are analogous to those involving calcite and CO_2 in skarns (Burt, 1972a). Replacement of limestone by fluorite can similarly be described in terms of the acidic anion-exchange operator $F_2(CO_3)_{-1} (=F_2O_{-1} - CO_2)$.

From the above discussion, it would appear that alteration processes can commonly be described in terms of neutralization of relatively acidic solutions by more basic wallrocks. Changes in composition of the wallrock itself can then be considered in terms of the addition of acidic ion exchange operators such as HK_{-1} (hydrolysis), $MgCa_{-1}$ (dolomitization), and $F_2(CO_3)_{-1}$ (fluorite replacement of limestone). The more acidic ions are added to the wallrock, and the more basic ions are leached in solution.

The exchange operator approach to acid-base reactions can also be used to formulate oxidation-reduction reactions. Acidic exchange operators, such as $Fe^{3+}(Fe^{2+})_{-1}$ (ferric iron is more acidic than ferrous iron), generally are also oxidizing. The cation-exchange operator above can be abbreviated as $(e^-)_{-1}$ or minus one electron. Similarly, the acidic anion-exchange operator $(SO_4)^{2-}(S_{-1}^{2-})$ can be written as $2O_2$, two oxygen molecules, and is obviously oxidizing. Thus the components e^- and O_2 are already, as mentioned, abbreviated forms of "exchange operators," and $(e^-)_{-1}$ or O_2 can be thought of as acidic as well as oxidizing.

Calculations of Alteration Equilibria

Calculations of equilibria between minerals and fluids require accurate thermochemical data for the phases involved. Unfortunately, despite major compilational efforts in recent years (Robie et al., 1978; Helgeson, 1969; Naumov et al., 1971; Stull and Prophet, 1971), basic data of adequate accuracy are still not readily available for many phases of geologic interest. Nevertheless, some of this data can be derived from experimentally determined phase equilibria (Fisher and Zen, 1971; Gordon, 1973; Helgeson et al., 1975) or can be approximated (Tardy and Garrels, 1974; Chen,

1975; Nriagu, 1975), but care must be taken in evaluating possible errors and their effects.

A related problem is that of phase characterization, both of minerals on which thermochemical and stability experiments are performed, and of phases that occur in nature. Particular problems involve grain size and degree of crystallinity, presence and activities of other components in solid solution, substitutional order-disorder (as in feldspars), "mixed-layering" of clay minerals, sorption of cations and anions, and so on.

Thermodynamic properties of fluids and dissolved species involved in hydrothermal alteration are more complex than those of solids and are beyond the scope of this article. The interested reader is referred to Helgeson (1969) and Helgeson and Kirkham (1974, 1976).

An alternative approach to experimental and thermodynamic methods is to use geologic observations to generate phase equilibrium diagrams. This approach has been used fruitfully by generations of petrologists and geochemists, among them C. E. Tilley, N. L. Bowen, J. B. Thompson, Jr., and R. M. Garrels. The approach is, in principle at least, applicable to systems on which thermodynamic and experimental data are completely lacking. The resulting diagrams are schematic only and lack precise numbers on the axes, but the deriver is at least assured that they will appear reasonable to the field geologist, and he can assure himself that they are consistent with thermodynamic principles.

The method (Burt, 1976) consists of the following steps: (1) choose the relevant system, as defined by appropriate components; (2) list all phases that might occur in this system; (3) list all incompatible combinations of these phases, as deduced from compositional restrictions, experimental data, field relations, or, in unfortunate circumstances, guesswork; (4) assume that all remaining reactions among phases are permitted. A computer program (REACTION: Finger and Burt, 1972) can then be used to calculate these reactions, which in turn can be used to generate the phase diagram. Its topology is already specified by the incompatibilities determined in step (3); these in turn reflect the topology of the rocks (i.e., which minerals never touch each other in nature). If P- and T-dependent reactions occur among the solids, various diagrams may be drawn for limited ranges of P and T. An example of such a diagram is Figure 5.13.

An advantage of this approach is that, over the span of geologic time, nature manages to avoid or minimize many problems of metastability that plague the experimentalist. On the other hand, nature does not "quench" her experiments, so that interpretations of the results must account for cooling processes.

Solid solution in minerals extends their field of stability and therefore must also be considered. For this reason, incompatibilities of minerals in natural samples, rather than compatibilities, are used as the basis of the method. Thus the fact that two minerals commonly touch each other in nature does not necessarily indicate that the end members are compatible, but if they never touch each other, despite natural solid solution, then one can be reasonably certain that they are incompatible in the model system. Some end members may even be unstable under all conditions, but solid solution with other components allows them to be stable in nature (thus there is no stable iron end member of the dolomite-ankerite series).

Presentation of Mineral Equilibria

A wide variety of variables, including pressure, temperature, mole fraction, molality, Eh, pH, fugacity, activity, and chemical potential are used by different workers in plotting diagrams of mineral equilibria. These sets of variables might appear to have little in common, but the opposite is true. The phase rule, as well as the specific stability relations, must be obeyed in all diagrams, resulting in equivalent topologies (configurations of divariant fields, univariant lines, and invariant points) on diagrams that involve quite different variables.

Figure 5.5 shows some stability relationships in the system Al_2O_3-K_2O-Na_2O-SiO_2-HCl-H_2O. In Figure 5.5a, stability of solid phases is shown, using mole fractions of Al_2O_3, $KAlSi_3O_8$, and $NaAlSi_3O_8$ as end members, and with quartz and aqueous solution present. This diagram is useful in depicting solid phases, including their solid solutions, but gives no information on composition of coexisting aqueous solutions. Figure 5.5b shows the same data on only two axes involving molar ratios of K_2O and Na_2O to Al_2O_3.

Figure 5.5c shows the same information in terms of $\log a_{Na_2O}$ vs $\log a_{K_2O}$. At the boundary between two phases such as pyrophyllite and muscovite, addition of K_2O to pyrophyllite forms muscovite plus quartz:

$$3Al_2Si_4O_{10}(OH)_2 + K_2O \rightleftharpoons 2KAl_3Si_3O_{10}(OH)_2 + H_2O + 6SiO_2 \tag{5.10}$$

If the pyrophyllite, muscovite, quartz, and water are all present as essentially pure end members, the activity of K_2O is a function only of temperature. Because of solid solution of sodium in muscovite, a_{musc} is less than unity at relatively high a_{Na_2O}, so the boundary curves to lower values of a_{K_2O} with increasing a_{Na_2O}.

The same relations are shown in Figure 5.5d, using the aqueous variables a_{Na^+}/a_{H^+} versus a_{K^+}/a_{H^+} as coordinates. The similarity between a_{K^+}/a_{H^+} and a_{K_2O} arises from the relationship

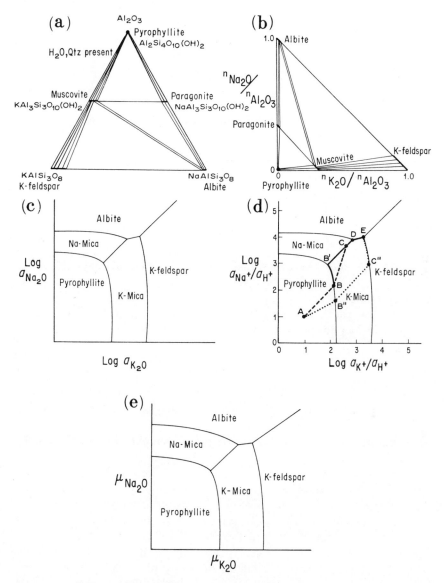

Fig. 5.5 Schematic stability relations in the system K_2O-Na_2O-Al_2O_3-SiO_2-H_2O-HCl at 400°C and 1 kb. Pyrophyllite is metastable. (a) Triangular mole fraction diagram, showing solid assemblages with quartz present. (b) Similar plot of molar Na_2O/Al_2O_3 vs. K_2O/Al_2O_3. (c) Stability of phases as a function of a_{Na_2O} vs a_{K_2O}. (d) Stability of phases as a function of $\log a_{Na^+}/a_{H^+}$ vs $\log a_{K^+}/a_{H^+}$. See text for discussion of the paths of solution composition during reaction of solution A with a mixture of feldspars. (e) Stability as a function of μ_{Na_2O} vs μ_{K_2O}. Figures based on data in Helgeson (1974), Meyer and Hemley (1967), and Montoya and Hemley (1974).

$$K_2O + 2H^+ \rightleftharpoons H_2O + 2K^+ \qquad (5.11)$$

$$K = \frac{a_{H_2O} \cdot a_{K^+}^2}{a_{K_2O} \cdot a_{H^+}^2}$$

Since a_{H_2O} is approximately unity in the aqueous solution, then

$$a_{K_2O} = \frac{1}{K} \cdot \frac{a_{K^+}^2}{a_{H^+}^2}$$

and

$$\log a_{K_2O} = 2 \log (a_{K^+}/a_{H^+}) - \log K.$$

The ratio a_{K^+}/a_{H^+} can be estimated from experimental data (Montoya and Hemley, 1975), so that the results of experiments can be plotted quantitatively on Figure 5.5d.

Figure 5.5e shows similar relations plotted on axes of μ_{K_2O} vs. μ_{Na_2O}. Because activity and chemical potential are related by

$$\mu_{K_2O} = \mu_{K_2O}^\circ + RT \ln a_{K_2O} = \mu_{K_2O}^\circ + 2.3 \, RT \log a_{K_2O} \qquad (5.12)$$

Figures 5.5c, 5.5d, and 5.5e are essentially identical except for the factor $2.3 \, RT$ and shifts of the origin. An analogous diagram can be plotted using $\mu_{HK_{-1}}$ and $\mu_{HNa_{-1}}$ (see Figure 5.9).

Complicated phase diagrams can be greatly simplified if certain phases (usually the vapor or fluid phase or common gangue minerals such as quartz or calcite) are assumed to be present "in excess." For graphic purposes, this procedure is usually referred to as "projecting through" the phase or phases in question. Projecting through a pure phase present in excess corresponds to arbitrarily fixing one of the chemical potentials (activities) of the system—that of the phase itself chosen as a component. Analogously, one can simplify topologies of phase diagrams by arbitrarily fixing other thermodynamic variables, such as P or T, as is implicit in the "mineralogical phase rule" of V. M. Goldschmidt. One or more chemical variables can then be changed from "internally buffered," "inert," or "initial value" to "externally controlled," "mobile," or "boundary value" to further simplify the topology of the diagram (see the simple review of terminology in Vidale and Hewitt, 1973). Examples of these types of manipulations are schematically illustrated in Figure 5.6 and in other diagrams of the chapter.

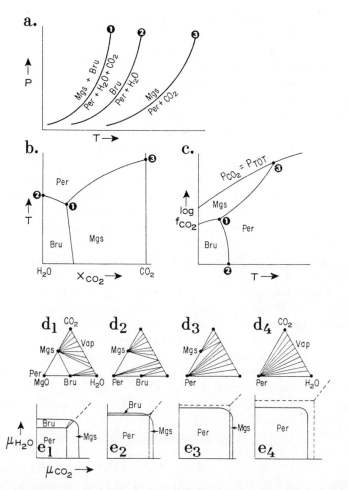

Fig. 5.6 Schematic stability relations in the system $MgO\text{-}CO_2H_2O$. (a) Pressure-temperature diagram showing univariant devolatilization reactions. (b) Isobaric temperature-X_{CO_2} relations; points 1, 2, and 3 are the intersection of curves 1, 2, and 3 of Figure 5.6a with this plane. (c) Isobaric temperature-log f_{CO_2} relations; points 1, 2 and 3 are those shown on Figure 5.6b. (d) Diagrammatic representation of solid-fluid compositions at equilibrium in the system. D_1, below curve 1; D_2, between curves 1 and 2; D_3, between curves 2 and 3; D_4, above curve 3. (e) Diagrammatic representation of stability in $\mu_{H_2O}\text{-}\mu_{CO_2}$ space. E_1–E_4 correspond to the conditions of D_1–D_4. Bru = brucite, Mgs = magnesite, Per = periclase.

Reaction Paths and Mass Transfer during Alteration

Compositional changes in a multicomponent solution interacting with a rock are complex, as indicated by the simple example below and by the more complicated examples discussed by Helgeson (1970b). The results depend on initial composition of the rock and solution, and on the manner in which the solution interacts with the rock.

Consider a fluid of composition A in the system K_2O-Na_2O-Al_2O_3-SiO_2-H_2O-HCl (Figure 5.5d), saturated with pyrophyllite and quartz. This idealized fluid contains 1 m NaCl and KCl and 0.1 m HCl, assumed for this example to be completely dissociated. The change in fluid composition during reaction with a "rock" of potassium-feldspar and albite depends on the ratio of amounts of the two feldspars decomposed and upon the nature of the solids and dissolved species formed. Assuming that the feldspars react in proportion to their relative abundance in the rock, that the newly formed solid phases are those in equilibrium with the fluid, and that reactions within the fluid are rapid compared to reactions with solids, the path of fluid composition is shown for two rocks, one with potassium-feldspar and albite in equal molar proportions (case I, path AB), and the other with twice as much potassium-feldspar as albite (case II, path AB″).

When the fluid reaches the potassium-mica boundary, additional assumptions must be made. If pyrophyllite formed along the path AB or AB″ remains in the system and reacts, the fluid composition moves along the potassium-mica-pyrophyllite boundary toward B′, all potassium released by decomposition of potassium-feldspar being used to convert pyrophyllite to muscovite. In case I, the composition reaches B′, where all pyrophyllite is consumed, after which it moves to D and then E, where fluid is in equilibrium with both feldspars. On the other hand, if the fluid leaves the initial pyrophyllite and reacts only with fresh rock to form potassium-mica, the path enters the potassium-mica field and follows ABCDE. For case II the analogous path is AB″C″E. Numerous other paths are possible for different "rocks" or processes of interaction with solids.

A quantitative method of modeling exchange reactions between hydrothermal fluids and wallrock has been developed by Helgeson (1970a), utilizing a computer routine. The complete path of the fluid and identity and amounts of solid phases decomposed and formed can be worked out by the computer, given sufficient thermodynamic data and appropriate initial assumptions. One convenient variable for monitoring the progress of alteration reactions is pH, which generally increases as acidic solutions are neutralized by wallrock. In a sense, the computer "titrates" the wallrock against the hydrothermal fluids, with results depending on assumptions about the conditions of the process.

STABILITY OF ALTERATION MINERALS

A large amount of experimental data has been obtained on stability of minerals at a wide range of temperatures, pressures, and chemical compositions of solid phases and fluids. Much of this data, especially that not involving melting, is of potential application in understanding conditions under which alteration takes place. A complete review of this data is not possible here, but some reactions selected for relevance to common types of alteration are shown graphically on Figures 5.7–5.13. For more complete summaries of experimental data on mineral stability, see Winkler (1974), Hewitt and Gilbert (1975), Kerrick (1974), Helgeson (1969), and Helgeson et al. (1969). Considerable care must be used in all quantitative applications of experimental data on mineral stability, because much published work does not represent true equilibrium, or the synthetic phases differ appreciably from natural minerals. If the past is any guide, some relations presented in Figures 5.7–5.13 will require modification on the basis of additional observations and experiments.

Dehydration Reactions

Figure 5.7 summarizes data on several dehydration reactions relevant to hydrothermal alteration. The curve for muscovite-quartz-andalusite-potassium-feldspar is one of the better established equilibria. For this reaction, the most divergent results appear to result from inability to recognize small amounts of reaction in the experiments. Other possible causes for error in this and other experimental studies include: (1) measurement of synthesis boundaries from metastable reactants, rather than reversing a reaction of stable phases, (2) unrecognized variability in the properties of phases, as suggested by Rosenberg (1974) for pyrophyllite, (3) effects of grain size and surface chemistry (Langmuir, 1971), and (4) supersaturation of the aqueous phase, as observed commonly for silica.

The effects of solution of other components on dehydration equilibria have been discussed by Barnes and Ernst (1963), Helgeson (1969), and Kerrick (1974). For the stability of muscovite and quartz, the equilibrium temperature for a solution with $a_{H_2O} = 0.9$ (equivalent to about 6 m or 26% NaCl) is calculated to be only approximately 10 °C lower than in pure water, using the procedure of Slaughter et al. (1975). A similar decrease in equilibrium temperature is calculated if P_f is 0.37 P_s, approximately the condition of hydrostatic fluid pressure and lithostatic solid pressure. These effects are small but would be increased for reactions releasing more moles of H_2O.

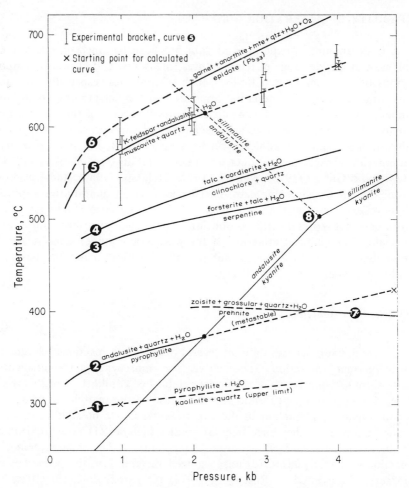

Fig. 5.7 Pressure-temperature relations of some dehydration equilibria and polymorphic transformations. Sources: curve 1, Reed and Hemley (1966); curve 2, Haas and Holdaway (1973); curve 3, Bowen and Tuttle (1949); curve 4, Chernosky (1975); curve 5, Chatterjee and Johannes (1974), plus experimental brackets indicating agreement of different investigators on this reaction (Althaus et al., 1970; Day, 1973; Evans, 1965; Kerrick, 1972; Fonarev and Ivanov, 1972); curve 6, Liou (1973); curve 7, Liou (1971); curve 8, Holdaway (1971). × indicates starting point for curve calculated by computer program of Slaughter et al. (1976). Dashed curves are metastable or of less precision than solid curves.

Anderson (1970) shows examples of calculations for dehydration equilibria using equilibrium constants.

Decarbonation Reactions

Many of the generalizations made for dehydration reactions also apply to decarbonation reactions. An additional experimental problem for decarbonation reactions is the slow reaction rate of experiments performed anhydrously (i.e., in pure CO_2), so that the trend in recent years has been to study decarbonation reactions (at moderate temperatures, at least) in H_2O-CO_2 mixtures. This method offers the additional advantage of allowing reactions involving OH-bearing minerals (e.g., tremolite) to be studied simultaneously. The technique, including many T-X_{CO_2} diagrams depicting stability of specific mineral assemblages, has been extensively reviewed by Kerrick (1974).

The calcium-aluminum silicates react with CO_2 to form calcite and aluminosilicates at low activities of CO_2, and thus furnish a useful guide to a_{CO_2} in altering fluids. Zeolites and related minerals, such as wairakite, laumontite, and prehnite, are converted to calcite plus clay or other aluminosilicates in H_2O-CO_2 mixtures with >0.01–0.03% CO_2 (Kerrick, 1974; Thompson, 1971). Serpentine, zoisite, and clinozoisite are also stable only at low contents of CO_2, but epidote is stable to higher values.

Hydrolysis Reactions

The stability of feldspars, micas, and clays is commonly controlled by hydrolysis, in which K^+, Na^+, Ca^{2+}, Mg^{2+}, and other cations are transferred from the mineral to solution, and H^+ enters the solid phases (Hemley and Jones, 1964). For instance, the stability of potassium-feldspar and muscovite at temperatures below about 300°C is limited by the following reactions:

$$\tfrac{3}{2}\ \underset{\text{K-spar}}{KAlSi_3O_8} + H^+ \rightleftharpoons \tfrac{1}{2}\ \underset{\text{muscovite}}{KAl_3Si_3O_{10}(OH)_2} + 3\ \underset{\text{quartz}}{SiO_2} + K^+ \quad (5.13)$$

$$\underset{\text{muscovite}}{KAl_3Si_3O_{10}(OH)_2} + H^+ + \tfrac{3}{2} H_2O \rightleftharpoons \tfrac{3}{2}\ \underset{\text{kaolinite}}{Al_2Si_2O_5(OH)_4} + K^+ \quad (5.14)$$

At higher temperatures, pyrophyllite and andalusite are the stable products instead of kaolinite and quartz, as illustrated in Figure 5.7. Experimental data on these and similar reactions are summarized by Montoya and Hemley (1975).

Calculations and graphic display of the stabilities in hydrolysis reactions are conveniently handled using the equilibrium constant, which for reaction (5.13) is

$$K = \frac{a_{\text{musc}}^{1/2} \cdot a_{\text{qtz}}^3 \cdot a_{K^+}}{a_{\text{ksp}}^{3/2} \cdot a_{H^+}}$$

If the solid phases and water are pure end members, the equilibrium constant reduces to $K = a_{K^+}/a_{H^+}$, which is a function only of temperature. Using this simplification, Figure 5.8 illustrates the stability of feldspars, micas, and aluminum silicates in the familiar plots of K/H, Na/H, and Ca/H^2 versus temperature.

The values of potassium and hydrogen used to obtain the K/H ratio in these diagrams are *not* activities of the ions at the high temperature of the experiment, but are derived from measured pH and total potassium in chloride solutions at room temperature. Because KCl and HCl occur predominantly as nonionized species in high-temperature solutions (Barnes and Ernst, 1963), and because HCl is nearly completely ionized at room temperature, the diagrams are actually plots of $(K^+ + KCl)/(H^+ + HCl)$. However, in the simple chloride system, ratios of total potassium to total hydrogen are closely related to the activity ratios of the ions at high temperature, as discussed by Shade (1974) and Montoya and Hemley (1975). In natural solutions containing significant amounts of other anions, relations will be more complex owing to formation of additional complex ions and nonionized solutes.

Both hydrolysis and hydration relations of minerals and fluids are schematically indicated on the μ_{H_2O}-$\mu_{HK_{-1}}$ diagram of Figure 5.9. Because increasing temperature leads to dehydration, reactions with increasing temperature are analogous to those with decreasing μ_{H_2O}. As discussed earlier, $\mu_{HK_{-1}}$ is analogous to log a_{H^+}/a_{K^+}. As a result, the reactions involving quartz (heavy lines) have a topology similar to the log K/H versus T diagram of Figure 5.8a. Figure 5.9 suggests that muscovite becomes unstable relative to potassium-feldspar and kaolinite at low temperature (130 ± 50°C according to Ivanov et al., 1974). This diagram also indicates relationships in quartz-free systems.

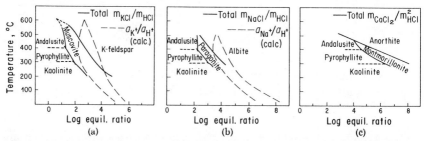

Fig. 5.8 (a) Total molality of K/H (solid line) and calculated $\alpha_{K^+}/\alpha_{H^+}$ (dashed line) for the system K$_2$O-Al$_2$O$_3$-SiO$_2$-H$_2$O-HCl after Montoya and Hemley (1975). (b) Total molality of Na/H (solid line) and calculated a_{Na^+}/a_{H^+} (dashed line) for the system Na$_2$O-Al$_2$O$_3$-SiO$_2$-H$_2$O-HCl with quartz present, at 1 kb, after Montoya and Hemley (1975). (c) m_{CaCl_2}/m^2_{HCl} for stability of anorthite, montmorillonite, and aluminum silicates at 1 kb with quartz present, after Hemley et al. (1971).

STABILITY OF ALTERATION MINERALS

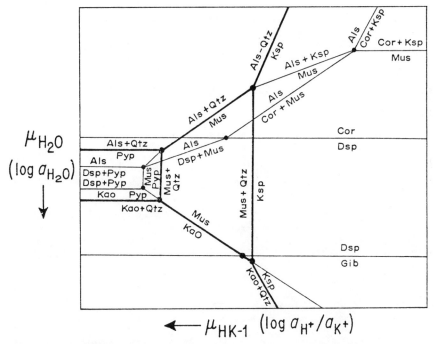

Fig. 5.9 Schematic stability of minerals in the system $K_2O\text{-}Al_2O_3\text{-}SiO_2\text{-}H_2O$ at low to moderate P and T under isothermal, isobaric conditions, in terms of $\mu_{HK-1}\text{-}\mu_{H_2O}$ (a_{K^+}/a_{H^+} vs a_{H_2O}). Heavy lines refer to quartz-bearing system; note similarity of topology to Figure 5.8a (after Burt, 1976). Als = aluminum silicate, Cor = corundum, Dsp = diaspore, Gib = gibbsite, Kao = kaolinite, Ksp = K-feldspar, Mus = muscovite, Pyp = pyrophyllite, Qtz = quartz.

Figures 5.10 and 5.11 illustrate the stability of phlogopite, chlorite, calcite, calcium-silicates, and alunite in terms of appropriate hydrolysis equilibria, based on experimental data and calculations from thermodynamic data. Note that stability of alunite depends on $a_{SO_4^{2-}}$ as well as ratios of cations.

Other Exchange Equilibria

Many minerals are solid solutions in which two or more components can be substituted in the same site. Data on such solid solutions can give information on the composition of the fluid coexisting with the mineral. An important example is the alkali feldspars, for which the exchange reaction can be written

$$KAlSi_3O_8 + Na^+ \rightleftharpoons NaAlSi_3O_8 + K^+ \tag{5.15}$$

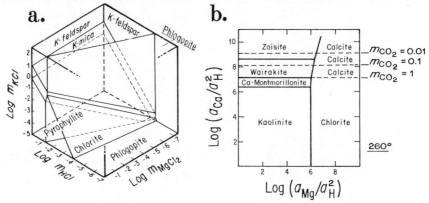

Fig. 5.10 (a) Preliminary diagram of some phase relations in the system K_2O-MgO-Al_2O_3-SiO_2 at 500°C and 1 kb in the presence of quartz (after Meyer and Hemley, 1967). (b) Stability diagram for calcium and magnesium minerals at 260°C, after Ellis (1971). The dashed lines indicate the limit of $a_{Ca^{2+}}/a_{H^+}^2$ in the presence of calcite at various values of m_{CO_2}.

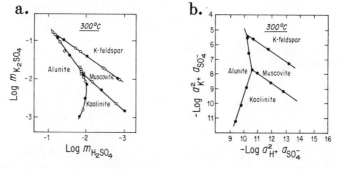

Fig. 5.11 Stability relations of alunite with quartz relative to muscovite, kaolinite, and potassium-feldspar at 300°C, after Hemley et al. (1969). (a) Experimental data for an H_2SO_4-K_2SO_4 fluid. (b) Activity products calculated from the experimental data.

The composition of coexisting alkali feldspar and solution at a series of temperatures has been summarized by Helgeson (1974). If temperature and pressure are known, the a_{K^+}/a_{Na^+} of coexisting fluids can be estimated from composition of the alkali feldspar. Waldbaum and Thompson (1969) discuss the thermodynamic character of solid solutions in the alkali feldspars and calculate activity coefficients for disordered sanidines (their Figure 3). Note that activities in feldspars with greater aluminum-silicon ordering or those containing H_3O^+ differ from those in disordered sanidines. Similar exchange equilibria exist for micas, plagioclases (Helgeson, 1974), and other minerals.

Exchange of fluorine and OH in muscovite, biotite, apatite, amphiboles, and other hydroxylated minerals can allow determination of HF/H_2O ratios in hydrothermal fluids. Munoz and Ludington (1974) have determined equilibrium constants for phlogopite, annite, siderophyllite, and muscovite for reactions of the type

$$\text{OH-mica} + HF \rightleftharpoons \text{F-mica} + H_2O \tag{5.16}$$

from which activity of HF in solution can be estimated from an analysis of fluorine and other constituents in the mineral.

Oxidation and Sulfidation Reactions

Elements undergoing oxidation-reduction reactions in hydrothermal solutions include iron, manganese, sulfur, carbon, and hydrogen. Minerals containing these elements can provide, from the valence state of these elements, an indication of the oxidation state of the fluids from which the minerals formed. Figure 5.12 shows limits of f_{O_2} for various minerals and mineral assemblages.

Combinations of oxidation with other types of reactions can be depicted on μ_{O_2}-μ_{CO_2} diagrams, like Figure 5.13, derived from natural assemblages (Burt, 1972a). Recent experimental studies (Gustafson, 1974; Liou, 1973) are consistent with this diagram. However, a bustamite-like phase of approximate composition $Ca_5FeSi_6O_{18}$ may be stable along the join $CaSiO_3$-$CaFeSi_2O_6$ at moderate temperatures (Shimazaki and Yamanaka, 1973). For many reactions on the diagram, the a_{O_2} and a_{CO_2} are known for specific temperatures (Figure 5.12), allowing conditions of other reactions to be estimated. Although accurate coordinates cannot yet be placed on these and similar diagrams, the topology and general relations are useful in studying zoning and other aspects of alteration. Of course, account must be taken of solid solution in natural minerals. Figure 5.13 is therefore only a *model* for more complex reactions that occur in nature—reactions that involve magnesium, aluminum, manganese, and other elements.

Because iron occurs in nearly all mineral assemblages, and because it commonly occurs in solid solutions with magnesium, its distribution is of special use in determining conditions of equilibration of hydrothermal assemblages. For instance, annite, the Fe^{2+} analogue of phlogopite, decomposes in an oxidizing environment by the reaction

$$KFe_3^{2+}AlSi_3O_{10}(OH)_2 + \tfrac{1}{2}O_2 = KAlSi_3O_8 + Fe_3O_4 + H_2O \tag{5.17}$$

Although pure annite is not reported in hydrothermally altered rocks, biotite containing annite in solid solution does occur. The stability of iron-

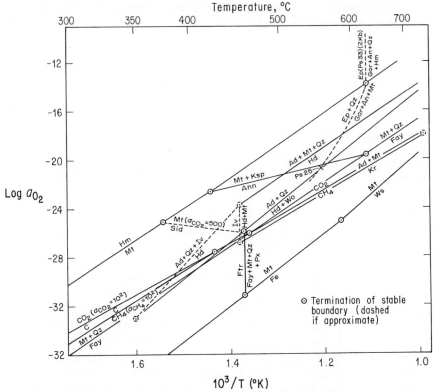

Fig. 5.12 Stability of iron minerals and other species in terms of log a_{O_2} vs $1/T$, based on data from Gustafson (1974), Liou (1973), Ernst (1966), Haas and Robie (1973), Huebner (1971), Robie et al. (1978), French (1971), and Wones and Eugster (1965). $P_{H_2O} = 1$ kb for reactions involving hydrous minerals (except epidote). Ad = andradite, An = anorthite, Ann = annite, Ep = epidote, Fay = fayalite, Fe = iron, Ftr = ferrotremolite, Gar = garnet, Hd = hedenbergite, Hm = hematite, Iv = ilvaite, Kr = kirschsteinite, Ksp = potassium feldspar, Mt = magnetite, Ps = pistacite, Px = pyroxene, Sid = siderite, Ws = wustite.

bearing biotites coexisting with potassium-feldspar and magnetite can be evaluated from the expression

$$k = \frac{a_{ksp} \cdot a_{mag} \cdot a_{H_2O}}{a_{ann} \cdot a_{O_2}^{1/2}}$$

as discussed by Wones and Eugster (1965) and Beane (1974). Both Fe^{3+} and Al^{3+} may substitute for Mg-Fe^{2+} in octahedral sites of biotites but do not form ideal solutions; thus estimation of activity of annite in biotite is discussed by these authors but not completely resolved. Beane (1974) sug-

STABILITY OF ALTERATION MINERALS

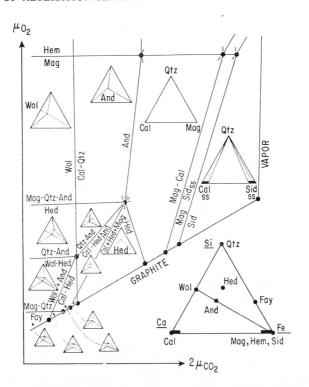

Fig. 5.13 Schematic isobaric, isothermal μ_{O_2}-$2\mu_{CO_2}$ diagram for the system Ca-Fe-Si-C-O, after Burt (1972a). The diagram expresses known mineral compatibilities and incompatibilities and (with adjustment for solid solution) can be applied to zoning sequences, or, for some reactions, the experimental data for specific temperatures and pressures can be used to quantitatively evaluate conditions of formation. The two lines for siderite-magnetite-calcite represent the equilibria of end members and of siderite-calcite solid solutions. Mag = magnetite, Qtz = quartz, And = andradite, Hed = hedenbergite, Wol = wollastonite, Cal = calcite, Fay = fayalite, Sid = siderite.

gests that the phlogopite-annite solution is nearly ideal, and that nonideal solution theory can be applied to Fe^{3+}-bearing solutions. The effect of aluminum in octahedral sites has been investigated by Rutherford (1973).

If Fe^{2+}, Fe^{3+}, and magnesium contents of biotites coexisting with potassium-feldspar + iron oxide or muscovite + quartz + iron oxide are known, the temperature and f_{O_2} of equilibrium can be estimated (Wones and Eugster, 1965; Beane, 1974). In addition, if both iron oxides and sulfides are present, an evaluation of f_{S_2} is possible from the reactions:

$$KFe_3^{2+}AlSi_3O_{10}(OH)_2 + Fe_3O_4 + 6S_2$$
$$\rightleftharpoons KAlSi_3O_8 + 6FeS_2 + H_2O + \tfrac{5}{2} O_2 \qquad (5.18)$$

$$KFe_3{}^{2+}AlSi_3O_{10}(OH)_2 + Fe_3O_4 + 3S_2$$
$$\rightleftharpoons KAlSi_3O_8 + 6FeS + H_2O + \tfrac{1}{2}O_2 \qquad (5.19)$$

The latter reaction has been studied experimentally by Hammarback and Lundqvist (1972), who used stoichiometry of pyrrhotite to determine f_{S_2}. If iron oxide is not present, an additional parameter (T, f_{O_2}, f_{S_2}) must be estimated from other data.

In principle, the approach described above can be used with any Fe^{2+}-bearing silicate mineral, such as an amphibole, pyroxene, chlorite, etc. Because iron-magnesium solid solution is approximately ideal in many silicates (Mueller, 1972), approximate evaluations of such equilibria can be calculated assuming ideal solution.

CLASSIFICATION OF ALTERATION

Because of the wide range in mineralogy, mineral abundance, texture, parent material, and other features of hydrothermal alteration, workers mapping and discussing alteration have generally simplified their observations by classifying the altered rocks into groups. Classifications useful in exploration and research studies are emphasized here, but the possibilities of other classifications should be kept in mind.

The simplest method of classification in the field is by a name reflecting the most abundant or most obvious mineral in the altered rock. Terms such as silicification, sericitization, argillization, and garnet alteration are used in this way. If alteration products are fine-grained, this terminology is frequently the best that can be accomplished in the field without microscopic study. Combinations of mineralogy, texture, and parent rock are the basis of terms like greisen, skarn, and jasperoid.

A different approach is to classify by dominant chemical changes during alteration. An altered rock with given proportions of minerals can be formed either by isochemical metamorphism of an appropriate parent or by metasomatism of a wide variety of parents. The type and degree of metasomatism are crucial in many cases, especially in attempting correlations with introduction of various ore minerals. One way of expressing the chemical changes is by terms such as hydrogen metasomatism (Hemley and Jones, 1964), potassium metasomatism, and fluorine metasomatism. However, in many cases, several elements are simultaneously introduced and removed to varying extents, so this approach becomes cumbersome in detail.

The utility of classifying hydrothermally altered aluminosilicate rock in terms of mineral assemblages was first pointed out by Creasey (1959), and his approach has been followed in various ways by all succeeding workers on porphyry copper deposits and similar ores in aluminosilicate rocks (Burnham, 1962; Hemley and Jones, 1964; Meyer and Hemley, 1967; Rose, 1970). The mineral assemblage developed during alteration reflects temperature, pressure, chemical composition of the hydrothermal fluid, chemical and mineralogical composition of the parent rock, and time available for equilibration.

Careful observation of the minerals and their textures is necessary to identify the proper mineral assemblage. The term equilibrium mineral assemblage refers to a group of minerals apparently formed at the same time and lacking any indication of disequilibrium, such as replacement or veining textures. In many such cases, the minerals probably formed at or close to equilibrium with each other and with the fluid, but disequilibrium or metastability in hydrothermal assemblages, such as those containing both kaolinite and potassium-feldspar, is common enough that caution must be exercised in equating coexistence with stable equilibrium.

A variety of terms for types of alteration (intermediate and advanced argillic, sericitic, potassic, propylitic, biotite-orthoclase) have been used by various workers to designate assemblages or groups of assemblages. Unfortunately, terms for alteration types are not clearly defined, are not used uniformly by different authors, or do not completely describe the rock. For instance, the term phyllic was originally proposed by Burnham (1962) to include both sericitic and biotitic alteration. However, Lowell and Guilbert (1970) and others have used phyllic in a more restricted manner to denote only strongly sericitized rocks. Many workers do not distinguish rocks containing sericite accompanied by either unaltered or hydrothermal potassium-feldspar from rocks with both feldspars altered to sericite, although the significance in terms of mineral assemblages is quite different.

For future work, a listing of the complete assemblage of minerals, preferably in approximate order of abundance and with a separation into temporal stages, if applicable, seems the best means of describing altered rocks. The list may be supplemented by data on changes in chemical composition or abundance of minerals to establish the extent or intensity of alteration. This complete description should be given in the text of any thorough alteration study, as is done in good studies of other metamorphic rocks. However, in aluminosilicate rocks, two abbreviated levels of terminology may be helpful for routine use. A listing of the alteration products of plagioclase, potassium-feldspar, and mafic minerals, in order of abundance, can form a first stage simplification. Persistence of potassium-feldspar or other primary phases without obvious alteration to a new phase would justify inclusion in the listing. As a second stage of simplification for general reference to the character of alteration, or for casual field use, terms ending in -ic or -ization may be used to indicate the most abundant or most obvious alteration products or alteration types. Examples are sericitic, biotitic, potassic, propylitic, sericitization, biotitization, etc. In view of the present variety of usages for these terms, their application in any specific sense is probably not justified, but an attempt should be made to use these terms consistently with previous usage.

As an example of the use of these three levels of terminology, the complete assemblage in the outer argillic zone of alteration around veins at Butte is quartz-montmorillonite-chlorite-potassium-feldspar-carbonate-pyrite $\pm$ kaolinite $\pm$ amorphous clay (Sales and Meyer, 1948). In condensed form, this can be expressed as montmorillonite-K-spar-chlorite or by the general descriptive term argillic.

With the less precise terms, such as propylitic and argillic alteration, an expression of intensity of metasomatism or the predominant elements added to the rock is useful as a second criterion to separate similar assemblages resulting from drastically different fluids. In such cases, designation as weak, moderate, or strong alteration is useful to indicate the extent to which relics of the original rock are preserved and the extent of metasomatic effects in the rocks.

Alteration Types in Aluminosilicate Rocks

Meyer and Hemley (1967) have classified alteration in most aluminosilicate rocks into propylitic, intermediate argillic, advanced argillic, sericitic, and potassic types. Typical mineral assemblages of these types are summarized in Figure 5.14. Essential features of the types may be summarized as follows:

Propylitic. Presence of epidote and/or chlorite, and a lack of appreciable cation metasomatism or leaching of alkalies or alkaline earths; H_2O, CO_2, and sulfur may be added.

Intermediate argillic. Presence of important amounts of kaolinite, montmorillonite, or amorphous clay, principally replacing plagioclase; sericite may accompany clays; potassium-feldspar unaltered or argillized; appreciable leaching of calcium, sodium, and magnesium.

Advanced argillic. All feldspar converted to dickite, kaolinite, pyrophyllite, diaspore, alunite, or other aluminum-rich phases.

Sericitic. Both potassium-feldspar and plagioclase converted to sericite, minor kaolinite may be present.

Potassic. Potassium-feldspar and/or biotite formed as an alteration of plagioclase or mafic minerals.

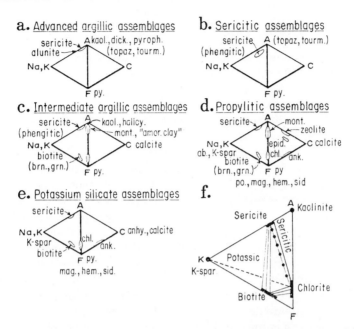

Fig. 5.14 (a–e) Common minerals occurring in the major types of alteration in aluminosilicate rocks; quartz usually present (after Meyer and Hemley, 1967). At higher temperatures, andalusite would occur in the advanced argillic type. (f) Suggested separation of potassic and sericitic types based on stable assemblages (after Rose, 1970). Dots on chlorite-sericite tieline indicate separation of types.

If sericitization of potassium-feldspar, as well as plagioclase, is made a criterion of sericitic alteration, as shown in the diagrams but not in the text of Meyer and Hemley (1967), sericitic alteration becomes identical to the quartz-sericite alteration of Rose (1970) and the phyllic alteration of Lowell and Guilbert (1970). Unfortunately, most workers have based names on abundance of minerals rather than mineral assemblages, so that sericitic, intermediate argillic, propylitic, and potassic alteration are not always distinguishable in their usage.

ALTERATION IN HOT SPRINGS AND EPITHERMAL ORE DEPOSITS

Investigations of alteration at hot springs and geothermal systems have defined at least two distinct environments and types of alteration. Beneath the water table, the hydrothermal fluid is usually near neutral or weakly alkaline, with chloride as the major anion. Alteration products in felsic rocks are feldspars, micas, zeolites, silica minerals, montmorillonite, and occasionally kaolinite. In contrast, at other hot springs and geothermal systems, near-surface fluids are acid, have high SO_4^{2-}/Cl^-, iron, magnesium, calcium, and aluminum, and relatively low sodium and potassium (Raymahashay, 1968; White et al., 1971). Alteration is of the advanced argillic type (kaolinite, alunite and other sulfates, sulfur, silica minerals, and zunyite).

In the first environment, alteration by hydrothermal fluids just beneath the water table at relatively low temperatures (up to approximately 150°C) commonly produces assemblages containing zeolites and clays, as at Wairakei and other hot springs in New Zealand, at Yellowstone Park, and in Iceland (Ellis, 1967). The alteration products are those expected for very low temperature "metamorphism" of felsic rocks, combined with some leaching of calcium, sodium, and magnesium. However, beneath this clay-zeolite zone, alteration to sericite, adularia or albite, along with chlorite, epidote, wairakite, and other zeolites, is common in potassium-bearing rocks.

Deep hot water in geothermal systems in felsic igneous and sedimentary rocks is usually approximately in equilibrium with quartz, potassium-feldspar, potassium-mica, albite, chlorite, pyrite, and epidote or zeolite (wairakite or laumontite) (Ellis, 1970). As this water approaches the surface, three main processes may change the character of alteration and mineral deposition: cooling, boiling, and condensation. Cooling results in supersaturation and possibly precipitation of quartz (or some other form of silica) and, in the higher temperature range, a tendency to form additional potassium-mica from feldspars. At temperatures below about 150°C (but dependent on the chemistry of the water), montmorillonite, mixed-layer illite, and mordenite apparently form as the result of cooling, although effects of mixing and dilution with shallow waters and condensation of volatiles from below may contribute to forming these products.

Boiling selectively removes CO_2, H_2S, and other volatile compounds from the liquid, thereby increasing pH and K/H. Potassium-feldspar, calcite, and perhaps wairakite or epidote form either by precipitation or by alteration of preexisting micas or plagioclases. In contrast, condensation of CO_2-bearing steam into hot water in equilibrium with feldspars and potassimum-mica has the reverse effect of lowered pH and formation of potassium-mica, waiarkite, and albite or montmorillonite.

In vapor-dominated hydrothermal systems, inferred by White et al. (1971) to exist at the Geysers, Larderello, and locally at Yellowstone Park and Steamboat Springs, vapor boiled from the water table of a deep body of hot water moves upward and condenses as it moves through cooler rock toward the surface. Oxidation of H_2S to H_2SO_4 (and native sulfur) gives rise to strong argillic alteration to clays, alunite, and other minerals of the advanced argillic suite. Chloride is low in such vapors and waters because of its low volatility in near-neutral waters; sulfate is the major anion, balanced by H^+, Ca^{2+}, Mg^{2+}, and other constitutents leached from the rock. The larger pores and channels are filled with vapor, but the smaller pores and channels are filled with condensate, at least some of which flows back downward to the water table.

At Steamboat Springs, material above the water table is a highly leached mass of opaline silica, apparently because of the strongly acid environment from oxidation of H_2S (Schoen et al., 1974). At the water table, alunite becomes abundant and is underlain by increasing amounts of kaolinite, gradually changing downward to normal montmorillonite-potassium-mica alteration. Apparently the hot water for a few tens of feet below the water table is appreciably affected by the descending acid condensate and high content of sulfate. Differences in rate and type of recharge and discharge, heat supply, permeability, rate of oxidation of H_2S, and other features may result in wide variability in such systems.

A third possible type of hot spring water is formed from absorption of volcanic H_2S, SO_2, CO_2, and other constituents by a deep ground water, as suggested by White (1957) and Ozawa et al. (1973). The latter authors describe extremely acid (pH 1.2-2.0) hot springs of moderate discharge which they attribute to absorption of SO_2- and HCl-rich gases at depth, and subsequent self-oxidation of SO_2 to H_2SO_4. Deep advanced argillic alteration at Butte, Cerro de Pasco, and other districts may have a similar origin.

Near-surface advanced argillic alteration, which has been described from the San Juan Mountains, Colorado; Steamboat Springs, Nevada; Marysvale, Utah; Goldfield, Nevada; and Cochiti, New Mexico (Meyer and Hemley, 1967); has also been termed solfataric or acid hot spring alteration. At Cochiti and Goldfield, the alteration around veins shows lateral zoning from propylitic at the fringes through argillic to sericitic to advanced argillic, with alunite accompanying the clay and sericite.

ALTERATION AROUND VEINS IN GRANITIC ROCKS

The earliest thorough descriptions of alteration and the most detailed considerations of the mechanism of hydrothermal alteration have been made of alteration around veins in granitic rocks, so characteristics and mechanisms of such alteration are worth examining in detail. Alteration around veins at Butte, as described by Sales and Meyer (1948, 1950), has already been summarized. An earlier classic study is that of Lovering (1941) on the tungsten veins of Boulder County, Colorado. Further work on veins in this region (Bonorino, 1959) has added details with more modern techniques.

In all these districts, the mineral assemblage changes progressively from fresh or propylitically altered rock through an argillic zone to a sericitized zone to the vein, although distinct differences occur in detail; for instance, Bonorino (1959) recognized six different zoning patterns in deposits of the Colorado Front Range. In many cases, the outer argillic zone contains montmorillonite and the inner argillic zone kaolinite.

At Butte, Sales and Meyer (1948, 1950) first demonstrated that the zones developed simultaneously by outward growth of several alteration fronts fed by a single type of hydrothermal solution flowing through the vein. Their evidence was the fact that all Main-Stage veins at Butte show the same arrangement of alteration zones, even small late veins cutting across early, more strongly mineralized veins. If the argillic and sericitic zones were formed by two chemically different fluids at different times, one would expect to find examples of fractures open only during one of the two stages, or examples of one type cutting across the other, but no such examples have been reported. Veins enclosed by only one type of alteration are found only where the inner alteration envelopes have coalesced between veins to eliminate the outer zones. In addition, within a given segment of the district, relative sizes of the alteration zones are similar from vein to vein, although regular vertical and lateral trends in relative width are observed (Meyer and Hemley, 1967), and total width of alteration around a given vein is greatest in the Central Zone in the vicinity of the 2800 level.

The importance of diffusion in transferring material within the alteration envelope is also demonstrated at Butte. Calcium, sodium, and magnesium are depleted from altered rocks of the sericitic and argillic zones but are not detectably enriched in adjacent zones of fresh rock, even where these rocks occur as residual unaltered "islands" completely surrounded by altered rock. Therefore, calcium, sodium, and magnesium must have diffused toward the veins and have been carried away by the flowing solution at the same time that H^+, S, and CO_2 diffused outward from the vein to form sericite, clays, pyrite, and carbonate. Although transfer by fluid flow clearly occurred in the veins, and minor flow may have occurred in the wallrock, the two-directional movement of material in forming the alteration envelopes requires diffusion to predominate across most of the thickness of the alteration envelope, a distance of perhaps as much as 10 m near some large veins.

At temperatures of 200–300°C, the assemblage potassium-feldspar-kaolinite should react to form muscovite-quartz, as indicated in Figure 5.10a, but this assemblage is common at Butte and elsewhere. The expected reaction is

$$\text{KAlSi}_3\text{O}_8 + \text{Al}_2\text{Si}_2\text{O}_5(\text{OH})_4 \rightleftharpoons \text{KAl}_3\text{Si}_3\text{O}_{10}(\text{OH})_2 + 2\text{SiO}_2 + \text{H}_2\text{O}$$
K-feldspar   kaolinite   muscovite

(5.20)

for which

$$K = \frac{a_{\text{mus}} a_{\text{SiO}_2}^2 a_{\text{H}_2\text{O}}}{a_{\text{Kf}} a_{\text{kaol}}}$$

This reaction goes to the right in 1–2 m chloride solutions at 1–2 kbar pressure (Figure 5.10a) and, if quartz is assigned unit activity of SiO_2, has K of about $10^{1.6}$ at 300°, based on combining equations (5.13) and (5.14). Potassium feldspar and kaolinite might coexist if activities of the natural phases differ enough from the standard states. Based on data summarized by Holland and Malinin (Chapter 9), solubility of amorphous silica is approximately three times that of quartz at 300°C, and hydrothermal solutions in some hot springs are in approximate equilibrium with amorphous silica (Fournier and Rowe, 1966), suggesting that a_{SiO_2} could reach about 3 relative to quartz. Fournier (1967, p. 222) suggests that a_{SiO_2} may reach 3.4 at 300°C. The activity of KAlSi_3O_8 in an alkali feldspar with 50–100% KAlSi_3O_8 at 300°C is 1 ± 0.1 (Helgeson, 1974). Activity of kaolinite may be somewhat less than unity owing to fine grain size and poor crystallinity, but an activity of less than 0.5 seems unlikely. Combination of these factors leaves muscovite and amorphous silica stable by about $10^{0.2}$ in log K. Alternatively, kaolinite-potassium-feldspar assemblages may form at temperatures lower than 300°C. At such temperatures, kaolinite-potassium-feldspar is more nearly stable. Other alternatives are that the association results from supersaturation of the fluid with sericite, local inhomogeneities in the fluid, or metastable persistence of potassium-feldspar. Appreciable concentration gradients between the solution within altering plagioclase crystals and the solution adjacent to orthoclase are possible at Butte, but in view of the intergrowths of kaolinite and potassium-feldspar observed by Bonorino (1959) in the argillic zone of the Front Range area, such gradients are not a general explanation.

Adjacent to Main-Stage veins at Butte, potassium-feldspar is stable in fresh rock. It is altered to sericite in the sericitic zone and to kaolinite or other aluminum silicates immediately adjacent to the veins in advanced argillic alteration in parts of the Central Zone. The potassium/hydrogen ratio of the fluid causing the alteration must have been lowest near the vein and increased away from it in a manner such that, within zones, the appropriate mineral assemblages were stable (or nearly stable), and the concentration gradient between alteration fronts was adequate to supply reactants and remove the products of the alteration reactions. Applying Fick's first law, $J = -D(dc/dx)$, where J is the flux (mass of the constituent passing through unit area in unit time), D is the diffusion coefficient, c is concentration, and x is distance, it can be seen that, assuming D is constant throughout the alteration envelope, the flux is proportional to the concentration gradient. Because constituents are being consumed or produced in alteration reactions at the zone boundaries, the flux on the side toward the vein must be greater than on the side away from the vein, and the concentration gradients must be correspondingly steeper toward the vein. Plagioclase decomposes to montmorillonite in the outer argillic zone, to kaolinite in the inner argillic zone, and is converted to sericite and locally back to kaolinite adjacent to the vein. A progressive increase in Na^+/H^+ and $Ca^{2+}/(H^+)^2$ away from the vein is thus indicated. Depletion of silica from the argillic zone despite the apparent stability of quartz suggests that the alteration to clay results in supersaturation of silica in this zone (barring complications such as "uphill diffusion").

Advanced argillic alteration as a zone bordering veins in the Central Zone extends to depths of at least 3800 feet at Butte and contains traces of alunite. Origin of this alteration in a simple vapor-dominated near-surface system does not seem reasonable. The presence of alunite suggests that the fluid contained appreciable concentrations of sulfate. Meyer et al. (1968) suggest that ionization of H_2SO_4 on cooling of the upward-flowing solutions, combined with dissociation of HCl and a lack of neutralizing feldspar in the strongly sericitized wallrock, allowed development of the advanced argillic alteration in a hypogene environment. Sulfate concentrations on the order of 10^{-2} m and pH values of 2-3 are required to produce alunite at temperatures of approximately 300°C (Figure 5.13). Figure 5.9 suggests a temperature of 300-350°C for the abundant pyrophyllite.

ALTERATION AT PORPHYRY COPPER DEPOSITS

At porphyry copper deposits, the mineralogy and character of hydrothermal alteration is typically zoned laterally over hundreds to thousands of feet (Figure 5.15), commonly exhibiting a crudely concentric pattern of alteration types (Lowell and Guilbert, 1970; Rose, 1970). In some districts, vertical zoning of alteration is also documented. However, closer study indicates that the peripheral zones of sericitic and argillic alteration are at least partly later in age than the central potassic alteration and that a wide variety of geometric relations exist.

In a considerable number of the described deposits, the central zone exhibits predominantly potassic alteration, in some cases centered on a por-

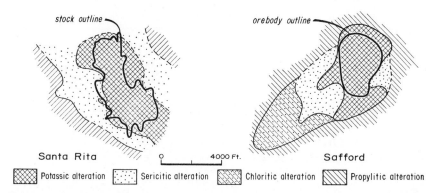

Fig. 5.15 Distribution of major alteration types at Santa Rita, New Mexico and Safford, Arizona (after Rose, 1970).

phyritic intrusive. The potassic zone is at least partly surrounded by a zone of predominantly sericitic alteration, grading outward to predominantly propylitic alteration (Lowell and Guilbert, 1970; Rose, 1970). Argillic alteration of plagioclase occurs within the potassic zone in some deposits (Nielsen, 1968; Fournier, 1967) and as a separate zone outside the sericitic zone in others. On detailed examination, local areas of one type of alteration commonly occur in other zones. A relatively unaltered core zone is enclosed by the potassic zone or lies beneath it at some deposits. The central zone of potassic alteration usually contains the best hypogene ore. Propylitic alteration generally has affected at least some of the surrounding rocks. The San Manuel-Kalamazoo deposit in Arizona forms the best-described example of this pattern of alteration (Figure 5.16a).

At other deposits, such as El Salvador in Chile (Gustafson and Hunt, 1975) and Yerington, Nevada (M. Einaudi, personal communication), sericitic alteration cuts across and largely overlies the other zones and is later than most potassic alteration (Figure 5.16b). Early andalusite-K-spar and andalusite-sericite assemblages also tend to overlie potassic and propylitic assemblages at El Salvador, and late advanced argillic alteration (pyrophyllite, diaspore, alunite) occurs in a still shallower zone.

In addition to potassic, sericitic, and propylitic alteration discussed above, considerable alteration at porphyry copper deposits is formed by supergene processes related to weathering and enrichment near the surface. Some kaolinite in and near the zone of chalcocite enrichment at Santa Rita has an isotopic composition consistent with formation at near-surface temperatures from normal surface waters of the area (Taylor, Chapter 6). Some features of sericite in leached cappings at porphyry copper deposits suggest an origin by supergene processes (Rose, 1970). In

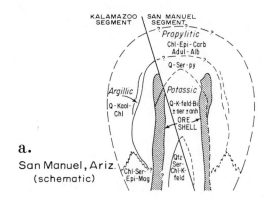

a. San Manuel, Ariz. (schematic)

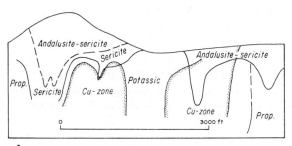

b. El Salvador, Chile (schematic)

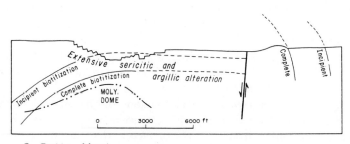

c. Butte, Mont.

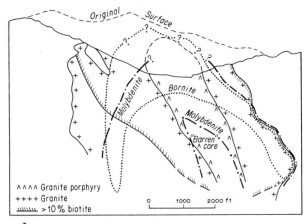

d. Bingham, Utah

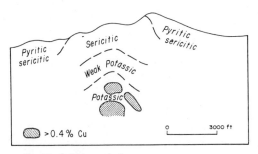

e. Red Mtn., Ariz.

Fig. 5.16 Vertical distribution of alteration and mineralization at porphyry copper and related deposits, after Lowell and Guilbert (1970), Gustafson and Hunt (1975), Corn (1975), John (1975), and Meyer et al. (1968).

most cases paragenetic evidence for the relative age of clays is ambiguous or misleading, and distinction of clays of different origin is difficult. Because kaolinite is not stable with quartz above about 300°C, but fluid inclusions indicate appreciably higher temperatures for most quartz-sulfide veins, the kaolinite and perhaps some montmorillonite are probably late retrogressive features, or may be supergene.

Mineralogy and Conditions in the Potassic Zone

In relatively felsic rocks, such as granitic rocks, potash-rich orthoclase (Or_{85-90}) typically replaces plagioclase and occurs in small veinlets with quartz and sulfides. Microcline, rather than orthoclase, is reported at a few deposits (Jones, 1974; Hollister et al., 1974). Biotite, greenish sericite, anhydrite, siderite, and sulfides are common associated minerals.

In more mafic rocks, such as andesites and diorites, biotitization of mafic minerals is the major evidence of potassic alteration. Chlorite, epidote, albite, and actinolite are found in some potassically altered dioritic rocks. Andesites usually alter to a nearly black rock with fine-grained, disseminated, brown biotite. Coarser-grained igneous rocks generally preserve some textural evidence of the original mafic minerals of the rock, but biotite may replace plagioclase and other minerals, as well as mafic minerals. The mica is generally richer in magnesium in the altered rock than igneous biotite. For instance, Moore and Czamanske (1973) find average Mg/(Mg + Fe) ratios of 0.76 for hydrothermal biotite at Bingham compared to 0.64 in unaltered igneous rocks. Beane (1974) calculates temperatures of 350–550°C from the composition of hydrothermal biotite coexisting with potassium-feldspar and magnetite at Santa Rita, New Mex., and estimates f_{O_2}, f_{S_2}, and other chemical parameters.

Small to moderate amounts of sericite commonly accompany alteration of plagioclase to orthoclase, indicating that K/H is on the sericite-K-spar boundary. Plagioclase in biotitized rocks may be unaltered, albitized, or converted partly or wholly to montmorillonite ± kaolinite, anhydrite, or biotite. Chlorite coexists with orthoclase at some deposits (Ajo) to the exclusion of biotite or as a partial alteration of earlier biotite. The assemblage chlorite + potassium-feldspar is restricted to the lower greenschist facies (Winkler, 1974), suggesting relatively lower temperatures for such alteration than in deposits containing muscovite + biotite (but the reaction is clearly dependent on Mg/Fe ratio). At Ray, Arizona, and other deposits, epidote occurs with hydrothermal potassium-feldspar in the central zone of a porphyry copper deposit. Andalusite with potassium-feldspar at El Salvador (Gustafson and Hunt, 1975) and Butte (Brimhall, 1977) suggests temperatures above about 550°C.

Concentrations of some chemical constituents in fluids forming porphyry copper deposits can be estimated from the stability of orthoclase, plus information from fluid inclusions. At 400°C, near the lower end of the range of filling temperatures found for fluid inclusions associated with potassic alteration at porphyry copper deposits (Roedder, Chapter 14), the K/H of fluids in equilibrium with orthoclase is 500 or greater. Most of the fluid inclusions contain NaCl crystals, indicating greater than 26% or 6 m NaCl. At 400°C, the Na/K of fluids equilibrated with orthoclase and plagioclase is 6.2 (Meyer and Hemley, 1967). From these data, a potassium content of approximately 1 m can be estimated, giving H^+ + HCl approximately $10^{-2.7}$ m. Because HCl is only weakly ionized under these conditions, the pH is considerably higher (4.1, according to calculations of Montoya and Hemley, 1975), but as the fluid cools, this acid will dissociate and cause hydrolytic alteration.

The oxides and sulfides in the potassic zone also help define the conditions of alteration and ore deposition. Chalcopyrite and molybdenite are generally accompanied by either pyrite or bornite, as well as rutile and occasionally hematite (Gustafson and Hunt, 1975; Field et al., 1974; Guilbert and Lowell, 1974) or magnetite (Phillips et al., 1974). Pyrrhotite is rare but present in a few atypical porphyry copper deposits (Middle Fork, Patton et al., 1973) and in skarns associated with porphyry copper deposits at Ely and Battle Mountain, Nevada. Calcite, siderite, dolomite or ankerite, and anhydrite are present in many potassic zones and apatite in some potassic zones (Gustafson and Hunt, 1975; Sillitoe, 1973). Presence of pyrite + chalcopyrite, accompanied by anhydrite and locally hematite, and rarity of pyrrhotite, magnetite, and pyrite-bornite, suggest relatively high f_{O_2} and f_{S_2} and, at temperatures of 500–600°C, an appreciable proportion of oxidized sulfur in the fluid. However, at 300°C, reduced sulfur appears to predominate.

Sericitic Zone

Nonsilicates associated with the pervasive quartz, sericite, pyrite, and locally kaolinite include calcite and other carbonates, anhydrite (Gustafson and Hunt, 1975; Corn, 1975), rutile, and apatite (Lowell and Guilbert, 1970). At nearly all deposits, pyrite, largely in veinlets, is most abundant in the sericitic zone, amounting to 5–20% of the rock; and chalcopyrite is less abundant than pyrite, especially in the outer part of the zone (Rose, 1970). In some cases, as at Ely and Santa Rita, sericitic envelopes around veins are bordered by argillic alteration (Fournier, 1967), but in the central part of the sericitic zone, the sericitic envelopes around veins have usually coalesced to eliminate argillic envelopes. Veins and fractured zones with strong sericitic alteration cut through the potassic zone and extend out into the propylitic zone at some deposits, making the boundaries of the sericitic zone very irregular in detail. The cross-cutting relations suggest that sericitic alteration either followed the other types (Gustafson and Hunt, 1975) or persisted longer.

Propylitic and Other Types of Alteration

Alteration to chlorite and epidote tends to be stronger near the inner edge of the propylitized zone but grades outward to merge with normal deuteric, metamorphic, and diagenetic phenomena. At some deposits, propylitic alteration affects a zone several miles in width outside the sericitic zone, but at others, is barely noticeable. In tonalite at the Big Bug district in Arizona, Sturdevant (1975) reports zoning within the propylitic zone from an inner chlorite-calcite-clinozoisite-sericite-magnetite-pyrite assemblage to a chlorite-rich zone and then an outer epidote-rich zone, but the latter having the assemblage chlorite-epidote-actinolite-sericite-magnetite. Along a traverse across these zones, chlorite becomes more iron-rich outward, changing in atomic ratio of $Fe/(Fe + Mg)$ from 0.42 at the inner edge of the propylitic zone to 0.56 at the outer edge.

Other types of alteration at porphyry copper deposits include strong chloritic alteration with abundant pyrite, which appears to proxy in part for sericite-pyrite alteration in some deposits in andesites and related rocks (Rose, 1970; Hollister et al., 1975). Actinolite has been recognized as a major component of the inner propylitic zone at Bingham (John, 1975) and the Big Bug district, Arizona (Sturdevant, 1975). Strong silicification and quartz veining are present in the central zone of some deposits (Ajo, Valley Copper, Santa Rita). Pyrophyllite, along with diaspore, alunite, and other aluminous minerals, forms a prominent part of the alteration at El Salvador.

Sequence of Veins and Alteration

In general, early veins contain quartz, potassium-feldspar, biotite, molybdenite, bornite, chalcopyrite, and, in several cases, magnetite, commonly accompanied by potassic alteration or little alteration, indicating a fluid compatible with magmatic minerals. A stockwork of barren quartz veins is commonly recorded as an early phase. Late veins, at most deposits, are pyrite-rich with sericitic and argillic alteration envelopes. Careful study of the age relations of igneous bodies, veins, and alteration envelopes is essential in improving our understanding of porphyry copper deposits, because it is clear that a complex sequence of intrusive and mineralizing events has produced the present deposits.

Vertical Zoning of Alteration

At San Manuel-Kalamazoo, Arizona, which has been rotated nearly on its side to expose at least 8000 ft of original vertical extent, the strongest alteration to sericite and pyrite is near the top of the known ore (Figure 5.16a). At depth, sericite and pyrite decrease in abundance, and chlorite and magnetite increase. In the potassic alteration zone, the assemblage quartz-K-spar-biotite-sericite-sulfides is found in a shallower zone, but quartz-K-spar-chlorite-sericite-sulfides-magnetite in the deeper zone.

At Red Mountain, Arizona (Corn, 1975), a sericitic zone with quartz-sericite-pyrite and minor alunite, accompanied by enargite-pyrite veins, grades downward over several thousand feet into weak potassic alteration with quartz-chlorite-biotite-pyrite-chalcopyrite-magnetite and then into stronger potassic alteration with quartz-K-spar-biotite-anhydrite-sericite-chalcopyrite-magnetite (Figure 5.16c). A wide halo of propylitic alteration surrounds the sericitic and potassic alterations. The best grades of copper accompany the potassic alteration.

At the Henderson porphyry molybdenite deposit in Colorado, the ore body is characterized by strong silicification, quartz veining, and potassic alteration at a depth of 3000 ft below the surface (Wallace et al., 1978). Above the ore are, successively, a quartz-topaz-magnetite zone, a quartz-sericite-pyrite zone, a spessartite-bearing zone, and an argillic zone with unaltered potassium-feldspar. A wide propylitic halo surrounds the strong alteration. A greisen zone and an argillic zone lie beneath the ore.

At Butte, early biotitization affected a deep dome-shaped zone, and molybdenite is concentrated beneath this zone (Brimhall, 1977). Main-Stage veins with their sericitic and argillic alteration envelopes pervasively alter the rock in and above the center of the biotite- and molybdenite-rich zone (Figure 5.16c). At Bingham, a core of weak alteration and low sulfides is surrounded by and overlain by a zone of abundant molybdenite (John, 1975). Biotitization tends to increase upward (Figure 5.16d) but is partly controlled by the rock type. At El Salvador, a zone of late strong sericitic alteration and pyritization overlies and surrounds a core of earlier pervasive potassic alteration with chalcopyrite and bornite (Gustafson and Hunt, 1975). Advanced argillic alteration occurs in the upper part of the sericitic zone. At Valley Copper, a deep zone of feldspathic alteration is overlain by argillic and sericitic alteration (Jones, 1974).

Gustafson and Hunt (1975) suggest, based on the paragenesis plus isotopic data and fluid inclusions, that the early potassic alteration and accompanying copper-molybdenum sulfides are of primarily magmatic hydrothermal origin. Later circulating ground waters heated by the intrusive body deposit the abundant pyrite and sericite and redeposit some copper sulfides. Most porphyry deposits show these two types and stages of alteration but to varying extents and with varying geometric relations.

High salinity of fluid inclusions, accompanied by vapor-rich inclusions (Roedder, Chapter 14), suggests that boiling was occurring in the potassic and sericitic zones, thereby creating potassic alteration, as at the hot springs discussed above. The extent and timing of mixing magmatic and meteoric waters remain to be worked out. Differences in permeability of country rock, structural controls, intrusive history, and timing of the two types of waters, in combination with mixing of waters at many of the best-zoned deposits, seem capable of explaining variations from one district to another.

ALTERATION IN SKARN DEPOSITS

"Contact," "contact metamorphic," "contact pneumatolytic," "contact metasomatic," "igneous metamorphic," "hydrothermal metamorphic," and "pyrometasomatic" are all terms that have been applied to the skarn type of replacement deposit in limestone or dolomite. The characteristic feature of this type of deposit is not the metals mined, nor the proximity to an intrusive, but rather the gangue, a zoned, coarse-grained, generally iron-rich mixture of calcium or magnesium silicates termed "skarn" or "tactite." "Skarn," being the older term, is generally preferred, especially outside the United States.

The word "skarn" itself is commonly used in two connotations. During progressive regional metamorphism, local diffusion and reaction between carbonate beds and pelitic schist or another suitable rock may produce relatively small "bimetasomatic diffusion" or "reaction" skarns that have been termed "calc-silicate bands" by Vidale (1969). Much more extensive "replacement skarns" that form a type of wallrock alteration in ore deposits are discussed here. Unzoned calc-silicate hornfels may be associated with the skarn but generally does not share the iron- and manganese-rich minerals and zoning of skarns. If the original wallrock was impure enough, a hornfels can grade into a skarn as the intrusive or ore-bearing zone is approached, although in pure carbonate wall rocks, the skarn boundary is typically very sharp. A distinct wollastonite zone on the outer part of a zoned skarn is, of course, considered part of the skarn.

Structures

Most skarn deposits belong to one of the following three types, depending on their structural relation to an intrusion (Figure 5.17): (1) skarn around intrusion, (2) intrusion around skarn, or (3) no intrusion exposed. Most skarns belong to the first type, that is, skarns replacing carbonates near relatively small granitic plutons or dikes (which themselves may or may not be mineralized). In the second type, skarns replace xenoliths or roof pendants in large plutons, as in some of the tungsten skarns of the Bishop district, California (Bateman, 1965) or the copper skarns of the Alder Creek (White, Knob, MacKay) district, Idaho (Nelson and Ross, 1968). The third type corresponds to skarn along veins, as in the Linchburg

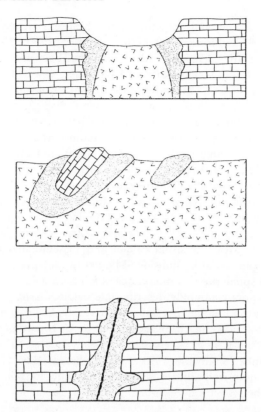

Fig. 5.17 Relations of skarn to intrusive. (a) Skarn around intrusive. (b) Intrusive around skarn. (c) Skarn without intrusive.

mine, New Mexico (Titley, 1961). As indicated on Figure 5.17, the first two types commonly have distinctive topographic expressions.

Magnesian and Calcic Skarns

Skarns that replace dolomite tend to be magnesian (forsterite, serpentine, talc, actinolite, humite group minerals, etc.), and commonly are mined for magnetite, as in the Eagle Mountain district, southern California (Dubois and Brummett, 1968), or exceptionally, for magnesium, as in the Gabbs magnesite-brucite deposit, Nevada (Schilling, 1968). Skarns replacing pure limestone tend to be more calcic, with iron entering calcium-iron or calcium-manganese silicates (andraditic garnet, hedenbergitic or johannsenitic pyroxene, ilvaite, ferroactinolite, etc.). Dolomitic and argillic limestones yield more complex mineralogies.

Calcic skarns commonly are mined for tungsten (scheelite), as in the Pine Creek mine, California (Bateman, 1965; Gray et al., 1968), or copper, as in the Twin Buttes district, southern Arizona (Lacy and Titley, 1962), or at Santa Rita, New Mexico (Rose and Baltosser, 1966). Calcic skarns are occasionally also of economic importance as sources of iron or molybdenum, or of lead and zinc, as in the Hanover (Central) district, New Mexico (Hernon and Jones, 1968), where iron-rich sphalerite is characteristically associated with manganiferous hedenbergite, johannsenite, and ilvaite.

Depending on original stratigraphy, magnesian and calcic skarns may be interbedded. Perry (1969), for example, describes interbedded forsteritic and andraditic skarns from the Christmas Copper mine, Arizona. In detail, the magnesian skarns consist of forsterite (largely replaced by serpentine), magnetite, and anhydrite, with variable talc and sulfides (chalcopyrite, pyrite, bornite, and sphalerite). The calcic skarns consist of fractured andraditic garnet, with interstitial calcite and disseminated bornite and/or chalcopyrite, and variable diopside. Magnetite and pyrite occur near the intrusive, and sphalerite is common near the limestone.

Russian authors commonly describe late calcic skarns superimposed on magnesian skarns ("apomagnesian calcic skarns," etc.), but this feature has, to our knowledge, not been recognized by U. S. authors.

The magnesian type of skarn is exhaustively treated in two recent Russian monographs (Shabynin, 1973; 1974), whereas Burt (1972a) similarly treats the calcium-iron silicate type of skarn and its relation to mineral stabilities in the system Ca-Fe-Si-C-O-S-H-F-W-Cu-Zn.

Zoning

Possibly the earliest recognition of a silicate zoning sequence in skarns was by G. Vom Rath (1868, p. 343-345), who described a generalized zoning pattern quartz porphyry dike/ilvaite/hedenbergite/marble in the ancient copper-lead-zinc mine of Temperino, near Campiglia Marittima, Italy. In places a zone of epidote-rich endoskarn replaces the porphyry next to the ilvaite, and elsewhere a zone of quartz is developed between the hedenbergite and the limestone. A similar zoning pattern is still visible at Temperino (Bartholomé and Evrard, 1970), with a magnetite zone up to a meter wide developed between ilvaite and porphyry in the deeper levels of the mine. Most of the copper (as chalcopyrite) is localized in the ilvaite zone (up to 5 to 10 m wide), whereas galena and sphalerite are localized in the hedenbergite zone (up to 30 m wide). Because the replaced rock was monomineralic limestone, perhaps it is not surprising that the skarn zones are nearly monomineralic. Minor amounts of other minerals (quartz, fluorite, calcite, pyrite, pyrrhotite, arsenopyrite, and andradite) also occur, mainly in the ilvaite zone. The zoning sequence at Temperino can be modeled on a μ_{Ca}-μ_{Fe}-μ_{Si} diagram (Figure 5.4c, after Burt, 1972a) similar to Figure 5.4b. The striking concentric and radiating structure developed in the fibrous hedenbergitic clinopyroxene from Temperino is typical of pyroxenes from many calcium-iron-silicon skarn deposits, as is the fact that the MnO content of Temperino clinopyroxene increases from 1 to 15% as the marble is approached (Bartholomé and Evrard, 1970).

Fukuchi (1907) described a number of similar zoning sequences in Japanese skarns, and his descriptions have been supplemented by numerous later Japanese studies (Watanabe, 1960, Ito, 1962; Tokunuga, 1965; Shimazaki, 1969). U.S. authors, in general, have been slow to recognize silicate zoning and instead have concentrated on the sequence of ore deposition.

The variety of generalized zoning sequences that may be developed in calcium-iron-silicon exoskarns has recently been summarized (Burt, 1974). The three common calc-silicates garnet, clinopyroxene, and wollastonite are usually zoned in the same order in which they are listed. The outer wollastonite zone, where present, is sometimes barren of ore (except locally, gold) possibly because wollastonite tends to be incompatible with iron-rich ore-bearing solutions, according to andradite-producing reactions such as

$$6CaSiO_3 + 5CuFeS_2 + 3H_2O \rightleftharpoons 2Ca_3Fe_2Si_3O_{12} + Cu_5FeS_4 + 3H_2S$$
wollastonite   chalcopyrite   andradite   bornite

(5.21)

Under H_2S-rich conditions, the reaction goes to the left, and chalcopyrite ore occurs with wollastonite. (Note that this H_2S-buffering reaction can also be written as an SO_{-1} buffer, using the exchange operator "shorthand").

It should be emphasized that, as observed in the field, skarn zones are not really monomineralic, due to the presence of preskarn contact metamorphic minerals, ore minerals, and retrogressive skarn alteration and breakdown products. The most important controls on zone development are structure (fracture systems) and stratigraphy (chemistry of the beds being replaced), but given these, the gross distribution of minerals commonly reflects chemical gradients.

Volume Relations

On the basis of bedding thickness measurements in replaced and nearby unreplaced carbonate rocks in the Bingham district, Utah, Lindgren (1912,

1925) proposed that the formation of skarn and other metasomatism proceeds with little or no change in the volume of the replaced rock (although the porosity might change). This hypothesis came to be called "Lindgren's Law," or "the constant volume hypothesis" (Ames, 1961). It had earlier been recognized that decarbonation during contact metamorphism must result in a considerable loss of volume (Barrell, 1902; Cooper, 1957).

The constant volume hypothesis, neglecting possible changes in porosity, has been immensely popular with geologists who wish to calculate quantitatively the gains and losses (exchanges of material between solution and wallrock) involved in skarn formation. Without the assumption that volume is preserved, or an analogous assumption that some other parameter, such as moles of titanium, moles of aluminum, or the titanium/aluminum molar ratio, is preserved, such calculations are impossible. If nothing is conserved during skarn formation, of course, the calculations become quantitatively meaningless, although they may still be "representative" of the replacement process.

A complicating factor is that if volume rather than pressure is the thermodynamic variable controlling replacement, then replacement reactions should be rewritten to balance molar volumes, rather than merely molar quantities of reacting minerals (Lindgren, 1912; Ridge, 1949). The conventional "phase rule" must then be modified to allow for this unconventional restraint (Korzhinskii, 1970; Thompson, 1970).

On the other hand, if pressure solution surfaces (stylolites), commonly observed in metamorphosed carbonate rocks adjacent to skarn, can form in response to volume losses caused by their lateral replacement, then bedding thickness measurements (Lindgren, 1924; Schmitt, 1939) become useless as indicators of volume preservation. Recent careful measurements of bedding attitudes and thicknesses have in fact indicated that slumping can occur over hydrothermal ore bodies in carbonate rocks (Hewitt, 1968; Perry, 1969). "Lindgren's Law" should therefore be regarded as no more than a hypothesis.

Genesis

Numerous studies suggest the following as stages in the evolution of a "typical" skarn deposit:

1. Shallow intrusion of a granitic magma at 900–700°C into carbonate sediments. The magma can also be mafic, as at Cornwall, Pennsylvania (Lapham and Gray, 1973). In some cases, it might be intruded following a wave of skarn-forming solutions, although direct evidence for this sequence is generally lacking.

2. Contact metamorphism at 700–500 °C (with concomitant crystallization of the magma). The magma may react with dolomite, forming "skarns of the magmatic stage," but it will generally undergo little reaction with limestone (Pertsev, 1974). Impurities in the marble react to give light-colored calc-silicates plus CO_2 and H_2O, provided that there is a passage for escape of these volatiles. Loss of volatiles causes a net volume loss and/or increase in the porosity of the rock, preparing the gound for later skarn formation and ore deposition.
3. Metasomatism and iron-rich skarn formation at 600–400 °C or lower. Magmatic crystallization at depth may provide the initial skarn-forming fluid, with important late contributions being made by meteoric waters (Burnham, Chapter 3; Taylor, Chapter 6). The properties of this fluid appear to fluctuate, possibly leading to the distinctive compositional growth zoning usually observed in anisotropic garnets of skarns. With time, the fluid seems to become progressively enriched in sulfur and the ore metals. Skarn formation proceeds outward into the carbonate wallrock (yielding exoskarn) and, from calcium acquired by the fluid, into the solidified intrusion (yielding endoskarn). Diffusion gradients may be set up next to fractures, the intrusive contact, preskarn dikes, and sedimentary contacts, leading to an orderly arrangement of skarn alteration zones, with the inner zones continually replacing the outer ones. This process eventually ceases as temperatures drop, and skarn destruction begins. If the skarn-forming fluid arrives at too low a temperature or too high a pressure, the fluid instead forms replacement bodies or magnetite, siderite, jasperoidal silica, and sulfides (Titley, 1968). At many intrusive contacts, the fluid never arrives at all, and no skarn or other ore deposit occurs.
4. Superposition of oxides and sulfides at 500–300 °C and lower (in part concomitant with late skarn formation). Scheelite and magnetite commonly appear to be earlier than sulfides. A preferential association of ore minerals with certain skarn minerals or zones suggests textural or chemical controls on ore deposition.
5. Late hydrothermal alteration (destruction of early skarn) at 400–200 °C or lower (concomitant with continued ore deposition). During this state, garnet may be altered to calcite, quartz, specular hematite, pyrite, epidote, chlorite, or nontronite. Clinopyroxene is commonly extensively altered to minerals, including calcite, fluorite, quartz, oxides and sulfides, ilvaite, babingtonite, amphibole, chlorite, serpentine, and rhodonite (or other minerals). Wollastonite may alter to calcite or fluorite plus quartz, or to xonotlite or stevensite. In some skarns, boron-rich minerals such as axinite or datolite occur; others are characterized by scapolite. In very late veinlets, apophyllite and zeolites may occur.

Relation to Ore

The fact that copper ore is almost always restricted to garnetiferous skarn zones, but that much garnet is barren in a given deposit, was an early observation (Keyes, 1909). The tendency of copper ore to be concentrated near the limestone side of the garnet skarns was recognized and well documented by Umpleby (1916). Although these observations were made with regard to copper ore in andraditic skarns, they also hold for lead-zinc ore in hedenbergitic skarns (Burt, 1972a). This enrichment near the "marble line" is presumably due to the neutralization of acidic hydrothermal fluids as they approach limestone.

The previously discussed tendency of ore to be restricted to skarn, and to particular zones in the skarn (already noted in several examples), is presumably due to the following factors: (1) ground preparation (volume decrease, increase in vugginess or brittleness, etc.) associated with contact metamorphism and skarn formation, (2) the fact that late ore-bearing solutions must, in general, utilize the same plumbing system involved in earlier skarn formation, (3) coprecipitation of certain skarn minerals with certain ore minerals, (4) textural or chemical controls on later ore deposition by early skarn minerals ("buffering"). The dominant controls on skarn formation and ore deposition are, of course, the structure and stratigraphy of the deposit.

The tendency for wollastonite to occur only with iron-poor ore minerals, except at high H_2S-fugacities, has already been cited. Another chemical relationship is the ability of the $CaFeSi_2O_6$ component in clinopyroxene to buffer O_2 and S_2 fugacities at low levels by reactions such as the following (Burt, 1972a):

$$9[CaFeSi_2O_6] + 2O_2 \rightleftharpoons 3Ca_3Fe_2Si_3O_{12} + 9SiO_2 + Fe_3O_4 \quad (5.22)$$
$$\text{clinopyroxene} \qquad\qquad\qquad \text{andradite} \qquad \text{quartz} \quad \text{magnetite}$$

$$3[CaFeSi_2O_6] + S_2 \rightleftharpoons Ca_3Fe_2Si_3O_{12} + 3SiO_2 + FeS_2 \quad (5.23)$$
$$\text{clinopyroxene} \qquad\qquad \text{andradite} \qquad \text{quartz} \quad \text{pyrite}$$

Garnet coexists with clinopyroxene in a large number of skarn deposits. The andradite and hedenbergite end members coexist over only a limited range of oxygen fugacities (Figure 5.14), but Al^{3+}/Fe^{3+} and $Mg^{2+} - Mn^{2+}/Fe^{2+}$ substitutions allow the two minerals to coexist over a wide range of oxygen fugacities in nature.

Zharikov (1970) has correlated the relative iron contents in coexisting garnet and clinopyroxene in skarns with the metals mined. He found that skarn deposits of tungsten and/or molybdenum are characterized by hedenbergitic clinopyroxene coexisting with grossularitic garnet (indicative of reducing and/or acid conditions), and the same finding was made by Shimazaki (1974). In iron and lead-zinc deposits, the iron contents of the pyroxene and garnet commonly increase together (more intermediate or neutral conditions). In U. S. copper deposits, andradite commonly coexists with diopsidic pyroxene (relatively oxidizing or basic conditions).

The exchange equilibrium, unfortunately, does not depend entirely on oxygen fugacity/acidity but also on calcium and silica activities:

$$[Ca_3Fe_2Si_3O_{12}] + [SiO_2] \rightleftharpoons 2[CaFeSi_2O_6] + [CaO] + \tfrac{1}{2}O_2 \quad (5.24)$$
$$\text{in garnet} \qquad\qquad\qquad \text{in clinopyroxene}$$

Strictly speaking, at least two other phases must coexist for garnet/clinopyroxene compositions to reflect only oxygen fugacities; for instance:

$$[Ca_3Fe_2Si_3O_{12}] + 2SiO_2 \rightleftharpoons 2[CaFeSi_2O_6] + CaSiO_3 + \tfrac{1}{2}O_2$$
$$\text{in garnet} \quad\ \text{quartz} \quad\ \text{in clinopyroxene} \quad \text{wollastonite}$$
$$(5.25)$$

Zharikov (1970) assumed that activities of CaO and SiO_2 in skarns probably are relatively constant or at least are externally controlled.

Relation to Other Types of Ore Deposits

The skarn type of alteration commonly is spatially and almost certainly genetically associated with the porphyry copper type of alteration described above. In fact, to our knowledge, it is *always* developed if carbonates (limestones or dolomite) occur near a copper-bearing porphyry. Examples (Bingham, Utah; Santa Rita, New Mexico; Gaspe, Quebec) are too numerous to cite individually.

Note, however, that the opposite assumption cannot be made—most intrusions associated with skarn appear to be barren of economic mineralization (Burt, 1972a). This relation probably reflects a wide range of geologic conditions (P, T, degree of fracturing, type of intrusion, etc.), permitting formation of skarn and a smaller range of conditions leading to formation of a porphyry copper deposit, in the strict sense of the term.

Relative to skarn-type iron deposits, Park (1972) has extensively summarized what little is known of the relation of skarns to the deposits of magnetite-apatite-amphibole that occur around the periphery of the Pacific. Many of these deposits do seem distinct from the "classical" skarn deposits of tungsten, copper, and zinc that occur further inland.

Finally, skarns can be spatially and genetically related to the acid greisen (tin-tungsten-molybdenum-beryllium-bismuth-lithium-fluorine) type of mineralization. Near acid intrusions, carbonate wallrock may be massively altered to a fluorite-rich greisen, commonly with a distinctively banded appearance (e.g., the fluorite-magnetite-hematite-helvite "ribbon rock" from Iron Mountain, New Mexico, Jahns, 1944). Other notable examples occur in the tin and beryllium deposits of the western Seward Peninsula, Alaska (Sainsbury, 1968). Here tin ores occur in association with tourmaline and topaz-bearing greisen, which in places has been affected by late kaolinization and sericitization, as well as in association with skarns, which in places are locally enriched in B minerals (paigeite or vonsenite, hulsite, and danburite). The beryllium ores (helvite, chrysoberl, euclase, bertrandite, phenakite, and beryl) occur replacing carbonates with fluorite, magnetite, diaspore, tourmaline, "white mica" (ephesite), and other minerals. Similar deposits have been described from the Soviet Union (Govorov, 1968).

A low-temperature equivalent of the fluorite-rich greisen appears to be the Spor Mountain beryllium district, Utah (Shawe, 1968), where bertrandite is associated with fluorite and silica replacements of carbonate nodules in water-laid tuff, which elsewhere is enriched in lithium-bearing montmorillonite and/or with potassium-feldspar.

ALTERATION IN GREISEN DEPOSITS

"Greisen" is a term little used by U. S. mining geologists, probably because the "greisen" or "pneumatolytic" type of tin-tungsten-beryllium-molybdenum deposit is rare in the continental United States. The two examples that come immediately to mind are small [Irish Creek, Virginia, (Glass et al., 1958); Silver Mine, Missouri, (Singewald and Milton, 1929)]. Other examples from Colorado include the Lake George beryllium deposits (Hawley, 1969) and, in part, the Climax molybdenum deposit.

Canada has some slightly better examples (Mulligan, 1974) including Brunswick tin mines and the Burnt Hill wolframite deposit, both in New Brunswick. The best known examples, of course, are found in southwest England and the Erzegebrige, Central Europe.

Recent general summaries of greisens are found in Stemprok (1974). The use of phase relations between topaz and beryllium minerals as indicators of relative fluorine activities in greisens has been proposed by Burt (1975). Except for lithium-, boron-, and fluoride-rich minerals (zinnwaldite, tourmaline, topaz, and fluorite), alteration features of greisens in granite are similar to those already described in granitic rocks. An interesting variant is that in fluorine-rich peralkaline granites albitization can be combined with greisenization, yielding "apogranites" enriched in beryllium, niobium, tantalum, and zirconium (Borodin and Pavlenko, 1974). A notable accessory may be cryolite, Na_3AlF_6, as in the albite-riebeckite granites of northern Nigeria (Bowden and Turner, 1974).

The fluorite-rich greisens developed in carbonate rocks have already been mentioned. Beryllium- and tin-bearing skarn minerals (e.g., $Ca_2BeSi_2O_7$, gugiaite, and $CaSnSiO_5$, malayaite) are unstable in the greisen environment with respect to assemblages of fluorite, quartz, and silicates or oxides [e.g., phenakite, Be_2SiO_4, or cassiterite, SnO_2 (Burt, 1975)]. More complex reactions also involving the acidic anion exchange operator F_2O_{-1} can be written to explain the occurrence of tungsten minerals in skarns and greisens. For example, in the reaction (Burt, 1972a);

$$Fe_3O_4 + 3CaWO_4 + 3F_2O_{-1} \rightleftharpoons 3CaF_2 + 3FeWO_4 + \tfrac{1}{2}O_2$$
magnetite scheelite   fluorite   ferberite

the left-hand assemblage appears stable in skarns and the right-hand assemblage in greisens.

ACKNOWLEDGMENTS

We are greatly indebted to Marco Einaudi, Richard L. Nielsen, and Derrill M. Kerrick for very helpful reviews of the text.

REFERENCES

Althaus, E., E. Karotke, and K. H. Nitsch (1970) An experimental reexamination of the upper stability limit of muscovite plus quartz: *N. Jh. Mineral. Monatsh.*, **1970**, 325-336.

Ames, L. L. (1961) Volume relationships during replacement reactions: *Econ. Geol.*, **56**, 1438-1445.

Anderson, G. M. (1970) Some thermodynamics of dehydration equilibria: *Am. J. Sci.*, **269**, 392-401.

Barnes, H. L. and W. G. Ernst (1963) Ideality and ionization in hydrothermal fluids: the system $MgO-H_2O-NaOH$: *Am. J. Sci.*, **261**, 129-150.

Barrell, J. (1902) Physical effects of contact metamorphism: *Am. J. Sci.*, **13**, 279-296.

Bartholomé, P. and F. Dimanche (1967) On the paragenesis of ilvaite in Italian skarns: *Ann. Soc. Geol. Belg.*, **90**(5), 17-48, (B533-564).

Bartholomé, P. and P. Evrard (1970) On the genesis of the zoned skarn complex at Temperino, Tuscany: in *Problems of Hydrothermal Ore Deposition*, Int. Union Geol. Sci., Ser. A, No. 2, p. 53-57.

Bateman, P. C. (1965) Geology and tungsten mineralization of the Bishop district, California: *U. S. Geol. Surv., Prof. Paper 470.*

Beane, R. E. (1974) Biotite stability in the porphyry copper environment: *Econ. Geol.,* **69,** 241-256.

Bonorino, F. G. (1959) Hydrothermal alteration in the Front Range mineral belt, Colorado: *Bull. Geol. Soc. Am.,* **70,** 53-90.

Borodin, L. S. and A. S. Pavlenko (1974) The role of metasomatic processes in the formation of alkaline rocks: in *The Alkaline Rocks,* H. Sørensen, ed., New York: Wiley, p. 515-534.

Bowden, P. and D. C. Turner (1974) Peralkaline and associated ring complexes in the Nigeria-Niger Province, West Africa: in *The Alkaline Rocks,* H. Sørensen, ed., New York: Wiley, p. 330-351.

Bowen, N. L. and O. F. Tuttle (1949) The system MgO-SiO_2-H_2O: *Geol. Soc. Am. Bull.,* **60,** 439-460.

Brady, J. B. (1975) Chemical components and diffusion: *Am. J. Sci.,* **275,** 1073-1088.

Brimhall, G. H., Jr. (1977) Early fracture-controlled mineralization at Butte, Montana: *Econ. Geol.,* **72,** 37-59.

Burnham, C. W. (1962) Facies and types of hydrothermal alteration: *Econ. Geol.,* **57,** 768-784.

Burt, D. M. (1972a) Mineralogy and geochemistry of Ca-Fe-Si skarn deposits: Unpublished Ph.D. thesis, Harvard University.

Burt, D. M. (1972b) The facies of some Ca-Fe-Si skarns in Japan: *24th Int. Geol. Cong., Montreal, Sect. 2, Petrology,* 284-288.

Burt, D. M. (1974) Metasomatic zoning in Ca-Fe-Si exoskarns: in *Geochemical Transport and Kinetics,* A. W. Hofmann et al., eds., Washington, D. C.: Carnegie Institution Wash., Pub. 634, p. 287-293.

Burt, D. M. (1975) Beryllium mineral stabilities in the model system CaO-BeO-SiO_2-P_2O_5-F_2O_{-1}, and the breakdown of beryl: *Econ. Geol.,* **70,** 1279-1292.

Burt, D. M. (1976) Hydrolysis equilibria in the system K_2O-Al_2O_3-SiO_2-H_2O-Cl_2O_{-1}: Comments on topology: *Econ. Geol.,* **71,** 665-671.

Chatterjee, N. D. and W. Johannes (1974) Thermal stability and standard thermodynamic properties of synthetic $2M_1$-muscovite: *Contr. Min. Pet.,* **48,** 89-114.

Chen, C. H. (1975) A method of estimation of standard free energies of formation of silicate minerals at 298.15°K: *Am. J. Sci.,* **275,** 801-817.

Chernosky, J. V. (1975) The stability of the assemblage clinochlore + quartz at low pressure (abst.): *EOS,* **56,** 466.

Cooper, A. R., Jr. (1974) Vector space treatment of multicomponent diffusion: in *Geochemical Transport and Kinetics,* A. W. Hofmann et al., eds., Washington, D.C.: Carnegie Institution Wash. Pub. 634, p. 15-30.

Cooper, J. R. (1957) Metamorphism and volume losses in carbonate rocks near Johnson Comp, Cochise County, Arizona: *Geol. Soc. Am. Bull.,* **68,** 577-610.

Corn, R. M. (1975) Alteration-mineralization zoning, Red Mountain, Arizona: *Econ. Geol.,* **70,** 1437-1447.

Creasey, S. C. (1959) Some phase relations in hydrothermally altered rock of porphyry copper deposits: *Econ. Geol.,* **54,** 351-373.

Day, H. W. (1973) The high temperature stability of muscovite plus quartz: *Am. Mineral.,* **58,** 255-262.

Dubois, R. L. and R. W. Brummet (1968) Geology of the Eagle Mtn. mine area: in *Ore Deposits of the United States 1933-1967,* J. D. Ridge, ed., New York: AIME, p. 1592-1606.

REFERENCES

Ellis, A. J. (1967) The chemistry of some explored geothermal systems: in *Geochemistry of Hydrothermal Ore Deposits*, H. L. Barnes, ed., New York: Holt, Rinehart and Winston, 465-514.

Ellis, A. J. (1970) Quantitative interpretation of chemical characteristics of hydrothermal systems: *Geothermics*, special issue **2**, 516-528.

Ellis, A. J. (1971) Magnesium ion concentrations in the presence of magnesium chlorite, calcite, carbon dioxide, quartz: *Am. J. Sci.*, **271**, 481-489.

Ernst, W. G. (1966) Synthesis and stability relations of ferrotremolite: *Am. J. Sci.*, **264**, 37-65.

Evans, B. W. (1965) Application of a reaction rate method to the breakdown equilibria of muscovite and muscovite plus quartz: *Am. J. Sci.*, **263**, 647-667.

Field, C. W., M. B. Jones, and W. R. Bruce (1974) Porphyry copper molybdenum deposits of the Pacific Northwest: *Am. Inst. Min. Eng. Trans.*, **255**, 9-22.

Finger, L. W. and D. M. Burt (1972) REACTION, a Fortran IV computer program to balance chemical reactions: *Carnegie Institution Wash. Yearbook* **71**, 616-620.

Fisher, G. W. (1973) Nonequilibrium thermodynamics as a model for diffusion-controlled metamorphic processes: *Am. J. Sci.*, **273**, 897-924.

Fisher, G. W. and D. Elliott (1974) Criteria for quasi-steady diffusion and local equilibrium in metamorphism: in *Geochemical Transport and Kinetics*, A. W. Hofmann et al., eds., Washington, D.C.: Carnegie Institution Wash. Pub. 634, 231-241.

Fisher, J. R. and E. Zen (1971) Thermochemical calculations from hydrothermal phase equilibrium data and the free energy of H_2O: *Am. J. Sci.*, **270**, 297-314.

Fletcher, R. C. and A. W. Hofmann (1974) Simple models of diffusion and combined diffusion-infiltration metasomatism: in *Geochemical Transport and Kinetics*, A. W. Hofmann et al., eds., Washington, D.C.: Carnegie Institution Wash. Pub. 634, 243-259.

Fonarev, V. I. and I. P. Ivanov (1972), Equilibria of the dehydration reactions of pyrophyllite and muscovite at $P_{tot} = P_{H_2O} = 1000$ to 8000 Kg/Cm^2: *Dokl. Akad. Nauk SSSR*, **205**, 218-219.

Fournier, R. O. (1967) The porphyry copper deposit exposed in the Liberty open-pit mine near Ely, Nevada: Part I. Syngenetic formation; Part II. The formation of hydrothermal alteration zones: *Econ. Geol.*, **62**, 57-81, 207-227.

Fournier, R. O. and J. J. Rowe (1966) Estimation of underground temperatures from the silica content of water from hot springs and wet-stream wells: *Am. J. Sci.*, **264**, 685-697.

Frantz, J. D. and H. K. Mao (1974) Metasomatic zoning resulting from intergranular diffusion. A theoretical model: in *Carnegie Institution Wash. Yearbook* **73**, 384-392.

Frantz, J. D. and P. M. Mao (1975) Bimetasomatism resulting from intergranular diffusion: Multimineralic zone sequences: in *Carnegie Institution Wash. Yearbook* **74**, 417-424.

Frantz, J. D. and A. Weisbrod (1974) Infiltration metasomatism in the system K_2O-SiO_2-Al_2O_3-H_2O-HCl: in *Geochemical Transport and Kinetics*, A. W. Hofmann et al., eds., Washington, D.C.: Carnegie Institution Wash. Pub. 634, 261-271.

French, B. M. (1971) Stability relations of siderite in the system Fe-C-O: *Am. J. Sci.*, **271**, 37-78.

Fukuchi, N. (1907) Mineral paragenesis in the contact metamorphic deposits found in Japan: *Beitrage zur Mineralogie von Japan*, herausgegeb. von T. Wada, No. 3, 88-89.

Glass, J. J., A. H. Koschmann, and J. S. Vhay (1958) Minerals of the cassiterite-bearing veins at Irish Creek, Virginia, and their paragenetic relations: *Econ. Geol.*, **53**, 65-84.

Gordon, T. M. (1973) Determination of internally consistent thermodynamic data from phase equilibrium experiments: *J. Geol.*, **81**, 199-208.

Govorov, I. N. (1968) Rare metal greisens in carbonate rocks: in *Geochemistry and Mineralogy of Rare-Element Deposits, Vol. 3, Genetic Types of Rare-Element Deposits,* Jersualem: Israel Prog. Sci. Transl., 138-166.

Gray, R. F., V. J. Hoffman, R. J. Bagan, and H. L. McKinley (1968) Bishop Tungsten district, California: in *Ore Deposits of the United States, 1933-1967,* J. D. Ridge, ed., New York: AIME, 1531-1554.

Guilbert, J. M. and J. D. Lowell (1974) Variations in zoning patterns in porphyry copper deposits: *Can. Min. Metall. Bull.,* **67**(742), 99-109.

Gustafson, L. B. and J. P. Hunt (1975) The porphyry copper deposit at El Salvador, Chile: *Econ. Geol.,* **70,** 857-912.

Gustafson, W. I. (1974) The stability of andradite, hedenbergite and related minerals in the system Ca-Fe-Si-O-H: *J. Petrol.,* **15,** 455-496.

Haas, H. and M. J. Holdaway (1973) Equilibria in the system Al_2O_3-SiO_2-H_2O involving the stability limit of pyrophyllite and thermodynamic data on pyrophyllite: *Am. J. Sci.,* **273,** 449-464.

Haas, J. L. and R. A. Robie (1973) Thermodynamic data for wustite, $Fe_{0.947}O$, magnetite, Fe_3O_4, and hematite, Fe_2O_3 (abst.): *EOS,* **54,** 483.

Hammarback, S. and B. Lundqvist (1972) The hydrothermal stability of annite in the presence of sulfur: *Geol. For. Stockholm Forhand.,* **94,** 549-564.

Hawley, C. C. (1969) Geology and beryllium deposits of the Lake George (or Badger Flats) beryllium area, Park and Jefferson Counties, Colorado: *U. S. Geol. Survey, Prof. Paper 608-A.*

Helgeson, H. C. (1968) Evaluation of irreversible reactions in geochemical processes involving minerals and aqueous solutions: I. Thermodynamic relations: *Geochim. Cosmochim. Acta,* **32,** 853-877.

Helgeson, H. C. (1969) Thermodynamics of hydrothermal systems at elevated temperatures and pressures: *Am. J. Sci.,* **267,** 729-804.

Helgeson, H. C. (1970a) A chemical and thermodynamic model of ore deposition in hydrothermal systems: *Min. Soc. Am., Spec. Paper 3,* 155-186.

Helgeson, H. C. (1970b) Reaction rates in hydrothermal flow systems: *Econ. Geol.,* **65,** 299-303.

Helgeson, H. C. (1974) Chemical interaction of feldspars and aqueous solutions: in *The Feldspars,* W. S. MacKenzie and J. Zussman, eds., Manchester: Manchester University Press, 184-215.

Helgeson, H. C., T. H. Brown, and R. H. Leeper (1969) *Handbook of Theoretical Activity Diagrams Depicting Chemical Equilibria in Geological Systems Involving an Aqueous Phase at 1 atm. and 0-300°C,* San Francisco: Freeman, Cooper.

Helgeson, H. C. and D. H. Kirkham (1974, 1976) Theoretical prediction of the thermodynamic behavior of aqueous electrolytes at high pressures and temperatures: Parts I and II: *Am. J. Science,* **274,** 1089-1261; Part III, **276,** 97-240.

Helgeson, H. C., H. W. Nesbitt, and J. Delany (1975) Summary and critique of the thermodynamic properties of rock-forming minerals: *Geol. Soc. Am., Abstr. Programs,* **7,** 1108-1109.

Hemley, J. J., P. B. Hostetler, A. J. Gude, and W. T. Mountjoy (1969) Some stability relations of alunite: *Econ. Geol.,* **64,** 599-612.

Hemley, J. J. and W. R. Jones (1964) Chemical aspects of hydrothermal alteration with emphasis on hydrogen metasomatism: *Econ. Geol.,* **59,** 538-569.

Hemley, J. J., J. W. Montoya, A. Nigrini, and H. A. Vincent (1971) Some alteration reactions in the system CaO-Al$_2$O$_3$-SiO$_2$-H$_2$O: *Soc. Min. Geol. Japan, Special Issue 2,* 58-63.

Hernon, R. M. and W. R. Jones (1968) Ore deposits of the Central mining district, Grant County, New Mexico: in *Ore Deposits of the United States, 1933-1967,* J. D. Ridge, ed., New York: AIME, p. 1211-1238.

Hewitt, D. A. and M. C. Gilbert (1975) Experimental metamorphic petrology: *Rev. Geophys. Space Sci.,* **13,** 79-81, 120-128.

Hewitt, W. P. (1968) Geology and mineralization of the main mineral zone of the Santa Eulalia district, Chihuahua, Mexico: *Soc. Min. Eng. Trans.,* **241,** 228-260.

Hofmann, A. (1972, 1973) Chromatographic theory of infiltration metasomatism and its application to feldspars: *Am. J. Sci.,* **272,** 69-90; **273,** 960-964.

Holdaway, M. J. (1971) Stability of andalusite and the aluminum silicate phase diagram: *Am. J. Sci.,* **271,** 97-131.

Hollister, V. F., S. A. Anzalone, and D. H. Richter (1975) Porphyry copper deposits of southern Alaska and contiguous Yukon Territory: *Can. Min. Met. Bull.,* **68**(756), 104-112.

Hollister, V. F., R. R. Potter, and A. L. Barker (1974) Porphyry-type deposits of the Appalachian orogen: *Econ. Geol.,* **69,** 618-630.

Huebner, J. S. (1971) Buffering techniques for hydrostatic systems at elevated pressures: in *Research Techniques for High Pressure and High Temperature,* G. C. Ulmer, ed., New York: Springer-Verlag, 123-178.

Huheey, J. E. (1972) Inorganic chemistry: New York: Harper and Row.

Ito, K. (1962) Zoned skarn of the Fujigatani mine, Yamaguchi Prefecture: *Japan J. Geol. Geogr.,* **33,** 169-190.

Ivanov, I. P., O. N. Belyaevska, and V. Yu. Potekhin (1974) Improved diagram of hydrolysis and hydration equilibria in the open multisystem KCl-HCl-Al$_2$O$_3$-SiO$_2$-H$_2$O at P = 1000 Kg/cm^2 (in Russian): *Dokl. Akad. Nauk SSSR,* **219,** 715-717.

Jahns, R. H. (1944) "Ribbon rock," an unusual beryllium-bearing tactite: *Econ. Geol.,* **39,** 173-205.

John, E. C. (1975) Mineral zones of the Bingham District: in *Guidebook to the Bingham Mining District,* R. E. Bray and J. C. Wilson, eds., Bingham Canyon, Utah: Kennecott Copper Corp., 59-72.

Jones, M. B. (1974) Hydrothermal alteration and mineralization of the Valley Copper deposit, Highland Valley district, British Columbia: Unpublished Ph.D. thesis, Oregon State University.

Kerrick, D. M. (1972) Experimental determination of muscovite + quartz stability with P_{H_2O} < P_{total}: *Am. J. Sci.,* **272,** 946-958.

Kerrick, D. M. (1974) Review of metamorphic mixed volatile (H$_2$O-CO$_2$) equilibria: *Am. Mineral.,* **59,** 729-762.

Keyes, C. R. (1909) Garnet contact deposits of copper and the depths at which they are formed: *Econ. Geol.,* **4,** 365-372.

Korzhinskii, D. S. (1936) Mobility and inertness of components in metasomatism (in Russian): *Izv. Akad. Nauk SSSR,* ser. geol. **1,** 36-60.

Korzhinskii, D. S. (1956) The dependence of component activity on solution acidity and the sequence of reactions in post-magmatic processes (in Russian): *Geokhimiya,* **7,** 3-10.

Korzhinskii, D. S. (1959) *Physicochemical Basis of the Analysis of the Paragenesis of Minerals,* New York: Consultants Bureau.

Korzhinskii, D. S. (1970) *Theory of metasomatic zoning* (trans. by J. Agrell): London: Oxford University Press.

Lacy, W. C. and S. R. Titley (1962) Geological developments in the Twin Buttes district: Min. Congr. J., **48**(4), 62-64.

Langmuir, D. (1971) Particle size effect on the reaction goethite = hematite + water: *Am. J. Sci.*, **271**, 147-156.

Lapham, D. M. and C. Gray (1973) Geology and origin of the Triassic magnetite deposit and diabase at Cornwall, Pennsylvania: *Penna. Geol. Surv. (4th), Bulletin M56.*

Lindgren, W. (1912) The nature of replacement: *Econ. Geol.*, **7**, 521-535.

Lindgren, W. (1924) Contact metamorphism at Bingham, Utah: *Geol. Soc. Am. Bull.*, **35**, 507-534.

Lindgren, W. (1925) Metasomatism: *Geol. Soc. Am. Bull.*, **36**, 247-261.

Liou, J. G. (1971) Synthesis and stability relations of prehnite, $Ca_2Al_2Si_3O_{10}(OH)_2$: *Am. Mineral.*, **56**, 507-531.

Liou, J. G. (1973) Synthesis and stability relations of epidote $Ca_2Al_2FeSi_3O_{12}(OH)$: *J. Petrol.*, **14**, 381-413.

Lovering, T. S. (1941) The origin of the tungsten ores of Boulder County, Colorado: *Econ. Geol.*, **36**, 229-279.

Lovering, T. S. (1949) Rock alteration as a guide to ore—East Tintic district, Utah: *Econ. Geol., Monograph 1.*

Lovering, T. S. (1950) The geochemistry of argillic and related types of rock alteration: *Colo. School Mines Quart.*, **45**(1B), 231-260.

Lowell, J. D. and J. M. Guilbert (1970) Lateral and vertical alteration-mineralization zoning in porphyry copper deposits: *Econ. Geol.*, **65**, 373-408.

Meyer, C. et al. (1968) Ore deposits at Butte, Montana: in *Ore Deposits of the U. S., 1933-1967*, J. D. Ridge, ed., New York: Am. Inst. Min. Eng., p. 1373-1416.

Meyer, C. and J. J. Hemley (1967) Wall rock alteration: in *Geochemistry of Hydrothermal Ore Deposits*, H. L. Barnes, ed., New York: Holt, Rinehart and Winston, p. 166-235.

Montoya, J. W. and J. J. Hemley (1975) Activity relations and stabilities in alkali feldspar and mica alteration reactions: *Econ. Geol.*, **70**, 577-594.

Moore, W. and G. K. Czamanske (1973) Compositions of biotites from unaltered and altered monzonitic rocks in the Bingham mining district, Utah: *Econ. Geol.*, **68**, 269-280.

Mueller, R. F. (1972) Stability of biotite: *Am. Mineral.*, **57**, 300-315.

Mulligan, R. (1974) Geology of Canadian tin occurrences: *Can. Geol. Surv., Econ. Geol. Rept 28.*

Munoz, J. L. and S. D. Ludington (1974) Fluoride-hydroxyl exchange in biotite: *Am. J. Sci.*, **274**, 396-413.

Naumov, G. B., B. N. Ryzhenko, and I. L. Khodakovsky (1971) *Handbook of Thermodynamic Data*, Moscow: Atomizdat; translated by G. J. Soleimani, *Nat. Tech. Inf. Serv. #PB226-722,* 1974.

Nelson, W. H. and C. P. Ross (1968) Geology of part of the Alder Creek mining district, Custer County, Idaho: *U.S. Geol. Surv. Bull. 1252A.*

Nielsen, R. L. (1968) Hypogene texture and mineral zoning in a copper bearing granodiorite porphyry stock, Santa Rita, New Mexico: *Econ. Geol.*, **63**, 37-50.

Nriagu, J. O. (1975) Thermochemical approximations for clay minerals: *Am. Mineral.*, **60**, 834-839.

Ozawa, T., M. Kamada, M. Yoshida, and I. Sanemasa (1973) Genesis of acid hot spring: in *Proceedings Sympos. Hydrogeochemistry and Biogeochemistry,* Vol. 1, E. Ingerson, Ed., Washington, D.C.: The Clarke Co., 105-121.

Park, C. F. (1972) The iron ore deposits of the Pacific Basin: *Econ. Geol.,* **67,** 339-349.

Patton, T. C., A. R. Grant, and E. S. Cheney (1973) Hydrothermal alteration at the Middle Fork copper prospect, central Cascades, Washington: *Econ. Geol.,* **68,** 816-830.

Perry, D. V. (1969) Skarn genesis at the Christmas mine, Gila County, Arizona: *Econ. Geol.,* **64,** 255-270.

Pertsev, N. N. (1974) Skarns as magmatic and postmagmatic formations: *Int. Geol. Rev.,* **16,** 572-582.

Phillips, C. H., N. A. Gambell, and D. S. Fountain (1974) Hydrothermal alteration, mineralization, and zoning in the Ray deposit: *Econ. Geol.,* **69,** 1237-1250.

Raymahashay, B. C. (1968) A geochemical study of rock alteration by hot springs in the Paint Pot Hill area, Yellowstone Park: *Geochim. Cosmochim. Acta,* **32,** 499-522.

Reed, B. L. and J. J. Hemley (1966) Occurrence of pyrophyllite in the Kekiktuk conglomerate, Brooks Range, Alaska: *U.S. Geol. Survey, Prof. Paper 550-C,* C162-C166.

Ridge, J. D. (1949) Replacement and the equating of volume and weight: *J. Geol.,* **57,** 522-550.

Roberts, S. A. (1973) Pervasive early alteration in the Butte district, Montana: in *Guidebook for the Butte Field Meeting of the Society of Econ. Geol.,* R. N. Miller, ed., The Anaconda Co., Butte, Montana, HH1-HH8; abstract, *Econ. Geol.,* **68,** 909.

Robie, R. A., B. S. Hemingway and J. R. Fisher (1978) Thermodynamic properties of minerals and related substances at 298.15°K and 1 bar (10^5 pascals) pressure and at higher temperatures: *U.S. Geol. Sur. Bull.* 1452.

Rose, A. W. (1970) Zonal relations of wall rock alteration and sulfide distribution at porphyry copper deposits: *Econ. Geol.,* **65,** 920-936.

Rose, A. W. and W. W. Baltosser (1966) The porphyry copper deposit at Santa Rita, New Mexico: in *Geology of the Porphyry Copper Deposits,* S. R. Titley and C. L. Hicks, eds., Tucson: University of Arizona Press, p. 205-220.

Rosenberg, P. E. (1974) Pyrophyllite solid solutions in the system Al_2O_3-SiO_2H_2O: *Am. Mineral.,* **59,** 254-260.

Rutherford, M. J. (1973) The phase relations of aluminous iron biotites in the system $KAlSi_3O_8$-$KAlSiO_4$-Al_2O_3-Fe-O-H: *J. Petrol.,* **14,** 159-180.

Sainsbury, C. L. (1968) Tin and beryllium deposits of the central York Mountains, western Seward Peninsula, Alaska: in *Ore Deposits of the United States, 1933-1967,* ed by J. D. Ridge, New York: AIME, p. 1555-1573.

Sales, R. H. and C. Meyer (1948) Wallrock alteration at Butte, Montana: *Am. Inst. Min. Eng. Trans.,* **178,** 9-35.

Sales, R. H. and C. Meyer (1950) Interpretation of wallrock alteration at Butte, Montana: *Colo. Sch. Mines Quart.,* **45**(1B), 261-274.

Schoen, R., D. E. White, and J. J. Hemley (1974) Argillization by descending acid at Steamboat Springs, Nevada: *Clays and Clay Minerals,* **22,** 1-22.

Schilling, J. H. (1968) The Gabbs magnesite-brucite deposit, Nye County, Nevada: in *Ore Deposits of the United States, 1933-1967,* J. D. Ridge, ed., New York: AIME, p. 1607-1622.

Schmitt, H. A. (1939) The Pewabic mine: *Geol. Soc. Am. Bull.,* **50,** 777-818.

Shabynin, L. I. (1973) *Formation of Magnesian Skarns,* Moscow: Nauka.

Shabynin, L. I. (1974) *Ore Deposits in Magnesian Skarn Formation(s)*, Moscow; Nedra.

Shade, J. W. (1974) Hydrolysis reactions in the SiO_2-excess portion of the system K_2O-Al_2O_3-SiO_2-H_2O in chloride fluids at magmatic conditions: *Econ. Geol.*, **69**, 218-228.

Shawe, D. R. (1968) Geology of the Spor Mountain beryllium district, Utah: in *Ore Deposits of the United States, 1933-1967*, J. D. Ridge, ed., New York: AIME, p. 1148-1162.

Shimazaki, H. (1969) Pyrometasomatic copper and iron ore deposits of the Yaguki mine, Fukushima prefecture, Japan: *Tokyo Univ. Fac. Sci.*, **17**(2), 317-350.

Shimazaki, H. (1974) The ratios of Cu/Zn-Pb of pyrometasomatic deposits in Japan and their genetical implications: *Econ. Geol.*, **70**, 717-724.

Shimazaki, H. and T. Yamanaka (1973) Iron wollastonite from skarns and its stability relation in the $CaSiO_3$-$CaFeSi_2O_6$ join: *Geochem. J.*, **7**, 67-79.

Sillitoe, R. H. (1973) The tops and bottoms of porphyry copper deposits: *Econ. Geol.*, **68**, 799-815.

Singewald, J. T. and C. Milton (1929) Greisen and associated mineralization at Silvermine, Mo.: *Econ. Geol.*, **24**, 569-591.

Slaughter, J., D. M. Kerrick, and V. J. Wall (1976) APL computer programs for thermodynamic calculations in P-T-X_{CO_2} space: *Contr. Min. Petrol.*, **54**, 157-171.

Stemprok, M., ed. (1974) *Metallization Associated with Acid Magmatism, Vol. 1*, Prague: Geological Survey of Czechoslovakia.

Stull, D. R. and H. Prophet (1971) *JANAF Thermochemical Tables (2nd Ed.)*, Washington, D.C.: National Bureau of Standards.

Sturdevant, J. A. (1975) Relationships between fracturing, hydrothermal zoning and copper mineralization at the Big Bug pluton, Big Bug mining district, Yavapai County, Arizona: Unpublished Ph.D. thesis, Pennsylvania State University, University Park, Pa.

Tardy, Y. and R. M. Garrels (1974) A method of estimating the Gibbs energies of formation of layer silicates: *Geochim. Cosmochim. Acta*, **38**, 1101-1116.

Thompson, A. B. (1971) P_{CO_2} in low-grade metamorphism; zeolite, carbonate, clay mineral, prehnite relations in the system CaO-Al_2O_3-SiO_2-CO_2-H_2O: *Contrib. Min. Petrol.*, **33**, 145-161.

Thompson, J. B., Jr. (1955) The thermodynamic basis for the mineral facies concept: *Am. J. Sci.*, **253**, 65-103.

Thompson, J. B., Jr. (1959) Local equilibrium in metasomatic processes: in *Researches in Geochemistry*, P. H. Abelson, ed., New York: Wiley, p. 427-457.

Thompson, J. B. (1970) Geochemical reaction and open systems: *Geochim. Cosmochim. Acta*, **34**, 529-551.

Titley, S. R. (1961) Genesis and control of the Linchburg ore body, Socorro County, New Mexico: *Econ. Geol.*, **56**, 695-722.

Titley, S. R. (1963) Lateral zoning as a result of a monoascendent hydrothermal process in the Linchburg Mine, New Mexico: in *Problems of Postmagmatic Ore Deposits, Vol. 1*, Prague: Geol. Sur. of Czechoslovakia, 312-316.

Titley, S. R. (1968) Environment of deposition of the hydrothermal metamorphic (pyrometasomatic) deposits (abst.): *Econ. Geol.*, **63**, 699.

Tokunuga, M. (1965) On the zoned skarn including bustamite, ferroan johannsenite and manganoan hedenbergite from Nakatatsu mine, Fukui Prefecture, Japan: *Sci. Rept. Tokyo Kyoiki Daigaku, Ser. C*, **9**, 67-87.

Umpleby, J. B. (1916) The occurrence of ore on the limestone side of garnet zones: *Univ. Calif., Dept. Geol. Bull. 10*, 25-37.

REFERENCES

Van der Werf, C. A. (1961) *Acids, Bases and the Chemistry of the Covalent Bond*, New York: Van Nostrand.

Vidale, R. J. (1969) Metasomatism in a chemical gradient and the formation of calc-silicate bands: *Am. J. Sci.*, **267**, 857-874.

Vidale, R. J. and D. A. Hewitt (1973) Mobile components in the formation of calc-silicate bands: *Am. Mineral*, **58**, 991-997.

Vom Rath, G. (1868) Die Berge von Campiglia in der Toskanischen Maremme: *Zeitschr. Deutschen Geologischen Gesellschaft*, **20**, 307-364.

Waldbaum, D. R. and J. B. Thompson, Jr. (1969) Mixing properties of sanidine crystalline solutions: IV, Phase diagrams from equations of state: *Am. Mineral.*, **54**, 1274-1298.

Wallace, S. R., W. B. MacKenzie, R. G. Blair and N. K. Muncaster (1978) Geology of the Urad and Henderson molybdenite deposits, Clear Creek County, Colorado: *Econ. Geol.*, **73**, 325-368.

Watanabe, T. (1960) Characteristic features of ore deposits found in contact metamorphic aureoles in Japan: *Int. Geol. Rev.*, **2**, 946-966.

White, D. E. (1957) Thermal waters of volcanic origin: *Bull. Geol. Soc. Am.*, **68**, 1637-1658.

White, D. E., L. J. P. Muffler, and A. H. Truesdell (1971) Vapor-dominated hydrothermal systems compared with hot-water systems: *Econ. Geol.*, **66**, 75-97.

Winkler, H. G. F. (1974) *Petrogenesis of Metamorphic Rocks*, 3rd ed., New York: Springer-Verlag.

Wones, D. R. and H. P. Eugster (1965) Stability of biotite: Experiment, theory and application: *Am. Mineral.*, **50**, 1228-1273.

Zharikov, V. A. (1970) Skarns: *Int. Geol. Rev.*, **12**, 541-559, 619-647, 760-775.

6

Oxygen and Hydrogen Isotope Relationships in Hydrothermal Mineral Deposits*

HUGH P. TAYLOR, JR.
California Institute of Technology

Because H_2O is the dominant constituent of ore-forming fluids, a knowledge of its origin is fundamental to any theory of ore formation. The other materials in solution also provide important evidence, but variations in dissolved salts and gases tell us principally something about the P-T-pH-f_{O_2} history of the solution and about the types of rocks with which the fluid came into contact. The ultimate source of the H_2O can best be deciphered by studying some geochemical parameter based on the water molecules themselves. Stable isotope analyses provide just such a parameter, because natural waters of various origins exhibit systematic differences in their deuterium and ^{18}O contents.

The aim of this chapter is to discuss the basic principles of hydrogen and oxygen isotope geochemistry that bear on the problems of ore deposition and hydrothermal alteration. Although isotopic analyses can, in principle, pro-

*Contribution Number 2700 of the Division of Geological and Planetary Sciences, California Institute of Technology, Pasadena, California 91125.

ISOTOPIC NOTATION AND STANDARDS

vide determinations of the temperatures of formation of hydrothermal mineral assemblages, the prevalence of isotopic disequilibrium in such assemblages severely restricts this application. At present the most useful application lies in using D/H and $^{18}O/^{16}O$ analyses as indicators of the origin and history of the H_2O in hydrothermal fluids.

Since the publication of the first edition of this book, a very large amount of isotopic data has been published on a wide variety of ore deposits. Therefore this chapter should be regarded as a supplement to the earlier one (Taylor, 1967), rather than a revision, and we shall concentrate on actual case histories rather than on principles. Other recent reviews that should be consulted in this connection are those of White (1968a, 1974) and Taylor (1974a).

Two methods are utilized to determine $^{18}O/^{16}O$ and D/H ratios of natural hydrothermal fluids: (1) direct measurement of the fluid itself in a geothermal area or of fluid inclusions in the minerals of an ore deposit, and (2) isotopic analyses of minerals, calculation of temperatures of formation utilizing various geothermometers, and finally, calculation of D/H and $^{18}O/^{16}O$ ratios of waters in equilibrium with the assemblages at their temperatures of formation. There are problems involved in the application of either of the above techniques, particularly with respect to whether or not isotopic ratios are preserved during subsequent cooling of the minerals. The water that is present in fluid inclusions of oxygen-bearing minerals, for example, will undergo exchange with the host mineral during cooling, thus changing the $^{18}O/^{16}O$ ratio of the fluid. Even nonoxygen-bearing minerals may contain secondary fluid inclusions that are not representative of the original fluid from which the mineral formed, especially if the mineral is highly fractured or full of imperfections.

ISOTOPIC NOTATION AND STANDARDS

The isotope data are reported as δD or $\delta^{18}O$, where

$$\delta = \left(\frac{R_{sample}}{R_{standard}} - 1 \right) 1000$$

and R_{sample} is D/H or $^{18}O/^{16}O$ in the sample and $R_{standard}$ is the corresponding ratio for the standard. We are not concerned with *absolute* ratios, only with relative deviations from a standard material, and the most convenient standard for both oxygen and hydrogen is ocean water. A particular set of ocean water values, designated Standard Mean Ocean Water (SMOW) by Craig (1961) is the most common standard in present-day use. Thus, a δD value =

+10 would mean that the sample is 10‰ (10 parts per thousand) or 1% richer in deuterium or ^{18}O than SMOW. Negative numbers signify relative *depletions* in the heavy isotopes.

The accuracy of determination of δD is approximately an order of magnitude worse than for $δ^{18}O$ (± 1‰ vs ± 0.1‰). However, the natural variations in D/H are much greater than for $^{18}O/^{16}O$; hence, a 10‰ variation represents a very large $δ^{18}O$ change but only a small δD change.

Other terms in common use are $α$, the fractionation factor or isotopic partition coefficient for two chemical compounds (i.e., for two species A and B, $α_{AB} = R_A/R_B$), and $Δ_{AB} = 1000 \ln α_{AB}$. Because most $α$ values are close to unity, it also follows that $1000 \ln α_{AB} ≈ δ_A - δ_B$.

EQUILIBRIUM ISOTOPE FRACTIONATIONS

If an isotopic exchange reaction is written such that only one of a set of exchangeable atoms in each compound takes part, using as an example the quartz-H_2O system:

$$H_2^{18}O + \tfrac{1}{2} Si^{16}O_2 \rightleftharpoons H_2^{16}O + \tfrac{1}{2} Si^{18}O_2,$$

then it can be shown (Urey, 1947) that $α \equiv K$, the equilibrium constant for the reaction as written. Urey (1947) and Bigeleisen and Mayer (1947) showed that $\ln K$ for an isotopic exchange reaction involving perfect gases closely follows $1/T^2$ dependence, and this is the reason for presenting the data in Figures 6.1 and 6.2 in the form of plots of $1000 \ln α$ versus $10^6/T^2$.

Some calculated and experimental $^{18}O/^{16}O$ fractionation curves of geologic interest are given in Figure 6.1. More extended discussions of the general applicability of these curves and the problems involved in their use are given by Taylor (1974a), Bottinga and Javoy (1973, 1975), and Blattner (1975).

In spite of some difficulties yet to be resolved, the curves on Figure 6.1 can be used to reliably calculate the $δ^{18}O$ values of H_2O in equilibrium with a particular mineral, if the temperature of formation can be estimated by some other means. Pressure is not a factor, because the equilibrium isotopic fractionations are solely temperature dependent. The composition of the aqueous fluid may, however, be important. This problem is discussed by Taylor (1967, p. 131), and the recent work by Truesdell (1974) has shown that salinity variations may require that slight adjustments be made in calculated $δ^{18}O$ values of hydrothermal waters. However, in view of the other variables involved, these difficulties are of second order and will be ignored in the discussions below.

Some D/H fractionation curves of geological interest are given in Figure

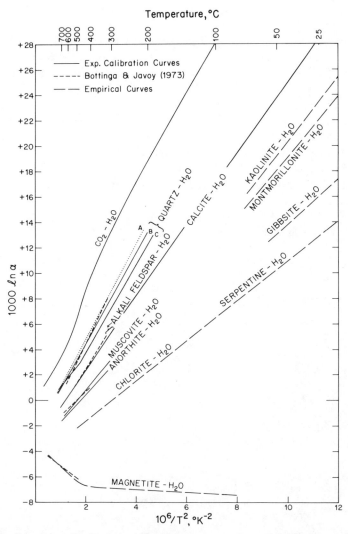

Fig. 6.1 Experimentally determined equilibrium oxygen isotope fractionation curves for various mineral-H_2O systems, all based on using $\alpha = 1.0412$ for calcite-H_2O instead of 1.0407: calcite-H_2O (O'Neil et al., 1969); quartz-H_2O (the dotted curve A is based on the quartz-feldspar curve of Blattner, 1975, recalculated to be compatible with the alkali feldspar-H_2O curves of O'Neil and Taylor, 1967; B = "partial" exchange and C = "complete" exchange experiments of Clayton et al., 1972); muscovite-H_2O (O'Neil and Taylor, 1969); and anorthite-H_2O (O'Neil and Taylor, 1967). Also shown is a calculated CO_2-H_2O curve (Bottinga, 1968), and some empirical curves based on natural assemblages; magnetite-H_2O (high-T portion = Anderson et al., 1971; low-T portion = Wenner and Taylor, 1971); serpentine-H_2O (Wenner and Taylor, 1971); kaolinite-H_2O (Savin and Epstein, 1970a); and gibbsite-H_2O (Lawrence and Taylor, 1971). The positions of some estimated curves for quartz, alkali feldspar, muscovite, and magnetite by Bottinga and Javoy (1973) are also indicated.

6.2. Note the enormously larger values of 1000 ln α in these systems compared to the $^{18}O/^{16}O$ systems.

The most important results of the experimental study by Suzuoki and Epstein (1976) are that the D/H fractionations among silicates are mainly a function of the magnesium, aluminum, and iron contents in the minerals. Water concentrates deuterium relative to all OH-bearing silicates, and the magnesium-rich and aluminum-rich minerals concentrate deuterium relative to iron-rich minerals. This helps explain why muscovite in natural mineral assemblages is invariably richer in deuterium than coexisting biotite, and why coexisting biotite and hornblende generally have similar δD values (they also generally have similar Mg/Fe ratios).

Above 400°C the various silicate-H_2O D/H fractionation curves determined by Suzuoki and Epstein (1976) form subparallel lines on a plot of 1000 ln α versus $10^6/T^2$. Below 400°C the positions of the curves are unknown, but if the low-temperature estimates of Savin and Epstein (1970a), Lawrence

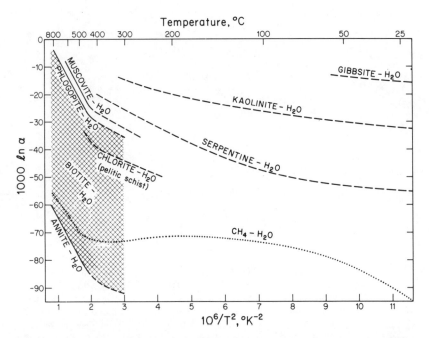

Fig. 6.2 Equilibrium hydrogen isotope fractionation curves for various mineral-H_2O systems. The solid lines represent data by Suzuoki and Epstein (1976). The dashed lines are empirical extrapolations based on natural assemblages (Savin and Epstein, 1970a; Lawrence and Taylor, 1971; Wenner and Taylor, 1973). The dotted line is the calculated CH_4-H_2O curve of Bottinga (1969). Note that below the critical point of H_2O, the curves are all based on values for *liquid* water.

and Taylor (1971), and Wenner and Taylor (1973) are reasonably accurate, all the hydrogen isotope fractionation curves must flatten out, as shown on Figure 6.2.

The calculated CH_4-H_2O equilibrium curve on Figure 6.2 indicates that at approximately 300°C, H_2O will be approximately 70‰ richer in deuterium than coexisting CH_4. This means that *if equilibrium is maintained* and a carbon-bearing hydrothermal fluid encounters a strongly reducing environment, some of the H_2O will react to form CH_4 and the latter will have a relatively low δD value. By material balance the remaining H_2O will have to become richer in deuterium. Therefore, redox reactions of this type represent possible means of changing the δD values of the H_2O in a natural hydrothermal fluid. However, to produce a $+20$‰ change in the H_2O, 30% of the H_2O in the hydrothermal fluid would have to react with carbon species to make methane. Large isotopic changes probably only occur in hydrothermal fluids that come into contact with systems containing abundant graphite (Eugster and Skippen, 1967), coal, or petroleum.

The reverse of the above situation occurs if a CH_4-bearing hydrothermal fluid encounters an oxidizing environment. Then the H_2O will become depleted in deuterium. Much larger effects are possible if H_2 gas, rather than CH_4, is involved, because the D/H fractionations between H_2 and H_2O may be 700–200‰ at 200–700°C (Bottinga, 1969). However, at the normal oxygen fugacities encountered in natural ore-forming solutions (e.g., see Meyer and Hemley, 1967; Barnes and Czamanske, 1967) only minuscule amounts of H_2 will be formed (Eugster and Skippen, 1967). This is particularly true if carbon is present in the system, because methane will then be an enormously more important constituent than H_2 gas. As shown on Figure 6.2, above 100°C the H_2O-CH_4 fractionation is never larger than 75‰ at equilibrium.

ISOTOPIC VARIATIONS IN NATURAL WATERS

Any naturally occurring water is potentially an ore-forming fluid if it becomes heated as a result of deep circulation in the crust or through interaction with a magma body, particularly if in the process it also becomes appreciably saline. Therefore, it is important to discuss the isotopic variations in the different kinds of waters that might conceivably be involved in ore deposition and hydrothermal alteration.

Meteoric Waters

The isotopic variations of H_2O in rain, snow, glacier ice, streams, lakes, rivers, and most low-temperature ground waters are extremely systematic;

the higher the latitude or elevation, the lower are the δD values and the $\delta^{18}O$ values of the waters. To a very close approximation all meteoric waters follow the equation (Craig, 1961):

$$\delta D \approx 8\delta^{18}O + 10 \text{ (in per mil)}$$

The linear relationship between δD and $\delta^{18}O$ arises because condensation of H_2O from the Earth's atmosphere is essentially an equilibrium process, and the D/H fractionation is proportional to the $^{18}O/^{16}O$ fractionation. For example, at 25°C, relative to liquid water, H_2O vapor is depleted in ^{18}O by approximately 9‰ and depleted in deuterium by 72‰. Both of these fractionations increase proportionally with decreasing temperature; this is essentially where the factor of eight arises in the slope of the meteoric H_2O line.

Thus, water condensed from atmospheric vapor in an air mass will be richer in ^{18}O and deuterium than the vapor, and by simple material-balance any subsequent precipitation from the same air mass must be lower in ^{18}O and deuterium than this initial condensate. As the air mass leaves the ocean and progresses across the continents, it becomes steadily lower in ^{18}O and deuterium. This leads to pronounced geographic and topographic isotopic effects on the rain and snow and the δD and $\delta^{18}O$ values are lower at the higher elevations or more northerly latitudes in the Northern Hemisphere. The average δD values of surface meteoric waters are consistent enough so that a rough isotopic contour map for the North American continent can be drawn, as shown in Figure 6.3. Note that a practically identical contour map could be drawn for $\delta^{18}O$ just by changing the δ values according to the aforementioned meteoric water equation (i.e., $\delta D = -90$ would be replaced by $\delta^{18}O = -12.5$, etc.).

Ocean Waters

The isotopic composition of present-day ocean water is exceedingly uniform at $\delta^{18}O = 0$ and $\delta D = 0$. Therefore, it is a useful isotopic standard (SMOW on Figure 6.4). Only in areas such as the Red Sea (Craig, 1966) where there has been appreciable evaporation and an increase in salinity are $\delta^{18}O$ values as high as $+2$ and δD values as high as $+11$. Also, if there is appreciable dilution with fresh waters, there can be small isotopic changes in the opposite direction, particularly in the Arctic and Antarctic.

The major isotopic problem concerning seawater is, how constant has its isotopic composition been through geologic time? We know that if we melted all the ice sheets in the world, the $\delta^{18}O$ value might become as low as -1 and the δD value as low as -10. At least throughout most of Phanerozoic time the isotopic composition of ocean water has probably fluctuated within those

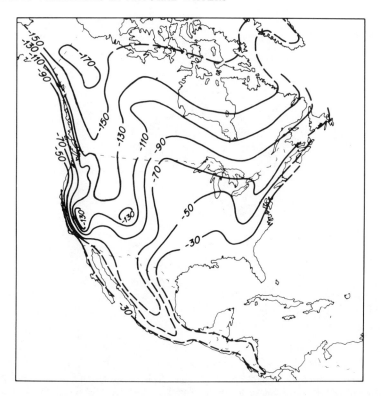

Fig. 6.3 Map of North America showing contours of the approximate average δD values of meteoric surface waters, based principally on the data of Friedman et al. (1964), Dansgaard (1964), and Hitchon and Krouse (1972).

limits, and paleotemperature studies on carbonate fossils indicate that the ocean has had a $\delta^{18}O \approx 0$ throughout the Mesozoic and Cenozoic.

The assumption that the isotopic composition of seawater has been essentially constant with time is important, obviously, if seawater is itself involved in processes of hydrothermal alteration or ore deposition. Equally important, however, is the fact that the isotopic composition of the oceans *controls* the isotopic composition of all meteoric waters as well. The intercept of the meteoric water line must shift if the isotopic composition of ocean water shifts in any direction other than parallel to the meteoric water line. We can be reasonably confident that the meteoric water line has been essentially in its present position for at least the last 150 million years ($\pm 1‰$ for $^{18}O/^{16}O$ and $\pm 10‰$ for D/H), but prior to that time the position of the meteoric water line is intimately tied into the problem of the isotopic evolution of ocean waters.

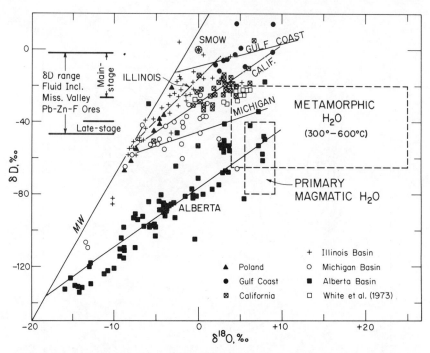

Fig. 6.4 Plot of δD vs. $\delta^{18}O$ for oil-field brines from North America (Clayton et al., 1966; Hitchon and Friedman, 1969; Kharaka et al., 1973) and Poland (Dowgiallo and Tongiorgi, 1972). Also shown are SMOW (standard mean ocean water), the calculated fields of primary magmatic and metamorphic waters (see text), some analyses of some possible modern metamorphic waters from the California Coast Ranges (White et al., 1973), the meteoric water line (MW) of Craig (1961), and the δD values of the Mississippi Valley ore fluids (Hall and Friedman, 1963).

Although evidence is beginning to build up that ocean waters have indeed been relatively constant in δD and $\delta^{18}O$ throughout most of geologic history, there are major uncertainties concerning the isotopic composition of ancient ocean waters. Conceivably, there may have been $\delta^{18}O$ depletions of as much as 20 per mil in the early Precambrian oceans (e.g., see Perry, 1967), but detailed discussion of these problems is beyond the scope of this chapter (see Taylor, 1974a, and Taylor and Magaritz, 1975, for a more extensive discussion).

Geothermal Waters

The $\delta^{18}O$ and δD variations in hot waters and steam from various geothermal areas throughout the world show that essentially all of the H_2O in these hot spring areas is of surface derivation (Craig, 1963). In almost all cases the hot water or steam shows a characteristic ^{18}O shift away from the meteoric water line to higher δ^{18}O values as a result of isotopic exchange with silicate and carbonate country rocks, all of which start out with δ^{18}O values higher than +5.5. The ^{18}O shift may be as small as 1 or 2‰ (e.g., in volcanic areas with a high water/rock ratio such as Wairakei, New Zealand) or as much as 15‰ (in very hot waters circulating through high-^{18}O sediments, such as in the Salton Sea geothermal brines).

In contrast to the ^{18}O/^{16}O ratios, the D/H values of the hot waters in these geothermal areas are *not* controlled by the exchange process; this is because the rocks contain so little initial hydrogen compared to the amounts of the H_2O involved. Instead, the δD values either remain constant, identical to the local meteoric waters, or are systematically enriched in deuterium along linear trends having a slope of approximately 3. The latter is due to nonequilibrium evaporation of steam at temperatures of 70-90°C (Craig, 1963).

"Connate" Formation Waters (Brines)

Oil-field brines at one time were thought to largely represent connate water or original trapped seawater. However, isotopic evidence obtained by Clayton et al. (1966) and Hitchon and Friedman (1969) show that meteoric ground waters are a major constituent of these brines in the mid-continent region of North America (Figure 6.4). These circulating groundwaters have increased in salinity either because of solution of evaporites and/or because of shale-membrane filtration (see Chapter 4). All sedimentary basins probably contain formation waters analogous to those found in wells from the midcontinent of North America. Such formation waters are probably the most common type of pore solution in deeply buried, unmetamorphosed sedimentary-rock sections.

Formation waters show a very wide range in δ^{18}O, δD, and salinity, but the waters within a given sedimentary basin are usually isotopically distinct. Just as with the surface meteoric waters, there is a general decrease in δD (and to a lesser extent δ^{18}O) as one moves to higher latitudes. For example, the Gulf Coast formation waters are appreciably richer in δD and δ^{18}O than the Alberta Basin waters (Figure 6.4). Within a given basin the highest δ^{18}O values are typically associated with the highest temperatures and salinities, but in contrast to the near-neutral, chloride-type geothermal waters, the δD values of the brines are not constant; instead, although they show a great deal of scatter, the δD values generally increase with salinity and δ^{18}O content. This is a result of mixing of the meteoric waters with true connate waters or waters of other origins, exchange with hydroxyl-bearing clay minerals in the rocks, to fractionation effects such as membrane-filtration, and/or to reactions involving petroleum hydrocarbons.

Metamorphic Waters

All the waters so far discussed represent materials for which we can obtain actual samples for isotopic analysis. However, for some other geologically important waters, isotopic data can only be indirectly obtained. Metamorphic pore waters are an example, because except for the possibility of making measurements on the fluid inclusions present in certain minerals, we must use calculations to obtain the isotopic composition of the H_2O coexisting in equilibrium with the rocks at the temperature and pressure at which metamorphism takes place.

Typically, regional metamorphic waters appear to have a relatively restricted range of δD (-20 to -65) but a wide range of $\delta^{18}O$ ($+5$ to $+25$), as shown in Figure 6.4. Because the isotopic fractionation factors are temperature-dependent, low-temperature metamorphic waters will typically have high δD and low $\delta^{18}O$, while the higher temperature waters will have lower δD and higher $\delta^{18}O$ within the approximate ranges quoted above. The general range of δD values in metamorphic minerals is almost identical to that exhibited by marine sedimentary rocks and altered volcanic rocks, and it is plausible that the metamorphic δD values are simply inherited from their precursor parent rocks. However, we also cannot rule out the introduction of large amounts of magmatic H_2O into the metamorphic environment, because the δD values of magmatic waters are very similar (see below).

Metamorphic muscovites are richer in deuterium than coexisting chlorites, which are in turn richer in deuterium than coexisting biotites and hornblendes. However, all of these metamorphic minerals typically have δD values in the narrow range -35 to $-85°/_{oo}$. The few samples known with $\delta D < -85$ are either iron-rich minerals or from contact metamorphic zones of epizonal intrusions that probably have interacted with heated meteoric waters.

The wide range of $\delta^{18}O$ in metamorphic waters is due to the fact that metamorphosed sedimentary and igneous rocks in large part retain their original $\delta^{18}O$ values during metamorphism. Shales, limestones, and cherts all tend to be very rich in ^{18}O ($\delta = +15$ to $+35$), whereas igneous rocks and sandstones, graywackes, arkoses, and volcanogenic sediments tend to be low in ^{18}O ($\delta = +5$ to $+13$).

There is a possibility that some metamorphic waters actually do make their way to the Earth's surface. White et al. (1973) have identified some isotopically unusual geothermal waters in the California Coast Ranges, which are a rare exception to the above-mentioned rule that hot spring waters are everywhere wholly of surface derivation. These anomalous waters are plotted on Figure 6.4, and they overlap the calculated field of metamorphic waters.

Magmatic Waters

Magmatic waters present some of the same problems referred to above concerning metamorphic waters. Nowhere, even in H_2O gas samples from active volcanoes, can we be absolutely certain that we are obtaining H_2O that has come directly from a pristine, uncontaminated magma. As before, we can, however, calculate the isotopic composition of the H_2O that would have coexisted with the various igneous minerals at magmatic temperatures. Inasmuch as crustal magmas only exist over a restricted temperature range of approximately 700–1100°C, and because most volcanic and plutonic igneous rocks typically have very uniform $\delta^{18}O$ values (+5.5 to +10.0) and δD values (−50 to −85) the "normal" isotopic range for magmatic H_2O (Figure 6.4) is much more restricted than the range for metamorphic H_2O.

True magmas with unusual oxygen isotopic compositions well outside the normal range do exist, however. Some of these rare examples are fresh volcanic flows on Iceland (Muehlenbachs et al., 1974) that have $\delta^{18}O$ values as low as +2, and at the other end of the spectrum unaltered flows from some Pleistocene volcanoes north of Rome, Italy, which have $\delta^{18}O$ values as high as +16 (Taylor and Turi, 1976). Any magmatic H_2O coming from such anomalous magmas would have $\delta^{18}O$ values outside the "normal" range.

The total δD range of *magmas* is at present not so well understood, although it is likely that most of the magmas in the world originally had δD values in the range −50 to −85. The term "primary magmatic water" (Figure 6.4) is thus defined somewhat arbitrarily as the calculated H_2O in equilibrium with these "normal" igneous rocks or magmas at $T \geq 700°C$. Note also that a δD value of −60 was obtained for the original H_2O in a carefully studied, recent, chilled, submarine basalt flow from Hawaii (Moore, 1970).

Nonetheless, quite a significant number of plutonic biotites and hornblendes are known with δD values ranging down to −170 (see compilation by Taylor, 1974a). These low δD values may largely be a result of postcrystallization exchange between the rocks and heated meteoric ground waters, but in most instances we cannot rule out the possibility that the δD values are a characteristic of the magmas themselves.

It is unfortunately very difficult at present to develop criteria to decide whether the low δD values are a magmatic phenomenon or are a result of later exchange. With oxygen isotopes this is feasible because the different minerals in an assemblage undergo ^{18}O exchange at vastly different rates (e.g., quartz versus feldspar). For example, most of the low-^{18}O igneous rocks throughout the world are *known* to have been produced by high-temperature interaction between the hot, solidified rocks and low-^{18}O meteoric groundwaters, because the quartz will typically have essentially a "normal" $\delta^{18}O$ value whereas the coexisting feldspar has been depleted in ^{18}O by several per mil (e.g., see Taylor and Forester, 1971).

Magma bodies emplaced at shallow levels in the Earth's crust may directly or indirectly interact with meteoric groundwaters (i.e., by direct influx of H_2O into the magma, or by stoping and subsequent exchange with, or assimilation of, hydrothermally altered roof rocks, see Burnham, Chapter 3, and Taylor, 1974b). Such a process definitely produced low-^{18}O magmas in some instances (Forester and Taylor, 1972; Friedman et al., 1974). Because of the small amounts of H_2O present in most magmas, such processes should much more readily produce a change in the δD value of a magma than in the $\delta^{18}O$ value. Therefore, water *derived from such a contaminated magma* may in some instances have a very low δD value, much lower than so-called "primary magmatic water".

Because of the similarity of D/H ratios in most igneous and metamorphic rocks, it is basically impossible to distinguish between high-temperature metamorphic waters and primary magmatic waters by means of δD values alone. Also, note that the δD values of most meteoric waters in temperate climates are indistinguishable from such deep-seated waters (see Figure 6.3). This similarity of δD in most igneous and metamorphic rocks might be ascribed to coincidence, but it is more likely because "primary magmatic water" is probably ultimately derived from the OH-bearing minerals of subducted sediments and altered volcanic rocks. If true "juvenile" water exists at all, it has not been identified, nor has its δD value been established. If sea floor spreading has proceeded at its present rate during a major fraction of geologic time, "primary magmatic water" may largely represent recycled marine clays, chlorites, and micas. The latter typically have $\delta D = -50$ to -85 because of exchange with ocean waters at low temperatures (see Figure 6.2 above, and Savin and Epstein, 1970b; Magaritz and Taylor, 1976a). These subducted materials may either be directly melted or they may be dehydrated at high temperatures in the upper mantle; in either case the H_2O would end up incorporated into magmas.

INTERACTIONS BETWEEN SURFACE-DERIVED WATERS AND IGNEOUS INTRUSIONS

In recent years it has become well established by means of oxygen and hydrogen isotope analyses that certain igneous intrusions have interacted on a very large scale with meteoric groundwaters. In favorable terranes, namely in highly jointed, permeable rocks, these intrusions act as gigantic "heat engines" that provide the energy necessary to promote a long-lived convective circulation of any mobile H_2O in the country rocks surrounding the igneous body. These systems probably represent the "fossil" equivalents of the deep portions of modern geothermal water systems such as occur at Wairakei, New Zealand; Steamboat Springs, Nevada; and Yellowstone Park, Wyoming (e.g., see Ellis, Chapter 13; Banwell, 1961; White, 1968b). The interaction and transport of large amounts of meteoric ground waters through the hot igneous rocks produces a depletion of ^{18}O in the igneous rocks and a corresponding ^{18}O enrichment or "^{18}O-shift" in the water.

Low-^{18}O igneous rocks produced by interaction with Mesozoic and Tertiary meteoric groundwaters have now been observed in the Skaergaard intrusion, in the San Juan volcanic field of Colorado, in the Scottish Hebrides, in Iceland, in the Western Cascade Range in Oregon, and in major portions of the Boulder batholith, the Idaho batholith, and the Coast Range batholith of British Columbia, as well as locally in the Southern California batholith (Taylor and Epstein, 1963, 1968; Taylor, 1968, 1971, 1973, 1974b; Taylor and Forester, 1971, 1973; Forester and Taylor, 1972; Sheppard and Taylor, 1974; Muehlenbachs et al., 1974; Magaritz and Taylor, 1976b). Precambrian examples are also known from 650-m.y.-old granite batholith of the Seychelles Islands (Taylor, 1974c), the 1450-m.y.-old St. Francois Mountains granite-rhyolite terrane of Missouri (Wenner and Taylor, 1976), and from some > 2.6-b.y.-old granites from Swaziland and Rhodesia (Taylor and Magaritz, 1975).

In most of the examples listed above, the meteoric-hydrothermal fluids were not obviously connected in any direct fashion with ore deposition. However, many such examples of meteoric-hydrothermal ore fluids are known (Chapter 15) and will be described below. The importance of the above studies with regard to this question is that they define the enormous scale and ubiquitous occurrence of such meteoric-hydrothermal convective circulation systems in nature. Given their abundance and demonstrated presence in the country rocks surrounding essentially all epizonal igneous intrusions as well as their presence within many of the more deeply seated batholiths, it is easy to understand why, in combination with a few other favorable factors, heated meteoric waters might well locally develop into ore-forming fluids.

The igneous complexes that are abnormally low in ^{18}O characteristically display the following geologic, petrologic, and isotopic features.

1. The most profound isotopic effects are found in intrusions emplaced into young, highly jointed, volcanic rocks that are particularly permeable to groundwater movement, notably in ring-dike intrusions emplaced into volcanic calderas.
2. In a given rock, the feldspars are commonly depleted in ^{18}O to a greater degree than the other coexisting minerals, and the feldspars very commonly show a "clouding" or turbidity.

3. The primary igneous pyroxenes, biotites, and hornblendes are usually partially altered to uralitic amphibole, chlorite, iron-titanium oxides, and/or epidote; locally, this process has gone to completion and only pseudomorphs of the primary mafic minerals remain.
4. Granophyric (micrographic) intergrowths of turbid alkali feldspar and quartz are ubiquitous.
5. Miarolitic cavities are locally present in the intrusives, and veins filled with quartz, alkali feldspar, epidote, chlorite, or sulfides are very common in both the intrusives and the surrounding country rocks.
6. The OH-bearing minerals usually have abnormally low δD values relative to "normal" igneous rocks.
7. The whole rock $\delta^{18}O$ values in the intrusions and the country rocks can usually be systematically contoured and the $\delta^{18}O$ contours are commonly centered on the intrusive body, forming concentric ovals that increase in $\delta^{18}O$ outward. "Normal" $\delta^{18}O$ values are commonly attained within approximately 1-3 stock diameters away from the intrusive contact, and the lowest $\delta^{18}O$ values are commonly found right at the contact ($\delta^{18}O$ commonly as low as -3 to -7).
8. Major influx of meteoric H_2O into the intrusion only occurs subsequent to solidification, after the body is sufficiently rigid to fracture.

Submarine hydrothermal systems produce some of the same types of effects as those described above, except that heated ocean waters are involved rather than heated meteoric waters, so the Na/K ratios are very high. The isotopic effects are also less because of the smaller isotopic contrast between ocean water and magmatic water. However, given favorable circumstances, such effects have been clearly documented in submarine dredge samples and in ophiolite complexes (Muehlenbachs and Clayton, 1972; Spooner et al., 1974; Magaritz and Taylor, 1974; Wenner and Taylor, 1973).

AMOUNTS OF METEORIC WATER INVOLVED IN HYDROTHERMAL SYSTEMS

We know the initial $\delta^{18}O$ values of unaltered basaltic and andesitic igneous rocks with a good deal of accuracy: $\approx +6.5 \pm 1 °/_{oo}$. If other rock types are involved, it is usually possible to estimate their original $\delta^{18}O$ values by analyzing rocks outside the hydrothermally altered area. Using deuterium/hydrogen analyses of the alteration mineral assemblages or of fluid inclusions, we can apply the meteoric water equation and calculate the initial $\delta^{18}O$ of the meteoric waters at any given locality. Given the above parameters together with some temperature information, it is then possible to calculate the total amounts of water involved in any meteoric-hydrothermal convective system, as follows:

$$w\delta^i_{H_2O} + r\delta^i_{rock} = w\delta^f_{H_2O} + r\delta^f_{rock}$$

where i is the initial value, f is the final value after exchange, w is the atom percent of meteoric water oxygen in the total system, and r is the atom percent of exchangeable rock oxygen in the bulk system.

In order to complete the calculation one must assume that $\delta^f_{H_2O}$ is determined by isotopic equilibration with the rocks at the temperature of hydrothermal alteration. Then:

$$\frac{w}{r} = \frac{\delta^f_{rock} - \delta^i_{rock}}{\delta^i_{H_2O} - (\delta^f_{rock} - \Delta)}$$

where $\Delta = \delta^f_{rock} - \delta^f_{H_2O}$. Note that for a given set of initial conditions, if w/r is a constant, then δ^f_{rock} is determined solely by Δ, which is only a function of temperature. Conversely, if the temperature is constant, δ^f_{rock} is controlled only by the w/r ratio.

If we make the reasonable approximation that in these systems δ_{rock} at equilibrium is equal to the $\delta^{18}O$ value of plagioclase (An_{30}), then we can utilize the feldspar-H_2O geothermometer to calculate Δ at any temperature [$\Delta \approx 2.68\,(10^6/T^2) - 3.53$, O'Neil and Taylor, 1967].

The above calculations give the w/r ratios integrated over the lifetime of the hydrothermal system, assuming continuous recirculation and re-equilibration of the H_2O. However, some of the heated water will be lost from the system, for example by escape to the surface. In the extreme open-system case in which the water makes only a single pass through the system, the integrated w/r ratio is given by the equation:

$$\frac{w}{r} = \ln\left[\frac{\delta^i_{H_2O} + \Delta - \delta^i_{rock}}{\delta^i_{H_2O} - (\delta^f_{rock} - \Delta)}\right]$$

For each of the two cases discussed above, we can calculate the w/r ratios as a function of $\delta^{18}O^f_{rock}$ and temperature; these two cases are compared for the specific parameters 500°C, 300°C, and $\delta^{18}O^i_{H_2O} = -14$ in Figure 6.5. In actual hydrothermal systems, the true 500°C curve would lie between the plotted 500°C curves on Figure 6.5, probably very close to the right-hand curve, as the single-pass system is a very unrealistic one. Note, however, that both types of calculations are strongly dependent on temperature and, in both cases, the lower the temperature the higher the required w/r ratio. Also, these models only give minimum values of w/r, because appreciable H_2O

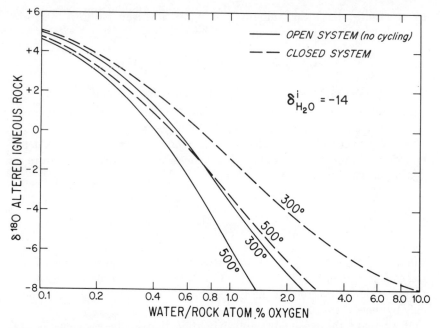

Fig. 6.5 Plot of $\delta^{18}O$ values of hydrothermally altered rocks vs. calculated water/rock ratio for the open system and closed system models discussed in the text, assuming initial values of $\delta^i_{rock} = +6.5$ and $\delta^i_{H_2O} = -14$. These values are very appropriate, for example, for the hydrothermal alteration effects observed in Western San Juan Mtns., Colorado and the western Nevada Au-Ag deposits.

may move through fractures in the rocks without exchanging (i.e., after the wall rocks next to the veins have become markedly depleted in ^{18}O).

The principal features illustrated by such calculations, as applied to the meteoric-hydrothermal systems discussed above, are that a wide range of water/rock ratios is required, from essentially zero up to values greater than 3 or 4. The average values at any one locality are seldom greater than one, however. Overall w/r values much higher than unity are also not compatible with the limited quantities of heat energy that are available to drive the convection systems. A simple calculation shows that even if the propylitically altered country rocks are only heated from about 50°C to an average temperature of 250°C, then with a specific heat of 0.3 cal/g approximately 60 cal/g of heat must be added throughout the alteration zone. With w/r ≈ 1, about 0.5 g of H_2O have to be heated for every gram of rock, thus increasing this to approximately 160 cal/g of altered rock. From a magma body initially intruded as a liquid at 950°C, the maximum amount of heat that can be obtained during crystallization and cooling to 300°C is only 275 cal/g (this includes about 80 cal/g for the latent heat of crystallization). Some heat is also available from the exothermic hydration reactions that occur during hydrothermal alteration, perhaps approximately 30 cal/g. Taking all this into account, a cylindrical stock of magma only contains enough heat to produce an alteration aureole approximately 0.4-0.5 stock diameters wide. While this is similar to the observed diameters of the alteration aureoles in the Scottish Hebrides (Taylor and Forester, 1971), much larger aureoles are found in many other examples of this type, notably in the Western Cascades and the Western San Juan Mountains (Taylor, 1971, 1974b).

In such cases, the stocks must either broaden in cross-sectional area with depth, or the alteration zones must be underlain by hidden intrusive bodies that do not outcrop. Another possibility is that the intrusions do not represent a single injection of magma, but instead are the result of episodic addition of new magma from below, as would be expected in a volcanic conduit to the surface (see Chapter 3). Inasmuch as most of these areas represent deeply eroded volcanic centers, the repeated intrusion and resupply of new magma that are known to take place in such localities probably removes any heat-balance problem.

"EPITHERMAL" ORE DEPOSITS IN VOLCANIC TERRANES

As far as ore deposits are concerned, the closest analogues to the types of meteoric-hydrothermal systems described above are the so-called "epithermal" or near-surface gold-silver and base metal deposits associated with volcanic centers. Indeed, the mineralogic and isotopic effects are virtually identical in all such terranes, irrespective of whether or not the volcanic centers are directly associated with major ore deposits. In addition, two present-day hot spring systems, Steamboat Springs, Nevada, and Broadlands, New Zealand, are presently depositing gold, silver, and base-metal sulfides (White et al., 1964; Browne, 1971).

Western Nevada Gold-Silver Deposits

The isotopic compositions of the hydrothermal fluids associated with several gold-silver districts in Nevada are shown on Figure 6.6. Each of these vein-type ore deposits occurs in an extensive area of low-^{18}O propylitically altered volcanic rocks. All of the hydrothermal vein minerals such as quartz and adularia that are associated with ore deposition also have remarkably low δ^{18}O values. At Tonopah, both pre-ore and post-ore vein quartz samples show a very narrow range of δ^{18}O from -5.1 to $-1.8^{\circ}/_{\circ\circ}$, and even the very latest stage, tiny, euhedral quartz crystals that line vugs have δ^{18}O $= -3.2$

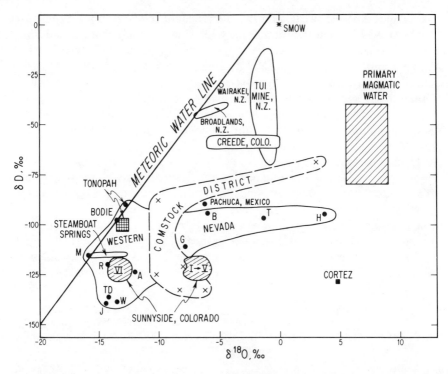

Fig. 6.6 Plot of δD of fluid inclusions vs. calculated $\delta^{18}O$ values of hydrothermal fluids in various "epithermal" ore deposits (mainly Au-Ag deposits in volcanic rocks). Solid dots and crosses (O'Neil and Silberman, 1974); Cortez (Rye et al., 1974a); ore stage periods I to V and VI at Sunnyside (Casadevall et al., 1974); Creede (Bethke et al., 1973); and Tui, New Zealand (Robinson, 1974). Also shown is the calculated ore fluid at Tonopah (Taylor, 1973) and some data on modern hot spring systems that have characteristics similar to these "epithermal" deposits (see White, 1974; Giggenbach, 1971; and Craig, 1963). Abbreviations: H = Humboldt, T = Tenmile, B = Bullfrog, G = Gilbert, A = Aurora, W = Wonder, TD = Trade Dollar, R = Rawhide, J = Jarbidge, M = Manhattan.

to +0.7 (Taylor, 1973). At Bodie, vein quartz has $\delta^{18}O = -3.2$ to -0.1 and adularia has $\delta^{18}O = -5.1$ to -3.0 (O'Neil et al., 1973). At the Comstock Lode, vein quartz has somewhat higher $\delta^{18}O = +0.2$ to $+3.2$ with one sample having $\delta^{18}O = +8.9$ (O'Neil and Silberman, 1974; Sugisaki and Jensen, 1971; Taylor, 1973). Various lines of evidence, including fluid-inclusion filling temperatures and isotopic geothermometers, suggest temperatures of approximately 200–300°C for these deposits (e.g., Nash, 1972). If these temperatures are valid, the $\delta^{18}O$ values of the quartz and adularia can be used to calculate the $\delta^{18}O$ values of the hydrothermal fluids (Figure 6.6). The D/H ratios are calculated from the δD values of the OH-bearing alteration minerals or determined directly from the fluid inclusions. For example, at Tonopah and Bodie the fluids must have had $\delta^{18}O \approx -13$ and $\delta D \approx -100$. This means that not only are these fluids derived from meteoric waters, they are practically pristine, unexchanged meteoric waters that apparently have not undergone the characteristic ^{18}O shift shown by most geothermal waters in hot-spring areas on Earth. In this sense they would be analogous to the Wairakei geothermal waters, and they must represent long-continued circulation through fractures where the wall-rocks had been previously depleted in ^{18}O.

The isotopic characteristics of these ore fluids imply that the solutions probably carried very low concentrations of the heavy metals and that extremely large amounts of water were involved. This is compatible with the chemical data on fluid inclusions from such deposits, which indicate only approximately 1-3% NaCl equivalent or less (Nash, 1972). At the Comstock Lode and Bodie there are outcrops of low-^{18}O intrusive stocks that were in part the source of the heat necessary to drive the meteoric-hydrothermal convection systems. However, at Tonopah and Goldfield there are no outcrops of such epizonal intrusions, and they must be inferred to exist at depth. Nolan (1935) postulated the existence of such a stock at Tonopah, and from the topography of the ore zone he indicated that it probably lay beneath the center of the district. The $\delta^{18}O$ data obtained on the altered volcanic rocks at Tonopah can be contoured, defining a central zone with $\delta^{18}O < -6$ that closely coincides with the apex of the productive ore zone as outlined by Nolan (1935). This low-^{18}O zone represents either the locus of the highest temperature solutions in the district, and/or the zone where the most H_2O was pumped through; in either case, it indicates the position of the high-temperature plume of upward-moving meteoric H_2O, which was presumably centered above the heat source (see Figure 26 in Taylor, 1974a).

In many ways, the most interesting of these western Nevada deposits is the Comstock Lode, although it is anomalous in that it contains the highest-^{18}O vein quartz and the fluid inclusions show a very wide range in δD (O'Neil and Silberman, 1974). A deep-level sample of high-grade "bonanza-type" silver ore contains high-^{18}O quartz ($\delta = +8.9$) with primary fluid inclusions ($T = 295°C$, 3.1% NaCl equivalent) that have $\delta D = -69$, which is 50-65‰ higher than the typical inclusions observed elsewhere in the Comstock district or in nearby deposits. As shown in Figure 6.6, these isotopic data can be explained by mixing between magmatic and meteoric waters in the deeper zones of the Comstock Lode, compatible with the relatively high calculated temperatures and salinities.

The ore fluids in three deposits on Figure 6.6, Humboldt, Tenmile (O'Neil and Silberman, 1974) and Cortez (Rye et al., 1974a), show much more pronounced ^{18}O shifts than the others. This can probably be explained by the fact that these three deposits all were emplaced into less permeable, high-^{18}O sedimentary rocks rather than volcanic rocks, so the water/rock ratios would be smaller and the exchanged hydrothermal fluids would also attain higher δ^{18}O.

San Juan Mountains, Colorado

The San Juan volcanic field is a very large pile of mid-Tertiary rhyolitic and andesitic volcanic rocks that have been extensively hydrothermally altered, particularly in the vicinity of epizonal intrusions emplaced along caldera ring-fractures (Lipman et al., 1970). This meteoric-hydrothermal alteration has produced extensive areas of low-^{18}O, low-deuterium rocks, but there are marked isotopic differences between the western San Juans in the vicinity of Ouray and Silverton (where the altered rocks commonly have δ^{18}O = -6 to 0 and δD = -150 to -130) and the eastern San Juans in the vicinity of Creede and Summitville. The rocks in the eastern San Juans typically show higher δ^{18}O and 35–60‰ higher δD values, probably reflecting climatic differences in the two regions, even though they are only about 100 km apart (Taylor, 1974b).

These regional isotopic differences also show up in the ore deposits. Data for the Sunnyside gold-silver-base metal deposit in the Silverton area and the Creede silver-lead-zinc-copper deposit in the eastern San Juan Mountains are shown in Figure 6.6. Both deposits show the characteristic ^{18}O-shift typical of modern meteoric-hydrothermal sytems, and at Sunnyside the early, higher temperature (260–330°C) mineralization of Periods I to V shows a 6‰ greater ^{18}O-shift than the late, lower-temperature (170–240°C) Period VI mineralization (Casadevall et al., 1974). At Creede, the δD of the ore fluid was about 60‰ higher than at Sunnyside, lying in the range of magmatic H$_2$O; however, the δ^{18}O$_{H_2O}$ values of -6 to 0 leave no doubt that these ore fluids were dominantly of meteoric origin (Bethke et al., 1973).

PORPHYRY COPPER AND MOLYBDENUM DEPOSITS

As indicated above, deep circulation and interaction between surface waters and epizonal intrusions is to be expected in volcanic terranes; *it is a common phenomenon*. Profound and extensive δ^{18}O effects are produced because the meteoric-water/rock ratios are very high. What happens, however, if the country rocks are sedimentary rocks or Precambrian crystalline rocks where the permeabilities are not so great? The δ^{18}O effects will be smaller, and they may be so small, given the natural isotopic variations in igneous rocks, that no clear-cut δ^{18}O changes may be discernible.

However, even if the $\delta^{18}O$ effects are very small, the δD values of fluid inclusions or of OH-bearing minerals will be affected by the influx of external waters, because the initial H_2O concentrations of most igneous rocks are so small.

Another problem exists with D/H ratios, however. This is the fact that the most common meteoric waters on Earth have δD values that are similar to primary magmatic waters. Over much of the North American continent, for example, the surface meteoric waters (Figure 6.3) and the saline formation waters (Figure 6.4) have δD values of -40 to -90; these values are too close to the range of magmatic values to provide any test of discrimination.

Because of the obstacles mentioned above, studies of intrusive bodies that involve a small water/rock ratio can only be definitive in certain geographic areas. Favorable regions would include submarine environments where $\delta D \approx 0$ or northerly regions such as Montana or British Columbia where δD_{H_2O} can be lower than -130. Areas of high elevation can in principle also be used, but generally it is difficult to ascertain the original surface elevation above a specific ore deposit. At least for Tertiary ore deposits, latitude is a much more definite parameter than original surface elevation.

A proper approach then might be to examine a certain class of ore deposits that occur over a wide geographic area. The isotopic composition of minerals formed by magmatic-hydrothermal solutions would be expected to be independent of latitude, whereas the minerals formed from meteoric-hydrothermal or "connate"-hydrothermal solutions should show the proper latitudinal dependence in δD (and perhaps to some extent in $\delta^{18}O$). The above approach was utilized by Sheppard et al. (1969; 1971) and Sheppard and Taylor (1974) in their studies of porphyry copper deposits in North America. Their isotopic data are reproduced in Figure 6.7, together with all other available isotopic data on OH-bearing minerals from hydrothermal ore deposits.

The δD values of hypogene clay minerals and sericites from the deposits shown on Figure 6.7 become lower as one moves north from Arizona and New Mexico (Santa Rita, Safford, Copper Creek, Mineral Park) to Utah, Colorado, and Nevada (Bingham, Ely, Gilman, Climax) to Montana, Idaho, and Washington (Butte, Wickes, San Poil, Ima). It is thus clear that the δD values of these minerals are a result of interaction with meteoric waters, the only question is whether or not they represent the original δD values of the sericites and clays or are a result of later exchange.

The possibility of later exchange cannot be totally ruled out, but in order to explain all of the $\delta D-\delta^{18}O$ relationships this would have to be a pervasive exchange with heated meteoric waters, not simply partial low-temperature exchange with groundwaters (see supergene discussion below).

The δD values of the hypogene clays and sericites at a given deposit such as

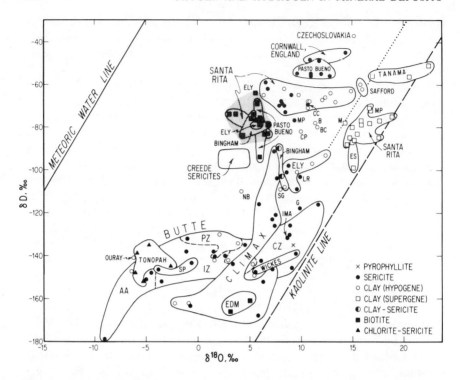

Fig. 6.7 Plot of δD vs $\delta^{18}O$ for most of the presently available analyses of OH-bearing minerals from ore deposits (Sheppard et al., 1969; 1971; Sheppard and Taylor, 1974; Hall et al., 1974; Landis and Rye, 1974); Bethke et al., 1973; and Taylor, 1973). The kaolinite line of Savin and Epstein (1970a) is shown for reference. The dotted line is drawn merely to emphasize the separation of supergene and hypogene clays. Abbreviations: CP = Cerro de Pasco, CC = Copper Creek, B = Bethlehem, M = Morenci, ES = El Salvador, MP = Mineral Park, LR = Lost River, Alaska, BC = Bond Creek, Alaska, G = Gilman, SG = St. George, NB = New Boston, and at Butte AA = Advanced Argillic alteration; PZ = Peripheral Zone, IZ = Intermediate Zone, CZ = Central Zone, EDM = Early Dark Micaceous alteration. The stippled pattern indicates the range of isotopic values in biotites from porphyry copper deposits.

Santa Rita are much more uniform than would be expected on the basis of variable degrees of partial exchange with such groundwaters. Therefore, the D/H data require that a geothermal H_2O circulation system was present at each deposit. In addition, the very low $\delta^{18}O$ values in sericites and clays from some of the northerly deposits (Butte, Climax) independently argue that these effects are not due to secondary exchange, as also does the similarity in δD values of sericites of vastly different grain size from the Climax deposit (Sheppard et al., 1971). Hall et al. (1974) also demonstrated very nicely that the sericites at Climax preserved their δD values even though nearby potassium-feldspars were drastically depleted in ^{18}O by exchange with heated groundwaters.

In striking contrast to the sericites and clays, the δD and δ^{18}O values of the hydrothermal biotites from porphyry copper deposits form a very tight grouping similar to "normal" igneous biotites and show no relationship to latitude (Figure 6.7). Biotites from the Panguna deposit, Bougainville (J. Ford, unpublished manuscript) and from the El Salvador deposit in Chile (Sheppard and Gustafson, 1976) also have this characteristic isotopic composition. The only biotites from ore deposits known to fall outside this narrow range are the very fine-grained EDM biotites from Butte and the biotites from the Copper Canyon porphyry deposit, Nevada (Batchelder and Blake, 1975). The EDM biotites at Butte probably exchanged with the low-δD, Main-Stage ore fluids at Butte, because the samples come from areas where there has been intense, subsequent Main-Stage mineralization (Sheppard and Taylor, 1974). It is doubtful, however, that this explanation can apply to the Copper Canyon body, because that deposit contains relatively minor sericitic and clay alteration. Instead, this might represent an example of low-δD magmatic water produced by prior interaction of the Copper Canyon magma body with meteoric H$_2$O.

The hydrothermal biotites in porphyry copper deposits thus in general appear to have formed from H$_2$O with an isotopic composition that coincides with the hypothetical field of primary magmatic waters, whereas the clays and sericites appear to have formed from heated meteoric waters of some type. The amounts of H$_2$O, however, are much less than in the low-^{18}O volcanic-intrusive terranes discussed previously. This implies that somewhat similar, but less extensive hydrothermal convective circulation systems are also established around the epizonal porphyry copper stocks, and that the potassium-silicate alteration assemblages in the cores of the stocks are in general probably not related to these external systems. A complicating factor, however, is that the coexisting quartz and potassium-feldspar are consistently out of isotopic equilibrium as a result of postdepositional ^{18}O exchange between the potassium-feldspars and aqueous solutions.

The only porphyry molybdenum deposit yet studied in any detail is Climax (Hall et al., 1974). It differs from typical porphyry copper deposits in its major ore mineralization (molybdenum and tungsten), its lack of hydrothermal biotite, and the presence of fluorite, muscovite, and topaz. Nonetheless, as shown by data both on fluid inclusions and on hydroxyl minerals (Figures 6.7 and 6.8), waters of at least two origins were present during the complex history of multiple intrusion and repeated ore mineralization of the Climax stock. An early-stage water with high salinity (up to 18%) and δD ≈ -90 was probably predominantly of magmatic derivation, and formed the Ceresco and Lower ore bodies; it was followed by a late-stage water with δD

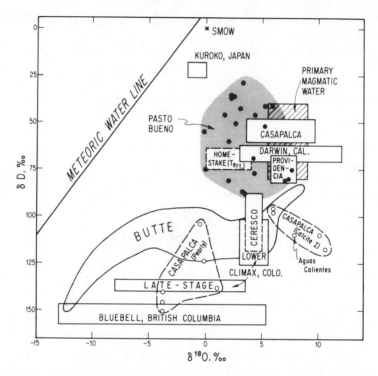

Fig. 6.8 Plot of the range of δD of fluid inclusions vs the calculated $\delta^{18}O$ values of hydrothermal fluids in a variety of ore deposits (Ohmoto and Rye, 1970; 1974; Rye, 1966, Rye et al., 1974b; Rye and Sawkins, 1974; Hall et al., 1974). The black dots and the heavy stippled pattern indicate the isotopic range in the Pasto Bueno veins (Landis and Rye, 1974), excluding the extreme late-stage meteoric-hydrothermal fluids, which have δD = −110 to −145. For the Homestake district, only the Tertiary mineralization is indicated (Rye and Rye, 1974). Also shown are the calculated Main-Stage fluids at Butte (Sheppard and Taylor, 1974).

≈ −120 to −140 that was predominantly of meteoric origin. The favorable geographic location of this stock insures a clear-cut distinction between the two types of fluids. The late-stage fluids were produced in several pulses, were responsible for the major silicification, and had salinities of 1-6%. The intermediate stages of molybdenum mineralization are attributed to mixed meteoric-magmatic fluids with salinities and δD values intermediate between the end-member values.

Thus, a basically similar genetic model can be developed to explain the isotopic relationships in both porphyry copper and molybdenum deposits. Magmatic-hydrothermal solutions under lithostatic pressure are formed during the late stages of crystallization in the upper, interior portions of a porphyry stock. Outside the stock, a meteoric-hydrothermal circulation is established under hydrostatic pressures that are probably only ⅓ as high as the pressures inside the stock. These external waters might be either meteoric groundwaters or saline formation waters, because the latter also display the proper isotopic dependence on latitude (Figure 6.4). Both hydrothermal systems are simultaneously present early in the history of the stock, but the external system will persist even after the magmatic-hydrothermal system fades away. With time, as the "heat engine" of the stock cools off, the external hydrothermal system collapses onto the hydrothermally altered rocks formed by the internal system; the argillic and sericite-pyrite alteration zones will then be locally superimposed upon the potassium-feldspar-biotite alteration zones or upon the fresh intrusive. The major locus of chalcopyrite deposition in the porphyry copper deposits tends to be along the boundary between the two different types of hydrothermal systems (see Lowell and Guilbert, 1970).

The above model envisages that although the H_2O has at least two major sources, the source of the sulfur, copper, and other heavy metals in these systems is basically the porphyry stock itself, with perhaps some contribution of materials leached from the surrounding country rocks. The size and intensity of the external hydrothermal system varies considerably in known porphyry copper deposits, from "wet" types (e.g., Bingham, Morenci) having high pyrite-to-chalcopyrite ratios and surrounded by enormous halos of pyrite-sericite-quartz, to "dry" deposits (e.g., Bethlehem) with relatively minor sericite-pyrite zones (Lowell and Guilbert, 1970). These variations in size of the external system are to be expected because the permeabilities and amounts of H_2O available in the country rocks must be quite variable.

SUPERGENE ALTERATION OF PORPHYRY COPPER DEPOSITS

Given favorable circumstances, stable isotope techniques can clearly distinguish between supergene and hypogene mineral assemblages. Clay-rich soils in the United States all plot very close to a curve labeled "kaolinite-line" on Figure 6.7 (Lawrence and Taylor, 1971); this line represents the locus of isotopic data points obtained in pure kaolinites from weathering zones (Savin and Epstein, 1970a). These data imply that kaolinites and montmorillonites formed during weathering are in approximate isotopic equilibrium with their coexisting meteoric waters, and that the fractionation factors are such that these clays are about 27‰ enriched in ^{18}O and about 30‰ depleted in deuterium relative to the water from which they formed (see Figures 6.1 and 6.2).

We should therefore expect supergene clay minerals to plot in the vicinity of the "kaolinite line" if they form in equilibrium with meteoric H_2O at

Earth-surface temperatures. Sheppard et al. (1969) showed this to be the case, if allowance is made for the fact that temperatures of supergene deposition may range up to 50-60°C because of the large amount of heat produced by oxidation of pyrite during the production of the acid supergene solutions. All of the presently available isotopic analyses of clay minerals from ore deposits are plotted on Figure 6.7, and based on geologic relations, the probable supergene clays are distinguished by a different symbol from the hypogene clays. Note that all of these supergene clays either plot on the "kaolinite line" or slightly to the left of it; the shift to the left is readily explained by slightly higher temperatures of formation than are involved in surface weathering.

At Santa Rita, which is the most extensively studied of these deposits, the deeper clay samples (below 5000-ft elevation) have $\delta^{18}O$ = 6.4-14.9 and δD = −62 to −71, whereas the shallow clay samples have $\delta^{18}O$ = 14.5-18.9 and δD = −71 to −88. On a δD—$\delta^{18}O$ diagram, such as Figure 6.7, there is a clear-cut gap between the shallow supergene samples and the deep hypogene samples.

MISSISSIPPI VALLEY-TYPE LEAD-ZINC-FLUORIDE DEPOSITS

Fluid inclusion, D/H, and $^{18}O/^{16}O$ studies of vein minerals and wallrocks from several Mississippi Valley deposits have demonstrated rather conclusively that these lead-zinc-fluorine vein deposits in Paleozoic sediments were formed from circulating saline formation waters at temperatures of approximately 70-180°C (Hall and Friedman, 1963; 1969; Pinckney and Rye, 1972; Roedder, 1972; Roedder et al., 1963; see review by Heyl et al., 1974). The salinities, chemical compositions, and the range of δD in the fluid inclusions are all remarkably similar to those observed in modern oil-field brines in the nearby Illinois basin (Figure 6.4) as discussed in detail in Chapter 4.

The $\delta^{18}O$ values of the ore fluids in a number of different localities have been calculated from $\delta^{18}O$ measurements on calcite, dolomite, and quartz, and these are also very similar to the $\delta^{18}O$ values observed in the oil-field brines. For example, at Cave-In-Rock the fluids had $\delta^{18}O$ = +1.4 to +4.7 (Pinckney and Rye, 1972), and in the Tri-State and Upper Mississippi Valley districts, the $\delta^{18}O_{H_2O}$ values in the late stages of mineralization were −3 and +0.4 to −1.6, respectively. The late-stage, lower-^{18}O fluids also tend to have lower salinities and δD values (down to δD = −45, see Figure 6.4). Exactly similar effects are also seen in the present-day oil-field brines, compatible with the hypothesis that the ore fluids mixed with relatively dilute meteoric groundwaters near the end stages of ore deposition (Hall and Friedman, 1963).

VOLCANOGENIC MASSIVE SULFIDE DEPOSITS

At present, very little $^{18}O/^{16}O$ and deuterium/hydrogen data are available on massive sulfide ore deposits. However, most recent models for these types of deposits envisage the operation of submarine hydrothermal systems, in which ocean water is probably a major constituent. The hot, metal-rich Red Sea brine may be such an ore fluid (see Craig, 1966). All isotopic studies of hydrothermal alteration in ophiolite complexes, for example, conclude that sea water was the major component of these hydrothermal solutions (Magaritz and Taylor, 1976a; Spooner et al., 1974). In addition, the ore fluids associated with the massive pyrite-chalcopyrite deposits in pillow lavas of the Troodos ophiolite on Cyprus also appear to be dominantly ocean water (S. M. F. Sheppard, personal communication). It is very important to extend these isotopic studies to Precambrian volcanogenic massive sulfide bodies, both because of their intrinsic interest, and because they might provide us with the best available estimates of the isotopic evolution of the Precambrian oceans.

Ohmoto and Rye (1974) and Sakai and Matsubaya (1974) have obtained some isotope data on the Kuroko massive iron-copper-lead-zinc sulfide deposits of Japan, a type of ore body that is generally thought to form in conjunction with island-arc type volcanism on the sea floor. The range of δD (-26 to -18) and $\delta^{18}O$ (-1.6 to -0.3) in H_2O of fluid inclusions from pyrite and chalcopyrite in the Kuroko ores are only slightly different from ocean water (Figure 6.8). This strongly suggests that seawater was the major constituent of these ore fluids. In particular, if the continental ice sheets were much reduced in size at the time of Miocene ore deposition, the discrepancy is reduced to approximately 10–15‰ in δD.

Sakai and Matsubaya (1974) have identified some modern coastal thermal waters in Japan that have isotopic characteristics very similar to the Kuroko ore fields. The slight lowering of δD in the ore fluids relative to ocean water can probably be explained by small additions of meteoric water to the submarine hydrothermal convective circulation systems, or to high-temperature exchange between the seawater and OH-bearing minerals in the altered volcanic rocks (Ohmoto and Rye, 1974).

METAMORPHIC ORE DEPOSITS

Ore deposits of regional metamorphic origin in general ought to be characterized by isotopic similarities between the ore body and the adjoining metamorphic host rocks. This is because regional metamorphism typically involves a lengthy period of time, during which there is a tendency toward isotopic homogenization. Such isotopic homogenization on at least a local scale has been demonstrated in almost every $^{18}O/^{16}O$ study of regional metamorphism (e.g., Taylor et al., 1963; Garlick and Epstein, 1967; Taylor, 1969; Taylor and Coleman, 1968). Typically, segregation vein minerals formed during moderate-to-high-grade metamorphism have $\delta^{18}O$ and δD values similar to their wallrocks, indicating isotopic equilibrium was attained. The only known exceptions are found in extremely low-grade metamorphic or diagenetic environments (e.g., see Magaritz and Taylor, 1976a).

Any type of ore deposit, of course, conceivably might end up involved in a later regional metamorphic episode. However, the deposits most likely to be involved in regional metamorphism are the various types of syngenetic, stratiform ore deposits that were originally formed at the time of deposition of their sedimentary or volcanic host rocks. Two such deposits have recently been identified and studied isotopically, the Ducktown, Tennessee copper-zinc deposits (Addy and Ypma, 1973) and the Homestake, South Dakota, gold deposit (Rye and Rye, 1974). At Ducktown, the δD values of OH-bearing minerals in the ores are identical to those in the host rocks (-68 to -77 for biotite, -62 to -69 for chlorite, -49 to -54 for muscovite). The $\delta^{18}O$ values are also similar to those in the immediately adjacent wallrock, but are approximately 2 ‰ depleted in ^{18}O relative to the more distant country rocks, compatible either with higher temperatures or slightly less ^{18}O-rich fluids in the ore zones. These fluids apparently had $\delta D \approx -30$ to -40 and $\delta^{18}O \approx +7$ to $+8$.

The Homestake gold deposit was for a long time considered to be an epigenetic, hydrothermal vein deposit of probable Tertiary age (see Noble, 1950). However, Rye and Rye (1974), utilizing a variety of geochemical techniques, conclude that it was actually a syngenetic deposit indigenous to the Homestake formation, and that the ore veins were formed in dilatant zones during a high-grade Precambrian regional metamorphic event. The δD data obtained on H_2O from fluid inclusions in the Precambrian vein minerals show a very wide range ($\delta D = -56$ to -112). This might at first appear to be incompatible with a metamorphic origin; however, because of the high concentrations of CH_4 in these fluid inclusions, and because of the probability of retrograde exchange between the CH_4 and coexisting H_2O, Rye and Rye (1974) concluded that the ore fluid very likely had a relatively uniform δD in the "normal" metamorphic range. Thus, it could have been in isotopic equilibrium with the OH-bearing minearls in the host rocks (which have $\delta D = -76, -78$).

The $\delta^{18}O$ values of Precambrian vein quartz in the ore bodies correlate very well with the type of host rock ($\delta^{18}O > 13.8$ in carbonate host rocks and < 13.8 in silicate host rocks). In addition, the vein quartz isotopic relationships in the mine are similar to those observed in regional metamorphic segregation veins outside the district. Some ore veins of definite Tertiary age do exist in the Black Hills, however, and the $\delta^{18}O$ values of quartz in these veins are distinctive and independent of wallrock type (the $\delta^{18}O$ and δD values of these Tertiary ore fluids may indicate a meteoric-hydrothermal origin, see Figure 6.8). All the isotopic data are thus compatible with a metamorphic origin for the Precambrian ore veins.

OTHER HYDROTHERMAL ORE DEPOSITS

The types of ore deposits described above can be classified into groups in rather systematic fashion, and distinct types of H_2O seem to characterize each kind of deposit. However, many other hydrothermal ore deposits seem to defy simple classification, and also, as will be reviewed below, a variety of waters were apparently present in the ore fluids at different times. The calculated isotopic compositions of the hydrothermal fluids in several such deposits are shown in Figure 6.8.

One difficulty with most of these deposits is that the only kinds of waters that are truly isotopically distinctive are meteoric H_2O and ocean H_2O. Hot, ^{18}O-shifted, saline formation waters, metamorphic waters, and magmatic waters all overlap in isotopic composition (Figure 6.4), and the only characteristic that might allow distinctions to be made is the fact that primary magmatic water falls into a much more narrow range of δD and $\delta^{18}O$ than the others. For example, this is the only isotopic reason why it is possible to claim that magmatic H_2O, rather than heated saline formation water, is responsible for the biotite-potassium-feldspar alteration in the cores of porphyry copper deposits. If a study were done solely on an individual porphyry copper deposit, this claim could not be definitive, but by considering a number of such bodies, the narrow isotopic range of the biotites would seem to be conclusive (Figure 6.7).

Most of the ore deposits shown in Figure 6.8 represent base-metal replacement veins in a wide variety of host rocks, and most are associated with granitic stocks that represent a plausible source of magmatic water, as well as of the heat energy necessary to drive a meteoric convective circulation system. In fact, as shown in Figure 6.8, in several of the deposits there is clear-cut evidence for the presence of meteoric water in the hydrothermal fluids, but the meteoric water commonly becomes abundant only in the late stages of ore deposition, associated with declining temperature and salinity in the ore fluid.

Providencia and Darwin are the only deposits shown in Figure 6.8 for which there is a lack of conclusive isotopic evidence for participation of heated meteoric waters in the hydrothermal fluids. However, both of these deposits occur in geographic areas where the δD values of meteoric and magmatic waters were probably similar at the time of ore deposition, and both also occur in high-^{18}O limestone country rocks, where the heated meteoric waters would be expected to have undergone very marked "^{18}O shifts." In such situations all isotopic evidence for an original meteoric-water origin could have been lost. The isotopic data shown in Figure 6.8 are nonetheless compatible with magmatic H$_2$O being the sole source of all of the ore fluids at Darwin and Providencia, as well as for all of the main-stage ore fluids at Casapalca (Rye et al., 1974b; Rye, 1966; Rye and Sawkins, 1974). In particular, the remarkably tight clustering of δD and δ^{18}O values at Providencia, over a wide range of salinity (5-20 wt. % NaCl equivalent) and temperature (200-400°C), is difficult to explain by any model involving ^{18}O shifts in a saline formation water or meteoric ground water.

The presence of heated saline formation waters in the ore fluids cannot be ruled out at Casapalca, but the main-stage ore fluid probably could not have formed from ^{18}O-shifted meteoric-hydrothermal waters. The local meteoric waters at the time of formation of this 5-10-m.y.-old, high-elevation deposit are known to have been much lower in δD than the main-stage ore fluid, as evidenced by the late influx of such meteoric water into the vein systems (Rye and Sawkins, 1974). In Figure 6.8, the isotopic compositions of the early ore fluids are compared with those of the waters that precipitated the post-ore calcite I and pearly calcite. Only minor amounts of ore in the Casapalca district were apparently deposited from such low-deuterium meteoric-hydrothermal fluids (e.g., the Aguas Calientes ore body). One problem, however, has not been fully resolved, namely whether the initial groundwaters at Casapalca might have formed prior to extensive uplift of the Andes; during such an uplift the local meteoric waters would have become steadily more depleted in δD with time, making it difficult to totally discount the presence of meteoric H$_2$O in the early ore fluids at Casapalca.

In striking contrast to Providencia, Casapalca, and Darwin, meteoric-hydrothermal waters were clearly heavily involved in ore deposition at the other three well-studied hydrothermal vein deposits shown on Figure 6.8 (Bluebell, Butte, and Pasto Bueno). There is also evidence for mixing of two or more waters in each of these three deposits, and because of the intrinsic interest in how such mixing processes might affect the deposition of the ore minerals, these three deposits are each described in a little more detail.

Bluebell, British Columbia

The Bluebell lead-zinc deposit, formed by replacement of limestones just east of the Nelson batholith, lies in a geographic location that is very favorable for distinguishing between primary magmatic and meteoric waters.

OTHER HYDROTHERMAL ORE DEPOSITS

The early massive ores are not suitable for detailed study, but Ohmoto and Rye (1970) showed that during the late-stage, vug period of ore deposition (representing the final 10% of the hydrothermal minerals) a decline in temperature from 450 to 320°C was accompanied by a decrease in salinity (10 to 3 wt. % NaCl equivalent) and $\delta^{18}O_{H_2O}$ (+5 to −13), but that the δD_{H_2O} was practically constant at −152 ±5 (Figure 6.8). These changes were accompanied by a large increase in the proportions of carbonate minerals and pyrite at the expense of quartz and the other sulfide minerals. At the extreme end stage of deposition, $\delta^{18}O_{H_2O}$ again increased to approximately +6, but the temperatures and salinities remained low.

The enormous fluctuations in $\delta^{18}O_{H_2O}$ at Bluebell are all the more remarkable in view of the constancy of δD_{H_2O}; the low δD value proves that this water is ultimately of meteoric derivation. Ohmoto and Rye (1970) conclude that mixing of two different kinds of meteoric waters is probably required to produce the $\delta^{18}O_{H_2O}$ variations; one fluid was apparently hotter, more saline, and better equilibrated with the country rocks, and the other was cooler, more dilute, and less exchanged. In view of the high temperatures involved (450°C or greater), and in light of recent isotopic work showing that many of the batholithic igneous rocks of British Columbia have interacted at high temperatures with low-δD meteoric waters (Magaritz and Taylor, 1976b), we probably cannot absolutely rule out the possibility that the early, high-^{18}O, saline ore fluids came out of a magma body that had previously incorporated meteoric water. A strong argument against this, however, is the uniformity of δD_{H_2O} at Bluebell. If the meteoric H_2O had actually entered and later come out of a magma, the likelihood of mixing with original magmatic water makes it doubtful that it would retain the same δD value as the heated, low-^{18}O meteoric water in the country rocks. Also, the most plausible manner in which the upper portions of a magma body are contaminated with meteoric waters is by stoping or assimilation of hydrothermally altered roof rocks (see Taylor, 1974b); the D/H ratios of such rocks are typically at least 30–40 ‰ lower than the meteoric-hydrothermal waters they exchanged with, so any such "magmatic" waters might conceivably have even lower δD values than the local meteoric waters.

Butte, Montana

The giant, early Tertiary Butte copper-zinc-manganese district is composed of myriads of ore veins that have been repeatedly fractured and rehealed and are emplaced almost entirely into a rather homogeneous quartz monzonite (see Meyer et al., 1968). Garlick and Epstein (1966), Sheppard et al. (1969; 1971), and Sheppard and Taylor (1974) studied the isotopic relationships at Butte, and some of the results are shown in Figures 6.7 and 6.8.

The Main-Stage Butte veins are all enveloped by prominent sericite-clay alteration halos that were formed at the time of ore deposition. This alteration is most intense in the Central Zone (CZ on Figure 6.7) where the veins are largely copper-bearing, and is less pervasive in the zinc-manganese-lead-rich Intermediate Zone (IZ) and Peripheral Zone (PZ). The Main-Stage fluids underwent drastic changes in their chemical characteristics through wallrock reaction, but Meyer et al. (1968) conclude that the original character of the ore solutions is best indicated by the intense Advanced Argillic alteration (AA on Figure 6.7) seen immediately adjacent to certain deep veins. At lower levels in the mine the Main-Stage veins can be observed to cut a type of potassium feldspar-biotite mineralization, termed EDM or pre-Main-Stage, that is similar to the potassium-silicate alteration zones of porphyry copper-molybdenum deposits.

At Butte, the pre-Main-Stage minerals are all very uniform in ^{18}O and identical to the values in the fresh Butte quartz monzonite pluton ($\delta^{18}O$ of quartz $= +9$ to $+10$); this is compatible with a derivation from primary magmatic waters, just as in the case of the porphyry copper deposits. However, the $\delta^{18}O$ values of Main-Stage quartz vary enormously, from -2 to $+13$, as do the $\delta^{18}O$ values of the associated hypogene clays and sericites (see Figure 6.7). Meyer et al. (1968) suggest temperatures of approximately 300°C for Main-Stage mineralization, and this estimate was used in calculating the isotopic compositions of the Butte fluids shown on Figure 6.8 (Sheppard and Taylor, 1974). If we exclude the two samples with extreme isotopic compositions, the calculated δD_{H_2O} values are quite constant at -110 ± 15, in spite of the fact that $\delta^{18}O_{H_2O}$ varies from -10 to $+6$. These waters must all represent heated meteoric groundwaters that have undergone extensive ^{18}O shifts. These calculated early Tertiary meteoric waters are 20–30 ‰ higher in δD than the present-day waters, suggesting a warmer climate in the past (see Figure 6.3).

The fine-grained EDM alteration biotites also have very low δD values (Figure 6.7), even though normal "primary magmatic" δD values are found in the fresh Butte quartz monzonite outside the mining district; this is probably a result of later D/H exchange between the EDM biotites and the Main-Stage ore fluids. Thus, although there is no hydrogen isotope evidence for anything but meteoric water in any of the Butte hydrothermal fluids, the $\delta^{18}O$ values and the similarities to porphyry copper-molybdenum deposits both argue that the pre-Main-Stage mineralization may have been a magmatic-hydrothermal event.

All of the high-$\delta^{18}O$ values associated with Main-Stage mineralization are from the altered Butte quartz monzonite at shallow levels in the Central Zone (Berkeley pit area). This indicates either lower temperatures or lower water/rock ratios. Both effects are likely in this area of closely spaced fractures and pervasive alteration. In the veins, particularly in the Peripheral Zone where the fractures are more widely spaced and the alteration is weaker, the effective water/rock ratios should be much higher; thus, the $\delta^{18}O_{H_2O}$ values are lower.

The Advanced Argillic assemblages in the deep Central Zone have the lowest $\delta^{18}O$ values found at Butte, indicating they formed from the meteoric-hydrothermal fluids that had undergone the smallest "^{18}O shift" and hence had exchanged least with the wallrocks. The isotope data provide strong support for the model of Meyer et al. (1968) regarding the evolution of the Butte hydrothermal fluids. In their model it is only after pervasive sericitization has sufficiently reduced the buffering power of the wallrocks that the original characteristics of the hydrothermal fluids can become apparent; these Advanced Argillic-type ore fluids had relatively low pH and produced high sulfur/metal mineral assemblages (chalcocite-digenite-covellite-enargite-pyrite, and associated dickite, pyrophyllite, and vug sericite). In a meteoric-hydrothermal convective system the solutions that have undergone the least interaction and exchange with the country rocks will have the lowest $^{18}O/^{16}O$ ratios, so with this model it is reasonable that the Advanced Argillic mineral assemblages should have the lowest $\delta^{18}O$ values, as shown on Figure 6.7.

Sheppard and Taylor (1974) thus conclude that the Main-Stage veins were formed by meteoric-hydrothermal fluids whose $\delta^{18}O_{H_2O}$ values varied widely in time and space throughout the Butte ore deposit, depending on the detailed history of wallrock reactions; the $\delta^{18}O$ variation is remarkably large for an ore deposit in a homogeneous igneous host rock. The meteoric-hydrothermal fluids apparently circulated to depths of at least 4–5 km (a 2-km range in vertical distance is observed just in the present mine workings) where they presumably interacted with a porphyry intrusion that provided the heat energy to drive the convective circulation and also may have provided the source of the heavy metals that were then carried upward and deposited in the Main-Stage veins. The necessary permeability for the hydrothermal circulation was provided by the intensive and repeated fracturing of the Butte quartz monzonite.

Pasto Bueno, Peru

The Pasto Bueno tungsten-copper-lead-silver deposit, studied by Landis and Rye (1974), lies 400 km NW of Casapalca along the axis of the Andes; both deposits are of similar age, and both lie at an elevation of about 4000 m. The ores at Pasto Bueno, however, are more clearly related to an igneous body, occurring in near-vertical quartz vein systems that cut across the upper contact of a 9.5-m.y.-old quartz monzonite stock and pass into a contact-metamorphosed shale-quartzite sequence. The veins are superimposed upon a concentric zonal hydrothermal alteration pattern in the stock that is analogous to that found in porphyry copper intrusions, but which also includes a period of fluoride-rich greisenization. The wallrocks next to the veins are intensely altered to sericite + clay ± pyrite ± fluorite.

Figure 6.7 shows the δD and $\delta^{18}O$ values of the alteration biotites and sericites. The biotites fit perfectly into the range of isotopic values exhibited by porphyry copper biotites, and Landis and Rye (1974) conclude that the early stages of hydrothermal activity at Pasto Bueno involved only magmatic waters. The sericites, however, have higher δD and $\delta^{18}O$ values than those in any of the porphyry copper deposits shown on Figure 6.7, and assuming $T \approx 300°C$ they must have formed from waters with $\delta D \approx -25$ and $\delta^{18}O \approx +7$ to 9 (based on the curves shown in Figures 6.1 and 6.2). These calculated sericite waters are very similar to the isotopically heaviest waters observed in the fluid inclusions in the veins (Figure 6.9), but they are significantly higher

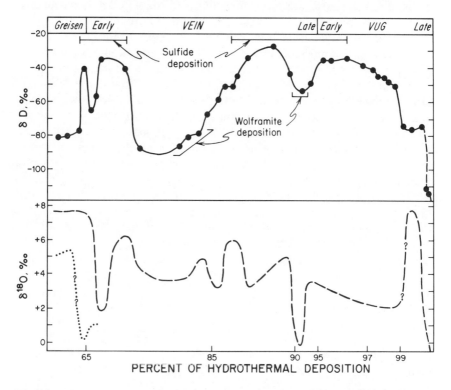

Fig. 6.9 Plot of changes in the δD in fluid inclusions and in the calculated $\delta^{18}O$ values of the hydrothermal fluids as a function of time in the Pasto Bueno tungsten-base metal deposit, Peru (modified after Landis and Rye, 1974). The dashed $\delta^{18}O$ curve is somewhat generalized and the dotted $\delta^{18}O$ curve is based on secondary inclusion filling temperatures.

CONCLUSIONS

in δD than so-called primary magmatic H_2O. These δD_{H_2O} waters are similar to those of the brines that formed the Mississippi Valley ore deposits (Figure 6.4); thus the isotopic compositions of these Pasto Bueno fluids are compatible with their being "^{18}O-shifted" saline formation waters or metamorphic waters derived from the country rocks surrounding the quartz monzonite stock.

During most of the period of vein deposition at Pasto Bueno, the temperatures and salinities fluctuated between 200 and 300°C and 11 and 17 wt. % NaCl equivalent, respectively. As indicated on Figure 6.9, low-deuterium, low-^{18}O meteoric waters clearly became a major component of these vein systems during the end stages of hydrothermal activity when the temperatures dropped below 200°C and the salinities to less than 2 wt. %. More interesting, however, is the fact that the major periods of wolframite deposition also correlate with dips in the δD and $\delta^{18}O$ curves on Figure 6.9, and with lowered salinity and temperature as well, indicating that the tungsten mineralization is somehow "triggered" by the influx of relatively dilute meteoric H_2O into the hydrothermal vein systems. Note also that the earliest period of meteoric H_2O influx corresponds to a major gap in the sulfide mineral paragenesis. It is not yet certain, however, whether the tungsten is carried in by the meteoric waters, or whether the wolframite precipitation is simply brought about by the abrupt fluctuations in temperature, salinity, or other chemical variables that accompany the mixing process (Landis and Rye, 1974).

Therefore, although other models are possible, the most plausible sequence of events at Pasto Bueno was (1) early magmatic-hydrothermal biotite alteration and greisenization; (2) influx of a saline connate-metamorphic water into the cooling stocks during the vein-forming stage, presumably in response to the establishment of a convective circulation in the country rocks; (3) subsequently, this saline ore fluid periodically mixed with an isotopically light meteoric water that also must have entered the hot, crystalline stock through a deep fracture system, causing the δD and $\delta^{18}O$ values of the vein fluids to fluctuate between the approximate limits -25 to -95 and $+7$ to 0, respectively; (4) at the extreme end stages of deposition, influx of a very light meteoric-hydrothermal water ($\delta D = -144$), much lighter than present-day local meteoric water, which has $\delta D \approx -95$ (Landis and Rye, 1974).

CONCLUSIONS

The usefulness of hydrogen and oxygen isotope measurements in solving problems of hydrothermal alteration and ore deposition has been demonstrated in a number of detailed studies of ore deposits during the past decade. Remarkably, ore deposits have been shown to have formed from almost every kind of water one can imagine in the Earth's crust. These include ocean waters, "connate" saline formation waters, pristine meteoric waters, "^{18}O-shifted" meteoric waters, metamorphic waters, and magmatic waters, as well as mixtures of several of the above. Mixing of waters from different sources in fact seems to be one of the most characteristic features of ore deposition and hydrothermal alteration in a number of localities.

It seems clear that important isotopic studies will continue to be carried out in the future. Each ore deposit seems to constitute a unique problem, however, making it difficult to generalize regarding the origin and evolution of the hydrothermal ore fluid(s); this is particularly true of most hydrothermal vein deposits.

Aside from continuing studies of individual ore deposits and laboratory measurements of equilibrium isotopic fractionation factors, much more work needs to be done on oxygen and hydrogen isotopic variations in magmas, particularly with regard to identifying low-^{18}O and low-deuterium magmas that have interacted with meteoric waters. The upper portions of certain magma chambers may become saturated with H_2O by incorporation of meteoric water from the country rocks, and such water-rich magmas might later give off large amounts of deuterium-depleted "magmatic"-hydrothermal fluids that conceivably could evolve into potential ore-forming solutions. Such deposits probably then ought to be classed as magmatic-hydrothermal even though the water is of meteoric origin! The sorting out of these problems may be of critical importance in fully understanding the generation of postmagmatic ore deposits, in spite of the basic difficulties in separating original magmatic isotopic effects from those later superimposed upon the hot, crystalline igneous rocks.

REFERENCES

Addy, S. K. and P. J. M. Ypma (1973) Metamorphism of sulfide deposits and the metamorphogenic circulation of pure water at Ducktown, Tennessee [abst.]: *Geol. Soc. Am., Ann. Mtg., Abst. with Prog.*, **5**, 529–530.

Anderson, A. T., Jr., R. N. Clayton, and T. K. Mayeda (1971) Oxygen isotope thermometry of mafic igneous rocks: *J. Geol.*, **79**, 715–729.

Banwell, C. J. (1961) Geothermal drillholes—Physical investigations: *U. N. Conf. New Sources of Energy, Paper G53*.

Barnes, H. L. and G. K. Czamanske (1967) Solubilities and transport of ore minerals: in *Geochemistry of Hydrothermal Ore Deposits*, H. L. Barnes, ed., New York: Holt, Rinehart and Winston, 334–381.

Batchelder, J. N. and D. W. Blake (1975) Geochemical variations in the Copper Canyon porphyry copper deposits, Lander County, Nevada [abst.]: *Geol. Soc. Am., Ann Mtg., Abst. Prog.*, **7**, 992–993.

REFERENCES

Bethke, P. M., R. O. Rye, and P. B. Barton, Jr. (1973) Hydrogen, oxygen, and sulfur isotope compositions of ore fluids in the Creede district, Mineral County, Colorado [abst.]: *Geol. Soc. Am., Ann. Mtg., Abst. with Prog.*, **5,** 549.

Bigeleisen, J. and M. G. Mayer (1947) Calculation of equilibrium constants for isotopic exchange reactions: *J. Chem. Phys.*, **72,** 800-808.

Blattner, P. (1975) Oxygen isotopic composition of fissure-grown quartz, adularia, and calcite from Broadlands geothermal field, New Zealand, with an appendix on quartz-K-feldspar-calcite-muscovite oxygen isotope geothermometers: *Am. J. Sci*, **275,** 785-800.

Bottinga, Y. (1968) Calculation of fractionation factors for carbon and oxygen isotopic exchange in the system calcite-carbon dioxide-water: *J. Phys. Chem.*, **72,** 800-808.

_____ (1969) Calculated fractionation factors for carbon and hydrogen isotope exchange in the system calcite-carbon-dioxide-graphite-methane-hydrogen-water vapor: *Geochim. Cosmochim. Acta*, **33,** 49-64.

_____ and M. Javoy (1973) Comments on oxygen isotope geothermometry: *Earth Planet. Sci. Lett.*, **20,** 250-265.

_____ and _____ (1975) Oxygen isotope partitioning among the minerals in igneous and metamorphic rocks: *Rev. Geophys. Space Phys.*, **13,** 401-418.

Browne, P. R. L. (1971) Mineralization in the Broadlands geothermal field, Taupo volcanic zone, New Zealand: *Soc. Min. Geol. Jap., Proc. IMA-IAGOD Mtg. Japan, Special Issue* **2,** 64-75.

Casadevall, T., H. Ohmoto, and R. O. Rye (1974) Sunnyside Mine, San Juan County, Colorado: Results of mineralogic, fluid-inclusion, and stable-isotope studies [abst.]: *Geol. Soc. Am., Ann. Mtg., Abst. Prog.*, **6,** 684.

Clayton, R. N., I. Friedman, D. L. Graf, T. K. Mayeda, W. F. Meents, and N. F. Shimp (1966) The origin of saline formation waters: I. Isotopic composition: *J. Geophys. Res.*, **71,** 3869-3882.

_____, J. R. O'Neil, and T. Mayeda (1972) Oxygen isotope exchange between quartz and water: *J. Geophys. Res.*, **77,** 3057-3067.

Craig, H. (1961) Isotopic variations in meteoric waters: *Science*, **133,** 1702-1703.

_____ (1963) The isotopic geochemistry of water and carbon in geothermal areas: in *Nuclear Geology on Geothermal Areas,* E. Tongiorgi, ed., Spoleto: Pisa, Consiglio Nazionale della Richerche, Laboratorio de Geologia Nucleare, 17-53.

_____ (1966) Isotopic composition and origin of the Red Sea and Salton Sea geothermal brines: *Science*, **154,** 1544-1548.

Dansgaard, W. (1964) Stable isotopes in precipitation: *Tellus*, **16,** 436-468.

Dowgiallo, J. and E. Tongiorgi (1972) The isotopic composition of oxygen and hydrogen in dome brines from the Mesozoic in northwest Poland: *Geothermics*, **1,** 67-69.

Eugster, H. P. and G. B. Skippen (1967) Igneous and metamorphic reactions involving gas equilibria: in *Researches in Geochemistry,* P. H. Abelson, ed., Vol. 2, New York: Wiley, 492-520.

Forester, R. W. and H. P. Taylor (1972) Oxygen and hydrogen isotope data on the interaction of meteoric ground waters with a gabbro-diorite stock, San Juan Mountains, Colorado: *Int. Geol. Congr., 24th, Montreal, Sec. 10, Geochemistry*, 254-263.

Friedman, I., A. C. Redfield, B. Schoen, and J. Harris (1964) The variation in the deuterium content of natural waters in the hydrologic cycle: *Rev. Geophys.*, **2,** 177-224.

_____, P. W. Lipman, J. D. Obradovich, J. D. Gleason, and R. L. Christiansen (1974) Meteoric water in magmas: *Science*, **184,** 1069-1072.

Garlick, D. G. and S. Epstein (1966) The isotopic composition of oxygen and carbon in hydrothermal minerals at Butte, Montana: *Econ. Geol.*, **61**, 1325-1335.

―― and ―― (1967) Oxygen isotope ratios in coexisting minerals of regionally metamorphosed rocks: *Geochim. Cosmochim. Acta*, **31**, 181-214.

Giggenbach, W. (1971) Isotopic composition of waters of the Broadlands geothermal field: *N. Z. J. Sci.*, **14**, 959-970.

Hall, W. and I. Friedman (1963) Composition of fluid inclusions, Cave-in-Rock fluorite district, Illinois, and Upper Mississippi Valley zinc-lead district: *Econ. Geol.*, **58**, 886-911.

―― and ―― (1969) Oxygen and carbon isotopic composition of ore and host rock of selected Mississippi Valley deposits: *U. S. Geol. Surv.*, *Prof. Paper 650-C*, C140-C148.

――, ――, and J. T. Nash (1974) Fluid inclusion and light stable isotope study of the Climax molybdenum deposits, Colorado: *Econ. Geol.*, **69**, 884-901.

Heyl, A. V., G. P. Landis, and R. E. Zartman (1974) Isotopic evidence for the origin of Mississippi Valley-type mineral deposits: A Review: *Econ. Geol.* **69**, 992-1006.

Hitchon, B. and I. Friedman (1969) Geochemistry and origin of formation waters in the western Canada sedimentary basin—I. Stable isotopes of hydrogen and oyxgen: *Geochim. Cosmochim. Acta*, **33**, 1321-1349.

―― and H. R. Krouse (1972) Hydrogeochemistry of surface waters of the Mackenzie River drainage basin Canada, III. Stable isotopes of oxygen, carbon and sulfur: *Geochim. Cosmochim. Acta*, **36**, 1337-1358.

Kharaka, Y. K., F. A. F. Berry, and I. Friedman (1973) Isotopic composition of oil-field brines from Kettleman North Dome, California, and their geologic implications: *Geochim. Cosmochim. Acta*, **37**, 1899-1903.

Landis, G. P. and R. O. Rye (1974) Geologic fluid inclusion, and stable isotope studies of the Pasto Bueno tungsten-base metal ore deposit, northern Peru: *Econ. Geol.*, **69**, 1025-1059.

Lawrence, J. R. and H. P. Taylor (1971) Deuterium and oxygen-correlation: Clay minerals and hydroxides in Quaternary soils compared to meteoric waters: *Geochim. Cosmochim. Acta*, **35**, 993-1003.

Lipman, P. W., T. A. Steven, and H. H. Mehnert (1970) Volcanic history of the San Juan Mountains, Colorado, as indicated by potassium-argon dating: *Bull. Geol. Soc. Am.*, **81**, 2329-2352.

Lowell, J. D. and J. M. Guilbert (1970) Lateral and vertical alteration-mineralization zoning in porphyry ore deposits: *Econ. Geol.*, **65**, 373-408.

Magaritz, M. and H. P. Taylor, Jr. (1974) Oxygen and hydrogen isotope studies of serpentinization in the Troodos ophiolite complex, Cyprus: *Earth Planet. Sci. Lett.*, **23**, 8-14.

―― and ―― (1976a) Oxygen, hydrogen and carbon isotope studies of the Franciscan formation, Coast Ranges, California: *Geochim. Cosmochim. Acta* **40**, 215-234.

―― and ―― (1976b) $^{18}O/^{16}O$ and D/H studies of igneous and sedimentary rocks along a 500-km traverse across the Coast Range Batholith into Central British Columbia at latitudes 54-55°N: *Can. J. Earth Sci.*, **13**, 1514-1536.

Meyer, C. and J. J. Hemley (1967) Wall rock alteration: in *Geochemistry of Hydrothermal Ore Deposits*, H. L. Barnes, ed., New York: Holt, Rinehart, and Winston, 166-235.

――, E. P. Shea, C. C. Goddard, and staff (1968) Ore deposits at Butte, Montana: in *Ore Deposits of the United States, 1933-1967*, J. D. Ridge, ed., New York: Amer. Inst. Min. Met. Petrol. Eng., 1373-1416.

Moore, J. G. (1970) Water content of basalt erupted on the ocean floor: *Contr. Miner. Petrol.*, **28**, 272-279.

REFERENCES

Muehlenbachs, K., A. T. Anderson, and G. E. Sigvaldason (1974) Low-O^{18} basalts from Iceland: *Geochim. Cosmochim. Acta*, **38**, 577-588.

____ and R. N. Clayton (1972) Oxygen isotope studies of fresh and weathered submarine basalts: *Can. J. Earth Sci.*, **9**, 172-184.

Nash, J. T. (1972) Fluid inclusion studies of some gold deposits in Nevada: *U.S. Geol. Surv., Prof. Paper 800-C*, C15-C19.

Noble, J. A. (1950) Ore mineralization in the Homestake gold mine, Lead, South Dakota: *Bull. Geol. Soc. Am.*, **61**, 221-252.

Nolan, T. B. (1935) The underground geology of the Tonopah mining district, Nevada: *Nevada Univ. Bull.*, **29**, No. 5.

Ohmoto, H. and R. O. Rye (1970) The Bluebell Mine, British Columbia. I. Mineralogy, paragenesis, fluid inclusions, and the isotopes of hydrogen, oxygen, and carbon: *Econ. Geol.*, **65**, 417-437.

____ and ____ (1974) Hydrogen and oxygen isotopic compositions of fluid inclusions in the Kuroko deposits, Japan: *Econ. Geol.*, **69**, 947-953.

O'Neil, J. R., R. N. Clayton, and T. Mayeda (1969) Oxygen isotope fractionation in divalent metal carbonates: *J. Chem. Phys.*, **51**, 5547-5558.

____ and M. L. Silberman (1974) Stable isotope relations in epithermal Au-Ag deposits: *Econ. Geol.*, **69**, 902-909.

____, M. L. Silberman, B. P. Fabbi, and C. W. Chesterman (1973) Stable isotope and chemical relations during mineralization in the Bodie mining district, Mono County, California: *Econ. Geol.*, **68**, 765-784.

____ and H. P. Taylor, Jr. (1967) The oxygen isotope and cation exchange chemistry of feldspars: *Am. Mineral.*, **52**, 1414-1437.

____ and ____ (1969) Oxygen isotope fractionation between muscovite and water: *J. Geophys. Res.*, **74**, 6012-6022.

Perry, E. G. (1967) The oxygen isotope chemistry of ancient cherts: *Earth Planet. Sci. Lett.*, **3**, 62-66.

Pinckney, D. M. and R. O. Rye (1972) Variation of O^{18}/O^{16}, C^{13}/C^{12}, texture and mineralogy in altered limestone in the Hill mine, Cave-in-Rock District, Illinois: *Econ. Geol.*, **67**, 1-18.

Robinson, B. W. (1974) The origin of mineralization at the Tui Mine, Te Aroha, New Zealand, in the light of stable isotope studies: *Econ. Geol.*, **69**, 910-925.

Roedder, E. (1972) Data of geochemistry, chapter JJ, Composition of fluid inclusions: *U. S. Geol. Surv., Prof. Paper 440-JJ*.

____, B. Ingram, and W. E. Hall (1963) Studies of fluid inclusions III: Extraction and quantitative analysis of inclusions in the milligram range: *Econ. Geol.*, **58**, 353-374.

Rye, D. M. and R. O. Rye (1974) Homestake gold mine, South Dakota: I. Stable isotope studies: *Econ. Geol.*, **69**, 293-317.

Rye, R. O. (1966) The carbon, hydrogen, and oxygen isotopic composition of the hydrothermal fluids responsible for the lead-zinc deposits at Providencia, Zacatecas, Mexico: *Econ. Geol.*, **61**, 1399-1427.

____, B. R. Doe, and J. D. Wells (1974a) Stable isotope and lead isotope study of the Cortez, Nevada, gold deposit and surrounding area: *J. Res. U.S. Geol. Surv.*, **2**, 13-23.

____, W. E. Hall, and H. Ohmoto (1974b) Carbon, hydrogen, oxygen, and sulfur isotope study of the Darwin lead-silver-zinc deposit, southern California: *Econ. Geol.*, **69**, 468-481.

____ and F. J. Sawkins (1974) Fluid inclusion and stable isotope studies on the Casapalca Ag-Pb-Zn-Cu deposit, central Andes, Peru: *Econ. Geol.*, **69**, 181-205.

Sakai, H. and O. Matsubaya (1974) Isotopic geochemistry of the thermal waters of Japan and its bearing on the Kuroko ore solutions: *Econ. Geol.*, **69,** 974-991.

Savin, S. M. and S. Epstein (1970a) The oxygen and hydrogen isotope geochemistry of clay minerals: *Geochim. Cosmochim. Acta*, **34,** 43-64.

___ and ___ (1970b) The oxygen and hydrogen isotope geochemistry of ocean sediments and shales: *Geochim. Cosmochim. Acta*, **34,** 43-64.

Sheppard, S. M. F. and L. Gustafson (1976) Oxygen and hydrogen isotopes in the porphyry copper deposit at El Salvador, Chile: *Econ. Geol.*, **71,** 1549-1559.

___, R. L. Nielsen, and H. P. Taylor (1969) Oxygen and hydrogen isotope ratios of clay minerals from porphyry copper deposits: *Econ. Geol.*, **64,** 755-777.

___, ___, and ___ (1971) Hydrogen and oxygen isotope ratios in minerals from porphyry copper deposits: *Econ. Geol.*, **66,** 515-542.

___ and H. P. Taylor, Jr. (1974) Hydrogen and oxygen isotope evidence for the origin of water in the Boulder batholith and the Butte ore deposits, Montana: *Econ. Geol.*, **69,** 926-946.

Spooner, E. T. C., R. D. Beckinsale, W. S. Fyfe, and J. D. Smewing (1974) O^{18} enriched ophiolitic metabasic rocks from E. Liguria (Italy), Pindos (Greece) and Troodos (Cyprus): *Contrib. Min. Petrol.*, **47,** 41-62.

Sugisaki, R. and M. L. Jensen (1971) Oxygen isotopic studies of silicate minerals with special reference to hydrothermal deposits: *Geochem. J.*, **5,** 7-21.

Suzuoki, T. and S. Epstein (1976) Hydrogen isotope fractionation between OH-bearing minerals and water: *Geochim. Cosmochim. Acta*, **40,** 1229-1240.

Taylor, H. P., Jr. (1967) Oxygen isotope studies of hydrothermal mineral deposits: in *Geochemistry of Hydrothermal Ore Deposits*, H. L. Barnes, ed., New York: Holt, Rinehart and Winston, 109-142.

___ (1968) The oxygen isotope geochemistry of igneous rocks: *Contr. Mineral. Petrol.*, **19,** 1-71.

___ (1969) Oxygen isotope studies of anorthosites with special reference to the origin of bodies in the Adirondack Mountains, New York: in *Origin of Anorthosites*, *N. Y. State Museum Science Service Memo* **18,** 111-134.

___ (1971) Oxygen isotope evidence for large-scale interaction between meteoric ground waters and Tertiary granodiorite intrusions, western Cascade Range, Oregon: *J. Geophys. Res.*, **76,** 7855-7874.

___ (1973) O^{18}/O^{16} evidence for meteoric-hydrothermal alteration and ore deposition in the Tonopah, Comstock Lode, and Goldfield mining districts, Nevada: *Econ. Geol.*, **68,** 747-764.

___ (1974a) The application of oxygen and hydrogen isotope studies to problems of hydrothermal alteration and ore deposition: *Econ. Geol.*, **69,** 843-883.

___ (1974b) Oxygen and hydrogen isotope evidence for large-scale circulation and interaction between ground waters and igneous intrusions, with particular reference to the San Juan volcanic field, Colorado: in *Geochemical Transport and Kinetics*, A. W. Hofmann, B. J. Giletti, H. S. Yoder, Jr. and R. A. Yund, eds., Washington, D.C.: Carnegie Institution Wash., 299-324.

___ (1974c) A low-^{18}O, late Precambrian granite batholith in the Seychelles Islands, Indian Ocean: Evidence for formation of ^{18}O-depleted magmas and interaction with ancient meteoric ground waters [abst.]: *Geol. Soc. Am., Ann. Mtg., Abst. Prog.*, **6,** 981-982.

___, A. L. Albee, and S. Epstein (1963) O^{18}/O^{16} ratios of coexisting minerals in three assemblages of kyanite-zone pelitic schist: *J. Geol.*, **71,** 513-522.

_____ and R. G. Coleman (1968) O^{18}/O^{16} ratios of coexisting minerals in glaucophane-bearing metamorphic rocks: *Bull. Geol. Soc. Am.*, **79**, 1727-1756.

_____ and S. Epstein (1963) O^{18}/O^{16} ratios in rocks and coexisting minerals of the Skaergaard intrusion: *J. Petrol.*, **4**, 51-74.

_____ and _____ (1968) Hydrogen isotope evidence for influx of meteoric ground water into shallow igneous intrusives [abst.]: *Geol. Soc. Am., Ann. Mtg., Abst. Prog.*, 294.

_____ and R. W. Forester (1971) Low-O^{18} igneous rocks from the intrusive complexes of Skye, Mull, and Ardnamurchan, western Scotland: *J. Petrol.*, **12**, 465-497.

_____ and _____ (1973) An oxygen and hydrogen isotope study of the Skaergaard intrusion and its country rocks [abst.]: *Am. Geophys. Union Trans.*, **54**, 500.

_____ and M. Magaritz (1975) Oxygen and hydrogen isotope studies of 2.6-3.4-b.y. old granites from the Barberton Mountain Land, Swaziland, and the Rhodesian craton, Southern Africa [abst.]: *Geol. Soc. Am., Ann. Mtg., Abst. Prog.*, **7**, 1293.

_____ and B. Turi (1976) High-^{18}O igneous rocks from the Tuscan magmatic province, Italy: *Contrib. Min. Petrol.*, **55**, 31-54.

Truesdell, A. H. (1974) Oxygen isotope activities and concentrations in aqueous salt solution at elevated temperatures: consequence for isotope geochemistry: *Earth Planet. Sci. Lett.*, **23**, 387-396.

Urey, H. C. (1947) The thermodynamic properties of isotopic substances: *J. Chem. Soc.*, 562-581.

Wenner, D. B. and H. P. Taylor, Jr. (1971) Temperatures of serpentinization of ultramafic rocks based on O^{18}/O^{16} fractionation between coexisting serpentine and magnetite: *Contr. Min. Petrol.*, **32**, 165-185.

_____ and _____ (1973) Oxygen and hydrogen isotope studies of the serpentinization of ultramafic rocks in oceanic environments and continental ophiolite complexes: *Am. J. Sci.*, **273**, 207-239.

_____ and _____ (1976) Oxygen and hydrogen isotope studies of a Precambrian granite-rhyolite terrane, St. Francois Mtns., S.E. Missouri: *Bull. Geol. Soc. Am.*, **87**, 1587-1598.

White, D. E. (1968a) Environments of generation of some base-metal ore deposits: *Econ. Geol.*, **63**, 301-335.

_____ (1968b) Hydrology, activity, and heat flow of the Steamboat Springs thermal system, Washoe County, Nevada: *U.S. Geol. Surv., Prof. Paper 458-C*, C1-C109.

_____ (1974) Diverse origins of hydrothermal ore fluids: *Econ. Geol.*, **69**, 954-973.

_____, I. Barnes, and J. R. O'Neil (1973) Thermal and mineral waters of non-meteoric origin, California coast ranges: *Bull. Geol. Soc. Am.*, **84**, 574-560.

_____, G. A. Thompson, and C. H. Sandberg (1964) Rocks, structure, and geologic history of Steamboat Springs thermal area, Washoe County Nevada: *U.S. Geol. Surv., Prof. Paper 458-B*, B1-B63.

7

Sulfide Mineral Stabilities

PAUL B. BARTON, JR.
U. S. Geological Survey

BRIAN J. SKINNER
Yale University

During the first decades of this century studies of ore minerals centered on identification of the minerals and the interpretation of their complex textural intergrowths. The vast body of observational data led to attempts to recreate observed assemblages in the laboratory in the hope that this would help define the variables controlling mineral formation. The last 20 years have seen great advances in our understanding of how and why sulfide minerals behave the way they do; most minerals have now been synthesized, and their relative stabilities at least approximately determined. The high hopes that laboratory studies would resolve all questions have not yet been fulfilled, partly because the behavior of ore minerals is much more complex than had at first been supposed, and particularly, because our perspective of the problems to be attacked has broadened so rapidly. The future continues to hold bright prospects for studies of sulfide minerals.

The greatest promise now seems to lie in a reexamination and reinterpretation of complex mineral assemblages and textures in the light of experimental studies, making full use of the modern instruments that have revolutionized ore petrology.

LIMITATIONS OF CHEMICAL AND PHYSICAL CONDITIONS

We are concerned with the conditions and processes by which sulfide mineral assemblages form. Our discussions are, therefore, concentrated on those chemical systems and minerals that are potentially useful to this end.

Useful Chemical Systems

We consider only groupings of elements that occur in the same geochemical environment and that can be meaningfully studied as subsystems. For example, the system iron-sulfur is of great importance because iron-sulfide minerals are widespread, whereas the system calcium-sulfur is of negligible importance to ore deposits. Calcium sulfide does not occur in the crust because its stability field is preempted by the formation of simple compounds containing calcium and sulfur with other common elements.

The number of elements known to combine with sulfur in natural environments is limited. Because of their geochemical and crystallochemical similarities, it is common practice to consider all the condensed native elements and alloys, sulfide, arsenide, antimonide, bismuthinide, selenide, telluride, and sulfo-salt minerals as constituting one family of compounds, the "sulfide" minerals.

Useful Temperature-Pressure Range

Sulfide mineral deposits occur in crustal rocks and broad limits can therefore be placed on temperature and pressure ranges. The extreme temperature range of interest is from 0°C to roughly 1200°C; the pressure range is from 1 bar to 12 kb, the estimated pressure at the base of the continental crust. Indications from the mineralogy of veins and host rocks suggest that hydrothermal ore deposits, the subject of this book, tend to form in the upper rather than the lower parts of the crust and hence that pressures less than 2 or 3 kb and temperatures of less than 700°C are typical.

SELECTION OF SIGNIFICANT NATURAL SYSTEMS

Application of controlled laboratory data to natural assemblages requires us to present the assemblage in a format amenable to quantitative interpretation. Physical parameters, such as temperature and pressure, cannot be evaluated independently of chemical variables. Recognition of interdependent variables is essential both for design of laboratory experiments and reduction of an assemblage to its simplest form. But such reduction can only take place by realizing which variables are truly independent and must be considered, and which are dependent and hence can be ignored. The way to achieve this goal is through the phase rule, a fundamental statement of chemical thermodynamics relating the number of stable phases in an equilibrium assemblage, ϕ, to the number of independent components, c, and the number of independent degrees of freedom, v, in the well-known formula $\phi = c - v + 2$. This formula assumes that the system being considered is uniform with respect to such exotic parameters as gravitational, magnetic, and electrostatic fields. Usually these fields can be treated as constants, or are so weak over the region of interest that they can be neglected. This conclusion might not be valid, however, if we consider very strong magnetic fields, because magnetic phases would be stabilized relative to nonmagnetic ones. Each additional condition of state requires modification of the phase rule. Thus, if a magnetic potential is also considered, the statement is $\phi = c - v + 3$.

Limits can be assigned to the degrees of freedom and number of components in such a way as to delineate *the least complicated representative subsystem* within a larger and more general system. Because a successful application of laboratory data to natural assemblages is our principal reason for making laboratory studies, it is useful to spell out some of the fundamental, but all too frequently overlooked, steps in the reasoning behind the application.

Restriction of Degrees of Freedom

A system possesses v degrees of freedom when at least v independent variables are required to describe completely the state of the system at equilibrium. The independent variables may be either physical, such as temperature and pressure, or they may be chemical. For example, whereas the state of pure water may be completely described in terms of physical variables, a multicomponent solution requires evaluation of chemical data, too, before its state is completely specified because each independent compositional variable of a phase will give rise to another degree of freedom.

The phase rule is independent of the amount of a phase; it is concerned simply with the number of phases present. At equilibrium the maximum number of phases that may stably coexist, equal to the number of components plus 2, occurs when there are zero degrees of freedom, meaning that no variables beyond the identity of phases and number components need be stated to specify uniquely the state of the system. A classic example of an invariant situation for a one-component system is the triple point of H_2O where water + ice + water vapor coexist at a unique temperature and pressure. Systems with 0, 1, 2, 3, or more degrees of freedom are described as invariant, univariant, divariant, trivariant, and so forth.

Mineral assemblages that occur widely must be stable over a range of both temperature and pressure (that is, the assemblages are *at least* divariant); two degrees of freedom are thereby preempted. Under this circumstance the phase rule reduces to $\phi = c$, a relation first pointed out by V. M. Goldschmidt and known as the mineralogic phase rule. It is obvious that an assemblage described only as obeying the mineralogic phase rule is not useful, in itself, as a fixed-point geothermometer or geobarometer, because under such divariant conditions both temperature and pressure can be independently varied without modifying the mineral assemblage. As frequently used, the mineralogic phase rule treats the phases present as having essentially fixed compositions. However, a more realistic approach may be taken by realizing that many solids are capable of continuous variations in their compositions. These compositional variables may be used to further restrict the degrees of freedom, a circumstance that, as discussed later in this chapter, does allow us to use mineral assemblages as indicators of conditions, such as temperature and pressure, under which minerals grow.

As an example of the application of the phase rule, let us consider the senary system lead-zinc-iron-cadmium-manganese-sulfur. To make this senary system invariant solely in terms of the phases present requires that eight phases coexist. Conceivably these phases could be galena, sphalerite, pyrrhotite, pyrite, greenockite, alabandite, sulfur-rich liquid, and vapor. This is a very unlikely assemblage to find in nature. A common circumstance is to find only galena + sphalerite + pyrite, an assemblage that in the framework of the senary system allows five degrees of freedom. In this case, however, additional components are present as solid solutions in all three phases: cadmium in sphalerite and galena, manganese in pyrite and sphalerite, and iron in sphalerite. We can completely describe the state of the system by specifying the compositions of the galena, sphalerite, and pyrite, thereby consuming the remaining five independent degrees of freedom and reducing the system to invariance. It is possible, at least in theory, to perform the necessary experiments to determine the compositional variations of galena, sphalerite, and pyrite as functions of temperature and pressure. Thus, again in theory, we would have a sliding-scale geothermobarometer that would be suitable for recording temperature and pressure (and, if adequately calibrated, the chemical potentials of the components as well) over a wide range of conditions. This type of reasoning is fundamental to the development of all sliding-scale indicators of equilibration conditions, of which the sphalerite, the pyrrhotite, and the trace-element-distribution indicators, discussed later, are examples.

Compare the usefulness of the sliding-scale indicators with the eight-phase invariant assemblage discussed previously. The invariant point could, if necessary, be located by suitable measurements. It would define a unique temperature and pressure, but the possibility of any geologic environment perching exactly on it would be exceedingly remote. This discussion is not meant to discourage the study of invariant points. To the contrary, they are exceedingly useful in providing limits and they are, in many cases, the only information that some studies of chemical systems yield. We wish to emphasize, however, that it is the sliding-scale indicators that must ultimately be used to investigate general physicochemical parameters of geologic processes.

We have already pointed out that specific chemical data may be necessary to completely define the state of a system. The example of pure water being described by the physical variables entails an assumption fundamental to our discussion. Water must be contained in a vessel, say of gold, before an experiment can be performed. A correct specification of the state of the system would contain a term stating that the solubility of gold in water is negligible. (The actual statement would be that the mole fraction of the component gold in the phase water is so small as to be effectively zero within the range of measurement capability.) Throughout the physical range of our experimental demonstration this would be true and hence the chemical term involving the presence of the gold would be constant and equal to zero. It can therefore be ignored. The same reasoning is true for the sulfide experiments performed in inert silica-glass containers.

Let us now consider the common mineral assemblage pyrite + pyrrhotite + sphalerite + galena + quartz with each mineral containing inclusions filled with a (CO_2 + K + Na + Ca + Cl)-bearing aqueous fluid. The most general system is Fe-Pb-Zn-S-Si-H-C-K-Na-Ca-Cl-O. Consideration of all possible phase relations in a 12-component system is practically impossible. Fortunately, quartz and the components in the fluid inclusions have essentially zero solubility in the sulfide phases, and hence can be dropped from consideration. The quartz and the fluid inclusions are equivalent to an inert container. Thus we can reduce the components necessary to specify the chemical subsystems for detailed study if we choose to do so. Thus, sphalerite, pyrite, and pyrrhotite do not dissolve appreciable amounts of lead, so their interrelationships can be studied in terms of the subsystem iron-zinc-sulfur. Iron enters the sphalerite phase, so the simplest subsystem within which to study the sphalerite is iron-zinc-sulfur. But no zinc enters either the pyrite or pyrrhotite phases and, hence, one can specify their state completely in terms of the iron-sulfur system.

If an additional element, such as nickel, were present in solid solution in appreciable amounts in either the pyrite or pyrrhotite, iron-sulfur could no longer be used as the simplest subsystem. The pyrite + pyrrhotite relations must then be considered in terms of the subsystem iron-nickel-sulfur and the sphalerite + pyrite + pyrrhotite relations in terms of the subsystem iron-nickel-zinc-sulfur.

The commonly stated belief that water and other fluxing agents present during mineral deposition can affect the equilibrium state but not appear as components in the minerals is incorrect. Whereas the *position of equilibrium* cannot be affected by fugitive components, the *rate* at which the equilibrium state is approached may very well be affected by extraneous components acting as fluxes. In some circumstances the very attainment of the equilibrium state may even depend on the presence of a suitable mineralizer, or catalyst, to increase reaction rates; nevertheless, the final equilibrium state will not be changed in the least. The argument concerning applicability of sulfide-phase equilibria studies conducted under dry, laboratory conditions to an aqueous, natural environment is reduced to a discussion of kinetics and the attainment of equilibrium. It does not involve different positions of equilibrium.

Mobile Components: Mineral Systems as Buffers and Indicators

The concept of mobile components, as proposed by Korzhinskii (1959), Thompson (1955, 1959) and others, merits discussion. It is implicit in the mineralogic phase rule that the pressure and temperature of the rocks are externally controlled. The assumption is generally valid because there are few reactions that consume or supply sufficient heat or have sufficiently large volume changes to control (or buffer) the temperature or pressure of their environment.

In addition to temperature and pressure being controlled by the external environment, and hence preempting two degrees of freedom, the chemical potential (or fugacity, or activity) of any component within the subsystem may also be controlled externally. We must then, in a manner analogous to specification of pressure and temperature, specify the chemical potential of this component within the subsystem before we can specify the state of the subsystem. This consumes another degree of freedom for each chemical potential controlled.

We observe most rocks or ores long after they have been formed, and part of the environment in which they formed is no longer available for examination. Korzhinskii (1959) proposed that the concept of an open system be used to handle this common geologic case, for which it is difficult to define the now nonexistent complete (or closed) system within which the assemblage formed. This is essentially a way of looking at the environment from the point of view of the remaining subsystem (that is, mineral assemblage) realizing that the mineral assemblage at a given point may have been fixed by an externally controlled medium such as a moving fluid or pervasive atmosphere. If the activity of a component in a mineral assemblage is completely controlled by its activity in the external environment and changes by transfer of matter across the boundary of the subsystem in response to any change in the external environment, the subsystem is open with respect to this component; the component itself is said to be "mobile". Clearly, if the activity of the mobile component should change from time to time in an isothermal system, as it may well do, the mineral assemblages, or compositions of minerals within assemblages, will change. The maximum number of phases able to coexist in equilibrium is reduced by one for each degree of freedom preempted by a mobile component. Accordingly, a detailed study of the sequence of assemblages produced should, under ideal circumstances, tell us whether or not one or more components are mobile within a specific subsystem.

Korzhinskii (1959), Thompson (1955), and Zen (1963) have divided components into two classes:

1. Components whose activities are not controlled by the mineral assemblages in which they occur, but are controlled by some larger, external environment. Such components have been given the names "perfectly mobile", "active", and "boundary value" components by the three authors, and are here termed simply "mobile". Remember that the mobility of a component gives rise to another degree of freedom, and that another but usually more complex system can always be selected within which the component is not mobile. For example, during deposition of an iron oxide from a well-stirred lake, the oxygen fugacity in the lake water and in the mineral that precipitates from it, is controlled by the vast reservoir of the atmosphere. With respect to the system lake + precipitate, oxygen is a mobile component, or to put it another way, the system selected is open to oxygen. If the system chosen were lake + precipitate + atmosphere, it would contain many more components but would be a closed system with respect to oxygen.

2. Components whose activities are buffered by the mineral assemblage in which they occur. Such components have been called "inert", "inactive", and "initial value" components. Buffering may come about in either of two ways: through a *fixed-point buffer* or a *sliding-scale buffer*. The maximum number of phases permitted by the mineralogic phase rule occurs with a fixed-point buffer. An example is the assemblage pyrite + pyrrhotite. At any temperature and pressure, the fugacity of sulfur is rigidly fixed so long as pyrite and pyrrhotite coexist at equilibrium. Therefore, addition of sulfur to the system will not change the fugacity

of sulfur or the compositions of the pyrite and pyrrhotite, *but* it will change the relative amounts of pyrite and pyrrhotite. Another important example of a fixed-point buffer is the case where a phase such as galena has the same composition as the component, PbS. The presence of galena buffers the activity of PbS at unity. This buffering effect by phases of fixed composition in equilibrium with an ore-forming fluid forms the basis of the surficial-equilibrium arguments presented by Sims and Barton (1961) and by Barton et al. (1963).

With sliding-scale buffers, phases of variable composition respond to a change in the fugacity of a component by changing their compositions. Although the fugacity of the component is buffered by the system, the position of the buffer will change in response to a change in the amount of the component. An example of such a case is pyrrhotite (in the absence of either iron or pyrite), which will change its composition in response to a changing amount of sulfur in the system. Once a second phase, either iron or pyrite, begins to form from the pyrrhotite, a fixed-point buffer is established.

The effectiveness of any assemblage as a buffer depends on its abundance and on the demands placed on it. Thus, a large body of pyrite + pyrrhotite at a constant temperature and pressure may well be an effective sulfur buffer because of its ability to consume or supply large amounts of sulfur at constant fugacity. Similarly, a large body of hematite + magnetite may act as an effective oxygen buffer. However, most assemblages that could be effective sulfur buffers, such as silver + argentite, orpiment + realgar, tennantite + enargite, and chalcopyrite + bornite + pyrite, usually occur in such small volumes that their buffering capacity is negligible. If we assume that an ore fluid contains appreciable quantities of sulfur relative to the small amounts fixed in the sulfide assemblage, it is probable that the observed assemblages record incomplete reaction as conditions changed from the stability field of one mineral to the stability field of the other. In this sense, we may use such assemblages as *indicators* of previous conditions, recognizing that, although they were not sufficiently abundant to control the changing conditions, they do indicate the direction of the change. Indicator assemblages, like the previously discussed buffer assemblages, may be of the fixed-point or sliding-scale types. Mineralogic and paragenetic studies suggest that indicators are far more common than buffers. Perhaps the most common and useful sliding-scale indicator is the FeS content of sphalerite in equilibrium with pyrite. The variation in FeS content of sphalerite reflects, at constant temperature and pressure, the fugacity of sulfur through the relation

$$2 \text{ FeS (in sphalerite)} + S_2 \rightleftharpoons 2 \text{ FeS}_2 \text{ (pyrite)}$$

Once formed, an ore deposit may become chemically isolated from its surroundings and may react internally (isochemical metamorphism) in response to changing temperature or pressure. The bulk composition of the equilibrating system then determines the mineral assemblage of the final ore. The term "equilibrating system" is emphasized because it is entirely possible for some sulfide minerals to react internally, or with each other, while others in the same ore remain in the state in which they were initially deposited (see Barton et al., 1963).

Requirements of Equilibrium

The successful application of phase-equilibrium studies requires natural assemblages that have attained equilibrium. Yet the attainment of equilibrium among a group of minerals can rarely be completely demonstrated. Because the equilibrium state is one from which a system has no spontaneous tendency to change, a disequilibrium state usually may be discerned, whereas the equilibrium state cannot. We must, therefore, argue the case for equilibrium on evidence, largely textural, gained jointly from laboratory and field studies, and we may also test for equilibrium by noting the degree of concordance in temperature and pressure given by the distributions of different chemical and isotopic components between coexisting minerals.

The attainment of equilibrium by solid-state reactions in many, but not all, hydrothermal ore-forming environments can be accepted for the copper-rich minerals in the copper-iron-sulfur system, all the copper-silver-sulfides, the lead sulfosalts and other similar compounds. Their reaction rates are exceedingly fast—so fast, in fact, that some reactions proceed rapidly even at room temperature. Energy barriers preventing the attainment of equilibrium in these reactions are so small that it is most unlikely that a disequilibrium state could be maintained at temperatures of more than 100 or 150°C for even a few years (see Figure 7.1).

Growth zoning in crystals attests to (1) fluctuations in the environment during crystal growth, (2) the ability of the growing crystal to respond to the changing environment, and (3) to the ability of the mineral to resist internal postdepositional modifications. Growth zoning is the key to recognition of fluctuations in the ore-forming environment. The origin of very delicately banded sphalerites from Creede, Colorado, was discussed by Barton et al. (1963, 1977), and they concluded that it arose from distinct changes in a depositing fluid that was always close to chemical equilibrium with the surfaces of the growing crystals. Reactions between solids and liquids usually are many orders of magnitude faster than purely solid-state reactions. The surface of a growing crystal can thus be in equilibrium with

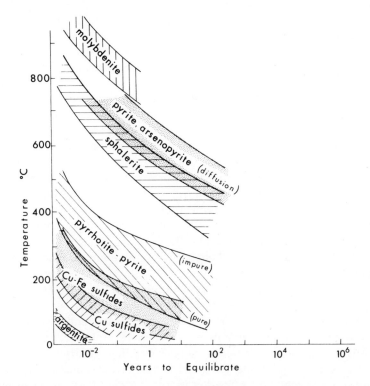

Fig. 7.1 Equilibration times for various sulfides involved in solid-state reactions. The field widths represent differing rates in different reactions as well as changes in rates due to compositional differences in a phase and a great deal of experimental uncertainty. Pyrite shows large variation in rates of reaction depending on whether it is in a redox reaction with pyrrhotite or a diffusion-controlled reaction such as the homogenization of $(Fe, Ni)S_2$ solid solutions. Yund and Hall (1970) have shown a wide range in pyrrhotite equilibration rates depending on the impurity content.

its external environment (a solid-liquid equilibrium), while it has virtually no ability to equilibrate with the previously deposited zones beneath it (a much slower solid-state equilibration).

There are numerous examples, besides the obvious case of zoned crystals, where disequilibrium states have been preserved. Inhomogeneous areas can be found within individual growth zones in some crystals. Marcasite, a polymorph of pyrite, may precipitate metastably at low temperature, or opal may deposit from natural waters and be preserved because of the exceedingly slow transformation to quartz, the stable mineral, via a series of intermediate, metastable states. Later in this chapter we show how a common metastable effect, supersaturation with respect to pyrite, can lead to some apparently paradoxic sulfide mineral associations.

Preservation of a Former Equilibrium State: The Quenching Problem

Preservation, or reconstruction, of an equilibrium state, despite later retrograde effects, is essential if we are to apply laboratory studies to mineral occurrences. Retrograde effects that obliterate evidence of the initial equilibrium will, at best, allow only the determination of some intermediate step or steps in the retrograde process. Unfortunately, the systems that are best studied in the laboratory are those that equilibrate most rapidly; such systems also reequilibrate most rapidly and are of dubious value in reconstructing the environment of deposition. Certain phases, such as the high-temperature forms of Cu_2S, the high-temperature form of $Ni_{3\pm x}S_2$, high quartz and the cubic form of Ag_2S cannot be quenched under any circumstances, although low-temperature pseudomorphs indicate their former presence.

The phases that equilibrate least rapidly are the phases that most strongly resist any retrograde effects. These phases, such as molybdenite, sphalerite, pyrite, arsenopyrite, hematite, magnetite, and wolframite, all have less metallic characteristics than the more rapidly reacting phases. Minerals with the lowest degree of metallic bonding, generally expressed as a low frequency of short metal—metal bonds, are the most useful for our purposes. This rough generalization is true not only from metal to metal, but also from one sulfidation state to another for the same element. Thus, reactions involving chalcocite (Cu_2S) are faster than those involving covellite (CuS); reactions involving pyrrhotite ($Fe_{1-x}S$) are faster than those involving pyrite (FeS_2). Solid-state reaction rates of sulfide can be grouped to give order-of-magnitude comparisons between them (Figure 7.1).

The rapid decrease in reaction rates at lower temperatures argues for the maximum preservation of initial-state conditions in low-temperature deposits that have not been affected by later metamorphism. It also indicates the difficulty to be expected in deciphering the initial state of high-temperature deposits. The chances for preservation of any initial-state relations in sulfides formed by magmatic segregation are almost nonexistent.

METHODS OF EXAMINING AND STUDYING ORE MINERALS

The most convenient way to study sulfide minerals is by optical examination. Most sulfides are opaque in the visible range and must be examined by reflected light. Preparation of reflecting surfaces, methods of examination, measurement of optical parameters, hardness determination, and criteria for discrimination have been covered in detail by several authors (see, for instance, Cameron, 1961; Ramdohr, 1969; Uytenbogaardt and Burke, 1971; and Galopin and Henry, 1972).

Reflected-light properties are generally less sensitive to slight compositional changes than are transmitted-light properties. Small compositional differences, as in zoned crystals, may be enhanced by etching techniques, but where transmitted light may be employed, as in sphalerite, much more detailed information can be obtained. In a few instances minerals are transparent to wavelengths in the nonvisible regions. Scanlon (1963), for example, monitored minute compositional variations in PbS by use of infrared absorption.

X-ray diffraction techniques lack the sensitivity of optical means for detecting trace quantities of phases, but in many cases X-ray identification is more certain than optical identification. X-ray studies can never provide the textural information gained from optical examination; however, single-crystal X-ray techniques can yield information on the relative crystallographic orientation of intergrown phases (for example, see Francis et al., 1976). X-ray and optical methods should always be considered complementary.

Optical and X-ray diffraction measurements record, indirectly, compositional and structural changes. Because a single optical or X-ray measurement can fix only a single compositional variable, it follows that such measurements can only be used to measure compositions in binary phases, such as the composition of hexagonal pyrrhotites (Arnold and Reichen, 1962). Where the number of compositional variables exceeds the number of independent optical or X-ray measurements, the results remain ambiguous. Unfortunately this is usually the case, so direct means of analysis must be used.

Electrical properties of sulfides are also indirect records of compositional variations and have been extensively investigated. Wide scatter in measurements has, unfortunately, precluded use of electrical properties as either identification or compositional measurement tools. Two electrical properties with possibly temperature-sensitive characteristics have been proposed for use in geothermometry. Smith (1947) suggested that the thermoelectric potential of pyrite was temperature-dependent, reflecting a temperature control of the number of defects in the crystal. However, Hayasa and Otuska (1952), and others, were not able to corroborate Smith's observations, so the method is of doubtful validity. Sato (1965) pointed out that many sulfides are semiconductors because their compositions depart slightly from strictly stoichiometric proportions. An electrochemical cell composed of two simultaneously deposited sulfides separated by a suitable electrolyte produces an electromotive force that is a function of temperature and should pass through zero at the temperature at which the minerals were originally in equilibrium. Sato demonstrated the feasibility of such measurements on synthetic preparations, and on natural oxides, but application to natural sulfides has not yet been accomplished.

An entirely different geothermometric scheme was suggested by Smith (1953, p. 110-111), and the theory supporting his suggestion was elaborated by Rosenfeld and Chase (see Rosenfeld, 1969). The scheme is based on the difference in thermal expansion between minerals or between different compositional zones of the same mineral. As temperatures change after crystallization, elastic strains are introduced in and around solid inclusions. If not annealed out by later thermal events, these strains can be observed as anomalously birefringent halos. One might then determine the zero strain condition by heating to an appropriate temperature while making optical observations. To our knowledge this relatively simple technique has not yet been attempted for ore mineral assemblages, but it appears promising for sphalerite and fluorite.

Bulk analyses performed on minerals from disaggregated ores, even with perfect separation, yield only an average composition from zoned crystals. But minerals that retain growth zoning are most useful in interpreting processes of ore formation. Thus, techniques for analyzing zoned mineral grains *in situ* provide data that can be interpreted in terms of the mineral paragenesis.

The electron-probe microanalyzer has revolutionized studies of minerals. Dozens of new minerals that might have otherwise remained as unidentifiable traces have been described thanks to the microprobe. Even more importantly, analysis of compositional details of well-known minerals has dramatically improved our ability to interpret natural environments (for example, see Barton et al., 1977). The facility to determine multicomponent compositional variables in individual phases without first having to separate the phases has provided a quantum jump in the accuracy and rapidity of progress in experimental studies dealing with mineral stabilities and has yielded the first significant advances into sulfide systems having more than three components (for examples, see Luce et al., 1977; Wiggins and Craig, 1975). The careful documentation of diffusion gradients in natural and synthetic samples offers promise of definitive understanding of reaction rates. The electron-probe microanalyzer magnetically focuses a beam of electrons on a spot as small as 1 μm in diameter; the characteristic X rays emitted by the irradiated elements can be resolved and their intensities measured. The technique is nondestructive and is useful for most elements found in the sulfide minerals. Under optimum conditions an accuracy of 1% or better of the amount present can be attained.

Another microanalytic technique, the ion probe, is more specialized and less developed than the electron probe. It focuses a beam of accelerated heavy ions onto a surface under study, and this causes secondary ions to be released from the surface of the bombarded specimen. The secondary ions so released are analyzed by mass spectroscopic techniques. It may soon become feasible to study isotopic distributions and even trace elements, across mineral surfaces *in situ*. Somewhat coarser in its spatial resolution than the electron-probe microanalyzer, the ion probe should eventually be refined to bombard an area approximately 4 μm in diameter.

Additional new techniques, as yet not widely employed in studies of sulfide minerals, deserve mention. One is a laser beam microprobe, first described by Maxwell (1963). A narrow, high-intensity laser beam impinges on a polished surface and causes the irradiated material to be completely vaporized. The vapor so formed is analyzed by standard spectroscopic techniques. The method is not limited by a reduction in sensitivity for elements of low atomic number, which is an inherent limitation in the microprobe analyzers using X radiation. The drawback of the method is the destruction of the material analyzed. Moreover, because the beam excavates a crater into the sample, there is uncertainty of the exact identity of the sample vaporized. The main advantage the method may have over the X-ray microprobe analyzer lies in its greater sensitivity at very low concentrations of elements and in the possibility that trace-element distributions may be studied *in situ*.

Another new technique is the scanning electron microscope. Employing a sweeping beam of electrons impinged on a natural or broken surface, the technique allows high-resolution electron-optical studies of unpolished surfaces far below visual resolution. The use of energy-dispersive counting techniques permits quantitative analysis of specific areas scanned.

High-resolution transmission electron microscopy is proving very useful in characterizing superstructures and vacancy distributions in materials whose basic structure is already known, as typified by the studies of Nakazawa et al. (1975) and Pierce and Buseck (1976).

METHODS OF STUDYING PHASE RELATIONS

Direct Determination

Direct determination of phase relations can be separated into three successively more difficult and useful classes: (1) determination of the way phase fields change as functions of pressure, temperature, and composition, (2) determination of compositional limits for individual minerals, and (3) determination of the tie lines between two or more coexisting solid or fluid solutions. Particularly useful data arise from the circumstance where, as temperature or pressure change, one assemblage of phases replaces another, producing invariant points such as those listed in Table 7-1 (See page 299). Determination of tie lines, however, provides the essential data for developing both sliding-scale and fixed-point buffers. Extensive reviews of experimental techniques have been prepared by Kullerud (1971) and Scott (1974, 1975).

Historically, phase relations have been determined by the appearance-of-phase method, in which boundaries of specific mineral assemblages are located by examination of a series of experiments of differing composition. Most studies are conducted by holding charges of known composition at a fixed temperature for sufficient time to ensure equilibration, then quenching the charges, and examining the solid phases by optical, X-ray, electrical, magnetic, and chemical techniques. *In situ* measurements of thermal effects (e.g., differential thermal analysis) or electrical conductivity (Potter and Barnes, 1978) have been used to locate reaction points.

Older work was often conducted in open containers with or without protective atmospheres. Results are sometimes useful for our purposes, especially when vapor pressures are low and when reactions with air have not occurred. More reliable results are obtained, however, when contaminating atmospheres and vapor losses are eliminated. This is done by using glass tubes into which reactants are sealed under vacuum. Because the walls of the tube are rigid, the volume is fixed and the pressure inside is the vapor pressure of the phases in the vessel. There are two disadvantages to the rigid-tube method: (1) only vapor-bearing assemblages (known as condensed assemblages because the total pressure is the vapor pressure of the solids and liquids) can be studied and (2) the actual vapor pressure is not measured.

Although the effect of total pressure is minor on sulfide systems, the partial pressure (or the more convenient thermochemical parameters activity, and fugacity) of sulfur is very important. The activity of sulfur can be measured in several ways as described by Kubaschewski et al. (1967), Barton and Toulmin (1964), Schneeberg (1973), and others.

1. The most-used method is a partial reduction of sulfides by hydrogen resulting in an $H_2 + H_2S$ gas mixture which can be analyzed easily for H_2S. The activity of S_2 is calculated from the reaction $2H_2(g) + S_2(g)$

$\rightleftharpoons 2H_2S(g)$. Alternatively, predetermined mixtures of $H_2 + H_2S$ are reacted and the state of the sample determined by weighing, optical examination, X-ray study, and similar techniques.

2. Dew-point techniques are simple, but limited to high-sulfur pressure ($>10^{-5}$ atm). There are two ways of using dew points. One uses a sulfur reservoir at a temperature below that of the reactants. This fixes the sulfur pressure in the apparatus. The other measures the temperature of the coldest spot in the apparatus, the temperature of the spot being adjusted to a point at which sulfur just condenses. Both methods require a correction from $P_{S_{total}}$, which includes $P_{S_8} + P_{S_7} + P_{S_6} + P_{S_5} + P_{S_4} + P_{S_3} + P_{S_2} + P_S$, to P_{S_2}. A discussion of the polymerization of sulfur is presented later in this chapter.

3. Total-pressure measurements can be made with a silica-spiral manometer or "spoon-gauge". The Rodebush or trap-door manometer (Bog and Rosenqvist, 1959) can also give good results.

4. The electrum-tarnish method developed by Barton and Toulmin (1964) uses the temperature at which a sulfide tarnish forms on silver-gold alloys. The method is useful for geologic studies because it allows study of sulfidation equilibrium to much lower temperatures than some of the other methods. The use of copper-gold alloys could extend the range of activity of S_2 to very low values, and similar methods might even be applied to the determination of activities of oxygen, selenium, and tellurium.

5. A pyrrhotite indicator method was developed by Toulmin and Barton (1964). At constant temperature, the composition of pyrrhotite is a function of the activity of S_2; the composition of pyrrhotite, in turn, can easily be measured by X-ray diffraction. The vapor of the assemblage whose sulfur activity is to be determined is permitted to equilibrate with pyrrhotite, the run is quenched, the pyrrhotite composition measured, and the activity of S_2 at the temperature of the run determined. Similar schemes, using other nonstoichiometric compounds of easily measured composition (e.g., $Ni_{(1-x)}Te$) can certainly be developed.

6. Electrochemical cells using solid or liquid electrolytes can give highly precise results; provided that care is taken to avoid electronic conduction and spurious reactions, the results can also be highly accurate. R. W. Potter (personal communication, 1976) determined the stabilities of several copper sulfides to within a few tens of calories per gram atom copper at temperatures of 100°C and below, using aqueous $CuSO_4$ solutions. Schneeberg (1973) followed Wagner (1953) and Sato (1971) in the use of solid AgI as an electrolyte to measure sulfur activities for sev-

eral univariant assemblages to over 400°C, but he expressed reservations about some of his results. Regardless of a few unfavorable results reported to date, the electrical potential techniques appear to offer the best hope for highly accurate measurements of sulfide stabilities in the future.

To study a reaction where a vapor phase is not present, it is necessary to use an inert, but malleable, reaction vessel. Reactants are placed inside and a pressure greater than the vapor pressure is applied to the outside causing the malleable tube to collapse, thus eliminating the vapor and transmitting the pressure to the reactants. Materials used for collapsible reaction vessels are platinum, gold, or a plastic such as polyethylene.

A combination of rigid and collapsible tube techniques allows study of the entire geologically important range of pressure and temperature. Nevertheless, problems still remain as to whether or not runs actually reach equilibrium and whether a high-temperature assemblage reequilibrates during quench. In practice, an equilibrium state is considered attained if we always produce the same products regardless of the choice of starting materials and the sequence of pressure and temperature changes to which we subject the charge. This is generally referred to as the ability to reverse a reaction.

Recognition of reaction during quench requires examination of the sample at the temperature and pressure of the run (as by high-temperature X-ray diffraction) or detection of thermal effects during quench by differential thermal analysis or by precise time-temperature studies.

Fluxes can be used to induce sluggish reactions to take place in reasonable lengths of time. Water and such diverse materials as molten alkali halides, glycerine, and alkali chloride solutions have been used, but it must always be shown that the role of the flux is strictly that of a catalyst. The combination of a suitable flux and a controlled thermal gradient can greatly enhance the rate of reaction and crystal growth (see Scott, 1974, 1975, and Pamplin, 1975). One hazard to this otherwise very attractive technique is that the nonequilibrium temperatures may lead to a nonequilibrium product (note our later discussion of the iron-zinc-sulfur system).

Indirect Determination

Many reactions are too sluggish to be studied by direct methods within the pressure-temperature range of interest. Such reactions may, however, be examined at high temperature and the equilibria so determined extrapolated to low temperature with a high degree of precision, provided that accurate heat-content data are available for all of the phases involved. Such a thermodynamic treatment does not guarantee, of course, that an unsuspected low-temperature phase will not render the projection meaningless. Some complex reactions such as the partial sulfidation of a dunite to a pyroxenite containing magnetite + pyrrhotite can even be calculated directly without recourse to any experiment when appropriate free-energy data are available.

Thermochemical data are available from many sources, the most useful sources for our purposes are Kubaschewski et al. (1967) and Mills (1974). Additional data are available from the compilations by Clark (1966), Coughlin (1954), Freeman (1962), Holland (1959, 1965), Kelley (1960, 1962), Kelley and King (1961), King et al. (1973), Pearson (1958), Richardson and Jeffes (1952), Robie and Waldbaum (1968), Rossini et al. (1952, and subsequent revisions of the NBS Tech. Note 270 series), Stull and Prophet (1971 with supplements in 1974 and 1975), Hultgren et al. (1973a,b), and Naumov et al. (1974).

To illustrate the graphic extrapolation of high-temperature equilibria let us consider the invariant assemblage pyrite + pyrrhotite + bismuth + bismuthinite (plus vapor). Mineral assemblages indicate that such an invariant assemblage does exist; the problem is to find what the temperature is because direct experimentation has failed to do so. The univariant curves for

$$\text{pyrrhotite} + S_2(g) \rightleftharpoons \text{pyrite}$$

and

$$\text{bismuth} + S_2(g) \rightleftharpoons \text{bismuthinite}$$

are known (Figure 7.23). Projecting these curves to an intersection indicates that the temperature of the invariant point is 213°C (just off-scale in Figure 7.23), a temperature at which it would take an impractically long time to obtain equilibrium in the laboratory because of the slow reaction rate of pyrite.

As an analytic example of the use of free-energy data, consider the portion of the copper-lead-sulfur system where galena, chalcocite, crystalline lead, and crystalline copper are found, each of essentially fixed composition. No solid solution or ternary-compound formation is considered. We can calculate the invariant point where these phases (plus vapor) coexist from the free-energy equations given in Table 7-2 (See page 337).

$$2\text{ Pb} + S_2(g) \rightleftharpoons 2\text{ PbS}; \Delta G° = -77,449 + 40.81\ T\ (\pm 1000)\text{ cal}$$

$$4\text{ Cu} + S_2(g) \rightleftharpoons 2\text{ Cu}_2\text{S}; \Delta G° = -63,945 + 17.22\ T\ (\pm 1000)\text{ cal}$$

subtracting,

$$2\,Pb + 2\,Cu_2S \rightleftharpoons 4\,Cu + 2\,PbS;\ \Delta G° = -13{,}504 + 23.59\,T\,(\pm 1400)\ \text{cal}$$

At equilibrium $\Delta G_R = 0$ and T must be 572K (299°C). Above the invariant temperature the reaction goes to the left and below, to the right. Considering that the uncertainty in the equation for the invariant point is approximately 1400 cal, the uncertainty in T is $\pm 60°$. Nevertheless, the calculated value of $299 \pm 60°C$ agrees with an experimental determination of $271 \pm 5°C$ by Craig and Kullerud (1968).

Thermodynamic calculations can be powerful aids when used to elucidate the qualitative aspects of phase diagrams, but the quantitative aspects are subject to considerable uncertainty. Conversely, experimentally determined invariant points provide a very sensitive way by which we can refine thermodynamic parameters.

Two reservations must be raised concerning the use of thermodynamic data. First, it is essential that the data used apply to the phase being considered. Some older texts, for example, do not distinguish digenite from chalcocite. Second, data given for end-member compositions require adjustment when applied to solid solutions. The following generalizations may aid in making approximate calculations:

1. In dilute solutions the activity of the component representing the solvent end member tends to behave ideally, that is, to follow Raoult's law. The activity of the solvent end member is equal to its mole fraction.
2. In dilute solutions the activity of the component representing the solute end member is proportional to its mole fraction, but the proportionality constant (Henry's law constant) in general is not unity.
3. Generally, the wider the solid solution, the greater the compositional range covered by the dilute-solution rules (1) and (2). For example, the activity of FeS in the (Fe, Zn)S solid solution does not depart drastically from Henry's law behavior even with more than 50% FeS in solid solution.
4. Substitutional solid solutions (which may occur in phases also displaying interstitial and omission phenomena), such as Pb(S, Se), (Zn, Fe, Cd, Mn)S, (Fe, Ni)S_2, (Fe, Ni)$_{1-x}$S, $Cu_{12}(As, Sb)_4S_{13}$, or $(Pb_2, AgBi)S_2$ are more likely to follow the dilute-solution rules over an appreciable compositional range than are the interstitial and omission types of solid solutions.

Several examples of the relation between activity and mole fraction of solid solutions are shown in Figure 7.2.

Table 7-1 Invariant points from condensed systems of possible interest to the geothermometry of ore deposits.

The footnoted remarks for this table are combined with those for Table 7-2 and are found following it. Compositions are given in atomic proportions, but mineral formulae are simplified where necessary to conserve space. Solid solutions are indicated by s.s. The uncertainties are those assigned by the authors and are therefore of somewhat uneven precision. Many of the references cited contain additional data on the melting relations of specific systems; most of the melting relations included here merely place upper limits on stability fields of phases of interest. The adjectives "low", "intermediate", and "high" designate relative thermal stabilities of polymorphs, except that they are omitted for low-temperature phases when such phases are well-defined minerals.

System	Low-temperature assemblage	High-temperature assemblage	Temperature °C	References/remarks
As	crystal	melt	818	$P = 29$ bars, Klement et al. (1963); $dT/dP = 4°/kb$
Bi	crystal	melt	271.5	Klement et al. (1963); $dT/dP = -7.6°/kb$
S	orthorhombic	monoclinic	95.4	Hultgren et al. (1973a); $dT/dP = 42°/kb$
	monoclinic	melt	115.2	Hultgren et al. (1973a); $dT/dP = 27°/kb$
	melt	vapor	444.7	Hultgren et al. (1973a); $P = 1$ atm
Sb	crystal	melt	630	Kubaschewski et al. (1967); $dT/dP = 0.4°/kb$; Klement et al. (1963)
Se	crystal	melt	220	Kubaschewski et al. (1967)
Te	crystal	melt	450	Kubaschewski et al. (1967)
Ag-As	silver + arsenic	huntilite Ag$_9$As	456	Hall and Yund (1964)
	huntilite	silver + melt	595	Hall and Yund (1964)

Table 7-1. (Continued)

System	Low-temperature assemblage	High-temperature assemblage	Temperature °C	References/remarks
Ag-Hg	moschellandsbergite Ag_2Hg_3 $Ag_{55}Hg_{45}$	$Ag_{55}Hg_{45}$ + Hg-rich melt	127	Hansen and Anderko (1958)
		Hg-rich melt + silver s.s. (Ag_5Hg_3)	276	Hansen and Anderko (1958)
Ag-S	acanthite (+ silver) Ag_2S	low argentite Ag_2S	176.5 ± 0.5	Hansen and Anderko (1958),[a] $dT/dP = 1.6°/kb$; Clark and Rapoport (1970)
	acanthite (+ sulfur melt) Ag_2S	low argentite Ag_2S	177.8 ± 0.7	Hansen and Anderko (1958)
	low argentite (+ silver) Ag_2S	high argentite Ag_2S	586 ± 3	Hansen and Anderko (1958)
	low argentite Ag_2S	high argentite (+ sulfur melt) Ag_2S	622 ± 3	Hansen and Anderko (1958)
	high argentite + silver Ag_2S	melt $(Ag_{68}S_{32})$	804 ± 2	Hansen and Anderko (1958)
	high argentite + sulfur melt Ag_2S	melt $(Ag_{64}S_{36})$	740 ± 2	Hansen and Anderko (1958)
	high argentite Ag_2S	melt	838	Hansen and Anderko (1958)
Ag-Sb	artimony + dyscrasite $(Ag_{73.4}Sb_{26.6})$	melt $(Ag_{59}Sb_{41})$	485	Hansen and Anderko (1958)
	dyscrasite $(Ag_{81.9}Sb_{18.1})$	allargentum + melt $(Ag_{83.7}Sb_{16.3})$ $(Ag_{75}Sb_{25})$	558	Hansen and Anderko (1958)
	allargentum $(Ag_{91.2}Sb_{8.8})$	silver + melt $(Ag_{92.8}Sb_{7.2})$ $(Ag_{83.1}Sb_{16.9})$	702.5	Hansen and Anderko (1958)

System	Phases	T (°C)	Reference
Ag-Se	naumannite Ag_2Se ⇌ high naumannite	133 ± 1	Shunk (1969); $dT/dP = 6.0°/kb$; Clark and Rapoport (1970)
	high naumannite + Ag_2Se ⇌ melt ($Ag_{55.5}Se_{44.5}$) selenium melt	616	Hansen and Anderko (1958)
	high naumannite Ag_2Se ⇌ melt	897 ± 3	Elliott (1965)
Ag-Te	hessite + stutzite Ag_2Te Ag_5Te_3 ⇌ low $Ag_{1.9}Te$	120 ± 15	Kracek et al. (1966)
	empressite $AgTe$ ⇌ stutzite + tellurium	170	o, Cabri (1965b)
	hessite Ag_2Te ⇌ int. hessite Ag_5Te_3	145 ± 3	Kracek et al. (1966)
	low $Ag_{1.9}Te$ ⇌ high $Ag_{1.9}Te$	178	Kracek et al. (1966)
	stutzite (+ $Ag_{1.9}Te$) ⇌ high stutzite Ag_5Te_3	265 ± 15	Kracek et al. (1966)
	stutzite Ag_5Te_3 ⇌ high stutzite (+ tellurium) Ag_5Te_3	295 ± 10	Kracek et al. (1966)
	high stutzite + tellurium Ag_5Te_3 ⇌ melt ($Ag_{33}Te_{67}$)	353 ± 0.5	Kracek et al. (1966)
	high stutzite Ag_5Te_3 ⇌ high $Ag_{1.9}Te$ + melt ($Ag_{44.4}Te_{55.6}$)	420 ± 5	Kracek et al. (1966)
	high $Ag_{1.9}Te$ ⇌ int. hessite + melt Ag_2Te ($Ag_{48.7}Te_{51.3}$)	460 ± 5	Kracek et al. (1966)
	int. hessite + melt ⇌ high hessite Ag_2Te ($Ag_{59.7}Te_{40.3}$) Ag_2Te	689 ± 5	Kracek et al. (1966)
	int. hessite ⇌ high hessite + silver Ag_2Te Ag_2Te	802 ± 3	Kracek et al. (1966)

Table 7-1. (*Continued*)

System	Low-temperature assemblage	High-temperature assemblage	Temperature °C	References/remarks
	high hessite + silver Ag_2Te	melt $(Ag_{87.5}Te_{12.5})$	869 ± 3	Kracek et al. (1966)
	high hessite + melt Ag_2Te $(Ag_{87}Te_{13})$	melt $(Ag_{69.7}Te_{30.3})$	881 ± 3	Kracek et al. (1966)
	high hessite Ag_2Te	melt	960	Kracek et al. (1966)
As-Au	arsenic + gold	melt $(As_{43}Au_{57})$	655	Hansen and Anderko (1958)[b]
As-Cu	α-domeykite Cu_3As	algodonite + β-domeykite $Cu_{5.2}As$ $Cu_{2.7}As$	90 ± 10	Skinner and Luce (1971a)
	algodonite Cu_8As	arsenian copper + β-domeykite $Cu_{19}As$ $Cu_{2.7}As$	300 ± 20	Skinner and Luce (1971a)
	β-domeykite + arsenic $Cu_{2.7}As$	koutekite $Cu_{19}As_8$	350 ± 10	Maske and Skinner (1971)
	koutekite + arsenic $Cu_{19}As_8$	melt $(As_{46}Cu_{54})$	600	Elliott (1965)
	koutekite $Cu_{19}As_8$	melt + β-domeykite $(As_{34}Cu_{66})$ $Cu_{2.7}As$	709	Elliott (1965)
	β-domeykite $Cu_{2.7}As$	melt	827	Elliott (1965)
	β-domeykite + copper $Cu_{2.7}As$	melt $(As_{18}Cu_{82})$	685	Elliott (1965)
As-Fe	loellingite $FeAs_2$	melt	1016 ± 8	Shunk (1969)

System	Phases		T (°C)	Reference
As-Ni	dienerite Ni_3As (?)	$Ni_{5-x}As_2$ + maucherite $Ni_{11}As_8$	<200	Yund (1961)[c]
	pararammelsbergite (+ niccolite) $NiAs_2$	rammelsbergite $Ni_{1\pm x}As$ $Ni_{1-x}As_2$	590 ± 10	Yund (1961)[d]
	pararammelsbergite $NiAs_2$	rammelsbergite (+ arsenic) $Ni_{1-x}As_2$	598 ± 10	Yund (1961)[d]
	rammelsbergite + arsenic $Ni_{1-x}As_2$	melt $(As_{94}Ni_6)$	783 ± 5	Yund (1961)
	$Ni_{5-x}As_2$ + maucherite $Ni_{11}As_8$	melt $(As_{38}Ni_{62})$	818 ± 5	Yund (1961)
	maucherite $Ni_{11}As_8$	niccolite + melt $Ni_{1\pm x}As$ $(As_{40}Ni_{60})$	830 ± 5	Yund (1961)
	rammelsbergite + niccolite $Ni_{1-x}As_2$ $Ni_{1\pm x}As$	melt $(As_{57}Ni_{43})$	852 ± 3	Yund (1961)
	nickel + $Ni_{5-x}As_2$	melt $(As_{22}Ni_{78})$	895 ± 5	Yund (1961)
	niccolite $Ni_{1\pm x}As$	melt	962 ± 3	Yund (1961)
	$Ni_{5-x}As_2$	melt	~998	Yund (1961)
	rammelsbergite $Ni_{1-x}As_2$	melt	>1040	Yund (1961)
As-Pt	platinum + sperrylite $PtAs_2$	melt $(As_{28}Pt_{72})$	597	Heyding and Calvert (1961)
	sperrylite $PtAs_2$	melt	>1400	Hansen and Anderko (1958)
As-S	low orpiment As_2S_3	high orpiment As_2S_3	170	Kirkinskii et al. (1967); $dT/dP = 14°/kb$
	low realgar AsS	high realgar AsS	265 ± 5	Hall and Yund (1964)

Table 7-1. (*Continued*)

System	Low-temperature assemblage	High-temperature assemblage	Temperature °C	References/remarks
	realgar + arsenic AsS	melt ($As_{54}S_{46}$)	281 ± 5	Hall and Yund (1964)
	realgar AsS	melt	307 ± 5	Hall and Yund (1964)
	orpiment As_2S_3	melt	315 ± 5	Hall and Yund (1964)
	arsenic + melt ($As_{83}S_{17}$)	melt ($As_{97}S_3$)	797 ± 5	Barton (1969)
				Duranusite (As_4S), dimorphite (As_3S_4), AsS_2, and As_2S_5 have no known stability fields.
As-Sb	arsenic-antimony solid solution	melt ($As_{25.5}Sb_{74.5}$)	612	Skinner (1965)
	arsenic s.s. + arsenic-antimony s.s.	complete solid solution	<300	Skinner (1965)
	arsenic-antimony s.s. + antimony s.s.	complete solid solution	<300	Skinner (1965)
Au-Bi	maldonite + bismuth Au_2Bi	melt ($Au_{17.2}Bi_{82.8}$)	241 ± 1	Shunk (1969)
	maldonite Au_2Bi	melt ($Au_{33.1}Bi_{66.9}$) + gold	371 ± 2	Shunk (1969)
	gold + bismuth	maldonite Au_2Bi	113	Calc'd. from Hultgren et al. (1973b); agrees with Ramdohr's comment (1969).

System	Assemblage		Temp. (°C)	Reference
Au-Cu	$Au_{3\pm x}Cu$ ($Au_{66}Cu_{34}$)	$Au_{1\pm x}Cu$ + gold-copper s.s. ($Au_{62}Cu_{38}$) ($Au_{67}Cu_{33}$)	~240	Shunk (1969)[e]
	$AuCu_{3\pm x}$ + $Au_{1-x}Cu$ ($Au_{35}Cu_{65}$) ($Au_{38}Cu_{62}$)	gold-copper s.s. ($Au_{36}Cu_{64}$)	285	Hansen and Anderko (1958)
	low AuCu	int. AuCu	385	Hultgren et al. (1973b)
	$AuCu_3$	gold-copper s.s.	390	Hultgren et al. (1973b)
	int. AuCu	gold-copper s.s.	410	Hultgren et al. (1973b)
Au-Sb	aurostibite + gold	melt	360	Hansen and Anderko (1958)
	$AuSb_2$	($Au_{65}Sb_{35}$)		
	aurostibite $AuSb_2$	melt (essentially congruent)	460	Hansen and Anderko (1958)
Au-Te	calaverite + tellurium $AuTe_2$	melt ($Au_{12}Te_{88}$)	416	Hansen and Anderko (1958)
	calaverite $AuTe_2$	high calaverite	435	Cabri (1965)
	high calaverite + gold $AuTe_2$	$AuTe_2$ melt ($Au_{47}Te_{53}$)	447	Hansen and Anderko (1958)
	calaverite $AuTe_2$	melt	464 ± 3	Cabri (1965)
	montbrayite $Au_{1.89}Ag_{0.03}Pb_{0.04}Sb_{0.07} \cdot Te_{2.88}Bi_{0.12}$	melt	410 ± 5	Bachechi (1972)[f]
Bi-S	bismuthinite + bismuth Bi_2S_3	melt ($Bi_{99.5}S_{0.5}$)	270	Hansen and Anderko (1958); Happ and Davey (1971)
	bismuthinite + sulfur melt Bi_2S_3	melt ($Bi_{35}S_{65}$)	715 ± 10	Glatz and Meikleham (1963)
	bismuthinite Bi_2S_3	melt	760 ± 5	Van Hook (1960)
Bi-Te	bismuth + hedleyite Bi_7Te_3	melt ($Bi_{97.6}Te_{2.4}$)	266	Elliott (1965)

Table 7-1. (*Continued*)

System	Low-temperature assemblage	High-temperature assemblage	Temperature °C	References/remarks
	hedleyite Bi_7Te_3	melt + $Bi_{2+x}Te$ ($Bi_{94}Te_6$)	312	Elliott (1965)
	tellurobismuthite + tellurium Bi_2Te_3	melt ($Bi_{10}Te_{90}$)	413	Elliott (1965)
	$Bi_{2+x}Te$	melt + wehrlite ($Bi_{72}Te_{28}$) $Bi_{1\pm x}Te$	420	Elliott (1965)
	wehrlite $Bi_{1\pm x}Te$	melt + tellurobismuthite ($Bi_{56}Te_{44}$) Bi_2Te_3	540	Elliott (1965)
	tellurobismuthite Bi_2Te_3	melt	588.5 ± 1	Shunk (1969)
Cd-S	hawleyite CdS	greenockite CdS	unknown, but must be at low temperature if hawleyite is at all stable.	
	greenockite CdS	melt	1475 ± 15	Shunk (1969)
Co-S	linneaite + cobalt pentlandite Co_3S_4 Co_9S_8	$Co_{1-x}S$	~460	Hansen and Anderko (1958)
	linneaite Co_3S_4	cattierite + $Co_{1-x}S$ CoS_2	670	Leegaard and Rosenqvist (1964)
	cobalt + cobalt pentlandite Co_9S_8	Co_4S_3	785	Elliott (1965)
	cobalt pentlandite Co_9S_8	$Co_4S_3 + Co_{1-x}S$	830	Elliott (1965)
	cattierite CoS_2	$Co_{1-x}S$ + sulfur melt	~970	estimated from Figure 7.23

System	Phase / Assemblage	Temp (°C)	Reference
Co-Se	trogtalite $CoSe_2$	~960	Elliott (1965)
	melt + $Co_{1-x}Se$		Hastite ($CoSe_2$), bornhardite (Co_3Se_4), and freboldite ($CoSe$) have no known stability fields.
Cu-S	anilite $Cu_{1.75}S$	75 ± 3	R. W. Potter (1976, personal communication)
	high digenite s.s. + covellite $Cu_{64.6}S_{35.4}$ CuS		
	djurleite + chalcocite $Cu_{1.96}S$ Cu_2S	90 ± 2	R. W. Potter (1976, personal communication)
	hex. chalcocite $Cu_{2-x}S$		
	djurleite $Cu_{1.96}S$	93 ± 2	R. W. Potter (1976, personal communication)
	hex. chalcocite + high digenite $Cu_{2-x}S$ $Cu_{64.9}S_{35.1}$		
	blaubleibender covellite $Cu_{1+x}S$	157 ± 3	Moh (1971)[g, h]
	covellite + digenite CuS $Cu_{63.75}S_{36.25}$		
	chalcocite Cu_2S	103 ± 2	Roseboom (1966)
	hex. chalcocite (+ copper)		
	hex. chalcocite Cu_2S	435 ± 10	Roseboom (1966)
	high digenite (+ copper) Cu_2S		
	covellite CuS	507 ± 3	Kullerud (1965)
	high digenite + sulfur melt $Cu_{63.4}S_{36.6}$		
	covellite CuS	647, 18 kbar	Taylor and Kullerud (1972)
	CuS_2 + high digenite		
	high digenite + sulfur melt $Cu_{64}S_{36}$	813	Jensen (1947)
	melt		
	high digenite $Cu_{67.1}S$	1129	Hansen and Anderko (1958)
	melt		
Cu-Se	klockmannite $CuSe$	60 ± 5	Bernardini et al. (1972)
	high klockmannite $CuSe$		
	athabascaite Cu_5Se_4	<100	Harris et al. (1970)
	umangite + klockmannite Cu_3Se_2 $CuSe$		

Table 7-1. (*Continued*)

System	Low-temperature assemblage	High-temperature assemblage	Temperature °C	References/remarks
	bellidoite Cu_2Se	berzelianite (+ copper) $Cu_{2-x}Se$	138 ± 5	Bernardini et al. (1972)
	umangite Cu_3Se_2	berzelianite + high klockmannite $Cu_{2-x}Se$ $\quad$ CuSe	143 ± 5	Bernardini et al. (1972)
	$CuSe_2$	high klockmannite + selenium melt	343 ± 5	Bernardini et al. (1972)[i]
	klockmannite CuSe	berzelianite + selenium melt $Cu_{2-x}Se$	384 ± 5	Bernardini et al. (1972)
	berzelianite + selenium melt $Cu_{2-x}Se$	melt $(Cu_{49}Se_{51})$	523 ± 5	Bernardini et al. (1972)
Cu-Te	vulcanite + tellurium CuTe	melt $(Cu_{29}Te_{71})$	340	Hansen and Anderko (1958)
	vulcanite CuTe	rickardite + melt Cu_4Te_3 $\quad (Cu_{33}Te_{67})$	367	Hansen and Anderko (1958)
	rickardite Cu_4Te_3	Cu_9Te_5 + melt $(Cu_{52}Te_{48})$	630	Hansen and Anderko (1958)[j]
	Cu_9Te_5	$Cu_{2-x}Te$ + melt $(Cu_{55}Te_{45})$	727	Hansen and Anderko (1958)[j]
	$Cu_{2-x}Te$	melt	1125	Hansen and Anderko (1958)
Fe-O	magnetite + iron Fe_3O_4	wüstite $Fe_{1-x}O$	560	Muan and Osborne (1965)
Fe-S	smythite Fe_3S_4	high smythite FeS	75	Taylor (1970a)[g, k]

System	Assemblage	Composition	T (°C)	Reference
	troilite FeS	pyrrhotite (1C) $Fe_{1-x}S$	140 ± 4	See text and Figure 7.8, Scott (1974)
	monoclinic pyrrhotite (4C) Fe_7S_8	pyrite + pyrrhotite (NA) $FeS_2 \quad Fe_{1-x}S$	251 ± 3	Rising (1973)
	pyrrhotite (NA) + pyrite $Fe_{1-x}S \quad FeS_2$	pyrrhotite (MC) $Fe_{1-x}S$	262	Scott (1974)
	pyrrhotite (MC) $Fe_{1-x}S$	pyrrhotite (1C) + pyrite $Fe_{1-x}S \quad FeS_2$	308	Scott (1974)
	pyrrhotite (1C) $Fe_{1-x}S$	pyrrhotite (1C) $Fe_{1-x}S$	320 ± 5	Scott (1974), magnetic transition
	pyrite FeS_2	pyrrhotite + sulfur melt $Fe_{1-x}S$	742 ± 1	Barton (1969), Kullerud and Yoder (1959)
	pyrrhotite + iron FeS	melt ($Fe_{56}S_{44}$)	988	Hansen and Anderko (1958)
	pyrrhotite + sulfur melt $Fe_{1-x}S$	melt ($Fe_{35.6}S_{64.4}$)	1092 ± 3	Arnold (1971)
	pyrrhotite $Fe_{1-x}S$	melt ($Fe_{48}S_{52}$)	1190	Hansen and Anderko (1958)
Fe-Se	tetragonal FeSe $FeSe_2$	hexagonal FeSe	335	Troften and Kullerud (1961)
	ferroselite $Fe_{1-x}Se$	$Fe_{1-x}Se$ + selenium melt	585	Troften and Kullerud (1961)
	$Fe_{1-x}Se$	melt ($Fe_{47}Se_{53}$)	1070 ± 5	Troften and Kullerud (1961)
Fe-Te	frohbergite $FeTe_2$	Fe_2Te_3 + tellurium melt ?	655	Elliott (1965)[c]
Hg-S	cinnabar s.s. + sulfur melt ($Hg_{49.2}S_{50.8}$)	metacinnabar s.s. ($Hg_{46.4}S_{53.6}$)	315 ± 3	Potter and Barnes (1978)
	cinnabar HgS	metacinnabar + mercury melt HgS	345 ± 2	Potter and Barnes (1978)

Table 7-1. (*Continued*)

System	Low-temperature assemblage	High-temperature assemblage	Temperature °C	References/remarks
	metacinnabar s.s. + sulfur melt ($Hg_{48.6}S_{51.4}$)	hypocinnabar s.s. ($Hg_{48.6}S_{51.4}$)	470 ± 2	Potter and Barnes (1978)
	metacinnabar s.s. ($Hg_{49.2}S_{50.8}$)	hypocinnabar s.s. + mercury ($Hg_{49.2}S_{50.8}$)	481 ± 2	Potter and Barnes (1978)
	hypocinnabar s.s. + sulfur melt ($Hg_{49.0}S_{51.0}$)	melt	788 ± 2	Potter and Barnes (1978)
	hypocinnabar s.s. + mercury ($Hg_{49.1}S_{50.9}$)	melt	804 ± 3	Potter and Barnes (1978)
	hypocinnabar s.s. ($Hg_{48.8}S_{51.2}$)	melt	820 ± 3	Potter and Barnes (1978)
Hg-Se	tiemannite HgSe	melt	795 ± 5	Elliott (1965)
Hg-Te	coloradoite HgTe	melt	670 ± 20	Elliott (1965), Shunk (1969)[c]
Mn-S	hauerite MnS_2	alabandite + sulfur melt MnS	423	Skinner (unpublished)
	alabandite	melt	1610 ± 10	Hansen and Anderko (1958)
Mo-S	Mo_3S_4	$molybdenite + molybdenum$ MoS_2	468	Chevrel et al. (1974)[a, l]
	$molybdenite + molybdenum$ MoS_2	Mo_2S_3	673 ± 10?	Pouillard and Perrot (1975)
Ni-S	polydymite + millerite Ni_3S_4 NiS	$Ni_{1-x}S$	282 ± 5	Kullerud and Yund (1962)

	Assemblage		Temp (°C)	Reference
	polydymite Ni_3S_4	vaesite + $Ni_{1-x}S$ NiS_2	356 ± 3	Kullerud and Yund (1962)
	millerite NiS	$Ni_{1-x}S$	379 ± 3	Kullerud and Yund (1962)
	godlevskite Ni_7S_6	high godlevskite Ni_7S_6	400 ± 3	Kullerud and Yund (1962)
	heazlewoodite + high godlevskite Ni_3S_2 Ni_7S_6	$Ni_{3\pm x}S_2$	524 ± 3	Kullerud and Yund (1962)
	heazlewoodite + nickel Ni_3S_2	$Ni_{3\pm x}S$	533 ± 3	Kullerud and Yund (1962)
	heazlewoodite Ni_3S_2	$Ni_{3\pm x}S_2$	556 ± 3	Kullerud and Yund (1962)
	high godlevskite Ni_7S_6	$Ni_{3\pm x}S_2 + Ni_{1-x}S$	573 ± 3	Kullerud and Yund (1962)
	nickel + $Ni_{3\pm x}S_2$	melt ($Ni_{66.6}S_{33.4}$)	635 ± 3	Kullerud and Yund (1962)
	$Ni_{3\pm x}S_2$	melt + $Ni_{1-x}S$ ($Ni_{44}S_{56}$)	806 ± 3	Kullerud and Yund (1962)
	$Ni_{1-x}S$ + vaesite NiS_2	melt ($Ni_{47}S_{53}$)	993 ± 3	Arnold and Malik (1975)
	$Ni_{1-x}S$	melt ($Ni_{48}S_{52}$)	999 ± 3	Arnold and Malik (1975)
	vaesite NiS_2	melt + sulfur melt ($Ni_{37}S_{63}$)	1022 ± 3	Arnold and Malik (1975)
Ni-Sb	nisbite $NiSb_2$	breithauptite + melt $Ni_{1\pm x}Sb$ (Ni_3Sb_{97})	621 ± 5	Cabri et al. (1970a); Hansen and Anderko (1958)
Ni-Se	makinenite NiSe	Ni_3Se_2 + sederholmite $Ni_{1-x}Se$	375	Elliott (1965)

Table 7-1. (*Continued*)

System	Low-temperature assemblage	High-temperature assemblage	Temperature °C	References/remarks
	wilkmanite Ni_3Se_4	sederholmite + penroseite $Ni_{1-x}Se$ $NiSe_2$	300–400	Komarek and Wessely (1972)
	penroseite $NiSe_2$	sederholmite + selenium melt $Ni_{1-x}Se$	853	Komarek and Wessely (1972)
	sederholmite + selenium melt $Ni_{1-x}Se$	melt $(Ni_{32}Se_{68})$	856	Komarek and Wessely (1972)
	sederholmite $Ni_{1-x}Se$	melt $(Ni_{46.5}Se_{53.5})$	959	Komarek and Wessely (1972)
				Kullerudite ($NiSe_2$) and trustedite (Ni_3Se_4) have no known stability fields.
Ni-Te	melonite $Ni_{1-x}Te$	melt $(Ni_{44}Te_{46})$	900.5	Klepp and Komarek (1972)
O-Sn	romarchite SnO	Sn_3O_4 + tin melt	720 ± 20	Moh (1974)
	Sn_3O_4	cassiterite + tin melt SnO_2	~475	Drabek and Stemprok (1974)
	cassiterite SnO_2	melt	1630 ± 5	Schneider (1963)
Pb-S	galena PbS	melt	1127 ± 5	Van Hook (1960)
Pb-Se	clausthalite PbSe	melt	1082 ± 2	Shunk (1969)
Pb-Te	altaite + tellurium PbTe	melt $(Pb_{14}Te_{86})$	~405	Hansen and Anderko (1958)

System				Temperature	Reference
S-Sb	altaite		melt	924 ± 0.5	Elliott (1965)
	PbTe				
	stibnite + sulfur melt		melt	496 ± 5	Barton (1971)
	Sb_2S_3		$(S_{65}Sb_{35})$		
	stibnite + antimony		melt	518 ± 5	Barton (1971)
	Sb_2S_3		$(S_{57}Sb_{43})$		
	stibnite		melt	556 ± 5	Barton (1971)
	Sb_2S_3				
	antimony + melt		melt	615	Hansen and Anderko (1958)
	$(S_{55}Sb_{45})$		(S_6Sb_{94})		
S-Sn	herzenbergite (+ ottemannite)		high herzenbergite	597 ± 1	Moh (1974)
	SnS	Sn_2S_3	$Sn_{1-x}S$		
	herzenbergite		high herzenbergite (+ tin melt)	599 ± 1	Moh (1974)
	SnS		$Sn_{1-x}S$		
	ottemannite (+ berndtite)		int. ottemannite	661	Moh (1974)
	Sn_2S_3	SnS_2	$Sn_{2-x}S_3$		
	ottemannite		int. ottemannite (+ $Sn_{1-x}S$)	675 ± 1	Moh (1974)
	Sn_2S_3		$Sn_{2-x}S_3$		
	berndtite (+ int. ottemannite)		high berndtite	680 ± 2	Moh (1974)
	SnS_2	$Sn_{2-x}S_3$	SnS_{2-x}		
	berndtite		high berndtite (+ sulfur melt)	691 ± 1	Moh (1974)
	SnS_2		SnS_{2-x}		
	int. ottemannite (+ high berndtite)		high ottemannite	710 ± 1	Moh (1974)
	$Sn_{2-x}S_3$	SnS_{2-x}	$Sn_{2\pm x}S_3$		
	int. ottemannite		high ottemannite +		
	$Sn_{2-x}S_3$		high herzenbergite		
			$Sn_{2\pm x}S_3$ $Sn_{1-x}S$		
	high ottemannite		high berndtite + melt	760 ± 2	Moh (1974)$^{m, n}$
	$Sn_{2-x}S_3$		SnS_{2-x} $(S_{56}Sn_{44})$		

Table 7-1. (*Continued*)

System	Low-temperature assemblage	High-temperature assemblage	Temperature °C	References/remarks
	high berndtite SnS_{2-x}	melt	867 ± 2	Moh (1974)
	high herzenbergite $Sn_{1-x}S$	melt	879 ± 2	Moh (1974)
S-Tl	carlinite Tl_2S	melt	449 ± 5	Radtke and Dickson (1975)
S-V	patronite VS_4	V_2S_3 + sulfur melt	~470	extrapolated from Biltz and Kocher (1939)[c]
S-W	tungstenite WS_2	melt	1800	Elliott (1965)
S-Zn	sphalerite (+ zinc melt) ZnS	wurtzite	1013	Moh (1975a); see discussion in text.
	wurtzite ZnS	wurtzite (+ sulfur melt) ZnS	1031	Moh (1975a)
	wurtzite ZnS	melt	1830 ± 20	Elliott (1965)
	sphalerite ZnS	wurtzite + melt (ZnS)	1670 at 40 kbar	Sharp (1969)
Se-Zn	stilleite ZnSe	hex. ZnSe	"high temp."	Elliott (1965)[c]
	hex. ZnSe	melt	1515 ± 20	Elliott (1965)
Ag-As-Cu	kutinaite ($Cu_{2.07}Ag_{0.84}As$)	breakdown product(s)	>400	Hak et al. (1970)[c]
Ag-As-S	xanthoconite Ag_3AsS_3	proustite Ag_3AsS_3	192 ± 10	Hall (1966)

312

System	Reaction	T (°C)	Reference
	trechmannite + orpiment → melt	280 ± 5	Hall (1966)
	AgAsS$_2$ As$_2$S$_3$ (Ag$_{1.6}$As$_{39}$S$_{59.4}$)		
	trechmannite → smithite	320 ± 5	Hall (1966)
	AgAsS$_2$ AgAsS$_2$		
	arsenostephanite → argentite + proustite	362 ± 5	Hall (1966)[c]
	Ag$_5$AsS$_4$ Ag$_2$S Ag$_3$AsS$_3$		
	smithite → high smithite	415 ± 5	Hall (1966)
	AgAsS$_2$ AgAsS$_2$		
	high smithite → melt	421 ± 2	Roland (1970)
	AgAsS$_2$		
	proustite + arsenic → melt	452 ± 2	Roland (1970)
	Ag$_3$AsS$_3$ (Ag$_{38.5}$As$_{23}$S$_{38.5}$)		
	proustite → melt	495 ± 1	Roland (1970)
	Ag$_3$AsS$_3$		
	arsenic billingsleyite → melt	571 ± 4	Roland (1970)
	Ag$_7$AsS$_6$		
	argentite + arsenic → melt	624 ± 3	Roland (1970)
	Ag$_2$S (Ag$_{22}$As$_{33}$S$_{45}$)		
Ag-Au-S	low Ag$_3$AuS$_2$ → high Ag$_3$AuS$_2$	183	Smit et al. (1970)
	low AgAuS → high AgAuS	307	Smit et al. (1970)
	AgAuS → melt (congruent?)	730	Smit et al. (1970)[c]
	Ag$_3$AuS$_2$ → melt	755	Tavernier et al. (1967)
Ag-Au-Se	fischesserite → high fischesserite	270	Smit et al. (1970)
	Ag$_3$AuSe$_2$ Ag$_3$AuSe$_2$		
Ag-Au-Te	petzite → Ag$_{11±x}$Au$_{1+x}$Te$_6$ + hessite	~50	Cabri (1965b); Kelly and Goddard (1969)
	Ag$_3$AuTe$_2$ Ag$_2$Te		
	petzite → int. petzite	210 ± 10	Cabri (1965b)
	Ag$_3$AuTe$_2$ Ag$_3$AuTe$_2$		

Table 7-1. (*Continued*)

System	Low-temperature assemblage	High-temperature assemblage	Temperature °C	References/remarks
	electrum + calaverite + int. petzite	melt ($Ag_{21}Au_{38}Te_{41}$)	304 ± 10	Cabri (1965b)
	(Ag, Au) AuTe$_2$ Ag$_3$AuTe$_2$ int. petzite Ag$_3$AuTe$_2$	high petzite Ag$_3$AuTe$_2$	319 ± 15	Cabri (1965b)
	tellurium + high stutzite + sylvanite Ag$_5$Te$_3$ AgAuTe$_2$	melt ($Ag_{35}Au_4Te_{61}$)	330	Cabri (1965b)
	sylvanite AgAuTe$_2$	krennerite + melt (Au, Ag)Te$_2$ Au-poor	354 ± 5	Cabri (1965b)
	krennerite (Au, Ag)Te$_2$	calaverite + melt AuTe$_2$ Au-poor	382 ± 5	Cabri (1965b)
	$Ag_{11\pm x}Au_{1\mp x}Te_6$	high hessite-high petzite s.s. Ag_3(Ag, Au)Te$_2$	~415	Cabri (1965b)
Ag-Bi-S	matildite AgBiS$_2$	high matildite AgBiS$_2$	195 ± 5	Craig (1967)
	silver + bismuth + matildite AgBiS$_2$	melt ($Ag_4Bi_{194}S_2$)	259 ± 2	Craig (1967)
	silver + high matildite AgBiS$_2$	argentite + melt Ag$_2$S ($Ag_{16}Bi_{82}S_2$)	343 ± 3	Craig (1967)
	argentite + high matildite Ag$_2$S AgBiS$_2$	melt	615 ± 2	Craig (1967)
	pavonite + sulfur melt AgBi$_3$S$_5$	melt ($Ag_{10}Bi_{30}S_{60}$)	676 ± 3	Craig (1967)

System	Assemblage		T (°C)	Reference
	pavonite $AgBi_3S_5$	melt + high matildite $AgBiS_2$	732 ± 4	Craig (1967)
	high matildite $AgBiS_2$	melt	801 ± 4	Craig (1967)
Ag-Cu-S	chalcocite + stromeyerite Cu_2S AgCuS	hex. chalcocite s.s. $Ag_{0.24}Cu_{1.76}S$	67 ± 2	Skinner (1966)
	stromeyerite + mckinstryite AgCuS $Ag_{1.2}Cu_{0.8}S$	hex. chalcocite s.s. $Ag_{1.04}Cu_{0.96}S$	90.4 ± 0.5	Skinner (1966)
	stromeyerite AgCuS	hex. chalcocite s.s. AgCuS	93.3 ± 0.7	Skinner (1966)
	mckinstryite $Ag_{1.2}Cu_{0.8}S$	hex. chalcocite s.s. + jalpaite AgCuS $Ag_{3+x}Cu_{1-x}S_2$	94.4 ± 1.5	Skinner (1966)
	acanthite + jalpaite Ag_2S $Ag_{3+x}Cu_{1-x}S_2$	argentite s.s. $Ag_{1.7}Cu_{0.3}S$	106 ± 1.5	Skinner (1966)
	hex. chalcocite + jalpaite AgCuS $Ag_{3+x}Cu_{1-x}S_2$	argentite s.s. $Ag_{1.48}Cu_{0.52}S$	115 ± 0.2	Skinner (1966)
	jalpaite $Ag_{3+x}Cu_{1-x}S_2$	argentite s.s.	117 ± 2	Skinner (1966)
	hex. chalcocite s.s. + argentite s.s. AgCuS $Ag_{1.36}Cu_{0.64}S$	digenite-high argentite s.s. $Ag_{1.1}Cu_{0.9}S$	119	Skinner (1966)
	covellite + argentite s.s. CuS	sulfur melt + digenite-high argentite s.s.	325 ± 25	Skinner (1966)
Ag-Fe-S	sternbergite $AgFe_2S_3$	pyrite + pyrrhotite + silver FeS_2 $Fe_{1-x}S$	<150	Czamanske (1969)[o]
	argentopyrite $AgFe_2S_3$	pyrite + pyrrhotite + silver FeS_2 $Fe_{1-x}S$	<150	Czamanske (1969)[o]

Table 7-1. (*Continued*)

System	Low-temperature assemblage	High-temperature assemblage	Temperature °C	References/remarks
	silver + pyrite FeS_2	argentite + pyrrhotite Ag_2S $Fe_{1-x}S$	248 ± 8	Taylor (1970b)
	argentite + pyrite + pyrrhotite Ag_2S FeS_2 $Fe_{1-x}S$	Ag_2S-rich melt	532 ± 2	Taylor (1970b)
	argentite + pyrite Ag_2S FeS_2	sulfur melt + Ag_2S-rich melt	607 ± 2	Taylor (1970b)
	argentite + pyrrhotite Ag_2S $Fe_{1-x}S$	silver + Ag_2S-rich melt	622 ± 2	Taylor (1970b)
Ag-Ge-S	argentite Ag_8GeS_6	high argyrodite Ag_8GeS_6	223	Pistorius and Gorochov (1970); $dT/dP = 5.4°/kb$
	high argyrodite	melt	955	Gorochov (1968)
Ag-Pb-S	argentite + galena + sulfur melt Ag_2S PbS	melt $(Ag_{21}Pb_{34}S_{45})$	528 ± 3	Craig (1967)
	argentite + galena Ag_2S PbS	melt $(Ag_{21}Pb_{34}S_{45})$	605 ± 5	Craig (1967)
	silver + galena PbS	melt + melt $(Ag_{60}Pb_{40})$ $(Ag_{55}Pb_9S_{36})$	784 ± 2	Craig (1967)
Ag-S-Sb	pyrostilpnite Ag_3SbS_3	pyrargyrite Ag_3SbS_3	192 ± 5	Keighin and Honea (1969)
	stephanite Ag_5SbS_4	pyrargyrite + argentite Ag_3SbS_3 Ag_2S	197 ± 5	Keighin and Honea (1969)[o]

System	Phase 1	Composition 1	Phase 2	Composition 2	T (°C)	Reference
	argentite + allargentum	Ag_2S $Ag_{10-x}Sb$	pyrargyrite + silver	Ag_3SbS_3	200-300	Keighin and Honea (1969)[c]
	miargyrite	$AgSbS_2$	high miargyrite	$AgSbS_2$	380	Keighin and Honea (1969)[p]
	pyrargyrite	Ag_3SbS_3	melt		485	Keighin and Honea (1969)
	cubic miargyrite	$AgSbS_2$	melt		519	Keighin and Honea (1969)
	Sb-billingsleyite	Ag_7SbS_6	melt		475	Keighin and Honea (1969)[q]
Ag-S-Sn	canfieldite	Ag_8SnS_6	high canfieldite	Ag_8SnS_6	172	Pistorius and Gorochov (1970); $dT/dP = 4.9°/\text{kb}$
	high canfieldite		melt		839	Gorochov (1968)
As-Co-S	cobaltite + safflorite	$CoAsS$ $CoAs_2$	alloclasite	$CoAs_{1+x}S_{1-x}$	800	Maurel and Picot (1974)
	anisotropic cobaltite	$CoAs_{1-x}S_{1+s}$	isotropic cobaltite	$CoAsS$	825 ± 25	Bayliss (1969)
	cobaltite	$CoAs_{1-x}S_{1+s}$	incongruent melting		>1000	Maurel and Picot (1974)
As-Cu-S	luzonite	Cu_3AsS_4	enargite	Cu_3AsS_4	<25?	Bernardini et al. (1972)
	sinnerite + lautite	$Cu_6As_4S_9$ $CuAsS$		$Cu_{24}As_{12}S_{31}$ + AsS-rich melt	379	Maske and Skinner (1971)
	enargite + sinnerite	Cu_3AsS_4 $Cu_6As_4S_9$		$Cu_{24}As_{12}S_{31}$ + AsS-rich melt	456	Maske and Skinner (1971)
	$Cu_{24}As_{12}S_{31}$ + lautite	$CuAsS$	tennantite + AsS-rich melt	$Cu_{12+x}As_4S_{13}$	474	Maske and Skinner (1971)
	sinnerite	$Cu_6As_4S_9$		$Cu_{24}As_{12}S_{31}$ + AsS-rich melt	489	Maske and Skinner (1971)

Table 7-1. (Continued)

System	Low-temperature assemblage	High-temperature assemblage	Temperature °C	References/remarks
	enargite + covellite Cu_3AsS_4 CuS	high digenite + S-rich melt Cu_9S_5	507	Maske and Skinner (1971)
	high digenite + lautite Cu_9S_5 CuAsS	tennantite + arsenic $Cu_{12+x}As_4S_{13}$	571	Maske and Skinner (1971)
	enargite + $Cu_{24}As_{12}S_{31}$ Cu_3AsS_4	tennantite + melt $Cu_{12-x}As_4S_{13}$	573	Maske and Skinner (1971)
	lautite CuAsS	arsenic + tennantite + melt $Cu_{12+x}As_4S_{13}$	574	Maske and Skinner (1971)
	$Cu_{24}As_{12}S_{31}$	tennantite + melt $Cu_{12+x}As_4S_{13}$	578	Maske and Skinner (1971)
	tennantite + arsenic $Cu_{12+x}As_4S_{13}$	high digenite + melt Cu_9S_5	598	Maske and Skinner (1971)
	tennantite + high digenite $Cu_{12+x}As_4S_{13}$ Cu_9S_5	enargite + melt Cu_3AsS_4	654	Maske and Skinner (1971)
	tennantite $Cu_{12.31}As_4S_{13}$	melt	665	Maske and Skinner (1971)
	enargite Cu_3AsS_4	melt	671 ± 1	Maske and Skinner (1971)
As-Fe-S	pyrite + arsenic FeS_2	arsenopyrite + AsS-rich melt FeAsS	363 ± 50	Barton (1969)
	pyrite + arsenopyrite FeS_2 FeAsS	pyrrhotite + AsS-rich melt $Fe_{1-x}S$	491 ± 12	Clark (1960)
	arsenopyrite + arsenic FeAsS	loellingite + AsS-rich melt $FeAs_2$	688 ± 3	Clark (1960)

System	Phases	Composition	T (°C)	Reference
	arsenopyrite	loellingite + pyrrhotite + AsS-rich melt	702 ± 3	Clark (1960)
	FeAsS			
	loellingite + AsS-rich melt	pyrrhotite + arsenic melt	825 ± 15	Barton (1969)
	FeAs$_2$	FeAs$_2$ Fe$_{1-x}$S		
As-Ni-S	niccolite + Ni$_{1-x}$(S, As)	Fe$_{1-x}$S complete solid solution	595 ± 5	Yund (1962)
	Ni$_{1-x}$(As, S)			
	gersdorffite	decomposition products	>700	Yund (1962)[c]
	NiAsS			
As-Pb-S	sartorite	baumhauerite + melt	305	Chang and Bever (1973)
	PbAs$_2$S$_4$	Pb$_3$As$_4$S$_9$ (As$_{39.7}$Pb$_{0.4}$S$_{59.9}$)		
	gratonite	jordanite	<250	Roland (1968)[o]
	Pb$_9$As$_4$S$_{15}$	Pb$_9$As$_4$S$_{15}$		
	baumhauerite	rathite II + melt	458	Chang and Bever (1973)
	Pb$_3$As$_4$S$_9$	Pb$_{19}$As$_{26}$S$_{58}$ (As$_{29.6}$Pb$_{14.3}$S$_{57.1}$)		
	rathite II	dufrenoysite + melt	474	Chang and Bever (1973)
	Pb$_{19}$As$_{26}$S$_{58}$	Pb$_2$As$_2$S$_5$ (As$_{25.8}$Pb$_{17.8}$S$_{56.4}$)		
	dufrenoysite	melt	525 ± 40	Chang and Bever (1973)[c]
	Pb$_2$As$_2$S$_5$			
	jordanite	galena + melt	549 ± 3	Roland (1968)
	Pb$_9$As$_4$S$_{15}$	PbS (As$_{18}$Pb$_{27}$S$_{55}$)		
As-S-Sb	getchellite	melt	345 ± 10	Weissberg (1965)
	AsSbS$_3$			
	AsSb$_2$S$_2$	melt (congruent?)	535 ± 5	Craig et al. (1974)
As-S-Tl	lorandite	melt	~ 300	Radtke et al. (1974)
	TlAsS$_2$			

Table 7-1. (Continued)

System	Low-temperature assemblage	High-temperature assemblage	Temperature °C	References/remarks
Au-Bi-Te	Bi_5Te_3 + bismuth + maldonite Au_2Bi	melt $(Au_{14}Bi_{84.5}Te_{1.5})$	235	Gather and Blachnik (1974)
	tellurobismuthite + gold + calaverite $AuTe_2$	melt $(Au_{37}Bi_{10}Te_{53})$	402	Gather and Blachnik (1974)
	Bi_2Te_3			
	tellurobismuthite + tellurium + calaverite	melt $(Au_{10.5}Bi_{7.5}Te_{82})$	383	Gather and Blachnik (1974)
	Bi_2Te_3 $AuTe_2$			
Bi-Cu-S	emplectite $CuBiS_2$	cuprobismutite + wittichenite $Cu_{10}Bi_{12}S_{23}$ Cu_3BiS_3	290 ± 10	Buhlmann (1971)
	hex. (?) chalcocite + wittichenite Cu_2S Cu_3BiS_3	Cu_9BiS_6	~400	Sugaki et al. (1972)
	low wittichenite Cu_3BiS_3	high wittichenite Cu_3BiS_3	442 ± 10	Buhlmann (1971)
	cuprobismutite + $CuBi_3S_5$ $Cu_{10}Bi_{12}S_{23}$	$Cu_3BiS_5S_9$	442 ± 3	Sugaki and Shima (1972)
	cuprobismutite $Cu_{10}Bi_{12}S_{23}$	$Cu_3Bi_5S_9$ + high wittichenite Cu_3BiS_3	474 ± 5	Sugaki and Shima (1972)
	$Cu_8Bi_8S_{19}$	$Cu_3Bi_5S_9$ + ternary melt + sulfur melt (?)	498 ± 5	Sugaki et al. (1972)
	high wittichenite Cu_3BiS_3	Cu_9BiS_6 s.s. + melt $(Bi_{17}Cu_{39}S_{44})$	527 ± 5	Sugaki and Shima (1972)

System	Assemblage	Temperature (°C)	Reference
Bi-Fe-S	pyrite + bismuth pyrrhotite + bismuthinite FeS_2 $Fe_{1-x}S$ Bi_2S_3	213 ± 25	See text.
	bismuthinite + Bi-rich melt + pyrrhotite Bi_2S_3 $FeBi_4S_7$ $Fe_{1-x}S$	608 ± 5	Craig et al. (1971)
	$FeBi_4S_7$ melt	713 ± 3	Craig et al. (1971)
Bi-Ni-S	parkerite melt $Bi_2Ni_3S_2$	686	DuPreez (1944)
Bi-Pb-S	lillianite + bismuthinite galenobismutite $Pb_8Bi_6S_{17}$ Bi_2S_3 $PbBi_2S_4$	410 ± 10	Hoda and Chang (1975)[c, r]
	cosalite lillianite + galenobismutite $Pb_2Bi_2S_5$ $Pb_8Bi_6S_{17}$ $PbBi_2S_4$	425 ± 25	Craig (1967)[c]
	galena + lillianite heyrovskyite + sulfur melt PbS $Pb_8Bi_6S_{17}$ $(Pb, Bi)_6Bi_2S_9$	473 ± 18	Salanci and Moh (1969)
	galenobismutite + bismuthinite $PbBi_2S_4$ Bi_2S_3 $PbBi_4S_7$	680 ± 5	Craig (1967)
	galenobismutite + melt $PbBi_2S_4$ $(Bi_{34}Pb_7S_{59})$	730 ± 5	Craig (1967)
	lillianite + melt $Pb_8Bi_6S_{17}$ $(Bi_{32}Pb_{10}S_{58})$	750 ± 3	Craig (1967)
	heyrovskyite + melt $(Pb, Bi)_6Bi_2S_9$ $(Bi_{22}Pb_{23}S_{55})$	816 ± 6	Craig (1967)
	galena + melt PbS $(Bi_{21}Pb_{24}S_{55})$	829 ± 6	Craig (1967)
Bi-S-Sb	bismuthinite s.s. + stibnite s.s. $(Bi, Sb)_2S_3$ $(Sb, Bi)_2S_3$ complete s.s.	< 200	Springer and Laflamme (1971)
Bi-S-Te	β-tetradymite γ-tetradymite + melt $Bi_{14}Te_{15}S_6$ $Bi_4Te_{3+x}S_{3-x}$ $(Bi_{40}S_4Te_{56})$	581 ± 3	Glatz (1967)[c]

Table 7-1. (Continued)

System	Low-temperature assemblage	High-temperature assemblage	Temperature °C	References/remarks
	γ-tetradymite $Bi_4Te_{3+x}S_{3-x}$	melt	630 ± 3	Glatz (1967)
Cl-Hg-S	corderoite $Hg_3Cl_2S_2$	breakdown products undefined	340	Carlson (1967)
Co-Cu-S	digenite + vaesite Cu_9S_5 NiS_2	carrollite + sulfur melt $CuCo_2S_4$	~550	Craig and Higgins (1973)
Co-Fe-S	cattierite s.s. + pyrite s.s. $(Co, Fe)S_2$ $(Fe, Co)S_2$	complete s.s.	600–700	Klemm (1965)
Co-Ni-S	vaesite s.s. + cattierite s.s. $(Ni, Co)S_2$ $(Co, Ni)S_2$	complete solid solution	400–500	Klemm (1965)
Co-S-Sb	paracostibite CoSbS	$Co_{1-x}S$ + 2 melts	876 ± 5	Cabri et al. (1970a)
Cu-Fe-S	bornite + pyrite Cu_5FeS_4 FeS_2	high digenite + chalcopyrite $(Cu, Fe)_{9-x}S_5$ $CuFeS_2$	~100	c, s
	low digenite $(Cu, Fe)_{9-x}S_5$	bornite + high digenite Cu_5FeS_4 $(Cu_2S)_{5+x}(Fe_2S_3)_{1-x}$	120 ± 70	c, s
	sulfur-rich bornite Cu_5FeS_{4+x}	bornite + chalcopyrite + Cu_5FeS_4 $CuFeS_2$ high digenite $(Cu_2S)_{5+x}(Fe_2S_3)_{1-x}$	140 ± 10	Yund and Kullerud (1966) c, g
	mooihoekite $Cu_9Fe_9S_{16}$	high mooihoekite $Cu_9Fe_9S_{16}$	~167	Cabri (1973)
	talnakhite $Cu_9Fe_8S_{16}$	int. talnakhite $Cu_9Fe_8S_{16}$	~186	Cabri (1973)

Assemblage	Formula	Product	T (°C)	Reference
bornite + high digenite	Cu$_5$FeS$_4$ (Cu$_2$S)$_{5+x}$(Fe$_2$S$_3$)$_{1-x}$	high bornite	~190	Brett (1963)
bornite + chalcopyrite	Cu$_5$FeS$_4$ CuFeS$_2$	high bornite (Cu$_2$S)$_{5\pm x}$(Fe$_2$S$_3$)$_{1\mp x}$	195 ± 5	Brett (1963)
haycockite	Cu$_4$Fe$_5$S$_8$	intermediate s.s. (Cu, Fe)$_{1+x}$S	<200	Cabri (1973)[o]
bornite + hex. chalcocite	Cu$_5$FeS$_4$ Cu$_2$S	high bornite (Cu$_2$S)$_{5\pm x}$(Fe$_2$S$_3$)$_{1\mp x}$	200 ± 10	Brett (1963)
chalcopyrite + covellite	CuFeS$_2$ CuS	idaite Cu$_{5.5}$FeS$_{6.5}$	<200	Sugaki (1957)[g,o]
cubanite	CuFe$_2$S$_3$	intermediate s.s. (Cu, Fe)$_{1+x}$S	205 ± 5	Cabri et al. (1973)
covellite + pyrite + high bornite	CuS FeS$_2$ (Cu$_2$S)$_{5\pm x}$(Fe$_2$S$_3$)$_{1\mp x}$	idaite Cu$_{5.5}$FeS$_{6.5}$	227 ± 50	calculated, Table 7.2
bornite	Cu$_5$FeS$_4$	high bornite (Cu$_2$S)$_{5\pm x}$(Fe$_2$S$_3$)$_{1\mp x}$	228 ± 5	Morimoto and Kullerud (1961)[t]
int. talnakhite	Cu$_9$Fe$_8$S$_{16}$	high talnakhite Cu$_9$Fe$_8$S$_{16}$	~230	Cabri (1973)
mooihoekite	Cu$_9$Fe$_9$S$_{16}$	intermediate s.s. (Cu, Fe)$_{1+x}$S	~236	Cabri (1973)
high digenite + high bornite	(Cu$_2$S)$_{5\pm x}$(Fe$_2$S$_3$)$_{1-x}$ (Cu$_2$S)$_{5\pm x}$(Fe$_2$S$_3$)$_{1\pm x}$	complete s.s. (Cu$_2$S)$_{5\pm x}$(Fe$_2$S$_3$)$_{1\mp x}$	290 ± 10	Brett (1963)
pyrrhotite + chalcopyrite	Fe$_{1-x}$S CuFeS$_2$	intermediate s.s. + pyrite (Cu, Fe)$_{1+x}$S FeS$_2$	334 ± 17	Yund and Kullerud (1966)

Table 7-1. (*Continued*)

System	Low-temperature assemblage	High-temperature assemblage	Temperature °C	References/remarks
	copper + intermediate s.s. $(Cu, Fe)_{1+x}S$	high bornite + $(Cu_2S)_{5\pm x}(Fe_2S_3)_{1\mp x}$ pyrrhotite $Fe_{1-x}S$	<300	Barton (1973)[c]
	pyrite + covellite FeS_2 CuS	idaite + sulfur melt $Cu_{5.5}FeS_{6.5}$	434	Roseboom and Kullerud (1958)
	copper + pyrrhotite $Fe_{1-x}S$	iron + high bornite $(Cu_2S)_{5\pm x}(Fe_2S_3)_{1\mp x}$	475 ± 5	Yund and Kullerud (1966)
	covellite + idaite CuS $Cu_{5.5}FeS_{6.5}$	high bornite + sulfur melt $(Cu_2S)_{5\pm x}(Fe_2S_3)_{1\mp x}$	482	Roseboom and Kullerud (1958)
	idaite $Cu_{5.5}FeS_{6.5}$	pyrite + sulfur melt + FeS_2 high bornite $(Cu_2S)_{5\pm x}(Fe_2S_3)_{1\mp x}$	501	Roseboom and Kullerud (1958)
	chalcopyrite + $CuFeS_2$ high bornite $(Cu_2S)_{5\pm x}(Fe_2S_3)_{1\mp x}$	pyrite + intermediate s.s. FeS_2 $(Cu, Fe)_{1+x}S$	532 ± 10	Barton (1973)
	chalcopyrite $CuFeS_2$	pyrite + intermediate s.s. FeS_2 $(Cu, Fe)_{1+x}S$	547 ± 5	Yund and Kullerud (1966)
	pyrite + high bornite FeS_2 $(Cu_2S)_{5\pm x}(Fe_2S_3)_{1\mp x}$	sulfur melt + intermediate s.s. $(Cu, Fe)_{1+x}S$	568	Roseboom and Kullerud (1958)
	high talnakhite $Cu_9Fe_8S_{16}$	intermediate s.s. $(Cu, Fe)_{1+x}S$	~520	Cabri (1973)

System	Assemblage	T (°C)	Reference
	pyrite + intermediate s.s. FeS_2 $(Cu, Fe)_{1+x}S$	739	Roseboom and Kullerud (1958)
	intermediate s.s. (S-rich; $(Cu, Fe)_{1+x}S$, $Cu = Fe$)	880	Dutrizac (1976)
Cu-Ni-S	villamaninite	vaesite + covellite + sulfur melt	503	Kullerud et al. (1969)
	$CuNi_2S_6$	NiS_2 CuS		
Cu-Pb-S	copper + galena	lead + hex. chalcocite	271 ± 5	Craig and Kullerud (1968)
	PbS	Cu_2S		
	galena + hex. chalcocite + lead melt	$Cu_{14}Pb_2S_{9-x}$	486 ± 4	Craig and Kullerud (1968)
	PbS Cu_2S			
	galena + digenite + sulfur melt	melt $(Cu_{37}Pb_{14}S_{49})$	508 ± 2	Craig and Kullerud (1968)
	PbS Cu_9S_5 $Cu_{14}Pb_2S_{9-x}$	digenite + melt Cu_9S_5	528 ± 2	Craig and Kullerud (1968)
Cu-S-Sb		complete s.s.	95	Tatsuka and Morimoto (1973)[u]
	Cu-rich tetrahedrite + Cu-poor tetrahedrite	$Cu_{12.88}Sb_{4.09}S_{13}$ $Cu_{12.59}Sb_{4.03}S_{13}$	125 ± 5 122 ± 3	Skinner et al. (1972)[u] Skinner et al. (1972)[q]
	skinnerite Cu_3SbS_3	high skinnerite Cu_3SbS_3		
	pseudotetrahedrite Cu_3SbS_3	tetrahedrite + antimony $Cu_{12+x}Sb_{4+y}S_{13}$	350 ± 5	Tatsuka and Morimoto (1973)
	chalcostibite + antimony + $CuSbS_2$ tetrahedrite $Cu_{12+x}Sb_{4+y}S_{13}$	high skinnerite Cu_3SbS_3	360 ± 2	Skinner et al. (1972); Tatsuka and Morimoto (1973)
	tetrahedrite $Cu_{12+x}Sb_{4+y}S_{13}$	+ antimony	high skinnerite + digenite Cu_3SbS_3 Cu_9S_5	436 ± 2	Skinner et al. (1972)

Table 7-1. (*Continued*)

System	Low-temperature assemblage	High-temperature assemblage	Temperature °C	References/remarks
	chalcostibite + stibnite $CuSbS_2$ Sb_2S_3	antimony + melt	476.5 ± 2	Skinner et al. (1972)
	famatinite + stibnite Cu_3SbS_4 Sb_2S_3	chalcostibite + melt $CuSbS_2$	476.5 ± 2	Skinner et al. (1972)
	tetrahedrite + $Cu_{12+x}Sb_{4+y}S_{13}$ chalcostibite $CuSbS_2$	famatinite + stibnite Cu_3SbS_4 Sb_2S_3	522 ± 2	Skinner et al. (1972)
	tetrahedrite $Cu_{12+x}Sb_{4+y}S_{13}$	digenite + famatinite + Cu_9S_5 Cu_3SbS_4 high skinnerite Cu_3SbS_3	543 ± 2	Skinner et al. (1972)
	chalcostibite $CuSbS_2$	melt	553 ± 2	Skinner et al. (1972)
	high skinnerite Cu_3SbS_3	melt	607.5 ± 3	Skinner et al. (1972)
	famatinite Cu_3SbS_4	melt	627 ± 2	Skinner et al. (1972)
Fe-Mo-S	pyrite + sulfur melt + molybdenite MoS_2	melt ($\sim Fe_{20}Mo_{10}S_{70}$)	726	Kullerud (1967)
	FeS_2 pyrite + molybdenite	melt + pyrrhotite ($\sim Fe_{20}Mo_{10}S_{70}$)	732	Kullerud (1967)

System	Assemblage	Temperature	Reference
Fe-Ni-S	pyrite + pentlandite + millerite FeS_2 $(Ni, Fe)_9S_8$ NiS	monosulfide s.s. $(Fe_{16}Ni_{32}S_{52})$	$\ll 200$	Craig (1973), see Figure 7.18
	pyrite + millerite FeS_2 NiS	violarite + monosulfide s.s. $(Ni, Fe)_3S_4$ $(Fe_{12}Ni_{35}S_{53})$	<200	Craig (1973)
	pyrite + pentlandite FeS_2 $(Ni, Fe)_9S_8$	monosulfide s.s. + monosulfide s.s. $(Fe_{26}Ni_{22}S_{52})$ $(Fe_{37}Ni_{10}S_{53})$	212 ± 13	Craig (1973)
	monosulfide s.s. + monosulfide s.s.	monosulfide s.s. $(Fe_{21}Ni_{26}S_{53})$	225 ± 25	Craig (1973)
	monosulfide s.s. + monosulfide s.s.	monosulfide s.s. $(Fe_{27}Ni_{19}S_{54})$	262 ± 13	Craig (1973)
	violarite $Fe_{1-x}Ni_{2+x}S_4$	monosulfide s.s. + pyrite + vaesite $(Fe, Ni)_{1-x}S$ FeS_2 NiS_2	461 ± 3	Craig (1968)
	pentlandite + vaesite $(Ni, Fe)_9S_8$	Fe-rich monosulfide s.s. + $Ni_{3\pm x}S_2$	610 ± 2	Kullerud (1963)
	pyrite + vaesite $(Fe, Ni)S_2$ $(Ni, Fe)S_2$	monosulfide s.s. + sulfur melt $(Fe, Ni)_{1-x}S$	729 ± 3	Clark and Kullerud (1963)
Fe-Pb-S	galena + pyrite + sulfur melt PbS FeS_2	melt $(Fe_{17}Pb_{21}S_{62})$	716 ± 2	Brett and Kullerud (1967)
	galena + pyrite PbS FeS_2	pyrrhotite + melt $Fe_{1-x}S$	719 ± 1	Brett and Kullerud (1967)

Table 7-1. (Continued)

System	Low-temperature assemblage	High-temperature assemblage	Temperature °C	References/remarks
Fe-S-Sb	gudmundite $FeSbS$	pyrrhotite + antimony + $FeSb_2$ $Fe_{1-x}S$	280 ± 10	Clark (1966b)[a]
	stibnite + sulfur melt + pyrite Sb_2S_3 FeS_2	melt $(Fe_1Sb_{36}S_{63})$	494 ± 5	Barton (1971)
	stibnite + antimony + berthierite Sb_2S_3 $FeSb_2S_4$	melt $(Fe_2Sb_{39}S_{59})$	516 ± 5	Barton (1971)
	berthierite + antimony $FeSb_2S_4$	pyrrhotite + melt $Fe_{1-x}S$ $(Fe_2Sb_{39}S_{59})$	530	Barton (1971)
	pyrite + berthierite FeS_2 $FeSb_2S_4$	pyrrhotite + stibnite $Fe_{1-x}S$ Sb_2S_3	532 ± 5	Barton (1971)
	pyrite + stibnite FeS_2 Sb_2S_3	pyrrhotite + melt $Fe_{1-x}S$ $(Fe_1Sb_{37}S_{62})$	545 ± 5	Barton (1971)
	pyrrhotite + stibnite $Fe_{1-x}S$ Sb_2S_3	berthierite + melt $FeSb_2S_4$ $(Fe_1Sb_{38}S_{61})$	548 ± 5	Barton (1971)
	berthierite $FeSb_2S_4$	pyrrhotite + melt $Fe_{1-x}S$ $(Fe_2Sb_{41}S_{57})$	563 ± 5	Barton (1971)
Fe-S-Tl	raguinite $TlFeS_2$	melt	659	Wandji and Kom (1972)
Fe-S-Sn	pyrite + ottemannite FeS_2 Sn_2S_3	pyrrhotite + berndtite $Fe_{1-x}S$ SnS_2	508 ± 12	Moh (1974)
	pyrite + herzenbergite FeS_2 SnS	pyrrhotite + ottemannite $Fe_{1-x}S$ Sn_2S_3	300 ± 200	[c]

System	Phases		State	Temp. (°C)	Reference
Hg-S-Sb	livingstonite		melt	453 ± 3	Craig (1970)
	$HgSb_4S_8$				
	stibnite + metacinnabar		melt	>450	Craig (1970)
	Sb_2S_3	HgS	($Hg_{11}S_{58}Sb_{31}$)		
Ni-Pb-S	shandite		melt	792	DuPreez (1944)
	$Ni_3Pb_2S_2$				
Ni-S-Sb	ullmannite		melt	752 ± 6	Williams and Kullerud (1970)
	NiSbS				
Pb-S-Sb	lead + antimony + galena		melt	240 ± 3	Craig et al. (1973)
			($Pb_{78}S_6Sb_{16}$)		
	zinckenite + antimony + stibnite		melt	480 ± 5	Craig et al. (1973)
	$PbSb_2S_4$	Sb_2S_3			
	zinckenite + boulangerite		robinsonite	318	Craig et al. (1973)[c]
	$PbSb_2S_4$ $Pb_5Sb_4S_{11}$		$Pb_6Sb_{10}S_{21}$		
	zinckenite + stibnite		melt	523 ± 3	Craig et al. (1973)
	$PbSb_2S_4$ Sb_2S_3		($Pb_5S_{59}Sb_{36}$)		
	zinckenite		robinsonite + melt	545 ± 5	Craig et al. (1973)
	$PbSb_2S_4$		$Pb_6Sb_{10}S_{21}$ ($Pb_{11}S_{57}Sb_{32}$)		
	robinsonite + boulangerite		$Pb_3Sb_4S_9$	576	Chang and Bever (1973)
	$Pb_6Sb_{10}S_{21}$ $Pb_5Sb_4S_{11}$				
	robinsonite		$Pb_3Sb_4S_9$ + melt	582 ± 2	Craig et al. (1973)
	$Pb_6Sb_{10}S_{21}$		($Pb_{20}S_{56}Sb_{24}$)		
	boulangerite		$Pb_3Sb_4S_9$ + melt	638 ± 2	Craig et al. (1973)
	$Pb_5Sb_4S_{11}$		($Pb_{23}S_{55}Sb_{22}$)		
Pb-S-Se	galena + clausthalite		complete s.s.	<300	Wright et al. (1965)
	Pb(S, Se) Pb(Se, S)				
Pb-S-Te	galena + altaite		complete s.s.	805	Darrow et al. (1966)
	Pb(S, Te) Pb(Te, S)				

Table 7-1. (Continued)

System	Low-temperature assemblage	High-temperature assemblage	Temperature °C	References/remarks
Pb-Se-Te	clausthalite + altaite Pb(Se, Te) Pb(Te, Se)	complete s.s.		Steininger (1970)
Zn-S-Se	sphalerite + stilleite Zn(S, Se) Zn(Se, S)	complete s.s.	<300	Wright et al. (1965)
Ag-As-Cu-S	arsenpolybasite $Ag_{15.4-x}Cu_{0.6+x}As_2S_{11}$	high arsenpolybasite	365 ± 5	Hall (1967)
	high arsenpolybasite $Ag_{14.7-x}Cu_{1.3+x}As_2S_{11}$	incongruent melting	~450	Hall (1967)
	pearceite $(Ag, Cu)_{16}As_2S_{11}$	incongruent melting	520	Hall (1967)
Ag-As-S-Sb	proustite + $Ag_3(As, Sb)S_3$ pyrargyrite $Ag_3(Sb, As)S_3$	complete s.s.	<300	Toulmin (1963)
Ag-Bi-Pb-S	galena + $(AgBi, Pb_2)S_2$ PbS	$(AgBi, Pb_2)S_2$ s.s.	215 ± 15	Craig (1967)
	schirmerite $PbAg_4Bi_4S_9$		220	Godovikov (1972)
Ag-Bi-S-Sb	$Ag(Bi, Sb)S_2 + Ag(Sb, Bi)S_2$	complete s.s.	380	Chen and Chang (1971)
Ag-Cu-Hg-S	balkanite $Cu_9Ag_5HgS_8$	higher temperature polymorphs	90 140 450	Atanassov and Kirov (1973)
	highest balkanite $Cu_9Ag_5HgS_8$	melt	690	Atanassov and Kirov (1973)

System	Assemblage	Products	T (°C)	Reference
Ag-Cu-S-Sb	polybasite $Ag_{15.4-x}Cu_{0.6+x}Sb_2S_{11}$	incongruent melting	~400	Hall (1967)
	antimonpearceite $Ag_{14.3-x}Cu_{1.7+x}Sb_2S_{11}$	incongruent melting	~480	Hall (1967)
Ag-Fe-Ni-S	pentlandite + argentite $(Ni, Fe)_9S_8$ Ag_2S	monosulfide s.s. + silver + $(Fe, Ni)_{1-x}S$ argentian pentlandite $(Ni, Fe)_{8+x}Ag_{1-x}S_8$	358	Mandziuk and Scott (1975)[c,v]
	argentian pentlandite $(Ni, Fe)_{8+x}Ag_{1-x}S_8$	breakdown products undefined	455	Mandziuk and Scott (1975)
Ag-Pb-S-Sb	diaphorite $Pb_2Ag_3Sb_3S_8$	isometric s.s.	350 ± 10	Hoda and Chang (1975)
	frieslebenite $AgPbSbS_3$	isometric s.s.	~340	Hoda and Chang (1975)
	galena + andorite PbS $AgPbSb_3S_6$	boulangerite + (AgSb, Pb_2)S_2 $Pb_5Sb_4S_{11}$	350 ± 50	Hoda and Chang (1975)
				ramdohrite ($PbAgSb_3S_6$), fizilyite ($Pb_5Ag_2Sb_8S_{18}$), and owyheeite ($Pb_5Ag_2Sb_6S_{15}$) are not stable at 300°C (Hoda and Chang, 1975).
As-Cu-Fe-S	arsenopyrite + chalcopyrite FeAsS $CuFeS_2$	tennantite + pyrite $Cu_{10}Fe_2As_4S_{13}$ FeS_2	400 ± 50 ?	Sugaki (1957)[c]
As-Cu-S-Sb	tennantite + chalcostibite $Cu_{12}As_4S_{13}$ $CuSbS_2$	high skinnerite + Cu_3SbS_3 $Cu_4(As, Sb)_2S_5$	392 ± 35	Luce et al. (1977)
	enargite + stibnite Cu_3AsS_4 Sb_2S_3	As-rich famatinite + melt + chalcostibite $Cu_3(As, Sb)S_4$ $CuSbS_2$	438 ± 12	Luce et al. (1977)

Table 7-1. (*Continued*)

System	Low-temperature assemblage	High-temperature assemblage	Temperature °C	References/remarks
As-Pb-S-Sb	chalcostibite + $AsSb_2S_2$ $CuSbS_2$	alloy + melt (As, Sb)	<476	Luce et al. (1977)[c, o]
	madocite $Pb_{17}(Sb, As)_{16}S_{41}$	breakdown products undefined	>400	Wallia and Chang (1973)
	veenite-dufrenoysite $Pb_2(Sb, As)_2S_5$	breakdown products undefined	>400	Wallia and Chang (1973)
	guettardite $Pb_9(As, Sb)_{16}S_{33}$	breakdown products undefined	>400	Wallia and Chang (1973)
	sorbyite $Pb_{17}(Sb, As)_{22}S_{50}$	breakdown products undefined	<400	Wallia and Chang (1973)
	sterryite $Pb_{12}(Sb, As)_{10}S_{27}$	breakdown products undefined	<400	Wallia and Chang (1973)
	twinnite $Pb(Sb, As)_2S_4$	breakdown products undefined	<400	Wallia and Chang (1973)
	geochronite Pb_2SbAsS_8	breakdown products undefined	<400	Wallia and Chang (1973)
Bi-Cu-Fe-S	$Cu_{8.4}Fe_{1.2}Bi_{10.8}S_{22}$	bornite + $Cu_3Bi_5S_9$ + chalcopyrite $CuFeS_2$	525	Sugaki et al. (1972)
Bi-Cu-Pb-S	bismuthinite + aikenite Bi_2S_3 $CuBiPbS_3$	Cu_5FeS_4 complete s.s.	<300	Springer (1971)
	aikenite $CuBiPbS_3$	incongruent melting	540 ± 5	Springer (1971)
				Gladite ($CuPbBi_5S_9$), hammarite ($Cu_2Pb_2Bi_4S_9$), and

				lindstromite (CuPbBi$_3$S$_6$) are not stable at 300°C or above (Springer, 1971).
Cu-Fe-Ge-S	briartite Cu$_2$FeGeS$_4$	high briartite	636 ± 5	Moh (1975a)
	high briartite Cu$_2$FeGeS$_4$	melt	990 ± 2	Moh (1975a)
Cu-Fe-Ni-S				Craig and Kullerud (1969)^w
Cu-Fe-Pb-S	betekhtinite Pb$_2$(Cu, Fe)$_{21}$S$_{15}$	galena + high digenite PbS (Cu$_2$S)$_{5+x}$(Fe$_2$S$_3$)$_{1-x}$	<150	Slavskaya et al. (1963)^o
Cu-Fe-S-Sn	mawsonite Cu$_6$Fe$_2$SnS$_8$	stannoidite + bornite + Cu$_5$FeS$_4$ chalcopyrite (+ pyrite?) CuFeS$_2$ FeS$_2$	390	Lee et al. (1975)^c
	chalcopyrite + stannite CuFeS$_2$ Cu$_2$FeSnS$_4$	intermediate s.s. (Cu, Fe, Sn)$_{1+x}$S	462 ± 5	Moh (1975a)
	intermediate s.s. + stannite (Cu, Fe, Sn)$_{1+x}$S Cu$_2$FeSnS$_4$	high stannite (Cu, Fe, Sn)S	610 ± 10	Moh (1975a)
	stannite Cu$_2$FeSnS$_4$	high stannite	706 ± 5	Moh (1975a)
	stannoidite Cu$_8$Fe$_3$Sn$_2$S$_{12}$	high stannite + bornite + Cu$_5$FeS$_4$ int. s.s. (+ S melt?) (Cu, Fe, Sn)$_{1+x}$S	830	Lee et al. (1975)^c
Cu-Fe-S-Zn	sphalerite + chalcopyrite (Zn, Fe)S CuFeS$_2$	intermediate s.s. (Cu, Fe, Zn)$_{1+x}$S	500 ± 10	Moh (1975a), Czamanske (1974)

Table 7-1. (*Continued*)

System	Low-temperature assemblage	High-temperature assemblage	Temperature °C	References/remarks
Cu-Pb-S-Sb	bournonite $PbCuSbS_3$	incongruent melting	522 ± 3	Frumar et al. (1973)
Cu-S-Sn-Zn	kesterite Cu_2ZnSnS_4	melt + sphalerite ZnS	991 ± 2	Moh (1975a)
Fe-Pb-S-Zn	galena + sphalerite + PbS (Zn, Fe)S pyrrhotite $Fe_{1-x}S$	melt $(Fe_{24.9}Pb_{17.8}Zn_{6.3}S_{50\pm?})$	820	Avetisyan and Gnatyshenko (1956)
As-Co-Fe-S-Sb	loellingite + paracostibite $FeAs_2$ $CoSbS$	pyrrhotite + (Co, Fe)As_2 + antimony $Fe_{1-x}S$	445 ± 10	Cabri et al. (1970b)
Cu-Fe-S-Sn-Zn	chalcopyrite + kesterite $CuFeS_2$ Cu_2ZnSnS_4	intermediate s.s. (Cu, Fe, Sn, Zn)$_{1+x}$S	490 ± 7	Moh (1975a)
Fe-Pb-S-Sb-Sn	cylindrite $FePb_3Sn_4Sb_2S_{14}$	breakdown products undefined	>500	Sachdev and Chang (1975)
	incaite $Fe_2(Pb, Ag)_9Sn_9Sb_5S_{33}$	breakdown products undefined	>500	Sachdev and Chang (1975)[x]
	"franckeite" $FePb_5Sn_3Sb_2S_{14}$	breakdown products undefined	>500	Sachdev and Chang (1975)[c]

Table 7-2 Selected univariant equilibria for defining conditions of ore formation

Each equation is written for 1 mole of chalcogenide; the free-energy change gives the chemical potential of S_2, Se_2, Te_2, or O_2 gas, at 1 atm (which is the standard state used throughout this chapter), for most reactions. A few reactions, denoted by *, are given with the condensed components in a standard state that differs significantly from their state for the equilibrium reaction because these are useful in consideration of solid solutions. Where there is possible ambiguity regarding whether or not the free-energy change includes terms from the activities of condensed components, the # symbol indicates that the value given is solely the chemical potential of S_2. The standard free-energy change is given as the right-hand side of the equation, $\Delta G° = A + BT$; "A" corresponds to the enthalpy change, $\Delta H°$, and "B" to the negative of the entropy change, $-\Delta S°$.

The accuracy of the equilibrium constants has been estimated on the basis of the method used for measurement, the fit of the linear equations to the data, and the magnitude of scatter from different measurements. A root-mean-square value is assigned to independent sources of error. The uncertainty is shown by "A" where within 1 kcal, "B" for 1-2 kcal, "C" for 2-5 kcal and "D" for 5+ kcal. Some data are included even though they are of very poor quality because they represent potentially useful equilibria that deserve further attention.

Formulae are simplified to conserve space, and they do not necessarily conform precisely to those given for the same phases in Table 7-1, as, for example, for bornite, which is written alternatively as Cu_5FeS_4 and $(Cu_2S)_{5\pm x}(Fe_2S_3)_{1\mp x}$.

System	Reaction	Free-energy change in calories for T in K	Accuracy	Temperature range (°C)	References/remarks
S	S_2 (v) = 2 S (s)	$-30,579 + 38.91 T$	A	25–115	King et al. (1973)[n, m]
	S_2 = 2 S (l)	$-29,334 + 35.65 T$	A	115–160	King et al. (1973)[m]
	S_2 = 2 S (l)	$-27,601 + 31.14 T$	A	160–950	Rau et al. (1973a, 1973b)[m]
Se	Se_2 (v) = 2 Se (s)	$-33,037 + 37.13 T$	A	25–220	Mills (1974)[m]
	Se_2 = 2 Se (l)	$-28,944 + 28.89 T$	A	220–527	Mills (1974)[m]
Te	Te_2 (v) = 2 Te (s)	$-37,443 + 35.77 T$	A	25–450	Mills (1974)[m]
	Te_2 = 2 Te (l)	$-25,508 + 19.35 T$	A	450–1027	Mills (1974)[m]
H-S	$2 H_2 + S_2 = 2 H_2S$	$-42,332 + 22.64 T$	A	25–1027	Mills (1974)[m]

Table 7-2. (*Continued*)

System	Reaction	Free-energy change in calories for T in K	Accuracy	Temperature range (°C)	References/remarks
Ag-S	$4\,Ag + S_2 = 2\,Ag_2S$	$-44{,}800 + 22.10\,T$	A	25–176	Kubaschewski et al. (1967)[a]
	$4\,Ag + S_2 = 2\,Ag_2S$	$-41{,}980 + 16.53\,T$	A	176–804	Kubaschewski et al. (1967)[n]
Ag-Se	$4\,Ag + Se_2 = 2\,Ag_2Se$	$-53{,}850 + 26.39\,T$	A	25–133	Mills (1974)[m]
	$4\,Ag + Se_2 = 2\,Ag_2Se$	$-48{,}710 + 14.27\,T$	A	133–727	Mills (1974)[m]
Ag-Te	$4\,Ag + Te_2 = 2\,Ag_2Te$	$-55{,}592 + 29.27\,T$	A	25–145	Mills (1974)[m]
	$4\,Ag + Te_2 = 2\,Ag_2Te$	$-50{,}087 + 16.10\,T$	A	145–527	Mills (1974)[m]
As-S	$2\,As + S_2 = 2\,AsS$	$-48{,}716 + 42.41\,T$	C	25–281	Craig and Barton (1973)
	$2\,As + S_2 = 2\,(As, S)\,(l)$	$-46{,}100 + 37.4\,T$	C	281–450	Barton (1969)[y]
	$4\,AsS + S_2 = 2\,As_2S_3$	$-40{,}568 + 34.28\,T$	C	25–175	Craig and Barton (1973)
Au-S	$4\,Au + S_2 = 2\,Au_2S$	$-12{,}500 + 24\,T$	C	25–?	Cugnac-Pailliotet and Pouradier (1972)[z]
Au-Te	$Au + Te_2 = AuTe_2$	$-42{,}022 + 37.15\,T$	A	25–447	Mills (1974)[m]
Bi-S	$4/3\,Bi + S_2 = 2/3\,Bi_2S_3$	$-52{,}121 + 36.20\,T$	A	25–271	Craig and Barton (1973)
	$4/3\,Bi + S_2 = 2/3\,Bi_2S_3$	$-55{,}600 + 42.60\,T$	A	271–450	Craig and Barton (1973)[v]
Cd-S	$2\,Cd + S_2 = 2\,CdS$	$-101{,}870 + 45.14\,T$	A	25–321	Mills (1974)[m]
	$2\,Cd + S_2 = 2\,CdS$	$-103{,}697 + 48.26\,T$	A	321–767	Mills (1974)[m]
Co-S	$9/4\,Co + S_2 = 1/4\,Co_9S_8$	$-80{,}978 + 40.85\,T$	B	25–834	Mills (1974)[m]
	$1/2\,Co_9S_8 + S_2 = 3/2\,Co_3S_4$	$-51{,}450 + 34.04\,T$	B	25–460	Rosenqvist (1954)[aa]
	$6\,CoS + S_2 = 2\,Co_3S_4$	$-90{,}900 + 87.86\,T$	A	460–670	Leegaard and Rosenqvist (1964)
	$Co_3S_4 + S_2 = 3\,CoS_2$	$-57{,}660 + 52.62\,T$	A	25–670	Leegaard and Rosenqvist (1964)

System	Reaction			Reference
	$2\,CoS + S_2 = 2\,CoS_2$	$-68{,}740 + 64.37\,T$	A	670–950	Rosenqvist (1954)
Cu-S	$4\,Cu + S_2 = 2\,Cu_2S$	$-69{,}030 + 30.69\,T$	A	25–103	bb
	$4\,Cu + S_2 = 2\,Cu_2S$	$-63{,}945 + 17.22\,T$	A	103–435	bb
	$4\,Cu + S_2 = 2\,Cu_2S$	$-62{,}109 + 14.63\,T$	A	435–1067	bb
	$18\,Cu_2S + S_2 = 4\,Cu_9S_5$		B	93–435	see Figure 7.22
	$2.2\,Cu + S_2 = 2\,Cu_{1.1}S$	$-57{,}505 + 38.01\,T$	A	25–100	blaubleibender covellite I$^{g.\,bb}$
	$2.8\,Cu + S_2 = 2\,Cu_{1.4}S$	$-59{,}929 + 34.19\,T$	A	25–100	blaubleibender covellite II$^{g.\,bb}$
	$3.53\,Cu + S_2 = 2\,Cu_{1.765}S$	$-67{,}311 + 36.57\,T$	A	25–75	low digenite$^{g.\,bb}$
	$2.67\,Cu_{1.750}S + S_2 = 4.67\,CuS$	$-41{,}405 + 41.72\,T$	A	25–75	bb
	$17.54\,Cu_{1.9495}S + S_2 = 19.54\,Cu_{1.750}S$	$-50{,}228 + 59.16\,T$	A	25–70	bb
	$77.21\,Cu_2S + S_2 = 79.21\,Cu_{1.9495}S$	$-60{,}674 + 86.69\,T$	A	25–90	bb
	$2\,Cu + S_2 = 2\,CuS$	$-54{,}897 + 34.66\,T^{*}$	A	25–507	bb
	$2\,Cu_2S + S_2 = 4\,CuS$	$-44{,}224 + 52.35\,T^{\#}$	B	75–507	Barton (1973)^{dd}
Fe-S	$2\,Fe + S_2 = 2\,FeS$	$-74{,}320 + 31.18\,T$	A	25–140	Richardson and Jeffes (1952)
	$2\,Fe + S_2 = 2\,FeS$	$-71{,}820 + 25.12\,T$	A	140–988	Richardson and Jeffes (1952)^n
	$2\,FeS + S_2 = 2\,FeS_2$	$-67{,}033 + 66.87\,T^{\#}$	A at 500° B at 250°	250–500	see Figure 7.9^{dd}
	$Fe + S_2 = FeS_2$	$-71{,}280 + 47.08\,T$	B	25–742	Toulmin and Barton (1964)
	$FeS_2\,(\text{pyrite}) = FeS_2\,(\text{marcasite})$	$+1000$	A	25–400	Grønvold and Westrum (1976)
Fe-O	$1.894\,Fe + O_2 = 2\,Fe_{0.947}O$	$-126{,}368 + 30.94\,T$	A	560–1200	Haas and Robie (1973)^m
	$\tfrac{3}{2}\,Fe + O_2 = \tfrac{1}{2}\,Fe_3O_4$	$-131{,}699 + 36.93\,T$	A	25–1200	Haas and Robie (1973)$^{m,\,ee}$
	$4\,Fe_3O_4 + O_2 = 6\,Fe_2O_3$	$-116{,}783 + 66.49\,T$	A	25–1200	Haas and Robie (1973)^m
Hg-S	$2\,Hg + S_2 = 2\,HgS\,(\text{cinnabar})$	$-57{,}140 + 50.8\,T$	A	25–345	Craig and Barton (1973)
	$2\,Hg + S_2 = 2\,HgS\,(\text{metacinnabar})$	$-55{,}970 + 49.0\,T$	A	345–480	Craig and Barton (1973)

Table 7-2. (Continued)

System	Reaction	Free-energy change in calories for T in K	Accuracy	Temperature range (°C)	References/remarks
Hg-Se	$2\,Hg + Se_2 = 2\,HgSe$	$-53{,}394 + 44.31\,T$	D	25–600	Mills (1974)[m]
Hg-Te	$2\,Hg + Te_2 = 2\,HgTe$	$-52{,}548 + 41.41\,T$	C	25–600	Mills (1974)[m]
MnO	$6\,MnO + O_2 = 2\,Mn_3O_4$	$-117{,}501 + 61.22\,T$	A	25–1000	Huebner and Sato (1970)
	$4\,Mn_3O_4 + O_2 = 6\,Mn_2O_3$	$-42{,}393 + 33.58\,T$	A	25–1000	Huebner and Sato (1970)
	$2\,Mn_2O_3 + O_2 = 4\,MnO_2$	$-40{,}311 + 50.97\,T$	A	25–700	Huebner and Sato (1970)
Mn-S	$2\,Mn + S_2 = 2\,MnS$	$-132{,}276 + 29.83\,T$	A	25–700	Mills (1974)[m]
	$2\,MnS + S_2 = 2\,MnS_2$	$-37{,}352 + 45.15\,T$	B	25–423	Mills (1974), Skinner (unpublished)[m]
Mo-S	$Mo + S_2 = MoS_2$	$-84{,}700 + 35.27\,T$	B	25–719	Pouillard and Perrot (1975)
	$4/3\,Mo + S_2 = 2/3\,Mo_2S_3$	$-83{,}640 + 34.15\,T$	B	719–1000	Pouillard and Perrot (1975)
	$2\,Mo_2S_3 + S_2 = 4\,MoS_2$	$-87{,}880 + 38.63\,T$	B	719–1000	Pouillard and Perrot (1975)
Ni-S	$3\,Ni + S_2 = Ni_3S_2$	$-79{,}240 + 39.01\,T$	B	25–533	Rosenqvist (1954), see Figure 7.21.
	$3\,Ni + S_2 = Ni_3S_2$	$-67{,}400 + 24.28\,T\#$	B	533–635	Rosenqvist (1954)[dd]
	$7/2\,Ni_3S_2 + S_2 = 3/2\,Ni_7S_6$	$-49{,}260 + 19.91\,T$	B	397–524	Rosenqvist (1954)
	$2\,Ni_3S_2 + S_2 = 6\,NiS$	$-76{,}360 + 57.21\,T$	B	573–806	Rosenqvist (1954)
	$2\,Ni_7S_6 + S_2 = 14\,NiS$	$-52{,}950 + 29.56\,T$	B	400–573	Rosenqvist (1954)
	$2\,NiS + S_2 = 2\,NiS_2$	$-49{,}880 + 46.67\,T\#$	B	356–1022	Leegaard and Rosenqvist (1964)[dd]
	$6\,NiS + S_2 = 2\,Ni_3S_4$	$-58{,}550 + 59.67\,T$	C	25–282	Calculated Rosenqvist (1954), Leegaard and Rosenqvist (1964), and Mills (1974).
	$6\,NiS + S_2 = 2\,Ni_3S_4$	$-62{,}400 + 66.60\,T$	C	282–356	

					Calculated Rosenqvist (1954), Leegaard and Rosenqvist (1964), and Mills (1974).
	$Ni_3S_4 + S_2 = 3 NiS_2$	$-43,620 + 36.71\,T$	C	25–356	
Pb-S	$2\,Pb + S_2 = 2\,PbS$	$-77,449 + 40.81\,T$	A	25–327	Mills (1974)^m
	$2\,Pb + S_2 = 2\,PbS$	$-78,595 + 42.82\,T$	A	327–900	Mills (1974)^m
Pb-Se	$2\,Pb + Se_2 = 2\,PbSe$	$-80,907 + 39.71\,T$	A	25–327	Mills (1974)^m
	$2\,Pb + Se_2 = 2\,PbSe$	$-82,483 + 42.44\,T$	A	327–900	Mills (1974)^m
Pb-Te	$2\,Pb + Te_2 = 2\,PbTe$	$-70,766 + 39.26\,T$	A	25–327	Mills (1974)^m
	$2\,Pb + Te_2 = 2\,PbTe$	$-72,157 + 41.59\,T$	A	327–700	Mills (1974)^m
Pt-S	$2\,Pt + S_2 = 2\,PtS$	$-66,100 + 43.6\,T$	C	700–1200	Kubaschewski et al. (1967)
	$2\,PtS + S_2 = 2\,PtS_2$	$-43,750 + 43.8\,T$	C	400–800	Kubaschewski et al. (1967)
Sb-S	$^4/_3\,Sb + S_2 = {^2/_3}\,Sb_2S_3$	$-55,430 + 40.1\,T$	B	25–518	Barton (1971)
Sn-O	$Sn + O_2 = SnO_2$	$-138,727 + 48.52\,T$	A	25–232	Robie and Waldbaum (1968)^m
	$Sn + O_2 = SnO_2$	$-139,124 + 49.71\,T$	A	232–1000	Robie and Waldbaum (1968)^m
Sn-S	$2\,Sn + S_2 = 2\,SnS$	$-82,186 + 41.37\,T$	A	25–232	Mills (1974)^m
	$2\,Sn + S_2 = 2\,SnS$	$-84,571 + 46.22\,T$	A	232–599	Mills (1974)^m
	$2\,Sn + S_2 = 2\,SnS$	$-82,318 + 43.60\,T$	A	599–860	Mills (1974)^m
	$4\,SnS + S_2 = 2\,Sn_2S_3$	$-56,000 + 47.78\,T$	C	25–599	ff
	$2\,Sn_2S_3 + S_2 = 4\,SnS_2$	$-53,200 + 49.07\,T$	C	25–760	^{n}, gg
V-S	$^2/_5\,V_2S_3 + S_2 = {^4/_5}\,VS_4$	$-36,600 + 39.35\,T$	C	25–470	Kubaschewski and Catterall (1956)
W-S	$W + S_2 = WS_2$	$-91,777 + 44.31\,T$	C	25–900	Mills (1974)^m

Table 7-2. (Continued)

System	Reaction	Free-energy change in calories for T in K	Accuracy	Temperature range (°C)	References/remarks
Zn-S	$2\,Zn + S_2 = 2\,ZnS$	$-128,471 + 45.73\,T$	A	25-420	Mills (1974)[m]
	$2\,Zn + S_2 = 2\,ZnS$	$-131,095 + 49.48\,T$	A	420-900	Mills (1974)[m]
As-Fe-S	$2\,Fe_2As + S_2 = 2\,FeS + 2\,FeAs$	$-78,000 + 34.4\,T$	C	400-800	Barton (1969)
	$4\,FeAs + S_2 = 2\,FeS + 2\,FeAs_2$	$-75,900 + 44.4\,T\#$	B	25-850	Barton (1969)[dd]
	$2\,FeAs_2 + 2\,FeS + S_2 = 4\,FeAsS$	$-69,900 + 55.6\,T\#$	A	25-650	Barton (1969),[dd] see Figure 7.23
	$2\,FeAs_2 + S_2 = 2\,FeAsS + 2\,As$	$-61,800 + 49.0\,T$	C	25-688	Barton (1969)
	$FeAs_2 + S_2 = FeS + (As, S)\,liquid$	$-30,700 + 17.1\,T\#$	B	702-825	Barton (1969)[dd]
	$2\,FeAsS + S_2 = 2\,FeS + 2\,(As, S)\,liquid$	$-23,600 + 9.8\,T\#$	B	491-702	Barton (1969)[dd]
	$FeAsS + S_2 = FeS_2 + (As, S)\,liquid$	$-53,300 + 48.7\,T\#$	C	363-491	Barton (1969)[dd]
	$2\,FeAsS + S_2 = 2\,FeS_2 + 2\,As$	$-61,600 + 61.7\,T$	C	25-363	Barton (1969)
Au-Bi-S	$4/3\,Au_2Bi + S_2 = 8/3\,Au + 2/3\,Bi_2S_3$	$-54,145 + 41.44\,T$	A	113-371	Hultgren et al. (1973b)
Au-Sb-S	$2/3\,AuSb_2 + S_2 = 2/3\,Au + 2/3\,Sb_2S_3$	$-53,900 + 39.9\,T$	C	25-360	Barton (1971)
As-Cu-S	$2/3\,Cu_{12}As_4S_{13} + S_2 = 8/3\,Cu_3AsS_4$	$-41,576 + 34.91\,T$	D	25-665	Craig and Barton (1973)[c]
Co-Cu-S	$3.6\,CuCo_2S_4 + S_2 = 0.4\,Cu_9S_5 + 7.2\,CoS_2$	$-44,657 + 51.84\,T$	B	25-550	Craig and Higgins (1973)
Cu-Fe-S	$5/2\,Cu + 1/2\,Fe + S_2 = 1/2\,Cu_5FeS_4$	$-73,809 + 30.18\,T$	A	25-228	Barton (unpublished)[g]
	$5/2\,Cu + 1/2\,Fe + S_2 = 1/2\,Cu_5FeS_4$	$-70,974 + 24.33\,T$	A	228-600	Barton (unpublished)[g]
	$Cu + Fe + S_2 = CuFeS_2$	$-75,928 + 36.70\,T$	A	25-547	Barton (unpublished)[g]
	$5\,CuFeS_2 + S_2 = Cu_5FeS_4 + 4\,FeS_2$	$-50,638 + 57.47\,T\#$	A	230-532	Schneeberg (1973)[dd]
	$2\,Cu_{5.5}FeS_{6.5} + S_2 = 2\,FeS_2 + 11\,CuS$	$-45,012 + 56.02\,T\#$	B	227-434	[dd, hh]
	$.96\,Cu_5FeS_4 + S_2 = .87\,Cu_{5.5}FeS_{6.5} + .09\,FeS_2$	$-41,820 + 49.64\,T\#$	B	227-501	[dd, hh]
	$Cu_5FeS_4 + .5\,CuS + S_2 = Cu_{5.5}FeS_{6.5}$	$-42,334 + 50.67\,T\#$	B	227-482	[dd, hh]
	$2/3\,Cu_5FeS_4 + S_2 = 2/3\,FeS_2 + 10/3\,CuS$	$-43,740 + 53.48\,T\#$	B	25-227	[dd, hh]
	$2\,Cu_5FeS_4 + 8\,FeS + S_2 = 10\,CuFeS_{1.8}$	$-103,270 + 77.58\,T\#$	C	360-700	Barton (1973)[dd]

System	Reaction			Reference
Cu-S-Sb	$2/3\ Cu_{12}Sb_4S_{13} + S_2 = 8/3\ Cu_3SbS_4$	$-41,577 + 39.85\ T$	D	Craig and Barton (1973)[c] ii
Fe-O-S	$3\ Fe_3O_4 + S_2 = FeS_2 + 4\ Fe_2O_3$	$-60,726 + 65.61\ T$	B	25-613
	$8/13\ Fe_2O_3 + S_2 = 10/13\ FeS_2 + 6/13\ FeSO_4$	$-42,200 + 44.7\ T$	C	25-613
	$2/5\ Fe_3O_4 + S_2 = 4/5\ FeS_2 + 2/5\ FeSO_4$	$-44,670 + 47.5\ T$	C	613-740
Fe-S-Sb	$2\ Fe + 2\ Sb + S_2 = 2\ FeSbS$	$-79,400 + 38.8\ T$	C	25-280 Barton (1971)[g]
	$1/2\ Fe + Sb + S_2 = 1/2\ FeSb_2S_4$	$-69,200 + 39.8\ T$	C	25-563 Barton (1971)[g]
	$2\ FeSb_2 + S_2 = 2\ FeS + 4\ Sb$	$-64,500 + 25.2\ T$	B	280-620 Barton (1971)[n]
	$2\ FeSb_2S_4 + S_2 = 2\ FeS_2 + 2\ Sb_2S_3$	$-57,450 + 55.5\ T$	C	25-532 Barton (1971)
	$4\ FeSb + S_2 = 2\ FeS + FeSb_2$	$-69,500 + 25\ T$	C	138-530 Barton (1971)
Fe-S-Sb	$2/3\ FeS + 4/3\ Sb + S_2 = 2/3\ FeSb_2S_4$	$-59,700 + 44.9\ T\#$	C	138-530 Barton (1971)[n, dd]
Fe-O-Pb-S	$3/8\ PbS + 1/2\ Fe_2O_3 + S_2 = FeS_2 + 3/8\ PbSO_4$	$-46,740 + 47.27\ T$	A	25-492 ii
	$1/3\ PbS + 1/3\ Fe_3O_4 + S_2 = FeS_2 + 1/3\ PbSO_4$	$-49,112 + 50.37\ T$	A	492-700 ii
Fe-O-S-Si	$4\ Fe_2SiO_4 + S_2 = 2\ FeS + 2\ Fe_3O_4 + 4\ SiO_2$	$-56,335 + 32.63\ T\#$	C	138-900 n, dd, ii
Fe-O-S-Ti	$4\ FeTiO_3 + S_2 = FeS_2 + Fe_3O_4 + 4\ TiO_2$	$-56,342 + 52.29\ T$	B	25-460 ii
Sulfosalts	Approximations for many reactions available in Craig and Barton (1973).			

[a] Jeannot et al. (1970) report Ag_4S stable below 450°C, but this is so contrary to mineralogic and experimental observation that it is omitted here.
[b] No binary compounds noted at least down to 500°C (Barton, unpublished).
[c] More work needed on this point.
[d] 1 wt. % S in rammelsbergite lowers the transition by 115 ± 35°C (Yund, 1962).
[e] The ordered gold-copper phases have not been described as minerals, but there is every reason to anticipate their eventual discovery.
[f] Bachechi (1972) notes that to stabilize montebrayite "impurities" are required, the most probable of which is Sb inasmuch as montbrayite does not appear in the Ag-Au-Te, Au-Bi-Te, or Au-Pb-Te systems.
[g] This assemblage does not constitute a stable equilibrium.
[h] R. W. Potter (personal communication, 1976) has shown that there are two distinct phases (see Table 7.2) known as blaubleibender coveliite, neither of which is ever stable relative to covellite + digenite or covellite + anilite.
[i] $CuSe_2$ is also $CuSe_2$ but with the pyrite structure. Krutaite is also $CuSe_2$ but with the marcasite structure; it is probably metastable except at high pressure (Hull and Hullinger, 1968).

NOTES TO TABLES 7-1 AND 7-2 Continued.

j Cu_9Te_5 may be weissite or a polymorph.

k The inversion of smythite at 75°C is based on a discontinuity in the cell volume vs temperature curve. There is also some suggestion that smythite does not form above 75°C (see Taylor, 1970a).

l $MoS_{3\pm x}$ decomposes to molybdenite + sulfur at 200–400°C; it is probably always metastable (Rode and Lebedev, 1961).

m Linear free-energy expression fit to tabulated free-energy values.

n Solid-state transition neglected.

o This is the maximum probable temperature, the reversible temperature may be much lower.

p Pb, Se, and Te in solid solution stabilize the high-temperature polymorph.

q Extensive mutual substitution of As and Sb.

r The Bi_2S_3—PbS join is complicated by slight departures from stoichiometry.

s Possibly not a valid invariant point, but it is included here for consistency with Figure 7.17.

t Sugaki and Shima (1975) report two transitions at 200 and 270°C; Pankratz and King (1970) report two transitions at 212 and 267°C; the need for additional work is obvious.

u The crest of the two-tetrahedrite field is not determined, although it must be not far above 125°C; the two-tetrahedrite field disappears with the introduction of small amounts of As, Fe, or Zn (Tatsuka and Morimoto, 1973).

v It is not clear whether pentlandite and argentian pentlandite actually form a complete solid solution. This interpretation of Mandziuk and Scott (1975) assumes that they do not.

w Complex melting relationships are described by Craig and Kullerud (1969).

x The requirement for Ag is unknown.

y Nonlinear above about 450°C; see Figure 7.23.

z Solubility measured at 25°C, entropy estimated.

aa Lower-temperature extrapolation has larger uncertainty.

bb Calculated using data of R. W. Potter (personal communication, 1976) and the equations for S_2 in Table 7-2.

cc This equation is for phases in their standard states; the equilibrium constant expression requires activities for solid components.

dd This equation gives the chemical potential of S_2.

ee Metastable above 560°C.

ff Calculated using Mills' entropy data assuming that herzenbergite + pyrite + pyrrhotite + ottemannite (+ vapor) constitute a stable assemblage at 300°C. This treatment is similar to that of U. Petersen cited by Drabek and Stemprok (1974).

gg Calculated using Mills' entropy data and Moh's (1974) invariant point at 508°C for pyrite + pyrrhotite + ottemannite + pyrrhotite (+ vapor).

hh Calculated from data of Schneeberg (1973), Roseboom and Kullerud (1958), and Pankratz and King (1970).

ii Calculated from data of Toulmin and Barton (1964) and Haas and Robie (1973).

jj Calculated from data of Toulmin and Barton (1964), Haas and Robie (1973), Mills (1974), and Robie and Waldbaum (1968).

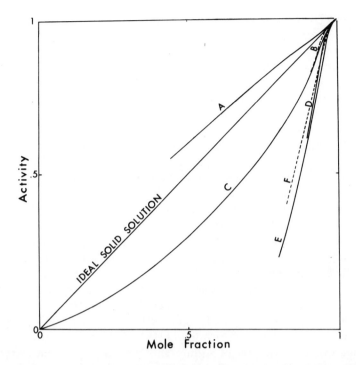

Fig. 7.2 Relation between activity and mole fraction for several solid solutions. The solid solutions of PbSe in Pb (S, Se) and ZnSe in Zn (S, Se) apparently lie close to the ideal solid solution, at least above 600°C (Bethke and Barton, 1971) and Scott et al. (1974) have shown that the (Fe, Ni)$_{1-x}$S solid solution is also ideal with respect to metal-metal substitution at high temperature. Curve A is for ZnS in (Zn, Fe)S (sphalerite) at 850°C from Fleet (1975); the previous curve by Barton and Skinner (1967) had a miscalculation in sign of the departure from ideal (first pointed out to us by B. Austria in 1969). Curve B (dashed) is for FeAsS in Fe(As, S)$_2$ at 500°C from Kretschmar and Scott (1976). Curve C is for Ag$_2$S in (Ag, Cu)$_2$S at 655°C from Perrot and Jeannot (1971). Curve D is for Cu$_2$S in (Cu, $\square$)$_2$ at 300°C from Barton's (1973) calculations from the data of Rau (1967). Curve E is for CuFeS$_2$ in (CuFe $\square_2$, Cu$_2$Fe$_2$)S$_2$ at 600°C from Barton (1973). Curve F (dashed to distinguish it from D) is for FeS in (Fe, $\square$)S at 700°C from Toulmin and Barton (1964). The $\square$ symbolizes a metal vacancy.

PRESENTATION OF DATA

Phase relations can be presented in many ways depending on the method of measurement or the need to emphasize a certain parameter. We shall discuss several methods briefly and point out the interrelationships. Further details may be found in standard texts such as those by Darken and Gurry (1953), Garrels and Christ (1965), Kubaschewski et al. (1967), Lewis and Randall (1961), and Ricci (1951).

For illustration we use the most important binary sulfide system, Fe-S. Whereas the same methods of presentation may be used for systems containing more components, the additional complexities soon exceed the capability of two-dimensional diagrams and limitations on the degrees of freedom that must be imposed for graphic presentation. Certain diagrams in common use, notably the activity-activity, Eh-pH, and activity of O_2-pH diagrams, are ideally suited for representing special multicomponent relations and have been developed for the ternary case, using iron-sulfur-oxygen as an example.

Data for the iron-sulfur binary were taken from references listed in Tables 7-1 and 7-2. The phases considered are iron (ir), pyrrhotite (po), pyrite (py), rhombic sulfur (su), liquid sulfur (L_1)[1] containing very little iron, vapor (V), and an iron-rich liquid (L_2). The α-γ-δ transitions in iron, the monoclinic polymorph of sulfur, and the low-temperature pyrrhotite transitions have been omitted for clarity. The phase relations themselves are largely self-explanatory; moreover, questionable points are omitted and discussed later in the section on specific sulfide systems.

The Temperature—Composition (or T-X) Diagram

Figure 7.3a is drawn for a 1-atm isobaric section, and the L_1 and V fields are exaggerated for clarity. Because the diagram is isobaric, vapor is stable only in the fields labeled. The T-X isobaric section is particularly useful in metallurgic studies where the total pressure is held constant, usually at 1 atm. For example, where $P_t = 1$ atm in the roasting of pyritic ores, pyrite will break down at just below 700°C rather than at 742°C as shown in Figure 7.3b.

The Condensed Temperature—Composition Diagram

Because of experimental difficulties, isobaric relations such as Figure 7.3a are rarely used in sulfide-system studies. More commonly, a diagram

---

[1] L_1 is referred to as a liquid although some diagrams extend to supercritical temperatures.

PRESENTATION OF DATA

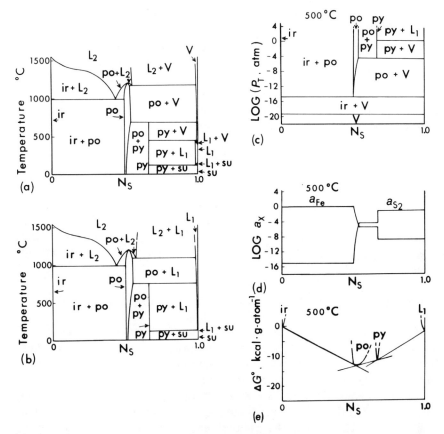

Fig. 7.3 Various ways of representing the phase relations in the binary system Fe-S as a function of mole fraction of sulfur, N_S. ir = iron, po = pyrrhotite, py = pyrite, su = rhombic sulfur, L = liquid, V = vapor. (a) isobaric (1 atm), temperature-composition (or T-X) diagram; (b) condensed temperature-composition diagram; (c) total-pressure-composition (or P_T-X) diagram; (d) activity-composition diagram; (e) free-energy-composition diagram.

is used in which vapor is a stable phase throughout, the pressure being fixed by the vapor pressures of the condensed phases present (Figure 7.3b). These phase relations are determined in the common evacuated-glass-tube techniques. They are always polybaric. Pressures may vary from negligible to a few tens of atmospheres, but the condensed phases are relatively incompressible, and, as shown in Figures 7.3c, 7.4a–c the effects of total pressure are negligible. Where vapor phases are involved in reactions, however, pressure is very important because of the high compressibility of gases. Failure to recognize this effect can lead to serious misinterpretation of experimental data and result in misuse of the derived diagram (see the example discussed by Barton and Toulmin, 1966).

The Total-Pressure—Composition (or P_t-X) Diagram

This diagram is infrequently used for depicting isothermal relations. Figure 7.3c is drawn for a 500°C isotherm. Total pressure, P_t, is plotted logarithmically to compress the pressure axis and expand the interesting low-pressure region. Although pressure effects take place in the vapor-bearing part of the diagram below 1 atm, the high-pressure range up to 1000 atm shows little effect by total pressure[2] on the solid and liquid assemblage.

The Activity—Composition Diagram

The effect of composition on the activity of a given component can be studied by holding both temperature and pressure constant. Figure 7.3d is drawn for a 500°C isotherm and 1 atm isobar, showing the activity of iron (a_{Fe}) and the activity of sulfur (a_{S_2}). The variables are related through equations of the sort

$$2\ Fe(s) + S_2(g) \rightleftharpoons 2\ FeS(s) \quad \text{for which} \quad K = a_{FeS}^2 \cdot a_{Fe}^{-2} \cdot a_{S_2}^{-1}$$

Because the standard state for sulfur is the ideal diatomic gas at one atmosphere, a_{S_2} is numerically equal to the pressure of S_2 in atmospheres. The diagram is calculated for total pressure of 1 atm. It would not change perceptibly, however, if the total pressure were 100 atm.

Activity is constant across an invariant (two components, two phases, T and P fixed), two-phase field. Changes in activity take place across the one-phase fields. Because the composition of the system is dependent on a_{S_2} (or on a_{Fe}), the composition and activities cannot be independent. There are, therefore, no two-dimensional fields on this diagram and all conditions fall along the curves shown. This type of diagram is useful in the evaluation of thermochemical parameters but is otherwise less applicable for our purposes than alternative types of presentation.

The Free-Energy—Composition Diagram

In Figure 7.3e the standard free energy of formation is plotted against composition, and the diagram is again isothermal and isobaric. Our example is given for the 500°C isotherm and 1-atm isobar.

---

[2] A common current practice is to record pressure in bars (1 bar = 10^5 Pa = 10^6 dynes/cm^2 = 0.987 atm). The difference between bars and atmospheres may be significant in some experimental work but there is no significant difference in geologic application.

One-phase fields are represented by curved segments; two-phase fields are represented by lines tangent to the one-phase fields. All equilibrium compositions must represent the minimum (most negative) free energy possible. The points of tangent contact give the compositions of phases in two-phase assemblages. Phases such as iron, pyrite, and sulfur vapor, having very narrow, stable compositional limits, have steep, narrow-bottomed, free-energy curves, and the curves rise to very high $\Delta G°$ values within a very small compositional range. Conversely, pyrrhotite has an appreciable range of stable compositions and the trough of the pyrrhotite free-energy curve is quite broad. Although not visible at the scale of the Figure 7.3e, the slopes of all free-energy curves must approach infinity as $N \to 1.0$.

If any tangent to the equilibrium curve is projected to 0 and 100% S, the intersections with the side lines indicate respectively the partial molal free energy (or chemical potential, μ) of Fe and $\frac{1}{2}$ S_2. Likewise, we could calculate $\mu_{1/2\,Fes}$ from the intercept with the 50% S composition, and so on.

We may easily convert information from the free-energy diagram to the activity of S_2 diagram through the relation $\mu_{S_2} = \mu°_{S_2} - RT \ln a_{S_2}$. Because of our choice of a standard state, $\mu°_{S_2} = 0$; therefore, we may write $\mu_{S_2} = -4.5756\, T \log a_{S_2}$.

The Total-Pressure—Temperature (or P_t-T) Diagram

This diagram (Figure 7.4a, b) is for plotting the effect of T and P_t (P_t = total pressure) on specific isochemical reactions. We have shown three univariant curves that have been measured or extrapolated to high pressure.

Two of the curves in Figure 7.4a are for reactions that do not involve vapor and, because of the relatively incompressible nature of the solid and liquid phases, the curves are shifted by less than 100°C by as much as 3000 atm pressure. On the other hand, the boiling curve for sulfur is very pressure dependent, as expected from the compressibility of sulfur vapor. The slope of such curves at a given point may be calculated from the Clausius-Clapeyron equation

$$\frac{dP_t}{dT} = \frac{\Delta H}{T \Delta V}$$

A refined treatment of this equation could allow for the variation of ΔH and ΔV with T and P_t; however, such refinements are not usually warranted for geologic processes occurring in the upper half of the crust because most P_t-T curves are sensibly straight lines, indicating little or no change in ΔH or ΔV. This plot is frequently made to present data from collapsible-tube runs. The low-pressure, or vapor, region of the diagram

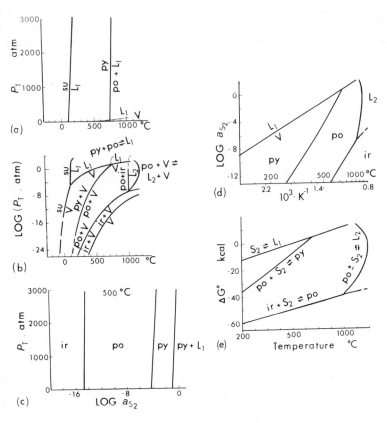

Fig. 7.4 Additional ways of representing phase relations in the system Fe-S. ir = iron, po = pyrrhotite, py = pyrite, su = rhombic sulfur, L = liquid, V = vapor. (a) and (b) total-pressure-temperature (P_T-T) diagrams; (c) total-pressure-activity of S_2 diagram; (d) activity of S_2-temperature diagram; (e) free-energy-temperature diagram. In (c), (d), and (e) the region having a_{S_2} or $\Delta G^\circ_{S_2}$ greater than that along the L_1-vapor curve [= sulfur condensation curve in (d) and (e)] is unattainable.

remains obscure because of the detail compressed into a small area. This deficiency can be corrected by plotting pressure on a log scale as in Figure 7.4b.

P-T curves represent only compositions compatible with the phase assemblages stable along a particular curve. For example, the apparent intersection of the po + ir + L_2 curve with the po + L_1 + V curve at about 1000°C has no significance, for these represent two different bulk compositions whose phases do not coexist. Four invariant points are known and correspond to the horizontal lines in Figure 7.3b.

The points for po + py + L_1 + V and po + ir + L_2 + V are shown by the intersections of the four univariant curves for each point. The invariant point for sulfur (su + L_1 + V) is not measurably influenced by the presence of pyrite and so is marked only by the curves from the sulfur unary. The fourth invariant point is indicated in Figure 7.3b but is not shown in Figure 7.4b.

The Total-Pressure—Activity of S_2 Diagram

Figure 7.4c is again constructed for a 500°C isotherm. The activity of S_2 is plotted on a log scale to present a large range of values more clearly. Sulfidation reactions can be written so that the equilibrium constant has the form $K = a_{S_2} \cdot \Pi a_i$. For most mineralogically significant reactions the activities of all products and reactants other than S_2 are unity, and therefore $K = a_{S_2}$. In a few reactions, such as that representing the pyrrhotite-pyrite univariant curve, the activities of end-member components of solid solutions may be far from unity and must be included in the evaluation of the equilibrium constant. We can then apply the expression

$$\frac{d \ln K}{dP_t} = \frac{-\Delta V}{RT}$$

to solve for the variation of a_{S_2} with pressure. Because a_{S_2} is a function of composition, fields as well as univariant curves may be labeled in this type of diagram. The diagram is not particularly informative except to show that total pressure is rather ineffective in influencing changes in a_{S_2}.

The Activity of S_2—Temperature Diagram

The calculations on which this diagram (Figure 7.4d) is based were made with data obtained from studies of the activity of S_2 from condensed equilibria, but the information thus derived may apply to vapor-free conditions as well. We emphasize the use of this diagram as a useful adjunct to the P_t-T and condensed T-X diagrams. A summary of available univariant equilibria pertinent to sulfide mineralogy is given in Figures 7.21, 7.22, 7.23, and in Table 7-2. The choice of plotting log a_{S_2} against the reciprocal of the absolute temperature is made because it permits most univariant sulfide reactions to be represented by linear equations; only when the activities of components within the condensed phases deviate from uniform values do the curves become appreciably nonlinear. The univariant curves for congruently melting phases must pass through temperature maxima, as shown for pyrrhotite, at which point the stoichiometric coefficient for S_2 in the balanced equation changes sign. One major advantage to this type of diagram is that extrapolation to other temperatures may reveal potentially intersecting curves that will reflect additional changes in the phase assemblage. Graphic and analytic examples of this type of forecasting were given earlier in this chapter.

The Free-Energy—Temperature Diagram

Figure 7.4e is topologically similar to Figure 7.4d. Diagrams of this type have been constructed by Richardson and Jeffes (1948, 1952) and Richardson et al. (1950) for a large number of oxides and sulfides. The diagram is related to the preceding one through the equation $\Delta G° = -RT \ln K$; and, because the reactions may usually be written so that S_2 is the only product or reactant not in its standard state (see discussion of Figure 7.4c), $\Delta G°$ (cal) $= -RT \ln a_{S_2} = -4.5756\ T \log a_{S_2}$.

Multicomponent Systems

Composition is commonly represented in triangles and tetrahedra for three- and four-component systems respectively, and isothermal sections are drawn for either isobaric or condensed conditions.

Certain of the diagrammatic representations discussed are well suited for presenting specific relations in multicomponent systems. Thus, the a_{S_2}-T diagram (Figure 7.4d) is excellent for comparing specific sulfidation reactions among a variety of metals.

A satisfactory method of presenting data can usually be selected from among the methods discussed by limiting the number of variables. Thus, a T-X diagram can be drawn with the restriction that all phases considered be in equilibrium with one or more universally present phases. This is accomplished geometrically by projecting the diagram from one composition of the ubiquitous phase, as discussed by Thompson (1957). An example would be the presentation of part of the copper-iron-zinc-sulfur system by projecting it from pyrite. In this example, the usable bulk compositions are necessarily limited to those in which pyrite is a stable phase.

Activity—Activity Diagram

A particularly useful type of phase diagram is that in which the activities of two components are used as coordinates. In Fig. 7.5a oxygen is added to the iron-sulfur system, enabling us to calculate the stability fields for magnetite and hematite in addition to those for pyrrhotite, pyrite, and sulfur. The details of construction of such diagrams are given by Garrels and Christ (1965). A comprehensive superposed series of diagrams for a

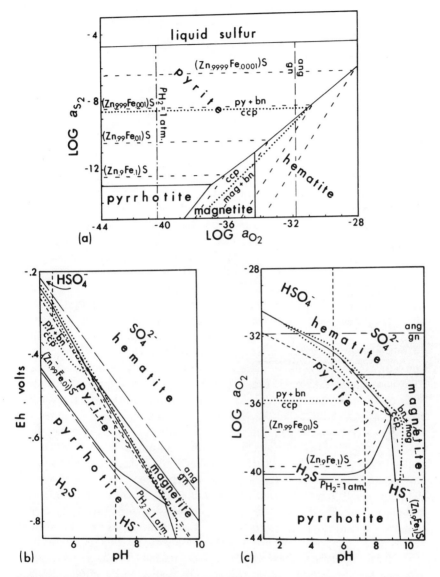

Fig. 7.5 Commonly employed methods of representing noninterfering, multicomponent equilibria. The same type of line is used to indicate each sort of reaction in each figure, but it is impractical to plot the full set of lines on each diagram. All of these diagrams are calculated for 250°C and an H$_2$O pressure of 40 bars. Abbreviations: py = pyrite; gn = galena; ang = anglesite; ccp = chalcopyrite; bn = bornite; mag = magnetite. The stability field for ferrous sulfate would appear in (a), but has not been included; it would be off the diagrams at low pH in (b) and (c).

large variety of ore-forming phases at several temperatures is given by Holland (1959, 1965). An exceptionally intricate discussion of the copper-iron-arsenic-sulfur system in which three activities are plotted simultaneously is presented by Gustafson (1963). If a pH is assigned, activities of aqueous sulfur and metal aqueous species may be plotted also; Barton et al. (1977) use such a diagram to discuss the environment of ore deposition at Creede, Colorado.

Eh-pH Diagram

The methods and data for constructing Eh-pH diagrams (Figure 7.5b) are given by Pourbaix (1966) and by Garrels and Christ (1965). The diagrams provide a useful way to summarize the aqueous-solution chemistry of a given system and provide a frame within which to compare redox and hydrolysis reactions of different metals.

Eh-pH diagrams can show different aspects of the system depending on the assumptions used in the calculation. In Figure 7.5b we assumed (1) $a_{H_2O} = 1.0$, (2) $T = 250°C$, (3) $P = 40$ bars, (4) the sum of the activities of all aqueous sulfur-bearing species is 0.02 molal, (5) the system is everywhere saturated with an iron-bearing phase. Actually, Figures 7.5b and c consist of two superposed diagrams, one for the sulfur ionic equilibria (closely spaced, short-dashed lines) and the other for the crystalline copper-iron-lead-zinc-oxygen-sulfur phases. The assumption of unit activity of H_2O is reasonable through the central part of the diagram, but in order for a solution to have very high or low pH, strong acids or bases must be present in appreciable concentration. These constituents will lower the activity of water, and so will appreciable concentrations of sulfur- and iron-bearing ionic species, but the effect on the diagram is small. The data on which to base construction of Eh-pH diagrams at high temperature are limited and of uneven reliability. H. C. Helgeson and coworkers (see Helgeson, 1969) have extended the available, geologically significant information to 300°C. If the sum of the aqueous sulfur-bearing species is changed, the positions of some of the field boundaries will be shifted. $FeSO_4$ does not appear because it is stable only at very high concentrations or acidities (negative pH values); moreover, its field would be usurped by hydrated ferrous sulfates. Other compounds, that is, ferric sulfates or basic sulfates, may also have stable fields, but these are omitted. We could have assumed a maximum value for total iron in solution [as an alternative to assumption (5) above] in which case the fields of the iron compounds might not have extended to the left margin but instead might have terminated against fields of ferric and ferrous ions.

A very useful aspect of the Eh-pH diagram is that, at modest temperatures, both the Eh and pH of natural and laboratory solutions can easily be measured directly. However, the cramming of many reactions into narrow regions, which, because they are inclined to the axes, cannot be expanded conveniently to show detail, causes the Eh-pH diagram to be less informative than the activity of O_2-pH diagram.

Activity of O_2—pH Diagram

This diagram (Figure 7.5c) is topologically identical to the Eh-pH diagram, but, for our purpose, it is much easier to read. The construction of such diagrams is discussed by Barnes and Kullerud (1961) and in Chapter 8 of the first edition of this volume. The activity of O_2 is related to Eh through the following half-cell reaction:

$$2\,H_2O = 4\,H^+ + 4\,e^- + O_2$$

Since $Eh = -\mathcal{E} = -\mathcal{E}° + \dfrac{RT}{n\mathfrak{F}} \ln K$, and since $\mathcal{E}° = -\Delta G°/n\mathfrak{F}$,

$$Eh = \frac{\Delta G°_{H_2O}}{2\mathfrak{F}} + \frac{4.5756\,T}{4\mathfrak{F}} \log a_{O_2} - \frac{4.5756\,T}{\mathfrak{F}} pH$$

The same symbols are used for each assemblage in Figure 7.5a-c; obviously c is much easier to read than b.

Note in both Figures 7.5b and 7.5c that when the boundaries between assemblages of solid phases cross the O_2-pH buffers provided by the aqueous sulfur species the boundaries are rounded; this curvature is a consequence of the significant changes in the activities of the aqueous buffer species (which are sliding-scale buffers) in the vicinity of the equivalence curves. If a fixed-point buffer is involved, the boundaries between phase assemblages will be straight lines with sharp corners where they change direction, as in Figure 7.5a.

Relative Importance of Variables in Controlling Sulfide Phase Assemblages

The important variables controlling sulfide mineral assemblages for a given metallic composition are clearly activity of S_2 and temperature. Total pressure plays a very minor role. For this reason we have many potential indicators of temperature and of activity of S_2, but very few geobarometers.

The most informative types of phase diagram for sulfide ores are the condensed T-X type (Figure 7.3b) and the a_{S_2}-T type (Figure 7.4d). If we had to consider reactions with a dense vapor, or perhaps also the crystallization paths for sulfide liquids, then total pressure would be important as well.

SPECIFIC SULFIDE SYSTEMS

We make no pretense of a comprehensive coverage of all sulfide phase diagrams, but in this section we emphasize those of most general importance. Several useful compilations are available. A summary of binary systems is presented by Hansen and Anderko (1958) with supplements by Elliott (1965) and Shunk (1969). Compilations of phase diagrams are found in Levin et al. (1964, 1969), Levin and McMurdie (1975), and Chukrov et al. (1974). Reviews of geologically significant systems have been made by Kullerud (1964, 1970) and more recently by Craig and Scott (1974), Shuey (1975), Vaughan and Craig (1978), and Moh (1974, 1975a,b).

We have supplemented the discussion of individual systems with an extensive tabulation of invariant points (Table 7-1) and univariant sulfidation equilibria (Table 7-2 and Figures 7.21, 7.22, and 7.23).

The reservoir of thermochemical data is constantly being enlarged and the fundamental base (that is, the stabilities of reference materials) is constantly being improved; therefore, we expect that further revisions of the data presented will continue as we achieve greater accuracy and precision. Certainly there are some dramatic modifications and additions to the 1967 compilation (Barton and Skinner), but a more fundamental modification is on the horizon. Haas and Fisher (1976) have devised a convenient, computer-based, data-evaluation scheme that uses all data available (equilibrium constants, heats of reactions, high and low temperature heat capacities, cell potentials, invariant points, vapor pressures, etc.) to arrive at the "best" set of values for all of the thermodynamic parameters. We would like to be in a position to do this for sulfides, but have not done so because of the magnitude of the effort (all of the original experimental data need to be reprocessed for each sulfide) and because sulfur itself presents a massive roadblock in that the existing "best" data are not internally consistent with respect to the heat content of liquid sulfur and speciation in the vapor.

Sulfur

Although phase relations involving sulfur are of little direct use in estimating the conditions of ore formation, sulfur is the first and most critical brick in the foundation for a study of sulfide stabilities. The reference state in all of our computations is pure sulfur. Some authors (e.g., Barton and Toulmin, 1964) have used the ideal diatomic gas at one atmosphere; many others (Kelley, 1962; Robie and Waldbaum, 1968) have used pure sulfur in whatever state (solid, liquid, or vapor) happens to be stable at 1-atm pressure and the temperature of interest; still others (Haas and Potter,

1974) have proposed a single reference state, crystalline sulfur at 25°C and 1-atm pressure. Each convention has disadvantages as well as merit, and confusion among different reference states is an ever-present trap for the unwary. We prefer to continue the use of the ideal diatomic gas at 1 atm and the temperature of interest because a single state for sulfur means that the standard equations for the stabilities of sulfides do not have to have breaks in slope at the melting and boiling points of sulfur.

Our choice of the reference state immediately brings us face to face with one of the major problems in the physical chemistry of sulfides: what is the stability of S_2 gas? We have chosen S_2 over any other of the at least 22 (Cocke et al., 1976) sulfur polymers because S_2 is dominant over so much of the chemical environment that we wish to consider (see Figure 7.7), yet there is no constraint that would preclude using, for example, S_7. Experiments designed to determine the stability of sulfides all, in one way or another, come down to a "relative humidity" of solid or liquid sulfur, or to an H_2S/H_2 ratio, or a cell potential that is interpreted in terms of S_2.

We have selected the quantitative model of Rau and others (1973b; see also Meyer, 1976) for the polymerization of sulfur vapor as being the most accurate over the entire temperature range, because they included corrections for nonideality of the sulfur species in the high-temperature, high-pressure region. When we use the total sulfur pressure data of Rau et al. (1973a) above 500°C and of West and Menzies (1930) below that temperature, a linear equation, $\log a_{S_2} = 6.8052 - 6.0323/10^{-3}\,T$ with a coefficient of determination of 0.9996 (standard error of estimate—0.047), fits the nonlinear array of points from 200 to 950°C to better than 0.10 log a_{S_2} unit and will be used throughout this chapter; *however,* refinements in the position of invariant points where the univariant curves for various reactions intersect liquid sulfur have not been made because the modifications should also include the recomputation of the entire array of data for each reaction. Such recomputations might involve dew point experimental results or hydrogen-reduction equilibria or cell potential data and would be far too extensive an undertaking for consideration in this chapter. The computations of other workers noted in Figure 7.6 agree for the activity of S_2 in saturated vapor to within 0.3 log units or better over the range 200-950°C, and thus, while the discrepancies may appear large, they are tolerably small for most geologic applications.

The change in structure of liquid sulfur near 160°C from S_n chains to puckered S_8 rings on cooling and the crystallization first of monoclinic and then orthorhombic sulfur at 115 and 95°C, respectively, precludes the use of a simple a_{S_2}-T equation at low temperature; two additional equations are therefore proposed in Table 7-2 for temperatures below 160°C.

Sulfur vapor is complex. At total sulfur pressures lower than approxi-

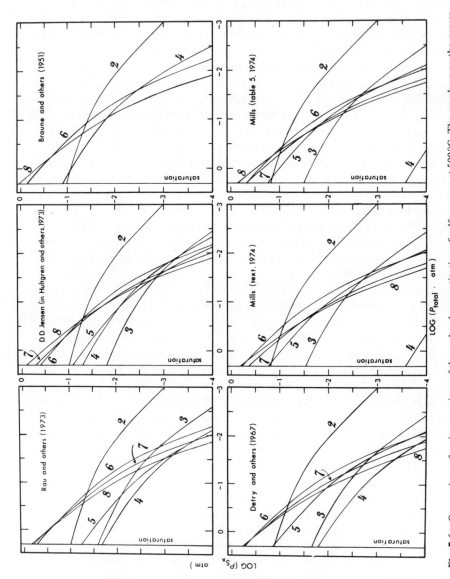

Fig. 7.6 Comparison of various versions of the molecular constitution of sulfur vapor at 500°C. The numbers on the curves indicate the number of sulfur atoms per molecule. The S_1 species has a maximum partial pressure of $10^{-7.17}$ atm for sulfur saturation at this temperature (Mills, 1974).

mately that compatible with the troilite + iron buffer the dominant sulfur vapor species is the monomer, S. As total sulfur pressure is increased the S_2 species becomes dominant, then at pressures approaching the condensation of liquid sulfur S_6, S_7 (in a limited temperature range), and then S_8 become successively the most abundant sulfur species. Also present in the relatively high-pressure region are other sulfur species, principally S_3, S_4, and S_5, and some studies have shown the existence of even higher polymers in the vapor, although they never become the most abundant species. A summary of sulfur vapor polymerization (assembled from the literature by J. L. Haas, Jr.) is given in Figures 7.6 and 7.7 wherein the partial pres-

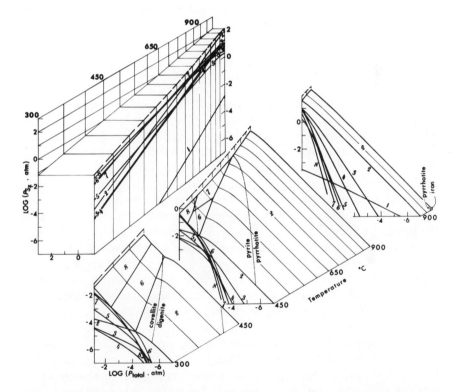

Fig. 7.7 Exploded isometric drawing of the covariation of the partial pressure of individual sulfur species for pure sulfur as functions of temperature and total pressure. The results of Rau et al. (1973b) are used for all sulfur species except S_1 for which the data of Mills (1974) are used. The univariant curves for covellite + digenite, pyrite + pyrrhotite, and pyrrhotite + iron are plotted to show why S_2 is the most convenient reference state for sulfur. The italic numbers within the diagram indicate the number of sulfur atoms per molecule for the gaseous species. The heavy, dashed curve is for total sulfur vapor pressure along the boiling curve. Except for the P_{S_x}-T surface on which the speciation of sulfur molecules in boiling sulfur is indicated, the entire upper left block represents unattainable conditions.

sures of the dominant sulfur species are shown as functions of total pressure and temperature. Note in Figure 7.6 that there are very large discrepancies in the relative stabilities of species such as S_4, whereas S_2, S_6, and S_8 are more consistent. Berkowitz (1965), Meyer (1976), and Rau et al. (1973b) present excellent reviews of the subject.

Estimates of the molecular composition of sulfur vapor are derived from three types of measurements: (1) pressure-volume-temperature studies that determine the average number of atoms per molecule from measurements of the pressure-temperature-volume relationships (Braune et al., 1951; Rau et al., 1973b), (2) mass spectrometric studies of the molecular constituents (Detry et al., 1967), and (3) calculations based on spectroscopic measurements that evaluate entropies for S_2, S_6, and S_8. Although the calculations have areas of agreement, discrepancies (See Figure 7.6) of several origins remain.

The System Iron-Sulfur

The most interesting and most studied binary sulfide system, both in geology and in metallurgy, is iron-sulfur. The system is shown in Figure 7.8. One or more forms of iron sulfide are found in most ore deposits and the iron sulfides are, by a wide margin, the most abundant of the sulfides; in fact, pyrite and pyrrhotite are the only sulfides sufficiently abundant to be called rock-forming minerals. The majority of multicomponent systems (other than those involving sulfosalts) studies to date are built on and around iron-sulfur as one of the bounding binaries. Moreover, the use of the pyrrhotite indicator for a_{S_2} (Toulmin and Barton, 1964) has provided thermodynamic information on many, diverse sulfide systems. Extensive discussions of iron sulfides have been made by Ward (1970) and Power and Fine (1976). S. D. Scott (in Craig and Scott, 1974) presents a comprehensive synthesis of the phase relations and mineralogy based principally on the studies of R. A. Arnold, N. Morimoto, H. Nakazawa, and S. A. Kissin.

The stability of the various members of the pyrrhotite family is not yet well understood; extensive metastability is known to occur, and this, together with the tendency for some researchers to rely too heavily on insensitive diffractometer X-ray methods for phase characterization, has contributed to a confused state of knowledge of the low-temperature phase relations. The phase relations above 300°C are straightforward and generally well accepted; but below 300°C there is a plethora of phases and much uncertainty regarding the condition of true equilibrium. In addition to the 10 binary iron-sulfur phases that may possibly have low-pressure stability fields (shown in Figure 7.8), there are at least 12 other iron sulfides, all of which we believe to be metastable or unstable: Fe_3S (Osaka and

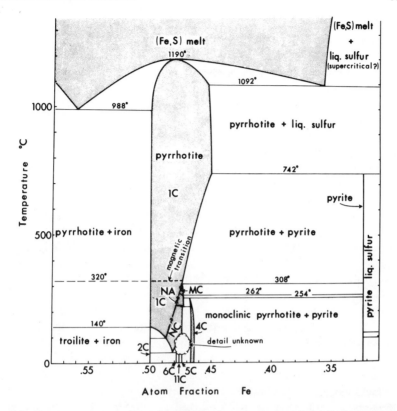

Fig. 7.8 Temperature-composition diagram for the central part of the Fe-S system. The designations of the nine structure types of pyrrhotite discovered to date are indicated using the nomenclature of Nakazawa and Morimoto (1971). The diagram is constructed from information given by Hansen and Anderko (1958), Arnold (1962, 1971), Scott (in Craig and Scott, 1974), Toulmin and Barton (1964), Yund and Hall (1968), Kullerud and Yoder (1959), Barton (1969), Nakazawa and Morimoto (1971), and Morimoto et al. (1975a,b). The question marks indicate uncertainty whether the transitions are first order (and therefore should show a two-phase area) or are of higher order with a continuous phase field. Some phases such as monoclinic pyrrhotite may be metastable, see text. One-phase fields are shaded.

Nakazawa, 1976), iron-pentlandite (Nakazawa et al., 1973), mackinawite (Berner, 1967), high-pressure FeS (Taylor and Mao, 1970), iron-sphalerite (De Medicis, 1970; Takeno et al., 1970), iron-wurtzite (Skinner, unpublished), anomalous pyrrhotite (Clark, 1966a), greigite (Berner, 1967), smythite (Berner, 1967), gamma Fe_2S_3 (Yamaguchi and Wada, 1973), monoclinic or tetragonal Fe_2S_3 (Schrader and Pietzsch, 1969), and marcasite (Rising, 1973; Grønvold and Westrum, 1976). The metastable phase relations should not be discarded lightly; they can be reproducible, and may be potentially even more useful than stable ones in deciphering some aspects of geologic history, particularly where low temperatures are involved.

Nakazawa and Morimoto (1971) have identified superstructure types having nonintegral multiples of the subcell. The nonintegral pyrrhotites, MC, NA, and NC, have fairly well-defined regions of existence in Figure 7.8 but the nature of their field boundaries is obscure. Some alternatives have been shown by Nakazawa and Morimoto (1971), but it has not yet been demonstrated whether the transitions (which are orderings of vacancies) are first order; if they are of higher-than-first order, there may be continuous transitions rather than two-phase regions between the 1C, MC, NA, and NC phases. The completely ordered low-temperature phases, $Fe_{11}S_{12}$ (6C), $Fe_{10}S_{11}$ (11C), Fe_9S_{10} (5C), and Fe_7S_8 (4C, or monoclinic) are shown to be stable in Figure 7.8, but it is not known whether or not this is correct. An alternative, showing monoclinic pyrrhotite as a metastable phase (in agreement with Hall and Yund, 1966) is given in Figure 7.11b.

Smythite is a member of the pyrrhotite family and its composition seems to range from Fe_3S_4 to at least $Fe_{3.25}S_4$ (Taylor, 1970a); whereas greigite is a sulfospinel whose composition falls in the interval between Fe_3S_4 and Fe_2S_3 in a solid solution similar to that of magnetite-maghemite. Taylor also suggests an inversion in smythite occurring, for $Fe_{3.25}S_4$ composition, at $75 \pm 15°C$.

The composition of high-temperature (1C) pyrrhotite and its quench products can be accurately determined by X-ray powder diffraction measurements; as summarized by Yund and Hall (1969) the interrelationship between atom percent iron and the (102) spacing is

$$X = 45.212 \pm 72.86(d_{102} - 2.0400) + 3.115(d_{102} - 2.0400)^2$$

Barton (unpublished) has demonstrated further that lead, zinc, silver, arsenic, antimony, bismuth, molybdenum, tungsten, tin, silicon, titanium, and oxygen enter pyrrhotite in such small amounts that their presence does not affect the determination of composition by X-ray. Appreciable quantities of nickel, cobalt, manganese, or selenium can enter the structure, and their presence in quantity does invalidate the above X-ray method. Skinner and Luce (1971a) showed that manganese expands the (102) spacing appreciably, but Arnold and Reichen (1962) demonstrated that, if nickel + cobalt + copper is less than 0.6% (by weight), there is no measurable effect on the (102) spacing. Although copper is known to enter pyrrhotite and to shift the (102) spacing (Yund and Kullerud, 1966), no significant spacing shift occurs below approximately 550°C for pyrrhotite in equilibrium with pyrite (Barton, unpublished).

The solvus along which high-temperature pyrrhotite coexists with pyrite has been carefully determined (Arnold, 1962; Yund and Hall, 1969) down to 290°C. This temperature range crosses a magnetic transititon in the high-temperature pyrrhotite field at about 325°C without regard for composition. Although the magnetic transition (neglected in Table 7-2) slightly modifies the thermodynamic properties of the high-temperature pyrrhotite, the phase remains stable down to 98 ± 4°C (Nakazawa and Morimoto, 1971). Immediately following Arnold's (1962) work the pyrite-pyrrhotite solvus was extensively applied as a geothermometer; later recognition of the many low-temperature complexities, plus the fact that pyrrhotite reacts rapidly and thus destroys most of its high-temperature heritage, negates application of this otherwise attractive technique.

Pyrite does not depart measurably from stoichiometric FeS_2 (Kullerud and Yoder, 1959) and melts incongruently to pyrrhotite + almost pure liquid sulfur at 742 ± 1°C (Barton, 1969). In the absence of vapor the effect of pressure is to raise the decomposition temperature by 14°C/kb (Kullerud and Yoder, 1959; Sharp, 1969). The invariant point at 742°C is not shifted measurably by the presence of lead, zinc, molybdenum or bismuth (Barton, unpublished), and even copper, which is soluble in pyrrhotite to the extent of almost 3 atom % (Yund and Kullerud, 1966) lowers the point only to 739°C (Roseboom and Kullerud, 1958).

Figures 7.9 and 7.10, which are based on the work of Toulmin and Barton (1964) and modified slightly using the refinements of Scott and Barnes (1971), show, respectively, the variation in composition of pyrrhotite and the activity of FeS (relative to stoichiometric FeS at the same temperature) as functions of temperature and activity of sulfur. This work provides a useful starting point for the investigation of several more complex systems. For example, the use of pyrrhotite as an indicator of sulfur activity in laboratory experiments has permitted the determination of several useful reactions in the systems iron-arsenic-sulfur, iron-antimony-sulfur, iron-bismuth-sulfur, and copper-iron-sulfur; and the knowledge of the activity of FeS has led to a much fuller understanding of the significance of the iron content of sphalerite in the system iron-zinc-sulfur.

Pyrite Disequilibrium

Pyrite has proven to be an extremely recalcitrant phase to study in the laboratory because it tends to remain inert while other phases (e.g., pyrrhotite, chalcopyrite, galena) react. In particular, pyrite can be very difficult to nucleate (see Springer et al., 1964). The nucleation problem is known to be shared by arsenopyrite (Kretschmar and Scott, 1976; Clark, 1960) and may be expected to be especially acute with other refractory sulfides

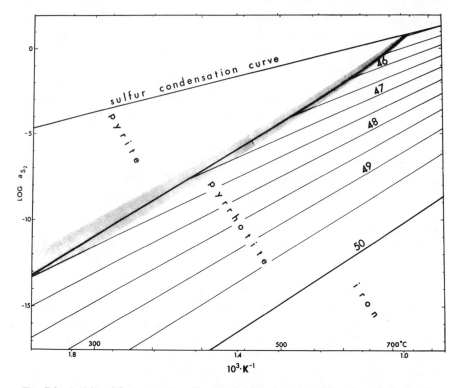

Fig. 7.9 Activity of S_2-temperature diagram showing the composition of pyrrhotite in atom percent iron according to Toulmin and Barton (1964). The shaded area is a generous estimate of the uncertainty in locating the pyrite + pyrrhotite univariant curve. The magnetic transition in pyrrhotite in the vicinity of 320°C has been ignored in this figure. See discussion in text.

(the iron-cobalt-nickel tri- and di-arsenides, nickel and cobalt disulfides, etc.) and perhaps with phases having complex structures (e.g., cylindrite). Nevertheless, the vast majority of kinetic problems in sulfide mineral assemblages concerns pyrite, and we will discuss pyrite as a model for other disequilibria.

A convenient way to quantify the disequilibrium of pyrite is to construct a diagram in which the activity of FeS_2 is a variable, as in Figure 7.11. By plotting the stability fields of the various copper-iron sulfides, it is easily seen that the common assemblage covellite + chalcopyrite is metastable and owes its existence to supersaturation with respect to pyrite. It is probable that some iron-sulfur phases such as marcasite, smythite, greigite, and some of the sulfur-rich pyrrhotite family originate only in the pyrite-supersaturated region, a point made previously by Berner (1967). If marcasite is,

SPECIFIC SULFIDE SYSTEMS 363

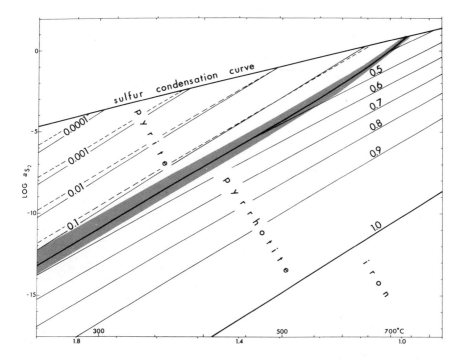

Fig. 7.10 Activity of S_2-temperature diagram for the Fe-S system showing the activity of FeS in the pyrrhotite and pyrite fields. The shaded area shows a generous estimate of the uncertainty in the location of the pyrite + pyrrhotite curve. Curves in the pyrrhotite field are from Toulmin and Barton (1964). The solid lines in the pyrite field are calculated from the free energy of pyrite given in Table 7-2; the dashed lines are from the free energy of pyrite given by Scott and Barnes (1971); see discussion in text.

in fact, slightly iron deficient with respect to pyrite, the marcasite curve should be rotated clockwise slightly, but there is no evidence that marcasite is ever more stable than pyrite, and thus the entire marcasite curve probably remains in the pyrite-supersaturated region (Rising, 1973). In the higher temperature part of the range illustrated by Figure 7.11a, the covellite + chalcopyrite assemblage is preempted by a field of idaite, which, below approximately 223°C, is metastable with respect to covellite + digenite + pyrite. As the temperature falls further, the metastable idaite field shrinks to a line coincident with the covellite + chalcopyrite boundary and then disappears at approximately 200°C, according to the results of Sugaki, 1957; thus, on cooling, idaite becomes metastable with respect to pyrite + digenite + covellite, then with respect to another metastable assemblage, covellite + chalcopyrite; there is little wonder that idaite is not a common mineral.

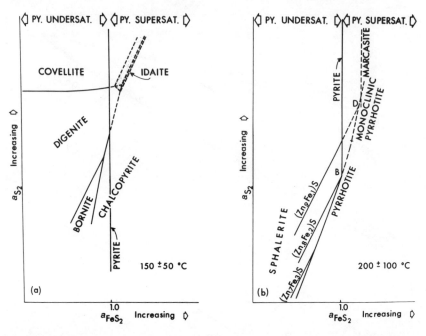

Fig. 7.11 Schematic isothermal diagrams of activity of S_2 vs activity of FeS_2 showing the influence of pyrite metastability on phase relations. (a) The system Cu-Fe-S showing the location of the covellite + chalcopyrite assemblage in the pyrite-supersaturated region. The hypothetical field (metastable) for idaite is shaded. See text. (b) The system Fe-Zn-S showing how a range of sphalerite compositions may originate in pyrrhotite-bearing assemblages if supersaturation in pyrite is permitted. Typical sphalerite isopleths and hypothetical curves for metastable monoclinic pyrrhotite and marcasite are also shown. Point B refers to the stable assemblage pyrite + pyrrhotite + sphalerite noted in Figures 7.13 and 7.14; point D is a typical metastable sphalerite + monoclinic pyrrhotite solvus.

Careful studies of some ores that have been involved in very slow cooling (Sulitjelma, Norway, by K. L. Rai, personal communication, 1975; Sullivan, B. C., by Ethier et al., 1976) reveal the puzzling situation in which sphalerite from pyrrhotite-bearing, pyrite-poor assemblages contains *less* FeS in solid solution than sphalerite from pyrite + pyrrhotite assemblages, a fact impossible to reconcile with the equilibrium diagram for the iron-zinc-sulfur system. This observation can be explained by assuming that sphalerite is still able to adjust its composition by reaction with pyrrhotite in conditions under which pyrite is no longer able to nucleate, although it may continue to grow on previously existing nuclei. Thus pyrite + pyrrhotite + sphalerite assemblages on cooling remain closer to the equilibrium position (point B in Figure 7.11b), while nearby pyrite-free assemblages undergoing identical physical conditions of reaction move into the pyrite-supersaturated field as temperature falls (point D in Figure 7.11b).

A quantitative measure of the degree of supersaturation is not easily obtained, but it appears that less than one log a_{FeS_2} unit would be sufficient to produce the results shown in Figure 7.11. Grønvold and Westrum (1976) show that marcasite is about a kilocalorie less stable than pyrite so that, at 200°C, a supersaturation with respect to pyrite of 0.5 log unit could permit marcasite to form. By comparison, Fournier (1973) shows that amorphous silica at approximately 75°C is one log a_{SiO_2} unit supersaturated with respect to quartz, so the degree of supersaturation we propose is not without precedent.

Supersaturation-related phenomena are not restricted to natural systems, but are frequently observed (although not always recognized as such) in the laboratory. The wide range in the reported maximum stability temperature for monoclinic pyrrhotite is probably due to the problem of pyrite-supersaturated experiments; and, in our opinion, adequate demonstration of equilibrium between pyrite and monoclinic pyrrhotite, at *any* temperature, has yet to be published. Sugaki et al. (1975) nicely document idaite + chalcopyrite assemblages at 300 and 350°C, yet digenite-bornite + pyrite clearly is the assemblage compatible with the thermodynamic data. Metastable equilibria may well be even more useful than stable equilibria for the interpretation of many geologic processes that occur at lower temperatures.

The System Iron-Zinc-Sulfur

The most important ternary system for our purposes is iron-zinc-sulfur, because it encompasses the iron-sulfur binary and the highly informative mineral, sphalerite. Sphalerite promises to be the most useful mineral for unraveling the environment of ore deposition because it has extremely desirable quenching properties, is capable of a wide range of substitution of iron for zinc, is widely distributed in natural environments, and is nearly ideal for both optical and X-ray examination techniques. The iron-zinc-sulfur system is a comparatively simple ternary system as is evident from the isothermal diagrams given in Figure 7.12. The following discussion is based predominantly on an extensive experimental study by Barton and Toulmin (1966) modified to include the more recent studies of Boorman (1967) and Scott and Barnes (1971).

The galvanized-iron alloy phases occurring along the iron-zinc binary are of little or no interest to us, being separated from the mineralogically interesting part of the diagram by the nearly binary join, Fe-ZnS. The alloy phases will not be discussed further. Sphalerite, wurtzite, and their mixed-

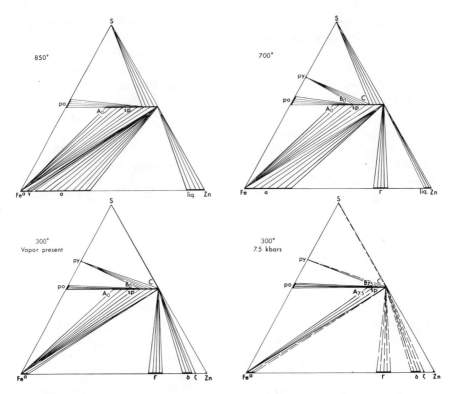

Fig. 7.12 Isothermal phase relations in the system Fe-Zn-S. One-phase regions are shown as heavy bars. Vapor is present throughout the first three diagrams, but the diagram in the lower right has points A and B located for a pressure of 7.5 kb (the remainder of the diagram is shown as if pressure had no effect, although the tielines are dashed to suggest uncertainty). Abbreviations: sp = sphalerite; py = pyrite; po = pyrrhotite. Points A, B, and C correspond to similarly labeled curves in Figures 7.13 and 7.14.

layer polytypes are the only crystalline phases stable along the zinc-sulfur binary join, and the extensions of their phase fields into the ternary region as iron substitutes for zinc in their structures generate the only stable crystalline single-phase regions within the ternary.

Sphalerite departs from a 1:1 metal/sulfur ratio by less than 0.1 atom %. Pyrrhotite dissolves less than 0.1% ZnS and pyrite appears to dissolve even less ZnS than pyrrhotite. These factors simplify experimental work in the iron-zinc-sulfur ternary, and it has been possible to determine the pyrrhotite composition by X-ray diffraction methods, to determine the sphalerite composition by X-ray diffraction or by electron microprobe, to calculate the activities of S_2 and FeS from the pyrrhotite composition, and thereby, to work out the complete thermochemistry and extent of the sphalerite field.

In the zinc-sulfur binary system there is an unresolved difference of opinion as to the true equilibrium relationships between sphalerite, the numerous ZnS polytypes, and wurtzite. The authors' experience, the available thermodynamic data, and a significant fraction of the dozens of other references, indicate that wurtzite is stable only at high temperatures. Allen and Crenshaw (1912) give 1020°C as the transition temperature and Moh (1975a) has refined this to 1031°C for sulfur-rich zinc sulfide and 1013°C for zinc-rich sulfide. On the other hand, there is a sizable body of data that has been interpreted to support the view advocated by Scott and Barnes (1972) and Scott (in Craig and Scott, 1974) that low activities of sulfur result in nonstoichiometry of ZnS and that, as a consequence, wurtzite is stable down to room temperature. Stoichiometry certainly does affect the positions of phase transitions in general, but the range in composition of zinc sulfide appears too small to have such a large effect on a transition; moreover, reversed experiments at approximately 900°C (Skinner and Barton, 1960) show that sphalerite can be in equilibrium with either free sulfur or free zinc thus precluding nonstoichiometry as the critical control for wurtzite-sphalerite stability. It seems probable (to us, but not to Scott or Barnes) that the thermal gradients associated with all of the low-temperature, wurtzite-yielding experiments permitted kinetic factors of crystal nucleation and growth to sidestep the thermodynamic requirement that, at constant temperature, a stable phase cannot convert spontaneously to a less stable one. Whether the polytypes do or do not have true stability fields is an open question; nevertheless, they occur frequently and in great diversity, both in the laboratory and in nature. We suggest that most natural occurrences of wurtzite or polytypes thus originate as metastable precipitates, perhaps promoted, but not truly stabilized, by "impurities" such as CdS, MnS, or ZnO in solid solution; possible exceptions are (Fe, Zn)S in meteorites, accessory or xenolithic minerals in rapidly quenched igneous rocks, or the presence of high concentrations (more than 10 mole %) of CdS or MnS in solid solution, which drastically lower the transition temperatures.

Because the sphalerite and wurtzite compositions fall essentially on the FeS-ZnS join, the compositions of sphalerites involved in the ternary phase relations can be conveniently represented on a composition-temperature plane along the FeS-ZnS join, as shown in Figure 7.13, even though the compositions of the coexisting iron sulfides do not fall on the join. The maximum FeS-content of sphalerite occurs where the activity of FeS is a maximum. This happens where the pyrrhotite coexisting with sphalerite is stoichiometric FeS. The true FeS-ZnS binary solvus as shown by curve A is

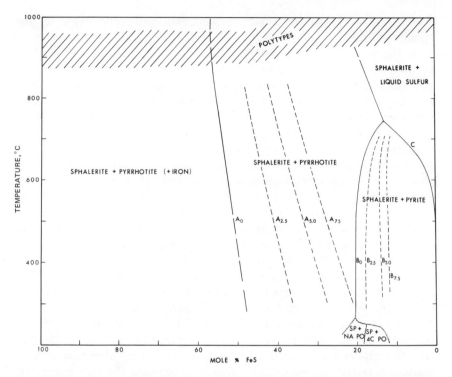

Fig. 7.13 Composition of sphalerite in the Fe-Zn-S system projected into the FeS-ZnS plane for several total pressures. Curves A, B, and C correspond to similarly labeled points in the isotherms in Figure 7.12. The subscripts indicate the pressure in kilobars taken from the experiments and calculations of Schwartz et al. (1975), Scott (1973), Scott and Barnes (1971), and Barton and Toulmin (1966); the O subscript indicated vapor-present experimental conditions. Speculative, low-temperature relations involving sphalerite (SP), "ordered pyrrhotite" (NC PO) and monoclinic pyrrhotite (4C PO), taken from Scott and Kissin (1973), are shown by the dotted lines. The MC and NA pyrrhotites also deserve representation here, but are not included in Scott and Kissin's diagram.

very steep and is sensibly linear with temperature, passing through 56 ± 1 mole % FeS at 850°C and 52 ± 2 mole % FeS at 580°C.

The very steep slope of curve A_0 (Figure 7.13) indicates that the sphalerite solvus in the FeS-ZnS binary is of little or no use as a geothermometer, being relatively insensitive to temperature. Also, the similar steep slope of the wurtzite solvus and its slight compositional displacement from that of sphalerite, indicates that the difference between the cubic and hexagonal packings of the sulfur atoms affects the FeS solubility in (Zn, Fe)S very little. For this reason, it is unlikely that regions of random hexagonal packing in sphalerite crystals, as found by Smith (1955), cause any significant alterations in the sphalerite phase fields.

Curve B_0 of Figure 7.13 projected from the ternary onto the FeS-ZnS binary is the composition curve for sphalerites in equilibrium with pyrite + pyrrhotite. Curve C is similarly a ternary projection and is the composition curve for sphalerites in equilibrium with pyrite + liquid sulfur. Curves B_0 and C each define isobarically univariant assemblages. The assemblage sphalerite + pyrite + liquid sulfur is rare in nature, but it does serve as a limit for pyrite-sphalerite assemblages.

Sphalerite + pyrrhotite + pyrite (curve B_0) is a common assemblage in ores, and several studies (e.g., Boorman, 1967; Scott and Barnes, 1971; Scott and Kissin, 1973) have attempted to refine its position at lower temperatures (below 580°C) than Barton and Toulmin (1966) were able to attain equilibrium. It is now clear that there is little change in sphalerite composition along this curve from approximately 550°C down to perhaps 300°C, or even lower, due to a fortuitous, mutually compensating combination of circumstances: the very slight variation in the activity coefficient for FeS in sphalerite with temperature and composition ($\gamma = 2.4 \pm 0.1$ for FeS contents up to 20-30%); the variation in the composition of pyrrhotite in equilibrium with pyrite (Figure 7.9); and the variation in the activity of FeS in pyrrhotite with temperature and pyrrhotite composition (see Figure 7.10). Scott and Kissin (1973) have attempted to determine the curves relating sphalerite composition to the apparently metastable univariant reactions involving smythite and monoclinic pyrrhotite at low temperatures (Figure 7.13), but uncertainties in the FeS phase diagram and severe experimental difficulties have precluded definitive results.

Sphalerite compositions in the divariant assemblages pyrite + sphalerite and pyrrhotite + sphalerite are useful in that they provide a continuous series of sliding-scale univariant curves covering the entire range of a_{S_2} versus temperature of interest in ore deposits. In the pyrite field, the sphalerite composition, $X_{(py)}$, is calculated quite rigorously from the relation FeS (in sphalerite) + $\frac{1}{2} S_2$ = FeS_2. Solving for sphalerite composition [using 2.4 as the activity coefficient for FeS in sphalerite, the data of Toulmin and Barton (1964) for FeS_2 and of Richardson and Jeffes (1952) for FeS] yields $\log X_{(py)} = 7.16 - 7730 \cdot T^{-1} - \frac{1}{2} \log a_{S_2}$. A reasonable alternative is that of Scott and Barnes (1971) who obtained $\log X_{(py)} = 6.65 - 7340 \cdot T^{-1} - \frac{1}{2} \log a_{S_2}$ by using 2.3 for the activity coefficient of FeS in sphalerite, their own revised value for FeS_2, and Robie and Waldbaum's (1968) for FeS. The two equations are in agreement within experimental uncertainties in the range 400-700°C (see Figure 7.14), but they diverge by approximately $10^{0.5}$ at 250°C and by $10^{1.0}$ at 100°C. This divergence is a good illustration of the danger of extrapolating even "consistent" experimental data.

Scott and Barnes (1971) have made a statistical treatment of the experimental data relating the mole percent FeS in sphalerite in equilibrium with

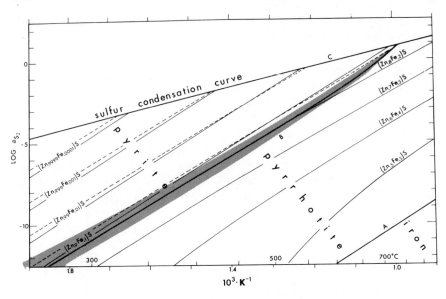

Fig. 7.14 Activity of S_2-composition diagram for the system Fe-Zn-S, showing the composition of sphalerite in equilibrium with an iron-bearing phase. Sphalerite isopleths within the pyrrhotite field are from Scott and Barnes (1971). Dashed lines in the pyrite field are sphalerite isopleths from the equation of Scott and Barnes (see text); the solid lines are from Barton and Toulmin (1966). The uncertainty in the position of the pyrite + pyrrhotite univariant curve is shown as a shaded area. The labeled curves correspond to similarly labeled curves and points in Figures 7.12 and 7.13.

pyrrhotite, $X_{(po)}$, to T (Kelvin), and activity of sulfur, a_{S_2}. They obtained the following relation:

$$X_{(po)} = 72.26695 - 15900.5 \cdot T^{-1} + 0.01448 \log a_{S_2} - 0.38918 \cdot 10^8 \cdot T^{-2}$$
$$- 7205.5 \cdot T^{-1} \cdot \log a_{S_2} - 0.34486 (\log a_{S_2})^2$$

This equation reproduces the experimental data with a standard deviation of 1.7 mole % FeS (favorably comparable with the uncertainties of the individual points themselves) and has been used to construct Figure 7.14. Unfortunately, the equation does not give reasonable extrapolations to temperatures below approximately 350°C or for log a_{S_2} values below approximately −12, nor does it combine with either of the equations for the FeS content of sphalerite within the pyrite field to accurately reproduce the univariant curve for the pyrite + pyrrhotite (+S_2 vapor) assemblage. Thus, an area of uncertainty (shaded in Figure 7.14) remains in the position of the pyrite + pyrrhotite curve in terms of a_{S_2}, even though we do know the composition of pyrrhotite and sphalerite along the curve fairly well. We have continued to use our previous values for iron-sulfur in Table 7-2 because the alternatives have not yet been proven clearly superior. A comprehensive reevaluation of the data is obviously warranted, but it must extend all the way to a reevaluation of the polymerization of sulfur vapor (see Figures 7.6 and 7.7) and is far too complex for treatment here. The discrepancies are generally within the uncertainties of the various experimental methods and in no way discount the use of the iron content of sphalerite as one of the most valuable guides to the chemical environment of ore deposition.

Because the iron content of sphalerite remains nearly constant along the pyrite + pyrrhotite curve, B_0, below 550°C and changes only slightly with temperature for curve A_0 (Figure 7.13), neither of the curves offers much promise as a geothermometer. However, FeS undergoes a large decrease in volume as it changes from the sphalerite to the pyrrhotite structure, which means that curves A_0 and B_0 are sensitive to pressure and thereby constitute potential geobarometers. Barton and Toulmin (1966) calculated the effects of pressure, and Scott and Barnes (1971) and Scott (1973) refined the calculation and demonstrated experimentally that both curves do indeed provide useful geobarometers (Figure 7.13). Schwarcz et al. (1975) have applied curve A to meteorites, and numerous applications of curve B have been made (Scott, 1976). The sphalerite geobarometer is very attractive from the point of view of precision and potential wide applicability, but we caution the would-be user to take great pains to demonstrate that both iron and pyrrhotite for curve A, or pyrite and pyrrhotite for curve B, were truly in equilibrium with the sphalerite analyzed and that postequilibration changes have not modified the sphalerite composition or the mineral assemblage. Except for large amounts of "impurities" in solid solution, as may occur in the pyrrhotite of some nickel-rich ores, reequilibration is a far more serious problem than impurities. If the temperatures of equilibration are low (less than, perhaps much less than, 300°C), curve B probably tends to lower iron contents (Scott and Kissin, 1973) and a false impression of high pressure might be anticipated. The low-temperature region is also subject to pyrite supersaturation, as discussed under the iron-sulfur system, which also would give an illusion of high pressure.

Scott and Barnes (1971) have described iron-rich patches in sphalerite grown hydrothermally below 450°C. Similar patches are noted in some natural sphalerite, but not all color in natural sphalerite is due to iron. The patches apparently owe their origin to sector zoning wherein certain crystallographic planes accept different quantities of "impurity" and the growth rate is too rapid for solid-state diffusion to homogenize or equilibrate the compositions of the sectors originating from different crystal faces; thus the metastable compositions become frozen in.

Wiggins and Craig (1975) have examined the copper-iron-zinc-sulfur system at high temperature and found up to 6.5 mole % CuS in sphalerite at 800°C (in the most sulfur-rich assemblage); the amount of copper in sphalerite decreases sharply with temperature. This observation is in agreement with the results of Moh (1975a) and Toulmin (1960). The cuprian sphalerite and the possible zinc analogue of idaite (Cu_3ZnS_4) described by Clark and Sillitoe (1970) have not proven stable in higher temperature laboratory studies and are, therefore, tentatively designated metastable.

The System Copper-Sulfur

The copper-sulfur system is characterized by extremely rapid reaction rates for topotactic reactions among members of the chalcocite-digenite family at low temperature, yet for some reconstructive reactions there appear to be sizable kinetic barriers. This produces not only phase changes with slight changes in composition, temperature, and pressure but, also, a capacity to form and retain metastable phases. As the temperature is raised above 100°C the multiplicity and preserved metastability of the phases diminishes; yet even above 400°C, the reconstructive inversion between hexagonal chalcocite and high chalcocite is so sluggish that the transition temperature for Cu_2S composition has been established only approximately despite considerable efforts to measure it more accurately.

Except for certain low-temperature, primarily supergene, processes, the copper-sulfur system is geologically important only as a bounding system for the intensely studied, and highly complex copper-iron-sulfur system.

Our present understanding of the copper-sulfur system (Figure 7.15) relies heavily on the detailed studies of Roseboom (1966), Rau (1967), Morimoto and Koto (1970), and Potter (1974, 1975, and personal communication, 1976). The high-temperature part of the diagram is dominated by the field of high digenite, a face-centered cubic phase having the capacity for a wide range of substitution in the copper-sulfur binary (and for other metals as well, e.g., silver and iron). From the studies of Rau (1967) and Roseboom (1966), Barton (1973) has calculated activity of Cu_2S for the high-digenite solid solution. Natural systems usually contain significant amounts of iron that can enter high digenite, forming a solid solution that extends beyond bornite, so that neither high digenite, nor its low-temperature descendants, are commonly encountered as primary phases in ores. In contrast to high digenite, covellite has a fixed, nearly impurity-free, composition $Cu_{1.000\pm0.0005}S$ at low temperature (according to Potter, personal communication, 1975). The only other binary phases known to be stable at high temperature are hexagonal chalcocite (below 435°C) and CuS_2, a pyrite-type phase known to be stable only at high pressures (Taylor and Kullerud, 1972).

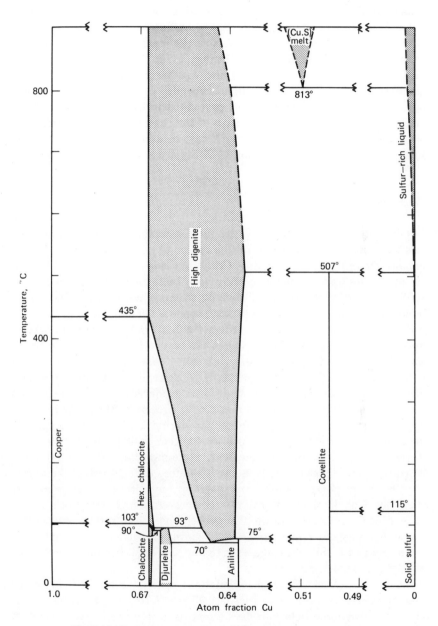

Fig. 7.15 Temperature-composition diagram for the system Cu-S.

In contrast with the simple phase relations at high temperatures, the copper-sulfur binary is exceedingly complex below 100°C. Using a combination of a sensitive electrical potential technique with microprobe analysis and Guinier X-ray techniques that enable the measurement and characterization of subtle phase differences, Potter (1974, 1975, and personal communication, 1976) has established the equilibrium diagram and shown quantitatively that both types of blaubleibender covellite (Moh, 1971), low digenite, tetragonal chalcocite, and protodjurleite are metastable at 25°C and 1 atm. The various copper sulfides of the chalcocite-digenite family are not easy to distinguish optically, by X-ray, or even by casual microprobe study, but criteria for identification by X-ray are given by Potter and Evans (1976).

The System Copper-Iron-Sulfur

Copper-iron-sulfur is probably the most complicated and certainly the most intensively studied ternary sulfide system of mineralogic interest, yet it remains one of the most confusing. Craig (in Craig and Scott, 1974) lists 18 ternary phases and two more recently have been suggested (Filimonova et al., 1974); most of them are well-defined minerals and fully half have been found since 1960 when the microprobe became a mineralogic tool.

The copper-iron-sulfur system offers the best and the worst in several aspects of sulfide studies. Some phases are easy to identify, whereas others can scarcely be distinguished from their kin. Some phases react so slowly in the laboratory that it is impractical to determine equilibrium assemblages directly, but other phases cannot be quenched. Tremendous ranges in compositions of solid solutions might offer precise geothermometers, but they fail to preserve those compositions during cooling. The topotactic nucleation and growth of some phases together with possible metastable ranges of composition of solid solutions, make the problems of determining the equilibrium phase relations in the laboratory difficult and make uncertain the application of equilibrium diagrams (even if we did know exactly what they look like) to natural assemblages. Although temperatures for many specific reactions within the copper-iron-sulfur system are known (Table 7-1), a great many more remain uncertain; indeed, the configuration of the phase diagram (Figures 7.16 and 7.17) still has many problems, especially at low temperature.

The main features of the phase relations can be described in terms of three principal solid solution series (see Figure 7.16). The first, the extention of the pyrrhotite field into the ternary, is quite limited; even at 700°C Yund and Kullerud (1966) found a maximum of only 3 atom % copper in pyrrhotite. The other two fields, that of bornite and that of the chalcopyrite family, are located in the vicinity of the join Cu_2S-Fe_2S_3; that

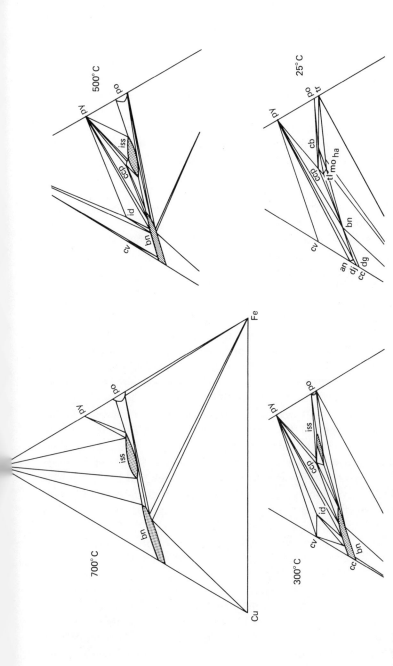

Fig. 7.16 Isothermal, condensed phase relations for the Cu-Fe-S system. The shaded areas are one-phase fields. The heavier-lined, sulfur-facing sides of the bornite, intermediate solid solution, and pyrrhotite phase fields are the surface onto which other phase fields are projected in Figure 7.17. Abbreviations: cv = covellite; cc = chalcocite; dg = digenite; dj = djurleite; an = anilite; id = idaite; bn = bornite; ccp = chalcopyrite; tl = talnakhite; mo = mooihoekite; ha = haycockite; cb = cubanite; po = pyrrhotite (several varieties included as a single point in the 25°C isotherm); tr = troilite; py = pyrite; iss = intermediate solid solution.

375

they represent cuprous-ferric phases with some cupric and ferrous substitution is supported by Mössbauer measurements on bornite, chalcopyrite, "cubic chalcopyrite", and cubanite (Marfunin and Mkrtchyan, 1967). The basic structural unit of all of the bornite-digenite and chalcopyrite-family crystal structures is that of the sphalerite-like, face-centered-cubic, sulfur subcell, 5.2–5.6 Å on an edge, with tetrahedrally coordinated metal atoms; slight distortions of the subcell, stacking of several subcells, shifts of the metal atoms from tetrahedral to nearly triangular coordination, and different kinds of metal ordering distinguishes the various phases. In some phases, for example cubanite and chalcocite, the sulfur framework has the hexagonal-closest-packed arrangement similar to wurtzite. There must also be a significant bonding difference between bornite, chalcopyrite, and the intermediate solid solution inasmuch as they do not form complete solid solutions at high temperatures despite close similarity in subcell size and extensive solid solution within the phases.

Figure 7.17 is an attempt to summarize many of the aspects of the sulfur-rich parts of the copper-iron-sulfur system, but it must be emphasized that this diagram is somewhat speculative. The projection used is similar to that used in Figure 7.13 for sphalerite in the iron-zinc-sulfur system, but the copper-iron-sulfur system is much more complex. Data are from a variety of sources, all of which are cited elsewhere in this chapter, except for an unpublished study of idaite by Shunzo Yui (personal communication, 1974). Projected onto the sulfur-facing phase fields of the solid solutions are the stability fields of the more sulfur-rich phases than the solid solutions; these sulfur-rich phases are covellite, idaite, chalcopyrite, cubanite, pyrite, and liquid sulfur. The lines separating the fields are the traces of the corners of the three-phase (plus vapor) fields seen on the isothermal sections in Figure 7.16. A comparison between Figures 7.16 and 7.17 should clarify most questions about the construction of this projection. Fukuchilite has been omitted because it is not believed to be stable except at high pressure. X-bornite is omitted because it is believed to be metastable. The bornite inversion is given as 228°C (Morimoto and Kullerud, 1961), yet it is probable that there are, indeed, two transitions, at 200 and 270°C (Pankratz and King, 1970; Sugaki and Shima, 1965). The bornite inversion at 228°C is shown as if high-temperature bornite could not be more rich in sulfur than low-temperature bornite at the inversion temperature, an assumption that may not prove valid.

Below 532°C the iron-rich extremity of bornite is assumed to be that in equilibrium with chalcopyrite and intermediate solid solution (iss) and the copper-rich extremity of iss is assumed to be that in equilibrium with chalcopyrite and bornite. On the other hand, the iron-rich extremity of iss is drawn a few percent more iron-rich than that in equilibrium with pyrite, and the copper-rich extremity of pyrrhotite is more copper-rich than that

SPECIFIC SULFIDE SYSTEMS

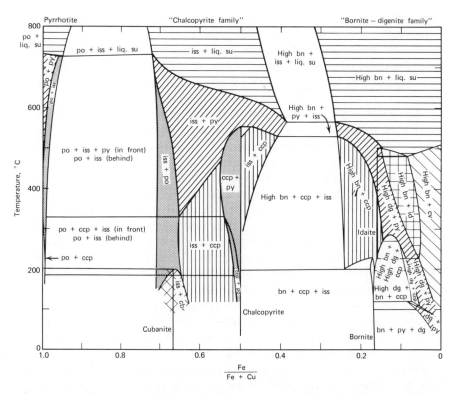

Fig. 7.17 Temperature-composition projection from sulfur of the sulfur-rich phase fields in the Cu-Fe-S system onto the sulfur-facing surfaces of the bornite, intermediate solid solution, and pyrrhotite phase fields. These solid solutions are variously patterned, the patterns identifying the sulfur-rich phase in equilibrium with the solid solution. Ccp (chalcopyrite) is heavily shaded and stands in front of iss (intermediate solid solution) as a phase showing up to 3 percent Fe/(Fe + Cu) variation in composition. Cb (cubanite) stands as a line in front of iss. Id (idaite) stands as a line in front of bn (bornite). Vapor is present in all assemblages. Other abbreviations: py = pyrite; po = pyrrhotite; dg = digenite; cv = covellite; su = sulfur. Tie lines involving complex, low-temperature phase relations along the Fe-S and Cu-S joins are omitted. The horizontal lines (isotherms) connect projected compositions of phases in equilibrium at invariant points; however, in the instances of invariant points for which one or more of the coexisting phases is cv, py, or liquid su (which are not among the phases whose composition is plotted) the tie line is terminated since it cannot be extended to a meaningful composition for py, cv, or liquid su. This diagram is highly speculative.

in equilibrium with pyrite. There is no requirement that tie lines between iss and pyrrhotite connect points of maximum mutual solid solution. Details associated with the numerous low-temperature phases on the copper-sulfur and iron-sulfur joins have been omitted, as have those involving talnakhite, because there are too few data to choose from among many possible topologies. A complexity similar to this is anticipated for the metal-facing surfaces of the solid solutions, but there are too few data to allow us to attempt a construction.

One of the interesting (although questionable) features of Figure 7.17 is the disappearance of a bornite + pyrite tie line below approximately 270°C (Yund and Kullerud, 1966, would place this at 228°C) and a reappearance of pyrite + bornite at low temperature. The reality of this circumstance is yet to be demonstrated, yet it is a reasonable construction. Unfortunately, the testing of this construction against natural mineral assemblages encounters the problem of pyrite metastability described in Figure 7.11a and elsewhere in this text.

The extensive high-temperature solid solutions shrink with decreasing temperature; cation ordering produces nearly stoichiometric phases at low temperature. On cooling below 547°C, chalcopyrite appears as an ordered phase, just to the sulfur-rich side of the intermediate solid solution. Chalcopyrite has an essentially fixed metal/sulfur ratio of 1 (Mukaiyama et al., 1968; MacLean et al., 1972), but recent studies by Karpenkov et al. (1974) and Sugaki et al. (1975) show that there can be a 2 or 3% substitution of iron for copper. Talnakhite ($Cu_9Fe_8S_{16}$), mooihoekite ($Cu_9Fe_9S_{16}$), haycockite ($Cu_4Fe_5S_8$), cubanite ($CuFe_2S_3$), and, if they are stable, the nickel-free analogues of the minerals described by Filimonova et al. (1974), $Cu_{17}(Fe, Ni)_{17}S_{32}$ and $Cu_{16}(Fe, Ni)_{19}S_{32}$, appear in relatively sulfur-deficient environments at low temperature but their phase relations are far from understood. One, or both, of the last two phases may be equivalent to the primitive-cubic synthetic phase described by Cabri (1973).

Fukuchilite and idaite are the only ternary phases that do not plot near the Cu_2S-Fe_2S_3 join. Fukuchilite and copper-rich pyrite, $(Cu, Fe)S_2$, have been reported from several localities and have been synthesized by Shimazaki and Clark (1970) who showed that they are not preserved above approximately 275°C, because they break down at a finite rate to covellite + pyrite + sulfur; they conclude that the fukuchilite-pyrite solid solution is metastable. Appreciable copper content in the $(Cu, Fe)S_2$ phase may be stable only at high pressure, and Barton (1970) has suggested that the highest stable copper contents in the pyrite phase may occur at low rather than high temperature.

Idaite, $Cu_{5.5}FeS_{6.5}$, is not a common mineral, although the assemblage covellite + chalcopyrite, which is chemically equivalent to idaite, is found frequently. The absence of idaite from chalcopyrite + covellite assemblages is explained by Schneeberg's (1973) experiments which yielded equilibrium constants showing that idaite should break down to digenite + covellite + pyrite below about 223°C, and Sugaki (1957), who noted that covellite and chalcopyrite did not react at 200°C in 12 hr, although they did react promptly at slightly higher temperatures. In ores the reluctance of pyrite to nucleate and grow permits idaite to decompose to chalcopyrite

+ covellite metastably at lower temperatures (see Figure 7.11a and the discussion of pyrite metastability).

Wiggins and Craig (1975) report up to 25 mole % ZnS in the intermediate solid solution saturated with sphalerite at 800°C and up to 4% at 500°C, and Moh (1975a) reports comparable results. Thus zinc might be expected to modify significantly the phase relations in the copper-iron-sulfur system at high temperatures, but most of it would be exsolved as sphalerite on slow cooling. The solid solution with ZnS so stabilizes the intermediate solid solution that chalcopyrite is incompatible with sphalerite down to just below 500°C.

The Copper-Iron-Nickel-Sulfur System

The copper-nickel-iron-sulfur system is of major importance to many magmatic-segregation deposits, but is much less significant for typical hydrothermal ores. However, because there are continuing controversies concerning the origin of certain deposits, we consider the quaternary briefly. For further details the definitive work of Kullerud et al. (1969) and Craig and Kullerud (1969) should be consulted.

The addition of large amounts of nickel to the copper-iron-sulfur system leads to the development of numerous nickel and nickel-iron sulfides (pentlandite, vaesite, bravoite, violarite, polydymite, etc.); there are no new quaternary phases and only pyrrhotite takes up large amounts of nickel. The rare mineral villamaninite, $CuNi_2S_6$, is the only ternary compound in the copper-nickel-sulfur system. The extension of the pyrrhotite solid solution into the quaternary is the most significant aspect of the system. There is complete solid solution between $Fe_{(1-x)}S$ and $Ni_{(1-x)}S$ down to less than 300°C, although pyrrhotite becomes nearly nickel-free and pyrite + pentlandite is clearly a stable assemblage at low temperature. The bulk composition of many magmatic segregation ores is clearly within the pyrrhotite solid solution, $(Fe, Ni, Cu)_{1-x}S$, at magmatic and high-grade metamorphic temperatures; the pentlandite and chalcopyrite-family phases exsolve as the temperature falls (Craig and Kullerud, 1969). The iron-nickel-sulfur system is shown in a series of isothermal diagrams in Figure 7.18; and the invariant points are given in Table 7-1. Most of the phases noted are compatible with chalcopyrite or other copper-iron sulfides.

The thermal stability of pentlandite is considerably enlarged by the substitution of cobalt for nickel and iron (Vaasjoki et al., 1974), but its stability is severely decreased by pressure. Thus, it is extremely unlikely that pentlandite will form as a liquidus phase in silicate magmas. The pentlandite structure accommodates silver as a phase having the composition $(Fe, Ni)_8AgS_8$, and argentian pentlandite is the principal residence for silver in many magmatic ores (Scott and Gasparrini, 1973).

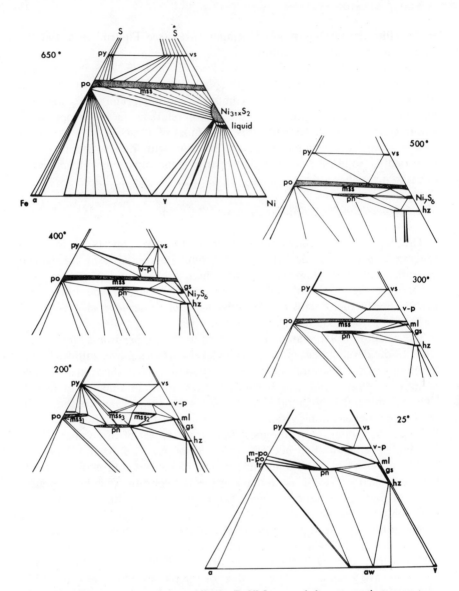

Fig. 7.18 Isothermal phase diagrams for the Fe-Ni-S system below magmatic temperatures. Vapor is present in all assemblages. To avoid confusion, tie lines are shown only for the 650°C isotherm. One-phase fields are shown as bars or are shaded. Abbreviations: py = pyrite, m-po = monoclinic pyrrhotite, h-po = hexagonal pyrrhotite (subtypes not distinguished); tr = troilite; mss = monosulfide solid solution; vs = vaesite; ml = millerite; v-p = violarite-polydymite; gs = godlevskite; hz = heazlewoodite; aw = awaruite; pn = pentlandite; α = alpha iron; γ = gamma nickel-iron. The low-temperature phase relations are speculative, and the compositions of some phases are poorly established, see Craig (1973) and references cited therein.

The Iron-Arsenic-Sulfur System

The iron-arsenic-sulfur system has had relatively little effort invested in it for several reasons: the ternary phase, arsenopyrite, is neither an ore of anything nor does it form in metallurgic mattes; thus the system is of marginal metallurgic interest. Arsenopyrite is not found in the wide variety of deposits that contain chalcopyrite or sphalerite. But arsenopyrite has a very desirable property; it is one of the most recalcitrant sulfide minerals in its participation in reactions in the laboratory; therefore, it does not reequilibrate at low temperatures. There have been three major studies of the ternary. Clark (1960) mapped out the system and proposed that the composition of arsenopyrite might be used as a geobarometer; his suggestion has been tried but not tested definitively. Barton (1969) evaluated data from a variety of sources to derive thermodynamic data for many of the binary and ternary phases. Kretschmar and Scott (1976) carefully examined the extent of, and phase relations around, the arsenopyrite field and concluded that the potential for use in geobarometry was small and that the composition of arsenopyrite in appropriately restricted assemblages could be useful in geothermometry, a suggestion that is just now beginning to be tested in the field. They refined the X-ray spacing curve to determine arsenopyrite, composition: atom % As $= 866.67 d_{131} - 1381.12$.

Figure 7.19 is taken from Kretschmar and Scott (1976). It shows the variation of the composition of arsenopyrite with mineral assemblages in the iron-arsenic-sulfur system. Arsenopyrite composition is sensibly along the FeS_2-$FeAs_2$ join, yet the drawing of the tie lines to arsenopyrite from pyrite and loellingite requires either that loellingite and pyrite can be slightly sulfur-rich or, more probably, that arsenopyrite is not precisely on the FeS_2-$FeAs_2$ join. Figure 7.2 shows that the activity of FeAsS in arsenopyrite solid solution at 500°C (from Kretschmar and Scott) has a moderate departure from ideal behavior.

Unfortunately, the experimental work with arsenopyrite is plagued by exactly the same property that makes it so attractive to the geologist, it is extremely difficult to get it to react. Kretschmar and Scott's extrapolation of compositions to temperatures below 400°C, particularly for the useful arsenopyrite + pyrite + pyrrhotite curve, are based on extremely short base lines; there is a great deal of room for error. Until an experimental breakthrough permits successful experimentation at lower temperatures, or field studies turn up material that can be unambiguously interpreted to provide natural "experiments," the arsenopyrite "geothermobarometer" will remain an interesting, but unproven, tool.

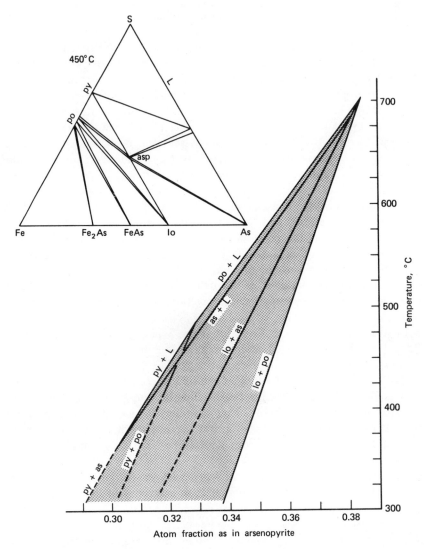

Fig. 7.19 Composition of arsenopyrite projected along the FeS_2-$FeAs_2$ join. All assemblages contain vapor. The total range in arsenopyrite composition is shaded; extrapolations beyond the experimental range are dashed. Data taken from Kretschmar and Scott (1976, *Canadian Mineral.*, *14*, 364–386). The inset shows the phase relations at 450 °C. Abbreviations: asp = arsenopyrite; py = pyrite; po = pyrrhotite; lo = loellingite; ar = arsenic; L = sulfur-arsenic liquid.

More Complex Sulfide Systems

Moh (1974, 1975a,b) has given an excellent summary (unfortunately, without the experimental data) of binary, ternary, and quaternary sections through the highly complex system copper-iron-zinc-antimony-sulfur (with or without germanium, lead, antimony, bismuth). Much of his discussion concerns the phase relations at high temperatures where very extensive, and in some cases complete, solid solution occurs among phases having the ZnS-type subcell (e.g., sphalerite—stannite). Most of the multicomponent phases melt in the surprisingly narrow 800–1000 °C temperature interval with only sphalerite (actually wurtzite) having liquidus temperatures greatly in excess of the others. As the temperature is lowered the amount of solid solution decreases markedly, particularly as the ordering of the ZnS-type subcells takes place so that below approximately 400 °C the "sphalerite superfamily" has many phases, each of which has relatively limited compositional ranges.

The detailed study of these high-temperature solid solutions does not have a great deal to offer directly to the understanding of most hydrothermal ores, but it should provide much of interest to those who would study the mechanisms and thermodynamics of solid solution and chemical-bond formation in sulfides.

Sulfosalts

Probably the most complex group of sulfide minerals is one called the sulfosalts. The name itself is not very informative; the compounds are more correctly called complex sulfides. To minimize confusion with names, we accept as sulfosalts only those compounds fitting the definition of Takeuchi and Sadanaga (1969). The group includes all compounds in which pyramid-shaped TS_3 groups are integral units in the mineral structure, T being restricted to the elements arsenic, antimony, or bismuth. Thus, enargite (Cu_3AsS_4) is not, but tennantite ($Cu_{12+x}As_{4+y}S_{13}$) is a sulfosalt. Most sulfosalts lie along simple joins such as $MS-T_2S_3$ or $M_2S-T_2S_3$ so that the usual valencies of the metals are retained.

Sulfosalts are of limited use for elucidating the physical conditions of ore formation. This is so for four reasons: first, most sulfosalts have undesirable quenching properties in that they react too quickly to preserve their heritage. They commonly have unquenchable high-temperature polymorphs and are often subject to disordering at relatively low temperatures. Second, although the sulfosalts are a widespread group of minerals, they are very diverse and many specific individuals occur infrequently.

Additionally, when sulfosalts do occur in an ore, they tend to be very minor constituents of the mineral assemblage. Third, the sulfosalts are a difficult group of minerals to examine, requiring painstaking and sophisticated mineralogic techniques for analysis and definitive identification. Fourth, they are often compositionally complex so that application of laboratory studies in simpler systems is unreliable.

Despite these drawbacks, one family of sulfosalts, tennantite-tetrahedrite, does hold a good deal of promise as an indicator of geologic environments. Tennantite and tetrahedrite are end members of a solid-solution series displaying complete diadochy of arsenic for antimony. They have a wide range in their stoichiometry with a general formula $Cu_{12+x}(As, Sb)_{4+y}S_{13}$ where x is 0 to 1.92 and y is -0.02 to 0.27 (Maske and Skinner, 1971, and Skinner et al., 1972). Tennantite-tetrahedrite reacts slowly in the laboratory, and preservation of zoned crystals (Yui, 1971) demonstrates that initial compositions may be retained in nature. Additionally, components such as FeS, ZnS, Ag_2S, and HgS can enter the structure in large amounts, offering the expectation that activities of these components might eventually be documented by tennantite-tetrahedrite compositions. Thus the wide composition field may eventually be useful, in the same manner that sphalerite is useful, in indicating conditions of formation.

Trace-Element Distribution

The distribution of a trace element between coexisting phases is commonly recognized as a means to reduce the degrees of freedom of a system and thereby to arrive at a unique value for the temperature (and perhaps pressure) of mineral equilibration. Experimentally determined distribution constants for sulfide systems are not abundant, but some are shown in Figure 7.20. There is predictably little pressure effect on these (in contrast with the pressure effect on the distribution for FeS between sphalerite and pyrrhotite shown in Figure 7.13) and neglect of pressure need not prove a serious barrier to application. However, the problem of sampling growth-zoned minerals to achieve concentrates of sufficient purity for meaningful analysis is a serious one.

The literature contains many references to attempts to use the concentration of a trace constituent *in a single phase* (such as the cadmium content of sphalerite or the cobalt content of pyrite) as a guide to temperature of mineralization. Although there may be valid geochemical reasons for some elemental patterns to be correlated with temperatures of mineralization, quantitative application of such guides is without theoretical support (see discussion by Barton, 1970).

SPECIFIC SULFIDE SYSTEMS

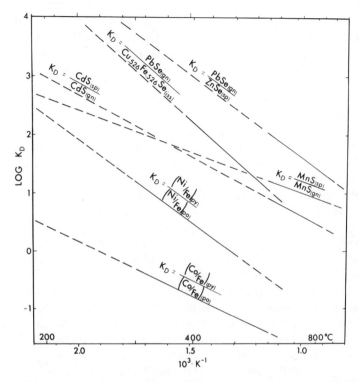

Fig. 7.20 Distribution coefficients for trace elements between coexisting phases as functions of temperature. The data for manganese, cadmium, and selenium and from Bethke and Barton (1971) and those for cobalt and nickel from Bezmen et al. (1975). The solid lines represent the temperature ranges of the experimental data; the dashes are linear extrapolations. All of the constants are in atomic proportions.

Condensed Invariant Points and Selected Univariant Equilibria

An extensive compilation of invariant points is presented in Table 7-1. Most of the assemblages are univariant, but a vapor phase, of composition dictated by the condensed phases, is also present in the evacuated silica tubes used for the laboratory studies. Thus the overall assemblages are invariant. Data are available for a few of the vapor-free univariant curves emanating from these points. Pressures of 2 or 3 kb displace the temperatures only slightly (see discussion by Barton, 1970) and, therefore, pressure is a relatively unimportant variable for our purposes.

Table 7-2 gives the free-energy changes for selected univariant equilibria for defining conditions of ore formation. These values may be added algebraically to evaluate reactions more complex than those considered in this chapter. In most instances the free-energy change, $\Delta G°$, equals the chemical potential of S_2, μS_2, which is related to the activity of S_2 through the relation $\Delta G° = \mu S_2 = -RT \ln a_{S_2}$. Many of the reactions are shown graphically in Figures 7.21, 7.22, and 7.23.

The Effect of Impurities

The effects on the invariant points and univariant curves of small amounts of additional components are highly varied, but they can be evaluated. The introduction of a new component without addition of a new phase adds a degree of freedom to the system; because minerals may contain many trace elements, we are forced to conclude that we shall never be able to use invariant points in minerals with Gibbsian thermodynamic rigor. Despite this fact, the overall effect of impurities may frequently be small enough to neglect. Were this not true we could not conduct meaningful equilibrium studies in the laboratory either, because completely pure starting materials cannot be obtained.

The entrance of a component into a phase will stabilize that phase relative to other phases in a binary system that take up less of the new component. Thus the field of the more impure phase expands relative to those of its neighbors. In an attempt to develop some guidelines for application we shall examine the ΔG-T diagram, Figure 7.24.

The solid lines represent a unary system having three phases, α, β, and γ, stable at successively lower temperatures. If a vapor phase is also present, transition points A and B will be invariant points. Consider now the addition of another component that is appreciably soluble only in β. The field of β will expand relative to α and γ, and the stability region of β will expand (dashed line). In the case of point A, the increased stability of β relative to α raises the invariant temperature only slightly to A'; on the other hand, invariant point B moves to B', a much larger temperature change.

An example of type A would be the slight shifts (usually only a few degrees) that other components make in the melting temperature of bismuth. This minor shift is associated with inversions having large enthalpies (or entropies) of transition such as those related to changes in bond type, coordination, or oxidation state. An example of type B is the drastic lowering of the cinnabar-metacinnabar inversion due to small amounts of nonstoichiometry (see Table 7-1) or FeS, ZnS, or HgSe in solid solution (Dickson and Tunnell, 1959) in the metacinnabar. The large shifts in inversion temperatures are associated with inversions having small enthalpies (or entropies) of transition such as displacive polymorphic transitions.

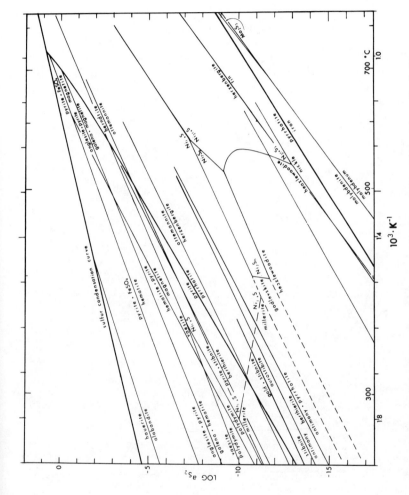

Fig. 7.21 Activity of S_2-temperature diagram for various sulfidation reactions. Dashed curves are schematic. Additional similar diagrams are found in Figures 7.22 and 7.23. Sources of data are indicated in Table 7-2.

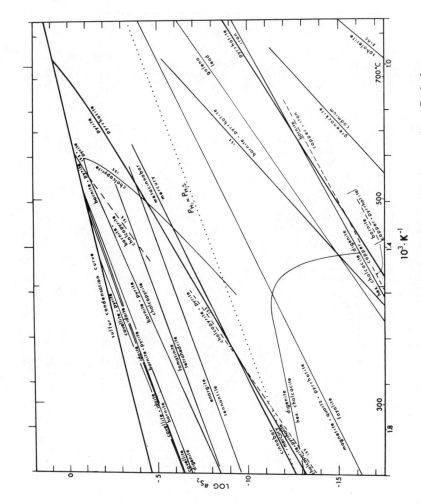

Fig. 7.22 Activity of S_2-temperature diagram for various sulfidation reactions. Dashed curves are schematic. Additional similar diagrams are found in Figures 7.21 and 7.23. Sources of data are indicated in Table 7-2.

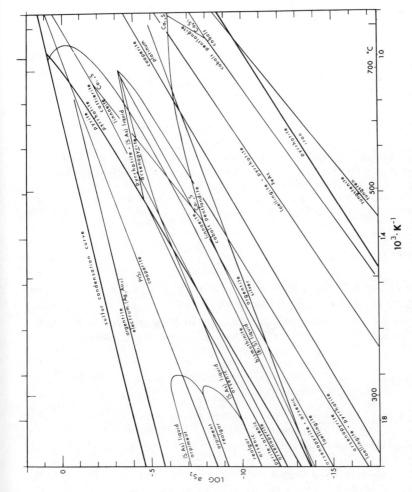

Fig. 7.23 Activity of S_2-temperature diagram for various sulfidation reactions. Dashed curves are schematic. Additional similar diagrams are found in Figures 7.21 and 7.22. Sources of data are given in Table 7-2.

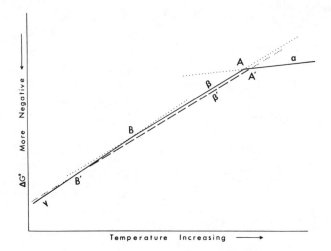

Fig. 7.24 Schematic diagram of free energy versus temperature showing the effect of impurities on invariant points. The solid curves represent the stable fields of three hypothetical polymorphs, α, β, and γ; the dotted curves are their metastable extentions. In the presence of vapor, A and B are invariant points. The dashed β' curve is for the β phase containing an appreciable amount of impurity which thereby shifts the invariant points from A and B to A' and B'.

ACKNOWLEDGMENTS

Many of our colleagues have made useful suggestions for which we are very appreciative. We are especially indebted to J. L. Haas, Jr., for computations on the speciation of sulfur vapor and to H. L. Barnes, J. R. Craig, W. E. Hall, and S. D. Scott for constructive reviews of the manuscript.

REFERENCES

Allen, E. T. and J. L. Crenshaw (1912) The sulfides of zinc, cadmium and mercury; their crystallographic forms and genetic conditions: *Am. J. Sci.,* 4, **34,** 341-396.

Arnold, R. G. (1962) Equilibrium relations between pyrrhotite and pyrite from 325° to 743°C: *Econ. Geol.,* **57,** 72-90.

____ (1971) Evidence for liquid immiscibility in the system FeS-S: *Econ. Geol.,* **66,** 1121-1130.

____ and O. P. Malik (1975) The NiS-S system above 980°C—A revision: *Econ. Geol.,* **70,** 176-184.

____ and L. E. Reichen (1962) Measurement of the metal content of naturally occurring, metal-deficient, hexagonal pyrrhotite by an x-ray spacing method: *Am. Mineral.,* **47,** 105-111.

Atanassov, V. A. and G. N. Kirov (1973) Balkanite, $Cu_9Ag_5HgS_8$, a new mineral from the Sedmochislenitsi mine, Bulgaria: *Am. Mineral.,* **58,** 11-15.

REFERENCES

Avetisyan, K. K. and G. I. Gnatyshenko (1956) Thermal and metallographic study of the lead sulfide-zinc sulfide-iron sulfide system (in Russian): *Izv. Akad. Nauk, Kaz SSR, Ser. Gorn. Dela, Met. Stroimaterialov,* **6,** 11-25.

Bachechi, F. (1972) Synthesis and stability of montebrayite, Au_2Te_3: *Am. Mineral.,* **57,** 146-154.

Barnes, H. L. and G. Kullerud (1961) Equilibria in sulfur-containing aqueous solutions, in the system Fe-S-O, and their correlation during ore deposition: *Econ. Geol.,* **56,** 648-688.

Barton, P. B., Jr. (1969) Thermochemical study of the system Fe-As-S: *Geochim. Cosmochim. Acta,* **33,** 841-857.

——— (1970) Sulfide petrology: *Min. Soc. Am., Spec. Paper No. 3,* 187-198.

——— (1971) The Fe-Sb-S system: *Econ. Geol.,* **66,** 121-132.

——— (1973) Solid solutions in the system Cu-Fe-S. Part I: The Cu-S and CuFe-S joins: *Econ. Geol.,* **68,** 455-465.

———, P. M. Bethke, and E. Roedder (1977) Environment of ore deposition in the Creede mining district, San Juan Mountains, Colorado: III. Progress toward interpretation of the chemistry of the ore-forming fluid for the OH vein: *Econ. Geol.,* **72,** 1-25.

———, P. M. Bethke, and P. Toulmin, III (1963) Equilibrium in ore deposits: *Min. Soc. Am., Spec. Paper No. 1,* 171-185.

——— and B. J. Skinner (1967) Sulfide mineral stabilities: *in Geochemistry of Hydrothermal Ore Deposits,* H. L. Barnes, ed., New York: Holt, Rinehart, and Winston, 236-333.

———, and P. Toulmin, III (1964) The electrum-tarnish method for the determination of the fugacity of sulfur in laboratory sulfide systems: *Geochim. Cosmochim. Acta,* **28,** 619-640.

——— and ——— (1966) Phase relations involving sphalerite in the Fe-Zn-S system: *Econ. Geol.,* **61,** 815-849.

Bayliss, P. (1969) X-ray data, optical anisotropism, and thermal stability of cobaltite, gersdorfite, and ullmannite: *Mineral. Mag.,* **37,** 26-33.

Berkowitz, J. (1965) Molecular composition of sulfur vapor; in *Elemental Sulfur,* B. Meyer, ed., New York: Interscience, 125-159.

Bernardi, G. P., F. Corsini, and R. Trosti (1972) Nuove relazioni di fase nel sistema Cu-Se: *Periodico Mineralogia,* **XLI,** 565-586.

Bernardini, G. P., G. Tanelli, and R. Trosti (1972) Relazioni di fase nel sistema Cu_3AsS_4-Cu_3SbS_4: *Soc. Ital. Mineral. Petrol.—Milano,* **29,** 281-296.

Berner, R. A. (1967) Thermodynamic stability of sedimentary iron sulfides: *Am. J. Sci.,* **265,** 773-785.

Bethke, P. M. and P. B. Barton, Jr. (1971) Distribution of some minor elements between coexisting sulfide minerals: *Econ. Geol.,* **66,** 140-161.

Bezmen, N. I., V. I. Tikhomirova, and V. P. Kosogova (1975) Pyrite-pyrrhotite geothermometer: distribution of nickel and cobalt: *Geokhimiya,* **5,** 700-714.

Biltz, W. and A. Kocher (1939) Über das System Vanadium/Schwafel: *Zeits. Anorg. Allgem. Chemie,* **241,** 324-337.

Bog, S. and T. Rosenqvist (1959) A high-temperature manometer and the decomposition pressure of pyrites: *Trans. Faraday Soc.,* **55,** 1565-1569.

Boorman, R. S. (1967) Subsolidus studies in the $ZnS-FeS-FeS_2$ system: *Econ. Geol.,* **62,** 614-631.

Braune, H., S. Peter, and V. Neveling (1951) Die Dissoziation des Schwefeldampfes: *Zeits. Naturf.,* **62,** 32-37.

Brett, P. R. (1963) Experimental data from the system Cu-Fe-S and their bearing on exsolution textures and reaction rates in ores: PhD thesis, Harvard University.

―― and G. Kullerud (1967) The Fe-Pb-S system: *Econ. Geol.*, **62**, 354–369.

Buhlmann, E. (1971) Untersuchungen im System Bi_2S_3-Cu_2S und geologische Schlussfolgerungen: *N. Jb. Min. Mh.*, **3**, 137–141.

Cabri, L. J. (1965a) Discussion of "Empressite and stuetzite redefined" by R. M. Honea: *Am. Mineral.*, **50**, 795–801.

―― (1965b) Phase relations in the Au-Au-Te system and their mineralogical significance: *Econ. Geol.*, **60**, 1569–1606.

―― (1973) New data on phase relations in the Cu-Fe-S system: *Econ. Geol.*, **68**, 443–454.

――, S. R. Hall, J. T. Szymanski, and J. M. Stewart (1973) On the transformation of cubanite: *Can. Mineral.*, **12**, 33–38.

――, D. C. Harris, and J. M. Stewart (1970a) Paracostibite (CoSbS) and nisbite ($NiSb_2$), New minerals from the Red Lake Area, Ontario, Canada: *Can. Mineral.*, **10**, 232–245.

――, ――, and ―― (1970b) Costibite (CoSbS), A new mineral from Broken Hill, N. S. W. Australia: *Am. Mineral.*, **55**, 10–17.

Cameron, E. N. (1961) *Ore Microscopy*, New York: Wiley.

Carlson, E. H. (1967) The growth of HgS and $Hg_3S_2Cl_2$ single crystals by a vapor phase method: *J. Crystal Growth*, **1**, 271–277.

Chang, L. L. Y. and J. E. Bever (1973) Lead sulfosalt minerals: crystal structures, stability relations, and paragenesis: *Minerals Sci. Eng.*, **5**, 181–191.

Chen, T. T. and L. L. Chang (1971) Phase relations in the systems Ag_2S-Bi_2S_3-Sb_2S_3 and Cu_2S-Bi_2S_3-Sb_2S_3 (abst.): *Geol. Soc. Am., Prog. Abst.* **3**, 524.

Chevrel, R., M. Sergent, and J. Prigent (1974) Un nouveau sulfure de molybdène: Mo_3S_4. Préparation, propriétés et structure cristalline: *Minerals Res. Bull.*, **9**, 1487–1498.

Chukrov, F. V., I. A. Ostrovskii, and V. V. Lapin (1974) *Minerals, Reference Book, Phase Equilibrium Diagrams*: Izdatel "Nauka", Moscow, *I*, (in Russian).

Clark, A. H. (1966a) Stability field of monoclinic pyrrhotite: *Inst. Min. Metall.*, **75B**, 232–235.

―― (1966b) Heating experiments on gudmundite: *Mineral. Mag.*, **35**, 1123–1125.

―― and R. H. Sillitoe (1970) Cuprian sphalerite and a probable copper-zinc sulfide, Cachiyuyo de Llampos, Copiapo, Chile: *Am. Mineral.*, **55**, 1021–1025.

Clark, L. A. (1960) The Fe-As-S System: Phase relations and applications: *Econ. Geol.*, **55**, 1345–1381, 1631–1652.

―― and G. Kullerud (1963) The sulfur-rich portion of the Fe-Ni-S system: *Econ. Geol.*, **58**, 853–885.

Clark, J. B. and E. Rapoport (1970) Effect of pressure on solid-solid transitions in some silver and cuprous chalcogenides: *J. Phys. Chem. Solids*, **31**, 247–254.

Clark, S. P., Jr. (1966) Handbook of the Physical Constants-Revised Edition: *Geol. Soc. Am. Mem.*, **97**.

Cocke, D. L., G. Abend, and J. H. Block (1976) Mass spectrometric observation of large molecules from condensed sulfur: *J. Phys. Chem.*, **80**, 524–528.

Coughlin, J. P. (1954) Contributions to the data on theoretical metallurgy. XII Heats and free energies of formation of inorganic oxides: *U. S. Bureau Mines Bull.*, **542**.

Craig, J. R. (1967) Phase relations and mineral assemblages in the Ag-Bi-Pb-S system: *Mineral. Depos.*, **1**, 278–306.

REFERENCES

———— (1968) The Fe-Ni-S system, violarite stability relations: *Carnegie Institution Wash. Yearbook 66,* 434–436.

———— (1970) Livingstonite, $HgSb_4S_8$: Synthesis and stability: *Am. Mineral.,* **55,** 919–924.

———— (1973) Pyrite-pentlandite assemblages and other low temperature relations in the Fe-Ni-S system: *Am. J. Sci.,* **273-A,** 496–510.

———— and P. B. Barton, Jr. (1973) Thermochemical approximations for sulfosalts: *Econ. Geol.,* **68,** 493–506.

————, ————, and B. H. Sepenuk (1971) Experimental investigations in the Bi-Fe-S system (abst.): *Geol. Soc. Am. Abst. Prog.,* **3,** 305.

————, L. L. Y. Chang, and W. R. Lees (1973) Investigations in the Pb-Sb-S system: *Can. Mineral.,* **12,** 199–206.

———— and J. B. Higgins (1973) Thiospinels: The carrollite-linnaeite series (abst.): *Geol. Soc. Am., Abst. Prog.,* **5,** 586.

———— and G. Kullerud (1968) Phase relations and mineral assemblages in the copper-lead-sulphur system: *Am. Mineral.,* **53,** 145–161.

———— and ———— (1969) Phase relations in the Cu-Fe-Ni-S system and their application to magmatic ore deposits: *Econ. Geol., Monograph,* **4,** 344–358.

———— and S. D. Scott (1974) Sulfide phase equilibria: *in Min. Soc. Am. Short Course Notes: Sulfide Mineralogy,* P. H. Ribbe, ed., CS1–CS110.

————, B. J. Skinner, C. A. Francis, F. D. Luce, and E. Makovicky (1974) Phase relations in the As-Sb-S system (abst.): *Trans. Am. Geophys. Union,* **55,** 483.

Cugnac-Pailliotet, A. Mme. and J. Pouradier (1972) Propriétès thermocynamiques du sulfure d'or: *C. R. Acad. Sci. Paris Ser. C,* **275,** 551–553.

Czamanske, G. K. (1969) The stability of argentopyrite and sternbergite: *Econ. Geol.,* **64,** 459–461.

———— (1974) The FeS content of sphalerite along the chalcopyrite-pyrite-bornite sulfur fugacity buffer: *Econ. Geol.,* **69,** 1328–1334.

Darrow, M. S., W. H. White, and R. Roy (1966) Phase relations in the system PbS-PbTe: *Trans. Metall. Soc. AIME,* **236,** 654–658.

Darken, L. S. and R. W. Gurry (1953) *Physical Chemistry of Metals,* New York: McGraw-Hill.

DeMedicis, R. (1970) Cubic FeS, a metastable iron sulfide: *Science,* **170,** 1191.

Detry, D., J. Drowart, P. Goldfinger, H. Keller, and H. Rickert (1967) Zur Thermodynamik von Schwefeldampf: *Z. Physik. Chem. N. Folge,* **55,** 314–319.

Dickson, F. W. and G. Tunnell (1959) The stability relationships of cinnabar and metacinnabar: *Am. Mineral.,* **44,** 471–487.

Drabek, M. and M. Stemprok (1974) The system Sn-S-O and its geologic implications: *N. Jb. Mineral. Abh.,* **122,** 90–118.

DuPreez, J. W. (1944) A thermal investigation of the parkerite series: *Ann. Univ. Stellenbosch,* **22,** 97–104.

Dutrizac, J. E. (1976) Reactions in cubanite and chalcopyrite: *Can. Mineral.,* **14,** 172–181.

Elliott, R. P. (1965) *Constitution of Binary Alloys,* 1st supplement, New York: McGraw-Hill.

Ethier, V. G., F. A. Campbell, R. A. Both, and H. R. Krouse (1976) The Sullivan ore body: I—Geological setting, and estimates of temperature and pressure of metamorphism: *Econ. Geol.,* **71,** 1570–1588.

Filimonova, A. A., I. V. Nurav'eva, and T. L. Evstigneeva (1974) Minerals of the chalcopyrite group in the copper-nickel ores of the Norilsk deposit (in Russian): *Geologiya Rudnykh Mestorozhdenii,* 36-46.

Fleet, Michael E. (1975) Thermodynamic properties of (Zn, Fe)S solid solutions at 850°C: *Am. Mineral.,* **60,** 466-470.

Fournier, R. O. (1973) Silica in thermal waters: Laboratory and field investigations: in *Proceedings of International Symposium on Hydrogeochemistry and Biogeochemistry, Japan, 1970, Vol. 1, Hydrogeochemistry,* Washington, D. C.: J. W. Clark, 122-139.

Francis, C. A., M. E. Fleet, K. Misra, and J. R. Craig (1976) Orientation of exsolved pentlandite in natural and synthetic nickeliferous pyrrhotite: *Am. Mineral.,* **61,** 913-920.

Freeman, R. D. (1962) Thermodynamic properties of binary sulfides: *Research Foundation Rept.,* **60,** Oklahoma State University.

Frumar, M., T. Kala, and J. Horak (1973) Growth and some physical properties of semiconducting $CuPbSbS_3$ crystals: *J. Crystal Growth,* **20,** 239-244.

Galopin, R. and N. F. M. Henry (1972) *Microscopic Study of Opaque Ore Minerals,* Cambridge: Heffer.

Garrels, R. M. and C. L. Christ (1965) *Solutions, Minerals and Equilibria,* New York: Harper and Row.

Gather, B. and R. Blachnik (1974) The Gold-Bismuth-Tellurium System: *Z. Metallkunde,* **65,** 653-656.

Glatz, A. C. (1967) The Bi_2Te_3-Bi_2S_3 system and the synthesis of the mineral tetradymite: *Am. Mineral.,* **52,** 161-170.

____ and V. F. Meikleham (1963) The preparation and electrical properties of bismuth trisulfide: *J. Electrochem. Soc.,* **110,** 1231-1234.

Godovikov, A. A. (1972) *Bismuth Sulfosalts:* Moscow: Nauka (in Russian).

Gorochov, O. (1968) Les composés Ag_8MX_6 (M = Si, Ge, Sn and X = S, Se, Te): *Bull. Soc. Chim. France,* 2263-2275.

Grønvold, F. and E. F. Westrum, Jr. (1976) Heat capacities of iron disulfides, thermodynamics of marcasite from 5 to 700 K, pyrite from 350 to 770 K, and the transformation of marcasite to pyrite: *J. Chem. Thermo.,* **8,** 1039-1048.

Gustafson, L. B. (1963) Phase equilibriums in the system Cu-Fe-As-S: *Econ. Geol.,* **58,** 667-701.

Haas, J. L., Jr. and J. R. Fisher (1976) Simultaneous evaluation and correlation of thermodynamic data: *Am. J. Sci.,* **276,** 525-545.

____ and R. W. Potter, II (1974) Internally consistent thermodynamic functions for sulfides and sulfosalts obtained using P-T data for sulfur (abst): *Geol. Soc. Am. Abst. Prog.,* **6,** 770.

____ and R. A. Robie (1973) Thermodynamic data for wüstite, $Fe_{0.947}O$, magnetite, Fe_3O_4, and hematite, Fe_2O_3 (abst.): *Am. Geophys. Union Trans.,* **54,** 438.

Hak, J., Z. Johan, and B. J. Skinner (1970) Kutinite; a new copper-silver arsenide mineral from Cerny Dul, Czechoslovakia: *Am. Mineral.,* **55,** 1083-1087.

Hall, H. T. (1966) The systems Ag-Sb-S, Ag-As-S, and Ag-Bi-S: Phase relations and mineralogical significance: Ph.D. dissertation, Brown University.

____ (1967) The pearceite and polybasite series: *Am. Mineral.,* **52,** 1311-1321.

____ and R. A. Yund (1964) Equilibrium relations among some silver sulfosalts and arsenic sulfides (abst.): *Am. Geophys. Union Trans.,* **45,** 122.

____ and ____ (1966) Pyrrhotite phase relations below 325°C (abst.): *Econ. Geol.,* **61,** 1297.

REFERENCES

Hansen, M. and K. Anderko (1958) *Constitution of Binary Alloys*, New York: McGraw-Hill.

Happ, J. V. and T. R. A. Davey (1971) Solubility of sulfur in liquid bismuth: *Inst. Min. Metall. Trans. Sec. C,* **80,** C190-C191.

Harris, D. C., L. J. Cabri, and S. Kaiman (1970) Athabascaite: A new copper selenide mineral from Martin Lake, Saskatchewan, part 2: *Can. Mineral.,* **10,** 207-215.

Hayasa, K. and R. Otuska (1952) Study on pyrite, Part I. On the electrical properties of pyrite: *J. Geol. Soc. Japan,* **58,** 133-143.

Helgeson, H. C. (1969) Thermodynamics of hydrothermal systems at elevated temperatures and pressures: *Am. J. Sci.,* **267,** 729-804.

Heyding, R. D. and L. D. Calvert (1961) Arsenides of the transition metals. IV. A note on the platinum-metal arsenides: *Can. J. Chem.,* **39,** 955-957.

Hoda, S. N. and L. L. Y. Chang (1975) Phase relations in the systems $PbS-Ag_2S-Sb_2S_3$ and $PbS-Ag_2S-Bi_2S_3$: *Am. Mineral.,* **60,** 621-633.

Holland, H. D. (1959) Some applications of thermochemical data to problems of ore deposits. I. Stability relations among the oxides, sulfides, sulfates and carbonate of ore and gangue minerals: *Econ. Geol.,* **54,** 184-233.

_____ (1965) Some applications of thermochemical data to problems of ore deposits. II. Mineral assemblages and the compositions of ore-forming fluids: *Econ. Geol.,* **60,** 1101-1166.

Huebner, J. S. and M. Sato (1970) The oxygen fugacity-temperature relationships of manganese oxide and nickel oxide buffers: *Am. Mineral.,* **55,** 934-952.

Hull, G. W., Jr., and Hulliger, F. (1968) Superconductors—$CuSe_2$ the first marcasite type: *Nature,* **220,** 257-258.

Hultgren, R., P. D. Desai, D. T. Hawkins, M. Gleiser, K. K. Kelley, and D. D. Wagman (1973a) *Selected Values of the Thermodynamic Properties of the Elements,* Am. Soc. Metals.

_____, _____, _____, _____, and _____ (1973b) *Selected Values of the Thermodynamic Properties of Binary Alloys,* Am. Soc. Metals.

Jeannot, C. and P. Perrot (1970) Étude thermodynamique de la sulfuration de l'argent. Charactérisation du sous-sulfure Ag_4S: *Rev. Chim. Minerale,* **7,** 47-61.

Jensen, E. (1947) Melting relations of chalcocite: *Avhandl. Norske Vid.-Akad. Oslo I Mat.— Naturw. Klasse,* 1-14.

Johan, Z., P. Picot, and R. Pierrot (1972) La krutaite, $CuSe_2$, un nouveau mineral du groupe de la pyrite: *Bull. Soc. Franc. Min. Crist.,* **95,** 475-481.

Karpenkov, A. M., G. A. Mitenkov, N. S. Rudashevskii, G. N. Sokolova, and N. N. Shishkin (1974) An iron-enriched variety of chalcopyrite: Notes of the *All-Union Mineral. Soc.,* **103,** 601-605 (in Russian).

Keighin, C. W. and Honea, R. M. (1969) The system Ag-Sb-S from 600° to 200°C: *Mineral. Deposita,* **4,** 153-171.

Kelley, K. K. (1960) Contributions to the data on theoretical metallurgy. XIII. High temperature-heat content, heat-capacity and entropy data for the elements and inorganic compounds: *U. S. Bur. Mines Bull. 584.*

_____ (1962) Contributions to the data on theoretical metallurgy. XV. A reprint of Bulletins 383, 384, 393, and 406: *U. S. Bur. Mines Bull. 601.*

_____ and E. G. King (1961) Contributions to the data on theoretical metallurgy. XIV. Entropies of the elements and inorganic compounds: *U. S. Bur. Mines Bull. 592.*

Kelly, W. C. and E. N. Goddard (1969) Telluride ores of Boulder County, Colorado: *Geol. Soc. Am. Mem.,* **109.**

King, L. G., A. D. Mah, and L. B. Pankratz (1973) Thermodynamic properties of copper and its inorganic compounds: *INCRA Monograph II on the Metallurgy of Copper, International Copper Research Association.*

Kirkinskii, V. A., A. P. Ryaposov, V. G. Yakushev (1967) Phase diagram of arsenic trisulfide for pressures up to 20 kilobars: *Bull. Acad. Sci. USSR, Inorg. Mater.,* **3,** 1931-1933.

Klement, W., A. Jayaraman, and G. C. Kennedy (1963) Phase diagrams of arsenic, antimony and bismuth at pressures up to 70 kilobars: *Phys. Rev.,* **131,** 632-637.

Klemm, D. D. (1965) Synthesen und Analysen in den Dreickdiagrammen FeAsS-CoAsS-NiAsS und FeS_2-CoS_2-NiS_2: *N. Jh. Mineral. Abhl.,* **102,** 205-255.

Klepp, K. O. and K. L. Komarek (1972) Übergangsmetall-Chalkogensysteme, 3 Mitt.: Das System Nickel-Tellur: *Monatshefte Chemie,* **103,** 934-946.

Komarek, K. L. and K. Wessely (1972) Übergangsmetall-chalkogensysteme, 2 Mitt.: Die Systeme Nickel-Selen und Kobalt-Nickel-Selen: *Monatshefte Chemie,* **103,** 923-933.

Korzhinskii, D. S. (1959) *Physicochemical Basis of the Analysis of the Paragenesis of Minerals,* (English translation) New York: Consultants Bureau.

Kracek, F. C., C. J. Kasada, and L. J. Cabri (1966) Phase relations in the system silver-tellurium: *Am. Mineral.,* **51,** 14-28.

Kretschmar, U. and S. D. Scott (1976) Phase relations involving arsenopyrite in the system Fe-As-S and their application: *Can. Mineral.,* **14,** 364-386.

Kubaschewski, O. and J. A. Catterall (1956) *Thermochemical Data for Alloys,* London: Pergamon Press.

———, E. L. Evans, and C. B. Alcock (1967) *Metallurgical Thermochemistry,* 4th Ed., Oxford: Pergamon Press.

Kullerud, G. (1963) Thermal Stability of Pentlandite: *Can. Mineral.,* **7,** 353-366.

——— (1964) Review and evaluation of recent research on geologically significant sulfide-type systems: *Fortschr. Mineral.,* **41,** 221-270.

——— (1965) Covellite stability relations in the Cu-S system: *Freiberger Forschungshefte, C186,* 145-160.

——— (1967) The Fe-Mo-S system: *Carnegie Institution Wash. Yearbook 65,* 337-342.

——— (1970) Sulfide phase relations: *Mineral. Soc. Am., Spec. Paper No. 3,* 199-210.

——— (1971) Experimental techniques in dry sulfide research, in *Research Techniques for High Pressure and High Temperature,* G. C. Ulmer, ed., New York: Springer-Verlag, 288-315.

——— and H. S. Yoder (1959) Pyrite stability relations in the Fe-S system: *Econ. Geol.,* **54,** 533-572.

——— and R. A. Yund (1962) The Ni-S System Related Minerals: *J. Petrol.,* **3,** 126-175.

———, ———, and G. Moh (1969) Phase relations in the Cu-Fe-S, Cu-Ni-S, and Fe-Ni-S systems: *Econ. Geol., Monograph,* **4,** 323-343.

Lee, M., S. Takenouchi, and H. Imai (1975) Syntheses of stannoidite and mawsonite and their genesis in ore deposits: *Econ. Geol.,* **59,** 834-843.

Leegaard, T. and T. Rosenqvist (1964) Der Zersetzungsdruck und die höheren Sulfide von Kobalt und Nickel: *Zeits. Anorg. Allgem. Chemie,* **328,** 294-298.

Legendre, B. and C. Souleau (1972) Étude de système ternaire Au-Pb-Te, *Bulletin de la Societe Chimique de France, No. 2,* 473-479.

Levin, E. M. and H. F. McMurdie (1975) *Phase Diagrams for Ceramists, 1975 Supplement*: Am. Ceramic Soc.

———, C. R. Robbins, and H. F. McMurdie (1964) *Phase Diagrams for Ceramists*: Am. Ceramic Soc.

———, ———, and ——— (1969) *Phase Diagrams for Ceramists, 1969 Supplement*: Am. Ceramic Soc.

Lewis, G. N. and M. Randall (1961) *Thermodynamics*, revised by K. S. Pitzer and L. Brewer, New York: McGraw-Hill.

Luce, F. D., C. L. Tuttle, and B. J. Skinner (1977) Studies of sulfosalts of copper. V. Phases and phase relations in the system Cu-Sb-As-S between 350° and 500°C: *Econ. Geol.*, **72**, 271–289.

MacLean, W. H., L. J. Cabri, and J. E. Gill (1972) Exsolution products in heated chalcopyrite: *Can. J. Earth Sci.*, **9**, 1305–1317.

Mandziuk, Z. L. and S. D. Scott (1975) Synthesis, stability, and phase relations of argentian pentlandite, $(Fe, Ni)_{8+x}Ag_{1-x}S_8$ in the system Ag-Fe-Ni-S (abst.): *Geol. Soc. Am., Abst. Prog.*, **7**, 1187.

Marfunin, A. S., and A. R. Mkrtchyan (1967) Mössbauer spectra of Fe^{57} in sulfides: *Geochem. Int.*, **4**, 980–989.

Maske, S. and B. J. Skinner (1971) Studies of the sulfosalts of copper. I. Phases and phase relations in the system Cu-As-S: *Econ. Geol.*, **66**, 901–918.

Maurel, C. and P. Picot (1974) Stability of Alloclasite and Cobaltite in Systems Co-As-S and Co-Ni-As-S: *Bull. Soc. Franc. Minéral. Cristallogr.*, **97**, 251–256.

Maxwell, J. A. (1963) The laser as a tool in mineral identification: *Can. Mineral.*, **7**, 727–737.

Meyer, B. (1976) Element sulfur: *Chem. Rev.*, **76**, 367–388.

Mills, K. C. (1974) *Thermodynamic Data for Inorganic Sulfides, Selenides and Tellurides*, London: Butterworths.

Moh, G. H. (1971) Blue remaining covellite and its relations to phases in the sulfur rich portion of the copper-sulfur system at low temperatures: *Mineral. Soc. Japan, Spec. Paper No. 1*, 226–232.

——— (1974) Tin-containing mineral systems, Part I: The Sn-Fe-S-O system and mineral assemblages in ores: *Chemie Erde*, **33**, 243–275.

——— (1975a) Tin-containing mineral systems, Part II: Phase relations and mineral assemblages in the Cu-Fe-Zn-Sn-S system: *Chemie Erde*, **34**, 1–61.

——— (1975b) Tin-containing mineral systems, Part III: Phase equilibria within the Sn-Pb-Sb-Bi-Fe-S system in relation to mineral assemblages of the Bolivian type: *Chemie Erde*, **34**, 201–238.

Morimoto, N. and G. Kullerud (1961) Polymorphism in bornite: *Am. Mineral.*, **46**, 1270–1282.

———, A. Gyobu, H. Mukaiyama, and E. Izawa (1975a) Crystallography and stability of pyrrhotites: *Econ. Geol.*, **59**, 824–833.

———, ———, K. Tsukuma, and K. Koto (1975b) Superstructure and nonstoichiometry of intermediate pyrrhotite: *Am. Mineral.*, **60**, 240–248.

——— and K. Koto (1970) Phase relations of the Cu-S system at low temperatures: Stability of anilite: *Am. Mineral.*, **55**, 106–117.

Muan, A. and E. F. Osborn (1965) *Phase Equilibria among Oxides in Steelmaking*, Reading: Addison-Wesley.

Mukaiyama, H., E. Izawa, and Y. Nakae (1968) The chemical composition and the structure of natural chalcopyrite: *J. Mineral. Soc. Japan,* **9,** 81-91.

Nakazawa, H. and N. Morimoto (1971) Transformation mechanisms of pyrrhotite: *Mineral. Soc. Japan, Spec. Paper No. 1,* 52-55.

_____, _____, and E. Watanabe (1975) Direct observation of metal vacancies by high-resolution electron microscopy: 4C type pyrrhotite (Fe_7S_8): *Am. Mineral.,* **60,** 359-366.

_____, T. Osaka, and K. Sakaguchi (1973) Crystallography—A new form of iron sulphide prepared by vacuum deposition: *Nature, Phys. Sci.,* **242,** 13-14.

_____, _____, and N. Morimoto (1971) Phase relations and superstructures of pyrrhotite, $Fe_{1-x}S$: *Mat. Res. Bull.,* **6,** 345-358.

Naumov, G. B., B. N. Ryzhenko, and I. L. Khodakovsky (1974) *Handbook of Thermodynamic Data* (translated from the 1971 Russian original): Nat. Tech. Info. Service PB-226-722.

Osaka, T. and H. Nakazawa (1976) Cementite structure for Fe_3S: *Nature,* **259,** 109-110.

Pamplin, B. R. (1975) *Crystal Growth* (Int. series of monographs on the science of the solid state), Oxford: Pergamon Press.

Pankratz, L. B. and E. G. King (1970) High-temperature enthalpies and entropies of chalcopyrite and bornite: *U. S. Bur. Mines, Rept. Inv. No. 7435.*

Pearson, W. B. (1958) *Lattice Spacings and Structures of Metals and Alloys,* New York: Pergamon Press.

Perrot, P. and C. Jeannot (1971) Relations activité-composition dans les solutions solides Ag_2S-Cu_2S: *Rev. Chim. Minerale,* **8,** 87-98.

Pierce, L. and P. R. Buseck (1976) A comparison of bright field and dark field imaging of pyrrhotite structures: in *Electron Microscopy in Mineralogy,* H. R. Wenk et al., eds., Berlin: Springer-Verlag, 137-141.

Pistorius, C. W. F. T. and O. Gorochov (1970) Polymorphism and stability of the semiconducting series Ag_8MX_6 (M = Si, Ge, Sn, and X = S, Se, Te) to high pressures: *High Temp.—High Press.,* **2,** 31-42.

Potter, R. W., II (1974) The low temperature phase relations in the system Cu-S derived from an electrochemical investigation (abst.): *Geol. Soc. Am., Abst. Prog.,* **6,** 915-916.

_____ (1975) Metastable phase relations in the system Cu-S (abst.): *Geol. Soc. Am., Abst. Prog.,* **7,** 1232-1233.

_____ and H. L. Barnes (1978) Phase equilibria in the binary Hg-S: *Am. Mineral.,* **63,** 1143-1152.

_____ and H. T. Evans, Jr. (1976) Definitive x-ray powder data for covellite, manilite, djurelite, and chalcocite: *J. Res. U. S. Geol. Survey,* **4,** 205-218.

Pouillard, G. and P. Perrot (1975) Free enthalpy of formation and region of stability of molybdenum sulphides Mo_2S_3 and MoS_2: *Comptes Rendus,* **281,** 143.

Pourbaix, M. (1966) *Atlas of Electrochemical Equilibria in Aqueous Solutions,* Oxford: Pergamon Press.

Power, L. F. and H. A. Fine (1976) The iron-sulfur system: Part 1. The structures and physical properties of the compounds of the low temperature phase fields: *Minerals Sci. Eng.,* **8,** 106-128.

Radtke, A. S., C. M. Taylor, R. C. Erd, and F. W. Dickson (1974) Occurrence of lorandite, $TlAsS_2$, at the Carlin gold deposit, Nevada: *Econ. Geol.,* **69,** 121-123.

_____ and F. W. Dickson (1975) Carlinite, Tl_2S, a new mineral from Nevada: *Am. Mineral.,* **60,** 559-565.

REFERENCES

Ramdohr, P. (1969) *The Ore Minerals and their Intergrowths,* Oxford: Pergamon Press.

Rau, Hans (1967) Defect equilibria in cubic high temperature copper sulfide (digenite): *J. Phys. Chem. Solids,* **28,** 903-916.

Rau, H., T. R. N. Kutty, and J. R. F. Guedes de Carvalho (1973a) High temperature saturated vapor pressure of sulfur and the estimation of its critical quantities: *J. Chem. Thermo.,* **5,** 291-302.

____, ____, and ____ (1973b) Thermodynamics of sulphur vapour: *J. Chem. Thermo.,* **5,** 833-844.

Ricci, J. E. (1951) *The Phase Rule and Heterogeneous Equilibria,* New York: Van Nostrand.

Richardson, F. D. and J. H. E. Jeffes (1948) The thermodynamics of substances of interest in iron and steel making from 0 to 2400°C. I. Oxides: *J. Iron Steel Inst.,* **160,** 261-270.

____ and ____ (1952) The thermodynamics of substances of interest in iron and steel making. III. Sulfides: *J. Iron Steel Inst.,* **171,** 165-175.

____, ____, and G. Withers (1950) The thermodynamics of substances of interest in iron and steel making. II. Compounds between oxides: *J. Iron Steel Inst.,* **166,** 213-235.

Rising, B. (1973) Phase relations among pyrite, marcasite and pyrrhotite below 300°C: Unpublished Ph.D. thesis, Pennsylvania State University.

Robie, R. A. and D. R. Waldbaum (1968) Thermodynamic properties of minerals and related substances at 298.15°K (25°C) and one atmosphere (1.013 bars) and at higher temperatures: *U. S. Geol. Survey Bull. 1259.*

See also R. A. Robie, B. S. Hemingway, and J. R. Fisher (1978) revision under the same title, *U. S. Geol. Survey Bull. 1452.*

Rode, Y. Y. and B. A. Lebedev (1961) Physico-chemical study of molybdenum trisulfide and of the products resulting from its thermal decomposition: *Zhur. Neorg. Khim.,* **6,** 1189-1197.

Roland, G. W. (1968) The system Pb-As-S, composition and stability of jordanite: *Mineral. Depos.,* **3,** 249-260.

____ (1970) Phase relations below 575°C in the system Ag-As-S: *Econ. Geol.,* **65,** 241-252.

Roseboom, E. H., Jr. (1966) An investigation of the system Cu-S and some natural copper sulfides between 25° and 700°C: *Econ. Geol.,* **61,** 641-672.

____ and G. Kullerud (1958) The solidus in the system Cu-Fe-S between 400° and 800°C: *Carnegie Institution Wash. Yearbook 57,* 222-227.

Rosenfeld, J. N. (1969) Stress effects around quartz inclusions in almandine and the piezothermometry of coexisting aluminum silicates: *Am. J. Sci.,* **267,** 317-351.

Rosenqvist, T. (1954) A thermodynamic study of the iron, cobalt and nickel sulfides: *J. Iron Steel Inst.,* **176,** 37-57.

Rossini, F. D., D. D. Wagman, W. H. Evans, S. Levine, and I. Jaffee (1952) Selected values of chemical thermodynamic properties: *Natl. Bur. Standards Circ. No. 500,* 1299. (See also revisions as N.B.S. Tech. Notes 270-1 through 270-7 by various authors.)

Sachdev, S. C. and L. L. Y. Chang (1975) Phase relations in the system tin-antimony-lead sulfides and the synthesis of cylindrite and franckeite: *Econ. Geol.,* **70,** 1111-1122.

Salanci, B. and G. H. Moh (1969) Die experimentalle Untersuchung des pseudobinaren Schnittes $PbS-Bi_2S_3$ innerholb des Pb-Bi-S systems in Beziehung zu naturlichen Blei-Bismut-Sulfosaltzen: *N. Jh. Mineral Abhl.,* **112,** 63-95.

Sato, M. (1965) Electrochemical geothermometer: a possible new method of geothermometry with electro-conductive minerals: *Econ. Geol.,* **60,** 812-818.

____ (1971) Electrochemical measurements and control of oxygen fugacity and other gaseous fugacities with solid state sensors: in *Research Techniques for High Pressure and High Temperature,* G. C. Ulmer, ed., New York: Springer-Verlag, 43-100.

Scanlon, W. W. (1963) The physical properties of semiconducting sulfides, selenide and tellurides: *Mineral. Soc. Am., Spec. Paper No. 1,* 135-143.

Schneeberg, E. P. (1973) Sulfur fugacity measurements with the electrochemical cell Ag|AgI| $Ag_{2+x}S$, f_{S_2}: *Econ. Geol.,* **68,** 507-517.

Schneider, S. J. (1963) Compilation of the melting points of the metal oxides: *U. S. Bur. Standards, Monograph 68.*

Schrader, R. and C. Pietzch (1969) Über amorphes und kristallines Eisen (III)-sulfid: *Kristall Technik,* **4,** 385-397.

Schwarcz, H. P., S. D. Scott, and S. A. Kissin (1975) Pressures of formation of iron meteorites from sphalerite compositions: *Geochim. Cosmochim. Acta,* **39,** 1457-1466.

Scott, S. D. (1973) Experimental calibration of the sphalerite geobarometer: *Econ. Geol.,* **68,** 466-474.

____ (1974) Experimental methods in sulfide synthesis: in *Mineral. Soc. Am. Short Course Notes: Sulfide Mineralogy,* P. H. Ribbe, ed. S1-S38.

____ (1975) Hydrothermal synthesis of refractory sulfide minerals: *Fortschr. Miner.,* **52,** 185-195.

____ (1976) Application of the sphalerite geobarometer to regionally metamorphosed terraines: *Am. Mineral.,* **61,** 661-670.

____ and H. L. Barnes (1971) Sphalerite geothermometry and geobarometry: *Econ. Geol.,* **66,** 653-669.

____ and ____ (1972) Sphalerite-wurtzite equilibria and stoichiometry: *Geochim. Cosmochim. Acta,* **36,** 1275-1295.

____ and E. Gasparrini (1973) Argentian pentlandite, $(Fe, Ni)_8AgS_8$, from Bird River, Manitoba: *Can. Mineral.,* **12,** 165-168.

____ and S. A. Kissin (1973) Sphalerite composition in the Fe-Zn-S system below 300°C: *Econ. Geol.,* **68,** 475-479.

____, A. J. Naldrett, and E. Gasparrini (1974) Regular solution model for the $Fe_{1-x}S\text{-}Ni_{1-x}S$ solid solution (abst.): Collected Abstracts, *International Mineralogical Assoc. Ninth General Meeting,* 172.

Sharp, W. E. (1969) Melting curves of sphalerite, galena and pyrrhotite and the decomposition curve of pyrite between 30 and 65 kilobars: *J. Geophys. Res.,* **74,** 1645-1652.

Shimazaki, H. and L. A. Clark (1970) Synthetic FeS_2CuS_2 solid solution and fukuchilite-like minerals: *Can. Mineral.,* **10,** 648-664.

Shuey, R. T. (1975) *Semiconducting Ore Minerals,* New York: Elsevier.

Shunk, F. A. (1969) *Constitution of Binary Alloys,* New York: McGraw-Hill.

Sims, P. K. and P. B. Barton, Jr. (1961) Some aspects of the geochemistry of sphalerite, Central City District, Colorado: *Econ. Geol.,* **56,** 1211-1237.

Skinner, B. J. (1965) The system arsenic-antimony: *Econ. Geol.,* **60,** 228-239.

____ (1966) The system Cu-Ag-S: *Econ. Geol.,* **61,** 1-26.

____ and P. B. Barton, Jr. (1960) The substitution of oxygen for sulfur in wurtzite and sphalerite: *Am. Mineral.,* **45,** 612-625.

____ and Luce, F. D. (1971a) Stabilities and compositions of domeykite and algodonite: *Econ. Geol.,* **66,** 133-139.

REFERENCES

___ and ___ (1971b) Solid solutions of the type (Ca, Mg, Mn, Fe)S and their use as geothermometers for the enstatite chondrodites: *Am. Mineral.*, **56**, 1269-1296.

___, ___, and E. Makovicky (1972) Studies of the sulfosalts of copper. III. Phases and phase relations in the system Cu-Sb-S: *Econ. Geol.*, **67**, 924-938.

Slavskaya, A. I., V. E. Rudnichenko, and E. I. Bogoslovskaya (1963) Dissociation of betekhtinite during heating: *Geologiya Rudnykh Mestorozhdenii*, **5**, 97-100 (In Russian), abstract by E. A. Alexanderov in *Econ. Geol.*, **59**, 344.

Smit, T. J. M., E. Venema, J. Wiersma, and G. A. Wiegers (1970) Phase transitions in silver-gold chalcogenides: *J. Solid State Chemistry*, **2**, 309-312.

Smith, F. G. (1947) The pyrite geo-thermometer: *Econ. Geol.*, **42**, 515-523.

___ (1953) *Historical development of inclusion thermometry:* Toronto University Press.

___ (1955) Structure of zinc sulfide minerals: *Am. Mineral.*, **40**, 658-675.

Springer, G. (1971) The synthetic solid-solution series Bi_2S_3-$BiCuPbS_3$ (Bismuthinite-Aikinite): *Mineral Sci. Div., Mines Branch, Dept. of Energy, Mines and Resources, Ottawa, Canada*, 19-24.

___ and J. H. G. LaFlamme (1971) The system Bi_2S_3-Sb_2S_3: *Can. Mineral.*, **10**, 847-853.

Springer, G., D. Schachner-Korn, and J. V. P. Long (1964) Metastable solid solution relations in the system FeS_2-CoS_2: *Econ. Geol.*, **59**, 475-491.

Steininger, J. (1970) Phase diagram of the PbTe-PbSe pseudobinary system: *Metallurg. Trans.*, **1**, 2939-2941.

Stull, D. R. and H. Prophet (1971) JANAF Thermochemical Tables, Second Edition, Nat. Stand. Ref. Data Serv., *U. S. Nat. Bureau Stand.* **37**; see also supplements by M. W. Chase and others (1974 and 1975) *J. Phys. Chem. Ref. Data*, **3**, 311-480 and **4**, 1-175).

Sugaki, A. (1957) Thermal studies on the diffusion between some sulphide minerals in the solid phase: *Sci. Repts. Tohoku University, Ser. 3*, **5**, 95-112.

___ and H. Shima (1965) Synthetic sulfide minerals (II): *Memoirs Faculty of Engineering, Yamaguchi University*, **15**, 33-47.

___ and ___ (1972) Phase relations of the Cu_2S-Bi_2S_3 system: *Tech. Rept. Yamaguchi University*, **1**, 45-70.

___, ___, and A. Kitakaze (1972) Synthetic sulfide minerals (IV): *Tech. Rept. Yamaguchi University*, **1**, 71-85.

___, ___, and ___ (1974a) Synthesized lead bismuth sulfosalt minerals; heyrovskyite, lillianite and galenobismutite: *Tech. Rept. Yamaguchi University*, **1**, 359-368.

___, ___, and ___ (1974b) Synthetic phases in the PbS-Bi_2S_3 system; $PbBi_4S_7$ and $Pb_2Bi_2S_5$ (Synthetic sulfide minerals (VII)): *Tech. Rept. Yamaguchi University*, **1**, 369-373.

___, ___, ___, and H. Harada (1975) Isothermal phase relations in the system Cu-Fe-S under hydrothermal conditions at 350°C and 300°C: *Econ. Geol.*, **70**, 806-823.

Takeno, S., H. Zoka, and T. Niihara (1970) Metastable cubic iron sulfide—with special reference to mackinawite: *Am. Mineral.*, **55**, 1639-1649.

Takeuchi, Y. and R. Sadanaga (1969) Structural principles and classification of sulfosalts: *Zeit Krist*, **180**, 346-368.

Tatsuka, K. and N. Morimoto (1973) Composition variation and polymorphism of tetrahedrite in the Cu-Sb-S system below 400: *Am. Mineral.*, **58**, 425-434.

Tavernier, B. H., J. Vervecken, P. Meissen, and M. Baiwir (1967) Uber das thermische Verhalten von Silber- und Silber-Gold(I) Calkogeniden: *Z. Anorg. Allgem. Chem.*, **356**, 77-88.

Taylor, L. A. (1970a) Low-temperature phase relations in the Fe-S system: *Carnegie Institution Wash. Yearbook 68*, 259-270.

―― (1970b) The System Ag-Fe-S: Phase relations between 1200° and 700°C: *Metall. Trans.*, **1**, 2523-2529.

―― and G. Kullerud (1972) Phase equilibria associated with the stability of copper disulfide: *N. Jh. Mineral. Monatsch.*, **10**, 458-463.

―― and H. K. Mao (1970) A high-pressure polymorph of troilite, FeS: *Science*, **170**, 850-851.

Thompson, J. B., Jr. (1955) The thermodynamic basis for the mineral facies concept: *Am. J. Sci.*, **253**, 65-103.

―― (1957) The graphical analysis of minerals in peletic schists: *Am. Mineral.*, **42**, 842-858.

―― (1959) Local equilibrium in metasomatic processes: in *Researches in Geochemistry*, ed. P. H. Abelson, New York: Wiley, 427-457.

Toulmin, P., III (1960) Effect of Cu on sphalerite phase equilibria—A preliminary report (abst.): *Am. Geol. Soc. Bull.*, **71**, 1993.

―― (1963) Proustite-pyrargyrite solid solutions: *Am. Mineral.*, **48**, 725-736.

―― and P. B. Barton, Jr. (1964) A thermodynamic study of pyrite and pyrrhotite: *Geochim. Cosmochim. Acta*, **28**, 641-671.

Troften, P. and G. Kullerud (1961) The Fe-Se system: *Carnegie Institution Publ. Geophys. Lab. No. 1363*, 176.

Tuller, W. M. (1954) *The Sulfur Data Book*, New York: McGraw-Hill.

Uytenbogaardt, W. and E. A. J. Burke (1971) *Tables for Microscopic Identification of Ore Minerals*, 2nd Ed., Amsterdam: Elsevier.

Vaasjoki, O., A. Hakli, and M. Tontti (1974) The effect of cobalt on the thermal stability of pentlandite: *Econ. Geol.*, **69**, 549-551.

Van Hook, H. J. (1960) The ternary system Ag_2S-Bi_2S_3-PbS: *Econ. Geol.*, **55**, 759-788.

Vaughan, D. J. and J. R. Craig (1978) *Mineral Chemistry of the Metal Sulfides*, Cambridge, Cambridge Univ. Press.

Wagner, C. (1953) Investigations on silver sulfide: *J. Chem. Physics*, **21**, 1819-1827.

Wallia, D. S. and L. L. Y. Chang (1973) Investigations in the systems PbS-Sb_2S_3-As_2S_3 and PbS-Bi_2S_3-As_2S_3: *Can. Mineral.*, **12**, 113-119.

Wandji, R. and J. K. Kom (1972) Préparation et étude de la raguinite et de son homologue sélenié: $TlFeS_2$ ét $TlFeSe_2$: *C. R. Series C—Sciences Chimiques*, **275**, 813-816.

Ward, J. C. (1970) The structure and properties of some iron sulphides: *Rev. Pure. Appl. Chem.*, **20**, 175-206.

Weissberg, B. G. (1965) Getchellite, $AsSbS_3$, a new mineral from Humboldt County, Nevada: *Am. Mineral.*, **50**, 1817-1826.

West, W. A. and A. W. C. Menzies (1930) The vapor pressures of sulfur between 100° and 550° with related thermal data: *J. Phys. Chem.*, **33**, 1880-1892.

Wiggins, L. B. and J. R. Craig (1975) A reconnaissance investigation of chalcopyrite-sphalerite relationships in the Cu-Fe-Zn-S system (abst.): *Geol. Soc. Am., Abst. Prog.*, **7**, 1317.

Williams, K. L. and G. Kullerud (1970) The Ni-Sb-S system: *Carnegie Institution Wash. Yearbook 68*, 270-273.

Wright, H. D., W. M. Barnard, and J. B. Halbig (1965) Solid solutions in the systems ZnS-ZnSe and PbS-PbSe at 300°C and above: *Am. Mineral.*, **50**, 1802-1815.

REFERENCES

Yamaguchi, S. and H. Wada (1973) Magnetic iron sulfide of the γ-Al_2O_3 type: *J. Appl. Physics,* **44,** 1929.

Yui, S. (1971) Heterogenity within a single grain of minerals of the tennantite-tetrahedrite series: *Soc. Mining Geol. Japan, Spec. Issue 2,* 22-29.

Yund, R. A. (1961) Phase relations in the system Ni-As: *Econ. Geol.,* **56,** 1273-1296.

―― (1962) The system Ni-As-S, phase relations and mineralogical significance: *Am. J. Sci.,* **260,** 761-782.

―― and H. T. Hall (1968) The miscibility gap between FeS and $Fe_{1-x}S$: *Mater. Res. Bull.,* **3,** 779-784.

―― and ―― (1969) Hexagonal and monoclinic pyrrhotites: *Econ. Geol.,* **64,** 420-424.

―― and ―― (1970) Kinetics and mechanism of pyrite exsolution from pyrrhotite: *J. Petrol.,* **11,** 381-404.

―― and G. Kullerud (1966) Thermal stability of assemblages in the Cu-Fe-S system: *J. Petrol.,* **7,** 454-488.

Zen, E. (1963) Components, phases, and criteria of chemical equilibrium in rocks: *Am. J. Sci.,* **261,** 929-949.

8

Solubilities of Ore Minerals

HUBERT L. BARNES
The Pennsylvania State University

The geologic environments in which hydrothermal ore deposits are found appear, at least superficially, to be highly variable. Both localizing structures and host rocks differ significantly from one deposit or type of deposit to another. However, this variability does not extend to the dominant mineralogies of hydrothermal ores. Compared to the number of compounds that might conceivably occur, we find in ores only relatively few in quantity, and these are predominantly the very common sulfides or oxides. This means that a useful framework for understanding ore transport processes may be based initially on the solubility characteristics of only the more common ore minerals. For this reason, relatively little emphasis will be placed in this chapter on the solubilities of the less common minerals of ore deposits.

The solubilities of common ore minerals, as simple ions, have long been recognized to be inadequate for ore transport. This can be demonstrated for metallic sulfides by considering dissolution in neutral to acidic solutions where the dominant aqueous sulfide species is H_2S (see Figure 8.1).

$$MeS(s) + 2H^+ \rightleftharpoons Me^{2+} + H_2S(aq) \qquad (8.1)$$

Note that the solubility product is less useful for testing of ionic solubilities because S^{2-} is subordinate compared to H_2S in the near-neutral to acidic solutions characteristic of ore transport (see Chapter 5). We can calculate the activity product of Me^{2+} and aqueous H_2S for neutral solutions from the

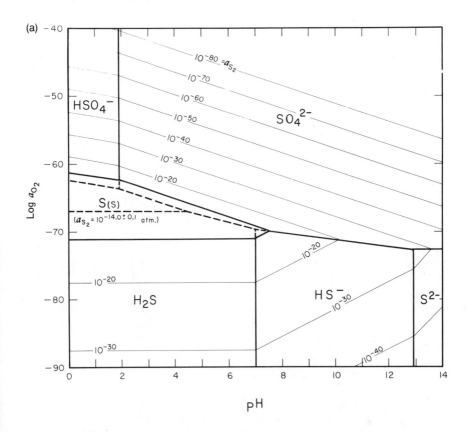
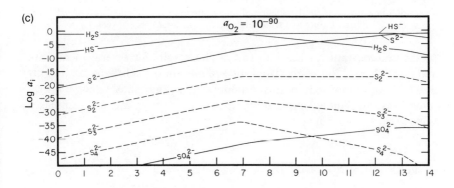

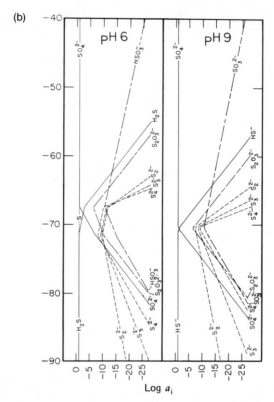

FIG. 8.1 Stabilities of aqueous, sulfur-containing species at $\Sigma S = 0.1$, 25°C. (a) Light, solid contours show a_{S_2}. Changes in the fields of predominance (heavy, solid lines) on decreasing ΣS to 0.001 are indicated by the heavy, dashed lines. (b) Variation with pH in activities of predominant and minor species under strongly reducing conditions at $a_{O_2} = 10^{-90}$. (c) Variation in activities of predominant and minor species with oxidation state at pH 6 and at pH 9. (Modified from Barnes and Kullerud, 1961, *Econ. Geol.*, **56**, 648-688.)

equilibrium constant for reaction (8.1). The results for several metals are given in Table 8-1. Representative values of a_{H_2S} for ore solutions are roughly 0.1–0.001 and, inserted in this product, show that most soluble metal sulfides have negligible simple solubilities. Apparently, aqueous complexes are required for ore transport.

COMPLEX-FORMING LIGANDS

The chemistries important to ore transport are restricted not only by the limited number of dominant ore minerals but also by the few ligands that are present in sufficient quantities to associate with metals in complexes under

Table 8-1 Activities of metal ions in equilibrium with sulfides

Data for the calculation have been taken from Naumov et al. (1974). pH_n represents the neutral pH at each temperature[a] and pH_{n-2} is two pH units acidic.

	Log $(a_{Me^{2+}} a_{H_2S})$					
	25°C		100°C		200°C	
Sulfide	pH_n	pH_{n-2}	pH_n	pH_{n-2}	pH_n	pH_{n-2}
FeS	−10.58	−6.58	−9.47	−5.47	−9.2	−5.2
NiS	−15.14	−11.14	−13.56	−9.56	−12.60	−8.6
ZnS	−19.03	−15.03	−16.48	−12.48	−14.61	−10.6
CdS	−21.30	−17.30	−18.03	−14.03	−15.7	−11.7
PbS	−22.17	−18.17	−18.19	−14.19	−15.40	−11.4
CuS	−29.96	−25.96	−25.15	−21.15	−21.43	−17.4
HgS	−46.8	−42.8	−38.2	−34.2	−31.4	−27.4

[a] Neutral pH's are 7.00 at 25°C, 6.12 at 100°C, and 5.69 at 200°C.

hydrothermal conditions. Clearly, the most important are Cl^-, HS^- or H_2S, and OH^-. At moderate temperatures, above about 250°C, NH_3 and F^- may also contribute to complexing and at lower temperatures $S_2O_3^{2-}$, S_x^{2-}, SO_3^{2-}, CN^-, SCN^-, organic ligands, and possibly various aqueous species of Se, Te, As, Sb, PO_4^{2-}, and Bi could be of local importance. Data on stabilities are available for relatively few of the complexes that might be formed by these ligands in reducing solutions at temperatures above 25-60°C. Nevertheless, the ore-carrying capacity of hydrothermal solutions is determined more by the activity of these ligands than by the abundances of the metals in host rocks that vary over a comparatively narrow range.

The availability of a ligand to form a metal-carrying aqueous complex depends upon its activity in solution. This activity is a function of (1) its total, analytic concentration, (2) temperature, (3) ionic strength of the solution, (4) extent of ion-pairing with other aqueous species, (5) acidity of the solution, and (6) the oxidation state of the solution. The expected range of values for these six variables changes considerably for different types of deposits. Among these, temperatures extend to above 600°C (although 100-400° is more common) and ionic strengths extend up to at least 6, based on fluid inclusion filling temperatures and saturation in halide daughter minerals (Roedder, Chapter 14). A brief review of the effects of these six variables on the activity of the most important ligands follows as a necessary basis for considering ore transport processes.

Sulfur-Containing Ligands

The thermodynamic stabilities of sulfur-containing aqueous species are shown in Figure 8.1 at 25°C, $\Sigma S = 0.1$, where ΣS equals the total activities of all aqueous sulfur species. Methods of calculating such diagrams were described in detail by Barnes and Kullerud (1961) and compared to Eh-pH diagrams by Barnes and Czamanske (1967). Complexes may be formed by reaction of metals with most of the aqueous species shown in Figure 8.1, including the subordinate species, polysulfide, thiosulfate, and sulfite. During oxidation of sulfides, these subordinate species may be produced in concentrations (Granger and Warren, 1969) that exceed those of the thermodynamically dominant sulfide and sulfate species. The thiosulfate and polysulfide species can persist metastably at significant concentrations for long periods, and, in the latter case, to above 200°C in near-neutral solutions (Hoffman, 1977; Giggenbach, 1974b). There may even be a small field near neutrality where polysulfide species are more stable than sulfides or sulfates at low temperatures (Giggenbach, 1974a). In contrast, sulfite species disproportionate relatively rapidly even at ambient temperatures (Avrahami and Golding, 1968).

The dominant complexing ligands among sulfur species are more likely to be HS^- or H_2S in hot, reducing hydrothermal solutions where equilibration is favored by time and high temperatures. Stronger complexing is caused by the bisulfide ionic field than by the dipole field of molecular H_2S. For this reason, it is in, *or close to*, the stability field of HS^- that bisulfide complexing is more likely to result in higher solubilities of the sulfides.

Figure 8.1 and 8.2 show that bisulfide ions are most stable in solutions that are both strongly reducing and alkaline. Note that there is still considerable uncertainty in the ionization constant for HS^- (Ramachandra Rao and Hepler, 1977) so that the pH of the HS^-/S^{2-} boundary is not well fixed, as emphasized by different choices for Figures 8.1 and 8.2. The stability field of HS^- lies between pHs equal to pK_1 for H_2S (between dotted curves on Figure 8.2 at the lower pHs and the poorly known pK_2, the upper dotted curve.) At higher temperatures, the bisulfide stability field deviates by as much as 3 pH units from neutrality (neutrality is shown by the heavy, solid curve of Figure 8.2) toward high alkalinity. Therefore, geologically improbable, very alkaline solutions are required above approximately 300°C for HS^- to become a dominant species. For this reason, bisulfide complexes may be important to ore transport only at temperatures below roughly 350°C.

Complexing with ore metals may also be depressed by competitive association of other cations with ligands, such as HS^-. Herr and Helz (1978) have shown that ion-pairing of HS^- with Na^+, Mg^{2+}, and Ca^{2+} is negligible up to about 150°C but that there is weak association to form NH_4HS. This is sur-

COMPLEX-FORMING LIGANDS

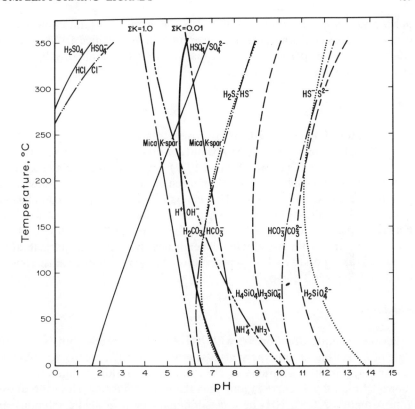

FIG. 8.2 The ionization of some aqueous acids and bases. The curves represent equal activities of aqueous species and separate their regions of predominance. The stability boundary for mica and K-feldspar in the presence of quartz is also shown for ΣK of 1.0 and 0.01. (J. D. Rimstidt assisted with the compilation of data and prepared the diagram.)

prising in view of the much stronger ligand field of the divalent cations. Their results, plus the stoichiometric activity coefficients for NaHS by Crerar and Barnes (1976), indicate that ion-pairing constants for NaHS and NaCl are very similar and not large. Consequently, ion-pairing with alkali cations probably doesn't significantly repress bisulfide complexing except at temperatures so high that prohibitively high pHs would be required anyway, as we have seen in Figure 8.2. There may be environments where $a_{NH_4^+}$ is great enough, approximately 0.6 m at 25°C, to decrease a_{HS^-} by 0.5x but these are probably not common as we shall see later in this chapter where NH_3 is examined as a complexing ligand.

Sulfate pairing with metal ions forms aqueous species less stable than those formed with alkali ions (Izatt et al., 1969; Reardon, 1975; and Shimizu et al., 1977). For this reason, we can neglect the contribution of Me-SO$_4$

pairing to the *solubilities* of ore minerals. Nevertheless, sulfate ion-pairing cannot be neglected when calculating equilibria at specified fixed values of ΣS, such as log a_{o_2}-pH diagrams, because the activities of alkali sulfate pairs can easily be a large fraction of ΣS.

Chloride Ions

As shown in Figure 8.2, HCl is effectively totally ionized by hydrothermal solutions below approximately 350°C but, at higher temperatures, it rapidly becomes ion-paired (Chou and Frantz, 1977). Similarly, it associates with the alkalies in progressively more stable molecular species under elevated temperatures (Naumov et al., 1974). This ion pairing is particularly complicated under these conditions because the degree of association varies considerably between aqueous $CaCl_2$, NaCl, and KCl (Johnson and Pytkowicz, 1978) and the concentrations of these alkalies depends largely on the feldspar alteration reactions taking place and the temperature. These reactions are discussed in detail in Chapters 3 and 5.

Ammonia

Because complexation occurs with NH_3, and only rarely with NH_4^+, the state of ionization of ammine species is critical for such complexes. Only in moderately alkaline solutions is NH_3 dominant at low temperatures, as shown by Figure 8.2 (curve of long plus three short dashes). However, above approximately 220°C, NH_3 becomes dominant even in acidic solutions; in addition, there is little association to form NH_4Cl (Hitch and Mesmer, 1976). For these reasons, hot hydrothermal solutions may contain ammine complexes in concentrations beyond those expected from lower temperature equilibria.

Organic Ligands

Although organic solids, liquids, and aqueous species have been noted in many low-temperature deposits (Saxby, 1976), especially Mississippi Valley-type deposits, little has been done to characterize these constituents (McLimans, 1977). This area of ignorance cannot be dismissed as inconsequential because there are many strong metal-organic complexes, such as with porphyrins, amino acids, carboxylates, and "humic" acids, that occur in nature (see, for example, Gardner, 1974). The upper thermal stability limits of these complexes are largely unknown, but it is thought that they can transport some components of Mississippi Valley-type ores (Giordano, 1978). Apparently, above about 300°C, organic complexes are dissociated and are unlikely to be important in high-temperature ore solutions. A major need remains for research to determine the types of organic constituents present in fluid inclusions and clays, particularly in low-temperature ore deposits, and the stabilities of related aqueous complexes.

CONCENTRATIONS IN ORE SOLUTIONS

Ore solubilities obviously depend not only on the types of complexing ligands present in hydrothermal solutions but also on their concentrations. Furthermore, we need an estimation of the concentrations of metals that must have accounted for ore transport if the contributions of each of the possible complexes is to be evaluated quantitatively. Estimates from diverse evidence indicate concentration ranges often for both ligands and metals.

The compositions of hydrothermal solutions thought to be responsible for the main stage of ore deposition in a variety of deposits have been compiled in Table 8-2. These compositions often are derived from analyses of fluid inclusions plus thermodynamic calculations based principally on the stabilities of observed mineral assemblages. For comparison, representative analyses of several modern metal-rich hydrothermal fluids are included in Table 8-2. Other analyses of similar fluids are given in Chapter 15. Comparison of the ranges of concentrations and their mean values suggest that the two types of fluids overlap sufficiently in composition to be taken as one population.

Ligand Concentrations

In experimental and thermodynamic investigations, it is convenient to begin from generally representative compositions of ore solutions. These can be obtained from Table 8-2 and associated data. In this table, Cl^- concentrations range 0.03–$7\ m$ and a typical value is about $1.0\ m$. Both total sulfate and sulfide vary greatly, from 10^{-10} to $0.3\ m$ and 10^{-4}–$10^{-1.4}\ m$ respectively. The sulfate to sulfide ratio may be either greater or less than one, depending upon the particular deposit considered. Typical mean values from this group might be $10^{-2}\ m$ for either total sulfate or total sulfide. However, evidence on compositions of solutions forming porphyry copper deposits presented in Chapters 3 and 10 suggests that the total content of sulfur-containing species lies between 0.2 and $2.0\ m$. Apparently, a representative value for these and the deposits of Table 8-2 around which research might center is $0.1\ m$ total sulfur-containing species. Carbonate concentrations extend from $10^{-3.4}$ to $1\ m$, but the analyses of geothermal fluids are distinctly lower than the results from ores; therefore, a representative value of $0.1\ m$ seems more reasonable.

In Table 8-2, there are no data on the concentrations of ammonia based on evidence from ores, but analyses from geothermal fluids range $10^{-3.9}$–$10^{-1.5}\ m$. Herr and Helz (1978) have examined similar analyses and select $10^{-2}\ m$ as a "plausible" concentration. A review of the ammonia concentrations of inclusion fluids from quartz or fluorite from various ore deposits by Volynets and Sushchevskaya (1972) also concludes that 10^{-2} m is typical. This value is consistent for the three sources and appears to be representative. Although Gunter (1973) has presented evidence that the N_2 content of hot springs (average approximately $10^{-3.6}$ m) in Yellowstone may originate from dissolved air and is not far below air-saturation (near $10^{-3.3}$ m) in that environment, the high or mean concentrations of geothermal and ore-forming fluids suggests that dissolved air is not the dominant source. Note that a representative value of 10^{-2} m is small compared to the average implicit in dissolving all atmospheric nitrogen in the oceans, 0.19 m (Bada and Miller, 1968).

The interpretations of alteration mineralogy in Chapter 5 suggest that hydrothermal solutions are normally weakly alkaline to somewhat acidic. The pH values of Table 8-2 lend further support to this range and suggest a pH averaging approximately one unit acidic, pH_{n-1}.

The above discussion leads to the following total composition as being probably the most common for hydrothermal ore-forming solutions: 1 m Cl^-, 0.1 m carbonate, 0.1 m in sulfur species, 0.01 m ammonia, and pH_{n-1}. Of course, solutions forming specific types of deposits may be systematically different from this average composition; this is discussed further according to the type of deposit. Relative *activities* will not necessarily follow these *concentrations* because of wide differences in the extent of ion-pairing among these components at various ionic strengths and temperatures. For example, the stronger pairing of sulfate relative to sulfide at high temperatures means that this solution may have a small ratio ($a_{HSO_4^-}/a_{H_2S}$), and be quite reducing although having total sulfate equal total sulfide concentration.

Below ore-grade mineralization may be the product of deposition from solutions of significantly lower ligand concentrations than the representative values given above. For example, in the detailed discussion of Broadlands, New Zealand in Chapter 15, a solution of $10^{-1.5}$ m chloride and $10^{-2.5}$ m sulfide was found to have deposited mineralization that is scientifically, but not economically interesting. In principal, too dilute a solution can cause a spatial retreat in the physical or chemical causes of deposition, for example, by transferring heat to decrease temperature gradients or by flushing away other mixing solutions. The effect of a retreat in the depositional interface is to disperse mineralization through large rock volumes and at sub-ore grades.

Metal Concentrations

In the first edition of this book, we reviewed several arguments suggesting that the concentrations of the common base and ferrous metals in ore-forming solutions must be at least 10 ppm (Barnes and Czamanske, 1967).

Table 8-2 Representative total (analytical) concentrations in hydrothermal solution

Location	ΣCl	ΣSO_4	Complex-Forming ΣS^{2-}
Ore Deposits[a]			
Creede, Colo.[b]	0.9–1.9	0.02–0.3	$10^{-3.8}$–$10^{-3.7}$
Darwin, Cal.[c]	>4.3	~0.01	—
Yatani, Japan[d]	0.1–0.2	10^{-6}–10^{-10}	~10^{-2}
Providencia, Mex.[e]	0.03–>4.3	<0.7	—
Echo Bay, N.W.T.[f]	~7	10^{-2}–10^{-3}	—
Kuroko Deposits, Japan[g]	0.4–1.5	—	<$10^{-1.7}$
Pasto Bueno, Peru[h]	0.4 to 3.5		
Tribag, Ontario[i]	—	<H_2S	10^{-2}–10^{-3}
Messina, S. Africa[i]	—		
S. W. Wisconsin, U.S.[j]	4.2–>5.1	—	>10^{-3}
Eureka, Col.[k]	0.1–0.6	$10^{-2.2}$	$10^{-1.4}$
Geothermal Fluids			
El Tatio, Chile[m]	0.2	—	$10^{-3.7}$
Imperial Valley[n]			
Salton Sea, Cal.	2.4	<$10^{-4.0}$	$10^{-3.7}$
Cerro Prieto, Mex.	0.3	$10^{-3.8}$	—
Cheleken, U.S.S.R.[o]	3.1	$10^{-2.8}$	$10^{-4.0}$
	2.9	$10^{-2.5}$	—
Central Miss. U.S.A.[p]	5.9	—	—
Eastern Kansas, U.S.A.[q]	6.0	$10^{-1.9}$	$10^{-2.9}$
	4.0	$10^{-2.9}$	$10^{-2.2}$

[a] Where sufficient data are available, the concentrations of the ore-depositing stage were selected.
[b] A vein deposit of Ag-Pb-Zn-Cu-Au in volcanics (Barton et al., 1977).
[c] A contact metasomatic Pb-Zn-Ag deposit in skarn (Rye et al., 1974).
[d] Pb-Zn and Au-Ag veins in volcanics (Hattori, 1975).
[e] Pb-Zn pipes in carbonates (Rye and Haffty, 1969).
[f] Vein deposit of U-Ni-Ag-Cu (Robinson and Ohmoto, 1973).
[g] Urabe (1974).
[h] Vein deposit of W-Cu-Zn-Pb (Landis and Rye, 1974; Norman, et al., 1976).
[i] Cu breccia pipe deposit (Norman, et al., 1976).
[j] Mississippi-Valley type Zn-Pb deposits (McLimans, 1977).
[k] Au-Ag-Pb-Zn-Cu-Cd veins in volcanics in stage 4 mineralization (Casadevall and Ohmoto, 1977).

Table 8-2. (*Continued*)

Species (molalities)		Metals (ppm)			pH at T°C	
ΣNH_3	ΣCO_3	Zn	Pb	Cu		
—	—	$10^{-2.8}$	$10^{-3.3}$ (Fe: 1)	$10^{-2.8}$	5.4	250
—	0.15 ± 0.06	<7700	—	740	4.8-6.7	350
—	—	—	—	—	~6	250
—	—	220-890	—	<70-530	—	300
—	1 to 0.1		(Ag: 0.1-1.)		4.2 ± .5	200
—	—	—	—	—	5.5 ± .5	250
					—	175-290
—	—				—	—
—	0.03-0.9	8700	~500	400	6.0 ± .3	150
—	0.01-0.2	>1-<1000	>1-<1000	(Au 10^{-2}-10^{-3})	4.3-5.9	300
$10^{-3.9}$	$10^{-1.1}$	—	—	—	—	263
					7	25
$10^{-1.5}$	$10^{-3.0}$	380	70	3	—	240
					6.1	25
—	$10^{-3.0}$	—	—	—	—	100
					7.9	25
—	$10^{-2.4}$	0.19	3.6	0.9	5.5	54
—	$10^{-3.4}$	2.3-4.7	3.6	0.8	5.4	80
$10^{-2.1}$		124	6.8			158
$10^{-3.0}$	$10^{-2.4}$	.06	1.7	.14	7.03	~20
$10^{-3.4}$	$10^{-2.2}$	.13	1.1	<.1	6.88	~20

[l] These are selected for their relatively high concentrations of metals and/or sulfide. Other examples are given by White in the first edition and in Chapters 13 and 15 of this edition.

[m] Cusicanqui et al. (1976).

[n] Wells sampled are Magmamax No. 1 near the Salton Sea and No. M-5 at Cerro Prieto (Phillips et al., 1977; Rimstidt and Barnes, 1977; and unpublished data).

[o] Analyses are of fluids from an upper, H_2S-rich aquifer and a lower, metal-rich aquifer. Metallic sulfides precipitate where they mix, either in surface tanks or by natural subsurface circulation. A general range of metal contents is 0.2-5.4 ppm Zn and 2-77 ppm Pb (Lebedov, 1967, 1972).

[p] Norphlet brine of the Pelahatchie oil Field (Carpenter et al., 1974).

[q] The upper analyses are for oil field brine from the Arbuckle Formation and the lower from the Viola Formation (McClure, 1973).

This minimum is reinforced by the concentrations of zinc, lead, and copper reported in Table 8-2 as well as in Chapter 14 for many fluid inclusions. The higher concentrations are more likely to be associated with ore deposition and are commonly in the range of tens to thousands (X0-X, 000) of parts per million. The only exceptionally low concentrations for deposits in Table 8-2, deduced by Barton et al. (1977) for Creede, are at variance with inclusion analyses for these ores of 1-25 ppm maximum copper by Tsui and Holland (1976) and the 100-1000 ppm copper by Czamanski et al. (1963). Another type of argument, based on dispersion from veins, has been used by Lavery and Barnes (1971) to approximate zinc concentrations of an ore solution for the Illinois-Wisconsin District. Because it depends on a stationary interface between the vein and adjacent wallrocks, the method is more accurate for narrow veins and gives a concentration of 200 ± 10 ppm zinc. The consensus is that ore-forming solutions carry X0-X, 000 ppm of the common metals for ores containing iron, zinc, lead, or copper and that a reasonable minimum concentration for ore transport remains 10 ppm (approximately 10^{-4} m).

For the rarer metals, ore solutions would be expected to be more dilute and the fragmentary evidence available is compatible with this supposition. Mercury-depositing hot springs contain 1 ppb mercury or more (see Chapter 15). In well BR-2 at Broadlands, gold concentration is 0.04 ppb but is 1-10 ppb in the Sunnyside Deposit, Eureka District, Colorado (Casadevall and Ohmoto, 1977). Silver is intermediate in concentration at Echo Bay, 0.1-1.0 ppm (Robinson and Ohmoto, 1973).

Where sulfidation of wallrocks is found, such as replacement of biotite with pyrite or pyrrhotite, this implies an excess of sulfide compared to metal concentrations of the ore solution. The metals with the lowest solubility products are those most likely to be found where this ratio falls below one.

SOLUBILITY DATA

A conceptual base has been prepared above for examining available information on the solubility of ore minerals. We have briefly reviewed the types of ligands capable of complexing metals in ore solutions, their availability for complexing (i.e., ionic state), and their probable concentrations. Minimum concentrations of importance to ore formation have also been approximated for complexes of these ligands with several metals. It now remains to be seen which complexes of these ligands and metals are most stable and if they can provide the required solubilities for the ore minerals in appropriate environments.

Quality of Data

When measuring solubilities, there are a minimum number of variables that must be measured or controlled to achieve a defined system that is thermodynamically interpretable. It is regrettable that in many published studies, this requirement has not been understood and the results of the considerable effort involved in such investigations become of only qualitative value.

The minimum number of determined variables to characterize a solubility experiment can be found simply by using the phase rule. Crerar (1974, Appendix X) has prepared a detailed explanation of this application. Intensive variables that might be determined include T, P_t, P_{H_2S}, P_{H_2O}, f_{O_2}, f_{S_2}, pH, or concentrations or activities of specific dissolved components. The minimum number of these is found from

$$f = c - p + 2 - r$$

where c is the number of components, p the number of phases, and r the number of restricted intensive parameters fixed initially. If we restrict the system to the aqueous phase, $p = 1$.

For a binary system, such as $ZnS-H_2O$, at a controlled T and measured P, then

$$f = 2 - 1 + 2 - 2 = 1$$

So long as the system can be kept closed to any loss of gas (H_2S) to a vapor, or gain of O_2 or other components (from the container walls), then only one variable need be fixed, such as the measured zinc concentration. In practice, there is normally some loss of H_2S to the vapor or sulfur to the container walls, the system becomes quaternary and two more intensive variables, such as f_{O_2} and P_{H_2S} or pH must be measured.

A second, more typical example may be a study of the solubilities of chalcopyrite + bornite + pyrite in $NaHS + H_2S + NaCl$ solutions. We find that $f = 3$ if again $p = 1$, $c = 7$, and $r = 5$ at controlled T and f_{S_2} (buffered by the sulfides), measured P_t, and concentrations of NaCl and NaHS fixed by the run composition. Selected variables may be P_{H_2S}, calculated from ($P_t - P_{H_2O}$), and measured Cu and Fe concentrations. Because there are several oxidation-reduction reactions involved, better resolution results if an f_{O_2} buffer is also present.

The particular intensive variables used should be those providing most resolution for species involved in any complexing reactions if the purpose of the study is the normal goal of identifying the stoichiometry and measuring the equilibrium constant of the dominant aqueous complex. Ideally, direct determination of more than the minimum number of intensive variables is desirable so that dual paths of calculation may be used for internal checking.

Equilibrium must be assured for variables to be treated as suggested above. For solubility measurements made by sampling reaction vessels (Barnes, 1971), it is easy to approach equilibrium from lower and higher temperatures and solubilities to effectively reverse the reactions to test for equilibration. Methods that depend upon weight loss of crystals involve an approach to equilibrium only from undersaturation and, consequently, are less reliable.

Once a set of parameters has been chosen and equilibration assured, then the remaining question is the accuracy within which each variable must be known to have useful resolution in determining either the stability of a complex or its contribution in models of ore transport. Anderson (1976) has recently described a Monte Carlo method for error analysis that allows the combining of multiple sources of error, such as that in solubility measurements, to assess a total error useful for our purposes. The method is based on making multiple calculations of a result using values with errors randomly distributed within each uncertainty range. He found that 50 iterations were sufficient to provide the net uncertainty when treating problems such as solubility calculations.

Stability of Complexes

A comprehensive review of halide, thio-, ammine, hydroxy-, and organic complexes of the metals of hydrothermal ore deposits would require several volumes of this size. Space constraints dictate, instead, that we review the stabilities of only those most important to hydrothermal ore genesis. For these reasons, those complexes stable only at relatively low temperature, below approximately 200°C, such as with S_x^{2-}, $S_2O_3^{2-}$, or organics, will not be examined here, particularly when critical compilations of their constants are available in convenient sources like Baes and Mesmer (1976) or Smith and Martell (1976). More useful here is an emphasis on higher temperature complexes, especially with chloride or bisulfide ions.

IRON COMPLEXES

There are still surprisingly few stability constants for chloride complexes of the metals that have been measured above 100°C. However, these include new data for several common ore metals and especially the ubiquitous iron.

Table 8-3 Stabilities of ferrous chloride complexes

Reaction	Log K						Reference
	25°C	200°C	250°C	300°C	350°C		
Aqueous Species							
(1) $Fe^{2+} + Cl^- \rightleftharpoons FeCl^+$	−0.51	1.58	2.34	2.33	—		Raman (1976), Crerar and Barnes (1976), Crerar et al. (1978)
(2) $Fe^{2+} + 2Cl^- \rightleftharpoons FeCl_2(aq)$	—	—	—	2.57	4.53		Crerar et al. (1978)
Pyrite							
(3) $FeS_2 + 2H^+ + nCl^- + H_2O(\ell)$ $\rightleftharpoons FeCl_n^{2-n} + 2H_2S(aq) + \tfrac{1}{2}O_2(g)$							
n = 1	—	−24.77	−19.47	−14.88	—		Crerar et al. (1978)
n = 2	—	—	—	−14.64	−9.37		Crerar et al. (1978)
Pyrrhotite							
(4) $FeS + 2H^+ + nCl^- \rightleftharpoons$ $FeCl_n^{2-n} + H_2S(aq)$							
n = 1	—	1.04	2.56	4.18	—		Crerar et al. (1978)
n = 2	—	—	—	4.42	7.51		Crerar et al. (1978)

Magnetite

(5) $Fe_3O_4 + 6H^+ + 3n\,Cl^- \rightleftharpoons$
    $3FeCl_n^{2-n} + 3H_2O(\ell) + \tfrac{1}{2}O_2(g)$

n = 1	—	−7.25	−2.43	3.04	—	Crerar et al. (1978)
n = 2	—	—	—	3.76	14.53	Crerar et al. (1978)

500–650°C, 2 kb

(6) $Fe_3O_4 + 6\,HCl(aq) + H_2(g) \rightleftharpoons$
    $3\,FeCl_2(aq) + 4\,H_2O(aq)$
    $\log K = (29410.6/T°K) - 25.90$ Chou and Eugster (1977)

Hematite

(7) $Fe_2O_3 + 4\,HCl(aq) + H_2(g) \rightleftharpoons$
    $2\,FeCl_2(aq) + 3\,H_2O(aq)$
    $\log K = (19420.2/T°K) - 16.497$ Chou and Eugster (1977)

Pyrite + Chalcopyrite + bornite

(8) $5/2\,CuFeS_2 + 2H^+ + Cl^- \rightleftharpoons$
    $FeS_2 + 1/2\,Cu_5FeS_4 + H_2S(aq) + FeCl^+$
    2.8, 5.0, 7.1, 8.4   Crerar and Barnes (1976)

(9) $5\,CuFeS_2 + 2H^+ + 1/2\,O_2(g) + Cl^- \rightleftharpoons$
    $3\,FeS_2 + Cu_5FeS_4 + H_2O(aq) + FeCl^+$
    26, 24, 24, 23   Crerar and Barnes (1976)

(10) $5/2\,CuFeS_2 + H^+ + 2\,O_2(g) + Cl^- \rightleftharpoons$
     $FeS_2 + 1/2\,Cu_5FeS_4 + HSO_4^- + FeCl^+$
     74, 67, 63, 59   Crerar and Barnes (1976)

Ferrous iron is dominant over ferric except in very abornmal, highly oxidizing, hydrothermal solutions. Increasing association with Cl^- is characteristic of Fe^{2+} at higher temperatures and a_{Cl^-}, as shown in Table 8-3, reactions (1) and (2). Crerar et al. (1978) find that the solubility of the assemblage pyrite + pyrrhotite + magnetite increases steeply above 250°C so that tens to thousands of parts per million of iron are stable in chloride complexes at 350°C in the pH range where muscovite is stable (Table 8-3, reactions 3, 4, and 5). These results are in agreement with those of Seward (1977) who also calculated that hydroxy complexes might become dominant under these conditions for the rare solution of negligible Cl^- concentration. However, Tremaine et al. (1977) have calculated that at 300°C and below, the solubility of magnetite in hydroxy complexes is less than 1 ppm within 2 pH units of neutrality. Geologically, the dominant iron species likely to be present changes with temperature from Fe^{2+} with $FeCl^+$, to $FeCl^+$ with $FeCl_2$, to $FeCl_2$ above roughly 300°C, depending on Cl^- concentration.

At 500–650°C, 2 kb, Chou and Eugster (1977) also found large solubilities of magnetite, but much less so of hematite, as $FeCl_2$ complexes (Table 8-3, reactions (6) and (7)). Apparently, this complex is sufficiently stable for hydrothermal transport of more than 10 ppm iron in weakly acidic solutions of 10^{-2} m or less sulfide at 300°C. Less acidity and higher sulfide concentrations may be present at higher temperatures. However, at lower temperatures ferrous chloride complexes become ineffective for moving iron in the quantities observed (see Seward, 1977, Figure 1) for example, such as in Mississippi Valley-type ores. Bisulfide complexes provide no alternative. Numerous experimental investigations have shown that they are insufficiently stable at temperatures up to at least 300°C (Crerar and Barnes, 1976; Seward, 1977). Organic, and perhaps ammine, complexes deserve investigation for the lower temperature deposits.

COPPER COMPLEXES

The solubility of copper at low temperatures is complicated by the large number of minerals that occur in nature and the various complexes that are stable. Detailed Eh-pH diagrams for the sulfides and oxides of copper present during supergene leaching have recently been calculated by Peters (1976; also in less detail for nickel, zinc, lead, molybdenum, and iron) and show the effects of ammine, chloride, and hydroxide complexes. There are about 40 distinct stability fields for combinations of a stable mineral plus a dominant aqueous species in each of his diagrams. Near-surface waters carry copper in several complexes according to Mann and Deutscher (1977), including $Cu(OH)_2$, $Cu(CO_3)_2^{2-}$, $CuSO_4$, and $CuCl^+$ as dominant species. The recent discovery of copper porphyrins in deep-sea sediments (Palmer and Baker,

1978) suggests that organic complexes should be added to their list. Rose (1976) has shown that transport by chloride complexes can account for the low-temperature formation of red-bed copper deposits without the additional contributions of these other complexes.

At higher temperatures in the more reduced environments of hydrothermal ores, this chemistry becomes much less complicated because the dominant aqueous species is cuprous (Crerar and Barnes, 1976; Snellgrove and Barnes, 1974) and not cupric as assumed by Romberger and Barnes (1970). The copper-containing sulfides also have relatively small solubility products (Barnes, 1975) so that in the presence of excess sulfide ion, only very strong ligands might compete successfully to complex copper. An alternative is for the undissociated sulfide to be included in the stoichiometry of the complex, as in the bisulfide complexes of Table 8-4. In anoxic marine waters, the bisulfide complexes have been calculated by Gardner (1974) to be the dominant inorganic aqueous copper species present.

Among the bisulfide complexes of Table 8-4, the dominant species changes with increasing temperature from $Cu(HS)_3^{2-}$ to $Cu(HS)_2^-$, and then $Cu(HS)_2(H_2S)^-$ with the crossover temperature depending upon ΣS (see Crerar and Barnes, 1976, Figure 6).

The simple cuprous chloride complex, CuCl, is apparently the dominant stoichiometry at all temperatures. Its stability at 200°C is readily evaluated by inverting the equilibrium constant for reaction (8) of Table 8-4

$$a_{Cl^-} \frac{a_{Cu^+}}{a_{CuCl}} = 10^{-2.2} \tag{8.2}$$

and noting that where $a_{Cu^+} = a_{CuCl}$, the ratio is unity and $a_{Cl^-} = 10^{-2.2}$. Consequently, any a_{Cl^-} greater than $10^{-2.2}$ makes CuCl dominant over Cu^+; since we have seen that hydrothermal solutions typically contain roughly 1.0 m Cl^-, the complex should normally be dominant over the ion.

The relative stabilities of the chloride and bisulfide complexes of copper have been compared in Crerar and Barnes (1976) and will be discussed here later in connection with porphyry copper deposits. The general conclusion for weakly acidic solutions is that CuCl predominates at 250-350°C up to ΣS = 0.1 to 1 but that bisulfide complexing becomes relatively more important below 250°C. For neutral solutions at $\Sigma S = 1.0$ and saturated with bornite + chalcopyrite at 200°C, the constants from Table 8-4 for reaction (2) tell us that approximately 70 ppm Cu is present in $Cu(HS)_3^-$ but that at higher temperatures, this complex becomes less important because of the divergence between $pK_{H,S}$ and neutral pH (Figure 8.2). Because of the increased stability of the chloride complex at temperatures above 250°C (see reaction 9 of Table 8-4), solubilities as CuCl may easily reach into thousands of parts per million copper.

Table 8-4 The stabilities of some copper complexes

Reaction	Log K				Reference
	200°C	250°C	300°C	350°C	
Bisulfide Complexes					
(1) $Cu_2S + 5 HS^- + H^+ \rightleftharpoons 2 Cu(HS)_3^{2-}$	$2.02 \pm .26$ ~-2.32	(22°C, $\mu = 2.1 - 4.4$) (75°C, $\mu = 2.1$)			Snellgrove and Barnes (1974 and unpublished data)
(2) $\frac{1}{4} Cu_5FeS_4 + HS^- + \frac{1}{2} H_2S(aq) \rightleftharpoons \frac{1}{4} CuFeS_2 + Cu(HS)_2^-$	-1.6	-1.5	-1.6	-1.6	Crerar and Barnes (1976)
(3) $\frac{1}{4} Cu_5FeS_4 + \frac{3}{2} H_2S(aq) \rightleftharpoons \frac{1}{4} CuFeS_2 + H^+ + Cu(HS)_2^-$	-8.7	-9.1	-9.6	$-10.$	Crerar and Barnes (1976)
(4) $Cu_5FeS_4 + 3 FeS_2 + \frac{3}{2} H_2O(aq) \rightleftharpoons 4 CuFeS_2 + \frac{3}{4} O_2(g) + H^+ + Cu(HS)_2^-$	-43	-38	-35	-33	Crerar and Barnes (1976)
(5) $\frac{1}{4} Cu_5FeS_4 + HS^- + \frac{3}{2} H_2S(aq) \rightleftharpoons \frac{1}{4} CuFeS_2 + Cu(HS)_2(H_2S)^-$	—	-3.8	-3.3	-3.1	Crerar and Barnes (1976)
(6) $\frac{1}{4} Cu_5FeS_4 + \frac{5}{2} H_2S (aq) \rightleftharpoons \frac{1}{4} CuFeS_2 + H^+ + Cu(HS)_2 (H_2S)^-$	—	-9.1	-9.3	-9.7	Crerar and Barnes (1976)
(7) $\frac{3}{2} Cu_5FeS_4 + 5 FeS_2 + \frac{5}{2} H_2O(aq) \rightleftharpoons \frac{13}{2} CuFeS_2 + \frac{5}{4} O_2(g) + H^+ + CuS(H_2S)_2^-$	—	-58	-52	-47	Crerar and Barnes (1976)
Chloride Complexes					
(8) $Cu^+ + Cl^- \rightleftharpoons CuCl$	$2.2 \pm .6$	$2.4 \pm .6$	$2.2 \pm .6$	$2.7 \pm .6$	Crerar and Barnes (1976)
(9) $\frac{1}{4} Cu_5FeS_4 + H^+ + Cl^- \rightleftharpoons \frac{1}{4} CuFeS_2 + \frac{1}{2} H_2S(aq) + CuCl$	-1.8	-0.08	1.3	2.6	Crerar and Barnes (1976)
(10) $CuFeS_2 + H^+ + \frac{1}{4} O_2(g) + Cl^- \rightleftharpoons FeS_2 + \frac{1}{2} H_2O(aq) + CuCl$	10	9.7	9.8	10	Crerar and Barnes (1976)
(11) $\frac{1}{4} Cu_5FeS_4 + \frac{1}{2} H^+ + O_2(g) + Cl^- \rightleftharpoons \frac{1}{4} CuFeS_2 + \frac{1}{2} HSO_4^- + CuCl$	34	31	29	28	Crerar and Barnes (1976)

ZINC AND CADMIUM COMPLEXES

Papers on the solubility of ZnS were reviewed in the first edition of this book (Barnes and Czamanske, 1967) and more recently by Barrett (1974). He derived approximate equilibrium constants from published data that were in gross agreement with his measurements in weakly to strongly acidic sulfide solutions. Assuming the following stoichiometries of the complexes, he calculated two constants

$$ZnS + 2H_2S(aq) \rightleftharpoons ZnS(H_2S)_2(aq)$$

$$\log K_{80°C} = -2.24 \tag{8.3}$$

$$ZnS + H_2S(aq) + HS^- \rightleftharpoons Zn(HS)_3^-$$

$$\log K_{80°C} = -1.89 \tag{8.4}$$

Because the sampling and analytic methods were suspect as discussed by Barrett (1974), the alternative constants given in Table 8-5 are preferred here. However, in weakly acidic solutions, the complex $ZnS \cdot (H_2S)_2(aq)$ of reaction (8.3) may be more stable than $ZnS \cdot H_2O$ (Table 8-5, reaction 2) at $m_{H_2S} > 0.01\ m$.

Because the chemistries of zinc and cadmium are similar, the constants for the sulfide complexes of both metals are shown for comparison in Table 8-5. The constants for reactions (1) and (3) in this Table imply that at 25°C, $Zn(HS)_4^{2-}$ exceeds the activity of $Zn(HS)_3^-$ only above $a_{HS^-} = 0.4$. If true, $Zn(HS)_4^{2-}$ becomes dominant only above $0.7\ m\ HS^-$, geologically unusual concentrations. In contrast, $Cd(HS)_4^{2-}$ is dominant above $a_{HS^-} = 10^{-2.12}$ and $Cd(HS)_3^-$ is always a minor species in comparison. Another possible inconsistency is the stoichiometry of the zinc and cadmium complexes in the H_2S-dominant pH and a_{O_2} region (Figure 8.1): $Cd(HS)_2$ versus $ZnS \cdot H_2O$ versus $ZnS(H_2S)_2$ of reaction (8.3).

The stability of the dominant zinc complex, $Zn(HS)_3^-$, is nearly independent of T (Barnes, 1960; Melent'yev et al., 1969), as shown also for $Cu(HS)_2^-$ in Table 8-4, reaction (2). There are no data available for Cd^{2+} complexes of this type at high temperatures.

The constants given in Table 8-5 for the formation of zinc chloride complexes are taken from the smoothed fit to available data by Helgeson (1969). The critical evaluation by Smith and Martell (1976) of more recent data as well as those available to Helgeson has led to a choice of exactly the same values for the four constants at 25°C. These values are in good agreement with the measurements at 200 and 400°C by Schulz (1974), these being the only experimental values available above 100°C.

At 25°C, the Cl^- complexes of Zn^{2+} are much more stable than those of

Table 8-5 The stabilities of some zinc and cadmium complexes

Reaction	Log K				References
Sulfide Complexes					
(1) $ZnS + H_2S\,(aq) + HS^- \rightleftharpoons Zn(HS)_3^-$	-3.0 ± 0.4 (25°C, $\mu = 1.0$)				Barnes (1975, Table 5)
	-2.9 ± 0.5 (100–200°C, $\mu = 1.0$)				Barnes (1975, Table 5) Barnes (1960, Figure 65),
(2) $ZnS + H_2O\,(aq) \rightleftharpoons ZnS \cdot H_2O\,(aq)$	$-5.87 \pm .01$ (25°C, $\mu = 1.0$)				Gubeli and Ste-Marie (1967)
(3) $ZnS + H_2S\,(aq) + 2HS^- \rightleftharpoons Zn(HS)_4^{2-}$	~ -2.6 (25°C, $\mu = 1.0$)				[a]
(4) $CdS + H_2S\,(aq) \rightleftharpoons Cd(HS)_2$	-4.57 (25°C, $\mu = 1.0$)				Ste-Marie et al. (1964)
(5) $CdS + H_2S\,(aq) + HS^- \rightleftharpoons Cd(HS)_3^-$	-2.69 (25°C, $\mu = 1.0$)				Ste-Marie et al. (1964)
(6) $CdS + H_2S\,(aq) + 2HS^- \rightleftharpoons Cd(HS)_4^{2-}$	-0.33 (25°C, $\mu = 1.0$)				Ste-Marie et al. (1964)
	25°C	100°C	200°C	300°C	
Chloride Complexes					
(7) $Zn^{2+} + Cl^- \rightleftharpoons ZnCl^+$	0.43	1.82	(3.9)	(6.0)	Helgeson (1969),[b] Smith and Martell (1976)
(8) $Zn^{2+} + 2Cl^- \rightleftharpoons ZnCl_2\,aq$	0.61	2.13	(4.4)	(6.9)	Helgeson (1969)[b]
(9) $Zn^{2+} + 3Cl^- \rightleftharpoons ZnCl_3^-$	0.53	2.23	(4.8)	(7.7)	Helgeson (1969)[b]
(10) $Zn^{2+} + 4Cl^- \rightleftharpoons ZnCl_4^{2-}$	0.20	2.14	(5.0)	(8.3)	Helgeson (1969)[b]
(11) $ZnCl_2(aq) + 2Cl^- \rightleftharpoons ZnCl_4^{2-}$	$+0.7 \pm 30\%$ [400°C, 1(?)kb, $\mu = 0$]				Schulz (1974)

[a] Stoichiometry assumed to be the same as that found for the cadmium complexes by Ste-Marie et al. (1964) and evaluated using the data shown in Barnes (1965, Figure 5).
[b] Values are for $\mu = 0$ and those in parentheses are more uncertain.

SOLUBILITY DATA

Cd^{2+} (Belousov and Alovyainikov, 1975) and $ZnCl_4^{2-}$ is stable at moderate a_{Cl^-}. However, Franck (1973, Figure 9) has shown that $ZnCl_2$ becomes progressively more dominant over other stoichiometries, especially $ZnCl^+$ and $ZnCl_3^-$, for temperatures increasing from 25 to 400°C. The trend toward less ionic and more molecular species in general is a consequence of the rapid decrease in the dielectric constant of water at higher temperatures. The constant for reaction (11) shows that $ZnCl_4^{2-}$ exceeds $ZnCl_2$ in activity at 400°C only if a_{Cl^-} is greater than 0.5. This activity may be recast in terms of NaCl concentration. Because NaCl is strongly associated at this temperature (ionization constant ~ 10^{-2}, depending on pressure), in excess of approximately 5 m NaCl must be present to form $ZnCl_4^{2-}$. For these reasons, the dominant complex of zinc in hydrothermal solutions is $ZnCl_2$.

LEAD COMPLEXES

There are available recent critical reviews of sulfide and chloride complexes of lead (Hamann and Anderson, 1978; Giordano, 1978; and Giordano and Barnes, 1979). They are in agreement with the constants given in Table 8-6 for the bisulfide complexes. We find that the dominant complex is $Pb(HS)_2$ at $\Sigma S < 0.1$ at 25°C, $\Sigma S < 10$ at 100°C, and in neutral to acidic solutions with $\Sigma S < 1.2$ at 200°C. At 300°C, $Pb(HS)_2H_2S$ should be dominant in neutral to acidic solutions at $\Sigma S < 1.0$. $Pb(HS)_3^-$ only becomes dominant at temperatures below about 50°C and in neutral to slightly alkaline solutions. The changes in the distribution of the lead complexes are shown in Figure 8.3.

In our experiments, PbS solubility increased with ΣS, P_{H_2S}, and especially temperature, reaching a maximum measured concentration of just 47 ppm PbS at 300°C, $P_{H_2S} = 75$ atm, and 0.5 m NaHS. Concentrations above 10 ppm were found only above approximately 200°C and 1.0 m NaHS. Therefore, lead sulfide complexes are not sufficiently stable to form low temperature, hydrothermal ore deposits, as also concluded by Hamann and Anderson (1978).

The constants for the chloride complexes in Table 8-6 may have uncertainties as large as an order of magnitude at 200 and 300°C because of poor agreement among the high temperature experimental values and Helgeson's earlier extrapolation. His approximated values are preferred here in the absence of better high temperature data (see the discussion by Giordano, 1978). In spite of their large uncertainties, these constants show a moderate stability for lead chloride complexes that increase significantly with temperature.

The extent of ion pairing of these complexes with other cations at 25°C is indicated by the constants in Table 8-6 for reactions (8) and (9). The activity

Table 8-6 Stabilities of lead complexes

Reaction	Log K^a			
	25°C	100°C	200°C	300°C
Sulfide Complexes				
(1) $PbS + H_2S(aq) + HS^- \rightleftharpoons Pb(HS)_3^{-\ b,c}$	-5.62	-5.49	—	—
(2) $PbS + H_2S(aq) \rightleftharpoons Pb(HS)_2(aq)^{b,c}$	(-7.6)	-4.97	-4.78	—
(3) $PbS + 2H_2S(aq) \rightleftharpoons PbS(H_2S)_2\ (aq)^{b,c}$	—	—	-4.88	-4.40
Chloride Complexes				
(4) $Pb^{2+} + Cl^- \rightleftharpoons PbCl^{+\ b,d,e}$	$1.59 \pm .02$	1.73	(2.1)	(3.0)
(5) $Pb^{2+} + 2Cl^- \rightleftharpoons PbCl_2(aq)^{b,e}$	$2.16 \pm .34$	2.04	(2.6)	(3.9)
(6) $Pb^{2+} + 3Cl^- \rightleftharpoons PbCl_3^{-\ b,e}$	$1.77 \pm .13$	2.13	(3.0)	(4.6)
(7) $Pb^{2+} + 4Cl^- \rightleftharpoons PbCl_4^{2-\ b,e}$	$1.1 \pm .4$	2.05	(3.2)	(5.3)

	K (25°, $\mu = 3.0$)				
	Na^+	K^+	Rb^+	Cs^+	NH_4^+
Ion Pairing of Chloride Complexes					
(8) $PbCl_3^- + X^+ \rightleftharpoons XPbCl_3(aq)^{b,f}$	0.09	0.49	0.62	0.82	0.1
(9) $PbCl_4^{2-} + X^+ \rightleftharpoons XPbCl_4^{-\ b,f}$	0.72	2.07	2.61	3.14	1.0

^{a}Values in parentheses are less certain.
^{b}Giordano (1978), values at $\mu = 0$ and with ± 0.2 uncertainty.
^{c}Giordano and Barnes (1979), values at $\mu = 0$ and with ± 0.2 uncertainty.
^{d}values at 25°C from Smith and Martell (1976).
^{e}Helgeson (1969).
^{f}Mironov et al. (1965).

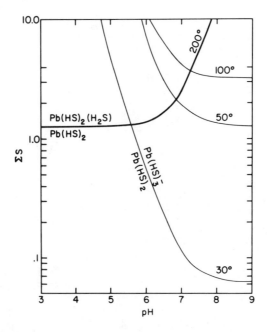

FIG. 8.3 The stoichiometries of lead sulfide complexes at different acidities, sulfide activities, and temperatures.

of the cation, where the complex is half ion-paired, is equal to K^{-1}. Ion pairs with the more abundant cations become dominant by reaction (8) only where $a > 1.6$ and in reaction (9) where $a > 0.5$. Reaction (8) may not be important at high temperatures because the constants for reactions (5), (6), and (7) suggest that $PbCl_3^-$ is apparently a minor species compared to $PbCl_2$ or $PbCl_4^{2-}$. The extent of reaction (9) at high temperatures deserves investigation.

MERCURY COMPLEXES

The constants for the bisulfide complexes of mercury are compiled in Table 8-7. The uncertainties in these values have not been established but they cannot be so large as to impugn an apparent stability that is remarkable for these complexes. The relative stabilities of these complexes are compared in Figure 8.4, which shows that approximately 1.0 ppm mercury dissolves in bisulfide complexes in equilibrium with metacinnabar at 25°C in near neutral solutions of $\Sigma S \sim 0.1$. The solubility increases approximately two times by 100°C and four to seven times by 200°C for the bisulfide complexes (Barnes et al., 1967) and by similar amounts for the sulfide complex as reported by Khodakovsky et al. (1975). These authors also suggest that vapor

Table 8-7 The stability of mercury bisulfide complexes

	Log K	T, °C	μ	Solid phase	References
(1) HgS + 2 H$_2$S (aq) ⇌ HgS (H$_2$S)$_2$ (aq)	−4.25	20	1.0	metacinnabar	Barnes et al. (1967)
	−4.31	20	0	cinnabar	Barnes et al. (1967)
	−3.0	100	0	cinnabar	Barnes et al. (1967)
(2) HgS + H$_2$S (aq) + HS$^-$ ⇌ Hg(HS)$_3^-$	−3.50	20	1.0	metacinnabar	Barnes et al. (1967)
	−3.59 ± 0.3	20	1.0	cinnabar	Barnes (1975)
	−3.3 ± 0.4	100	1.0	cinnabar	Barnes (1975)
(3) HgS + 2 HS$^-$ ⇌ HgS(HS)$_2^{2-}$	−3.51	20	1.0	metacinnabar	Barnes et al. (1967)
	−3.60	20	1.0	cinnabar	Barnes et al. (1967)
(4) HgS + S^{2-} ⇌ HgS$_2^{2-}$	0.57	20	1.0	metacinnabar	Barnes et al. (1967)
	0.48	20	1.0	cinnabar	Barnes et al. (1967)
(5) HgS + HS$^-$ + OH$^-$ → HgS$_2^{2-}$ + H$_2$O (aq)					

−Log K (μ = 0; for metacinnabar)

	25°C	100°C	150°C	200°C	250°C	
1 bar	0.31	0.49	0.57	0.63	0.69	Khodakovsky et al. (1975)
250 bar	0.38	0.53	0.61	0.64	0.71	
500 bar	0.44	0.58	0.68	0.69	0.74	

Temperature dependence (in K)
for cinnabar (μ = 0): $pK_5 = 1.185 - 260.32/T$
for metacinnabar (μ = 0): $pK_5 = 1.3565 - 366.42/T$

Pressure dependence (in bars and K)
for cinnabar (μ = 0): $pK_{5,p} = pK_5 + \left\{ \dfrac{(P-1) \; 20.79689 - 0.0206\,T}{2.303\,RT} \right\}$

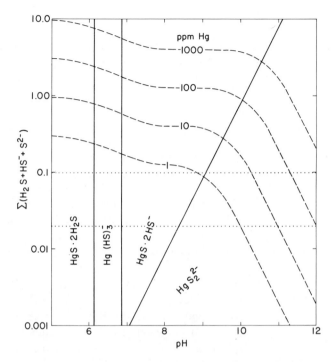

FIG. 8.4 The stoichiometries of mercury sulfide complexes at different acidities and sulfide activities at 20°C. The solubility contours are for metacinnabar and may be converted to those for cinnabar by multiplying by 0.8x. (Modified from Barnes, Romberger, and Stemprok, 1967, Figure 7.)

transport of mercury may be possible near 200°C, a possibility that was explored by Krauskopf (1964) with equivocal results.

Mercury readily forms other complexes besides those with aqueous sulfide ligands, including Cl^-, OH^-, and a variety of organic complexes (for example, see Baes and Mesmer, 1976). However, the extremely small solubility product of HgS, as shown in Table 8-1, means that mercury will bond first to any aqueous sulfide species in preference to these other ligands. Nriagu and Anderson (1970) have calculated that, in order for mercury chloride complexes to reach 1 ppm mercury at pH 4 or higher at 25°C, $\Sigma m_{H2S} + m_{HS^-}$ must be less than approximately 10^{-19}. For this reason, ligands other than chlorides are likely to be of importance for ore deposition. There remain a variety of organic complexes (Gardner, 1974) that may be sufficiently stable at 100–200°C in sulfide-containing solutions to dissolve at least 1 ppm mercury.

SILVER COMPLEXES

The chloride complexes of silver are strong as the particularly thorough investigation by Seward (1976) has demonstrated. His results, summarized in Table 8-8, also show that $AgCl_2^-$ is dominant over a progressively wider range in a_{Cl^-} with increasing temperatures. Similar studies are needed for ammine complexes of silver.

The sulfur-containing complexes of silver are poorly known. The review in the first edition (Barnes and Czamanske, 1967) included data showing Ag_2S solubilities near 1 ppm in a polysulfide complex, $Ag(HS)S_4^{2-}$, and a sulfide complex possible with a stoichiometry of $Ag_2S(H_2S)_2$. Schwarzenbach and Widmer (1966) suggest instead AgHS at pH < 4 and $Ag(HS)_2^-$ at pH 5-9 but because they only worked with solutions at a fixed 0.02 m sulfide concentration, their stoichiometries are not well established by this study alone. When compared with $Cu(HS)_2^-$ (see Table 8-4), for the relatively noble-acting Cu^+, the identical stoichiometries lend further support to their complex.

Although the composition and stabilities of the complexes of sulfides with silver are unknown at high temperatures, available solubility data (Melent'yev et al., 1969 and Sugaki, Scott, and Barnes, unpublished) show clearly that bisulfide complexes become very stable. Ag_2S solubility exceeds 10 ppm in near-neutral solutions at 20°C at $\Sigma S \geq 1.0$; it increases steeply above about 100°C and between 100°C and 160°C, the solubility in near-neutral solutions increases approximately fourfold.

GOLD COMPLEXES

There are many mechanisms that may affect the mobility of gold in oxidizing, near-surface environments, including complexes with Cl^-, Br^-, $S_2O_3^{2-}$, SCN^-, CN^-, organics, etc., and in colloids (Lakin et al., 1974; Goni et al. 1967); in addition complexes with S_x^{2-} or NH_3 are found in moderately reducing solutions (Skibsted and Bjerrum, 1974). In hydrothermal solutions, chloride and sulfide complexes are apparently dominant, although others with tellurium, selenium, bismuth, antimony, and/or arsenic are probably locally prominent judging from the occurrences of related minerals. Seward (1973) has investigated the stabilities of gold sulfide complexes at high temperature with the results shown in Table 8-9. The complex, $Au(HS)_2^-$, provides the maximum gold solubility at fixed ΣS and at pH = pK_{H_2S}. Therefore, this complex is dominant in near-neutral solutions at low temperatures but only in increasingly alkaline solutions at higher temperatures as shown by the left dotted curve of Figure 8.2. Solubilities resulting from this complex exceed 10 ppm Au ($10^{-5.3}$ m) above 250°C in near-neutral solutions with $m_{NaHS} > 0.15$; the highest value measured in

Table 8-8 The stabilities of silver chloride and sulfide complexes

Reaction	Log K						
	18°C	100°C	150°C	197°C	277°C	353°C	
Chloride Complexes[a]							
(1) $Ag^+ + Cl^- \rightleftharpoons AgCl\,(aq)$	3.48 +.03	2.88 ±.10	2.88 ±.09	2.87 ±.03	3.29 ±.09	4.26 ±.03	
(2) $Ag^+ + 2Cl^- \rightleftharpoons AgCl_2^-$	5.31 ±.03	4.46 ±.05	4.45 ±.04	4.57 ±.02	5.10 ±.06	6.60 ±.03	
(3) $Ag^+ + 3Cl^- \rightleftharpoons AgCl_3^{2-}$	5.44 ±.05	3.85 ±.08	3.73 ±.07	3.67 ±.04	—	—	
(4) $Ag^+ + 4Cl^- \rightleftharpoons AgCl_4^{3-}$	4.19 ±.07	1.94 ±.20	—	—			
	$\mu = 0.1$			$\mu = 1.0$			
Sulfide Complexes (at 25°C)[b]							
(5) $Ag^+ + HS^- \rightleftharpoons AgHS\,(aq)$		13.6		13.30			
(6) $Ag^+ + 2HS^- \rightleftharpoons Ag(HS)_2^-$		17.7		17.17			
(7) $2Ag(HS)_2^- \rightleftharpoons Ag_2S(HS)_2^{2-} + H_2S\,(aq)$		—		3.2			

[a] Seward (1976).
[b] Smith and Martell (1976).

Table 8-9 Stabilities of gold sulfide complexes

Reaction	Log K				Reference
	175°C	200°C	225°C	250°C	
(1) $2Au + 2HS^- \rightleftharpoons 2AuS^- + H_2(g)$		−11.2 (25°C, $\mu = 0$?)			Kakovskii and Tyurin (1962)[a]
(2) $Au + H_2S\,(aq) + HS^- \rightleftharpoons Au(HS)_2^- + \frac{1}{2}H_2(g)$	−1.29	−1.28	−1.22	−1.19	Seward (1973)[b]
(3) $2Au + H_2S\,(aq) + 2HS^- \rightleftharpoons Au_2S\,(HS)_2^{2-} + H_2(g)$	−2.14	−2.40	−2.50	−2.55	Seward (1973)[b]

[a] This stoichiometry is compatible with their data but it is not proven.
[b] For $\mu = 0.50$ and total sulfide molality of 0.50 at 1 kb.

these solutions was 546 ppm at 238°C, 1 kb, $m_{NaHS} = 0.50$, $m_{H_2S} = 2.4$, and $m_{NaCl} = 2.5$. Apparently the gold sulfide complexes are very stable.

Gold forms both aurous and auric chloride and hydroxychloride complexes, factors that have complicated the identification of their compositions and stabilities. Even at 25°C, they have not been reliably defined (Baes and Mesmer, 1976). According to Henley (1973), Au_2Cl_6 may be the principal aqueous complex but further work is obviously needed for confirmation. He finds in relatively oxidizing conditions above 500°C, 2 kb, with approximately 2 m KCl, that gold solubilities exceed 1000 ppm. This work demonstrates high stabilities of gold complexes, but stoichiometries are uncertain.

COMPLEXES OF OTHER METALS

There are too few published data on the solubilities and complexes of other ore metals at high temperatures to warrant extended discussion here. Reviews of current data are available for some of these metals. Sulfide complexing of arsenic and antimony is found in Chapter 8 of the first edition (Barnes and Czamanske, 1967, p. 364-365). The fragmentary information on hydrothermal solubilities of uranium minerals is examined by Rich et al. (1977, Chapters 3-6). Transport is favored by the uranyl state (U^{6+}), which readily is included in complexes with OH^-, CO_3^{2-}, SO_4^{2-}, F^-, and possibly PO_4^{3-} and deposition may result from reduction to relatively insoluble uranous (U^{4+}) minerals, especially uraninite, UO_2.

There is virtually no information published on the high temperature solubility of UO_2 in reduced systems containing NH_3, H_2S, and HS^-, although vein deposits of uraninite frequently contain marcasite and sulfarsenides and were deposited below 250°C.

There are some high-temperature measurements that have been reported for tin minerals. Because of the great stability of cassiterite (Moh, 1974; 1975a, b), it is comparatively far more common than other tin minerals and its solubility is of most consequence to tin ores. The typical association of these ores with fluorite or tourmaline implies that F^- is one of the complexing ligands. Sn^{4+} readily complexes with F^- and also OH^- or Cl^- or combinations thereof. There are few equilibrium constants measured at high temperatures for OH^- and/or F^- complexes of Sn^{4+} (Klintsova et al., 1975), but further work is needed on chloride complexes and on complexes of Sn^{2+} with H_2S, HS^-, NH_3, and Cl^-, because of its incorporation in stannite and other sulfides.

There are virtually no hydrothermal solubility data now available for tungsten (see Bryzgalin, 1976), the arsenides, selenides, tellurides, platinum group metals, and the sulfosalts in general.

PREDICTING SOLUBILITIES

A variety of structural (e.g., bonding, size, and charge determination) and thermodynamic and statistical thermodynamic methods have evolved that have achieved significant successes in predicting the stabilities of simple aqueous complexes, most often either by interpolation or extrapolation from known constants. A review of some of these methods of geochemical interest has been prepared by Langmuir (1979). It is beyond the scope of this chapter to discuss the nature of these methods except to encourage caution. For many of the complexes for which constants are presented in Tables 8-3-8-9, the uncertainties are unknown and, in some cases, even the stoichiometry is not firmly established. In addition, many of the complexes include either dual ligands (e.g., H_2S and HS^-) or cations. Generally the predictive methods are not sufficiently well developed to be used meaningfully for such systems.

Tables 8-3-8-9 demonstrate that normally a succession of complexes of different stoichiometries become dominant at higher temperatures in hydrothermal solutions, particularly between 25 and perhaps 300°C. To predict the temperatures of transition between complexes requires a resolution not yet attainable. Yet without knowing the composition of the dominant complexes, the reactions and processes potentially capable of causing ore deposition cannot be usefully evaluated.

ORE DEPOSITION

If we limit our discussion to the most common hydrothermal ore minerals, the sulfides, then there are in principle only a few types of processes that may cause precipitation. Because the ore sulfides are usually absent along the presumed flow channels where observed near ore bodies, the transporting fluid is very probably undersaturated up to the site of the deposit. What are the processes causing the solution to become saturated? The only possibilities that could deposit valuable concentrations are (1) cooling, (2) decrease in the activity(ies) of the complex-forming ligand(s), or (3) increase in $a_{S^{2-}}$. Reactions with these effects may be homogeneous or heterogeneous and involve either loss of gases or, typically, reaction with host rocks. If mixing with a second solution occurs, then addition of metals may be a fourth depositional process. Pressure decrease is not included because it has little effect locally on the ionic ore solution unless throttling and associated boiling take place. In this case, the change in composition may be included in effect as part of processes (2) and (3).

In detail, there are a myriad of reactions that may accompany and control these four types of depositional processes. This will become evident in the examination of mass transfer phenomena by Helgeson in Chapter 11. Although we recognize that the reactions accompanying precipitation form a complex, interdependent matrix, an understanding of their individual effects can best be achieved by conceptually isolating some representative reactions to note their consequences. This can be done most easily by grouping complexes on the basis of their ligands. Because complexing reactions typically equilibrate in milliseconds and rarely as slowly as in minutes even at 25°C (Cotton and Wilkinson, 1962, 548-559), rates of precipitation will be assumed here to be geologically instantaneous except where sulfate reduction is involved.

Deposition from Chloride Complexes

A representative reaction for deposition of sulfides from chloride-complexed metals, Me^{2+} in weakly acidic solutions can be written

$$MeCl_2(aq) + H_2S(aq) \rightarrow MeS + 2H^+ + 2Cl^- \qquad (8.5)$$

This reaction is written not in terms of HS^- or S^{2-} but in terms of H_2S, because it is the predominant sulfide species at pHs below neutral at low temperatures as well as in mildly alkaline solutions at temperatures above approximately 300°C. Equation (8.5) shows that either the degree of saturation or the extent of deposition of the sulfides will increase as a consequence of the following processes.

1. Increased H_2S concentration (in terms of $a_{S^{2-}}$) due to, for examples, sulfate reduction, reaction from organics, or mixing with sulfide solutions (e.g., at Cheleken, Lebedev 1972). Because the concentration of H_2S is typically low in equilibrium with Cl^- complexes (except above about 300°C or where strongly acid), this reaction goes virtually quantitatively to the right upon addition of H_2S. Note that very rapid precipitation and consequent tiny grain size results from mixing with sulfide solutions.
2. Increased pH caused by either reaction with carbonates or feldspars or boiling off acid gases. Boiling is also quantitatively effective as found by Nash (1973) for the lead-zinc-copper-silver-gold veins of the Park City District, Utah. He observed an erratic degree of fluid inclusion filling which is presumptive evidence for boiling found in many hydrothermal deposits (see Chapter 15).
3. Decreased chloride concentration (a_{Cl^-}), resulting from dilution by circulating meteoric waters or caused by reactions adding strong Cl^--ion-pairing cations like Ca^{2+}.
4. Decreased temperature (see Chapter 12).

The quantitative dependence of lead and zinc solubility on these parameters has been discussed by Anderson (1973) and of copper and iron solubilities by Crerar and Barnes (1976). The concentration dependencies can be expressed for the metals, Me by

$$\left[\frac{\partial (\text{Me})}{\partial a_{S^{2-}}}\right]_{P,T,\mu,a_{Cl^-},a_{H^+}\cdots}, \quad \left[\frac{\partial (\text{Me})}{\partial a_{H^+}}\right]_{P,T,\mu,a_{S^{2-}},a_{Cl^-}\cdots},$$

$$\left[\frac{\partial (\text{Me})}{\partial a_{Cl^-}}\right]_{P,T,\mu,a_{H^+},a_{S_2}\cdots}, \text{ and } \left[\frac{\partial (\text{Me})}{\partial T}\right]_{P,\mu,a_{Cl^-},a_{H^+},a_{S^{2-}}\cdots},$$

as given by Crerar and Barnes (1976). However, these partial differentials are functions of many variables and their evaluation must be at a fixed set of conditions out of many that are pertinent.

The relative chemical efficiencies of these processes in causing sulfide precipitation can be summarized for geologically feasible changes in conditions: very effective are processes (1) and (2), addition of sulfide or increase in pH; processes (3) and (4), dilution or cooling may be moderately effective; and pressure decrease without boiling is ineffective. The efficiency of dilution increases as the nth power, where n is the stoichiometric coefficient for Cl^- in the complex, so that it is more effective for $MeCl_4^{2-}$ and less so for $MeCl_2$. For $MeCl_2$ in an *initially sulfide-saturated solution*, 1:1 dilution by Cl^--free water causes deposition of 85 ± 10% of the metal in solution, other conditions remaining constant. A temperature decrease of 100°C generally deposits approximately 90% of the metal. On the other hand, for $\Delta pH = +1.0$ deposition is near 99%, an efficiency comparable with sulfide addition where the number of moles of sulfide precipitated nearly equals that added.

Although sulfate reduction has been suggested often as a cause for ore deposition from chloride complexes, there are reasons for doubting that it is a primary agent. Below 250°C, the rate of reduction becomes very slow even for powerful reductants like H_2 (see Chapter 10). Sakai (1977) finds some oxygen isotope exchange at lower temperatures with a half-time of reaction of 5.6 yr at neutrality; the rate increases markedly at lower pHs. According to Orr (1974, 1975) there may be some reduction in particularly organic-rich systems as low as 77–121°C but in the hundreds of hours. Microbial reduction occurs optimally at 30–45°C, far below hydrothermal temperatures. These rates do not imply that, in hydrothermal systems between 100 and 250°C, sulfate reduction should be common. Furthermore, many moles of reductant are necessary; $SO_4^{2-} + 4H_2 \rightarrow H_2S + 2OH^- + 2H_2O$. The large proportion necessary for reduction makes difficult the maintaining of an interface while ore-grade concentrations are being precipitated.

Deposition from Sulfide Complexes

Deposition from the sulfide complex of the most common stoichiometry can be written

$$Me(HS)_3^- \rightarrow MeS + HS^- + H_2S(aq) \qquad (8.6)$$

Clearly, any processes that reduce the product ($a_{HS^-} \cdot a_{H_2S}$) will either increase the degree of saturation in this complex or cause deposition. Also, for many metals, a decrease in temperature or pressure, especially where boiling takes place, may cause deposition. There is a large variation in the fraction of dissolved metals that might be expected to precipitate depending upon the processes affecting the activity product or upon changes in physical conditions.

Decreasing pressure may allow $H_2S(aq) \rightarrow H_2S(g)$ if a vapor phase separates. Only if throttling provides a large decrement in pressure could much sulfide deposition be caused by this process. Dilution is more effective in that an *initially saturated* solution would precipitate approximately half of its metals from $Me(HS)_3^-$ if diluted 1:1 with sulfide-free water and other conditions remain constant. Decreasing temperatures has little effect on the concentrations of $Zn(HS)_3^-$; on the other hand roughly 75% of the mercury, copper, silver, and lead in sulfide complexes in near-neutral solutions is deposited on cooling from 200 to 100°C. The temperature effect includes both changes in the extent of ion-pairing in associated reactions and decreases in stability of the complexes. Decreasing pH also changes the sulfide activity product of reaction (8.6) by

$$HS^- + H^+ \rightarrow H_2S(aq)$$

Because the complex $Me(HS)_3^-$ reaches maximum concentration where the pH = pK_{H_2S}, any pH decrease below this value causes deposition. About 90% of the complexed metal should deposit if the pH were to drop by 2.0 units from pK_{H_2S}. Most effective chemically in causing deposition is oxidation because it decreases both total sulfide concentration and pH (unless well buffered by the host rocks), for example, by

$$H_2S(aq) + 2O_2 \rightarrow H^+ + HSO_4^- \qquad (8.7)$$

A difficulty here is the quantity of the oxidant locally available. Air-saturated waters have approximately $10^{-3.6}$ m O_2 and $10^{-3.3}$ m N_2 either of which, depending on kinetics, may cause oxidation (Fisher et al., 1975).

Fluid inclusion gases also indicate that oxidation is commonly associated with deposition of hydrothermal ores. Norman et al. (1976) analyzed by mass spectrometer inclusions from several deposits and found H_2S contents typically of 0.01–0.001 m. SO_2 was less common and its distribution was explicable on the basis of an influx of oxidizing meteoric water during deposition. They conclude from this evidence that oxidation of hydrothermal solutions may be a critical factor in ore precipitation. Except for probable armoring that may impede reaction, hematite or goethite could also be oxidizing agents.

The chemical efficiencies of these processes are generally similar in order of importance but the effects in some cases are in opposite directions. In summary, solubilities are decreased for the two types of complexes by

Sulfide complexes: $-\Delta m_{S^{2-}}$, $+O_2$, $-\Delta pH$, dilution, $-\Delta T$, $-\Delta P_t$

Chloride complexes: $+\Delta m_{S^{2-}}$, $-O_2$, $+\Delta pH$, dilution, $-\Delta T$, $-\Delta P_t$.

Replacement

In deposits formed above roughly 250°C, ore textures give evidence that replacement is common, but that it is rare and limited in extent at lower temperatures. Nevertheless, there has been remarkably little research carried on that relates the host dissolution reactions with those of precipitation. Instead, somewhat more effort has been focused on diffusion and its influence on trace elements within minerals, rather than wholesale replacement and the infiltration metasomatism of ores. Replacement is a major area of ignorance in understanding hydrothermal ore genesis.

Replacement of carbonates is far more common than of other rock types in ore deposits. It apparently occurs where the ore solution first contacts the carbonate, as in some skarn deposits, or also after passing through some thickness of carbonates, as in the Hanover District, New Mexico. The case of deposition on first contact is very probably caused by acid neutralization but the second case is problematic. Textures also prove that there is a small decrease in volume associated with replacement, which means that possible linked reactions for dissolution plus precipitation are restricted to those with a small decrease in volume of reaction, $-\Delta V_R$.

Replacement of marble by sphalerite from a well-buffered brine has been carried out experimentally (D'Andrea, 1976; D'Andrea and Barnes, 1975). The solution contained 1.0 m Cl^-, 0.1 m CO_3^{2-}, and 0.001 m H_2S, which was initially equilibrated at 250-350°C with sphalerite, graphite, quartz, kaolinite, Na-montmorillonite, pyrite, and pyrrhotite. On contact, the carbonate was etched and replaced especially along grain boundaries by a

ORE DEPOSITION

cohesive assemblage of sphalerite and iron sulfides. The linked reactions were probably

$$CaCO_3 + 2H^+ \rightarrow Ca^{2+} + H_2CO_3 \quad (8.8)$$

$$ZnCl_2(aq) + H_2S(aq) \rightarrow ZnS + 2H^+ + 2Cl^- \quad (8.9)$$

A representative reaction path is shown by the arrow on Figure 8.5. The results differ from those found in ores in several respects. As might be expected for these rapid reactions and the steep chemical potential gradients generated in these experiments, the grain size of the deposited sulfides were only several tens of micrometers. Little replacement was found below the surface and the proportion dissolved appears excessive.

Another model for carbonate replacement has also been tested experimentally (Howd and Barnes, 1975). If a sulfide-rich ore solution were to flow through channels in marble—it is possible for it to do so without reaction—until it contacted a circulating meteoric water; then oxidation and sulfide deposition could occur. We arranged the experiments to have the oxidizing interface within the walls of a marble cylinder during reaction at

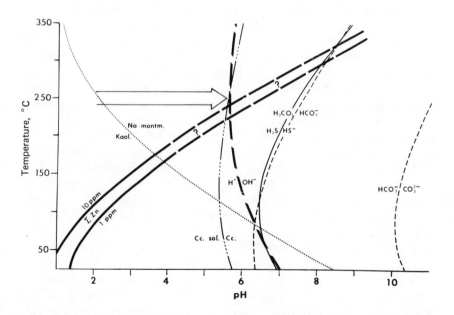

FIG. 8.5 Replacement of marble with ZnS from an NaCl-H$_2$S solution. The contours of zinc concentration were calculated for the initial solution prior to contact with the marble. The arrow represents the change in pH caused by replacement after contact with the marble. Calcite = Cc. (From D'Andrea, 1976.)

450°C and found pyrrhotite replacement (0.2 mm grains) at that site, although the 5.5 m NaHS solutions (pH 8.6 at 25°C) were also saturated with sphalerite.

Assuming that HS^- was dominant during replacement, the dominant reactions with limited O_2 present may be

$$\text{above 300°C: } HS^- + 2O_2 \rightarrow HSO_4^- \tag{8.10}$$

$$\text{below 300°C: } 2HS^- + 2O_2 \rightarrow SO_4^{2-} + H_2S \tag{8.11}$$

and with excess O_2 (and other oxidants) available

$$\text{above 300°C: } HS^- + 2O_2 \rightarrow HSO_4^- \tag{8.12}$$

$$\text{below 300°C: } HS^- + 2O_2 \rightarrow H^+ + SO_4^{2-} \tag{8.13}$$

Of these four reactions, only (8.13) releases protons. Alternatively, if H_2S were dominant, a similar set of reactions to equations (8.10)-(8.13) shows that protons are then released both under limited and excess oxidant and above and below 300°C. The released protons may dissolve carbonates in five of the eight combinations, the four where H_2S is dominant or the one where HS^- is dominant, $T < 300°C$, and O_2 is in excess. Although oxidation in all eight cases causes deposition from sulfide complexes, at the equilibrium concentrations of these complexes, there is little deposited per kilogram of solution. It is necessary, therefore, for near volume-for-volume replacement to alternate between carbonate-dissolving, sulfide-depositing conditions and conditions where only sulfide deposition takes place to fill any available open spaces. Feasible combinations are (1) an HS^- solution contacting excess O_2 alternating above and below 300°C; (2) an HS^- solution below 300°C alternating between a limited supply and an excess of O_2; and (3) a limited O_2 supply and pH alternating between HS^-- and H_2S-dominant regions.

These replacement experiments produced textures similar to those of ores and the model can satisfy the volume requirement. It is deficient in two respects, the requiring of very high pHs (and to some extent high ΣS) at high temperatures where replacement is abundant, and the need for an adequate supply of oxidant.

A third type of replacement is simple exchange of metals between a sulfide-deficient, metal-rich solution and sulfides of the host rock, such as at White Pine, Michigan. Typically, the exchange is between a chloride or sulfate solution and pyrite in the host rock. Maurel (1973) has produced such replacement at 300°C using 0.36 m Cl^- solutions of Pb^{2+} with ZnS and also Zn^{2+} with PbS. Nicely euhedral crystals were grown within the surface of either sulfide in these experiments. Nevertheless, the principal problem remains the replacement, not of another sulfide, but of a carbonate or silicate host rock.

Zoning

There have been two recent investigations of zoning in massive sulfide deposits (Large, 1977; Barnes, 1975). Both of these quantitatively correlate the observed distribution of metals with precipitation from chloride complexes. This sequence is largely fixed by the relative solubility products of the sulfides because the constants for the chloride complexes and the concentrations (abundances) of the metals do not vary over such a wide range.

The zoning typical of other hydrothermal deposits from earliest, and most central, outward is: molybdenite-arsenopyrite-pyrrhotite-pentlandite-stannite-chalcopyrite-sphalerite-tetrahedrite-galena-acanthite-gold tellurides-stibnite-cinnabar (Barnes, 1975). It correlates well with parageneses. No single deposit presents the entire idealized sequence. In addition, it is not inviolate but switching between adjacent minerals, such as chalcopyrite and sphalerite, happens where the relative concentrations of the metals in the deposit are reversed from normal.

At present, equilibrium constants for complexing in hydrothermal solutions are not known well enough for useful comparison of calculated relative saturation concentrations with zoning. A combined error in the constants for the metals from adjacent zones may reasonably be $10^{\pm 1.0}$. This translates into a 10-fold uncertainty in their relative solubilities, probably as much as their concentration differences at saturation, i.e., the inception of precipitation. For this reason, a quantitative understanding of the causes of zoning may require direct measurements of relative solubilities simultaneously from one system at pertinent P-T-X conditions.

Duration of the Mineralizing Process

The questions "how long did nature take to form ore bodies?" and "how much fluid was involved?" are related by the concentrations of ore solutions and the quantities of ore observed in deposits. We have known for some time the contents of many ore deposits and now have several lines of evidence pointing to ore solutions with hundreds to thousands of parts per million of the more common metals. There are presently developing some lines of evidence that suggest a time frame, as we shall see next.

A quantitative model was developed by Burnham in Chapter 3 that ties the genesis of porphyry copper deposits to the thermal evolution of granodioritic intrusives. Norton and Cathles (Chapter 12) and Cathles (1977) have investigated hydrodynamically and thermodynamically the cooling rates of plutons of appropriate sizes and depths. Together they provide a maximum interval within which ore genesis took place, approximately 0.1 m.y.

Because ore deposits are essentially extinct geothermal systems (see Chapter 13 and 15), their ages provide further clues to the longevity of hydrothermal processes. These are compiled together with some approximations from investigations of some ore deposits in Table 8-10. In general, 10^5–10^6 yr seems to be typical and the estimate for Creede is anomalously low.

Deposition of Porphyry Copper Ores

The petrologic and isotopic environments in which porphyry copper ore solutions originate have been examined in detail in Chapters 3 and 10, respectively. The solution is apparently 0.2–2.0 m in sulfur-containing species varying around 1:1 in sulfate to sulfide mole ratio. From fluid inclusion evidence, Nash (1976) concludes that the solutions precipitating most ores were less than approximately 2.4 m NaCl. Deposition took place at less than 350°C at pressures generally less than 500 bars and depths of about 1.8–3.0 km.

The solubilities of chalcopyrite, bornite, and pyrite have been determined for conditions including those of the above ore-forming environment by Crerar and Barnes (1976) and the resulting complexation constants are given in Tables 8-3 and 8-4. To use these to calculate copper and iron concentrations, it is necessary first to specify the a_{O_2} and pH of the ore-forming environment. This has been done using the stabilities of mineral assemblages of the ore zone with the results shown in Figure 8.6

The ore-depositing environment can be identified on Figure 8.6 in the following manner. Pyrrhotite and hematite (except in fluid inclusions) are absent, but pyrite is abundant, so conditions lie within the field labeled Py. In fact, bornite, chalcopyrite, and pyrite are each prominent in the ore assemblages indicating that conditions are close to the Bn + Py/Cp line. The pH is fixed generally in the muscovite (sericite) stability field near the feldspar limit, i.e., weakly acidic. Graphite is absent so conditions lie at a_{O_2} above this field. Barite and anhydrite are present either in trace or major amounts precipitated hydrothermally so that the solution must have reached saturation close to the "Barite and Anhydrite Insoluble" curve. Calcite is present in small veins in the ores so that the solution must have become saturated and been close to the "CaCO$_3$ Insoluble" curve. Only the presence of some siderite is inconsistent with the above indicators and this may well be due to the quality of pertinent thermodynamic data. The preferred region for deposition is a zone centered on the triangle including the dotted carbonate

Table 8-10 The duration of hydrothermal systems

Location	Duration[a] or Age (m.y.)	Basis	Reference
Yellowstone	1.9	Paleomagnetism of tuffs	Reynolds (1977)
Wairakei, N. Z.	0.5	Stratigraphic	Grindley (1965)
Steamboat Springs, Nev.	1-3	Isotopic?	White (1974)
Bodie, Calif.	1.5	K-Ar dates on alteration	Silberman et al. (1972)
Creede, Colo.	0.00X–0.0X	Hematite flake hydrodynamics	Barton et al. (1977)
Wisconsin-Illinois	0.25	Dispersion from ore body	Lavery and Barnes (1971)
Porphyry Cu Deposits	<0.1	Cooling of plutons	Cathles (1977)
Geothermal Systems	>(0.1–0.01)	Multiple	Ellis and Mahon (1977)
Kupferschiefer[b]	0.01	Average sedimentation rate	Wedepohl (1971)

[a] X represents any simple integer between 1 and 9.
[b] This 30-cm-thick black shale deposit is only included for comparison.

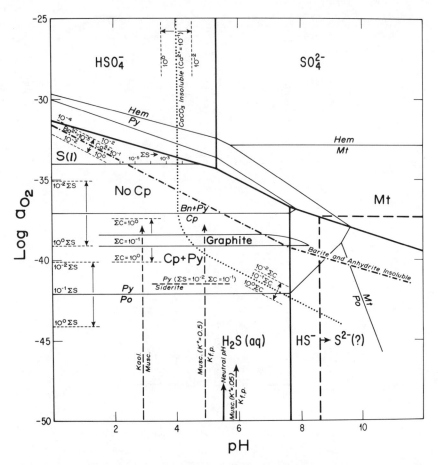

FIG. 8.6 The stabilities of minerals from porphyry copper deposits at 250°C. The solid boundaries and those with longer dashes represent activities respectively of $\Sigma S = 0.1$, $\Sigma C = 0.1$, $K^+ = 0.5$, $Ca^{2+} = 0.1$, and $Ba^{2+} = 0.001$. A change of $\pm$ 10x in activity is indicated by the light lines of shorter dashes. (Modified from Crerar and Barnes, *Econ. Geol.*, 71, 772-794.)

line, the dash-dot barite-anhydrite line, and the solid Bn + Py/Cp line. Although the diagram has been drawn for 250°C, only the values on the ordinate change materially at 350°, the upper temperature for the inception of ore precipitation.

The concentrations of copper and iron present in the preferred zone, dotted area of Figure 8.7, are shown by contours on this diagram. $CuCl$, $FeCl^+$, and Fe^{2+} make a significant contribution, but the sulfide complexes do not. At 350°C, concentrations of thousands of parts per million of both iron and copper are present, but at 250°C, virtually all of the copper and most of the

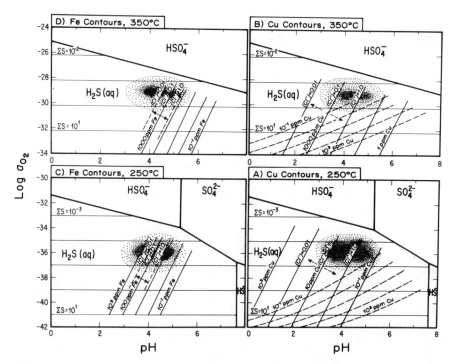

FIG. 8.7 The concentration of iron and copper in average porphyry copper ore solutions, stippled area, in equilibrium with chalcopyrite + bornite + pyrite. The copper concentrations are shown solid for chloride complexes and dashed for sulfide complexes. (From Crear and Barnes, *Econ. Geol.*, 71, 772-794.)

iron must have precipitated. A concentration in excess of approximately 1,000 ppm copper satisfies Rose's (1970) contention that lower concentrations require, both thermally and chemically, circulation of improbable quantities of solutions through porphyry copper bodies. It also agrees with the iron content of $(0.4-1.2) \times 10^4$ ppm found from daughter minerals in fluid inclusions in the ore zone at Bingham, Utah (Roedder, 1971).

Cathles (1977) has investigated the influence of convective circulation and of boiling on the thermal development of plutons and concludes that temperature gradients are an unlikely reason for precipitation, but the boiling seems more reasonable. Also effective would be an increase in pH accompanying the sericitization of the feldspars on dilution by circulating, low-salinity meteoric waters. Taylor (1974) has shown isotopically that deposition takes place at the interface between hydrothermal and meteoric waters. Depending on the local circumstances, cooling, dilution, and wallrock reactions are each probable contributors to ore deposition. Nevertheless, the frequency with which deposition temperatures of 350-250°C occur indicates that cooling is prominent, especially when one realizes that this is the interval where solubilities change from more than adequate to submarginal for ore transport.

Deposition of Mississippi-Valley-Type Deposits

The geochemical environment of deposition in the relatively uncomplicated Wisconsin deposits has been examined using fluid inclusion data, sphalerite stratigraphy, and sulfur isotopes by McLimans (1977) and from the viewpoint of lead transport by Giordano (1978). They also evaluated the thermodynamic stabilities of mineral assemblages in the ores as described for porphyry copper deposits above. The ores contain pyrite, marcasite, galena, sphalerite, barite, 2M-illite, microcline, quartz, and calcite. The solutions were 2.1–5.7 m Cl$^-$ and probably had an average $a_{Cl^-} \sim 2.0$ and total sulfur molality ~ 0.02. The filling of fluid inclusions in sphalerite indicates temperatures of 85–210°C (corrected), very similar to those in other desposits of this type (Roedder, 1976, Table IV). Pressure was low because burial could not have been more than a few hundred meters but not so low that the solutions boiled.

Representative conditions for deposition lie in the stipled area on Figure 8.8. Contours of lead concentrations were calculated for both sulfide and chloride complexes using the constants in Table 8-6 but they show remarkably little solubility in the preferred region. A cross-section through Figure 8.8, at any constant a_{O_2} below 10^{-56}, reveals more explicitly the dependence of solubility on pH and sulfide concentration (see Figure 8.9). Even at the higher temperature of 150°C, a minimum concentration of 10 ppm can be realized only in very sulfide-rich solutions ($> 5\ m$ H$_2$S + HS$^-$, pH 7) or in acidic sulfide-deficient solutions ($m_{Pb} > m_{H,S+HS-}$). Neither is geologically reasonable in view of two factors: sulfidation of iron in host rocks demonstrates an excess of sulfide present but inclusions contain far less than the minimum 5 m total sulfide for sulfide complexing. No inorganic complex is sufficiently stable to transport lead and H$_2$S or HS$^-$ simultaneously into these deposits. There are many organic complexes that might provide the needed mechanism for transport and these deserve investigation.

Instead of carrying lead and other metals in reduced sulfide solutions, Anderson (1973, 1975) has advocated that sulfate solutions could easily carry the metals. Deposition would require reduction or mixing with sulfide solutions. There are several mitigating arguments, however. Barite is much less soluble in sulfate than sulfide solutions yet it is carried in the ore solution and precipitates in the deposit, evidence for oxidation as a cause for deposition.

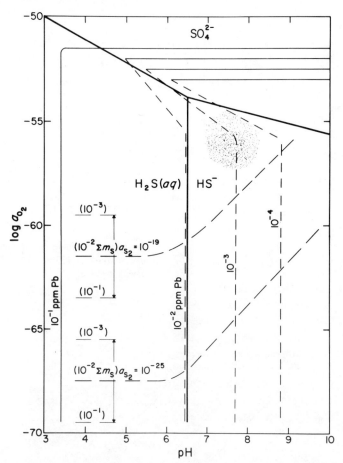

FIG. 8.8 Depositional conditions in the Illinois-Wisconsin District are shown as stipled. Calculated for 100°C, $a_{Cl^-} = 2.0$, and total concentration of sulfur-containing species of 0.01. Contours of a_{S_2} are based on the FeS contents of sphalerite. Contours of lead concentration are dashed where dominantly in sulfide complexes and solid were mostly chloride-complexed. (From Giordano, 1978.)

Reduction of sulfate at about 100°C is kinetically much too slow by either inorganic or organic agents. In addition sulfur isotopic evidence cited by Ohmoto and Rye in Chapter 10 shows that the sulfur of the solution must have been reduced prior to reaching the site of deposition.

The potential of inorganic complexes to contain in excess of 10 ppm zinc can also best be tested by a "V"-diagram. Figure 8.10 updates the original diagram of this type (Barnes, 1967) and adds a third ligand, NH_3. Because the zinc chloride complexes are considerably weaker than those for lead,

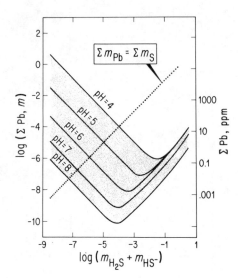

FIG. 8.9 The solubility of PbS in sulfide solutions at 150°C, $3m$ Cl$^-$. (From Giordano, 1978.)

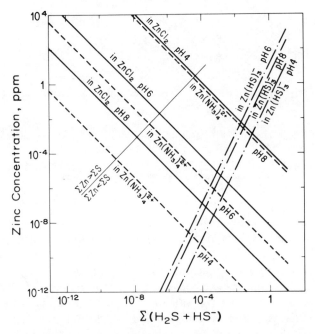

FIG. 8.10 The solubility of ZnS at 100°C at activities of Cl$^-$ and NH$_3$ both equal 1.0.

448

such species are even less viable. Ammine complexes are more stable than chloride complexes of zinc only in alkaline solutions and when ΣN approaches 1.0, an improbably high value. Again either complex requires $\Sigma Zn > \Sigma S$. Alternatively, the bisulfide complex, $Zn(HS)_3^-$ can dissolve 10 ppm zinc at pH6 ($\sim$ neutral) if $\Sigma(H_2S + HS^-) = 1.0$. There is also an additional solubility due to a second bisulfide complex of unproved stoichiometry (see Table 8-5) whose contribution is not included in Figure 8.10.

The combined solubility from the bisulfide complexes of zinc are shown in Figure 8.11. It was approximated by adding to the concentrations attributable to $Zn(HS)_3^-$, an empirically observed increment due to the second complex. The abscissa gives pHs at 25°C or, for higher temperature, this can be read as ΔpH with respect to the $pH = pK_{H_2S}$. The ordinate is also flexible in that the three curves are virtually immobile with respect to each other as the integers on this scale change or for temperatures to at least 200°C. This diagram shows $m_{Zn} \sim 10^{-2.8}$ or ~ 160 ppm zinc at $\Delta pH = 1.2$ (weakly alkaline) if $m_{HS^-} = 1.0$. Inasmuch as this solubility only exists under special and restricted conditions, for transport to persist through a variety of geologic environments, it may still be necessary that there be a contribution from organic complexes. It is interesting to note that McClure (1973) has found some oil field brines in Kansas, saturated in ZnS, to contain dominantly bisulfide complexes.

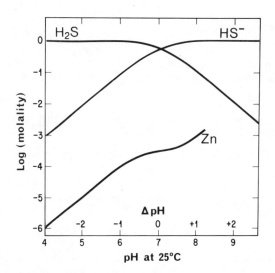

FIG. 8.11 The solubility of ZnS in bisulfide solutions up to 200°C. The adaptation of this figure to temperatures above 25°C and to other molalities of H_2S and HS^- is discussed in the text.

Deposition from either bisulfide or organic complexes could be caused by oxidation. Again the principal problem is the availability of oxidizing agents in the subsurface. The quantities of O_2 needed to precipitate zinc from a bisulfide solution is illustrated by Figure 8.12. If the solution were initially saturated at pH 8 (or 1 pH above $pK_{H,S}$ at higher temperatures), the 1 m HS^- solution would contain approximately 10^{-3} m zinc (~100 ppm), i.e., the upper right end of the curve. As the O_2 reacts, the zinc concentration falls toward the left until at 0.6 moles O_2 reacted, 90% of the initial concentration of zinc has been precipitated as ZnS. This quantity of O_2 is equivalent to the oxidizing capacity of $N_2 + O_2$ in 750 ℓ of water saturated with air at 20°C. On this basis, the dilution effect would certainly be more effective in causing deposition per kilogram of initial solution.

The texture of ores of this type prove that deposition was generally under quiescent conditions in Wisconsin. The common etching of sulfides after deposition is a consequence of a readily reversible complexing reaction. The regularity in thickness of individual segments of the sphalerite stratigraphy (McLimans, 1977) across the entire district negates mixing of metal-rich and sulfide-rich solutions as the cause of deposition. The constancy of inclusion filling temperatures throughout the district at each stage of deposition also refutes cooling as the cause. If mixing with circulating meteoric waters is the principal agent, it must have been remarkably regular throughout the district to produce nearly constant thickness of sulfides.

Deposition of Mercury Ores

Regardless of its concentration, chloride is chemically too feeble a ligand to affect the solubility of the very strongly associated HgS (Table 8-1). In-

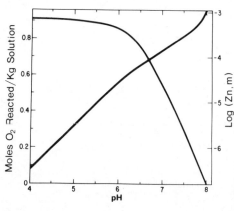

FIG. 8.12 The quantities of oxygen required to deposit various amounts of ZnS from a solution saturated at pH 8, 25°C, 1 m HS^-.

stead, only sulfide and perhaps some organic complexes are suffuciently stable in the presence of even trace concentrations of aqueous sulfides to account for transport at typical conditions for mercury ore formation, 100-200°C and 1-30 atm. (Barnes et al., 1967). Based on hot spring evidence, these ore solutions are probably neutral to weakly alkaline (I. Barnes et al., 1973 and Chapter 15). Some mercury-rich springs contain organic material with high mercury contents (e.g., 500 ppm in the Abbott Mercury Mine, North Central California) but others are nearly organic-free (I. Barnes et al., 1973), indicating that organic complexes are neither always present nor required for transporting mercury. Instead, where sufficiently accurate analytic data on the spring waters are available, such as at Wilbur Springs, North Central California, there is gross agreement between measured mercury contents of the spring waters and calculated contents in bisulfide complexes. Because the spring waters are cooling, the solutions are apparently somewhat supersaturated. It is clear that the dominant complex is not generally HgS_2^{2-} (Khodakovsky et al., 1975), but is one of the other bisulfide complexes shown on Figures 8.4 and 8.13 and Table 8-7.

Changes in conditions that could cause mercury deposition are shown by the concentration contours in Figure 8.13. As with other bisulfide complexes, these include increasing a_{O_2} or a_{H^+}, decreasing a_{HS^-}, and a_{H_2S}, as well as dilution or cooling. Both textural evidence and monitoring of the mercury contents of some cinnabar-containing hot springs shows that deposition of mercury ores is episodic (I. Barnes et al., 1973; Ellis, Chapter 13). It is also evident from Figure 8.13 that either native mercury or more commonly cinnabar may be deposited at equilibrium. For this reason, the occurrence of native mercury is not necessarily evidence of oxidation of cinnabar or metacinnabar.

Orpiment (As_2S_3), realgar (AsS), and stibnite (Sb_2S_3) are associated in mercury-containing hot springs and all form sulfide complexes of high solubilities, as shown by Dickson and co-workers (e.g., Weissberg et al., 1966; Weissberg et al., Chapter 15 of this volume; and the review by Barnes and Czamanske, 1967). The stoichiometries of their sulfide complexes are not yet determined for neutral to weakly alkaline solutions so that the causes for their precipitation cannot be discussed quantitatively except to note that they are apparently similar to those for HgS.

At temperatures above approximately 200°C, mercury is apparently mobile in steam even in the presence of excess sulfide species, particularly H_2S (aq), as noted by Ellis (Chapter 13) and Khodakovsky et al. (1975). In fact, Lawrence (1971) reports that cinnabar was actively deposited along with sulfur from H_2S-containing fumaroles where no liquid but only a vapor phase was present. Therefore, mercury transport is probably by aqueous sulfide complexes up to approximately 200°C, but vapor transport may be dominant at higher temperatures, at least in shallow ore deposits.

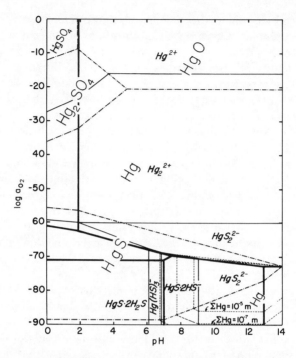

FIG. 8.13 The stabilities of mercury minerals at 25°C, $\Sigma S = 0.1$. The lines separate the following: heavy solid lines for sulfur species, light solid lines for mercury species, and the light dot-dash lines for minerals. Total mercury concentration is given by the light, dashed contours. (From Barnes, Romberger, and Stemprok, 1967, *Econ. Geol. 62*, 957-982, Figure 8.)

Deposition of Gold-Silver Ores

We are not yet ready to evaluate quantitatively the formation of gold-silver-tellurium ores because of the lack of experimental investigations of aqueous complexes with HTe^-, $HTeO_3^-$, for examples. Instead, only chloride and sulfide complexes in telluride-free deposits will be considered here.

Both chloride and sulfide complexes of gold are probably dominant in some ore-forming hydrothermal solutions, but at very different oxidation states—chloride complexes in relatively oxidizing and sulfide complexes in reducing solutions. A gap between the regions where these complexes are stable and where gold solubility is negligible is evident from the Eh-pH diagrams by Garrels and Christ (1965, Figures 7.33a and 7.33b). Although this gap narrows at higher temperatures, present stability data (Table 8-9) suggest that sulfide complexes are preferred in deposits formed both at less than approximately 300°C and where the mineralogy indicates reducing conditions, such as where pyrrhotite is present. On the other hand, for gold deposits formed at temperatures above approximately 400°C where the alteration reveals acidic conditions (e.g., boehmite or muscovite assemblages), and where relatively oxidizing conditions are indicated by oxide or sulfide minerals, for examples, chloride complexes were likely dominant during transport (Henley, 1973; Seward, 1973). Again for silver, it seems that at lower temperatures, sulfide complexes are more important, as for gold, but that above 300°C chloride complexes become dominant. It is possible that gold colloids also deposit some ore-grade concentrations (Henley, 1973), especially where ionic strengths are particularly low. However, a key problem for both gold and silver are the precise conditions of ΣS, ΣCl, pH, T, and a_{O_2} that demark the separation between the regions of dominance of chloride and sulfide complexing. Several examples of deposits follow below that indicate both types of complexes may form deposits of either silver or gold or both.

At Broadlands, New Zealand, 4×10^{-5} ppm gold, 4×10^{-3} m sulfide, and 0.033 m Cl^- occur in deep waters at 260°C, $pH_{260°C} = 6.1$ as summarized by Seward (1973). He calculated that only sulfide complexing could account for this gold concentration and that the solutions were, indeed, undersaturated under reduced conditions. Seward (1976) has also calculated the solubility of Ag_2S in the same Broadlands solution. Chloride complexes account for only 0.3-0.4 times the measured silver content of 0.25 ppb, suggesting that sulfide complexing may again be dominant, although more data are needed on silver sulfide complexes to be certain.

In contrast, his calculations of the concentrations of silver chloride complexes in the Salton Sea geothermal brine (see Chapters 13 and 15) prove that chloride complexes are dominant silver-carriers in these fluids at approximately 300°C. Robinson and Ohmoto (1973) reach the same conclusion for the Echo Bay uranium-nickel-silver-copper deposits, but for temperatures of only approximately 200°C. In the Yatani lead-zinc-gold-silver veins, Hattori (1975) concludes that the environment of deposition was 200-250°C, 0.1-0.2 m NaCl, 10^{-2}-10^{-3} m total sulfide species, $pH_{250°C} \sim 6.0$, and moderately reducing. She notes that most epithermal gold-silver veins in Japan have comparable geochemical characteristics and probably formed in a similar manner. Calculation of the concentrations of the gold chloride and bisulfide complexes shows the latter to be dominant. By analogy, the argentite probably was transported in bisulfide complexes. It is closely associated with gold in the veins.

Three springs in the Taupo volcanic zone of New Zealand contain precipitates rich in gold and silver (Weissberg, 1969 and Chapter 15). They are accompanied by high arsenic, antimony, mercury, and thallium. The maximum concentrations in the spring waters are 4×10^{-5} ppm gold, $6 \times$

10^{-4} ppm silver, 120 ppm total sulfide, 2900 ppm Cl^-, with weak alkalinity prior to surface oxidation. The high pH and sulfide/chloride ratio imply bisulfide complexing, although further data are required to be certain. These hot spring deposits closely resemble the disseminated type of gold ore found at Carlin, Nevada, and elsewhere (Roberts et al., 1971). Besides sinter, often cinnabar, realgar, and orpiment occur, and the gold is too fine grained to be readily visible.

Deposition of gold and of silver in ores occurs by those processes that we have already found to be characteristic of chloride or bisulfide complexes. In those deposits thought to have been formed by chloride-complexed metals, probable causes of deposition reported by the authors are either reduction both at Echo Bay (Robinson and Ohmoto, 1973) and for the common texture of gold replacing pyrrhotite (Henley, 1973), or cooling as in the Salton Sea brine (Seward, 1976). In precipitates from sulfide complexes, cooling, oxidation, and decreasing pH may all be involved in the New Zealand spring (Weissberg, 1969) but in the Yatani veins, oxidation seems to be primary cause (Hattori, 1975). Because the ore solution reduced minerals adjacent to the veins at Yellowknife, Northwest Territories (Kerrich, 1977), the inference may be reached that oxidation, and perhaps cooling, contributed significantly to deposition of gold in these veins.

ACKNOWLEDGMENTS

I appreciated the opportunity to review prior to publication a preprint from D. A. Crerar, Princeton University, a thesis by T. J. Barrett from G. A. Anderson, University of Toronto, and a thesis by K. R. Schulz from E. U. Franck, University of Karlsruhe. Several colleagues at Penn State assisted in the preparation of this chapter: H. Ohmoto and C. W. Burnham by discussions of processes, J. D. Rimstidt by compiling and evaluating data for Figure 8.2, and especially A. C. Lasaga who carefully reviewed the manuscript. Funding of the research on which this chapter is based has been provided by the Earth Sciences Section of the National Science Foundation under grants GA-36815X, EAR72-01713, and EAR77-13073.

REFERENCES

Anderson, G. M. (1973) The hydrothermal transport and deposition of galena and sphalerite near 100°C: *Econ. Geol.*, **68**, 480-492.

―― (1975) Precipitation of Mississippi Valley-type ores: *Econ. Geol.*, **70**, 937-942.

―― (1976) Error propagation by the Monte Carlo method in geochemical calculations: *Geochim. Cosmochim. Acta*, **40**, 1533-1538.

REFERENCES

Avrahami, M. and R. M. Golding (1968) The oxidation of the sulfide ion at very low concentrations in aqueous solutions: *J. Chem. Soc.*, 647-651.

Bada, J. L. and S. L. Miller (1968) Ammonium ion concentration in the primitive ocean: *Science*, **159**, 423-425.

Baes, C. F., Jr. and R. E. Mesmer (1976) *The Hydrolysis of Cations*, New York: Wiley.

Barnes, H. L. (1960) Ore solutions: *Carnegie Institution Yearbook* **59**, 137-141.

_____ (1965) Environmental limitations to mechanisms of ore transport: *Symposium, Problems of Postmagnetic Ore Deposition, Prague*, **2**, 316-326.

_____ (1967) Sphalerite solubility in ore solutions of the Illinois-Wisconsin District: *Econ. Geol., Monograph*, **3**, 326-332.

_____ (1971) Investigations in hydrothermal sulfide solutions: in *Research Techniques for High Pressure and High Temperature*, G. C. Ulmer, ed., New York: Springer-Verlag, Chapter 12, 317-355.

_____ (1975) Zoning of ore deposits: Types and causes: *Trans. Roy. Soc. Edinburgh*, **69**, 295-311.

_____ and G. K. Czamanske (1967) Solubilities and transport of ore minerals: *Geochemistry of Hydrothermal Ore Deposits*, 1st ed., by H. L. Barnes, ed., New York: Holt, Rinehart, and Winston, 334-381.

_____ and G. Kullerud (1961) Equilibria in sulfur-containing aqueous solutions, in the system Fe-S-O, and their correlation during ore deposition: *Econ. Geol.* **56**, 648-688.

_____ S. B. Romberger, and M. Stemprok (1967) Ore solution chemistry II. Solubility of HgS in sulfide solutions: *Econ. Geol.*, **62**, 957-982.

Barnes, I., M. E. Hinkle, J. B. Rapp, C. Heropoulis, and W. W. Vaughn (1973) Chemical composition of naturally occurring fluids in relation to mercury deposits in part of North-Central California: *U. S. Geol. Surv. Bull.*, *1382-A*.

Barrett, T. J. (1974) Solubility of galena and sphalerite in sodium chloride brines up to 95°C: Unpublished M.Sc. dissertation, Dept. of Geology, University of Toronto.

Barton, P. B. Jr., P. M. Bethke, and E. Roedder (1977) Environment of ore deposition in the Creede Mining District, San Juan Mountains, Colorado: Part III. Progress toward interpretation of the chemistry of the oreforming fluid for the OH Vein: *Econ. Geol.* **72**, 1-24.

Belousov, E. A. and A. A. Alovyainikov (1975) Determination of the stability constants of zinc and cadmium chloro-complexes in aqueous solutions by the distribution method: *Russ. J. Inorg. Chem.* (transl.), *20*, 803-804.

Bryzgalin, O. V. (1976) On the solubility of tungstic acid in aqueous salt solutions at high temperatures: *Geochem. Int.*, **13**, 155-159.

Carpenter, A. B., M. L. Trout, and E. E. Pickett (1974) Preliminary report on the origin and chemical evolution of lead- and zinc-rich oil field brines in Central Mississippi: *Econ. Geol.*, **69**, 1191-1206.

Casadevall, T. and H. Ohmoto (1977) Sunnyside Mine, Eureka Mining District, San Jaun County, Colorado: Geochemistry of gold and base metal ore deposition in a volcanic environment: *Econ. Geol.*, **72**, 1285-1320.

Cathles, L. M. (1977) An analysis of the cooling of intrusives by ground-water convection which includes boiling: *Econ. Geol.*, **72**, 804-826.

Chou, I-M. and H. P. Eugster (1977) Solubility of magnetite in supercritical chloride solutions: *Am. J. Sci.*, **277**, 1296-1314.

_____ and J. D. Frantz (1977) Recalibration of Ag + AgCl acid buffer at elevated pressures and temperatures: *Am. J. Sci.* **277**, 1067-1072.

Cotton, F. A. and G. Wilkinson (1962) *Advanced Inorganic Chemistry*, New York: Wiley.

Crerar, D. A. (1974) Solvation and deposition of chalcopyrite and chalcocite assemblages in hydrothermal solutions: PhD. dissertation, Dept. of Geological Sciences, The Pennsylvania State University.

―――― and H. L. Barnes (1976) Ore solution chemistry V. Solubilities of chalcopyrite and chalcocite assemblages in hydrothermal solution at 200° to 350°C: *Econ. Geol.*, **71**, 772-794.

―――― H. J. Susak, M. Borcsik, and S. Schwartz (1978) Solubility of the buffer assemblage pyrite + pyrrhotite + magnetite in NaCl solutions from 200° to 350°C: *Geochem. Cosmochim. Acta,* **42**, 1427-1439.

Cusicanqui, H., W. A. J. Mahon, and A. J. Ellis (1976) The geochemistry of the El Tatio Geothermal Field, Northern Chile: *Proc. Sec. United Nations Sympos. on the Development and Use of Geothermal Resources*, **1**, 703-711.

Czamanske, G. K., E. Roedder, and F. C. Burns (1963) Neutron activation analysis of fluid inclusions for copper, manganese, and zinc: *Science*, **140**, 401-403.

D'Andrea, R. F. (1976) Replacement of marble by ZnS in chloride solutions: Unpublished M.Sc. dissertation: Dept. of Geosciences, The Pennsylvania State University.

D'Andrea, R. F., Jr. and H. L. Barnes (1975) Experimental replacement of marble in chloride systems (abs.): *Geol. Soc. Am., Abst. Prog.*, **7**, 1044-1045.

Ellis, A. J. and W. A. J. Mahon (1977) *Chemistry and Geothermal Systems*, New York: Academic Press.

Fisher, J. R., J. L. Haas, Jr., and P. B. Barton, Jr. (1975) Nitrogen as an oxidant in hydrothermal systems (abst.): *Geol. Soc. Am., Abst. Prog.*, **7**, 1074.

Franck, E. U. (1973) Concentrated electrolyte solutions at high temperatures and pressures: *J. Solut. Chem.,* **2**, 339-356.

Garrels, R. M. and C. L. Christ (1965) *Solutions, Minerals, and Equilibria,* San Francisco: Freeman-Cooper.

Gardner, L. R. (1974) Organic versus inorganic trace metal complexes in sulfidic marine waters—some speculative calculations based on available stability constants: *Geochim. Cosmochim. Acta*, **38**, 1297-1302.

Giggenbach, W. F. (1974a) Equilibria involving polysulfide ions in aqueous sulfide solutions up to 240°: *Inorg. Chem.*, **13**, 1724-1730.

―――― (1974b) Kinetics of the polysulfide-thiosulfate disproportionation up to 240°: *Inorg. Chem.*, **13**, 1730-1733.

Giordano, T. H. (1978) Dissolution and precipitation of lead sulfide in hydrothermal solutions, and the point defect chemistry of galena: Unpublished Ph.D. dissertation, Dept. of Geological Sciences, The Pennsylvania State University.

―――― and H. L. Barnes (1979) Ore solution chemistry. VI. PbS solubility in bisulfide solutions to 300°C: *Econ. Geol.,* in press.

Goni, J., C. Guillemin, and C. Sarcia (1967) Geochimie de l'or exogene: *Mineral. Depos.*, **1**, 259-268.

Granger, H. C. and C. G. Warren (1969) Unstable sulfur compounds and the origin of roll-type uranium deposits: *Econ. Geol.*, **64**, 160-171.

Grindley, G. W. (1965) The geology, structure, and exploitation of the Wairakei geothermal field, Taupo, New Zealand: *N. Z. Geol. Surv. Bull.*, **75**.

Gubeli, A. O. and J. Ste-Marie (1967) Constantes de stabilite de thiocomplexes et produits de solubilite de sulfures de mataux. II Sulfure de zinc: *Can. J. Chem.*, **45**, 2101-2108.

Gunter, B. D. (1973) Aqueous phase-gaseous phase material balance studies of argon and nitrogen in hydrothermal features at Yellowstone National Park: *Geochim. Cosmochim. Acta*, **37**, 495-513.

Hamann, R. J. and G. M. Anderson (1978) Solubility of galena in sulfur-rich NaCl solutions: *Econ. Geol.*, **73**, 96-100.

Hattori, Keiko (1975) Geochemistry of ore deposition at the Yatani lead-zinc and gold-silver deposit, Japan: *Econ. Geol.*, **70**, 677-693.

Helgeson, H. C. (1969) Thermodynamics of hydrothermal systems at elevated temperatures and pressures: *Am. J. Sci.*, **267**, 729-804.

Henley, R. W. (1973) Solubility of gold in hydrothermal chloride solutions: *Chem. Geol.* **11**, 73-87.

Herr, F. L., Jr. and G. R. Helz (1978) On the possibility of bisulfide ion-pairs in natural brines and hydrothermal solutions: *Econ. Geol.*, **73**, 73-81.

Hitch, B. F. and R. E. Mesmer (1976) The ionization of aqueous ammonia to 300°C in KCl media: *J. Solut. Chem.*, **5**, 667-680.

Hoffman, M. R. (1977) Kinetics and mechanism of oxidation of hydrogen sulfide by hydrogen peroxide in acidic solution: *Environ. Sci. Tech.*, **11**, 61-66.

Howd, F. H. and H. L. Barnes (1975) Ore solution chemistry IV. Replacement of marble by sulfides at 450°C; *Econ. Geol.*, **70**, 968-981.

Izatt, R. M., D. Eatough, J. J. Christensen, and C. H. Bartholomew (1969) Calorimetrically determined log K, $\Delta H°$, and $\Delta S°$ values for the interaction of sulphate ion with several bi-and ter-valent metal ions: *J. Chem. Soc. Ser. A.*, 47-53.

Johnson, K. S. and R. M. Pytkowicz (1978) Ion association of Cl^- with H^+, Na^+, K^+, Ca^{2+} and Mg^{2+} in aqueous solutions at 25°C: *Am. J. Sci.*, **278**, 1428-1447.

Kakovskii, I. A. and N. G. Tyurin (1962) Behavior of gold in polysulfide solutions at elevated temperatures and pressures (in Russian): *Cvetnaya Metall.*, No. **2**, 104-111.

Kerrich, R., W. S. Fyfe, and I. Allison (1977) Iron reduction around gold-quartz veins, Yellowknife District, Northwest Territories, Canada: *Econ. Geol.*, **72**, 657-663.

Khodakovsky, I. L., M. Ya.Popova, and N. A. Ozerova (1975) On the role of sulfide complexes in the transport of mercury by hydrothermal solutions: (transl.) *Geochem. Int.*, **12**, 37-45.

Klintsova, A. I., V. L. Barsukov, T. P. Shemarykina, and I. L. Khodakovsky (1975) Experimental determination of stability constants of tetravalent tin hydroxyfluoride complexes: *Geochem. Int.*, **12**, 207-215.

Krauskopf, K. B. (1964) The possible role of volatile metal compounds in ore genesis: *Econ. Geol.* **59**, 22-45.

Lakin, H. W., G. C. Curtin, and A. E. Hubert (1974) Geochemistry of gold in the weathering cycle: *U. S. Geol. Surv. Bull. 1330*.

Landis, G. P. and R. O. Rye (1974) Geologic, fluid inclusion, and stable isotope studies of the Pasto Buena tungsten-base metal ore deposit, Northern Peru: *Econ. Geol.*, **69**, 1025-1059.

Langmuir, D. (1979) Techniques of estimating thermodynamic properties for some aqueous complexes of geochemical interest: in *Chemical Modelling-Speciation, Sorption, Solubility, and Kinetics in Aqueous Systems*, R. F. Gould, ed., Am. Chem. Soc. Sympos. Series, Washington, D. C., in press.

Large, R. R. (1977) Chemical evolution and zonation of massive sulfide deposits in volcanic terrains: *Econ. Geol.*, **72**, 549-572.

Lavery, N. G. and H. L. Barnes (1971) Zinc dispersion in the Wisconsin zinc-lead district: *Econ. Geol.*, **66**, 226-242.

Lawrence, E. F. (1971) Mercury mineralization at the Senator Fumaroles, Dixie Valley, Nevada (abst.): *Geol. Soc. Am.*, *Abst. Prog.*, **3**, 147.

Lebedev, L. M. (1967) Modern growth of sphalerite: *Doklady Akad. Nauk S.S.S.R.*, **175**, 920-923.

―――― (1972) Minerals of contemporary hydrotherms of Cheleken: *Geochem. Int.*, **9**, 485-504.

Mann, A. W. and R. I. Deutscher (1977) Solution geochemistry of copper in water containing carbonate, sulphate, and chloride ions: *Chem. Geol.*, **19**, 253-265.

Maurel, C. (1973) Mechanism of hydrothermal sphalerite = galena replacement at 300°C: *Econ. Geol.*, **68**, 665-670.

McClure, J. W. (1973) Base metal content of oil field brines in Eastern Kansas (abst.) *Geol. Soc. Am.*, *Abst. Prog.*, **5**, 336.

McLimans, R. K. (1977) Geological, fluid-inclusion, and stable isotope studies of the Upper Mississippi Valley Zinc-Lead District, Southwest Wisconsin: Unpublished Ph.D. dissertation, Dept. of Geological Sciences, The Pennsylvania State University.

Melent'yev, B. N., V. V. Ivanenko, and L. A. Pamfilova (1969) Solubility of some ore-forming sulfides under hydrothermal conditions: *Geochem. Int.*, **6**, 416-460.

Mironov, V. E., F. Ya. Kul'ba, and V. A. Fedorov (1965) Chloro-complexes of lead (II) and their interaction with alkali metal cations: *Russ. J. Inorg. Chem.* (trans.), **10**, 495-497.

Moh, G. H. (1974) Tin-containing mineral systems. Part I: The Sn-Fe-S-O system and mineral assemblages in ores: *Chem. Erde*, **33**, 243-275.

―――― (1975a) Tin-containing mineral systems. Part II: Phase relations and mineral assemblages in Cu-Fe-Zn-Sn-S system: *Chem. Erde*, **34**, 1-61.

―――― (1975b) Tin-containing mineral systems. Part III. Phase equilibria within the Sn-Pb-Sb-Bi-Fe-S system in relation to mineral assemblages of the Bolivian ore type: *Chem. Erde*, **34**, 201-238.

Nash, J. T. (1973) Geochemical studies in the Park City District: I: Ore fluids in the Mayflower Mine: *Econ. Geol.*, **68**, 34-51.

―――― (1976) Fluid inclusion petrology—Data from porphyry copper deposits and applications to exploration. *U. S. Geol. Surv. Prof. Pap. 907-D*.

Naumov, G. B., B. N. Ryzhenko, and I. L. Khodakovsky (1974) *Handbook of Thermodynamic Data* (translated from the 1971 Russian original): Nat. Tech. Info. Service PB-226-722.

Norman, D. I., G. P. Landis, and F. J. Sawkins (1976) H_2S and SO_2 detected in fluid inclusions (abst.): *Geol. Soc. Am.*, *Abst. Progr.*, **8**, 1031.

Nriagu, J. O. and G. M. Anderson (1970) Calculated solubilities of some base-metal sulphides in brine solutions: *Trans. Inst. Mining Metall.*, **79**, B208-212.

Orr, W. L. (1974) Changes in sulfur content and isotopic ratios of sulfur during petroleum maturation—study of Big Horn Basin Paleozoic oils: *Bull. Am. Assoc. Petrol. Geol.*, **50**, 2295-2318.

―――― (1975) Geologic and geochemical controls on the distribution of hydrogen sulfide in natural gas: *Proc. Seventh Int. Cong. Organic Geoch.*, Madrid, 571-597.

Palmer, S. E. and E. W. Baker (1978) Copper porphyrins in deep-sea sediments: A possible indicator of oxidized terrestial organic matter: *Scinece*, **201**, 49-51.

Peters, E. (1976) Direct leaching of sulfides: Chemistry and applications: *Metallurg. Trans.*, **7B**, 505-517.

Phillips, S. L., A. K. Mathur, and R. E. Doebler (1977) A study of brine treatment: *Lawrence Berkeley Laboratory Report LBL-6731.*

REFERENCES

Ramachandra Rao, S. and L. G. Hepler (1977) Equilibrium constants and thermodynamics of ionization of aqueous hydrogen sulfide: *Hydrometallurgy*, **2**, 293-299.

Raman, S. (1976) Use of the complexation equilibrium between Fe(II) and 2,2′-bipyridine or phenanthroline in the determination of the formation constant of $FeCl^+$: *J. Inorg. Nucl. Chem.* **38**, 1741-1742.

Reardon, E. J. (1975) Dissociation constants of some monovalent sulfate ion pairs at 25° from stoichiometric activity coefficients: *J. Phys. Chem.*, **79**, 422-425.

Reynolds, R. L. (1977) Paleomagnetism of welded tuffs of the Yellowstone group: *J. Geophys. Res.*, **82**, 3677-3693.

Rich, R. A., H. D. Holland, and U. Petersen (1977) *Hydrothermal Uranium Deposits*, New York: Elsevier.

Rimstidt, J. D. and H. L. Barnes (1977) Development of geothermal energy: Some problems: *Earth Min. Sci.*, **47**, 9-12.

Roberts, R. J., A. S. Radtke, and R. R. Coats (1971) Gold-bearing deposits in North-Central Nevada and Southwestern Idaho: *Econ. Geol.*, **66**, 14-33.

Robinson, B. W. and H. Ohmoto (1973) Mineralogy, fluid inclusions, and stable isotopes of the Echo Bay U-Ni-Ag-Cu Deposits, Northwest Territories, Canada: *Econ. Geol.*, **68**, 635-656.

Roedder, E. (1971) Fluid inclusion studies on the porphyry-type ore deposits at Bingham, Utah, Butte, Montana, and Climax, Colorado: *Econ. Geol.*, **66**, 98-120.

____ (1976) Fluid-inclusion evidence on the genesis of ores in sedimentary and volcanic rocks: in *Handbook of Strata-bound and Stratiform Ore Deposits*, K. H. Wolf, ed., Amsterdam: Elsevier.

Romberger, S. B. and H. L. Barnes (1970) One solution chemistry. III Solubility of CuS in sulfide solutions: *Econ. Geol.*, **65**, 901-919.

Rose, A. W. (1970) Zonal relations of wallrock alteration and sulfide distribution at porphyry copper deposits: *Econ. Geol.*, **65**, 920-936.

____ (1976) The effect of cuprous chloride complexes in the origin of red-bed copper and related deposits: *Econ. Geol.*, **71**, 1036-1048.

Rye, R. O. and J. Haffty (1969) Chemical composition of the hydrothermal fluids responsible for the Lead-Zinc Deposits at Providencia, Zacatecas, Mexico: *Econ. Geol.*, **64**, 629-643.

____ W. E. Hall, and H. Ohmoto (1974) Carbon, hydrogen, oxygen and sulfur isotope study of the Darwin lead-silver-zinc deposit, Southern California: *Econ. Geol.*, **69**, 468-481.

Sakai, H. (1977) Sulfate-water isotope thermometry applied to geothermal systems: *Geothermics*, **5**, 67-74.

Saxby, J. D. (1976) The significance of organic matter in ore genesis: in *Handbook of Strata-bound Stratiform Ore Deposits*, K. H. Wolf, ed., New York: Elsevier, vol. 2, Chapter 5, p. 111-133.

Schulz, K. R. (1974) Ramanspektroskopische Utersuchungen an wässerigen Zink-chloridlösungen bis 500°C and 4000 bar: Unpublished Ph.D. dissertation, Physics Faculty, University of Karlsruhe.

Schwarzenbach, G. and M. Widmer (1966) Die Löslichkeit von Metallsulfiden II. Silbersulfid (1): *Helvetica Chim. Acta*, **49**, 111-123.

Seward, T. M. (1973) Thio complexes of gold and the transport of gold in hydrothermal solutions: *Geochim. Cosmochim. Acta*, **37**, 379-399.

____ (1976) The stability of chloride complexes of silver in hydrothermal solutions up to 350°C: *Geochim. Cosmochim. Acta*, **40**, 1329-1341.

——— (1977) Solubility of coexisting pyrite and pyrrhotite in the system $NaHS$-H_2S-$NaCl$-H_2O at elevated temperature and pressure: in *Geochemistry 1977*, A. P. W. Hodder, ed., N. Z. Dept. Sci. Ind. Res., Bull. 218, Wellington.

Shimizu, K., N. Tsuchihashi, and Y. Kondo (1977) Effect of pressure on dissociation of ion pairs—$MnSO_4$, $CuSO_4$, and $ZnSO_4$: *Rev. Phys. Chem. Japan*, **47,** 80-89.

Silberman, M. L., C. W. Chesterman, F. J. Kleinhampl, and C. H. Gray, Jr. (1972) K-Ar ages of volcanic rocks and gold-bearing quartz-adularia veins in the Bodie mining district, Mono County, California: *Econ. Geol.*, **67,** 597-604.

Skibsted, L. H. and J. Bjerrum (1974) Studies on gold complexes. II. The equilibrium between gold (I) and gold (III) in the ammonia system and the standard potentials of the couples involving gold, diammine gold (I), and tetrammine gold (III): *Acta Chem. Scandinavia*, **A28** 764-770.

Smith, R. M. and A. E. Martell (1976) Critical Stability Constants. *Volume 4*: Inorganic Complexes, New York: Plenum Press; also *Volume 1*: *Amino Acids*, *Volume 2*: *Amines*, and *Volume 3*: *Other Organic Ligands*.

Snellgrove, R. A. and H. L. Barnes (1974) Low temperature solid phases and aqueous species in the copper sulfide system (abst.): *Trans. Am. Geophys. Union*, **55,** 484.

Ste-Marie, J., A. E. Torma, and A. O. Gübeli (1964) The stability of thiocomplexes and solubility products of metal sulphides: *Can. J. Chem.*, **42,** 662-668.

Taylor, H. P., Jr. (1974) The application of oxygen and hydrogen isotope studies to problems of hydrothermal alteration and ore deposition. *Econ. Geol.*, **69,** 843-883.

Tremaine, P. R., R. V. Massow, and G. R. Shierman (1977) A calculation of Gibbs free energies for ferrous ions and the solubility of magnetite in H_2O and D_2O to 300°C: *Thermochim. Acta*, **19,** 287-300.

Tsui, T. F. and H. D. Holland (1976) The Cu content of fluid inclusions in three epithermal ore deposits (abst.): *Trans. Am. Geophys. Union*, **57,** 1014.

Urabe, T. (1974) Mineralogical aspects of the Kuroko Deposits in Japan and their implications: *Mineral. Depos.*, **9,** 309-324.

Volynets, V. F. and T. M. Sushchevskaya (1972) Nitrogen in hydrothermal processes (ammonium in inclusion solutions): *Geochem. Int.*, **9,** 42-46.

Wedepohl, K. H. (1971) "Kupferschiefer" as a prototype of syngenetic sedimentary ore deposits: *Soc. Mining Geol. Japan, Spec. Issue* **3,** 268-273.

Weissberg, B. G. (1969) Gold-silver ore-grade precipitates from New Zealand thermal waters: *Econ. Geol.*, **64,** 95-108.

———., F. W. Dickson, and G. Tunell (1966) Solubility of orpiment (As_2S_3) in Na_2S-H_2O at 50-200°C and 100-1500 bars, with geological applications: *Geochim. Cosmochim. Acta*, **30,** 815-827.

White, D. E. (1967) Mercury and base-metal deposits with associated thermal and mineral meters: *Geochemistry of Hydrothermal Ore Deposits*, 1st ed., H. L. Barnes, ed., New York: Holt, Rinehart, and Winston, p. 575-631.

——— (1974) Diverse origins of hydrothermal ore fluids: *Econ. Geol.*, **69,** 954-973.

9

The Solubility and Occurrence of Non-Ore Minerals

HEINRICH D. HOLLAND
Harvard University

SERGEY D. MALININ
Vernadsky Institute, Moscow

Solution, deposition, and replacement of minerals are the dominant processes during the formation of hydrothermal ore deposits. Each of these processes depends on the solubility of minerals. In this chapter we deal with the solubility of non-ore minerals and with the application of solubility data toward the solution of problems concerning the physical and chemical evolution of hydrothermal ore deposits. We proceed from relatively simple to relatively more complex systems in the course of our discussion of mineral solubilities. Our first concern is with the system $NaCl-H_2O$, then with the solubility of quartz, the solubility of fluorite, that of the carbonates, and finally with the solubility of the alkali earth sulfates.

THE SYSTEM $NaCl-H_2O$

The analytical data for fluids in fluid inclusions in hydrothermal minerals invariably indicate that Na^+, Ca^{2+}, K^+, and Mg^{2+} are the dominant cations, and that Cl^-, HCO_3^-, and SO_4^{2-} are the dominant anions (see

Chapter 14). Among these ions Na$^+$ and Cl$^-$ are normally the most abundant, and many of the properties of hydrothermal fluids can be understood in terms of the properties of solutions in the system NaCl-H$_2$O. Fortunately this system has been studied quite intensively. Keevil (1942) showed that the saturation curve for aqueous sodium chloride solutions is continuous up to the melting point of NaCl. As shown in Figure 9.1, the solubility of NaCl in aqueous solutions in equilibrium with a vapor phase rises from 26.4 wt. % at 25°C to 77 wt. % at 650°C. The vapor pressure of these saturated solutions has been measured by Keevil (1942), Ölander and Liander (1950), Morey and Chen (1956), and by Sourirajan and Kennedy (1962). There is substantial agreement among the various sets of data. Figure 9.2 has been taken from Sourirajan and Kennedy's (1962) paper.

In most hydrothermal systems, the total pressure rises more steeply with increasing temperature than the vapor pressure of solutions saturated with NaCl. Boiling of hydrothermal solutions saturated with respect to NaCl occurs only in systems where the geothermal gradient is in excess of approximately 150°C/km, or where the vapor pressure of the hydrothermal solutions is enhanced by the presence of dissolved gases.

Fluid inclusions containing cubes of NaCl are rather common (see

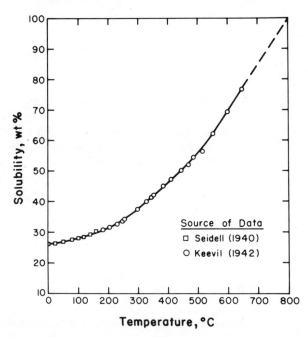

Fig. 9.1 The solubility of NaCl in water along the three-phase boundary vapor-liquid-NaCl (see Keevil, 1942).

THE SYSTEM NaCl-H₂O

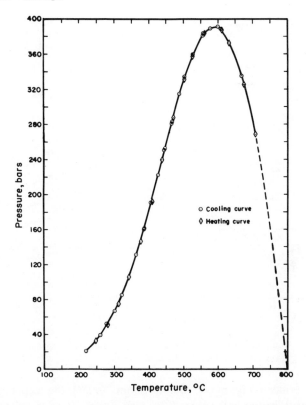

Fig. 9.2 Vapor pressure of liquids saturated with respect to NaCl in the system NaCl-H$_2$O as determined in cooling and heating runs. (From Sourirajan and Kennedy, 1962, *Am. J. Sci., 260,* 115–141.)

Chapter 14). However, these cubes normally dissolve on heating before the filling temperature of the inclusions is reached, suggesting that hydrothermal solutions are not normally saturated with respect to NaCl during ore deposition. The vapor pressure of unsaturated solutions is larger than that of a saturated solution at the same temperature. Below the critical temperature of water (374.1 °C), the vapor pressure rises with decreasing NaCl concentration toward the vapor pressure of water at that temperature. Above 374 °C, the vapor pressure rises with decreasing NaCl concentration to a maximum that lies on the critical curve of the system NaCl-H$_2$O. These relationships are shown in Figures 9.3 and 9.4 that have been taken from the paper by Sourirajan and Kennedy (1962). The pressure along the critical curve above 400 °C has been plotted in Figure 9.5 together with Tuttle and Bowen's (1958) data for the relationship between the water pressure and the temperature of the albite-orthoclase-quartz

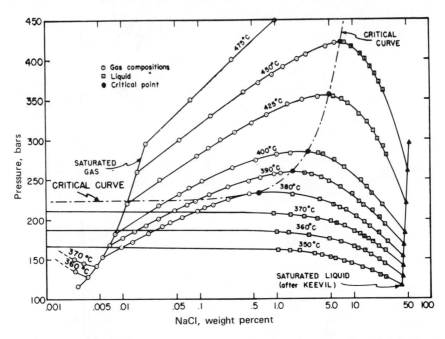

Fig. 9.3 The composition of coexisting gases and liquids in the sytem NaCl-H_2O between 350° and 450°C. (From Sourirajan and Kennedy, 1962, *Am. J. Sci.*, *260*, 115–141.)

cotectic (ternary) minimum. Below about 700°C the vapor pressure of water in equilibrium with melts of granitic composition is greater than that along the critical curve of the NaCl-H_2O system. Simple NaCl-H_2O mixtures separating from such a magma below 700°C should therefore be single-phase fluids. Such mixtures separating above 700°C could well be two-phase fluids. In such two-phase fluids, the NaCl concentration in the fluid of lower density could be considerably smaller than that of its higher density companion. The behavior of two-phase fluids in the system NaCl-H_2O during cooling depends on the variation of pressure and temperature during their rise and on the composition and relative proportion of the initial vapor and liquid phases. If both pressure and temperature drop linearly toward a surface pressure of 1 atm and a surface temperature of 100°C, then two-phase fluids will tend to homogenize on cooling. The cooling history of particular fluids can be worked out from the data in Figures 9.3 and 9.4.

In hydrothermal solutions the effect of solutes other than NaCl can rarely be neglected safely in discussing phase changes on cooling (see Chapters 3 and 4). It is reassuring then, to find that the system KCl-H_2O is very similar to the system NaCl-H_2O (Keevil, 1942; Benedict, 1939), and that small admixtures of KCl to NaCl in hydrothermal solutions serve

THE SOLUBILITY OF QUARTZ AND OTHER SiO$_2$ POLYMORPHS

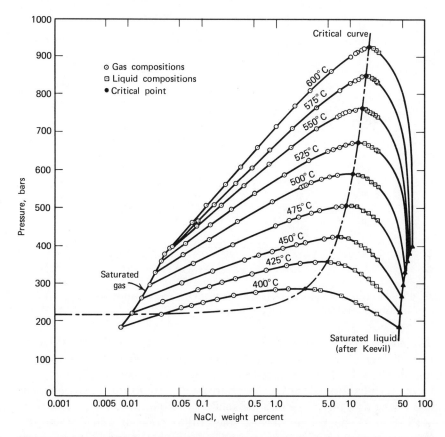

Fig. 9.4 The composition of coexisting gases and liquids in the system NaCl-H$_2$O between 400° and 600°C. (From Sourirajan and Kennedy, 1962, *Am. J. Sci.*, *260*, 115-141.)

mainly to lower the pressure slightly along the liquid-vapor-salt boundary (see also Wood, 1976). It seems almost certain that the addition of small amounts of CaCl$_2$ and MgCl$_2$ has similar, small effects. At high CaCl$_2$ and MgCl$_2$ concentrations, complications can arise in the form of hydrates of CaCl$_2$ and MgCl$_2$ and of hydrated double salts (see Linke, 1958), but fortunately the composition range in which these salts appear lies outside the composition range of normal hydrothermal solutions.

THE SOLUBILITY OF QUARTZ AND OTHER SiO$_2$ POLYMORPHS

The system SiO$_2$-H$_2$O is certainly the most thoroughly studied system involving a common gangue mineral of hydrothermal systems. The comprehensive studies of quartz solubility by Kennedy (1944, 1950) have been

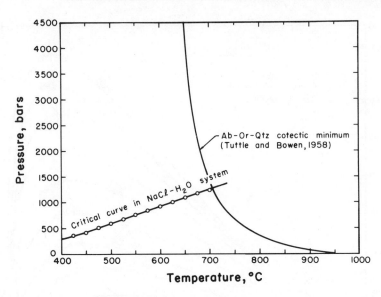

Fig. 9.5 The pressure and temperature of the critical curve in the system NaCl-H_2O and its relationship to the melting curve of granite.

followed in rapid succession by a series of important, and generally concordant sets of measurements. Kennedy's (1950) now classic diagram showing the solubility of quartz in water over a large portion of the hydrothermally interesting range of pressures and temperatures is shown in Figure 9.6. At a given pressure the solubility of quartz increases with temperature except in the region near the critical point of water, where the density of the solutions is quite small. The solubility of quartz is very much smaller than that of NaCl, and the critical curve of the system SiO_2-H_2O has the two end points which are characteristic of aqueous systems of slightly soluble salts.

Kennedy's low-temperature results were somewhat disturbing, as they seemed to indicate an unreasonably small solubility of quartz at and below 160°C. Since 1950, van Lier and others (1960), Siever (1962), and Morey et al. (1962) have shown that equilibrium was apparently not attained in Kennedy's (1950) runs below 200°C, and their data have been used in constructing the low-temperature portion of the summary diagram of Figure 9.7. Kennedy's (1950) data in the intermediate range of temperature and pressure were largely confirmed by Kitahara's (1960a) results and have been extended by Morey and Hesselgesser (1951), Khitarov (1956), Weill and Fyfe (1964), and Anderson and Burnham (1965). The agreement between solubility measurements at the same pressures and temperatures reported in these papers is usually very good. The upper limit of the solu-

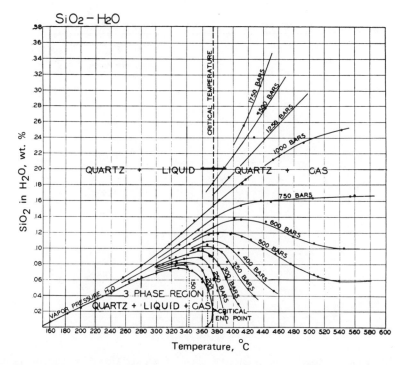

Fig. 9.6 The solubility of quartz in water at temperatures up to 560°C and pressures up to 1750 bars. (From Kennedy, 1950, *Econ. Geol.*, 45, 629-653.)

bility of SiO_2 in water vapor at pressures up to 9.7 kbar is determined by the composition of the vapor phase along the coexistence curves liquid-vapor-β-quartz and liquid-vapor-tridymite in the upper three-phase region. These curves have been determined by Kennedy et al. (1962), and their data have been used to plot the SiO_2-solubility curves above 900°C at 2, 3, 4, 5, 6, and 8 kbar. The solubility of the metastable polymorphs of SiO_2 is higher than that of quartz, but the differences decrease toward higher temperatures as shown in Figure 9.8.

The solubility of the SiO_2 polymorphs in aqueous solutions is virtually independent of the concentration of dissolved salts. This has been shown by Siever (1962), Greenberg and Price (1957), and Krauskopf (1956) at low temperatures, and by Kitahara (1960b) except in the vicinity of the lower critical end point. The solubility of silica also appears to be essentially independent of the pH of aqueous solutions in most of the geologically interesting range (Alexander et al., 1954).

Anderson and Burnham (1967) have shown that the solubility of quartz is decreased slightly by the addition of chlorides at 600°C and 3 kbar and

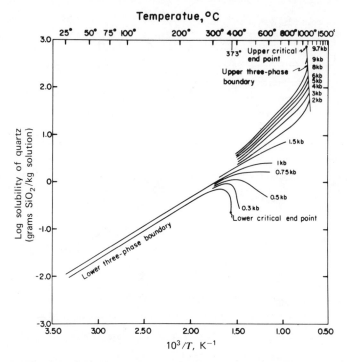

Fig. 9.7 Summary diagram for the solubility of quartz in water. Data from Kennedy (1950), Khitarov (1956), Kitahara (1960a), Morey and Hesselgesser (1951), Siever (1962), Morey, Fournier, and Rowe (1962), Weill and Fyfe (1964), Anderson and Burnham (1965), and Kennedy, Wasserburg, Heard, and Newton (1962). At the lower critical end point solid quartz is in equilibrium with a fluid phase saturated with quartz and consisting almost entirely of water; at the upper critical end point solid quartz is in equilibrium with a fluid phase saturated with quartz and consisting largely of SiO_2.

at 700°C and 4 kbar. Under these conditions the addition of NaOH increases the solubility of quartz by an amount directly proportional to the hydroxide concentration. Dickson (1966) and Learned et al. (1967) have reported the same effect of NaOH on the solubility of quartz at 250°C. They have also shown that quartz apparently reacts with Na_2S solutions between 100 and 250°C to produce solutions containing largely Na^+, HS^-, and $SiO_2 \cdot nH_2O \cdot OH^-$.

In solution dissolved silica appears to be present largely as one or more monomers of formula $SiO_2 \cdot nH_2O$ (Alexander et al., 1954). The lack of influence of the ionic strength of aqueous solutions on the solubility of the SiO_2 polymorphs is undoubtedly due to the uncharged nature of the $SiO_2 \cdot nH_2O$ species; the lack of influence of the pH in the acid and slightly alkaline range is due to the small value of the first ionization constant of these species. The value of the parameter n has been the subject of con-

THE SOLUBILITY OF QUARTZ AND OTHER SiO$_2$ POLYMORPHS

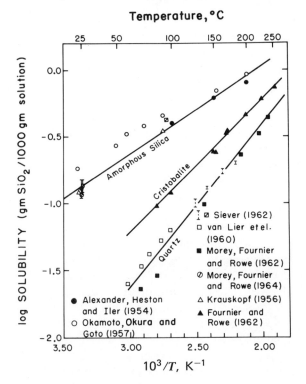

Fig. 9.8 The solubility of quartz, cristobalite, and amorphous silica in water between 25 and 250°C.

siderable discussion. It is reasonable to assume that in aqueous solutions Si^{4+} would tend to be in fourfold coordination with oxygen, and that the most likely formula for a neutral silica species would be Si(OH)$_4$, that is, SiO$_2 \cdot$2H$_2$O. However, both Weill and Fyfe (1964) and Anderson and Burnham (1965) find that their data cannot be interpreted readily if this is assumed to be the dominant form of dissolved silica in their experiments at high temperatures.

The near-independence of the solubility of the silica polymorphs on ionic strength and pH permits us to use the silica concentration as a geothermometer in geothermal systems (Arnórsson, 1975; Truesdell, 1976), and to discuss the solution and deposition of quartz in hydrothermal systems without undue worry concerning the effects of other dissolved species in solution. We can ask directly how much quartz is likely to be deposited during the passage of a solution through a vein or pipe in which there is a given temperature and pressure gradient, or conversely, what quantity of solution must have passed through a particular conduit to account for its quartz content (Smith, 1958). In most hydrothermal systems the thermal gradient is almost certainly equal to or greater than the normal gradient of 35°C/km. The high temperatures encountered at depth in wells in such areas as Wairakei, New Zealand, and the Salton Sea, California (see Chapter 13), suggest that thermal gradients during ore deposition may exceed 100°C/km. The near-surface pressure gradient in most wells is nearly hydrostatic. In deep wells the pressure gradient approaches the lithostatic gradient, and it seems likely that during ore deposition the pressure gradient lies between these two extreme limits. In a column of solution of density 1.0 g/cc, the rate of increase of pressure is nearly 100 atm/km. In a column of rock of specific gravity 3.0, the increase is nearly 300 atm/km. We can use the solubility data in Figures 9.6, 9.7, and 9.8 to define at least approximately the quantity of quartz deposited during the rise of hydrothermal solutions along a variety of possible geothermobars. The solubility of quartz along four limiting geothermobars between 15 and 700°C has been compiled in Table 9-1 (for a similar compilation see Table 12.3 in Smith, 1963). The solubility of quartz rises steadily with increasing temperature along three of the four geothermobars. Only when the thermal gradient is 100°C/km and the pressure gradient is 100 atm/km do we find a slight reversal in solubility between 400 and 600°C. In most vein systems the solubility of quartz almost certainly drops during the entire path of the hydrothermal solutions. The greatest amount of quartz deposition should take place at high temperatures. Between 600 and 700°C on the order of 0.5–14 g of quartz can be deposited per 1000 g of solution, whereas between

Table 9-1 Solubility of quartz in water along four geothermobars

	Solubility (g SiO_2/1000 g solution)			
Thermal Gradient:	35°C/km		100°C/km	
Pressure Gradient: Temperature (°C)	100 atm/km	300 atm/km	100 atm/km	300 atm/km
15	0.006	0.006	0.006	0.006
100	0.060	0.064	0.055	0.062
200	0.38	0.40	0.29	0.31
300	0.81	1.3	0.76	0.91
400	1.9	2.8	0.79	1.8
500	3.2	6.9	0.72	3.5
600	5.1	15	0.76	6.0
700	10.2	29	1.20	11

100 and 200°C only 0.2–0.3 g of quartz can be deposited from the same weight of solution.

The data in Table 9-1 are helpful in understanding quartz deposition in certain hydrothermal systems; however, quartz precipitation is often controlled by factors other than the shape of the solubility surface in the system SiO_2-H_2O. The formation of calc-silicates can remove so much dissolved silica from quartz-saturated hydrothermal fluids entering carbonate rocks, that quartz deposition is entirely prevented. On the other hand, hydrothermal solutions frequently liberate silica during reaction with silicate wall rocks (see Chapter 5). Since much of this silica is reprecipitated as vein quartz, the quantity of vein quartz deposited in hydrothermal systems can be enhanced greatly by wallrock reactions.

THE SOLUBILITY OF FLUORITE

Fluorite is one of the many sparingly soluble salts which occur in hydrothermal deposits. Its solubility is similar to that of quartz, but the response of its solubility to the presence of other salts is completely different. In aqueous solutions, fluorite is at least partially ionized. Its solubility is therefore strongly affected by the presence of other electrolytes, and the quantity of fluorite dissolved or precipitated is a function not only of temperature, pressure, and the ionic strength of a given hydrothermal solution but also of the pH and the ratio of the concentration of calcium and magnesium to that of fluoride ion in solution.

The numerous early measurements of the solubility of fluorite near room temperature have been summarized by Strübel (1965). The earliest systematic solubility measurements above 100°C were those of Booth and Bidwell (1950). Unfortunately, there is a great deal of scatter in their data as well as an anomalous solubility maximum just below the critical temperature of water. Ellis and Mahon (1964) repeated Booth and Bidwell's experiments, found a considerably smaller solubility at all temperatures between 100 and 350°C, and observed that above approximately 230°C, the calcium concentration in solution was less than one half the fluoride ion concentration. They suggest that $Ca(OH)_2$ formed in their reaction vessels but do not report actually observing portlandite at the end of any of their runs. The measurements of the solubility of $Ca(OH)_2$ by Blount and Dickson (1967) are in agreement with this conclusion. Strübel's (1965) values are in good agreement with those of Ellis and Mahon (1964).

In pure water the solubility of fluorite passes through a maximum near 100°C (see Figure 9.9) and approaches very small values near the critical temperature of water. Qualitatively this behavior is similar to that of the

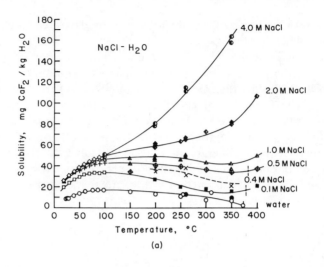

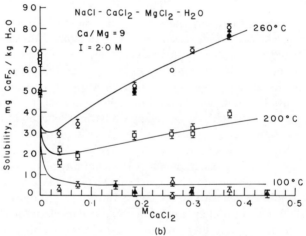

Fig. 9.9 (a) Solubility of fluorite as a function of temperature in NaCl solutions. The symbols represent sets of data for the following solutions: water—⊙ Strübel, ● Richardson and Holland; 0.1 M NaCl—□ Strübel, ■ Strübel and Schaefer; 0.4 M NaCl— × Richardson and Holland; 0.5 M NaCl— + Strübel, ⬙ Strübel and Schaefer; 1.0 M NaCl— △ Strübel, ▲ Strübel and Schaefer, ▼ Richardson and Holland; 2.0 M NaCl— ◇ Strübel, ◆ Strübel and Schaefer, ◆ Richardson and Holland; and 4.0 M NaCl— ◐ Richardson and Holland. The vertical dashes indicate the critical temperature for each solution. For detailed discussion, see Richardson and Holland (1979a). (b) Solubility of fluorite as a function of the concentration of $CaCl_2$ and $MgCl_2$ at a constant ionic strength of 2.0, a molar ratio of Ca/Mg of 9/1, and temperatures from 100 to 260°C. The position of the curves has been calculated to include the effects of the ion-pairs: NaF, Na_2F^+, CaF^+, and MgF^+. The symbols indicate data for: ○ 260°C, ● fluorite + anhydrite at 260°C, □ 200°C, △ 100°C, and ▲ KCl + $CaCl_2$ at 100°C. For detailed discussion, see Richardson and Holland (1979a).

solubility of quartz along the lower three-phase boundary, but the solubility maximum of quartz lies well above 100 °C. The solubility of fluorite increases with increasing pressure and NaCl concentration (see Figure 9.9a based on Strübel, 1965, Strübel and Schaefer, 1975, and Richardson, 1976). The most marked effect of NaCl is at low concentrations, and there is a possibility especially at low temperatures that the solubility of fluorite passes through a maximum with increasing NaCl concentration above 2 m. The solubility maximum with increasing temperature is shifted by the addition of NaCl, and it seems likely that at salt concentrations in excess of 1 m, the solubility of fluorite increases uniformly to high temperatures. Anikin and Shushkanov (1963) found this to be the case in solutions containing between 11 and 45 wt. % LiCl when the degree of filling of the autoclave was between 75 and 95%. The effect of NaCl on the solubility of fluorite in aqueous solutions is due largely to the decrease in the activity coefficient of Ca^{2+} and F^- with increasing ionic strength of the solutions and to the formation of the species NaF.

The addition of NaF to aqueous solutions decreases the solubility of fluorite to a degree that is quantitatively explained by the common ion effect (Richardson, 1976). Complexes of the type CaF_{2+n}^{n+} are therefore not particularly stable, and do not contribute significantly to the solubility of fluorite in aqueous solutions. On the other hand, Anikin and Shushkanov (1963) and Richardson and Holland (1979a) found that the solubility of fluorite is enhanced by the addition of $CaCl_2$ and $MgCl_2$ as shown in Figure 9.9b; it is therefore likely that the complexes CaF^+ and MgF^+ are important in the transport of fluoride ions in calcium-rich hydrothermal solutions (Richardson and Holland, 1979b; Richardson and Malinin, 1975).

The solubility of fluorite in acetic acid is less than in a NaCl solution of equivalent normality (Duparc et al., 1925). However, the solubility of fluorite in nitric acid is much greater than in an equivalent NaCl solution (Wilding and Rhodes, 1970). The rather low solubility in acetic acid is due to the small degree of ionization of this acid, while the enhanced solubility in nitric acid is largely due to the formation of the complex HF(aq). In HCl-NaCl solutions of a given ionic strength the solubility of fluorite increases with the proportion of HCl in the mixtures at least up to 260 °C; the magnitude of this increase is quantitatively explained by the presence of HF(aq) (Malinin, 1976, and Richardson and Holland, 1979b). The effect of HCl is small until the initial HCl concentration in solution exceeds approximately 10^{-3} m; when the initial HCl concentration exceeds 0.1 m, the solubility of fluorite is far greater than the solubility in pure NaCl solutions of comparable ionic strength.

Several mechanisms are available to explain the deposition of fluorite in hydrothermal systems (Richardson and Holland, 1979b). A decrease in solubility with decreasing temperature may well be the most important of these. In NaCl solutions the decrease in CaF_2 solubility is most pronounced below 150°C (see Figure 9.9a). In concentrated $CaCl_2$ solutions the solubility decreases rapidly even well above 150°C (Anikin and Shushkanov, 1963 and Richardson and Holland, 1979a).

Fluorite deposition can also be brought about by an increase in solution pH and by mixing solutions rich in Ca^{2+} with solutions rich in F^-. Since the ratio $m_{HF(aq)}/m_{F^-}$ exceeds unity only in rather acid solutions (pH ≤ ca. 3 at 260°C), an increase in pH in the flow direction of hydrothermal solutions is quantitatively important only for the deposition of fluorite from hydrothermal solutions that are initially rather acid.

THE SOLUBILITY OF ALKALINE EARTH CARBONATE MINERALS

A discussion of the solubility of the alkaline earth carbonate minerals is inherently more complicated than a discussion of the solubility of fluorite. The solubility of fluorite is insensitive to the pH of solutions over a considerable pH range and is not strongly affected by the fugacity of gases dissolved in hydrothermal fluids. On the other hand the solubility of carbonate minerals is strongly influenced by the pH and by the fugacity of CO_2 dissolved in hydrothermal fluids. An understanding of the relationship between the chemistry of hydrothermal solutions and the solubility of carbonate minerals proceeds best from a discussion of the system CO_2-H_2O, via the system CO_2-H_2O-NaCl, to the systems MCO_3-CO_2-H_2O and MCO_3-CO_2-NaCl-H_2O (where M stands for a divalent cation) and thence toward real hydrothermal solutions.

The System CO_2-H_2O

Figure 9.10 is a qualitative P-T-X diagram of the CO_2-H_2O system between 5 and 400°C for pressures up to 3500 atm (Tödheide and Franck, 1963). The front face shows the boiling-point curve of water, the rear face the boiling-point curve of CO_2. The critical curve is not continuous. The segment that starts at the critical point of CO_2 terminates almost immediately at the lower critical end point *LCEP*. The segment that starts at the critical point of water (C_{H_2O}) first moves rapidly toward lower temperatures (C_4 to C_7), becoming progressively steeper. At 266°C and 2450 bars (point C_8) the critical curve becomes vertical and at higher pressures it moves slightly toward higher temperatures (C_9 and C_{10}). A projection of Figure 9.10 on the P-X plane is shown in Figure 9.11. This is in substantial agreement with Takenouchi and Kennedy's (1964) results.

In hydrothermal solutions the mole fraction of CO_2 rarely exceeds 0.2. The

THE SOLUBILITY OF ALKALINE EARTH CARBONATE MINERALS

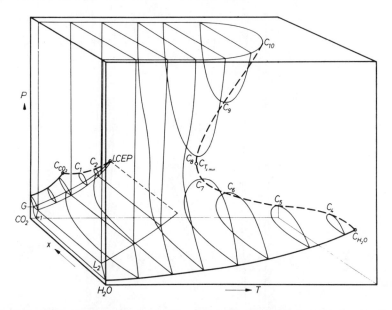

Fig. 9.10 Schematic representation of the two-phase region and the critical curve in the system CO_2-H_2O; dashed line = critical curve. (From Tödheide and Franck, 1963, *Zeit. Phys. Chem.*, **37**, 388–401.)

most interesting portion of Figure 9.10 therefore lies close to the front face. At low temperatures we are concerned largely with the solubility of CO_2 in water; at temperatures somewhat below the critical point of water we are concerned with the solubility of CO_2 in water and with critical phenomena; at temperatures above the critical point of water we are concerned primarily with the fugacity of CO_2 and of water in CO_2-H_2O mixtures.

For many purposes the representation of the distribution of a gas between a vapor and a solution phase is conveniently described in terms of the Henry's law coefficient, K, defined by the expression

$$K = \frac{f_c}{X} \tag{9.1}$$

where f_c is the fugacity of the gas and X the ratio of the moles of gas to the sum of moles of gas and water in solution. Ellis and Golding's (1963) summary diagram of the value of K for CO_2 in water mixtures and in NaCl solutions is shown in Figure 9.12. The maximum value of K near 125 °C represents a minimum in the solubility of CO_2 in water.

At high pressures the value of K is not constant, because the chemical

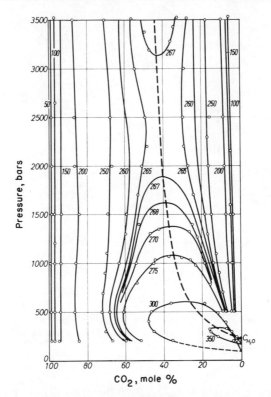

Fig. 9.11 Composition of vapor and liquid phases in the system CO_2-H_2O projected on to the *P-X* plane. Temperature in °C. (From Tödheide and Franck, 1963, *Zeit. Phys. Chem.*, **37**, 388-401.)

potential of CO_2 dissolved in water is no longer simply proportional to $\ln X_{CO_2}$. Malinin (1974) has shown that between 200 and 350°C and up to values of $X_{CO_2} = 0.14$ the fugacity of CO_2 in the vapor phase is related rather precisely to the mole fraction X_{CO_2} in the liquid phase, by the equation

$$\ln \frac{f_{CO_2}}{X_{CO_2}} \simeq \ln K^\circ + \frac{\overline{V}^L_{CO_2}(P_t - P_O)}{RT} + \frac{A(1 - X^2_{H_2O})}{RT} \qquad (9.2)$$

where

K° = value of the Henry's Law constant at low CO_2 pressures,
$\overline{V}^L_{CO_2}$ = partial molal volume of CO_2 in the liquid phase,
P_t = total pressure,
P_O = vapor pressure of water,
X_{H_2O} = mole fraction of water in the liquid phase.

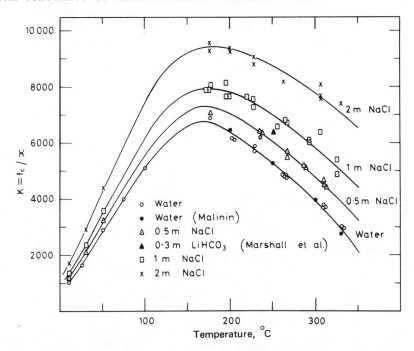

Fig. 9.12 Values of the Henry's law constant K for the solution of carbon dioxide in water and sodium chloride solutions. (From Ellis and Golding, 1963, *Am. J. Sci.*, *261*, 47-60.)

The term in $\overline{V}^L_{CO_2}$ is the pressure correction for $\mu^L_{CO_2}$; the third term is equal to $\ln \gamma^L_{CO_2}$, the logarithm of the activity coefficient of CO_2 in the liquid phase. As P_t approaches P_0 the second and third terms both approach zero, and equation (9.2) reduces to the form of equation (9.1).

At temperatures above 450°C, CO_2 and H_2O are completely miscible except at extremely high pressures. Franck and Tödheide (1959) have shown that interaction between the two components is slight, and only that to be expected in a mixture of polar and nonpolar molecules.

The System CO_2-H_2O-Salts

In a good deal of work on calcite solubility (see for instance Segnit et al., 1962) the function B, defined by the expression

$$m_{CO_2} = Bf_{CO_2} \tag{9.3}$$

has been used. B is obviously related to the reciprocal of K. The presence of salts depresses the solubility of gases in water (Prutton and Savage, 1945).

Often the depression follows the Setschenow equation

$$\log \frac{B_0}{B} = km \qquad (9.4)$$

where B_0 is the solubility of a gas in the pure solvent, B the solubility in a solution at salt concentration m, and k is a constant.

The salting-out effect of NaCl on the solubility of CO_2 in aqueous solutions is well shown in Figure 9.12 (Ellis and Golding, 1963). The Henry's law constant, K, for CO_2 in NaCl solutions is always greater than for CO_2 in pure water, and the ratio of the Henry's law constant for a solution of a given NaCl content to that in pure water increases somewhat with increasing temperature. However, the solubility of CO_2 in NaCl solutions is more than 40% of its solubility in pure water in all parts of the diagram. Ellis and Golding's (1963) results have been confirmed and extended to higher pressures and high salt concentrations by Takenouchi and Kennedy (1965) and by Malinin and Savelyeva (1972). Tiepel and Gubbins (1972) have related the values of $\bar{V}^L_{CO_2}$ in salt solutions to those in pure CO_2-H_2O mixtures; Malinin (1974) and Munjal and Stewart (1971) have successfully applied equation (9.2) to the analysis of the observed solubility of CO_2 in salt solutions.

The presence of NaCl in aqueous solutions markedly increases the temperature and pressure of their critical point (Sourirajan and Kennedy, 1962). The presence of CO_2 has the opposite effect (Tödheide and Franck, 1963). However, the effect of dissolved salts probably outweighs the effect of dissolved gases in most hydrothermal solutions, and it seems likely that the critical point of most hydrothermal solutions lies well above that of pure water.

The Solubility of Calcite

At any given temperature, the solubility of calcite in solutions in equilibrium with a vapor phase increases with increasing CO_2 pressure until $m_{CO_2} \approx 1$ mole/kg (Miller, 1952; Segnit et al., 1962; Malinin, 1963). This is shown in Figure 9.13 at a temperature of 150°C (Holland and Borcsik, 1965a). In solutions held at a constant total pressure, the solubility increases with increasing CO_2 concentration until $m_{CO_2} \approx 1$ mole/kg and then decreases toward higher CO_2 concentrations (Sharp and Kennedy, 1965; Berendsen, 1971; Malinin and Kanukov, 1971).

At any given CO_2 pressure in the vapor phase, the solubility of calcite decreases with increasing temperature as shown in Figure 9.14 (Ellis, 1959). Sharp and Kennedy (1965) report that the decrease in calcite solubility in solutions of a particular mole fraction of CO_2 decreases at least up to 600°C,

THE SOLUBILITY OF ALKALINE EARTH CARBONATE MINERALS

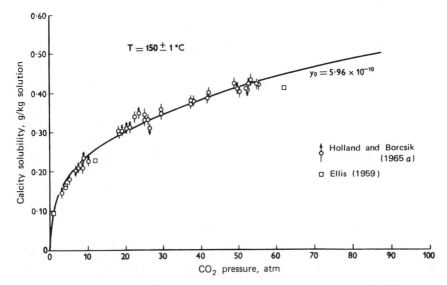

Fig. 9.13 The solubility of calcite in water as a function of the CO_2 pressure in the vapor phase at 150°C. (From Holland and Borcsik, 1965b, Symposium, *Problems of Postmagmatic Ore Deposition, Prague,* 2, 364-374.)

and Figure 9.15 shows the solubility surface of calcite in the system $CaCO_3$-CO_2-H_2O at temperatures up to 300°C and CO_2 pressures up to 70 atm.

At extremely low CO_2 pressures, especially in water cleansed of CO_2, the solubility of calcite is very small. It is almost constant up to 150°C and then falls steadily toward the critical point of water (Morey, 1962). The difference in the behavior of the solubility of calcite at high, from that at extremely low, CO_2 pressures can be understood in terms of the difference in the dominant ions in solution (see, for instance, Chapter 3 of Garrels and Christ, 1965).

In the range of CO_2 pressures most commonly encountered in hydrothermal systems the calcium ion concentration has been shown (Segnit, Holland, and Biscardi, 1962) to be related to the fugacity of CO_2 in the vapor phase by the expression

$$m_{Ca^{2+}}^3 = \frac{K_1' K_{Cal} B f_{CO_2}}{4 K_2 \gamma^3_{\pm Ca(HCO_3)_2}}$$

where

K_1' = first ionization constant of carbonic acid,
K_2 = second ionization constant of carbonic acid,
K_{Cal} = solubility product of calcite,

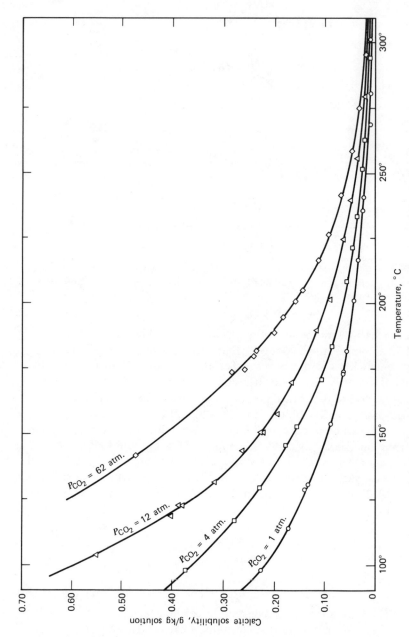

Fig. 9.14 The solubility of calcite in water up to 300°C at various partial pressures of carbon dioxide. (From Ellis, 1959, *Am. J. Sci.*, 257, 354–365.) The solubility values have been revised downward slightly by Ellis (1963).

THE SOLUBILITY OF ALKALINE EARTH CARBONATE MINERALS

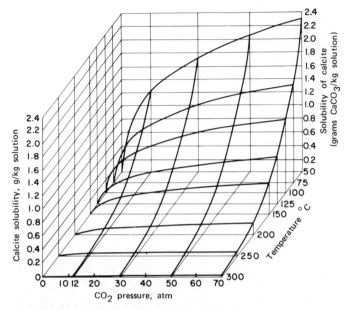

Fig. 9.15 The solubility surface of calcite in the system $CaCO_3$-$CO_2$$H_2O$ between 50° and 300°C and between 0 and 70 atm CO_2.

B = inverse Henry's law constant for CO_2 in water; moles CO_2/l of solution per atm,

f_{CO_2} = fugacity of CO_2,

$\gamma_{\pm Ca(HCO_3)_2}$ = mean activity coefficient of $Ca(HCO_3)_2$ in solution.

Calcite can be precipitated from solutions within the confines of Figure 9.15 by a decrease in total pressure at constant m_{CO_2} and by a decrease in m_{CO_2} due to boiling. However, hydrothermal solutions rising without boiling through a limestone terrain will tend to dissolve rather than to precipitate calcite, because the decrease in calcite solubility due to the decrease in total pressure is normally more than compensated by the increase in calcite solubility due to falling temperatures in the flow direction.

The addition of NaCl increases the solubility of calcite in hydrothermal solutions. The data in Figure 9.16 show that this increase is more pronounced at high than at low temperatures. Below 200°C the solubility of calcite passes through a maximum with increasing NaCl concentration; above 200°C, the calcite solubility probably increases to high NaCl concentrations. The effect of NaCl on the solubility of calcite is therefore similar to its effect on the solubility of fluorite, the sulfates (see below), and other sparingly soluble salts. The similarity is easily explained in terms

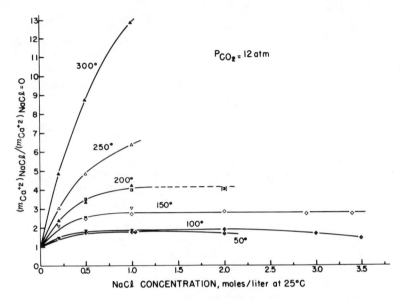

Fig. 9.16 The effect of NaCl on the solubility of calcite in water at a CO_2 pressure of 12 atm. (Data from Holland and Borcsik, 1965b, and Ellis, 1963.).

of Debye-Hückel theory; over a reasonably wide range of NaCl concentrations, the log of the calcite solubility is linearly related to the ionic strength, μ, by the expression $\sqrt{\mu}/1 + A_s\sqrt{\mu}$ where A_s is a weak function of temperature (Malinin, 1975). At least up to 275 °C, the increasingly positive effect of NaCl on the solubility of calcite is not sufficient to offset the decrease in calcite solubility with increasing temperature. Hydrothermal solutions in the system $CaCO_3$-CO_2-NaCl-H_2O will, therefore, tend to become undersaturated during their rise to the surface at temperatures below 275 °C. It seems likely, however, that at higher temperatures the solubility of calcite becomes prograde in concentrated NaCl solutions and that calcite deposition can take place from such solutions during simple cooling.

In dilute $CaCl_2$ solutions at 100 °C, the solubility of calcite is similar to that in water at the same CO_2 pressures (Holland and Borcsik, 1965a). At any given temperature, the effect of $CaCl_2$ will depend on the balance of two effects: a solubility depression due to the common ion effect and an enhancement due to the increase in the ionic strength of the solution (Malinin, 1963). The addition of $NaHCO_3$ should have similar effects. In most hydrothermal solutions, the concentration of calcium is greater than that of bicarbonate. Formally, such solutions can be classed as belonging to the NaCl-$CaCl_2$ type (Holland and Borcsik, 1965b). The solution and deposition of calcite from such solutions can be predicted roughly, at least near 100°C. In any complete theory for the solution and precipitation of calcite in hydrothermal systems, we cannot, however, neglect the effect of complexing, whose importance for the solubility of calcite in sea water has been well documented (see for instance Disteche, 1974, and Dyrssen and Wedborg, 1974).

In highly alkaline solutions, the solubility of calcite increases with temperature and with increasing CO_3^{2-} concentration. Malinin and Dernov-Pegarev's (1974) data in Figure 9.17 show that at 200, 300, and 400°C the solubility of calcite increases approximately with the third power of the CO_3^{2-} concentration, is nearly the same in Na_2CO_3 and K_2CO_3 solutions, and is unaffected by the addition of KCl. The slope of the lines in Figure 9.17 could be explained by the formation of carbonate complexes such as $Ca(CO_3)_4^{6-}$. Most natural hydrothermal solutions are probably slightly

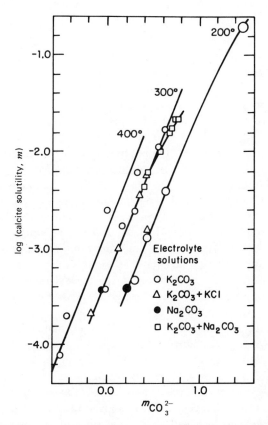

Fig. 9.17 The solubility of calcite in alkaline solutions (Malinin and Dernov-Pegarev, 1974).

acid or too slightly alkaline to be in the pH range where the solubility of calcite is strongly affected by carbonate complexing. The data in Figure 9.17 are, however, important for understanding the behavior of calcite in hydrothermal solutions related to carbonatites or to nonmarine evaporites.

The solution and precipitation of calcite in hydrothermal systems can be understood reasonably well in terms of the accumulated body of solubility data. The frequent formation of vugs and open space fillings in hydrothermal ore deposits in limestone terrains is probably due to the increase in the solubility of calcite with decreasing temperature in solutions within a wide range of chemical compositions. The deposition of calcite in hydrothermal systems can be caused by a variety of mechanisms. At high temperatures and high salinities, as well as in strongly alkaline solutions, the solubility of calcite probably decreases with falling temperature. However, boiling and changes in solution pH due to wallrock reactions are more likely mechanisms for calcite precipitation (see for instance, Ellis and Mahon, 1977).

The quantity of calcite deposited as a consequence of boiling depends on the composition of the boiling fluid. A complete removal of CO_2 from solutions within the confines of Figure 9.15 could produce a maximum precipitation of 2300 mg $CaCO_3$/kg of solution. A release of 70 atm of CO_2 at a temperature of 150°C would produce the precipitation of 450 mg $CaCO_3$/kg of solution. At 300°C 60 mg $CaCO_3$ would be precipitated by the complete release of 70 atm CO_2 pressure. A complete loss of CO_2 during boiling of hydrothermal solutions is unlikely; these figures are, therefore, maximum values for the precipitation of calcite from solutions in the system $CaCO_3$-CO_2-H_2O. On the other hand, hydrothermal solutions usually have a high ionic strength. The solubility of calcite will normally be higher in these solutions, and the quantity of calcite precipitated for a given CO_2 loss at a particular temperature will be greater than from solutions in the system $CaCO_3$-CO_2-H_2O. At 200°C, the amount of calcite deposited during boiling from a 1 m NaCl solution saturated with respect to $CaCO_3$ at 70 atm CO_2 is four times as great as from the corresponding NaCl-free solution. The precipitation of 100 mg calcite/kg of a hydrothermal solution due to boiling is therefore well within the realm of possibility at temperatures up to 300°C.

In silicate terrains, reactions involving H^+ metasomatism (see Chapter 5) may well exert a critical control on the deposition of calcite via their effect on the pH of the hydrothermal solutions. Many of the important reactions in alteration zones around hydrothermal veins involve the loss of H^+ from the solutions. Such a reduction in the H^+ concentration leads to an increase in the CO_3^{2-} concentration at a given total carbon species concentration, and can therefore lead to the precipitation of carbonate minerals. It seems likely that the carbonates described by Pinckney (1965) in veins in the northern part of the Boulder Batholith owe their origin to reactions of this type. Redox reactions can also lead to the precipitation of calcite. When sulfate is reduced to sulfide, H^+ is extracted from hydrothermal solutions because H_2S is a much weaker acid from H_2SO_4. The effect of SO_4^{2-} reduction on calcite precipitation is therefore similar to that of wallrock reactions involving H^+ metasomatism. Oxidation of sulfide to sulfate should, of course, have the opposite effect.

In many vein systems calcite in the wallrock has been replaced by sulfide minerals. Reactions of the type

$$CaCO_3 + Zn^{2+} + HS^- \rightleftharpoons ZnS + Ca^{2+} + HCO_3^- \tag{9.5}$$

have almost certainly played an important part in such replacement processes; the stability of sulfide and chloride complexes of the ore metals and the variation of these stabilities in the flow direction of the hydrothermal solutions determines in large part the extent of the replacement produced in particular hydrothermal settings (see Chapter 8).

The Solubility of Dolomite

Next to calcite, dolomite is undoubtedly the most important carbonate gangue mineral. The measurement of its solubility has turned out to be much more difficult than the measurement of the solubility of calcite. The rate of reaction of solutions with dolomite is unexpectedly slow even at 200°C, and a good many reported measurements of the solubility of dolomite are almost certainly too low because insufficient time was allowed for equilibrium to be reached. This must be the explanation for the anomalous data of Morey (1962) and of Garrels et al. (1960). A value near 2×10^{-17} for the solubility product of dolomite at 10-20°C was proposed by Hsu (1963) and by Holland et al. (1964) on the basis of the composition of water in limestone-dolomite aquifers and in caves. This value has been confirmed by Langmuir (1964). The solubility product of dolomite near room temperature is therefore nearly equal to the square of the solubility product of calcite at the same temperature.

In solutions saturated with respect to calcite and dolomite, equilibrium prevails in the reaction

$$2CaCO_3 + Mg^{2+} \rightleftharpoons CaMg(CO_3)_2 + Ca^{2+}$$

The equilibrium constant $K_{\text{Cal-Dol}}$ for this reaction,

$$K_{\text{Cal-Dol}} = \frac{a_{Ca^{2+}}}{a_{Mg^{2+}}}$$

has the value

$$K_{\text{Cal-Dol}} = \frac{K_{\text{Cal}}^2}{K_{\text{Dol}}},$$

where K_{Cal} is the solubility product of calcite and K_{Dol} the solubility product of dolomite. At room temperature the ratio $a_{\text{Ca}^{2+}}/a_{\text{Mg}^{2+}}$ in solutions saturated with respect to calcite and dolomite is near unity. Normal seawater, in which $a_{\text{Ca}^{2+}}/a_{\text{Mg}^{2+}}$ is 0.19 should therefore be able to dolomitize calcite. In fact, seawater concentrated by evaporation has been shown to be a dolomitizing solution until the ratio $m_{\text{Ca}^{2+}}/m_{\text{Mg}^{2+}}$ has risen to at least 0.50 (Kinsman, 1964, 1966). The value of this ratio in 2 m CaCl$_2$-MgCl$_2$ solutions in equilibrium with dolomite and calcite and with dolomite and magnesite has been studied by Rosenberg and Holland (1964) at temperatures between 260° and 425°C. Their results are shown in Figure 9.18. These data have been confirmed and extended by Usdowski (1964), Johannes (1966), Rosenberg et al. (1967) and Katz and Mathews (1977). In this temperature range, the ratio $m_{\text{Ca}^{2+}}/m_{\text{Mg}^{2+}}$ in solutions in equilibrium with calcite and dolomite is very much larger than unity. The solubility product of dolomite must, therefore, become progressively smaller than the square of the solubility product of calcite with increasing temperature. As the solubility product of calcite decreases steadily with increasing temperature, the solubility product of dolomite must decrease even more rapidly than that of calcite. Therefore dolomite, like calcite, can not be deposited from hydrothermal solu-

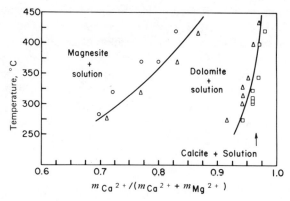

Fig. 9.18 The mole fraction $m_{\text{Ca}^{2+}}/(m_{\text{Ca}^{2+}} + m_{\text{Mg}^{2+}})$ in solutions in equilibrium with calcite + dolomite and dolomite + magnesite at temperatures between 275 and 420°C. Squares indicate runs in which dolomite was replaced by calcite; circles, dolomite replaced by magnesite; and triangles, calcite or magnesite replaced by dolomite. (From Rosenberg and Holland, 1964, *Science*, 145, 700–701.)

tions by simple cooling, but any of the mechanisms proposed for calcite precipitation can also be responsible for the precipitation of dolomite. Whether calcite, dolomite, or magnesite is precipitated by the operation of one of these mechanisms will depend essentially on the ratio of the activity of calcium to that of magnesium at the temperature of precipitation. From Figure 9.18, it is obvious that highly magnesium-rich solutions are not required to form dolomite or even to form magnesite at temperatures above 250°C.

The extensive dolomitization of limestone wallrock around many high-temperature hydrothermal ore bodies can be readily understood in terms of the data of Figure 9.18. A hydrothermal solution entering a limestone section at high temperatures might be expected to have a Ca/Mg ratio within the stability range of dolomite rather than of limestone. On entering a limestone section, such solutions will react with calcite until the Ca/Mg ratio has risen to the value at the dolomite-calcite boundary in Figure 9.18 at the temperature of the hydrothermal solution. Conversely, of course, a hydrothermal solution that is essentially devoid of magnesium on entering a dolomite section will tend to replace dolomite by calcite until the appropriate Ca/Mg ratio at the calcite-dolomite boundary is reached. At high temperatures, the potential amount of dolomitization per kilogram of solution, however, is much greater than the potential amount of dedolomitization, as the width of the dolomite field in Figure 9.18 is a good deal greater than that of calcite.

Consider, for instance, the behavior of a solution containing 1.0 mole Ca^{2+} and 0.10 mole Mg^{2+}/kg of solution entering a limestone at 400°C. The solution will replace calcite by dolomite until $m_{Ca^{2+}}/(m_{Ca^{2+}} + m_{Mg^{2+}})$ is about 0.97, that is, until $m_{Ca^{2+}}/m_{Mg^{2+}}$ is about 32. At equilibrium the solution will contain 1.068 moles Ca^{2+} and 0.032 moles Mg^{2+}. Each kilogram of solution will have produced 0.068 moles, or 12.5 g of dolomite from an initial 13.6 g of calcite. It is interesting that this quantity of dolomite is considerably larger, perhaps as much as an order of magnitude larger, than the quantity of quartz which might reasonably be expected to be deposited from 1 kg of solution in and around the conduit of such a hydrothermal system.

The Solubility of Strontianite

The solubility of strontianite ($SrCO_3$) has been studied by Helz and Holland (1965) between 50 and 200°C. Its solubility product at 25°C is 1/10 that of calcite. Toward higher temperatures the solubility of strontianite decreases (see Figure 9.19) but not as rapidly as that of calcite, so that above about 200°C the solubility of strontianite exceeds that of calcite.

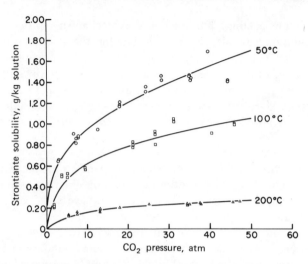

Fig. 9.19 The solubility of strontianite in water as a function of CO_2 pressure at 50°, 100°, and 200°C. (From Helz and Holland, 1965, *Geochim. Cosmochim. Acta*, **29**, 1303-1315.)

The same processes that can lead to the precipitation of calcite can also lead to the precipitation of strontianite. When strontianite and calcite are in equilibrium with the same solution, the ratio $a_{Sr^{2+}}/a_{Ca^{2+}}$ is defined approximately by the equation

$$SrCO_3 + Ca^{2+} \rightleftharpoons CaCO_3 + Sr^{2+}$$

$$K = \frac{a_{Sr^{2+}}}{a_{Ca^{2+}}} = \frac{K_{Str}}{K_{Cal}}$$

where K_{Str} is the solubility product of strontianite and K_{Cal} the solubility product of calcite. The relationship is only approximate, because each phase will contain some of the other in solid solution. In most underground waters the ratio $a_{Sr^{2+}}/a_{Ca^{2+}}$ is less than 0.01, that is, smaller than the ratio K_{Str}/K_{Cal}; calcite rather than strontianite is therefore the carbonate commonly precipitated in hydrothermal systems. Rather special processes must be at work in order to produce solutions with the abnormally high Sr/Ca ratio required for strontianite precipitation. Solutions passing through carbonate rocks containing celestite have been shown to have abnormally high Sr/Ca ratios (Feulner and Hubble, 1960), and it seems likely that the Sr/Ca ratio of solutions passing through sediments containing strontian barite or through aragonite which is being altered to calcite (Helz and Holland, 1965) can also reach the requisitely high values.

Most strontianite seems to have been precipitated at temperatures near or below 100°C. This is not surprising in view of the progressively higher value of the ratio $a_{Sr^{2+}}/a_{Ca^{2+}}$ which must prevail with increasing temperature in order to make strontianite stable with respect to calcite. Hydrothermal strontianite deposited at temperatures in excess of 200°C would hardly be expected, as values of the ratio $a_{Sr^{2+}}/a_{Ca^{2+}}$ in excess of unity are not at all likely.

The Solubility of Witherite

The solubility of witherite ($BaCO_3$) has been studied at 25°C by Garrels et al. (1960) and by Townley et al. (1937). The value of the solubility product of witherite calculated from the two sets of data is nearly the same. At higher temperatures the solubility of witherite has been studied by Malinin (1963), whose data have been summarized in Figure 9.20. The solubility of witherite in water in equilibrium with a vapor phase containing a given CO_2 pressure decreases with increasing temperature. At 25°C calcite and witherite have similar solubility products, but at 225°C the solubility product of calcite is more than an order of magnitude smaller than that of witherite. The solubility of witherite, like that of strontianite,

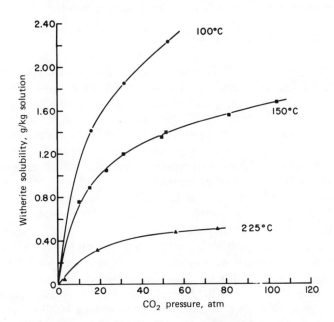

Fig. 9.20 The solubility of witherite in water as a function of the CO_2 pressure at 100° and 225°C. (After Malinin, 1963.)

decreases less rapidly than the solubility of calcite with increasing temperature.

In the absence of data for the solubility of alstonite and barytocalcite, both polymorphs of the double carbonate $BaCa(CO_3)_2$, the values of the ratio $a_{Ba^{2+}}/a_{Ca^{2+}}$ in solutions saturated with respect to minerals in the system $CaCO_3$-$BaCO_3$ remain unknown. However, the scarcity of the barium carbonates is undoubtedly due to the relatively high value of this activity ratio and to the low solubility of barite.

THE SOLUBILITY OF THE CALCIUM, STRONTIUM, AND BARIUM SULFATES

The solubility of the alkaline earth sulfate minerals in hydrothermal solutions behaves qualitatively somewhat like that of fluorite. In solution, their components exist largely as ions, and the quantity of each mineral that can be precipitated from a particular solution depends in large part on the temperature, pressure, and salinity, as well as on the ratio of the activity of the alkaline earth to that of sulfate in solution. The sulfate minerals are, however, treated separately, because their solution and precipitation also depends on the oxidation state of hydrothermal solutions.

The Solubility of Gypsum and Anhydrite

Along the join $CaSO_4$-H_2O, the phases anhydrite ($CaSO_4$), bassanite ($CaSO_4 \cdot 1/2H_2O$), and gypsum ($CaSO_4 \cdot 2H_2O$) have been synthesized. Bassanite is stable only at pressures in excess of 2 kb (Yamamoto and Kennedy, 1969). The problem of the stability relations between gypsum and anhydrite was a topic of continuing experimentation and debate until Hardie (1967) finally reversed the gypsum-anhydrite reaction, and showed that at a pressure of 1 atm these minerals are at equilibrium in water at 57°C. The equilibrium temperature increases only slightly with increasing total pressure (see Figure 9.21). In hydrothermal systems gypsum is therefore unlikely to form at temperatures above 70°C and is generally restricted to zones of secondary alteration.

The solubility of anhydrite in water decreases rapidly with increasing temperature and increases with increasing total pressure (see Figure 9.21). NaCl enhances the solubility of anhydrite, bassanite, and gypsum. Blount and Dickson's (1969) data in Figure 9.22 show that the solubility of anhydrite passes through a maximum with increasing NaCl concentration at temperatures up to approximately 175°C. At any given salinity the solubility passes through a minimum with increasing temperature. In 2 m NaCl solutions, the minimum solubility is probably near 400°C, in 6 m NaCl

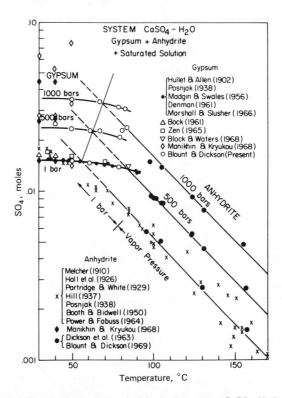

Fig. 9.21 The solubility of gypsum and anhydrite in the system CaSO$_4$-H$_2$O at 1 bar or the vapor pressure of the solution, at 500 bars, and at 1000 bars. The line connecting the intersection of the isobaric curves is the projection onto the composition-temperature plane of the curve separating the field of gypsum-liquid from that of anhydrite-liquid (Blount and Dickson, 1973).

solutions near 250°C. At temperatures below these minima, the solubility of anhydrite is retrograde; at higher temperatures it becomes prograde.

Marshall et al. (1964) have shown that the logarithm of the anhydrite solubility in NaCl solutions is proportional to $\mu^{1/2}/(1 + 1.5\mu^{1/2})$ at temperatures up to 200°C over a surprisingly wide range of the ionic strength, μ. As shown in Figure 9.23, the anhydrite solubility is well represented by the extended Debye-Hückel equation except at high ionic strengths. Subsequent studies (Marshall and Slusher, 1973, and Kalyanaraman et al., 1973a) have shown that the extended Debye-Hückel equation is equally useful for representing the solubility of anhydrite in NaNO$_3$, LiNO$_3$, and NaClO$_4$ solutions.

In most hydrothermal solutions, the calcium concentration is probably not equal to the sulfate concentration. Various studies of the solubility of anhydrite in solutions in which $m_{\text{Ca}^{2+}}/m_{\text{SO}_4^{2-}} \neq 1$ (Templeton and

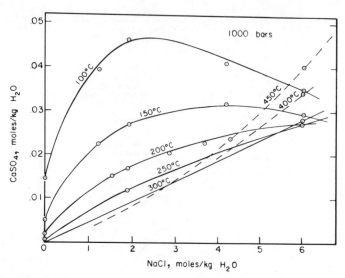

Fig. 9.22 The solubility of anhydrite in the system $CaSO_4$-$NaCl$-H_2O; dashed lines are based on data extrapolation (Blount and Dickson, 1969).

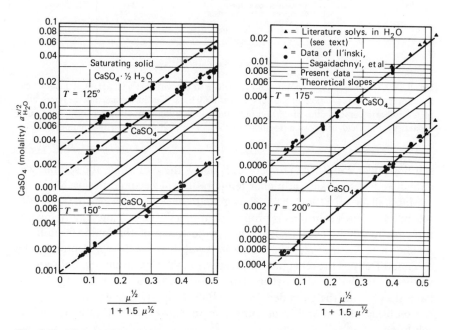

Fig. 9.23 The solubility function of bassanite at 125°C and of anhydrite at 125°, 150°, 175°, and 200°C in NaCl solutions. (From Marshall, Slusher, and Jones, 1964, *J. Chem. Eng. Data*, *1*, 187-191.)

Rodgers, 1967; Block and Waters, 1968; Kalyanaraman et al., 1973b) have shown that the addition of salts such as Na_2SO_4 depresses the solubility of anhydrite, and that complex species such as $CaSO_4(aq)$, $CaSO_4^{2+}$, and $Ca(SO_4)_2^{2-}$ are important carriers of sulfate in calcium-rich solutions and of calcium in sulfate-rich solutions.

The solubility of anhydrite is retrograde in essentially all hydrothermal solutions that participate in the formation of epithermal ore deposits. Boiling and an increase in the solution pH in the flow direction move solutions toward saturation with respect to anhydrite, but their effect is much less pronounced than for calcite. The increase in $m_{Ca^{2+}}$ due to wallrock reactions in the epithermal temperature range is apparently minor. The virtual absence of anhydrite as a gangue mineral in most epithermal ore deposits should therefore come as no surprise.

One group of epithermal deposits does, however, contain large quantities of anhydrite and of anhydrite that has been converted to gypsum. These are the Kuroko deposits (see for instance Tatsumi, 1970). Their origin is not completely understood, but geologic and isotopic evidence suggests that seawater penetrated into volcanic sequences, and that anhydrite was deposited when these solutions were heated near rhyodacite intrusions (Farrell, Holland, and Petersen, 1978 and Farrell, 1979).

If seawater is heated, saturation with respect to anhydrite is reached at approximately 110°C along the vapor pressure curve (Marshall and Slusher, 1968, and Glater and Schwartz, 1976). Above this temperature anhydrite begins to precipitate. Since $m_{Ca^{2+}}$ in seawater is 1.0×10^{-2} m and $m_{SO_4^{2-}}$ is 2.7×10^{-2} m, the precipitation of anhydrite is limited by the initial calcium concentration, and $m_{Ca^{2+}}/m_{SO_4^{2-}}$ decreases progressively during anhydrite precipitation. If seawater is heated in the presence of excess basalt or of other igneous rocks containing large quantities of calcium, calcium is released to the solution in such large quantities that virtually all of the initial seawater sulfate is precipitated as anhydrite. The final solution is saturated with respect to anhydrite, contains three to five times the initial concentration of calcium, and is virtually SO_4^{2-} free (Mottl et al., 1974; Hajash, 1975; Bischoff and Dickson, 1975; and Mottl and Holland, 1978). In the Reykjanes Peninsula of Iceland, seawater passes through hot basalts and emerges at the surface with a composition nearly identical to that predicted by laboratory experiments (Björnsson et al., 1972). Drill chips from this natural marine geothermal system contain several percent anhydrite.

Samples of water from thermal springs on the Galapagos Rift indicate that hydrothermal solutions at depth in this area have a composition very similar to that of the Reykjanes brines (Corliss et al., 1979).

Anhydrite is a fairly common mineral in high-temperature mineral deposits (Lindgren, 1933; Nagell, 1957) and has been found in particularly large quantities in many porphyry copper deposits. Since the ore-forming fluids in porphyry copper deposits are frequently very saline (see Chapter 14), the solubility of anhydrite in such fluids is probably prograde, and the deposition of anhydrite is easily explained. However, the prograde solubility of anhydrite may not be the most important contributor to its abundance in porphyry copper deposits. During cooling from 600 to 400°C, SO_2 disappears almost entirely from hydrothermal fluids (Holland, 1965). The reactions by which SO_2 is destroyed during the cooling of hydrothermal fluids depend on the oxidation state of the fluid at 600°C. In fluids which are in or near equilibrium with pyrite, magnetite, and hematite at 600°C, the removal of SO_2 can take place largely via the reaction

$$4SO_2 + 4H_2O \rightarrow H_2S + 3H_2SO_4 \tag{9.6}$$

H_2SO_4 formed by this reaction probably participates vigorously in wall-rock reactions. The alteration of plagioclase and of other calcium silicates can release large quantities of calcium. The solubility of anhydrite is therefore apt to be exceeded. Blount and Dickson's (1973) data indicate that the solubility of anhydrite in 5 m NaCl solutions at a pressure of 1 kb and 450°C is approximately 0.03 moles $CaSO_4$/kg H_2O, and the concentration product $m_{Ca^{2+}} \cdot m_{SO_4^{2-}}$ is 9×10^{-4} m^2. Even if $m_{Ca^{2+}}$ is only 0.1 m, $m_{SO_4^{2-}}$ need only be 9×10^{-3} m in order to saturate the solution with anhydrite. Much larger quantities of H_2SO_4 are apt to be generated during cooling by reaction (9.6), and the presence of large quantities of anhydrite in porphyry copper deposits is entirely expected.

The Solubility of Celestite

The behavior of celestite in aqueous solutions is similar to that of anhydrite. Its solubility is retrograde except at temperatures above 373°C and possibility below 20°C; it increases in the expected fashion with increasing NaCl concentration (Strübel, 1966) and with increasing total pressure (North, 1974). Strübel's (1966) data are shown in Figure 9.24. The solubility of celestite is smaller by approximately an order of magnitude than that of anhydrite. Lucchesi and Whitney (1962) have shown that at 0 and 25°C the solubility of celestite passes through a maximum with increasing NaCl concentration at a normality of 1.5 and 2.8 m respectively, and it is most likely that at higher temperatures the solubility maximum occurs at progressively higher NaCl concentrations.

In sediments, the deposition of celestite is frequently the result of the recrystallization of aragonite or of high-magnesian calcite in the presence of

SOLUBILITY OF CALCIUM, STRONTIUM, AND BARIUM SULFATES

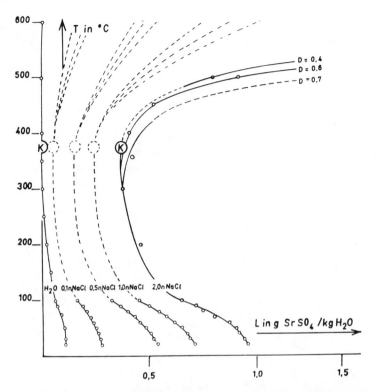

Fig. 9.24 The solubility of celestite in water and in NaCl solutions (Strübel, 1967).

solutions, such as seawater, containing fairly high concentrations of sulfate. In hydrothermal ore deposits, celestite is a relatively rare mineral. Its scarcity in epithermal deposits is almost certainly due to its retrograde solubility below 250°C. Seawater becomes saturated with respect to celestite on heating to approximately 110°C; celestite can therefore be expected to occur as a minor constituent, together with anhydrite, in hydrothermal systems where seawater has been heated above 110°C. Small amounts of celestite probably also accompany anhydrite in porphyry copper deposits unless the coprecipitation of Sr^{2+} with anhydrite prevents the appearance of a separate $SrSO_4$ phase.

The Solubility of Barite

Barite is the least soluble and most common of the alkaline earth sulfates in hydrothermal systems. Its solubility has been measured repeatedly, and many of the earlier measurements have been summarized by Strübel (1967) and by Gundlach et al. (1972). The most recent measurements are those of Blount (1977). They agree well with the measurements of Uchameyshvili et al. (1966) and are probably the most precise of the various somewhat divergent sets of data. In contrast to the solubility of anhydrite and celestite, the solubility of barite in water is prograde up to approximately 100°C (see Figure 9.25). In the presence of NaCl, the solubility maximum shifts to progressively higher temperatures, and at NaCl concentrations above 1 m the solubility of barite increases with temperature at least up to 300°C. Figure 9.25 shows that at any given temperature and NaCl concentration, the solubility increases with increasing pressure. Figure 9.26 shows that between 100 and 250°C the solubility of barite in-

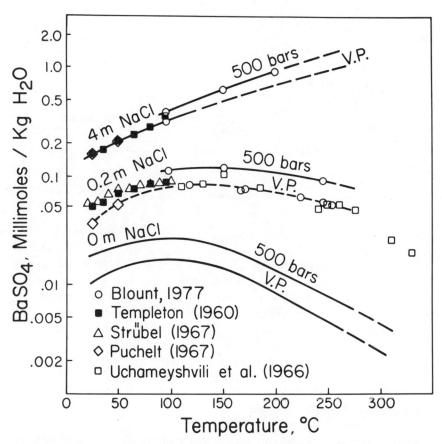

Fig. 9.25 The solubility of barite in water, 0.2 m NaCl, and 4 m NaCl along the vapor pressure (V.P.) curve and at a pressure of 500 bars (Blount, 1977).

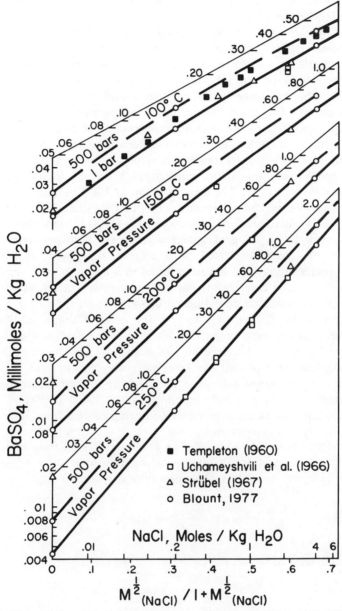

Fig. 9.26 The solubility of barite in NaCl solutions at 100°, 150, 200, and 250°C (Blount, 1977).

creases with increasing NaCl concentration as predicted by the extended Debye-Hückel equation.

Most hydrothermal solutions contain more than 1 mole NaCl (58.5 g)/kg of solution. Barite can therefore precipitate during the simple cooling of most hydrothermal solutions. The frequent occurrence of barite in epithermal deposits is in striking contrast with the rarity of anhydrite and celestite. The difference in the relative abundance of these minerals is readily explained by the difference in the temperature dependence of their solubility, and it seems likely that simple cooling accounts for the precipitation of much of the barite in epithermal ore deposits.

A 2 m NaCl solution saturated with respect to barite at 300°C along the vapor pressure curve contains approximately 160 mg $BaSO_4$/kg solution when $m_{Ba^{2+}} = m_{SO_4^{2-}}$. The same solution at 100°C contains only 56 mg $BaSO_4$/kg solution. Slow, simple cooling of 1 kg of such a solution from 300 to 100°C therefore produces some 0.1 g $BaSO_4$. From the same solution approximately 0.7 g of quartz would have been precipitated if the solution had been saturated with respect to quartz at 300°C. A barite/quartz ratio of 1/7 has been found in some epithermal ore deposits. However, in many deposits the barite/quartz ratio is much smaller. Silica release during wallrock alteration probably accounts for quartz deposition in excess of the amount expected on the basis of simple cooling alone. The barite/quartz ratio in systems where such "excess" quartz deposition is important should therefore be less than 1/7. On the other hand the amount of barite deposited can be considerably smaller than 0.1 g/kg of solution. In less saline solutions the solubility of barite decreases much less rapidly than in 2 m NaCl solutions, and below a salinity of approximately 0.8 m, no barite is precipitated during cooling from 300 to 100°C along the vapor pressure curve.

A smaller amount of barite will be precipitated even from highly saline solutions if the ratio $m_{Ba^{2+}}/m_{SO_4^{2-}}$ differs from unity. The differential of the solubility product K'_{Bar} with respect to temperature is

$$\frac{dK_{Bar}}{dT} = m_{Ba^{2+}}\frac{dm_{SO_4^{2-}}}{dT} + m_{SO_4^{2-}}\frac{dm_{Ba^{2+}}}{dT} \qquad (9.7)$$

In the precipitated barite, the ratio Ba^{2+}/SO_4^{2-} is unity; thus

$$\frac{dm_{Ba^{2+}}}{dT} = \frac{dm_{SO_4^{2-}}}{dT} \qquad (9.8)$$

and

$$\frac{dK_{Bar}}{dT} = (m_{Ba^{2+}} + m_{SO_4^{2-}})\frac{dm_{Ba^{2+}}}{dT} \qquad (9.9)$$

If we let r be defined as

$$r = m_{Ba^{2+}}/m_{SO_4^{2-}}$$

then

$$\frac{dm_{Ba^{2+}}}{dT} = \frac{r}{m_{Ba^{2+}}(1+r)}\frac{dK'_{Bar}}{dT} \qquad (9.10)$$

and since

$$K_{Bar} = m_{Ba^{2+}} \cdot m_{SO_4^{2-}} = \frac{m_{Ba^{2+}}^2}{r} \qquad (9.11)$$

it follows that

$$\frac{dm_{Ba^{2+}}}{dT} = \frac{r^{1/2}}{(K'_{Bar})^{1/2}(1+r)}\frac{dK'_{Bar}}{dT} \qquad (9.12)$$

The function $r^{1/2}/(1+r)$ has a maximum at $r = 1$, and decreases to 0 as r approaches 0 and infinity. The amount of barite precipitated during cooling through a given temperature interval therefore depends on the Ba^{2+}/SO_4^{2-} ratio in solution. Precise calculation of the magnitude of the ratio effect would require a knowledge of the stability of the various barium and sulfate complexes in hydrothermal solutions. Sulfate concentrations in epithermal solutions are generally less than 10^{-2} m since anhydrite would otherwise be a more common constituent of hydrothermal deposits. By the same token SO_4^{2-} in hydrothermal solutions is almost certainly greater than 10^{-4} m since witherite rather than barite would be the stable barium mineral at a pH of 7 and a CO_2 fugacity equal to or greater than 0.1 atm at lower sulfate concentrations (Holland, 1965).

Although these results suggest that sufficient barite can be precipitated from hydrothermal solutions simply by cooling, they do not by any means prove that this is the only or even the most important process controlling the precipitation of barite in hydrothermal systems. Hall and Friedman (1963) have shown that their data for the chemical and isotopic composition of fluids in fluid inclusions from the Kentucky-Illinois fluorspar district can best be interpreted by a mixing of at least two components: a component similar to that of the connate brines in the Illinois basin, and a hydrothermal component. Such mixing could lead to the precipitation of barite, particularly if the connate waters are rich in barium, a hypothesis that has been proposed by Sawkins (1966) for the origin of barite on the periphery of the Northern Pennine ore field in England.

If the solutions with which the hydrothermal components are mixed are highly oxidized, the sulfate concentration can be increased at the expense of the bisulfide concentration. However, in many epithermal deposits barite is associated with sulfide minerals that are unstable in highly oxidized environments. The mixing ratio for the two fluids might, therefore, have to be adjusted rather delicately to prevent the disappearance of sulfides such as pyrite and the appearance of oxides such as hematite from the mineral association.

SUMMARY

There is now a gratifyingly large body of experimental data dealing with the solubility of gangue minerals in hydrothermal solutions. With the exception of quartz, the common gangue minerals are largely ionized in hydrothermal solutions, and the salient features of their solubility can be understood in terms of Debye-Hückel theory. Complexing does play a role, however, particularly at temperatures above 200°C. Fortunately, data are available for the stability of many of the important complexes, so that the reaction paths of hydrothermal solutions can be calculated with a fair degree of confidence (see Chapter 11).

Several classes of mechanisms are available for precipitating nonmetallic minerals in hydrothermal systems. Any one gangue mineral may be precipitated in response to several or even all of these mechanisms, but one or two mechanisms are generally dominant.

Table 9-2 lists four categories of mechanisms. Simple cooling includes the effect of all of the processes that depend only on the chemistry and speciation of hydrothermal solutions during cooling. Boiling involves processes that accompany the separation of a single fluid into two or more fluids. Reactions with wallrocks and earlier vein minerals include cation exchange reactions and H^+ metasomatism (see Chapter 5) and reactions that affect the oxidation state of hydrothermal solutions. Mixing includes the interaction of shallow solutions (meteoric, connate, and seawater) with each other and with deep (metamorphic and magmatic hydrothermal) solutions.

Simple cooling is of importance primarily for minerals with prograde solubility. Quartz, fluorite, and barite fall in this group, as does anhydrite at temperatures greater than approximately 350°C. Anhydrite precipitation in meso- and hypothermal deposits may be related more to the loss of SO_2 by reaction with water during cooling rather than due to its prograde solubility in saline solutions at high temperatures.

Boiling can lead to the deposition of virtually all minerals, but it is probably of major significance for the carbonates, particularly calcite. In

Table 9-2 The important mechanisms for the precipitation of gangue minerals in hydrothermal systems

Mechanism of precipitation	Quartz	Fluorite	Calcite	Dolomite	Strontianite	Witherite	Anhydrite	Celestite	Barite
1. Simple Cooling									
$T > 350°C$	✓	✓	?				✓		✓
$T < 350°C$	✓	✓							
2. Simple Heating							✓		
3. Boiling			✓	✓			?		
4. Reaction with wallrocks and earlier vein minerals									
(a) cation-cation reactions (including H^+ metasomatism)	✓	✓	✓	✓	✓	✓	✓	✓	?
(b) redox reactions			?	?	?	?	?	?	?
5. Mixing of solutions		?	?	?	?	?	?	?	?

most of the geologically interesting range, the loss of CO_2 during boiling reduces the solubility of calcite, so that boiling can easily lead to calcite precipitation. The same is true for strontianite and witherite, but these minerals are rarely deposited at the high temperatures at which boiling is common.

Most wallrock reactions involve cation exchange, redox reactions, or both, and can have a profound effect on the precipitation of gangue minerals. Silica is frequently released and reappears as vein quartz. The loss of H^+ moves hydrothermal solutions toward saturation with respect to carbonates; concurrent gains of Ca^{2+}, Mg^{2+}, Na^+, K^+, and other cations reinforce this movement and increase the saturation of the solutions with respect to the sulfate minerals. Redox reactions generally involve one or more of the elements iron, carbon, and sulfur, and produce gains or losses of Fe^{2+} and Fe^{3+}, CH_4 and CO_2, and SO_4^{2-} and S^{2-}. These are of potential importance, particularly for the precipitation of carbonates and sulfates, and it is likely that at least some hydrothermal barite owes its origin to redox reactions.

The mixing of solutions is an attractive precipitating mechanism because nearly everything can be explained by the proper choice of ingredients. Several recent studies have strongly indicated the presence of more than one type of solution during the formation of individual ore deposits, and it is likely that mixing has played a role in the precipitation of nearly all gangue minerals in at least some ore deposits.

This summary shows that the experimental data bearing on the solubility of gangue minerals are sufficient to explain the occurrence of these minerals in hydrothermal ore deposits, but that they and the distribution of the gangue minerals are frequently not sufficient by themselves to define the precipitation mechanism or mechanisms in particular hydrothermal systems. Fortunately, the addition of evidence supplied by the composition of fluid inclusions and by the isotopic composition of elements in minerals and inclusion fluids can narrow the bounds of speculation considerably and has already led to compelling interpretations of particular hydrothermal systems.

REFERENCES

Alexander, G. B., W. M. Heston, and R. K. Iler (1954) The solubility of amorphous silica in water: *J. Phys. Chem.*, **58,** 453-455.

Anderson, G. M. and C. W. Burnham (1965) The solubility of quartz in supercritical water: *Am. J. Sci.*, **263,** 494-511.

―――― and ―――― (1967) Reactions of quartz and corundum with aqueous chloride and hydroxide solutions at high temperatures and pressures: *Am. J. Sci.*, **265,** 12-27.

Anikin, I. N. and A. D. Shushkanov (1963) The solubility of fluorite in aqueous solutions of electrolytes: *Kristallografiya*, **8,** 128-130 (pp. 98-100 in transl.).

REFERENCES

Arnorsson, S. (1975) Application of the silica geothermometer in low temperature hydrothermal areas in Iceland: *Am. J. Sci.*, **275**, 763-784.

Benedict, M. (1939) Properties of saturated aqueous solutions of potassium chloride at temperatures above 250°C: *J. Geol.*, **67**, 252-276.

Berendsen, P. (1971) The solubility of calcite in CO_2-H_2O solutions, from 100° to 300°C, 100 to 1000 bars, and 0 to 10 weight per cent CO_2, and geologic implications: Ph.D. thesis, University of California at Riverside.

Bischoff, J. L. and F. W. Dickson (1975) Seawater-basalt interation at 200°C and 500 bar: Implications for origin of sea floor heavy-metal deposits and regulation of sea-water chemistry: *Earth Planet. Sci. Letters*, **25**, 385-397.

Björnsson, S., S. Arnorsson, and J. Tomasson (1972) Economic evaluation of Reykjanes thermal brine area, Iceland: *Bull. Am. Assoc. Petrol. Geol.*, **56**, 2380-2391.

Block, J. and O. B. Waters, Jr. (1968) The $CaSO_4$-Na_2SO_4-$NaCl$-H_2O system at 25° to 100°C: *J. Chem. Eng. Data*, **13**, 336-344.

Blount, C. W. (1977) Barite solubilities and thermodynamic quantities up to 300°C and 1400 bars: *Am. Mineral.*, **62**, 942-957.

Blount, C. W. and F. W Dickson (1967) The solubility of portlandite ($Ca(OH)_2$) in water at temperatures to 257° and at pressures to 1500 bars (abst.): *Trans. Am. Geophys. Union*, **48**, 249-250.

──── and ──── (1969) Solubility of anhydrite ($CaSO_4$) in $NaCl$-H_2O from 100 to 450°C and 1 to 1000 bars: *Geochim. Cosmochim. Acta*, **33**, 227-245.

──── and ──── (1973) Gypsum-anhydrite equilibria in systems $CaSO_4$-H_2O and $CaSO_4$-$NaCl$-H_2O: *Am. Mineral.*, **58**, 323-331.

Booth, H. S. and R. M. Bidwell (1950) Solubilities of salts in water at high temperatures: *J. Am. Chem. Soc.*, **72**, 2567-2575.

Corliss, J. B., J. Dymond, L. I. Gordon, J. M. Edmond, R. P. von Herzen, R. D. Ballard, K. Green, D. Williams, A. Bainbridge, K. Crane, and T. H. van Andel (1979) Submarine thermal springs on the Galápagos Rift: *Science*, **203**, 1073-1082.

Dickson, F. W. (1966) Solubilities of metallic sulfides and quartz in hydrothermal sulfide solutions: *Bull. Volcanol.*, **29**, 605-628.

Disteche, A. (1974) The effect of pressure on dissociation constants and its temperature dependency: *The Sea, Vol. 5*, E. D. Goldberg, ed. New York: Wiley-Interscience, Chapter 2.

Duparc, L., P. Wenger, and G. Graz (1925) Etude de la solubilité du fluorure de calcium dans l'acide acétique: *Helv. Chim. Acta*, **8**, 280-284.

Dyrssen, D. and M. Wedborg (1974) Equilibrium calculations of the speciation of elements in sea water: in *The Sea, Vol. 5*, E. D. Goldberg, ed. New York: Wiley-Interscience, Chapter 5.

Ellis, A. J. (1959) The solubility of calcite in carbon dioxide solutions: *Am. J. Sci.*, **257**, 354-365.

──── (1963) The solubility of calcite in sodium chloride solutions at high temperatures: *Am. J. Sci.*, **261**, 259-267.

──── and R. M. Golding (1963) The solubility of carbon dioxide above 100°C in water and in sodium chloride solutions: *Am. J. Sci.*, **261**, 47-60.

──── and W. A. J. Mahon (1964) Natural hydrothermal systems and experimental hot-water rock interactions: *Geochim. Cosmochim. Acta*, **28**, 1323-1357.

──── and ──── (1977) *Chemistry and Geothermal Systems*, New York: Academic Press.

Farrell, C. W. (1979) *Strontium Isotopes of Kuroko Deposits;* Ph.D. Thesis, Harvard University.

———, H. D. Holland, and U. Petersen (1978) The isotopic composition of strontium in barites and anhydrites from Kuroko Deposits: *Mining Geology* (Japan), **28,** 281-291.

Feulner, A. J. and J. H. Hubble (1960) Occurrence of strontium in the surface and ground waters of Champaign County, Ohio: *Econ. Geol.,* **55,** 176-186.

Fournier, R. O. and J. J. Rowe (1962) The solubility of cristobalite along the three-phase curve, gas plus liquid plus cristobalite: *Am. Mineral.,* **47,** 897-902.

Franck, E. U. and K. Tödheide (1959) Thermische Eigenschaften überkritischer Mischungen von Kohlendioxyd und Wasser bis zu 750°C und 2000 atm.: *Zeit. Physik. Chem.,* N.F., **22,** 232-245.

Garrels, R. M. and C. L. Christ (1965) *Solutions, Minerals, and Equilibria,* New York: Harper & Row.

———, M. E. Thompson, and R. Siever (1960) Stability of some carbonates at 25° and one atmosphere total pressure: *Am. J. Sci.,* **258,** 402-418.

Glater, J. and J. Schwartz (1976) High-temperature solubility of calcium sulfate hemihydrate and anhydrite in natural seawater concentrates: *J. Chem. Eng. Data,* **21,** 47-52.

Greenberg, S. A. and E. W. Price (1957) The solubility of silica in solutions of electrolytes: *J. Phys. Chem.,* **61,** 1539-1541.

Gundlach, H., D. Stoppel, and G. Strübel (1972) Zur hydrothermalen Löslichkeit von Baryt: *N. Jb. Miner. Abh.,* **116,** 321-338.

Hajash, A., Jr. (1975) Hydrothermal processes along mid-ocean ridges: An experimental investigation: Ph.D. thesis, Texas A&M University.

Hall, W. E. and I. Friedman (1963) Composition of fluid inclusions, Cave-in-Rock fluorite district, Illinois, and upper Mississippi Valley zinc-lead district: *Econ. Geol.,* **58,** 886-911.

Hardie, L. A. (1967) Gypsum-anhydrite equilibrium at one atmosphere pressure: *Am. Mineral.,* **52,** 171-200.

Helz, G. R. and H. D. Holland (1965) The solubility and geologic occurrence of strontianite: *Geochim. Cosmochim. Acta,* **29,** 1303-1315.

Holland, H. D. (1965) Some applications of thermochemical data to problems of ore deposits. II. Mineral assemblages and the composition of ore-forming fluids: *Econ. Geol.,* **60,** 1101-1166.

——— and M. Borcsik (1965a) On the solution and deposition of calcite in hydrothermal systems: Symposium, *Problems of Postmagmatic Ore Deposition, Prague,* **2,** 364-374.

——— and ——— (1965b) On the solution of calcite by and the deposition of calcite from hydrothermal solutions: Symposium, *Problems of Postmagmatic Ore Deposition, Prague,* **2,** 418-421.

———, T. V. Kirsipu, J. S. Huebner, and U. M. Oxburgh (1964) On some aspects of the chemical evolution of cave waters: *J. Geol.,* **72,** 36-67.

Hsu, K. J. (1963) Solubility of dolomite and composition of Florida ground waters: *J. Hydrol.,* **1,** 288-310.

Johannes, W. (1966) Experimentelle Magnesitbildung aus Dolomit + $MgCl_2$: *Contr. Mineral. Petrol.,* **13,** 51-58.

Kalyanaraman, R., L. B. Yeatts, and W. L. Marshall (1973a) High-temperature Debye-Hückel correlated solubilities of calcium sulfate in aqueous sodium perchlorate solutions: *J. Chem. Thermodyn.,* **5,** 891-898.

———, ——— and ——— (1973b) Solubility of calcium sulfate and association equilibria in $CaSO_4$ + Na_2SO_4 + $NaClO_4$ + H_2O at 273 to 623°K: *J. Chem. Thermodyn.,* **5,** 899-909.

REFERENCES

Katz, A. and A. Matthews (1977) The dolomitization of $CaCO_3$: an experimental study at 252-295°C: *Geochim. Cosmochim. Acta,* **41,** 297-308.

Keevil, N. B. (1942) Vapor pressures of aqueous solutions at high temperatures: *J. Am. Chem. Soc.,* **64,** 841-850.

Kennedy, G. C. (1944) The hydrothermal solubility of silica: *Econ. Geol.,* **39,** 25-36.

_____ (1950) A portion of the system silica-water: *Econ. Geol.,* **45,** 629-653.

_____, G. J. Wasserburg, H. C. Heard, and R. C. Newton (1962) The upper three-phase region in the system SiO_2-H_2O: *Am. J. Sci.,* **260,** 501-521.

Khitarov, N. I. (1956) The 400° isotherm for the system H_2O-SiO_2 at pressures up to 2000 kg/cm^2: *Geochemistry,* 55-61 (transl.).

Kinsman, D. J. J. (1964) Recent carbonate sedimentation near Abu Dhabi, Trucial Coast, Persian Gulf: doctoral dissertation, University of London.

_____ (1966) Gypsum and anhydrite of Recent age, Trucial Coast, Persian Gulf: *Proceedings of the Second Salt Symposium,* Northern Ohio Geological Society.

Kitahara, S. (1960a) The solubility of quartz in water at high temperatures and high pressures: *Rev. Phys. Chem. Japan,* **30,** 109-114.

_____ (1960b) The solubility of quartz in the aqueous sodium chloride solution at high temperatures and high pressures: *Rev. Phys. Chem. Japan,* **30,** 115-121.

Krauskopf, K. B. (1956) Dissolution and precipitation of silica at low temperatures: *Geochim. Cosmochim. Acta,* **10,** 1-26.

Langmuir, D. (1964) Stability of carbonates in the system CaO-MgO-CO_2-H_2O: Ph.D. thesis, Harvard University.

Learned, R. L., F. W. Dickson, and G. Tunell (1967) The solubility of quartz in Na_2S and $NaOH$ solutions at elevated temperatures and 100 bars (abst.): *Trans. Am. Geophys. Union,* **48,** 249.

Lindgren, W. (1933) *Mineral Deposits,* 4th ed., New York: McGraw-Hill.

Linke, W. F. (1958) *Solubilities, Vol. 1, Inorganic and Metal-Organic Compounds,* 4th ed., New York: Van Nostrand.

Lucchesi, P. J. and E. D. Whitney (1962) Solubility of strontium sulphate in water and aqueous solutions of hydrogen chloride, sodium chloride, sulphuric acid and sodium sulphate by the radiotracer method: *J. Appl. Chem.,* **12,** 277-279.

Malinin, S. D. (1963) An experimental investigation of the solubility of calcite and witherite under hydrothermal conditions: *Geochemistry,* 650-667 (transl.).

_____ (1974) Questions concerning the thermodynamics of the H_2O-CO_2 system: *Geokhimiya,* **10,** 1523-1549.

_____ (1975) Rastvorimost $CaCO_3$ v uglekyskikh rastvorakh H_2O i $NaCl$ pri visokykh temperaturakh u davleniakh; Trudy IX po experimentalnoy i technicheskoy mineralogii i petrografi: *Irkutsk 1973,* Iz-vo ANSSR.

_____(1976) Solubility of fluorspar (CaF_2) in $NaCl$ and HCl solutions under hydrothermal conditions: *Geokhimiya,* No. 2, 223-228.

_____ and V. F. Dernov-Pegarev (1974) Investigation of calcite solubility in K_2CO_3 and Na_2CO_3 solutions at temperatures of 200-350°C: *Geokhimiya,* **3,** 454-462.

_____ and A. B. Kanukov (1971) The solubility of calcite in homogeneous H_2O-$NaCl$-CO_2 systems in the 200-600°C temperature interval: *Geochem. Int.,* **8,** 668-679.

_____ and N. I. Savelyeva (1972) Solubility of CO_2 in $NaCl$ and $CaCl_2$ solutions at 25, 50, and 75° and elevated CO_2 pressure: *Geochem. Int.,* **9,** 410-418.

Marshall, W. L. and R. Slusher (1973) Debye-Hückel correlated solubilities of calcium sulfate in water and in aqueous sodium nitrate and lithium nitrate solutions of molality

0 to 6 mol kg^{-1} and at temperatures from 398 to 623°K: *J. Chem. Thermodyn.,* **5,** 189-197.

____, ____, and E. V. Jones (1964) Aqueous systems at high temperature. XIV. Solubility and thermodynamic relationships for CaSO$_4$ in NaCl-H$_2$O solutions from 40° to 200°C, 0 to 4 molal NaCl: *J. Chem. Eng. Data,* **9,** 187-191.

Miller, J. P. (1952) A portion of the system calcium carbonate-carbon dioxide-water, with geological implications: *Am. J. Sci.,* **250,** 161-203.

Morey, G. W. (1962) The action of water on calcite, magnesite, and dolomite: *Am. Mineral.,* **47,** 1456-1460.

____ and W. T. Chen (1956) Pressure-temperature curves in some systems containing water and a salt: *J. Am. Chem. Soc.,* **78,** 4249-4252.

____, R. O. Fournier, and J. J. Rowe (1962) The solubility of quartz in water in the temperature interval from 25° to 300°C: *Geochim. Cosmochim. Acta,* **26,** 1029-1043.

____, ____, and ____ (1964) The solubility of amorphous silica at 25°C: *J. Geophys. Res.,* **69,** 1995-2002.

____ and J. M. Hesselgesser (1951) The solubility of some minerals in super-heated steam at high pressures: *Econ. Geol.,* **46,** 821-835.

Mottl, M., R. F. Corr, and H. D. Holland (1974) Chemical exchange between sea water and mid-ocean ridge basalt during hydrothermal alteration (abst.): *Geol. Soc. Am. GAAPBC,* **6,** 879-880.

____ and H. D. Holland (1978) Chemical exchange during hydrothermal alteration of basalt by seawater—I. Experimental results for major and minor components of seawater: *Geochim. Cosmochim. Acta,* **42,** 1103-1115.

Munjal, P. and P. B. Stewart (1971) Correlation equation for solubility of carbon dioxide in water, sea water, and sea water concentrates: *J. Chem. Eng. Data,* **16,** 170-172.

Nagell, R. H. (1957) Anhydrite complex of the Morococha district, Peru: *Econ. Geol.,* **52,** 632-644.

North, N. A. (1974) Pressure dependence of SrSO$_4$ solubility: *Geochim. Cosmochim. Acta,* **38,** 1075-1081.

Okamoto, G., T. Okura, and K. Goto (1957) Properties of silica in water: *Geochim. Cosmochim. Acta,* **12,** 123-132.

Ölander, A. and H. Liander (1950) The phase diagram of sodium chloride and steam above the critical point: *Acta. Chem. Scand.,* **4,** 1437-1445.

Pinckney, D. M. (1965) Veins in the northern part of the Boulder Batholith, Montana: Ph.D. thesis, Princeton University.

Prutton, C. F. and R. L. Savage (1945) The solubility of carbon dioxide in calcium chloride-water solutions at 75°, 100°, 120°, and high pressure: *J. Am. Chem. Soc.,* **67,** 1550-1554.

Richardson, C. K. (1976) The solubility of fluorite in hydrothermal solutions: Ph.D. thesis, Harvard University.

Richardson, C. K. and S. D. Malinin (1975) Solubility of fluorite and complexing in the system CaF$_2$-NaCl-HCl-H$_2$O (abst.): *Geol. Soc. Am. Abst. Prog.,* **7,** 1246.

____ and H. D. Holland (1979a) The solubility of fluorite in hydrothermal solutions, an experimental study: *Geochim. Cosmochim. Acta,* in press.

____ and ____ (1979b) Fluorite deposition in hydrothermal systems: *Geochim. Cosmochim. Acta,* in press.

Rosenberg, P. E. and H. D. Holland (1964) Calcite-dolomite-magnesite stability relations in solutions at elevated temperatures: *Science,* **145,** 700-701.

_____, D. M. Burt, and H. D. Holland (1967) Calcite-dolomite-magnesite stability relations in solutions: The effect of ionic strength: *Geochim. Cosmochim. Acta,* **31,** 391-396.

Sawkins, F. J. (1966) Ore genesis in the North Pennine ore-field in the light of fluid inclusion studies: *Econ. Geol.,* **61,** 385-401.

Segnit, E. R., H. D. Holland, and C. J. Biscardi (1962) The solubility of calcite in aqueous solutions—I. The solubility of calcite in water between 75° and 200° at CO_2 pressures up to 60 atm.: *Geochim. Cosmochim. Acta,* **26,** 1301-1331.

Sharp, W. E. and G. C. Kennedy (1965) The system $CaO\text{-}CO_2\text{-}H_2O$ in the two-phase region calcite and aqueous solution: *J. Geol.,* **73,** 391-403.

Siever, R. (1962) Silica solubility, 0°-200°C, and the diagenesis of siliceous sediments: *J. Geol.,* **70,** 127-150.

Smith, F. G. (1958) Transport and deposition of the nonsulphide vein minerals. VI. Quartz: *Can. Mineral.,* **6,** 210-221.

_____ (1963) *Physical Geochemistry,* Reading, Mass.: Addison-Wesley.

Sourirajan, S. and G. C. Kennedy (1962) The system $H_2O\text{-}NaCl$ at elevated temperatures and pressures: *Am. J. Sci.,* **260,** 115-141.

Strübel, G. (1965) Quantitative Untersuchungen über die hydrothermale Löslichkeit von Flusspat (CaF_2): *N. Jb. Mineral.,* No. 3, 83-95.

_____ (1966) Die hydrothermale Löslichkeit von Cölestin im System $SrSO_4\text{-}NaCl\text{-}H_2O$: *N. Jb. Mineral.,* No. 4, 99-108.

_____(1967) Zur Kenntnis und genetischen Bedeutung des Systems $BaSO_4\text{-}NaCl\text{-}H_2O$: *N. Jb. Mineral.* 7/8, 223-233.

_____ and B. Schaefer (1975) Experimentalle Untersuchngen zur hydrothermalen Löslichkeit von Fluorit im System $CaF_2\text{-}NaCl\text{-}H_2O$: in Geochemie der Lagerstättenbildung und -Prospektion GDMB-DMG Symposium, 21-23 February 1974, Karlsruhe.

Takenouchi, S. and G. C. Kennedy (1964) The binary system $H_2O\text{-}CO_2$ at high temperatures and pressures: *Am. J. Sci.,* **262,** 1055-1074.

_____ and _____ (1965) The solubility of carbon dioxide in NaCl solutions at high temperatures and pressures: *Am. J. Sci.,* **263,** 445-454.

Tatsumi, T. (1970) *Volcanism and Ore Genesis,* Tokyo: University of Tokyo Press.

Templeton, C. C. (1960) Solubility of barium sulfate in sodium chloride solutions from 25° to 95°C: *J. Chem. Eng. Data,* **5,** 514-516.

_____ and J. C. Rodgers (1967) Solubility of anhydrite in several aqueous salt solutions between 250° and 325°C: *J. Chem. Eng. Data,* **12,** 536-547.

Tiepel, E. W. and K. E. Gubbins (1972) Partial molal volumes of gases dissolved in electrolyte solutions: *J. Phys. Chem.,* **76,** 3044-3049.

Tödheide, K. and E. U. Franck (1963) Das Zweiphasengebiet und die kritische Kurve im System Kohlendioxid-Wasser bis zu Drucken von 3500 bar: *Zeit. Phys. Chem., N.F.,* **37,** 388-401.

Townley, R. W., W. B. Whitney, and W. A. Felsing (1937) The solubilities of barium and strontium carbonates in aqueous solutions of some alkali chlorides: *J. Am. Chem. Soc.,* **59,** 631-633.

Truesdell, A. H. (1976) Geochemical techniques in exploration: in *Proceedings 2nd U.N. Symp. Geothermal Research,* 53-79, U.S. Government Printing Office, Wash., D.C.

Tuttle, O. F. and N. L. Bowen (1958) Origin of granite in the light of experimental studies in the system $NaAlSi_3O_8\text{-}KAlSi_3O_8\text{-}SiO_2\text{-}H_2O$: *Geol. Soc. Am. Mem.,* **74.**

Uchameyshvili, N. Y., S. D. Malinin, and N. I. Khitarov (1966) Solubility of barite in concentrated chloride solutions of some metals at elevated temperatures in relation to problems of the genesis of barite deposits: *Geokhimiya,* 1193–1205.

Usdowski, H. E. (1964) Die Phasenbeziehungen der Systeme Ca^{2+}-Mg^{2+}-CO_3^{2-}-Cl_2^{2-}-SO_4^{2-}-H_2O und Na_2^{2+}-Ca^{2+}-Mg^{2+}-CO_3^{2-}-SO_4^{2-}-Cl_2^{2}-H_2O: *Nachr. Akad. Wiss. Göttingen, Math-Physik., II.,* **20,** 263–265.

van Lier, J. A., P. L. de Bruyn, and J. T. G. Overbeek (1960) The solubility of quartz: *J. Phys. Chem.,* **64,** 1675–1682.

Weill, D. F. and W. S. Fyfe (1964) The solubility of quartz in H_2O in the range 1000–4000 bars and 400–550°C: *Geochim. Cosmochim. Acta,* **28,** 1243–1255.

Wilding, M. W. and D. W. Rhodes (1970) Solubility isotherms for calcium fluoride in nitric acid solution: *J. Chem. Eng. Data,* **15,** 297–298.

Wood, J. R. (1976) Thermodynamics of brine-salt equilibria-II. The system NaCl-KCl-H_2O from 0 to 200°C: *Geochim. Cosmoschim. Acta,* **40,** 1211–1220.

Yamamoto, H. and G. C. Kennedy (1969) Stability relations in the system $CaSO_4$-H_2O at high temperatures and pressures: *Am. J. Sci.,* **267-A,** 550–557.

10

Isotopes of Sulfur and Carbon

HIROSHI OHMOTO
The Pennsylvania State University

ROBERT O. RYE
U. S. Geological Survey

Sulfur has four stable isotopes with the approximate natural abundance of $^{32}S = 95.02\%$, $^{33}S = 0.75\%$, $^{34}S = 4.2\%$ and $^{36}S = 0.017\%$ (MacNamara and Thode, 1950). Although some studies have been made on the variation of $^{36}S/^{32}S$ ratios in natural samples, the majority of sulfur isotopic studies deal with the variation of $^{34}S/^{32}S$ ratios.

The sulfur isotopic composition of a compound is usually expressed as a $\delta^{34}S$ value, which is defined as a per mil deviation of the $^{34}S/^{32}S$ ratio of the compound relative to that of the troilite phase of the Cañon Diablo meteorite ($^{34}S/^{32}S = 0.0450045$, Ault and Jensen, 1963):

$$\delta^{34}S_{sample} = \left[\frac{(^{34}S/^{32}S)_{sample}}{(^{34}S/^{32}S)_{standard}} - 1\right] \times 1000 \qquad (10.1)$$

Carbon has two stable isotopes with abundances of $^{12}C = 98.89\%$ and $^{13}C = 1.11\%$ (Craig, 1957). The carbon isotopic composition of a compound is similarly reported as a $\delta^{13}C$ value relative to the $^{13}C/^{12}C$ ratio of the Chicago PDB standard ($^{13}C/^{12}C = 0.0112372$, Craig, 1957).

The variation of sulfur and carbon isotopic compositions of naturally occurring materials are summarized in Figures 10.1 and 10.2, respectively. In general, sulfides in igneous rocks are isotopically similar to those in meteorites with average $\delta^{34}S$ close to 0‰, whereas seawater and sedimentary sulfates are enriched in the heavy isotope by approximately 10 to 30‰ depending on geologic age. Sedimentary sulfides have a wide range in $\delta^{34}S$ values (-70 to $+70$‰), but are typically depleted in the heavy isotope. The $\delta^{13}C$ values of igneous carbon are commonly around -5‰, whereas the sedimentary marine carbonates are close to 0‰ and the carbon associated with biologic processes is usually isotopically more negative than igneous carbon. The total range of $\delta^{34}S$ and $\delta^{13}C$ values in hydrothermal ore deposits is approximately 100‰ and 50‰, respectively. Even within a single ore deposit, $\delta^{34}S$ and $\delta^{13}C$ variations can be as much as 50‰.

Until recently the approach to sulfur and carbon isotopic studies on ore deposits has been to investigate the variability in $\delta^{34}S$ and $\delta^{13}C$ values and their deviation from "magmatic" values. Ore deposits in which sulfides have $\delta^{34}S$ values near 0‰ were interpreted to have formed from magmatic fluids. On the other hand, deposits in which $\delta^{34}S$ values were variable (e.g., more

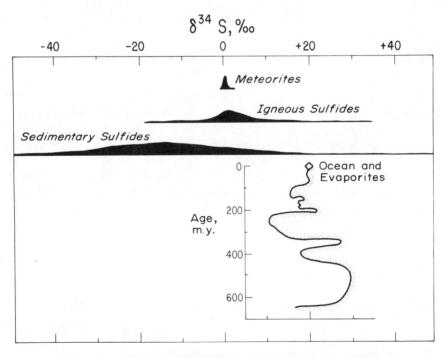

Fig. 10.1 Sulfur isotopic variation in nature.

ANALYTIC TECHNIQUES

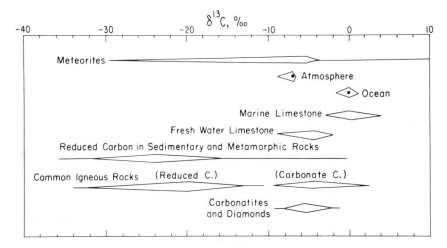

Fig. 10.2 Carbon isotopic variation in nature.

than 10‰) were interpreted as biogenic. Similarly, hydrothermal carbonates with $\delta^{13}C$ values near -5‰ were interpreted to have formed from magmatic fluids. However, with an increase in the knowledge of the mechanisms of isotopic fractionation in various natural conditions from recent theoretical studies, laboratory experiments, and detailed isotopic and geochemical studies on numerous ore deposits, such simple interpretations of sulfur and carbon isotopic data are gradually disappearing.

It is now realized that data on sulfur or carbon isotopic compositions of minerals from an ore deposit alone are usually insufficient to define the genesis of the deposit. However, when combined with detailed geologic and geochemical studies, sulfur and carbon isotopic data can provide information on (1) the temperature of mineralization, (2) the chemical conditions and mechanisms of ore deposition, and (3) the sources of sulfur and carbon in the ore-forming fluid. In this paper, the principles of interpreting such data on ore deposits are presented by examining the causes of sulfur and carbon isotopic variation in hydrothermal systems. The emphasis is placed on sulfur isotopes, because the principles of carbon isotope geochemistry are essentially identical to those of sulfur (Ohmoto, 1972) and because carbon is not as important an ore constituent as sulfur.

ANALYTIC TECHNIQUES

Sulfur from a compound may be converted to SF_6 for isotopic analyses (Puchelt et al., 1971). This technique is particularly suitable for measuring

$^{36}S/^{32}S$ ratios. Sulfur isotope measurements, however are almost always made on SO_2, whereas carbon isotope measurements are routinely made on CO_2 (e.g., Craig, 1957; Hulston, 1964 cited in Thode et al., 1971). A variety of chemical techniques are used in various laboratories to convert milligram quantities of almost any naturally occurring sulfur or carbon bearing compound to SO_2 or CO_2, respectively. Sulfur compounds that are not in a form suitable for direct conversion to SO_2 are usually first converted to Ag_2S or $BaSO_4$.

Sulfides

Sulfide minerals can be converted to SO_2 by reaction with an oxidizing agent, CuO, Cu_2O, V_2O_5, or a stream of O_2 at 900–1000°C. The CuO method developed by Grinenko (1962) is rapidly becoming popular because it is fast, simple, and quantitative and creates less CO_2 and SO_3 than the other techniques. Furthermore most sulfide minerals can be processed directly without prior conversion to Ag_2S.

Sulfates

There are three basic methods currently used to prepare SO_2 gas from sulfates and each has certain advantages. The method originally described by Rafter (1957) consists of converting sulfate samples to $BaSO_4$ followed by reduction with graphite at 900–1050°C to BaS and CO_2. The BaS is dissolved and reprecipitated as Ag_2S, which can be processed to produce SO_2. The principal advantage of the technique is that $\delta^{18}O$ measurements can also be made on the evolved CO_2 (Rafter and Mizutani, 1967).

Sulfates can also be reduced to H_2S by boiling in a mixture of HI, H_3PO_2, and HCl (Thode et al., 1961) or in Kiba reagent, Sn^{2+}-H_3PO_4 solution (Sasaki et al., 1976). The H_2S is precipitated as Ag_2S. The principal advantage of these techniques is that complex sulfur-bearing minerals like alunite and scapolite can be processed without prior chemical treatment, and that the total sulfur in rocks can be analyzed.

Finally, $BaSO_4$ may be thermally decomposed to SO_2 directly in the presence of quartz powder (Holt and Engelkemeir, 1970; Bailey and Smith, 1972). This technique is fast and simple and especially suitable for small samples, but care must be made to correct for ^{18}O interference from the quartz in the isotopic analyses.

Carbonates

CO_2 can be liberated from most carbonates by reaction with 100% phosphoric acid at 25°C (McCrea, 1950). Carbonates that react very slowly

SULFUR ISOTOPIC GEOTHERMOMETRY

in phosphoric acid can also be treated by thermal decomposition in BrF_5 systems (Sharma and Clayton, 1965).

Reduced Carbon

Graphite, diamonds, and organic carbon compounds may be converted to CO_2 by reaction with an oxidizing agent such as CuO, V_2O_5, or a stream of O_2 at 900–1100 °C (Craig, 1953).

SULFUR ISOTOPE GEOTHERMOMETRY

When isotopic equilibrium is attained between two sulfur-bearing compounds (e.g., H_2S and SO_4^{2-} in hydrothermal fluids, ZnS and PbS in ore deposits), a typical equilibrium relationship can be written as follows:

$$H_2{}^{34}S + {}^{32}SO_4{}^{2-} \rightleftharpoons H_2{}^{32}S + {}^{34}SO_4{}^{2-} \qquad (10.2)$$

In such equations where there is one exchangeable atom, the isotopic fractionation factor (α) can be shown to be related to the equilibrium constant (K). For example, in the above reaction

$$\alpha_{(SO_4^{2-}-H_2S)} = \frac{({}^{34}S/{}^{32}S)_{SO_4^{2-}}}{({}^{34}S/{}^{32}S)_{H_2S}} = K \qquad (10.3)$$

The difference in the $\delta^{34}S$ values between the two compounds (Δ) can be related to α

$$\begin{aligned}\Delta_{(SO_4^{2-}-H_2S)} &= \delta^{34}S_{SO_4^{2-}} - \delta^{34}S_{H_2S} \\ &= 1000\,(\alpha - 1) \times \left[1 + \frac{\delta^{34}S_{H_2S}}{1000}\right] \qquad (10.4) \\ &\simeq 1000(\alpha - 1) \simeq 1000 \ln \alpha\end{aligned}$$

The isotopic equilibrium constants are a function primarily of temperature, and for most reactions are approximately proportional to $1/T^2$ (K). The effect of pressure is negligible for most systems at pressures less than 10 kb. Therefore, when the isotopic equilibrium constants between specific compounds are accurately known as a function of temperature, we may define the temperature of equilibration among the compounds by measuring the difference in their $\delta^{34}S$ values.

Equilibrium Fractionation Factors

The isotopic equilibrium constants between various sulfur compounds may be estimated theoretically by statistical mechanics from data on the vibrational frequencies of the compounds. Such theoretical treatment was first put forward by Urey (1947), and has subsequently been refined by several authors (Tudge and Thode, 1950; Sakai, 1957, 1968; Thode et al., 1971). For example, for reaction (10.2), the equilibrium constant at 25°C has been calculated to be 1.071 by Tudge and Thode (1950) and 1.069 by Sakai (1968), which implies that under equilibrium conditions, SO_4^{2-} is enriched in ^{34}S over H_2S by approximately 70‰.

Under equilibrium conditions, the heavier isotope is enriched in compounds with stronger sulfur bonds, such that the order of ^{34}S enrichment among gaseous and aqueous species is $SO_4^{2-} > SO_3^{2-} > SO_2 > SCO > S_x \sim H_2S \sim HS^- > S^{2-}$. From the comparison of the bond strength among compounds of similar structure, the order of ^{34}S enrichment has also been estimated among some sulfate compounds and sulfide compounds by Sakai (1968) and among some sulfide minerals by Bachinski (1969).

A number of experimental studies have been carried out since approximately 1968 to define the equilibrium fractionation factors between various sulfur compounds. The order of ^{34}S enrichment among various sulfur compounds obtained in different experiments generally agrees with that estimated theoretically. However, the agreement in the fractionation factors suggested by different groups of researchers is not entirely satisfactory. For example, experimental fractionation factors between sphalerite and galena ($\Delta_{ZnS-PbS}$ value) vary from a low of $(0.63/T^2) \times 10^6$ (Grootenboer, 1969) to a high of $(0.90/T^2) \times 10^6$ (Kajiwara et al., 1969). For a $\Delta_{ZnS-PbS}$ value of 3.0‰, the temperatures calculated from the above equations differ by 90°C (185 vs 275°C). Such discrepancies in the suggested fractionation factors arise in part from the fact that each set of experiments was carried out, in most cases, at only a few selected temperature conditions and from the fact that the slope of the equation is strongly influenced by low-temperature data. Furthermore, because of the difficulty in attaining isotopic equilibrium, the low-temperature data have the largest uncertainty.

Instead of directly comparing the equations for the fractionation factors suggested by various investigators, we have critically examined all the available experimental raw data in terms of (a) attainment of equilibrium, (b) uncertainties in the measurements, (c) minimum or maximum fractionation factors when equilibrium was not attained, and (d) compatibility with the fractionation factors estimated from other sets of experiments. For example, the Δ_{FeS_2-PbS} values measured directly by Kajiwara and Krouse (1971) must be in agreement, within analytic uncertainties, with the Δ_{FeS_2-PbS} values calculated from the data on $\Delta_{HSO_4^--FeS_2}$ (Nakai, 1970), $\Delta_{HSO_4^--S(g)}$ (Robinson,

1973; Bahr, 1976) and $\Delta_{PbS-S(g)}$ (Grootenboer, 1969). Even for a system where very few direct measurements were made, reasonable estimates on the equilibrium fractionation factors may be made indirectly. For example, $\Delta_{HS^--H_2S}$ values can be estimated from the data on $\Delta_{ZnS-S(g)}$ (Grootenboer, 1969), $\Delta_{H_2S-S(g)}$ (Grinenko and Thode, 1970) and Δ_{ZnS-HS^-} (Kiyosu, 1973). The order of ^{34}S enrichment estimated for various sulfur compounds (Sakai, 1968; Bachinski, 1969) may also be used to estimate the fractionation factors where no experimental data are available. A summary of what we believe to be the best fractionation factors relative to H_2S is shown in Table 10-1 and Figure 10.3.

Experimental data on the isotopic fractionation factors between sulfate species are few, except for some measured Δ values of approximately 1.6‰

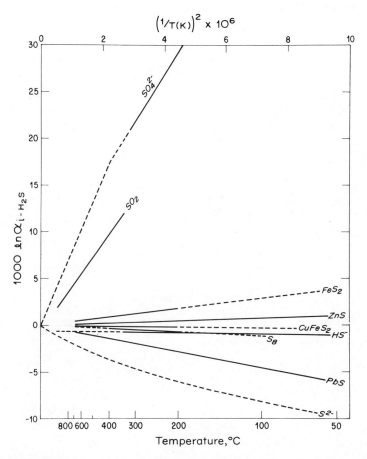

Fig. 10.3 Equilibrium isotopic fractionation factors among sulfur compounds relative to H_2S. Solid lines—experimentally determined. Dashed lines—extrapolated or theoretically calculated.

Table 10-1 Equilibrium isotopic fractionation factors of sulfur compounds with respect to H_2S

$$1000 \ln \alpha_{i-H_2S} = \frac{A}{T^2} \times 10^6 + \frac{B}{T} \times 10^3 + C$$

(T in K)

Compounds[e]	A	B	C	T(°C) Range
$CaSO_4$				
$CaSO_4 \cdot 2H_2O$	5.26		6.0 ± 0.5	200–350
$SrSO_4$				
$BaSO_4$[a]	(8.0)		(±1.0)	>400
Aqueous sulfates				
Sulfites[b]	(4.12)	(5.82)	(−5.0)	> 25
SO_2	4.70		−0.5 ± 0.5	350–1050
SCO[c]	(0.67)	(0.43)	(−1.15)	> 25
CaS[d]	(0.6 ± 0.1)			
SrS[d]	(0.6 ± 0.1)			
BaS[d]	(0.6 ± 0.1)			
MgS[d]	(0.5 ± 0.1)			
MoS_2[d]	(0.45 ± 0.1)			
FeS_2	0.40 ± 0.08			200–700
CoS_2[d]	(0.4 ± 0.1)			
NiS_2[d]	(0.4 ± 0.1)			
MnS_2[d]	(0.4 ± 0.1)			
ZnS	0.10 ± 0.05			50–705
MnS[d]	(0.10 ± 0.05)			
FeS	0.10 ± 0.05			200–600
CoS[d]	(0.10 ± 0.05)			
NiS[d]	(0.10 ± 0.05)			
$CuFe_2S_3$[d]	(0.05 ± 0.05)			
H_2S (g.aq.)	0			
$CuFeS_2$	−0.05 ± 0.08			200–600
$S (=S_8)$	−0.16		±0.5	200–400
HS^-	−0.06 ± 0.15		−0.6	50–350
Cu_5FeS_4[d]	(−0.25 ± 0.1)			
CdS[d]	(−0.4 ± 0.1)			
CuS[d]	(−0.4 ± 0.1)			
SnS[d]	(−0.45 ± 0.1)			
PbS	−0.63 ± 0.05			50–700
HgS[d]	(−0.7 ± 0.1)			
Cu_2S[d]	(−0.75 ± 0.1)			
Sb_2S_3[d]	(−0.75 ± 0.1)			
Ag_2S[d]	(−0.8 ± 0.1)			
S^{2-}[b]	(−0.21)	(−1.23)	(−1.23)	> 25

between anhydrite and aqueous sulfate at 25°C (Thode and Monster, 1965). In this paper, however, we have followed Sakai's (1968) suggestion and assumed that there is very little isotopic fractionation among coexisting sulfate species in hydrothermal conditions at temperatures above 100°C.

Evaluation of Sulfur Isotope Thermometers

Pairs of sulfur compounds in which equilibrium fractionation factors have the steepest slopes versus temperature are the most sensitive geothermometers. Suitable pairs include SO_2-H_2S in volcanic gases; SO_4^{2-}-H_2S in hot spring systems; sulfates-sulfides, pyrite-galena, sphalerite-galena, and pyrite-chalcopyrite in ore deposits. The sensitivity of each thermometer, which is inherited from a typical analytical uncertainty of $\pm 0.2‰$ for the Δ values, is listed in Table 10-2.

In addition to an adequate temperature dependence of the equilibrium fractionation factor, the successful application of sulfur isotope thermometry depends on the following conditions: (1) both mineral phases were formed in equilibrium, (2) no isotopic exchange took place between the mineral phases, or between a mineral phase and a fluid phase after the formation of the minerals (i.e., the original isotopic composition was frozen in), and (3) pure mineral phases were separated for isotopic analyses. Conversely, a comparison of isotopic temperatures with the temperatures estimated from other means (e.g., fluid inclusions) can provide information on the nature of equilibrium between the minerals, between minerals and sulfur species in ore-forming fluids, and between species in solution.

---

[a] Extrapolation above 400°C.
[b] Calculated from the partition coefficients of Sakai (1968) and Thode et al. (1971).
[c] Calculated from the partition coefficients of Tudge and Thode (1950) and Thode et al. (1971).
[d] Estimated values.
[e] Order of ^{34}S enrichment among sulfates are from Sakai (1968), and among sulfides from Bachinski (1969).

Experimental data used to estimate other fractionation factors are: Oana and Ishikawa (1966) on HSO_4^--S-H_2S; Grootenboer (1969) on FeS_2-ZnS-PbS-S; Kajiwara et al. (1969) on FeS_2-ZnS-PbS; Grinenko and Thode (1970) on SO_2-S-H_2S; Nakai (1970) on FeS_2-HSO_4^-; Schiller et al. (1970) on ZnS-HS-PbS; Kajiwara and Krouse (1971) on FeS_2-$CuFeS_2$-FeS-ZnS-PbS; Salomons (1971) on FeS_2-PbS-S; Thode et al. (1971) on SO_2-S-H_2S; Kiyosu (1973) on ZnS-HS$^-$-PbS; Czamanske and Rye (1974) on ZnS-PbS; Robinson (1973) on HSO_4^--S-H_2S; Bahr (1976) on HSO_4^--SO_4^{2-}-S-H_2S-HS$^-$.

Table 10-2 Sulfur isotopic thermometers

Mineral Pair	Equation (T in Kelvin; $\Delta = \delta^{34}S_A - \delta^{34}S_B$)	Uncertainties[a] 1	2
Sulfates-chalcopyrite	$T = \dfrac{2.85 \times 10^3}{(\Delta \pm 1)^{1/2}}$ ($T > 400°C$)	± 25[3]	± 5[3]
	$T = \dfrac{2.30 \times 10^3}{(\Delta - 6 \pm 0.5)^{1/2}}$ ($T < 350°C$)	± 10	± 5
Sulfates-pyrite	$T = \dfrac{2.76 \times 10^3}{(\Delta \pm 1)^{1/2}}$ ($T > 400°C$)	± 25[3]	± 5[3]
	$T = \dfrac{2.16 \times 10^3}{(\Delta - 6 \pm 0.5)^{1/2}}$ ($T < 350°C$)	± 10	± 5
Pyrite-galena	$T = \dfrac{(1.01 \pm 0.04) \times 10^3}{\Delta^{1/2}}$	± 25	± 20
Sphalerite (pyrrhotite)-galena	$T = \dfrac{(0.85 \pm 0.03) \times 10^3}{\Delta^{1/2}}$	± 20	± 25
Pyrite-chalcopyrite	$T = \dfrac{(0.67 \pm 0.04) \times 10^3}{\Delta^{1/2}}$	± 35	± 40
Pyrite-pyrrhotite (sphalerite)	$T = \dfrac{(0.55 \pm 0.04) \times 10^3}{\Delta^{1/2}}$	± 40	± 55

[a] 1 = uncertainty in the calculated temperature due to the uncertainty in the equation (at $T = 300°C$); 2 = uncertainty in the calculated temperature due to the analytical uncertainty of $\pm .2‰$ for Δ values (at $T = 300°C$); 3 = uncertainties in the calculated temperature at $T = 450°C$.

SULFIDE PAIRS

Some of the problems with determining temperatures of mineralization can be illustrated by comparing the measured fractionation factors between coexisting sphalerite-galena with measurements of the filling temperatures of the fluid inclusions in sphalerite from the same or similar samples from several deposits (Figure 10.4). The best fit line, based on a least squares analysis for all the data points, excluding the samples from Pine Point, is $\Delta_{\text{sp-gn}} = (0.74/T^2) \times 10^6$, which compares well with $\Delta_{\text{sp-gn}} = (0.73/T^2) \times$

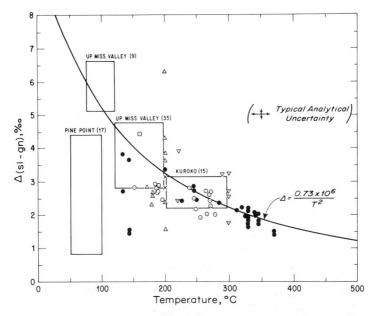

Fig. 10.4 Comparison of spalerite-galena isotopic fractionation factors with depositional temperatures determined by fluid inclusion studies. Equilibrium curve as based on data in Table 10-1. • Providencia, Creede, etc. (Czamanske and Rye, 1974), ○ Finlandia vein (Kamilli and Ohmoto, 1977), ▽ Sunnyside (Cadadevall and Ohmoto, 1977), X Echo Bay (Robinson and Ohmoto, 1973), Hansonburg (Allmandinger and Ohmoto, unpublished data), △ Pasto Bueno (Landis and Rye, 1974), □ Kuroko (Sasaki, 1974), Pine Point (Sasaki and Krouse, 1969), Upper Mississippi Valley (McLimans, 1977), () Numbers of pairs analyzed.

10^6 in Table 10-2, estimated from available experimental data. These data as well as the close agreement in the sphalerite-galena temperatures, pyrite-chalcopyrite temperatures, and pyrite-galena temperatures observed in several ore deposits (see Rye and Ohmoto, 1974 for the summary of the data) suggest that for a large number of sulfide samples from unmetamorphosed deposits isotopic equilibrium was attained during mineralization and that no change in the $\delta^{34}S$ values of the minerals took place after crystallization.

A significant number of sphalerite-galena pairs in Figure 10.4, however, show noticeable deviation from the equilibrium fractionation factors. Some of these "apparent disequilibrium" relationships may be due to impure mineral separates or to the selection of noncontemporaneous phases. For example, a sphalerite-galena pair formed in equilibrium at 145°C ($\Delta_{equil} = 4.0‰$) would show an apparent Δ value of 2.9‰ and an apparent temperature of 215°C, if each of the mineral separates contained 10 vol % of the other mineral phase. It is obviously imperative that mineral separates be examined very carefully for purity.

It is exceedingly difficult to sample mineral pairs so that each analyzed sample represents the same period of time in the history of the hydrothermal fluids (Rye, 1974). The usefulness of mineral pairs in geochemical studies depends to a large extent on the variability of the temperature and the $\delta^{34}S_{H_2S}$ value of the hydrothermal fluid during the period in which the minerals were precipitated. Even where a pair of minerals belongs to the same period of mineralization the $\delta^{34}S_{H_2S}$ and temperature of the fluids may have been so variable that it is difficult to get good temperature information. This problem has been studied for sphalerite-galena pairs at Providencia (Rye, 1974) and for pyrite-sulfide pairs at the Sunnyside mine (Casadevall and Ohmoto, 1977). At Sunnyside the only pairs that provided reasonable isotopic temperatures were those that grew in contact with each other, whereas the pairs obtained from what appeared to be the same growth zone often showed an apparent disequilibrium relationship. The reason that pyrite-chalcopyrite and pyrite-galena pairs often give abnormal sulfur isotopic temperatures may also be due in part to the fact that pyrite tends to precipitate over a much longer period of the paragenesis than chalcopyrite or galena, thus allowing less chance for the minerals to precipitate under identical conditions. On the other hand, even if the minerals are clearly not contemporaneous, $\delta^{34}S_{H_2S}$ and temperature may have been so uniform during deposition that useful temperatures may be obtained as was the case for some mineral pairs from the Darwin deposit (Rye et al., 1974).

Some mineral pairs in Figure 10.4 (e.g., Pine Point samples) may indeed indicate true disequilibrium relationships in which the isotopic fractionation factors between the dominant sulfide species in aqueous solutions (H_2S, HS^-) and the precipitating sulfide minerals were controlled by kinetic rather than by equilibrium isotopic effects. The difference between the apparent equilibrium state of the Upper Mississippi Valley deposits and the disequilibrium condition at Pine Point may reflect a difference in the mechanisms of metal and sulfur transportation in ore-forming fluids and in the mechanisms of mineral deposition for the two districts. This aspect will be discussed further in a later section.

$\delta^{34}S$ relationships among coexisting sulfide minerals in which later minerals were formed by replacement of earlier minerals (e.g., chalcopyrite after pyrite) do not always indicate equilibrium. The attainment of equilibrium among minerals during selective replacement probably depends upon whether replacement was by simple cation substitution or by complete breakdown, involving dissolution and reprecipitation.

Some suggestions have been made that reequilibration of sulfur isotopes takes place among coexisting sulfide minerals during metamorphism, and that $\delta^{34}S$ values of sulfide pairs in metamorphosed deposits provide a good estimate of the temperatures of metamorphism (see Lusk and Crocket, 1969).

However a closer examination of the isotopic data of many metamorphosed deposits suggests that the isotopic reequilibration usually was not complete at metamorphic conditions below upper amphibolite facies unless there was a change in mineralogy. The degree of isotopic reequilibration between coexisting mineral phases is dependent on the composition of metamorphic fluids as well as on temperature and pressure. Various data obtained during sulfur isotopic exchange experiments (e.g., Kajiwara and Krouse, 1971) suggest that (1) the rate of isotopic exchange reactions between coexisting mineral phases increases with increasing fugacity of sulfur, and with increasing concentration of aqueous sulfides in the exchange media, and that (2) the rate of isotopic exchange between a sulfide mineral and a gaseous or aqueous sulfur species increases from pyrite to sphalerite to galena. These data imply that isotopic exchange between sphalerite and galena may take place at a lower grade of metamorphism than that between pyrite and other minerals.

SULFIDE-SULFATE PAIRS

In hydrothermal ore deposits sulfate-sulfide pairs are often not contemporaneous. Further, they are sluggish to equilibrate at low temperatures as indicated in Figure 10.5. Temperatures of formation used in the figure are based on filling temperatures of fluid inclusions. At temperatures above approximately 300°C, the observed isotopic fractionation agrees reasonably well with equilibrium fractionation factors, suggesting that there was establishment of isotopic equilibrium between sulfates and sulfides. However, at temperatures below ~300°C, isotopic equilibrium was not always established between these pairs.

SULFUR ISOTOPIC VARIATION IN HYDROTHERMAL SYSTEMS

Sulfur in hydrothermal ore deposits, which is usually fixed as sulfide and/or sulfate minerals, has originated ultimately from (1) an igneous source and/or (2) a seawater source. The former includes (a) sulfur carried in magmatic fluids and (b) sulfur obtained by leaching of sulfur bearing minerals in igneous rocks.

Sulfur in seawater occurs as aqueous sulfates and may be incorporated into sulfide deposits through various paths. In environments near the earth's surface, *in situ* reduction of sulfate to H_2S by sulfur-reducing bacteria generates sedimentary sulfides. These sedimentary sulfides may then be leached by hydrothermal fluids, or they may be replaced by other sulfide minerals when encountered by metal-carrying brines. Reduction of aqueous sulfates to aqueous sulfides can also take place nonbacterially at elevated

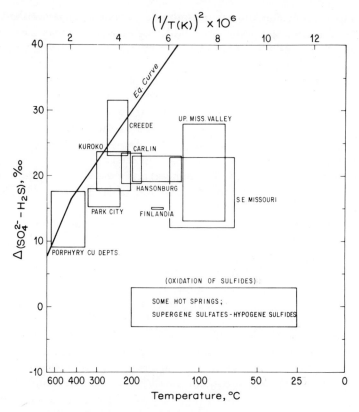

Fig. 10.5. Comparison of sulfate-sulfide isotopic fractionation factors with depositional temperatures. Data on porphyry copper deposits from Field (written commun., 1975); S. E. Missouri from Brown (1967); Carlin from Rye, Radtke and Dickson (unpublished data, 1976). For other deposits, see Figure 10.4 captions and Rye and Ohmoto (1974). Equilibrium curve is based on the equation in Table 10-1.

temperatures. Therefore, sulfate-bearing seawater, "sulfate-rich" connate water, or meteoric waters bearing sulfate dissolved from marine evaporites can evolve into sulfide-bearing hydrothermal fluids.

Fractionation of sulfur isotopes may occur in a variety of situations in the history of an ore fluid. Isotopic fractionation may take place (1) at the source of sulfur such as during separation of fluids from a magma, or during leaching of sulfides, (2) during the evolutionary history of hydrothermal fluids involving reduction of seawater sulfate, (3) during cooling of hydrothermal fluids, and (4) during precipitation or replacement of minerals. Isotopic fractionation may proceed while maintaining isotopic equilibrium among various sulfur-bearing compounds in fluids, melts, and minerals, or may proceed in a nonequilibrium manner. Such isotopic fractionation may take place in closed or in open systems. Observed $\delta^{34}S$ values of hydrothermal minerals, therefore, reflect a varied geochemical history of the sulfur in hydrothermal fluids, and the proper interpretation of the significance of the $\delta^{34}S$ values can be made only through understanding of the geology of the deposits and of the many processes of isotopic fractionation in hydrothermal systems.

Magmatic Systems

During the formation of magma by partial melting of mantle or lower crustal rocks, it is unlikely that isotopic fractionation takes place between sulfur in parental rock and sulfur in the melt (i.e., $\delta^{34}S_{melt} \simeq \delta^{34}S_{parental\ rock}$). Isotopic fractionation, however, may take place between sulfur in the melt and crystals, and between sulfur in the melt and separating fluids during the crystallization of magma.

According to recent experimental data on the solubility of sulfur in hydrous silicate melts, sulfur occurs in a melt predominantly as SH^-, much the same way that H_2O dissolves as OH^- in a melt (Burnham, Chapter 3). If we assume that SH^- in a melt behaves isotopically in a similar manner to aqueous HS^-, the isotopic relationship among melt-crystals-fluids can be quantitatively evaluated from the equilibrium fractionation factors summarized in Table 10-1.

MELT-CRYSTAL EQUILIBRIA

Primary sulfide minerals, which crystallize from a magma within the temperature range 750–1250°C, are commonly cupriferous pyrrhotite solid solutions. These minerals tend to enrich ^{34}S with respect to HS^-. Therefore, fractional crystallization of these sulfide minerals would decrease the $\delta^{34}S$ value of residual melts. However, because the $\Delta_{(sulfides-HS^-)}$ value at these high temperatures is less than 1‰, the decrease in the $\delta^{34}S$ value of residual melt would not exceed 2‰, even after the removal of 90% of the sulfur in the initial magma assuming a Rayleigh distillation process. Therefore, for practical purposes, we may assume that the $\delta^{34}S$ value of primary igneous sulfide minerals is essentially identical to that of the sulfur in the melt and also in the parent rock.

$\delta^{34}S$ values of primary sulfide minerals in mafic- to ultramafic igneous rocks are typically in the range $-1 - +2$‰ with an average of $+1.3$‰ (Smitheringale and Jensen, 1963; Schneider, 1970; Sasaki, 1970; Kanehira et al., 1973), suggesting similar values for the sulfur in the upper mantle. However, there are many mafic rocks that have abnormally high $\delta^{34}S$ such as $+5 - +17$‰ for the Muskox intrusion (Sasaki, 1969), $+11.0 - +15.5$‰ for the Duluth complex (Mainwaring and Naldrett, 1974), and $+8.6$‰ for

Grenfenburg basalt (Schneider, 1970), or abnormally low $\delta^{34}S$ values as $-13.4‰$ for the Talcott Basalt (Smitheringale and Jensen, 1963). The abnormal values of these igneous sulfides were probably caused by contamination from sedimentary sulfides or sulfates in the melt, rather than by initially inhomogeneous $\delta^{34}S$ values in the mantle.

Average $\delta^{34}S$ values of acidic igneous rocks and of the crust are less well defined. Values of $+10 \pm 5‰$ for average granitic rocks and $+6 \pm 2.3‰$ for average crustal rocks have been suggested by Holser and Kaplan (1966). These values were calculated, however, on the basis of relatively few published data on the granitic rocks, and, therefore, are strongly influenced by a few granitic rocks of extremely high $\delta^{34}S$ values (e.g., $+30‰$ for the Rice Lake batholith, Manitoba, Shima et al., 1963). If average crustal rocks had $\delta^{34}S$ values as high as $+6‰$, it would imply the (crust + ocean) system has a $\delta^{34}S$ average value of around $+7‰$, because the $\delta^{34}S$ of the ocean is $20‰$ and the mass ratio of sulfur in the crust to that in the ocean is approximately 9:1 (Ohmoto, in prep.). A value of $7‰$ is highly unlikely because there is no evidence that the formation of the crust from mantle materials at high temperatures resulted in large sulfur isotopic fractionation. Using more recent geochemical data on the sulfur content of various rock types (Ronov and Yaroshevsky, 1969) for calculations, we conclude that the average $\delta^{34}S$ values of acidic igneous rocks, as well as the crust, are $0 \pm 3‰$. These values are essentially identical to those of mantle rocks, and suggest that melts formed by partial melting of average crustal rocks have essentially the same $\delta^{34}S$ values as melts formed by partial melting of mantle rocks.

MELT-FLUID EQUILIBRIA

Sulfur species that are quantitatively important in high-temperature ($T \gtrsim 400°C$) hydrothermal fluids are H_2S and SO_2. From the consideration of free-energy data for various sulfur compounds, we can assume that under these temperature conditions other sulfur species, such as SO, SO_3, S_2, S_6, S_8, SCO, H_2SO_4, H_2SO_3, and ionic species, are always minor compared to H_2S or SO_2. Therefore, the $\delta^{34}S$ value of a hydrothermal fluid ($\delta^{34}S_{fluid}$) can be related to the $\delta^{34}S$ values of H_2S and SO_2 by the approximation

$$\delta^{34}S_{fluid} = \delta^{34}S_{H_2S} \cdot X_{H_2S} + \delta^{34}S_{SO_2} \cdot X_{SO_2}$$

$$= \delta^{34}S_{H_2S} + \Delta_{SO_2} \cdot \left(\frac{R}{1+R}\right), \qquad (10.5)$$

in which X_{H_2S} is the mole fraction of H_2S with respect to the total sulfur content in the fluid (i.e., $X_{H_2S} = m_{H_2S}/(m_{H_2S} + m_{SO_2})$, R is the mole ratio of SO_2/H_2S, and Δ_{SO_2} is the equilibrium enrichment factor between SO_2 and

H_2S (i.e., $\Delta_{SO_2} = \delta^{34}S_{SO_2} - \delta^{34}S_{H_2S}$). For the principles and terminology of isotopic mass balance in fluids, see the work of Ohmoto (1972).

When isotopic equilibrium is attained between sulfur in the fluid and sulfur in the melt and when SH^- is the only sulfur species present in the melt, equation (10.5) can also be written as

$$\delta^{34}S_{fluid} = \delta^{34}S_{melt} - \Delta_{HS^-} + \Delta_{SO_2} \cdot \left(\frac{R}{1+R}\right) \quad (10.6)$$

in which Δ_{HS^-} is the equilibrium enrichment factor between HS^- and H_2S. Because of the existence of significantly large isotopic fractionation factors between SO_2 and the reduced sulfur species, even at magmatic temperatures, equation (10.6) implies that the $\delta^{34}S$ value of magmatic fluids can be quite different from that of silicate magmas, and that the extent of the deviation in $\delta^{34}S$ values is dependent on the SO_2/H_2S ratio in the fluid, temperature, and the relative masses of fluid and magma at the time of fluid separation.

If we assume that fluids at high temperatures behave similarly to ideal gas mixtures, the mole ratio of SO_2/H_2S, R, can be computed as a function of T, P_f (total fluid pressure), and f_{O_2}:

$$R = \frac{m_{SO_2}}{m_{H_2S}} = \frac{X_{SO_2}}{X_{H_2S}} = \frac{[f_{SO_2}/P_f \cdot (\nu_{SO_2})]}{[f_{H_2S}/P_f \cdot (\nu_{H_2S})]}$$

$$= \frac{K \cdot \nu_{H_2S} \cdot (f_{O_2})^{3/2}}{P_f \cdot \nu_{H_2O} \cdot X_{H_2O} \cdot \nu_{SO_2}} \quad (10.7)$$

in which X_i is the true mole fraction of the species i in the mixture, for example,

$$X_{H_2S} = \frac{m_{H_2S}}{m_{H_2S} + m_{SO_2} + m_{H_2O} + m_{CO_2} + \cdots}$$

(ν_i) is the fugacity coefficient of the species i in the pure system at P_f and T, and K is the equilibrium constant for the following reaction

$$H_2S(g) + 3/2\, O_2(g) \rightleftharpoons H_2O(g) + SO_2(g). \quad (10.8)$$

Equilibrium constants for reaction (10.8) are summarized in Holland (1965), and the fugacity coefficients of H_2O and H_2S over a wide range of T and P are available from Burnham et al. (1969) and Ryzhenko and Volkov (1971), respectively. The fugacity coefficient of SO_2 can be estimated from critical gas data, together with the equation of the corresponding states (Ryzhenko and Volkov, 1971).

Deviation of $\delta^{34}S_{fluid}$ from $\delta^{34}S_{melt}$ as functions of T and f_{O_2} have been computed from equations (10.6) and (10.7) at the condition of $P_f \simeq P_{H_2O} = 1$ kb, and $X_{H_2O} \simeq 1.0$, and are shown in Figure 10.6. Also shown in Figure 10.6 are the $f_{O_2} - T$ conditions for basaltic rocks and for biotite-amphiboles in rhyolites and dacites (Carmichael et al., 1974). The latter may also be assumed to be the $f_{O_2} - T$ conditions of magmatic fluids derived from granitic melts.

Effects of changes in P_f and X_{H_2O} values on the positions of the SO_2/H_2S equal concentration boundary and of the $\delta^{34}S_{fluid}$ contours can easily be evaluated from equation (10.7). A decrease in the P_f value or in the X_{H_2O}, as may be due to a large increase in the concentration of CO_2, shifts the SO_2/H_2S boundary and the $\delta^{34}S$ contours to lower f_{O_2} values. However, for hydrothermal fluids responsible for the formation of ore deposits, it is unlikely that the P_f value at high temperatures lies outside the range 0.5–3 kb

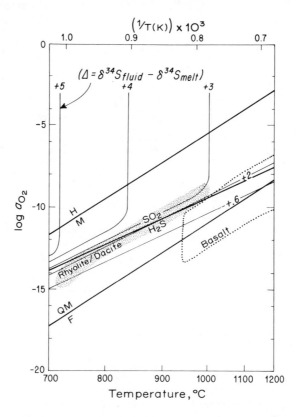

Fig. 10.6. Sulfur isotopic fractionation between magmatic fluids and hydrous silicate melts at $P_{H_2O} = 1$ kb. F = fayalite, Q = quartz, M = magnetite, H = hematite, SO_2/H_2S line is an equal concentration boundary.

and that X_{H_2O} is less than 0.8. Within these P_f and X_{H_2O} ranges, the positions of the SO_2/H_2S boundary and the $\delta^{34}S$ contours in $\log f_{O_2} - T$ space do not change more than $\pm 0.3 \log f_{O_2}$, units from those in Figure 10.6.

Figure 10.6 indicates that magmatic fluids derived from basic magmas at high temperatures and pressures are H_2S-rich. For example, fluid in equilibrium with pyrrhotite + magnetite + quartz + fayalite at $1000°C$ and $P_f = 1kb$ can be calculated to contain $0.4 m$ H_2S and $0.003 m$ SO_2 in 1000 g of H_2O. The $\delta^{34}S_{fluid}$ values of such magmatic fluids are essentially identical to those of the magmas (i.e., $\delta^{34}S_{fluid} = \delta^{34}S_{H_2S} \simeq \delta^{34}S_{melt}$). Fluids derived from granitic magmas, on the other hand, can be either H_2S-rich or SO_2-rich, depending on the oxidation state of the melt at the time of fluid separation.

The solubility of sulfur in hydrous melts decreases with decreasing f_{H_2S} (increasing f_{O_2}), because SO_2 is not very soluble in melts as shown by Burnham (Chapter 3). Therefore, essentially all the sulfur in the melt could be removed from the melt as SO_2 when the f_{O_2}-T conditions are above the SO_2/H_2S line in Figure 10.6. Under these conditions, $\delta^{34}S_{fluid} \simeq \delta^{34}S_{SO_2} \simeq \delta^{34}S_{melt}$. However, if the volume ratio of fluid to melt is small, SO_2 in the fluid may be buffered isotopically by the SH^- remaining in the melt, and $\delta^{34}S_{fluid}$ can become as much as $4\%_{oo}$ larger than $\delta^{34}S_{melt}$ (Figure 10.6). Also the oxidation state of melts may increase toward the later stages of crystallization due to diffusive loss of H_2. In this case, the later stage magmatic fluids would become SO_2-rich. This would be reflected in more or less continuously increasing values of $\delta^{34}S_{fluid}$ with evolution of the magmatic fluid phases.

FLUID-MINERAL EQUILIBRIA

Deviation of $\delta^{34}S_{H_2S}$ from $\delta^{34}S_{fluid}$ was calculated using equations 10.6 and 10.7 at temperatures in the range of $400-800°C$, and these results are shown in Figure 10.7. The oxidation state of ore-forming fluids responsible for high temperature ore deposits (including porphyry copper deposits, contact-type deposits, etc.) generally lies between or slightly above the magnetite-hematite buffer (pyrite + hematite stable), and the quartz-magnetite-fayalite buffer line (pyrrhotite stable). Under such f_{O_2} conditions of ore-formation, the $\delta^{34}S_{H_2S}$ value can be as much as $10\%_{oo}$ lower than that of $\delta^{34}S_{fluid}$ (Figure 10.7). Moreover, a change in T and/or f_{O_2} of a hydrothermal fluid, alone, can cause as much as $10\%_{oo}$ variation in the $\delta^{34}S$ value of hydrothermal minerals, even if $\delta^{34}S_{fluid}$ remains constant.

Figure 10.7 also indicates that (1) the variation in the $\delta^{34}S$ values of minerals in high-temperature deposits does not necessarily indicate multiple sources of sulfur, (2) to estimate $\delta^{34}S_{fluid}$ and the source of sulfur, f_{O_2} conditions as well as $\delta^{34}S_{mineral}$ must be determined, and (3) the f_{O_2} conditions in fluids may be estimated by examining the variability of $\delta^{34}S$ values of

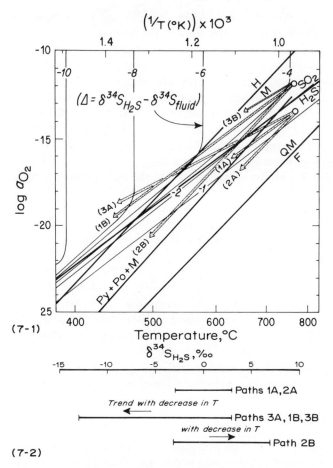

Fig. 10.7. Deviation of $\delta^{34}S_{H_2S}$ from $\delta^{34}S_{fluid}$ in high temperature hydrothermal fluids of $P_{H_2O} = 1$ kb. The lower diagrams (7-2) show trends and ranges in $\delta^{34}S_{H_2S}$ values produced through various cooling paths (1A, etc.) of magmatic fluids which were initially derived from melts of $\delta^{34}S = 0 \pm 3‰$ where the reservoir of sulfur in solution remains large compared to that precipitated.

minerals. For example, if a large number of sulfides from a given ore deposit show constant $\delta^{34}S$ values, the oxidation state can be assumed to have been below the SO_2/H_2S boundary or to have remained constant with respect to it. A large variation of $\delta^{34}S$ values may suggest that f_{O_2} was near or above the SO_2/H_2S line and either remained constant or increased with respect to it.

In the previous discussions on "melt-crystal" and "melt-fluid" equilibria, it was suggested that $\delta^{34}S_{melt}$ values of uncontaminated granitic magmas are likely to be between -3 and $+3‰$, and that the fluids derived from such melts will have $\delta^{34}S_{fluid}$ values in the range of -3 to $+7\%_o$, with the high $\delta^{34}S$ values ($+3$-$+7\%_o$) characterized by initially high oxidation states. These suggestions, together with the data in Figure 10.7 allow us to examine the possible range and trends in the $\delta^{34}S_{H_2S}$ values of fluids evolved during the cooling of magmatic fluids.

Several cooling paths may be hypothesized (see Figure 10.7): (1) The SO_2/H_2S ratio of the fluid may remain constant when the initial H_2 content was much less than that of H_2S or SO_2, and the oxidation state of the fluid is internally controlled. (2) The SO_2/H_2S ratio may continuously decrease due to redox reactions with ferromagnesian minerals in wallrocks

$$SO_2 + 6 \text{ "FeO"} + H_2O \rightarrow H_2S + 3 \text{ "Fe}_2O_3\text{"}$$

or with reduced carbon in wallrocks

$$SO_2 + 3/2 C + H_2O \rightarrow H_2S + 3/2 CO_2$$

(3) The SO_2/H_2S ratio may increase due to a continuous loss of hydrogen or due to redox reactions with ferric minerals in wallrocks, such as in red-bed sandstones. (4) The SO_2/H_2S ratio may increase when boiling of fluids takes place, because the solubility of SO_2 in an aqueous fluid is much higher than that of H_2S. (5) The SO_2/H_2S ratio may increase when reduced sulfur is removed as sulfide minerals, or decrease when oxidized sulfur is removed as sulfate minerals. Paths (1)-(3) may be followed at constant $\delta^{34}S_{fluid}$ because no sulfur is lost from the system. However, paths (4) and (5) may result in a change in $\delta^{34}S_{fluid}$. The magnitude of the change is dependent on the relative amount of sulfur removed from the system.

When the sulfur in the magmatic fluid is initially H_2S predominant, paths (1A) and (2A) of Figure 10.7 result in no change in $\delta^{34}S_{H_2S}$ and $\delta^{34}S_{H_2S} = \delta^{34}S_{fluid} = \delta^{34}S_{melt} = 0 \pm 3\%_o$. Path (3A) could cause a continuous decrease in $\delta^{34}S_{H_2S}$ which may become as much as $10\%_o$ smaller than $\delta^{34}S_{melt}$ when most of the H_2S is converted to SO_2 as the temperature drops to $\sim 400°C$. The $\delta^{34}S_{H_2S}$ values may range between $+3$ and $-13\%_o$ when the initial $\delta^{34}S_{melt}$ value was $0 \pm 3\%_o$.

When sulfur is initially SO_2 dominant, paths (1B) and (3B) will cause a continuous decrease in $\delta^{34}S_{H_2S}$ during cooling and $\delta^{34}S_{H_2S}$ may range between $+3$ and $-13\%_o$. Path (2B) may cause a continuous increase in $\delta^{34}S_{H_2S}$, that ranges from several per mil lower than $\delta^{34}S_{fluid}$ at high temperature and high f_{O_2} to the value essentially identical to that of $\delta^{34}S_{fluid}$ when most of the SO_2 is converted to H_2S. Because $\delta^{34}S_{fluid}$ can be as much as $4\%_o$ larger than $\delta^{34}S_{melt}$ for this type of magmatic fluid (see previous discussion in melt-fluid equilibria), path (2B) may result in $\delta^{34}S_{H_2S}$ values being as much as $4\%_o$ higher than the $\delta^{34}S_{melt}$ value at $T \simeq 400°C$.

Low-Temperature Equilibrium Systems

At temperatures below about 350°C and under low pressure conditions (i.e., near liquid/vapor, two-phase boundary of water), sulfur species in hydrothermal fluids can no longer be regarded as simple, ideal gas mixtures of H_2S and SO_2. The equilibrium

$$4H_2O(g) + 4SO_2(aq) \rightleftharpoons H_2S(aq) + 3H^+ + 3HSO_4^-$$

shifts to the right with decreasing temperatures (e.g., $\log K = -18.98$ at 350°C and -13.76 at 300°C according to the thermodynamic data compiled by Naumov et al., 1974). Sulfate species (e.g., HSO_4^-, $NaSO_4^-$) become the dominant oxidized sulfur species at temperatures below 400–350°C. The exact temperature at which $m_{SO_2}(aq) = m_{\Sigma SO_4^{2-}}$ depends on the values of f_{H_2O}, ΣS, f_{O_2}, pH, and salt concentrations of fluid.

Aqueous sulfur species that could become quantitatively significant in ore-forming fluids at temperatures below about 350°C can be classified into the following three groups: (1) $H_2S(aq)$, (2) HS^-, and (3) sulfates (SO_4^{2-}, HSO_4^-, KSO_4^-, $NaSO_4^-$, $CaSO_4$). $MgSO_4$ may become as important as $CaSO_4$ when the Mg content of the fluids is high. In the papers by Sakai (1968) and Ohmoto (1972), S^{2-} was also listed as an important species to consider for sulfur isotopic balance. However, recent data on the second dissociation constant of H_2S indicate that S^{2-} does not become an important sulfur species under normal pH conditions in hydrothermal processes (see Chapter 8). Sulfite species, polysulfide species, other sulfate species (e.g., H_2SO_4, K_2SO_4) and gaseous species (e.g., S_x) are always negligible in near-equilibrium hydrothermal solutions compared to one of the above three groups of sulfur species.

The isotopic fractionation between H_2S and HS^-, according to the summary in Table 10-1 is likely to be much smaller (i.e., Δ less than 1.0‰ at T above 100°C) than that previously estimated by Sakai (1968). Therefore, for practical purposes the Δ_{HS^-} value can be assumed to be 0. If we further assume that all the sulfate species have essentially the same isotopic composition, equation (4) in Ohmoto (1972) can be simplified as follows:

$$\delta^{34}S_{H_2S} = \delta^{34}S_{fluid} - \Delta_{SO_4^{2-}} \cdot \left(\frac{R'}{1 + R'}\right) \qquad (10.9)$$

in which R' is the mole ratio of $\Sigma SO_4 / \Sigma H_2S$ in the fluid:

$$R' = \frac{\Sigma SO_4}{\Sigma H_2S} = \frac{m_{SO_4^{2-}} + m_{HSO_4^-} + m_{KSO_4^-} + m_{NaSO_4^-} + m_{CaSO_4}}{m_{H_2S} + m_{HS^-}}$$

$$(10.10)$$

Values of R' can be computed as a function of T, pH, f_{O_2}, a_{K^+}, a_{Na^+}, and $a_{Ca^{2+}}$ using the following equation:

$$R' = \left\{ (f_{O_2}^{\,2} \cdot \gamma_{H_2S}) \cdot \left(\frac{1}{\gamma_{SO_4^{2-}}} + \frac{a_{H^+}}{K_{HSO_4^-} \cdot \gamma_{HSO_4^-}} + \frac{a_{K^+}}{K_{KSO_4^-} \cdot \gamma_{KSO_4^+}} + \right. \right.$$

$$\left. \left. \frac{a_{Na^+}}{K_{NaSO_4^-} \cdot \gamma_{NaSO_4^-}} + \frac{a_{Ca^{2+}}}{K_{CaSO_4} \cdot \gamma_{CaSO_4}} \right) \right\} \times \left\{ (K_{SO_4^{2-}} \cdot a_{H^+}^{+2}) \cdot \right.$$

$$\left. \left(1 + \frac{K_{H_2S} \cdot \gamma_{H_2S}}{a_{H^+} \cdot \gamma_{HS^-}} \right) \right\}^{-1} \tag{10.11}$$

in which K_{H_2S}, $K_{SO_4^{2-}}$, $K_{HSO_4^-}$, $K_{KSO_4^-}$, $K_{NaSO_4^-}$, and K_{CaSO_4} are respectively the equilibrium constants for the reactions

$$H_2S(aq) \rightleftharpoons H^+ + HS^- \tag{10.12}$$

$$2H^+ + SO_4^{2-} \rightleftharpoons H_2S(aq) + 2O_2(g) \tag{10.13}$$

$$HSO_4^- \rightleftharpoons H^+ + SO_4^{2-} \tag{10.14}$$

$$KSO_4^- \rightleftharpoons K^+ + SO_4^{2-} \tag{10.15}$$

$$NaSO_4^- \rightleftharpoons Na^+ + SO_4^{2-} \tag{10.16}$$

$$CaSO_4(aq) \rightleftharpoons Ca^{2+} + SO_4^{2-} \tag{10.17}$$

Equilibrium constants for these reactions and the activity coefficients for various species are summarized in Ohmoto (1972), Helgeson (1969), and discussed by Barnes (Chapter 8).

The $\delta^{34}S_{H_2S}$ relations can be projected onto a $\log f_{O_2}$-pH diagram (Figure 10.8) using equations (10.9) and (10.11) and can be compared with mineral stability relationships or can be projected on a $\log \Sigma SO_4/\Sigma H_2S - T$ diagram (Figure 10.9). Figures 10.8 and 10.9 indicate that a change in the redox state of the fluids can affect $\delta^{34}S_i$ values significantly at or near the SO_4^{2-}/H_2S boundary. In the region near the SO_4^{2-}/H_2S boundary, an increase of the SO_4^{2-}/H_2S ratio by an order of magnitude (i.e., an increase in f_{O_2} by 0.5 order of magnitude or in pH by 1 unit) can decrease the $\delta^{34}S_i$ value by more than 20‰ at temperatures approximately 250°C. The figures also show that the sulfide minerals formed in equilibrium with magnetite and/or with hematite will have $\delta^{34}S$ values much smaller than $\delta^{34}S_{fluid}$. Only in the H_2S-dominant regions, at low f_{O_2} and low pH, do $\delta^{34}S$ of sulfide minerals become nearly identical to $\delta^{34}S_{fluid}$. It follows that the composition of the

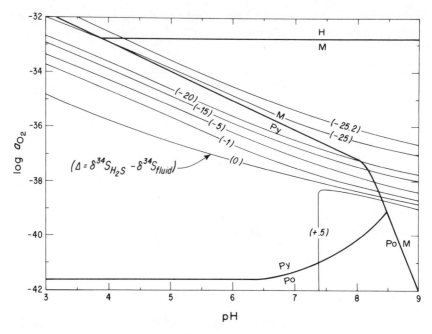

Fig. 10.8. Deviation of $\delta^{34}S_{H_2S}$ from $\delta^{34}S_{fluid}$ at 250°C under equilibrium conditions. Mineral boundaries are for $\mu = 1.0$, $\Delta S = 0.01\,m$, $m_{K^+} = 0.1$, $m_{Na^+} = .9$, and $m_{Ca^{2+}} = 0.01$.

fluids (pH, f_{O_2}, a_{K^+}, etc.) as well as temperature must be known to estimate the $\delta^{34}S_{fluid}$ from values of $\delta^{34}S_{mineral}$. Consequently, by examining the magnitude of changes of $\delta^{34}S_{mineral}$ values with respect to time and space, and by comparing them with other geochemical studies, one can estimate the physicochemical conditions of ore-forming processes.

Figure 10.9 also indicates that magmatic fluids with $\delta^{34}S_{fluid}$ between -3 and $+7‰$ are unlikely to produce sulfide minerals, such as pyrite or sphalerite, with $\delta^{34}S$ values larger than approximately $+8‰$ at temperatures above $\sim 200°C$, but may produce sulfide minerals with $\delta^{34}S$ values as low as $\sim -30‰$. Sulfur of seawater origin, including that from evaporites and connate water, would have $\delta^{34}S_{fluid}$ values between $+10$ and $+30‰$ (see Figure 10.1). Therefore, when the SO_4^{2-} in such fluids is reduced to H_2S (and/or HS^-) at temperatures above $\sim 250°C$ under equilibrium conditions sulfide minerals with $\delta^{34}S$ values as high as $\sim +30‰$ to as low as $\sim -15‰$ can precipitate (see Figure 10.6).

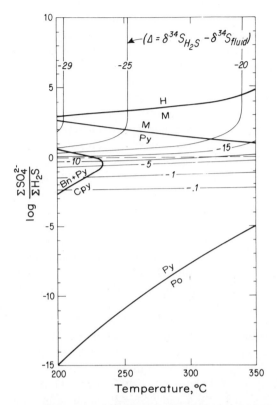

Fig. 10.9 Deviation of $\delta^{34}S_{H_2S}$ from $\delta^{34}S_{fluid}$ as a function of T and $\Sigma SO_4^{2-}/\Sigma H_2S$ ratio of fluid. Mineral boundaries are for pH = neutral, $\Sigma S = 0.01\ m$, u.c./$m_k^+ = 0.05$, $m_{Na^+} = 0.5$, $m_{Ca^{2+}} = 1.0$, $m_{mg^{2+}} = 0.25$. Bn = bornite, Cpy = chalcopyrite, Py = pyrite, Po = pyrrhotite, M = magnetite, H = hematite.

Nonequilibrium Systems

PROBLEMS OF SULFATE-SULFIDE EQUILIBRIA

The $\delta^{34}S$ variation of hydrothermal minerals in several ore deposits, which formed at temperatures in the range 350–200°C, including Darwin, Mogul, Echo Bay, and Kuroko (Rye and Ohmoto, 1974) can be explained as due to the variation in the SO_4^{2-}/H_2S ratio of their hydrothermal fluids together with the assumption of establishment of isotopic equilibrium between the oxidized and the reduced sulfur species in hydrothermal fluids. However, there are a number of ore deposits formed under similar temperature conditions where the variation in the SO_4^{2-}/H_2S ratio in the hydrothermal fluids did not result in a corresponding variation in the $\delta^{34}S$ values of precipitating sulfide minerals. Examples are the Creede and Park City deposits that are summarized in Rye and Ohmoto (1974). Observed isotopic fractionation factors between coexisting sulfate and sulfide minerals in many deposits also show a large deviation from equilibrium values at temperatures below $\sim 300°C$ (see Figure 10.5), with a general tendency for the observed fractionation factors to be smaller than the equilibrium values.

Temperature is obviously an important parameter in controlling the rate of isotopic exchange reactions between sulfate and sulfide species in solution. However, more important parameters appear to be pH, and/or the concentrations of sulfur species of intermediate valency state (e.g., +4, +2, 0, $-2/x$), which, in turn, are related to pH, f_{O_2}, and f_{S_2} of the fluids. For example, the rate of isotopic exchange between aqueous sulfate and sulfides in the presence of liquid sulfur (i.e., conditions of low pH, high f_{S_2}, and high concentrations of sulfur with intermediate valency state) is very fast and the equilibration takes place within a few days at temperatures as low as $\sim 200°C$ (Robinson, 1973; Bahr, 1976). However, in a system in which H_2S and SO_4^{2-} were both produced by thermal decomposition of $Na_2SO_3 \cdot 5H_2O$ at pH = 10-12 (no native sulfur present), no measurable isotopic exchange took place over a period of 2 weeks and at temperatures as high as 320°C (Bahr, 1976). Igumnov (1976) also suggested that the rate of isotopic exchange between aqueous sulfide and sulfate was inversely proportional to pH, and that the equilibrium was quickly established in acidic to neutral solutions at temperatures above 250-270°C. Typical data are 50% exchange in 99 hr at 272°C, pH = 5.4 and $\Sigma S \simeq 1\ m$. Therefore, the difference between equilibrium and nonequilibrium situations observed in ore deposits formed at 200-300°C may reflect the difference in the pH, f_{O_2}, and f_{S_2} conditions of ore-forming fluids.

Differences in the degree of attainment of isotopic equilibrium between sulfate and sulfide may also reflect differences in the environments where the change in the SO_4^{2-}/H_2S ratio of fluid took place. For example, suppose it takes one month to attain isotopic equilibrium between SO_4^{2-} and H_2S at 250°C, pH = 5, and $\Sigma S = 10^{-2}\ m$. If a change in the SO_4^{2-}/H_2S ratio took place in the plumbing system and it took longer than a month for the fluid to travel to the site of mineral deposition, isotopic equilibrium would be attained between sulfates and sulfides. On the other hand, if a change in the SO_4^{2-}/H_2S ratio took place at the site of mineral deposition and if the residence time of the fluid at the depositional site was less than one month, isotopic equilibrium would not be attained between sulfates and sulfides.

The $\delta^{34}S$ relationship between aqueous sulfates and sulfides in nonequilibrium systems (particularly at temperatures below $\sim 200°C$) depends on the type of chemical reactions that occurred during the evolutionary history of the fluids. If no chemical reaction took place between aqueous sulfides and sulfates, they could retain earlier equilibration temperatures through quenching of isotopic relationships. However, when one of the sulfur species was produced from the other by oxidation or reduction, or when both species were produced from another compound (e.g., dissolution of pyrite), the isotopic relationship will be controlled by kinetic isotopic effects.

KINETIC ISOTOPIC EFFECTS DURING OXIDATION OF SULFIDES

During nonequilibrium, unidirectional chemical reactions, the fractionation of sulfur isotopes arises from the fact that chemical reaction rates are mass dependent and that one isotopic species reacts more rapidly than another. In general, the molecules containing the lighter isotope will have the faster reaction rate. Consequently, the product tends to be enriched in the lighter isotope in simple systems.

For example, oxidation of sulfide to sulfate can be considered as two separate reactions with different rate constants:

$$H_2{}^{32}S \xrightarrow{k_1} {}^{32}SO_4{}^{2-} \qquad (10.18)$$

$$H_2{}^{34}S \xrightarrow{k_2} {}^{34}SO_4{}^{2-}. \qquad (10.19)$$

The ratio of the two rate constants k_1/k_2 is equal to the kinetic isotopic effect, and $(k_1 + k_2)$ is the overall rate of sulfate oxidation. A k_1/k_2 value of 1.005 implies that the $H_2{}^{32}S$ species reacts 5‰ faster than $H_2{}^{34}S$ and that $SO_4{}^{2-}$ produced at a given instance is depleted in ^{34}S by 5‰ with respect to the remaining H_2S. (The use of k_1/k_2 values in this discussion refers to the instantaneous enrichment factors regardless of the number of steps actually involved in the reactions and does not necessarily imply single step reactions.)

The kinetic isotopic effect occurs most effectively when the rupturing of a chemical bond including the isotopes in question is the rate-controlling step of the reaction. In the oxidation of sulfide, the isotopic effect is small because the oxidation rate is not controlled by M-S bond rupture but by some other processes, such as the rate of formation of an intermediate complex between H_2S and O_2 and/or H_2O. For both organic and inorganic oxidation of H_2S at low temperature, $k_1/k_2 = 1.000 \pm 0.003$ (Nakai and Jensen, 1964; Ohmoto, unpublished data). Therefore, sulfates produced by nonequilibrium oxidation of sulfides should show $\delta^{34}S$ values essentially identical to those of the initial sulfides. Such a relationship is observed between $SO_4{}^{2-}$ in mine water and sulfide minerals in ores, between $SO_4{}^{2-}$ and H_2S in some hot spring waters, and between supergene sulfates and hypogene sulfide minerals (see Figure 10.5). In fact, this isotopic relationship between supergene sulfate and hypogene sulfides, where $^{34}S_{sulfate} \simeq {}^{34}S_{sulfide}$, has been used to distinguish supergene sulfates from hypogene sulfate, which typically show a $^{34}S_{sulfate} \gg {}^{34}S_{sulfide}$ relationship (Field, 1966). It should be noted, however, that during low-temperature oxidation of sulfides, as may occur during the formation of sandstone-type uranium deposits, unstable sulfites and thiosulfates may be formed, which upon disproportionation have large sulfur isotope fractionations that can result in large variations of $\delta^{34}S$ values of the minerals (Granger and Warren, 1969; Warren, 1972).

Kinetic isotopic effects during the oxidation of fluids may also play an important role in the variation of $\delta^{34}S$ values of hypogene sulfate minerals. For example, suppose sulfate and sulfide species in hydrothermal fluids had initially attained isotopic equilibrium at 250°C and the $\delta^{34}S$ values are 25‰ for SO_4^{2-} and 0‰ for H_2S. If the H_2S is partially oxidized to SO_4^{2-} during or shortly before mineralization, the result would be very consistent $\delta^{34}S$ values of sulfide minerals at or near 0‰, whereas $\delta^{34}S$ values of sulfate minerals could vary widely between 25 and 0‰ depending on the degree of oxidation of the initial and final fluid (see Figure 10.10b). The wide variation in the $\delta^{34}S$ values of sulfate minerals, contrasting with very uniform $\delta^{34}S$ values of sulfide minerals observed in some hydrothermal deposits such as in the Creede, Colorado deposits, may be partly caused by such nonequilibrium kinetic isotopic effects. A similar type of $\delta^{34}S$ relationship, uniform $\delta^{34}S_{H_2S}$ values but variable $\delta^{34}S_{SO_4^{2-}}$ values, may also be produced by *mixing* of H_2S-rich solution (A) with SO_4^{2-}-rich solution (B) at temperatures below $\sim 200°C$ (see Figure 10.10c). This is because such mixing can be considered as two separate processes. One process is the formation of H_2S from SO_4^{2-} in solution B (reduction of solution B), and the other is the formation of SO_4^{2-} from H_2S in solution A (oxidation of solution A). The former process does not operate as easily as the latter process at temperatures below $\sim 200°C$, and it may result in a variable $\delta^{34}S_{SO_4^{2-}}$ that is a mixture of $\delta^{34}S_A$ and $\delta^{34}S_B$, whereas $\delta^{34}S_{H_2S}$ remains constant.

When there is isotopic disequilibrium between H_2S and SO_4^{2-} in solution, isotopic equilibrium is also unlikely to be established between mineral pairs involving pyrite or chalcopyrite (py-sp, py-po, py-gn, py-cpy) even when isotopic equilibrium is attained among sphalerite, pyrrhotite, and galena. This is because mechanisms of formation of pyrite and chalcopyrite under normal hydrothermal conditions from Fe^{2+} and Cu^+ complexes involve redox reactions that require both H_2S and SO_4^{2-}, such as:

$$4\ Fe^{2+} + 7\ H_2S + SO_4^{2-} \rightarrow 4\ FeS_2 + 4\ H_2O + 6\ H^+ \quad (10.20)$$

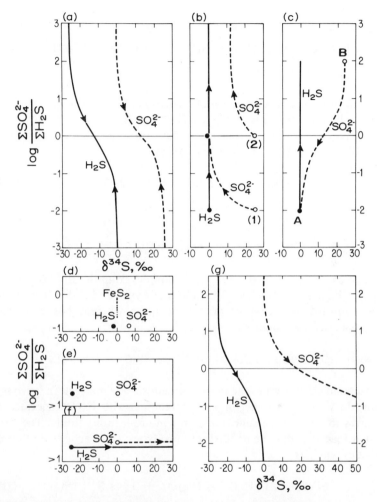

Fig. 10.10 Isotopic relationships between coexisting sulfate and sulfide species as a function of $\Sigma SO_4^{2-}/\Sigma H_2S$. (a) Equilibrium conditions at 250°C where $\delta^{34}S_{\Sigma S} = 0‰$. (b) Nonequilibrium oxidation. $\delta^{34}S_{H_2S}$ remains constant, but variation in $\delta^{34}S_{S_4^{2-}}$ depends on the initial SO_4^{2-}/H_2S ratio of fluids (e.g., 1 and 2). (c) Nonequilibrium mixing of H_2S-rich fluid (A) with SO_4^{2-}-rich and fluid (B) ($\delta^{34}S_{H_2S}$ remains constant, but $\delta^{34}S_{SO_4^{2-}}$ continuously approaches $\delta^{34}S_{H_2S}$ as a result of mixing). (d) Nonequilibrium leaching of pyrite (preliminary experimental data). (e) Nonequilibrium reduction of SO_4^{2-} in systems open to both SO_4^{2-} and H_2S ($\Delta_{SO_4^{2-}-H_2S}$, $\delta^{34}S_{H_2S}$ and $\delta^{34}S_{SO_4^{2-}}$ values remain constant). (f) Nonequilibrium reduction of SO_4^{2-} in systems closed to SO_4^{2-} but open to H_2S ($\Delta_{SO_4^{2-}-H_2S}$ remains constant, but $\delta^{34}S_{H_2S}$ and $\delta^{34}S_{SO_4^{2-}}$ values increase continuously). (g) Nonequilibrium reduction of SO_4^{2-} in systems closed to both SO_4^{2-} and H_2S ($\Delta_{SO_4^{2-}-H_2S}$, $\delta^{34}S_{H_2S}$ and $\delta^{34}S_{SO_4^{2-}}$ values increase continuously).

and

$$8\,Cu^+ + 8\,Fe^{2+} + 15\,H_2S + SO_4^{2-} \rightarrow 8\,CuFeS_2 + 4\,H_2O + 22\,H^+ \quad (10.21)$$

This is in contrast to the processes of formation of sphalerite, pyrrhotite, and galena, which require only H_2S (e.g., $Zn^{2+} + H_2S \rightarrow ZnS + 2H^+$). Pyrite and chalcopyrite formed in such disequilibrium systems may inherit the $\delta^{34}S$ values of H_2S and SO_4^{2-} in such manner as

$$\delta^{34}S_{py} = {}^{7}\!/_{8} \cdot \delta^{34}S_{H_2S} + {}^{1}\!/_{8} \cdot \delta^{34}S_{SO_4^{2-}} \quad (10.22)$$

and

$$\delta^{34}S_{cpy} = {}^{15}\!/_{16} \cdot \delta^{34}S_{H_2S} + {}^{1}\!/_{16} \cdot \delta^{34}S_{SO_4^{2-}}. \quad (10.23)$$

The best examples of such cases are in samples from the Illinois-Wisconsin deposits (Figure 10.14) where $\delta^{34}S$ values of pyrite and $BaSO_4$ continuously increase toward the later stages of mineralization, whereas those of sphalerite and galena decrease slightly.

KINETIC ISOTOPIC EFFECTS DURING LEACHING OF SULFIDES

Pyrite and chalcopyrite, the most common sulfide minerals in igneous and sedimentary rocks, may decompose by reaction with H_2O at elevated temperatures to become the source of sulfur in ore-forming fluids. The decomposition reactions may be divided into two types: (1) reactions that produce Fe-oxides, such as

$$6FeS_2 + 11\,H_2O \rightarrow 3Fe_2O_3 + 11\,H_2S + SO_2, \quad (10.24)$$

and (2) dissolution by acid solutions, such as

$$3FeS_2 + 2\,H_2O + 6H^+ \rightarrow 3Fe^{2+} + 5\,H_2S + SO_2 \quad (10.25)$$

or reversals of equations (10.20) and (10.21).

Reactions similar to (10.24) were investigated by Grinenko and Grinenko (1967) who observed that the $\delta^{34}S_{H_2S}$ values were between $+1$ and $+3‰$ with respect to the $\delta^{34}S_{py}$ values at temperatures of 350–550°C. Preliminary experimental data on the reversal reaction of (10.20) at 100°C and a reaction rate of 10^{-8} mole FeS_2/sec, indicate $\delta^{34}S_{H_2S}$ values of $-1.0‰$ with respect to the $\delta^{34}S_{py}$ value (Ohmoto and Archer, unpublished data). From mass balance calculations, this $\delta^{34}S_{H_2S}$ value suggests $\delta^{34}S_{SO_4^{2-}}$ and $\Delta_{(SO_4^{2-}-H_2S)}$ values of 7‰ and 8‰, respectively (see Figure 10.10d). Although more experimental studies are needed, these data suggest that hydrothermal fluids that obtained H_2S by decomposition of sulfides in igneous rocks would have $\delta^{34}S$ values nearly identical to those of magmatic fluids, and that the $\Delta_{SO_4^{2-}-H_2S}$ values in such fluids may initially be much smaller than those of the equilibrium values.

KINETIC ISOTOPIC EFFECTS DURING REDUCTION OF SULFATES (ROLE OF SEAWATER IN ORE GENESIS)

Under low-temperature ($T \lesssim 50\,°C$) geologic conditions, perhaps the only mechanism for the reduction of sulfate is by sulfate-reducing bacteria. The $\delta^{34}S$ distributions of biogenic H_2S and related sedimentary sulfides are controlled by (1) the value of the kinetic isotopic effect (k_1/k_2) and by (2) the open or closed nature of the system with respect to SO_4^{2-} and H_2S (or HS^-).

Experimental data on kinetic isotopic effects obtained during bacterial reduction of sulfate (Harrison and Thode, 1957; Nakai and Jensen, 1964; Kemp and Thode, 1968; Kaplan and Rittenberg, 1964) as well as sulfur isotopic data of sulfides in recent marine sediments (for summary, see Goldhaber and Kaplan, 1975) indicate that the kinetic isotopic effect increases with a decrease in the rate of reduction from $k_1/k_2 = 1.015\text{--}1.025$ at a rate of $\sim 10^{-0.5}$ mole $SO_4^{2-}/l \cdot sec$, a typical rate of sulfate reduction under laboratory conditions, to $k_1/k_2 >$ ca. 1.065 at a rate of $10^{-4.5}$ mole $SO_4^{2-}/l \cdot sec$. The latter value is close to the equilibrium value of 1.075 at $25\,°C$. Because the overall rate of sulfate reduction is dependent on the nature and amount of nutrient available for the sulfate-reducing bacteria, it can be expected that where the rate of sedimentation is fast as in near-shore environments or where there is unusual organic activity, the kinetic effect will be small, but where the rate of sedimentation is slow, as in off-shore environments, the kinetic effect will be large.

Since the open or closed nature of a system with respect to SO_4^{2-} and H_2S is dependent on the geologic environment, different $\delta^{34}S$ systematics of bacteriogenic sulfides can be expected in different environments as indicated below (see Schwarcz and Burnie, 1973, for a review on this subject). The following discussion assumes that most sulfate reduction occurs at the sediment-water interface and ignores sulfur isotope effects during diagenesis, which may not be accurate for some environments (Goldhaber and Kaplan, 1975).

Systems open to SO_4^{2-}. Environments where the rate of sulfate reduction is much slower than the rate of SO_4^{2-} supply can be considered as open to

SO_4^{2-}. Deep euxinic basins (e.g., Black Sea) may be typical examples where sulfate reduction occurs slowly at the bottom but where SO_4^{2-} from the overlying water mass is continuously supplied through diffusion. Schwarcz and Burnie (1973) suggested k_1/k_2 values of 1.040-1.060 for such environments. The $\delta^{34}S$ values of sulfide minerals formed in such environments are approximately 40-60‰ lower than the $\delta^{34}S$ values of seawater (see Figure 10.11). Because the $\delta^{34}S$ values of seawater varied between +10 and +30‰ through geologic time, the total range of the mean $\delta^{34}S$ values of sulfides formed in such environments may be from ~ −50 to −10‰.

The sulfur isotopic characteristics of the Kuperschiefer deposits with a range −50 to −30‰ and a maximum population occurring at −40‰ (which is approximately 50‰ lower than the $\delta^{34}S$ values of the Permian seawater) and of the Graptolithenschiefer deposits have been interpreted by Schwarcz and Burnie (1973) as the best examples of deposits that were formed in an SO_4^{2-}-open system, although a trend of increasing $\delta^{34}S$ values toward the top

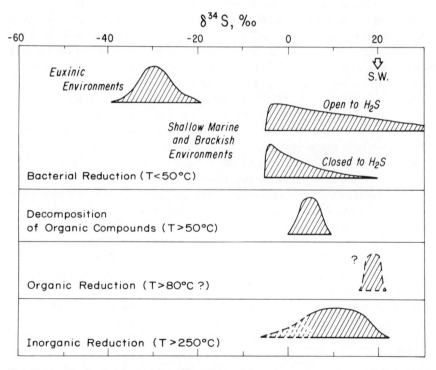

Fig. 10.11 Distribution pattern for $\delta^{34}S$ values of H_2S and sulfide minerals when sulfate of $\delta^{34}S = +20‰$ is reduced by various mechanisms. SW is sea water. See text for trends with time that are superimposed on these distributions.

of the sequence in the Kupferschiefer suggests the system was partially closed to SO_4^{2-}.

Systems closed to SO_4^{2-}. Environments where the rate of sulfate reduction is much faster than the rate of sulfate supply, including semiclosed shallow-marine and brackish-water environments, can be considered as systems closed to SO_4^{2-}. Schwarcz and Burnie (1973) have suggested k_1/k_2 values of approximately 1.025 as typical kinetic effects in such environments. These systems can further be divided into (1) systems open to H_2S (or HS^-) when aqueous sulfide is continuously removed from the system as sulfide minerals, and (2) systems closed to H_2S (HS^-) when the supply of metals to form sulfide minerals is limited with respect to H_2S, or when the formation of sulfide minerals is delayed.

The changes in the $\delta^{34}S$ values of SO_4^{2-} and of H_2S during the reduction of sulfate in systems closed to SO_4^{2-} can be calculated using the mathematical treatment of the Rayleigh distillation process. The $\delta^{34}S$ value of SO_4^{2-} at time t can be closely approximated as

$$\delta^{34}S_{SO_4^{2-}-(t)} = \delta^{34}S_{SO_4^{2-}-(o)} + 1000\ (F^{[1-(k_1/k_2)]} - 1), \quad (10.26)$$

where $\delta^{34}S_{SO_4^{2-}-(o)}$ is the original $\delta^{34}S$ value of SO_4^{2-}, and F is the fraction of SO_4^{2-} remaining at time t.

The $\delta^{34}S$ value of H_2S at time t in such a system open to H_2S (or HS^-) is

$$\delta^{34}S_{H_2S(t)} = \delta^{34}S_{SO_4^{2-}-(t)} - 1000\left(\frac{k_1}{k_2} - 1\right), \quad (10.27)$$

and in such a system closed to H_2S (or HS^-) is

$$\delta^{34}S_{H_2S(o-t)} = \frac{\delta^{34}S_{SO_4^{2-}-(o)} - \delta^{34}S_{SO_4^{2-}-(t)} \cdot F}{1 - F}. \quad (10.28)$$

Examples of the change in the $\delta^{34}S$ values for the two systems are shown in Figure 10.10*f* and 10.10*g*.

The distribution pattern of the $\delta^{34}S$ values of sulfide minerals formed in systems closed to SO_4^{2-} and having a k_1/k_2 value of 1.025 is also shown in Figure 10.11. Characteristic features are (1) spread out $\delta^{34}S$ distributions that are skewed toward positive values, (2) both minimum $\delta^{34}S$ values and the $\delta^{34}S$ values of the maximum population are approximately 25‰ lower than the original $\delta^{34}S_{SO_4^{2-}}$ and (3) $\delta^{34}S$ values increase continuously toward the later stages (i.e., toward the top of the sedimentary sequence). Although the later sulfide minerals may have $\delta^{34}S$ values close to $\delta^{34}S_{SO_4^{2-}-(o)}$, only a small number of samples in systems open to H_2S will have $\delta^{34}S$ values larger than the $\delta^{34}S_{SO_4^{2-}(o)}$ values. White Pine and the Zambian copper deposits have the $\delta^{34}S$ characteristics of sulfides formed in shallow marine or brackish environments (Schwarcz and Burnie, 1973).

The $\delta^{34}S$ data of these "sedimentary type" deposits, however, do not necessarily imply that all the metals in the deposits were fixed by the same process. It is quite possible that the majority of sulfur in the deposits was first fixed as iron sulfide by bacterial reduction of seawater sulfate, but later high-temperature brines were responsible for the formation of other heavy metal sulfides. For example, after examining the $\delta^{34}S$ relationship among pyrite, sphalerite, and galena at the McArthur River deposits, Williams and Rye (1974) have suggested that the majority of pyrite was formed syngenetically by bacterial reduction of seawater sulfate but that all the sphalerite and galena and some pyrite were formed at temperatures around 200°C by later hydrothermal fluids that obtained H_2S, iron, zinc, and lead through leaching of sedimentary units in lower sedimentary sequences.

Optimum temperatures for bacterial reduction of sulfate are between 30 and 45°C and rate of reduction is reduced drastically at higher temperatures (Orr, 1974). At temperatures above ~ 50°C, thermal decomposition of sulfur-bearing organic compounds (such as those in petroleum) produces H_2S. The $\delta^{34}S$ values of such H_2S are $15 \pm 5\%_0$ lower than those of seawater sulfates (Thode and Monster, 1965; Orr, 1974, 1975); see Figure 10.11.

Experimental studies on sulfate reduction by organic compounds (e.g., methane, xylene) at high temperatures have been carried out by Toland (1960), Orr (1974, 1975), and Drean (1978). Organic compounds alone did not produce a detectable quantity of H_2S from sulfate solutions at temperatures up to 325°C and for a period of up to 2 weeks (Drean, 1978). The rate of sulfate reduction was found, however, to be accelerated drastically when reduced sulfur (e.g., H_2S, NH_4SH) as well as organic compounds was initially present (Toland, 1960). For example, when initial concentrations of H_2S and SO_4^{2-} were in the order of $1\ m$, more than 50% of SO_4^{2-} was found to be reduced by organic compounds within a period of a few weeks at temperatures as low as ~ 175°C (Orr, personal communication).

The mechanisms of sulfate reduction by organic compounds and the role of H_2S (or other forms of reduced sulfur) are not well understood, and more experimental studies are needed. However, these experimental data together with the experimental study by Toland et al. (1958), which has shown that the sulfur of intermediate valency state (e.g., S, Na_2SO_3) is readily reduced to H_2S by organic compounds at temperatures as low as 180°C and within a period of less than a few hours, suggest that the process of organic sulfate reduction operates by the following steps: (1) reduction of sulfate by reduced sulfur to form sulfur of intermediate valency state, such as

$$2H^+ + SO_4^{2-} + 3H_2S \rightarrow 4S + 4H_2O \qquad (10.29)$$

and (2) reduction of sulfur of intermediate valency state by organic compounds, such as

$$CH_3C_6H_4CO_2H \text{ (toluic acid)} + 3S + 2H_2O \rightarrow \\ HO_2CC_6H_4CO_2H \text{ (phthalic acid)} + 3H_2S \qquad (10.30)$$

(see Toland et al., 1958). The rate of sulfate reduction, as seen in the above equations, is dependent on the pH and concentration of reduced sulfur in the fluid as well as on temperature. In geologic conditions, H_2S produced by decomposition of sulfur-bearing organic compounds may provide the necessary reduced sulfur to catalyze the organic reduction of sulfate.

From the temperature and H_2S-SO_4^{2-} concentration relationships in natural gas and oil, Orr (1974, 1975) has suggested that the reduction of aqueous sulfates by organic compounds occurs at temperatures above 80–120°C if sufficient time is allowed. This further suggests that some oil field brines that acquired sulfates from adjacent evaporite beds and underwent organic reduction can become H_2S-bearing ore-forming fluids, and that seawater and sulfate-bearing meteoric water can become sulfide-bearing fluids when circulated through organic-rich sedimentary sequences at temperatures above ~100°C.

The kinetic isotopic effects accompanied with organic reduction have not been experimentally investigated. However, regardless of the k_1/k_2 values, organic reduction appears to cause the complete conversion of SO_4^{2-} into H_2S within a relatively short time, so that the H_2S will have $\delta^{34}S$ values essentially identical to the source sulfates (see Figure 10.11). In fact, Orr (1974) has observed that H_2S produced by organic reduction in some natural gas and oil reservoirs has $\delta^{34}S$ values essentially identical to that of the associated evaporite beds.

During experiments of basalt-seawater interactions, SO_4^{2-} is reduced and pyrite and/or pyrrhotite formed at temperatures above 300°C (Mottl, 1976). Sulfide minerals were not found in the experiments at lower temperatures, but the SO_4^{2-}/H_2S ratio in the solution decreased significantly in the runs at 250°C after 9 months. These data indicate that SO_4^{2-} will be reduced to form H_2S at temperatures above ~250°C by reactions with Fe^{2+} components in rocks ($SO_4^{2-} + 8Fe^{2+} + 10H^+ \rightarrow H_2S + 8Fe^{3+} + 4H_2O$) and such a process is an important mechanism for the evolution of seawater into ore-forming fluids, particularly in environments where submarine volcanism is active and circulation of seawater through hot volcanic rocks is established. At temperatures above ~250°C isotopic equilibrium is probably established between SO_4^{2-} and H_2S in circulating fluid systems as discussed earlier on "Problems of Sulfate-Sulfide Equilibria", and seawater with a $\delta^{34}S$ value of $+20‰$ can produce H_2S of between $+20$ and $-5‰$ depending on the degree of reduction (see Figure 10.11). In fact, $\delta^{34}S$ values of eight samples of pyrite and pyrrhotite formed during basalt-seawater experiments were measured to be between $+8.0$ and $+20.7‰$ (Ohmoto et al., 1976), while some epigenetic pyrite samples from ocean floor basalts have been analyzed to show large $\delta^{34}S$ values, up to $\sim +23‰$ (Field et al., 1976).

OTHER PROCESSES OF ISOTOPIC FRACTIONATION

The preceeding discussions emphasized the effects of chemical processes on the isotopic fractionation of sulfur isotopes. However, mass dependent physical processes, such as diffusion and adsorption, can also produce sulfur isotopic fractionation.

Diffusion processes can be caused in nature by differences in osmotic pressure or in temperature. For example, in a standing column of solution under a temperature gradient, the lighter molecules migrate toward the warmer and the heavier molecules toward the cooler regions in the Soret effect. However, in hydrothermal systems involving ore-formation, isotopic fractionation due to diffusion may be neglected in most cases. This is because considerable turbulence and convective mixing of fluids is common and because most substances are transported by flow rather than by diffusion. Systematic variations of $\delta^{34}S$ values in disseminated sulfides adjacent to hydrothermal veins have sometimes been attributed to diffusion, but such variations are more likely the result of processes involving mass transport and chemical fractionation.

During the passage through clay-rich sediments both oxygen and hydrogen isotopic compositions of residual water may become heavier: a phenomenon commonly referred to as the shale membrane or microfiltration effect. A similar phenomenon has been experimentally demonstrated by Nriagu (1974) on sulfur isotopic composition of SO_4^{2-} in solutions. He observed that the isotopically lighter SO_4^{2-} was preferentially adsorbed into sediments and the remaining SO_4^{2-} in solution was enriched in ^{34}S by 0.9 to 6.0‰. The fractionation factor was dependent on the concentration of SO_4^{2-} in solution and on the amount of SO_4^{2-} adsorbed in sediments. This process could play an important role in altering the $\delta^{34}S$ value of SO_4^{2-} in formation waters, and also that of circulating seawater in sedimentary sequences.

SULFUR ISOTOPIC DATA FROM HYDROTHERMAL DEPOSITS

During the last 20 years, scores of ore deposits involving up to hundreds of sulfur isotope analyses have been studied by various investigators. Because it is impractical to present all of these data, we have selected only a small number of deposits to show general characteristics of the sulfur isotopic variation of ore deposits that were formed under different geologic settings and to illustrate how sulfur isotopic data can be used to obtain information on (1) the degree of equilibrium, temperature, f_{O_2}, and pH conditions of ore-forming fluids, (2) the mechanisms of mineralization, and (3) the source of sulfur.

Porphyry Copper Deposits

The sulfur isotopic data on sulfides and sulfates in porphyry copper deposits of the American Cordilleran are summarized in Figure 10.12. Although the type of presentation used in Figure 10.12 has the obvious limitation of not indicating time-space features, which are essential to an interpretation of the data of a specific ore deposit, it shows some general characteristics.

The majority of the $\delta^{34}S$ values of the sulfides and sulfates in these deposits fall between -3 and $+1‰$ and between $+8$ and $+15‰$, respectively. The sulfate-sulfide isotopic temperatures lie typically between 450 and 650°C. These values generally agree with temperatures estimated from other geochemical methods, and suggest establishment of isotopic equilibrium among the oxidized and the reduced species in fluids and in minerals. Although the exact $\delta^{34}S_{fluid}$ values at each deposit cannot be determined without the f_{O_2} information, they can be estimated to lie between -3 and $+9‰$ for the majority of these deposits. These $\delta^{34}S_{fluid}$ values suggest that the sulfur was derived largely from igneous sources, either as magmatic fluids or by dissolution of igneous sulfides. Low $\delta^{34}S$ values for sulfides in some deposits (e.g., Galore Creek) may mean that the oxidation state of the fluids was high or that they incorporated some sedimentary sulfides. Deposits that have high $\delta^{34}S$ values for both sulfides and sulfates, such as Morococha may indicate that some sulfur was derived from an evaporite source. (The host rocks at Morococha contain substantial sedimentary anhydrite-bearing units.)

Stratabound Sulfide Deposits in Volcanic Rocks

Stratabound massive sulfide deposits that occur in volcanic rocks are generally thought to be related to submarine volcanism and formed at or near the seawater-rock interface. The $\delta^{34}S$ values of sulfides in these deposits are typically positive except for the very late stages of mineralization in some deposits, like the Kuroko deposits. The $\delta^{34}S$ variation within a given deposit is usually slightly larger than that in the porphyry copper deposits (Figure 10.13).

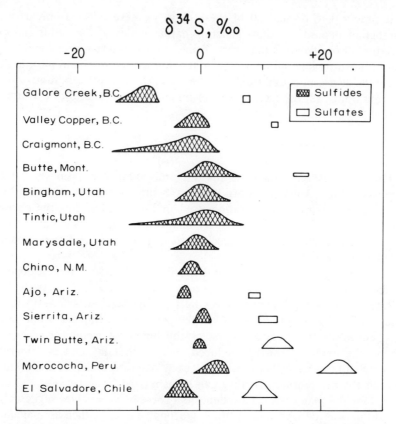

Fig. 10.12 Schematic presentation of sulfur isotopic data of some porphyry copper deposits (modified from unpublished data of Field, 1975).

Through examination of the isotopic data of 25 deposits belonging to this type, Sangster (1968) noted that the average $\delta^{34}S$ value of sulfides in each deposit is approximately 17‰ lower than the $\delta^{34}S$ value of the contemporaneous seawater, and suggested that the sulfides in this type of deposits were formed by mixture of magmatic sulfur and H_2S (or HS^-) produced by bacterial reduction of seawater sulfate. However, the host rocks in this type of ore deposits are generally too poor in organic carbon content and the temperatures of deposition were much too high (approximately 200-300°C for Kuroko, 120-350°C for Raul, and 50-220°C for the Red Sea), to allow bacterial activity at the site of ore deposition. The trend of decreasing $\delta^{34}S$ values of sulfides upward in the stratigraphic section often observed for some stratabound volcanogenic deposits (Kuroko, Cyprus, Raul) is also in constrast to the trend of increasing $\delta^{34}S$ values typically observed in "biogenic" deposits.

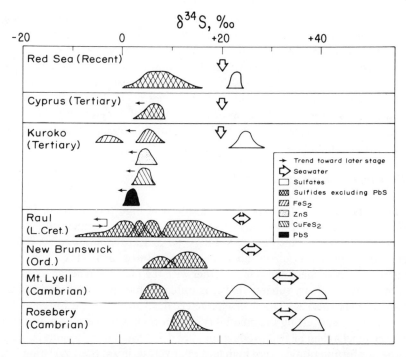

Fig. 10.13 Schematic presentation of sulfur isotopic data of some stratiform massive sulfide deposits associated with submarine volcanism. The sources of data on Cyprus, Kuroko, and New Brunswick are summarized in Rye and Ohmoto (1974). The data on Red Sea are from Kaplan et al. (1969) and Shanks and Bishoff (1975), on Raul deposits from Ripley and Ohmoto (1977), and on Mt. Lyell and Rosebery deposits from Solomon et al. (1969).

The sulfur isotopic data, the paragenetic sequence and the $\delta^{18}O$ and δD values of the Kuroko, Raul, and other volcanic-stratabound deposits can best be explained by a mechanism in which seawater became an ore-forming fluid by various degrees of reduction of its sulfate through reactions with Fe^{2+} components in hot volcanic rocks. For details of a model of the evolution of seawater into ore-forming fluids, see Ohmoto and Rye (1974) and Ripley and Ohmoto (1977). Such a proposed model, however, does not exclude the possibility that some sulfur from igneous sources was also present in the ore-forming fluids. In fact, the $\delta^{34}S$ data of the Raul deposit cannot be explained without a contribution of some sulfur from leaching of igneous sulfides (Ripley and Ohmoto, 1977).

The $\delta^{34}S$ values of sulfate minerals, when present in the volcanic stratabound ores, are larger than those of contemporaneous seawater, as would be expected when seawater sulfate is partially reduced at temperatures above approximately 250°C in equilibrium with aqueous sulfides. However, the sulfur isotope fractionations between coexisting sulfate and sulfide minerals are usually smaller than the equilibrium values at the temperatures of mineralization. For example, in the Kuroko ores, the values of $\Delta_{(SO_4^{2-}-H_2S)}$ are approximately 20‰, which is smaller than the equilibrium Δ value of 25‰ at 250°C. Such a relationship may reflect the slow rates of equilibration between aqueous sulfate and sulfide species below 300°C and/or some mixing of less reduced seawater at or near the site of mineralization. Neither process, however, affects the isotopic equilibria between aqueous sulfides and sulfide minerals.

Mississippi-Valley-Type Deposits

The Mississippi-Valley-type deposits shown in Figure 10.14 occur in carbonate host rocks and were formed under relatively low-temperature conditions (50-100°C for Pine Point, 50-140°C for S. E. Missouri, and 50-200°C for the Illinois-Wisconsin and the Hansonburg districts). The $\delta^{34}S$ characteristics of these deposits, together with the fact that igneous rocks are scarce in these areas, suggest that it is unlikely that the sulfur was derived from igneous sources. The $\delta^{34}S$ characteristics, however, are different for each deposit, suggesting the possibility that in each case the source of sulfur and the mechanism of H_2S production were different. The $\delta^{34}S$ values of H_2S in the ore-forming fluids were high and relatively uniform (i.e., variation less than 10‰) throughout the depositional stages at Pine Point, at the Illinois-Wisconsin district, and at Sardinia. They were high and highly variable (i.e., more than 20‰ variation) in the Southeast Missouri district and low and relatively uniform in the Hansonburg district.

Large positive $\delta^{34}S_{H_2S}$ values for some of these deposits suggest that the sulfur was derived from seawater or ancient evaporites that had undergone reduction by organic compounds. The difference in the degree of isotopic equilibrium between coexisting sphalerite and galena at the Wisconsin deposits (i.e., establishment of equilibrium at temperatures as low as 50°C) and those at Pine Point (i.e., disequilibrium relationships at 50-100°C) may indicate that at the Wisconsin deposits the reduction of SO_4^{2-} was completed before the ore-forming fluids reached the site of deposition, whereas reduction of SO_4^{2-} continued during the deposition of sulfides at Pine Point. [The host rocks at Pine Point are rich in bitumen and native sulfur (Sasaki and Krouse, 1969), the compounds which accelerate the rate of sulfate reduction.]

In the Southeast Missouri district, Brown (1967) observed a clear correlation between the $\delta^{34}S$ values of galenas and their lead isotopic compositions, such that the most radiogenic galenas have low $\delta^{34}S$ values while galenas with less radiogenic leads have $\delta^{34}S$ values similar to seawater. Such a trend is dif-

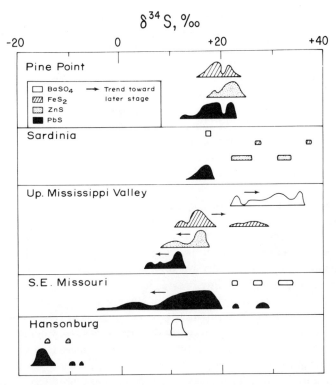

Fig. 10.14 Schematic presentation of sulfur isotopic data of some Mississippi Valley type deposits. Pine Point (Sasaki and Krouse, 1969), Sardinia (Jensen and Desau, 1966), Upper Mississippi Valley (McLimans, 1977), S. E. Missouri (Brown, 1967), and Hansonburg (Allmandinger and Ohmoto, unpublished data, 1976).

ficult to explain by a "biogenic" model, but easily explainable if metals and sulfur were derived from essentially two sources: (1) evaporites and (2) normal (sulfide-bearing) sedimentary rocks. The extractable lead in the normal sedimentary rocks in the district is highly radiogenic (Doe and Delevaux, 1972), whereas lead isotopic compositions of evaporites are probably similar to that of contemporaneous seawater and less radiogenic. Fluids that obtained sulfur from evaporites and underwent organic reduction would have had high $\delta^{34}S_{H_2S}$ values, whereas fluids that obtained sulfur by leaching of sedimentary sulfides would have had low $\delta^{34}S_{H_2S}$ values.

Nearly identical $\delta^{34}S$ values for H_2S and barite in the Southeast Missouri and Sardinia deposits may indicate that the sulfates in these deposits were formed by oxidation of H_2S (or HS^-) at the site of mineralization. However, variable $\delta^{34}S$ values for sulfate minerals occurring during the late stages of mineralization in the Illinois-Wisconsin district are likely to be produced by mixing of SO_4^{2-}-bearing fluids toward the later stage of mineralization (see Figure 10.10c). Such a mixing of fluid is also demonstrated by the change in the δD and $\delta^{18}O$ values of fluids (McLimans, 1977).

In the Hansonburg, New Mexico, district, the $\delta^{34}S$ values of barite in the ores are essentially identical to those of the Pennsylvanian anhydrite in the area, suggesting the source of sulfate in the ore-forming fluids was sedimentary anhydrite. The low $\delta^{34}S$ values for the H_2S in the ore-forming fluids may indicate that the sulfate was only partially reduced because of low organic content in the stratigraphic sections in the area, or alternatively the H_2S was largely derived by leaching of sedimentary sulfides.

Vein- and Replacement-Type Deposits

The $\delta^{34}S$ characteristics of vein- and replacement-type deposits that formed at temperatures of 350–140°C and have been studied in detail both geologically and geochemically are summarized in Rye and Ohmoto (1974). Many of the deposits, including Providencia and Pasto Bueno, are associated with calcalkaline intrusives, are usually lacking sulfate minerals, and have $\delta^{34}S$ values of sulfide minerals in a narrow range near 0‰. The sulfur in these ore-forming fluids appears to have been derived from igneous sources, and sulfides were precipitated in pH-f_{O_2} conditions where H_2S was the dominant sulfur species. However, when such fluids equilibrated with different host rocks, the pH and/or f_{O_2} may have increased and produced sulfides with lower $\delta^{34}S$ values such as probably occurred at the Mogul and Darwin deposits.

Sulfur from nonigneous sources may have been incorporated in some deposits, such as evaporite sulfur at the Sunnyside deposit, sedimentary sulfides at some Tertiary deposits in the Black Hills, both sedimentary sulfates and sulfides at the Bluebell deposit, and seawater or evaporite sulfates at the Echo Bay deposits.

CARBON ISOTOPE GEOTHERMOMETRY

Equilibrium isotopic fractionation factors among important carbon species in hydrothermal systems are summarized in Table 10-3 and Figure 10.15. Experimental data, particularly at temperatures above 50°C, are few. The only system that has been experimentally investigated over a wide temperature range is $HCO_3^- - CO_2$. For other systems, available data are mostly those estimated through theoretical computations by Bottinga (1968, 1969). Uncertainties in the values in Table 10-3 are probably within $\pm 10\%$

Table 10-3 Equilibrium isotopic fractionation factors of carbon compounds with respect to CO_2

$$1000 \ln \alpha_{i-CO_2} = \frac{A}{T^3} \times 10^8 + \frac{B}{T^2} \times 10^6 + \frac{C}{T} \times 10^3 + D$$

(T in °K)

Compounds	A	B	C	D	T(°C) Range
$CaMg(CO_3)_2$[a]	−8.914	8.737	−18.11	8.44	≤600
$CaCO_3$[b]	−8.914	8.557	−18.11	8.27	≤600
HCO_3^-[c]	0	−2.160	20.16	−35.7	≤290
CO_3^{2-}[d]	−8.361	8.196	−17.66	6.14	≤100
$H_2CO_3(ap)$[e]	0	0	0	0	≤350
CH_4[f]	4.194	−5.210	−8.93	4.36	≤700
CO[g]	0	−2.84	−17.56	9.1	≤330
$C(graphite)$[f]	−6.637	6.921	−22.89	9.32	≤700

[a] Based on empirical fractionation factors between dolomite and calcite of Sheppard and Schwarcz (1970).
[b] Polynomial fit of values calculated by Bottinga (1969).
[c] Polynomial fit of experimental data of Malinin et al. (1967), Deuser and Degens (1967), Wendt (1968), Emrich et al. (1970), Mook et al., (1974), and Vogel (1961).
[d] Polynomial fit of values calculated by Thode et al. (1965).
[e] Estimated in Ohmoto (1972) on the basis of $H_2CO_3(ap) = CO_2(aq) + H_2CO_3$.
[f] Polynomial fit of values calculated by Bottinga (1968).
[g] Polynomial fit of values calculated by Urey (1947).

for the $CaCO_3$-CO_2 system, and the CH_4-CO_2 system, as indicated from comparisons with a few experimental data on the $CaCO_3$-CO_2 system and with the measured fractionation factors between CH_4 and CO_2 in natural systems (see Bottinga, 1968, 1969). Uncertainties in the equations for graphite-CO_2, CO-CO_2, and CO_3^{2-}-CO_2 fractionations in Table 10-3 are difficult to evaluate because of the lack of either experimental or analytical data on natural samples.

Carbon isotopic fractionation factors among carbonate minerals have not been well studied. The magnitude of the fractionation factors, however, are probably similar to those of sulfur isotopic fractionation factors among sulfate minerals. The Δ values between carbonate minerals are probably less than 1.0 at temperatures above 200°C, and too insensitive to temperature to be useful in geothermometry even if the fractionation factors are experimentally determined in the future. For example, from the empirical equation for

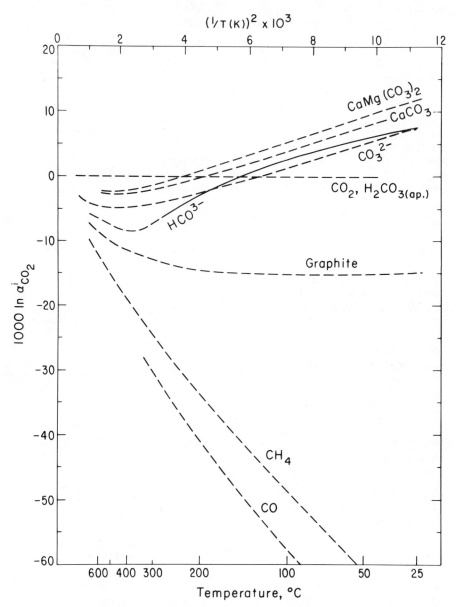

Fig. 10.15 Equilibrium isotopic fractionation factors among carbon compounds relative to $H_2CO_3(ap)$, where $H_2CO_3(ap) = H_2CO_3 + CO_2(aq)$.

the dolomite-calcite fractionation factors of Sheppard and Schwarcz (1970), the uncertainty of $\pm 0.2‰$ in the measurement of $\Delta_{Dol\text{-}Cal}$ values gives more than $\pm 80°C$ uncertainty in the calculated temperatures at approximately $300°C$.

The steep slope versus temperature of the equilibrium carbon isotopic fractionation factors for $CH_4\text{-}CO_2$ has been used to estimate temperatures of volcanic gases, hot spring systems, and natural gas fields. Calcite-graphite pairs may also be used as a sensitive geothermometer. However, application of this thermometer is limited to special cases, such as to metamorphic deposits, because most hydrothermal deposits do not contain primary graphite.

A potential geothermometer for hydrothermal ore deposits is calcite-CO_2. Comparing the $\delta^{13}C$ values of CO_2 in primary inclusion fluids in carbon-free host minerals (e.g., quartz, sulfides) with those of contemporaneously precipitated calcite, the temperature of mineralization should be determinable to within $\pm 20°C$ even at temperatures as high as $300°C$.

CARBON ISOTOPIC VARIATION IN HYDROTHERMAL SYSTEMS

Carbon in hydrothermal ore deposits is usually fixed in carbonate minerals. For the formation of carbonate minerals, fluids must contain oxidized carbon species (CO_2, H_2CO_3, HCO_3^-, and CO_3^{2-}). Oxidized carbon species in hydrothermal fluids may originate from (1) a magmatic source, from (2) oxidation of reduced carbon (organic compounds in sedimentary rocks, graphite in metamorphic and igneous rocks), and from (3) leaching of sedimentary carbonates. Each of these sources may contribute carbon of different $\delta^{13}C$ values to hydrothermal fluids. $\delta^{13}C$ values of CO_2 may also vary in hydrothermal systems due to changes in T, f_{O_2}, P, and pH of the fluids in a manner analogous to the $\delta^{34}S$ variations of H_2S. This is because there are large isotopic fractionation factors between oxidized carbon species and reduced carbon species (CH_4), and because the proportion of these species are affected by changes in the physicochemical parameters of the hydrothermal solutions.

Magmatic Systems

$\delta^{13}C$ values of carbonatites and of diamond in kimberlites are both similar and fall in most cases within a narrow range $-5 \pm 2‰$ (e.g., Deines and Gold, 1973). These values may be taken as the $\delta^{13}C$ of carbon in magmas formed by partial melting of average mantle rocks.

$\delta^{13}C$ values of carbon in magmas formed by partial melting of crustal rocks may be estimated from the average $\delta^{13}C$ value of crustal rocks. From the data on carbon content of igneous rocks (Hoefs, 1973) and of sedimentary and metamorphic rocks (Ronov and Yaroshevsky, 1969) and from the estimate of relative proportions of these rocks in the earth's crust (Ronov and Yaroshevsky, 1969), we can calculate that approximately 93% of the total carbon in the crustal system is fixed in sedimentary and metamorphic rocks, and approximately 7% in igneous rocks. Carbon in the atmosphere + hydrosphere + biosphere is less than 0.01% of the total carbon in the crustal system. The carbon reservoir in sedimentary and metasedimentary rocks, is approximately 22% reduced carbon (organic carbon and graphite) and about 78% oxidized carbon (carbonates). The average $\delta^{13}C$ values for these two types of carbon in sediments are approximately $-25\%_{oo}$ and $0\%_{oo}$, respectively (Schwarcz, 1969). Therefore, the average $\delta^{13}C$ value for sedimentary and metamorphic rocks can be calculated to be approximately $-5.5\%_{oo}$.

Carbon in granitic, mafic, and ultramafic rocks shows a much wider range in $\delta^{13}C$ values than the range in carbonatites. $\delta^{13}C$ values in these rocks are commonly between $+2$ and $-10\%_{oo}$ for carbonate and between -15 and $-30\%_{oo}$ for reduced carbon (for a summary of the data, see Feux and Baker, 1973). Because the proportion of carbonate carbon and reduced carbon varies greatly between samples, an estimate of the average $\delta^{13}C$ value for igneous rocks is difficult. However, even if extreme values are used in the computation, the average $\delta^{13}C$ value for the crustal rocks does not differ more than $\pm 1.5\%_{oo}$ from that of the average sedimentary and metamorphic rocks (i.e., approximately $-5.5\%_{oo}$). This suggests that melts formed from partial melting of average mantle rocks, of average sedimentary rocks, and of average crustal rocks, all would have $\delta^{13}C$ values around $-5\%_{oo}$.

In the presence of graphite, partial melting of pelitic rocks takes place at an oxidation state slightly below that of the quartz-fayalite-magnetite (QFM) buffer (Ohmoto and Kerrick, 1977). In such systems, carbon in the rocks could dissolve in melts as CO_2 and/or CH_4 depending on the original mineral assemblages and proportions of hydrous, carbonate, and oxide minerals in the rocks and also on the open or closed nature of the systems with respect to volatile components (Ohmoto and Kerrick, 1977). Consequently, the $\delta^{13}C$ values of melts formed by partial melting of graphitic rocks can become quite different from that of melts formed by partial melting of *average* sedimentary rocks.

Igneous rocks that are associated with base metal mineralization are, however, usually graphite-free, and the oxidation state is slightly above that of the QFM buffer (see Figure 10.7). For such an oxidation state, carbon in melts occurs predominantly as molecular CO_2 with minor HCO_3^- (Burnham, Chapter 3). The dominant carbon species in magmatic fluids under such an oxidation state is CO_2 (see Figure 10.16). Although there is no experimental data on the isotopic fractionation factor between CO_2 in silicate

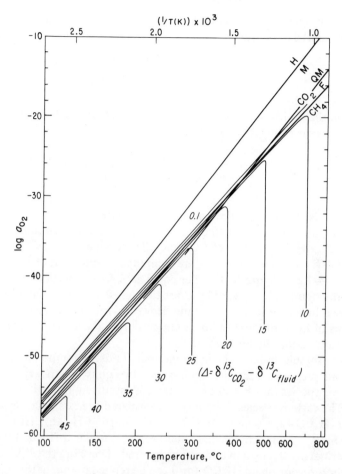

Fig. 10.16 Deviation of $\delta^{13}C_{CO_2}$ from $\delta^{13}C_{fluid}$ in hydrothermal fluids. At $T \geq 400°C$ $P_{H_2O} = 1$ kb, and at $T < 400°C$. P_{H_2O} values are those at liquid/vapor two phase boundary of salt solutions.

melts and $CO_2(g)$, we may assume that the isotopic fractionation factor between them is negligible. Then $\delta^{13}C_{magma} \cong \delta^{13}C_{magmatic\ fluid} \cong \delta^{13}C_{CO_2,(g)}$ for such magmatic systems.

During cooling of magmatic fluids, some CO_2 may be converted to CH_4. Other carbon species such as CO and COS are always much less abundant than CO_2 or CH_4. The proportion of CO_2 to CH_4 in fluids can be related to f_{O_2}, P_{H_2O}, and T, because of the following equilibrium

$$CH_4(g) + 2\ O_2(g) \rightleftharpoons CO_2(g) + 2\ H_2O(g) \qquad (10.31)$$

Then, the carbon isotopic balance in the fluid, by analogy to equations (10.7) and (10.9), can be written as

$$\delta^{13}C_{CO_2} = \delta^{13}C_{fluid} - \Delta_{CH_4} \cdot \left(\frac{1}{1 + R''}\right) \qquad (10.32)$$

Assuming the fluids are ideal mixtures of gases at high T and P, the CO_2/CH_4 molar ratio (R'') can be expressed as:

$$R'' = \frac{m_{CO_2}}{m_{CH_4}} = \frac{K \cdot (f_{O_2})^2}{(P_f \cdot \nu_{H_2O} \cdot X_{H_2O})^2} \cdot \left(\frac{\nu_{CH_4}}{\nu_{CO_2}}\right) \qquad (10.33)$$

in which K is the equilibrium constant for equation (10.31). Using the Δ values computed from Table 10-3, equilibrium constants in Holland (1965), and the fugacity coefficients of CO_2 and CH_4 summarized in Ohmoto and Kerrick (1977), changes in $\delta^{13}C_{CO_2}$ are computed as functions of T and f_{O_2} at $P_f \simeq P_{H_2O} = 1$ kb (Figure 10.16). The equal m_{CO_2}/m_{CH_4} line and $\delta^{13}C_{CO_2}$ contours in the figure shift upward by approximately 0.5 log f_{O_2} units when P_{H_2O} is increased by one order of magnitude. However, within the typical pressure conditions of hydrothermal fluids at temperatures above 400°C, their positions remain essentially unchanged.

Carbon isotopic balance in hydrothermal fluids at temperatures below 350°C and pressures near the liquid-vapor two-phase boundary of water, where ionic species may become important, is discussed in detail in Ohmoto (1972). The carbon species that must be considered in mass balance calculations are $H_2CO_3(ap)$ (equals $H_2CO_3 + CO_2(aq)$), $CH_4(aq)$, HCO_3^- and CO_3^{2-}; $m_{CO_3^{2-}}$ is much smaller than $m_{HCO_3^-}$ under pH conditions of hydrothermal ore deposition and can be ignored. Therefore the $\delta^{13}C$ value of CO_2 in the aqueous phase [$H_2CO_3(ap)$] can be related to that of the fluid and functions of T, f_{O_2}, and pH by the following equations;

$$\delta^{13}C_{H_2CO_3(ap)} = \delta^{13}C_{fluid} - \Delta_{HCO_3^-} \cdot \left(\frac{A}{1 + A + B}\right) - \Delta_{CH_4} \cdot \left(\frac{B}{1 + A + B}\right) \qquad (10.34)$$

in which

$$A = \frac{K_{H_2CO_3} \cdot \gamma_{H_2CO_3}}{a_{H^+} \cdot \gamma_{HCO_3^-}} \qquad (10.35)$$

and

$$B = \frac{a_{H_2O}}{K_{CH_4} \cdot (f_{O_2})^2} \tag{10.36}$$

and K_{CH_4} and $K_{H_2CO_3}$ are, respectively, the equilibrium constants for the following reactions:

$$CH_4(aq) + 2 O_2(g) \rightleftharpoons H_2CO_3(ap) + H_2O(aq) \tag{10.37}$$

and

$$H_2CO_3(ap) \rightleftharpoons H^+ + HCO_3^- \tag{10.38}$$

Using the $\Delta_{HCO_3^-}$ and Δ_{CH_4} values calculated from the equations in Table 10-3, and γ_i and $K_{H_2CO_3}$ and K_{CH_4} summarized in Ohmoto (1972), $\delta^{13}C$ values for $H_2CO_3(ap)$ are computed at pH conditions where $m_{H_2CO_3(ap)} > m_{HCO_3^-}$, and are shown in Figure 10.16.

The application of Figure 10.16 at temperatures below ~350°C depends on the assumption of isotopic equilibrium between CO_2 (or H_2CO_3) and CH_4. If the rate of isotopic exchange reactions between them is as slow as that in $SO_4^{2-} - H_2S$ systems, carbon isotopic equilibrium may not take place in hydrothermal systems at temperatures below approximately 200°C. Unfortunately, no experimental study has been carried out on the rate of carbon isotopic exchange reactions between CO_2 and CH_4. However, the fact that the $\Delta_{(CH_4-CO_2)}$ values give reasonable temperatures for both low temperature natural gas and hot spring reservoirs (see Bottinga, 1968) suggest that the rates are such that Figure 10.16 is applicable to most natural hydrothermal systems.

Figure 10.16 indicates that with decreasing f_{O_2} and decreasing CO_2/CH_4 ratio in the fluids, $\delta^{13}C$ values of oxidized carbon species in fluids increase, and that magmatic fluids with $\delta^{13}C_{fluid}$ value of $-5‰$ can produce $CaCO_3$ with $\delta^{13}C$ as high as $+20‰$. The figure also indicates that small variations in the CO_2/CH_4 ratio of fluids near the CO_2/CH_4 boundary can result in large variations in the $\delta^{13}C$ values of the precipitated carbonates. The best example of such an effect is seen among the carbonate minerals formed at approximately 200°C at the Renison Bell tin deposits in Tasmania in which the $\delta^{13}C$ values of carbonates vary between $+5$ and $-14‰$ and the CO_2/CH_4 ratios of the fluids as indicated by gas analyses of fluid inclusions vary between 1 and 0.5 (Patterson and Ohmoto, 1976).

When the pH of fluids above 200°C is near neutral or slightly acidic (i.e., conditions of typical hydrothermal fluids) the f_{O_2} values at $X_{oxid.\,C} = X_{red.\,C}$ are lower than those at $X_{oxid.\,S} = X_{red.\,S}$. Therefore, when sulfides and carbonate minerals are formed at temperatures above 200°C, it is unlikely that both sulfide and carbonate minerals will show large variation in isotopic compositions. Therefore, by examination of the variability in both $\delta^{34}S$ and $\delta^{13}C$ values of minerals in an ore deposit, one can quickly estimate f_{O_2} conditions of hydrothermal fluids. See Ohmoto (1972) and Rye and Ohmoto (1974) for detailed discussion.

Nonmagmatic Systems

CARBON FROM ORGANIC COMPOUNDS

The $\delta^{13}C$ value of atmospheric CO_2 is approximately $-7‰$. The kinetic isotopic fractionation accompanying photosynthesis is about 17-27‰, which is nearly the same order of magnitude as the kinetic isotopic effects associated with rapid reduction of sulfate. Therefore plants typically have $\delta^{13}C$ values between -24 and $-34‰$ (see Hoefs, 1973 for a summary of the data).

In marine water (pH = 8.1), atmospheric CO_2 dissolves mostly as HCO_3^-, which is approximately 7-10‰ enriched in ^{13}C with respect to the CO_2 and has a $\delta^{13}C$ of about 0‰. Because of the affect of dissolved atmospheric CO_2, marine organisms tend to be more enriched in ^{13}C than terrestrial plants, but they still possess very negative $\delta^{13}C$ values. Consequently, all organic components in sediments, coal, petroleum, and graphite typically have $\delta^{13}C$ values between -10 and $-35‰$ with a mean near $-25‰$.

Reduced carbon may become the source of carbon in hydrothermal fluids principally through two processes: (1) oxidation ($C + O_2 \rightarrow CO_2$), and (2) hydrolysis reactions (e.g., $2C + 2 H_2O \rightarrow CO_2 + CH_4$). The former is the most important mechanism at surface conditions. Both (1) and (2) are important at high-temperature metamorphic conditions. Inorganic oxidation of reduced carbon at surface conditions takes place irreversably. Although the kinetic isotopic effects during oxidation have not been studied experimentally, the isotopic composition of the ensuing CO_2 (or HCO_3^- depending on pH) is probably similar to that of organic carbon (i.e., $\delta^{13}C_{CO_2} = -10$ to $-35‰$). CO_2 produced by hydrolysis reactions at temperatures between 350 and 600°C would have $\delta^{13}C$ values between 3 and 12‰ heavier than that of graphite, depending on whether graphite is completely consumed and whether isotopic equilibrium is established between the remaining graphite and CO_2 produced. Thus, the combination of redox reactions and hydrolysis reactions together with the variation in $\delta^{13}C$ of organic carbon can result in a wide range of possibilities for the $\delta^{13}C_{CO_2}$ value of hydrothermal fluids, although values less than $-10‰$ are most likely.

CARBON FROM SEDIMENTARY CARBONATES

Most marine carbonates, regardless of the age of formation, have constant $\delta^{13}C$ values at $0 \pm 4‰$. Carbonate species in fresh water and carbonate minerals in fresh water environments however tend to be more negative and variable with $\delta^{13}C$ values between -2 to $-10‰$ (Schwarcz, 1969). This is because both CO_2 (and HCO_3^-) produced by the oxidation of organic carbon and atmospheric CO_2 are important constituents in fresh waters.

Carbonates can become the sources of CO_2 in ore-forming fluids through (1) dissolution reactions (e.g., $CaCO_3 + 2H^+ \rightarrow H_2CO_3 + Ca^{2+}$) and (2) decarbonation reactions (e.g., 3 dolomite + 4 quartz + $H_2O \rightarrow$ talc + 3 calcite + 3 CO_2). The former are the dominant reactions at surface temperature conditions, whereas both reactions are important at high temperatures. Dissolution reactions should produce HCO_3^- at surface conditions and CO_2 at high temperatures that are isotopically similar to the original carbonates $\delta^{13}C_{CO_2} \simeq 0$). Decarbonation reactions however may produce ^{13}C enriched CO_2 (Shieh and Taylor, 1969).

Carbon Isotopic Data of Hydrothermal Ore Deposits

Carbon isotopic data on hydrothermal ore deposits are much fewer than sulfur isotopic data. This is partly due to the fact that carbonate minerals usually form after sulfides in the paragenesis. The study of carbon isotopes in carbonates may or may not reveal the nature of ore-forming fluids responsible for sulfide mineralization.

Carbon isotopic data on carbonate minerals and CO_2 in fluid inclusions from hydrothermal ore deposits that have been subjected to systematic geochemical investigations are summarized in Figure 10.17. In many of the deposits shown in Figure 10.17, a trend of increasing $\delta^{13}C$ is observed for carbonate minerals formed in the later stages of mineralization, as first pointed out in Rye and Ohmoto (1974). Such a trend may be produced by (1) decreasing temperature, (2) decreasing CO_2/CH_4 ratios in the fluids, and/or (3) increasing contribution of CO_2 from other sources.

Within each of the deposits shown in Figure 10.17, temperatures of mineralization decreased toward the later stages of mineralization. However, the decrease in temperature alone cannot account for all the increase in $\delta^{13}C$ of carbonate minerals. For example, at Providencia, the maximum temperature change during the carbonate mineralization was from 300 to 200°C (Rye, 1966). A 100°C drop in temperature can account for only approximately 2 of the 5‰ increase in the $\delta^{13}C$ of the carbonates. A temperature decrease from near 100° to 25°C was possible at Pine Point

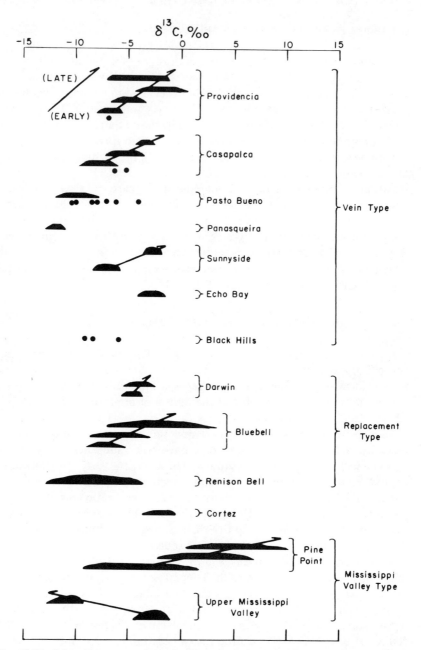

Fig. 10.17 Schematic presentation of carbon isotopic data of some hydrothermal ore deposits. Sunnyside (Casadevall and Ohmoto, 1977), Renison Bell (Patterson and Ohmoto, 1976), Upper Mississippi Valley (McLimans, 1976), Pine Point (Fritz, 1969). For other deposits, see Rye and Ohmoto (1974).

(Roedder, 1968). This drop in temperature can account for only approximately 7‰ of the more than 15‰ increase in the $\delta^{13}C$ of the carbonates.

At Pine Point, the abundance of organic carbon in the environment suggests that the increase in $\delta^{13}C$ of calcite during the later stages of mineralization was due to the increased ratio of CH_4/CO_2 at an oxidation state near the CH_4/CO_2 boundary (Ohmoto, 1972).

Mineralogic and fluid inclusion data at the Providencia, Casapalca, Bluebell, and Sunnyside deposits indicate that the f_{O_2}-pH conditions of hydrothermal fluids in the deposits were such that the dominant carbon species in the fluids was $H_2CO_3(ap)$. The temperatures of carbonate deposition was also higher than 200°C. Therefore, $\delta^{13}C_{CaCO_3} \cong \delta^{13}C_{H_2CO_3(ap)} \cong \delta^{13}C_{fluid}$. Increases in the $\delta^{13}C$ values of late-stage carbonate minerals in these deposits must indicate that the $\delta^{13}C_{fluid}$ increased during the later stage of mineralization. The most likely cause of this increase was the dissolution or decarbonation of limestones, because these deposits occur either in limestone host rocks or in regions where limestones occur.

The source of carbon for the earlier generation of carbonates in these deposits and other deposits that show $\delta^{13}C$ values between -5 and -10‰ is often difficult to define conclusively because, as was discussed earlier, such values can be produced from many sources.

ACKNOWLEDGMENTS

We should like to thank C. W. Burnham for the benefit of stimulating discussions and C. W. Field for permission to use unpublished data.

Critical reading of the manuscript and numerous suggestions by H. L. Barnes, C. W. Burnham, T. Casadevall, T. Drean, M. B. Goldhaber, D. M. Rye, H. Sakai, and A. Sakai are gratefully acknowledged. This research was supported by National Science Foundation Grants GA-31901 and EAR 76-03724.

REFERENCES

Ault, W. V. and M. L. Jensen (1963) Summary of sulfur isotope standards, in *Biogeochemistry of Sulfur Isotopes*, M. L. Jensen, ed., Natl. Sci. Found., Symposium Proc., Yale University.

Bachinski, D. J. (1969) Bond strength and sulfur isotope fractionation in coexisting sulfides: *Econ. Geol.*, **64**, 56-65.

Bahr, J. R. (1976) Sulfur isotopic fractionation between H_2S, S^0 and SO_4^{2-} in aqueous solutions and possible mechanisms controlling isotopic equilibrium in natural systems: Unpublished Masters Paper, Pennsylvania State University

Bailey, S. A. and J. W. Smith (1972) Improved method for the preparation of sulfur dioxide from barium sulfate for isotope ratio studies: *Anal. Chem.* **44**, 1542–1543.

Bottinga, Y. (1968) Calculation of fractionation factors for carbon and oxygen isotopic exchange in the system calcite-carbon dioxide-water: *J. Phys. Chem.*, **72**, 800–808.

—— (1969) Calculated fractionation factors for carbon and hydrogen isotope exchange in the system calcite-carbon dioxide-graphite-methane-hydrogen-water vapor: *Geochim. Cosmochim. Acta*, **33**, 49–64.

Brown, J. S. (1967) Isotopic zoning of lead and sulfur in S. E. Missouri, in *Genesis of Stratiform Lead-Zinc-Barite-Fluorite Deposits*, J. S. Brown, ed., *Econ. Geol. Monograph*, **3**, 410–426.

Burnham, C. W., J. R. Holloway, and N. F. Davis (1969) Thermodynamic properties of water to 1000°C and 10,000 bars: *Geol. Soc. Am. Spec. Paper* **132**.

Carmichael, I. S. E., F. J. Turner, and J. Verhoogen (1974) *Igneous Petrology*, New York: McGraw-Hill.

Casadevall, T. and H. Ohmoto (1977) Sunnyside mine, Eureka mining district, San Juan County, Colorado: Geochemistry of gold and base metal ore formation in the volcanic environment: *Econ. Geol.*, **72**, 1285–1320.

Craig, H. (1953) The geochemistry of the stable carbon isotopes: *Geochim. Cosmochim. Acta*, **3**, 53–92.

—— (1957) Isotopic standards for carbon and oxygen and correction factors for mass-spectrometric analysis of carbon dioxide: *Geochim. Cosmochim. Acta*, **12**, 133–149.

Czamanske, G. K., and R. O. Rye (1974) Experimentally determined sulfur isotope fractionation between sphalerite and galena in the temperature range 600°C to 275°C: *Econ. Geol.*, **69**, 17–25.

Deines, P. and D. P. Gold (1973) The isotopic composition of carbonatite and kimberlite carbonates and their bearing on the isotopic composition of deep-seated carbon: *Geochim. Cosmochim. Acta*, **37**, 1709–1733.

Deuser, W. G. and E. T. Degens (1967) Carbon isotope fractionation in the system CO_2(gas)-CO_2(aqueous)-HCO_3(aqueous): *Nature*, **215**, 1033–1035.

Doe, B. R. and M. H. Delevaux (1972) Source of lead in southeast Missouri galena ores: *Econ. Geol.*, **67**, 409–425.

Drean, T. (1978) Experimental study on sulfate reduction by methane, xylene, and iron in a temperature range of 175° to 350°C: Unpublished Masters Paper, Pennsylvania State University.

Emrich, K., D. H. Erhalt, and J. C. Vogel (1970) Carbon isotope fractionation during the precipitation of calcium carbonate: *Earth Planet Sci. Lett.*, **8**, 363–371.

Field, C. W. (1966) Sulfur isotopic method for discriminating between sulfates of hypogene and supergene origin: Econ. Geol., **61**, 1428–1435.

——, J. R. Dymond, G. R. Heath, J. B. Corliss, and E. J. Dasch (1976) Sulfur isotope reconnaissance of epigenetic pyrite in ocean-floor basalts, Leg 34 and elsewhere: in *Initial Reports of the Deep Sea Drilling Project, XXXIV*, R. S. Yeats, ed., Washington, D.C.: U. S. Govt. Printing Ofc., p. 381–384.

Fritz, P. (1969) The oxygen and carbon isotopic composition of carbonates from the Pine Point lead-zinc ore deposits: *Econ. Geol.*, **64**, 733–742.

Fuex, A. N. and Baker, D. R. (1973) Stable carbon isotopes in selected granitic, mafic, and ultramafic igneous rocks: *Geochim. Cosmochim. Acta*, **37**, 2509–2521.

Goldhaber, M. B. and I. R. Kaplan (1975) Controls and consequences of sulfate reduction rates in recent marine sediments: *Soil Sci.*, **119**, 42–55.

Granger, H. C. and C. G. Warren (1969) Unstable sulfur compounds and the origin of roll-type uranium deposits: *Econ. Geol.*, **64**, 160-71.

Grinenko, V. A. (1962) Preparation of sulfur dioxide for isotopic analysis: *Zeits. Neorgan. Khimii.*, **7**, 2478-2483.

____ and Grinenko, L. N. (1967) Fractionation of sulfur isotopes in high temperature decomposition of sulfides by water vapor: *Geokhimiya*, **9**, 1049-1055.

____, ____, and Zagryazhskaya (1969) Kinetic isotope effect in high temperature reduction of sulfate: *Geokhimiya*, **4**, 484-491.

____ and Thode, H. G. (1970) Sulfur isotope effects in volcanic gas mixtures: *Can. J. Earth Sci.*, **7**, 1402-1409.

Grootenboer, J. (1969) Equilibrium sulfur isotope fractionation between pyrite, sphalerite and galena in synthetic and natural assemblages: Unpublished Masters Thesis, McMaster University.

Harrison, A. G. and H. G. Thode (1957) The kinetic isotope effect in the chemical reduction of sulphate: *Trans. Faraday Soc.*, **53**, 1648-1651.

____ and ____ (1958) Mechanism of the bacterial reduction of sulphate from isotope fractionation studies: *Trans. Faraday Soc.*, **54**, 84-92.

Helgeson, H. C. (1969) Thermodynamics of hydrothermal systems at elevated temperatures and pressures: *Am. J. Sci.*, **267**, 729-804.

Hoefs, J. (1973) *Stable Isotope Geochemistry*, New York: Springer-Verlag.

Holland, H. D. (1965) Some applications of thermochemical data to problems of ore deposits. II. Mineral assemblages and the composition of ore-forming fluids: *Econ. Geol.*, **60**, 1101-1166.

Holser, W. T. and I. R. Kaplan (1966) Isotope geochemistry of sedimentary sulfates: *Chem. Geol.*, **1**, 93-135.

Holt, B. D. and A. G. Engelkemeir (1970) Thermal decomposition of barium sulfate to sulfur dioxide for mass spectrometer analysis: *Anal. Chem.*, **42**, 1451-1453.

Igumnov, S. A. (1976) Experimental study of isotope exchange between sulfide and sulfur in hydrothermal solution: *Geokhimiya*, **4**, 497-503.

Jensen, M. L. and G. Dessau (1966) Ore deposits of southwestern Sardinia and their sulfur isotopes: *Econ. Geol.*, **61**, 917-932.

Kajiwara, Y., H. R. Krouse, and A. Sasaki (1969) Experimental study of sulfur isotopic fractionation between coexistent sulfide minerals: *Earth Planet. Sci. Lett.*, **7**, 271-277.

____ and ____ (1971) Sulfur isotope partitioning in metallic sulfide systems: *Can. J. Earth Sci.*, **8**, 1397-1408.

Kamilli, R. J. and H. Ohmoto (1977) Paragenesis, zoning, fluid inclusion and isotopic studies of the Finlandia vein, Colqui district, Central Peru: *Econ. Geol.*, **72**, 950-982.

Kanehira, K., S. Yui, H. Sakai, and A. Sasaki (1973) Sulfide globules and sulfur isotope ratios in the abyssal tholeiite from the mid-Atlantic ridge near 30°N latitude: *Geochem. J.*, **7**, 89-96.

Kaplan, I. R. and S. C. Rittenberg (1964) Microbiological fractionation of sulfur isotopes: *J. Gen. Microbiol.*, **34**, 195-212.

____, Sweeney, R. E., and Nissenbaum, A. (1969) Sulfur isotope studies on Red Sea geothermal brines and sediments: in *Hot Brines and Recent Heavy Metal Deposits in the Red Sea*, E. T. Degens and D. A. Ross, eds., New York: Springer Verlag, p. 474-498.

Kemp, A. L. W., and H. G. Thode (1968) The mechanism of bacterial reduction of sulphate and of sulphite from isotopic fractionation studies: *Geochim. Cosmochim. Acta*, **32**, 71-91.

Kiyosu, Y. (1973) Sulfur isotopic fractionation among sphalerite, galena and sulfide ions: *Geochem. J.*, **7**, 191-199.

Landis, G. P. and R. O. Rye (1974) Geologic, fluid inclusion, and stable isotope studies of the Pasto Bueno tungsten-base metal ore deposit, Northern Peru: *Econ. Geol.*, **69**, 1025-1059.

Lusk, J. and J. H. Crocket (1969) Sulfur isotope fractionation in coexisting sulfides from the Health Steel B-1 ore body, New Brunswick, Canada: *Econ. Geol.*, **64**, 147-155.

Macnamara, J. and H. G. Thode (1950) Comparison of the isotopic constitution of terrestrial and meteoric sulfur: *Phys. Rev.*, **78**, 307-308.

Mainwaring, P. R. and A. J. Naldrett (1974) Genesis of Cu-Ni sulfides in the Duluth complex: *Geol. Soc. Am. Ann. Mtg., Abst. Prog.*, **6**, 854-855.

Malinin, S. D., O. I. Kropotava, and V. A. Grinenko (1967) Experimental determination of equilibrium constants for carbon isotope exchange in the system $CO_2(g)$-HCO_3(sol) under hydrothermal conditions: *Geochem. Int.*, **4**, 764-771.

_____ and N. I. Khitarov (1969) Reduction of sulfate sulfur by hydrothermal conditions: *Geochem. Int.*, **6**, 1020-1025.

McCrea, J. M. (1950) The isotopic chemistry of carbonates and a paleotemperature scale: *J. Chem. Phys.*, **18**, 849.

McLimans, R. K. (1977) Geologic, fluid inclusion, and stable isotope studies of the Upper Mississippi Valley zinc-lead district, southwest Wisconsin: Unpublished Ph.D. thesis, The Pennsylvania State University.

Mook, W. G., J. C. Boomerson, and W. H. Staverman (1974) Carbon isotope fractionation between dissolved bicarbonate and gaseous carbon dioxide: *Earth and Planet. Sci. Lett.*, **22**, 169-176.

Mottl, M. (1976) Chemical exchange between sea water and basalt during hydrothermal alteration of the oceanic crust: Unpublished Ph.D. thesis, Harvard University.

Nakai, N. (1970) Isotopic ratios of sulfide compounds on sulfide ore deposits: *Chikuykagaku*, **1**, 31-34 (in Japanese).

_____ and M. L. Jensen (1964) The kinetic isotope effect in the bacterial reduction and oxidation of sulfur: *Geochim. Cosmochim. Acta*, **28**, 1893-1912.

Naumov, G. B., B. N. Ryzhenko, and I. L. Khodakovsky (1974) *Handbook of Thermodynamic Data*, Washington, D.C. U.S. Department of Commerce, PB-226 722.

Nriagu, J. (1974) Fractionation of sulfur isotopes by sediment adsorption of sulfate: *Earth Planet. Sci. Lett.*, **22**, 366-370.

Oana, S. and H. Ishikawa (1966) Sulfur isotopic fractionation between sulfur and sulfuric acid in the hydrothermal solution of sulfur dioxide: *Geochem. J.*, **1**, 45-50.

Ohmoto, H. (1972) Systematics of sulfur and carbon isotopes in hydrothermal ore deposits: *Econ. Geol.*, **67**, 551-579.

_____, D. R. Cole, and M. J. Mottl (1976) Experimental basalt-seawater interaction: sulfur and oxygen isotope studies: *EOS*, **57**, 342.

_____ and D. M. Kerrick (1977) Devolatilization equilibria in graphitic systems: *Am. J. Sci.*, 1013-1044.

_____ and R. O. Rye (1974) Hydrogen and oxygen isotopic compositions of fluid inclusions in the Kuroko deposits, Japan: *Econ. Geol.*, **69**, 947-953.

Orr, W. L. (1974) Changes in sulfur content and isotopic ratios of sulfur during petroleum maturation—study of Big Horn basin paleozoic oils: *Bull. Am. Assoc. Petroleum Geologists*, **58**, 2295-2318.

_____ (1975) Geologic and geochemical controls on the distribution of hydrogen sulfide in natural gas: *Geol. Soc. Am. Ann. Mtg. Abst. Prog.*, 1220-1221.

Patterson, D. and H. Ohmoto (1976) Stable isotope and fluid inclusion studies at the Renison Bell cassiterite-sulfide deposits, Western Tasmania: *International Conference on Stable Isotopes*, New Zealand Abst., 52.

Puchelt, H., B. R. Sabels, and T. C. Hoering (1971) Preparation of sulfur hexafluoride for isotope geochemical analysis: *Geochim Cosmochim. Acta.*, **35**, 625.

Rafter, T. A. (1957) Sulphur isotopic variation in nature, part 2, A quantitative study of the reduction of barium sulphate by graphite for recovery of sulphate-sulphur for sulphur isotopic measurements: *N. Z. J. Sci. Technol.* Sect. B., **38**, 955-968.

_____ and Y. Mizutani (1967) Oxygen isotopic composition of sulphates, part 2, Preliminary results on oxygen isotopic variation in sulphates and relationship to their environment and to their $\delta^{34}S$ values: *N. Z. J. Sci.*, **10**, 816-840.

Ripley, E. M. and H. Ohmoto (1977) Mineralogic, sulfur isotope, and fluid inclusion studies of the stratabound copper deposits at the Raul Mine, Peru: *Econ. Geol.*, **72**, 1017-1041.

Robinson, B. W. (1973) Sulfur isotope equilibrium during sulfur hydrolysis at high temperatures: *Earth Planet. Sci. Lett.*, **18**, 443-450.

_____ and H. Ohmoto (1973) Mineralogy, fluid inclusions, and stable isotopes of the Echo Bay U-Ni-Ag-Cu deposits, Northwest Territories, Canada: *Econ. Geol.*, **68**, 635-656.

Roedder, E. (1968) Temperature, salinity and origin of the ore-forming fluids at Pine Point, Northwest Territories, Canada, from fluid inclusion studies: *Econ. Geol.*, **63**, 439-450.

Ronov, A. B. and Yaroshevsky, A. A. (1969) Chemical composition of the Earth's crust: in *The Earth's Crust and Upper Mantle*, P. J. Hart, ed., Washington, D.C.: Am. Geophys. Union, p. 37-57.

Rye, R. O. (1966) The carbon, hydrogen, and oxygen isotopic composition of the hydrothermal fluids responsible for the lead-zinc deposits at Providencia, Zacatecas, Mexico: *Econ. Geol.*, **61**, 1339-1427.

_____ (1974) A comparison of sphalerite-galena sulfur isotope temperatures with filling temperatures of fluid inclusions: *Econ. Geol.*, **69**, 26-32.

_____, W. E. Hall, and H. Ohmoto (1974) Carbon, hydrogen, oxygen, and sulfur isotope study of the Darwin lead-silver-zinc deposit, southern California: *Econ. Geol.*, **69**, 826-842.

_____ and H. Ohmoto (1974) Sulfur and carbon isotopes and ore genesis: A review: *Econ. Geol.*, **69**, 826-842.

Ryzhenko, B. N. and V. P. Volkov (1971) Fugacity coefficients of some gases in a broad range of temperatures and pressures: *Geochem. Int.*, 468-481.

Sakai, H. (1957) Fractionation of sulfur isotopes in nature: *Geochim. Cosmochim. Acta*, **12**, 150-169.

Sakai, H. (1968) Isotopic properties of sulfur compounds in hydrothermal processes: *Geochem. J.*, **2**, 29-49.

Salomons, W. (1971) Isotope fractionation between galena and pyrite and between pyrite and elemental sulfur: *Earth Planet. Sci. Lett.*, **11**, 236-238.

Sangster, D. F. (1968) Relative sulfur isotope abundance of ancient seas and stratabound sulphide deposits: *Proc. Geol. Assoc. Can.*, **19**, 79-86.

Sasaki, A. (1969) Sulfur isotope study of the Muscox intrusion, District of Mackenzie: *Geol. Survey Can.*, *Paper 68.*

―― (1970) Re-examination of sulfur isotopic composition of the earth's upper mantle (abst.): in *International Symposium of Hydrogeochemistry and Biogeochemistry*, Tokyo, Washington, D.C.: Clarke, p. 129.

―― (1970) Seawater sulfate as a possible determinant for sulfur isotopic compositions in some strata-bound sulfide ores: *Geochem. J.*, **4**, 41-51.

―― (1974) Isotopic data of Kuroko deposits: in *Geology of Kuroko Deposits*, S. Ishihara et al., ed., Tokyo: Soc. Mining Geologists of Japan, p. 389-397.

――, Y. Arikawa, and R. E. Folinsbee (1976) Application of tin (II)-strong phosphoric acid ("Kiba reagent") to the study of sulfur isotopes in rocks: *International Conference on Stable Isotopes*, New Zealand. Abst., 58.

Sasaki, A. and Kajiwara, Y. (1971) Evidence of isotopic exchange between seawater sulfate and some syngenetic sulfide ores: *Soc. Mining Geol. Japan*, *Spec. Issue* **3**, 289-294.

Sasaki, A. and H. R. Krouse (1969) Sulfur isotopes and the Pine Point lead-zinc mineralization: *Econ. Geol.*, **64**, 718-730.

Schiller, W. R., K. v. Gehlen, and H. Nielsen (1970) Hydrothermal exchange and fractionation of sulfur isotopes in synthesized ZnS and PpS: *Econ. Geol.*, **65**, 350-351.

Schneider, A. (1970) The sulfur isotope composition of basaltic rocks: *Contr. Mineral. Petrol.*, **25**, 95-124.

Schwarcz, H. P. (1969) The isotopes in nature, *Handbook of Geochemistry*, Vol. II/1, K. H. Wedelpohl, ed., Berlin: Springer-Verlag.

Schwarcz, H. P. and S. W. Burnie (1973) Influence of sedimentary environments on sulfur isotope ratios in clastic rocks: A review: *Mineral. Depos.*, **8**, 264-277.

Shanks, W. C. and J. L. Bishoff (1975) Sulfur isotopes and sulfide deposits in the Red Sea geothermal system: *Geol. Soc. Am. Ann. Mtg. Abst. Programs*, 1266.

Sharma, T. and R. N. Clayton (1965) Measurement of $^{18}O/^{16}O$ ratios of total oxygen of carbonates: *Geochim. Cosmochim. Acta*, **29**, 1347-1354.

Sheppard, S. M. F. and H. P. Schwarcz (1970) Fractionation of carbon and oxygen isotopes and magnesium between coexisting metamorphic calcite and dolomite: *Contr. Mineral. Petrol.*, **26**, 161-198.

Shieh, Y. N. and H. P Taylor, Jr. (1969) Carbon and hydrogen isotope studies of contact metamorphism in the Santa Rosa Range, Nevada and other areas: *Contrib. Mineral. Petrol.*, **20**, 306-356.

Shima, M., W. H. Gross, and H. G. Thode (1963) Sulfur isotope abundances in basic sills, differentiated granites and meteorites: *J. Geophys. Res.*, **68**, 2835-2846.

Smitheringale, W. G. and M. L. Jensen (1963) Sulfur isotopic composition of the Triassic igneous rocks of eastern United States: *Geochim. Cosmochim. Acta*, **27**, 1183-1207.

Solomon, M., T. A. Rafter, and M. L. Jensen (1969) Isotope studies on the Rosebery, Mount Farrell and Mount Lyell ores, Tasmania: *Mineral Depos.*, **4**, 172-199.

Thode, H. G., J. Monster and H. B. Dunford (1961) Sulfur isotope geochemistry: *Geochim. Cosmochim. Acta*, **25**, 159-174.

――, M. Shima, C. E. Rees, and K. V. Krishnamurty (1965) Carbon-13 isotope effects in systems containing carbon dioxide, bicarbonate, carbonate, and metal ions: *Can. J. Chem.*, **43**, 582-595.

――, C. B. Cragg, J. R. Hulston, C. E. Rees (1971) Sulfur isotope exchange between sulfur dioxide and hydrogen sulfide: *Geochim. Cosmochim. Acta*, **35**, 35-45.

REFERENCES

―― and J. Monster (1965) Sulfur isotope geochemistry of petroleum, evaporites, and ancient seas: *Am. Assoc. Petroleum Geol., Mem.* **4,** 367-317.

Toland, W. G. (1960) Oxidation of organic compounds with aqueous sulfate: *J. Am. Chem. Soc.*, **82,** 1911-1916.

――, D. L. Hagmann, J. B. Wilkes, and F. J. Brutschy (1958) Oxidation of organic compounds with aqueous base and sulfur: *J. Am. Chem. Soc.*, **80,** 5423-5427.

Tudge, A. P. and H. G. Thode (1950) Thermodynamic properties of isotopic compounds of sulfur: *Can. J. Res.*, **28B,** 567-578.

Urey, H. C. (1947) The thermodynamic properties of isotopic substances: *J. Chem. Soc.*, 562-581.

Vogel, J. C. (1961) Isotopic separation factors of carbon in the equilibrium system CO_2-HCO_3^--CO_3^{2-}, in *Summer Course on Nuclear Geology*, Pisa: Laboratorio di Geoligia Nucleare, p. 216-221.

Warren, C. G. (1972) Sulfur isotopes as a clue to the genetic chemistry of a roll-type uranium deposit: *Econ. Geol.*, **67,** 759-767.

Wendt, I. (1968) Fractionation of carbon isotopes and its temperature dependence in the system CO_2-gas-CO_2 in solution and HCO_3^--CO_2 in solution: *Earth Planet. Sci. Lett.*, **4,** 64-68.

Williams, N. and D. M. Rye (1974) Alternative interpretation of sulphur isotope ratios in the McArthur lead-zinc-silver deposit: *Nature*, **247,** 535-537.

11

Mass Transfer Among Minerals and Hydrothermal Solutions

HAROLD C. HELGESON
University of California, Berkeley

Ample field and laboratory evidence indicates that incongruent reactions among silicates and aqueous electrolyte solutions are responsible for deposition of sulfides in many hydrothermal ore deposits. The extent to which these reactions take place depends on the composition of the solution, the mineralogy of the system, and the rate at which the aqueous phase moves through the depositional environment. Fluid inclusion analyses indicate that hydrothermal solutions commonly contain relatively high concentrations of NaCl, KCl, and $CaCl_2$, with lesser concentrations of $MgCl_2$, $AlCl_3$, H_2S, H_2SO_4, CO_2, SiO_2, and the ore-forming metals. As a result of extensive experimental and theoretical investigation of departures from ideality, speciation, and the stoichiometric solubilities of minerals in such solutions, the chemical and thermodynamic properties of hydrothermal systems are reasonably well understood. However, because the components of hydrothermal solutions are rarely present in proportions corresponding to those in minerals, and because many silicates react irreversibly and incongruently with aqueous solutions, stoichiometric solubility studies yield little information about the *genesis* of ore deposits. Similarly, owing to technologic limitations and the multivariant nature of hydrothermal systems, experimental attempts to simulate ore-forming processes generally fail to define the chemical controls responsible for simultaneous alteration of host rocks and deposition of sulfides.

It has long been recognized that the association of sulfides with altered wallrocks is a manifestation of the chemical link between the alteration process and the precipitation of ore minerals in hydrothermal veins. Nevertheless, the chemical reactions responsible for the occurrence of one or another sulfide in a given depositional environment are still poorly understood. Similar ambiguities attend determination of the effect of diffusional transport of material from reaction fronts in alteration zones on the chemical consequences of bulk transport in adjoining fractures. Consequently, genetic implications of zonal patterns and the directional distribution of alteration minerals and associated sulfides remain obscure.

Recognition of subtle indicators of favorable environments for ore deposition depends to a large extent on the ability of the geologist to distinguish cause and effect in the geologic record. Mass transfer calculations indicate that precipitation of one or another sulfide may be caused indirectly by seemingly unrelated processes in different parts of the system. For example, the incongruent reaction of a calcium-poor, sulfate-rich solution with plagioclase to produce anhydrite may deplete the solution in SO_4^{2-}. As a consequence, the fugacity of oxygen is decreased significantly, which favors sulfide deposition. Similarly, precipitation of hydrothermal biotite, siderite, or chlorite in response to incongruent reaction of an aqueous solution with a wallrock may cause the Cu/Fe ratio in solution to increase sufficiently to precipitate chalcopyrite or bornite. Depending on the presence or absence of other iron-bearing minerals in the system, hydrothermal leaching and more than one generation of a given sulfide may also result from these reactions (Helgeson, 1970). All of the various ramifications and consequences of these and many other much more complicated sets of interdependent reaction couples have yet to be unraveled. Theoretical calculations of equilibrium and mass transfer among minerals and aqueous solutions with the aid of thermodynamics, fluid mechanics, chemical kinetics, transport theory, and computer technology afford a means to this end.

Accelerated interest in hydrothermal transport processes over the past decade has led to a number of different theoretical models of chemical transfer in hydrothermal systems. Most of these incorporate specific mechanisms of transport, such as the diffusional transfer models adopted by Korzhinskii (1970), Fisher (1970, 1973, 1975), Elliot (1973), Fisher and Elliot (1974), Wood (1974), Fletcher and Hofmann (1974), Weare et al., (1976), Frantz and Mao (1976), Joesten (1977), Brady (1977), and others.

Recent developments in the application of chromatographic transport theory to infiltration metasomatism (Korzhinskii, 1959, 1970, 1973; Hofmann, 1972, 1973; Fletcher and Hofmann, 1974; Frantz and Weisbrod, 1973) parallel those in diffusion theory. Other theoretical studies incorporate composite models of fluid flow and heat transfer (Norton, 1972, 1977, 1978, 1979; Norton and Knapp, 1977; Norton and Knight, 1977; Villas and Norton, 1977; Norton and Taylor, 1979; Norton and Cathles, Chapter 12; Cathles, 1976; Brownell et al., 1975; Cheng and Lau, 1974; Donaldson, 1970; Faust and Mercer, 1975; Lasseter and Witherspoon, 1975; Mercer and Pinder, 1974; Mercer et al., 1975; Wolery and Sleep, 1976; Straus and Schubert, 1977), many of which provide quantitatively for simultaneous diffusional flux and bulk transport as a function of space and time.

An alternate approach to those cited above, and one that is not restricted to a given mechanism of mass transfer nor explicit with respect to space and time, takes advantage of thermodynamic constraints imposed by local and partial equilibrium to calculate the extent to which components are redistributed among minerals and an aqueous solution as the solution reacts irreversibly with its mineralogic environment (Helgeson, 1967, 1968, 1970, 1972, 1974; Helgeson et al., 1969; Helgeson et al., 1970; Lafon and Mackenzie, 1971; Fritz and Tardy, 1976a, b; Droubi et al., 1976a, b; Sarazan et al., 1976; Fouillac et al., 1977; Wolery, 1978). The concepts involved in these calculations are reviewed below to encourage further application of the approach to the study of ore-forming processes.

CHEMICAL REACTIONS AND MASS TRANSFER

Mass transfer among minerals and aqueous solutions in geochemical processes can be represented by an array of chemical reactions representing reversible and irreversible changes in the distribution of components within or among the phases in a system. Regardless of whether the reactions are homogeneous or heterogeneous, each contributes to progress in the process, which leads ultimately to a state of overall equilibrium in the system; that is, a state of minimal Gibbs free energy at a given temperature and pressure. Of the many reactions that might occur in a given process, some are inhibited or favored kinetically, whereas others are precluded or promoted by thermodynamic constraints operating on or within the system. These constraints include constant temperature and/or pressure and fixed chemical potentials of one or more of the components in the system. The chemical potential of a component may be fixed by the presence of a phase with a composition corresponding to the component, by osmotic equilib-

CHEMICAL REACTIONS AND MASS TRANSFER

rium with an external reservoir, or simply by the presence of a large mass of the component in the aqueous solution relative to the mass of the component redistributed among the phases in the process. The paragenesis of reaction products and the relative degree to which components are redistributed among phases depends to a large extent on the relative magnitude of the chemical potentials of the components in the aqueous solution at the beginning of the process. These chemical potentials are determined by the initial solution composition and the distribution of species in the aqueous phase.

Chemical reactions can be viewed as symbolic representations of differential equations describing mass transfer of components or ionic species among phases. For example, the congruent dissolution of microcline in an acid aqueous phase in which no complexing occurs can be described by writing

$$KAlSi_3O_8 \text{ (microcline)} + 4H^+ \rightarrow K^+ + Al^{3+} + 3SiO_2(aq) + 2H_2O \tag{11.1}$$

which expresses the dissolution process in terms of the conservation of mass and charge required by the first law of thermodynamics, that is,

$$\sum_i \nu_{\epsilon,i} \frac{dn_{i,j}}{d\xi_j} = 0 \tag{11.2}$$

where $\nu_{\epsilon,i}$ refers to the number of moles of the ϵth element in one mole of the ith solid or aqueous species, $dn_{i,j}$ stands for the change in the number of moles of the species caused by the jth reaction, and ξ_j represents the progress variable for the reaction (De Donder, 1920; De Donder and Van Rysselberghe, 1936; Prigogine and Defay, 1954; Prigogine, 1955). The progress variable, ξ_j, which is a measure of the extent (or degree of advancement) of the jth reaction, is equal to the number of moles of a given reactant that has been converted into products by the reaction, divided by the stoichiometric coefficient of the reactant in the jth reaction. It follows that the differentials in equation (11.2) correspond to the stoichiometric coefficients in reaction (11.1). We can thus replace j by reaction (11.1) and write,

$$\frac{dn_{KAlSi_3O_8,(11.1)}}{d\xi_{(11.1)}} = -1 \tag{11.3}$$

$$\frac{dn_{H^+,(11.1)}}{d\xi_{(11.1)}} = -4 \tag{11.4}$$

$$\frac{dn_{K^+,(11.1)}}{d\xi_{(11.1)}} = 1 \tag{11.5}$$

$$\frac{dn_{Al^{3+},(11.1)}}{d\xi_{(11.1)}} = 1 \tag{11.6}$$

$$\frac{dn_{SiO_2,(11.1)}}{d\xi_{(11.1)}} = 3 \tag{11.7}$$

and

$$\frac{dn_{H_2O,(11.1)}}{d\xi_{(11.1)}} = 2 \tag{11.8}$$

The stoichiometric coefficients in reaction (11.1) are constants under all circumstances, but they represent the actual change in the number of moles of the subscripted species only as long as the reaction remains congruent and no significant complexing takes place in the aqueous phase. For example, as reaction (11.1) proceeds and H^+ is consumed, the solution pH must eventually increase. Consequently, at some stage of reaction progress, $Al(OH)^{2+}$ may form to a significant degree. Reaction (11.1) alone is then not adequate to describe the process, but rather some combination of reaction (11.1) and

$$Al^{3+} + OH^- \rightarrow Al(OH)^{2+} \tag{11.9}$$

for which

$$\frac{dn_{Al^{3+},(11.9)}}{d\xi_{(11.9)}} = \frac{dn_{OH^-,(11.9)}}{d\xi_{(11.9)}} = -\frac{dn_{Al(OH)^{2+},(11.9)}}{d\xi_{(11.9)}} = -1 \tag{11.10}$$

Under these conditions, the extent to which dissolved aluminum derived from the feldspar by the aqueous phase is distributed among Al^{3+} and $Al(OH)^{2+}$ is no longer *a priori* obvious. At each stage of reaction progress, the distribution of aluminum transferred to the solution is determined by $d\xi_{(11.9)}/d\xi_{(11.1)}$. This can be seen by taking advantage of equations (11.6) and (11.10) to describe the net change in the number of moles of Al^{3+} in the aqueous phase ($dn_{Al^{3+}}$), which can be expressed as

$$dn_{Al^{3+}} = dn_{Al^{3+},(11.1)} + dn_{Al^{3+},(11.9)} = d\xi_{(11.1)} - d\xi_{(11.9)} \tag{11.11}$$

CHEMICAL REACTIONS AND MASS TRANSFER

Hence,

$$\frac{dn_{Al^{3+}}}{d\xi_{(11.1)}} = 1 - \frac{d\xi_{(11.9)}}{d\xi_{(11.1)}} \qquad (11.12)$$

The quotient on the right side of equation (11.12) corresponds to the relative rate of the two reactions. If $d\xi_{(11.9)}/dt$ (where t stands for time) is smaller than $\partial\xi_{(11.1)}/dt$, $dn_{Al^{3+}}/\partial\xi_{(11.1)}$ is positive and the bulk of the aluminum transferred to the aqueous solution is present as Al^{3+}. If $d\xi_{(11.9)}/d\xi_{(11.1)}$ is $\ll 1$, $dn_{Al^{3+}}/d\xi_{(11.1)} \approx 1$. The distribution of aluminum in the aqueous phase is thus a function of the kinetics of the homogeneous reaction of Al^{3+} with OH^- to form $Al(OH)^{2+}$, relative to the kinetics of the heterogeneous reaction of microcline with the aqueous solution. Where the latter is slower (which is almost invariably the case), equilibrium among the aqueous species prevails and we can write,

$$Al^{3+} + OH^- \rightleftharpoons Al(OH)^{2+} \qquad (11.13)$$

which defines the conservation requirement for reversible mass transfer among $Al(OH)^{2+}$, Al^{3+}, and OH^- in solution.

Although activity coefficients of aqueous species change during the hydrolysis of feldspar in H_2O, the changes are slight and affect negligibly the extent of mass transfer in the process. Accordingly, if temperature and pressure, as well as the activity coefficients of the aqueous species are constant, it follows from the law of mass action (see below) that

$$\frac{d \ln n_{Al(OH)^{2+}}}{d\xi_{(11.1)}} = \frac{d \ln n_{Al^{3+}}}{d\xi_{(11.1)}} + \frac{d \ln n_{OH^-}}{d\xi_{(11.1)}} \qquad (11.14)$$

Conservation of mass and the constraint imposed by equation (11.14) specify implicitly the magnitude of $d\xi_{(11.9)}/d\xi_{(11.1)}$ at each stage of reaction progress. Consequently, the entire reaction process can be described in terms of the progress variable for reaction (11.1), which is the rate limiting reaction for the process as long as reaction (11.13) remains reversible.

As reaction (11.1) continues and the pH of the solution increases further, $Al(OH)_4^-$ may begin to form to a significant degree, which can be described by writing

$$Al^{3+} + 4OH^- \rightarrow Al(OH)_4^- \qquad (11.15)$$

If the rate of this reaction is also faster than that of reaction (11.1) and equivalent to that of reaction (11.13), the reversible mass transfer constraint imposed by

$$\frac{d\ln n_{\text{Al(OH)}_4^-}}{d\xi_{(11.1)}} = \frac{d\ln n_{\text{Al}^{3+}}}{d\xi_{(11.1)}} + 4\frac{d\ln n_{\text{OH}^-}}{d\xi_{(11.1)}} \quad (11.16)$$

becomes important (see below). Although the absolute rates of reaction (11.13) and the reversible analog of reaction (11.15) may not be strictly equivalent, most homogeneous reactions in an aqueous solution are so much faster than heterogeneous reactions that their rates can be regarded as infinite and equal in the context of geologic time. Equation (11.16) corresponds to the analog of equation (11.14) for the reversible dissociation of Al(OH)_4^-. Note that equation (11.16) obviates the need to consider explicitly $d\xi_{(11.15)}/d\xi_{(11.1)}$.

In addition to Al^{3+}, Al(OH)^{2+}, H^+, OH^-, and Al(OH)_4^-, other aqueous species may become important during the reaction of feldspar with an aqueous phase. The degree to which these various species form determines to a large extent the ultimate equilibrium composition of the solution. If and when overall equilibrium is achieved we can write

$$\text{KAlSi}_3\text{O}_8 \text{ (microcline)} + 4\text{H}^+ \rightleftharpoons \text{K}^+ + \text{Al}^{3+} + 3\text{SiO}_2(\text{aq}) + 2\text{H}_2\text{O} \quad (11.17)$$

or

$$\text{KAlSi}_3\text{O}_8 \text{ (microcline)} + 3\text{H}^+ \rightleftharpoons \text{K}^+ + \text{Al(OH)}^{2+} + 3\text{SiO}_2(\text{aq}) + \text{H}_2\text{O} \quad (11.18)$$

and

$$\text{KAlSi}_3\text{O}_8 \text{ (microcline)} + 2\text{H}_2\text{O} \rightleftharpoons \text{K}^+ + \text{Al(OH)}_4^- + 3\text{SiO}_2(\text{aq}) \quad (11.19)$$

as well as other reversible reactions. Each of these must hold simultaneously at equilibrium, regardless of the presence of other minerals produced by the process. All mass transfer caused by slight perturbations of the equilibrium state is then reversible and the system is in a state of lowest Gibbs free energy characterized by products and ratios of the activities of aqueous species.

Values of $dn_{\text{Al}^{3+}}/d\xi_{(11.1)}$, $dn_{\text{Al(OH)}^{2+}}/d\xi_{(11.1)}$, $dn_{\text{Al(OH)}_4^-}/d\xi_{(11.1)}$ and the corresponding differentials for other aqueous species formed during the hydrolysis of feldspar can be computed by simultaneous consideration of equations describing reversible mass transfer and conservation of mass in the process. The equations and algorithms used to carry out such calculations are discussed in detail below. However, at this point in the discussion it is of interest to consider the results of such calculations in the context of chemical affinity and the precipitation of incongruent reaction products caused by the hydrolysis of feldspar.

Changes in the concentrations of aqueous species caused by the hydrolysis of microcline in H_2O at 25°C and 1 bar are depicted in Figure 11.1, which was generated from mass-transfer calculations reported elsewhere (Helgeson et al., 1969). The reaction path corresponding to ABCDE in Figure 11.1 is shown on the activity diagram in Figure 11.2. It can be seen in the idealized case represented in Figure 11.1 that the reaction of microcline with the aqueous phase leads to the sequential appearance of gibbsite, kaolinite, and muscovite, which precipitate in response to the changing activities of aqueous species in the hydrolytic process. In other cases, reaction of microcline with aqueous solutions may result in a different sequence of reaction products, which may include metastable phases. Note that the reaction paths labeled $A'B'C'D'$ and $A'B'D'$ in Figure 11.2, which were generated from calculations representing microcline reacting alternately with a static ($A'B'C'D'$) and flowing ($A'B'D'$) solution containing given initial concentrations of HCl and $SiO_2(aq)$, fail to encounter the muscovite stability field. Higher initial concentrations of $SiO_2(aq)$ may cause the appearance of kaolinite (instead of gibbsite) as the first phase to precipitate. Incongruent reactions of minerals with solutions having different initial compositions may thus lead to different zonal patterns of reaction products in geochemical processes. The identity and paragenesis of the minerals produced by the reactions depend on the relative changes in the chemical affinities of hydrolysis reactions for all potential reaction products during the process.

CHEMICAL AFFINITY AND REACTION PROGRESS

Chemical affinity can be regarded as a measure of reaction potential. The chemical affinity of the jth reaction among a given set of minerals, gases, and/or aqueous species at constant pressure and temperature ($\mathbf{A}_j$) can be expressed as a function of reaction progress by writing (De Donder, 1920; De Donder and Van Rysselberghe, 1936)

$$\mathbf{A}_j = -\left(\frac{\partial G}{\partial \xi_j}\right)_{P,T,\xi_k} = -\sum_i \mu_i \hat{n}_{i,j} \qquad (11.20)$$

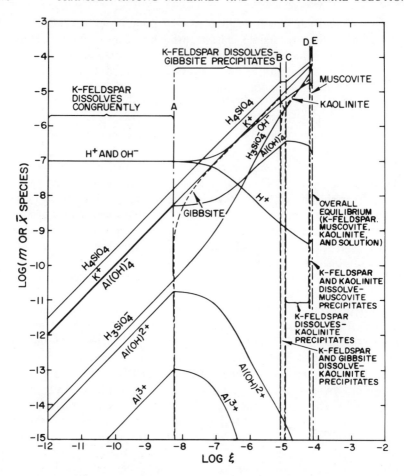

Fig. 11.1 Molalities (m) of species (———) in the aqueous phase and moles of minerals (kg $H_2O)^{-1}$ ($\bar{x}$) produced and destroyed (- - - -) during the hydrolysis of potassium feldspar at 25 °C and 1 bar (after Helgeson et al., 1969; Helgeson, 1974, reproduced with permission from Pergamon Press and Manchester University Press). The curves shown in the figure were generated from mass transfer calculations. The letters correspond to points ABCDE on the reaction path in Figure 11.2.

where G stands for the Gibbs free energy of the system, ξ_j again represents the progress variable for the jth reaction, μ_i refers to the chemical potential of the ith species in the system ($i = 1, 2, \ldots, i$) and $\hat{n}_{i,j}$ corresponds to the stoichiometric reaction coefficient of the species in the reaction, which is positive for products and negative for reactants. The stoichiometric reaction coefficient can be expressed as

CHEMICAL AFFINITY AND REACTION PROGRESS

$$\hat{n}_{i,j} \equiv \frac{dn_{i,j}}{d\xi_j} \tag{11.21}$$

where $dn_{i,j}$ refers to the change in the number of moles of the ith species caused by the jth reaction. In contrast to ξ_j, which has units of moles, $\hat{n}_{i,j}$ is dimensionless. Note that equations (11.3)–(11.8) are specific statements of equation (11.21) for reaction (11.1). As emphasized above, stoichiometric reaction coefficients are constants determined solely by constraints imposed by equation (11.2), which can now be written as

$$\sum_i \nu_{\epsilon,i} \hat{n}_{i,j} = 0 \tag{11.22}$$

where $\nu_{\epsilon,i}$ again stands for the number of moles of the ϵth element in one mole of the ith species. Note that the subscript j in equation (11.20) refers

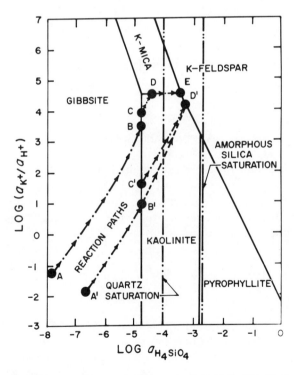

Fig. 11.2 Equilibrium activity diagram for the system $K_2O\text{-}Al_2O_3\text{-}SiO_2\text{-}H_2O$ at 25°C and 1 bar (after Helgeson, 1974 and reproduced with permission from Manchester University Press). Reaction path ABCDE corresponds to that generated by the reaction represented in Figure 11.1.

only to reactions for which $\hat{n}_{i,j}$ can be assessed unambiguously with the aid of equation (11.22). All others are composites of these reactions.

The chemical affinity of the jth reaction can also be expressed as a function of activity by taking account of

$$\mu_i = \mu_i^0 + RT \ln a_i \qquad (11.23)$$

where a_i stands for the activity of the ith species in the system. Multiplying equation (11.23) by $\hat{n}_{i,j}$ and summing across all values of i permits equation (11.20) to be written as

$$\mathbf{A}_j = - \sum_i \mu_i^0 \hat{n}_{i,j} - RT \sum_i \hat{n}_{i,j} \ln a_i \qquad (11.24)$$

which can be combined with

$$\sum_i \mu_i^0 \hat{n}_{i,j} = \Delta G_{r,j}^0 = -RT \ln K_j \qquad (11.25)$$

and

$$Q_j = \prod_i a_i^{\hat{n}_{i,j}} \qquad (11.26)$$

to give

$$\mathbf{A}_j = RT \ln(K_j/Q_j) \qquad (11.27)$$

where μ_i^0 stands for the chemical potential of the ith species in its standard state, $\Delta G_{r,j}^0$ refers to the standard molal Gibbs free energy of the jth reaction, and K_j and Q_j represent the equilibrium constant and activity product for the reaction, respectively. The derivative of equation (11.27) with respect to ξ_j appears as

$$\frac{d\mathbf{A}_j}{d\xi_j} = -RT \frac{d \ln Q_j}{d\xi_j} = -RT \sum_i \hat{n}_{i,j} \frac{d \ln a_i}{d\xi_j} \qquad (11.28)$$

As equilibrium is approached, $Q_j \rightarrow K_j$ and $\mathbf{A}_j \rightarrow d\mathbf{A}_j/d\xi_j \rightarrow 0$. When the jth reaction becomes reversible,

$$\mathbf{A}_j = \frac{d\mathbf{A}_j}{d\xi_j} = 0 \qquad (11.29)$$

Partial and/or local equilibrium (see below) requires equation (11.29) to hold for one or more of the simultaneous reactions occurring in a given geochemical process, but overall equilibrium is achieved only when $A_j = 0$ for all such reactions. Note that the second identity in equation (11.29) is not in itself a sufficient criterion for equilibrium. For example, chemical potential constraints or changes in solution composition and the distribution of species in the aqueous phase arising from precipitation or dissolution of minerals in geochemical processes may cause the derivatives of the chemical affinities of reactions involving other minerals to approach zero at an intermediate stage of reaction progress. Nevertheless, the affinities of these reactions may be much greater than zero. A case in which this behavior occurs is shown in Figure 11.3, where it can be seen that $dA_j/d\xi_j$ for the hydrolysis of high sanidine during reaction of a granitic rock with a hydrothermal solution abruptly changes to ~zero after the appearance of calcite as a reaction product, despite the fact that A_j for the hydrolysis of high sanidine is >0. It is thus necessary, but not sufficient that $dA_j/d\xi = 0$ at equilibrium, but *both* necessary and sufficient that $A_j = 0$ for all values of j at equilibrium.

Chemical affinity is commonly used in geochemistry as a measure of the relative extent to which one or another mineral is out of equilibrium with a given aqueous solution. However, ranking minerals according to the

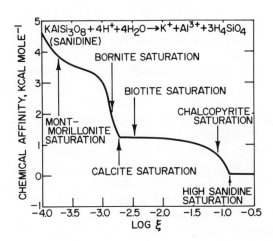

Fig. 11.3 Chemical affinity of the reaction shown above as a function of progress in the irreversible reaction of a granitic rock with a hydrothermal solution at 200 °C and 15 bars. The curve was generated from calculations of the kind reported by Helgeson (1970) using thermodynamic data taken form Helgeson (1969).

chemical affinities of their hydrolysis reactions affords little indication of the order in which they will equilibrate with the aqueous phase. The relative magnitude of these chemical affinities takes no account of the differential influence of homogeneous and heterogeneous reaction coupling on the extent to which they change as a function of reaction progress. Reaction coupling arises from ion association and the appearance of intermediate reaction products, which affect the change in the chemical affinities of the hydrolysis reactions for all other potential reaction products. This can be deduced by combining two statements of equation (11.28) for $j = 1$ and $j = 2$ representing two simultaneous heterogeneous reactions to give

$$\frac{d\mathbf{A}_1}{d\mathbf{A}_2} = \frac{d \ln Q_1}{d \ln Q_2} = \frac{\sum_i \hat{n}_{i,1} \dfrac{d \ln a_i}{d\xi_1}}{\sum_i \hat{n}_{i,2} \dfrac{d \ln a_i}{d\xi_2}} \qquad (11.30)$$

Because all of the species in the two reactions represented by $j = 1$ and $j = 2$ in equation (11.30) are not necessarily the same, and because changes in the activities of the species depend differentially on homogeneous reactions taking place in the process, the order of relative change in the chemical affinities of the reactions may be drastically different from the order of relative affinities at any given stage of the process. This dependence is manifested in part by the curves representing chemical affinities in Figure 11.4. Note that in the early stages of the reaction of microcline with the aqueous solution, the order of increasing chemical affinity of the hydrolysis reactions for the minerals corresponds to gibbsite, amorphous silica, kaolinite, microcline, potassium montmorillonite, muscovite. In contrast, the order at $\log \xi = -8$ is amorphous silica, kaolinite, potassium montmorillonite, muscovite, microcline. However, it can be seen that only two of these minerals actually appear as reaction products (gibbsite and kaolinite) before microcline equilibrates with the aqueous phase.

Although De Donder (1920) assigned a separate progress variable to each reaction that takes place in a given system, it is advantageous to introduce the concept of an *overall* progress variable (ξ) for a geochemical process. The overall progress variable refers to the rate-limiting reaction in the process, which in most cases corresponds to the irreversible reaction of the aqueous phase with one of the minerals in the reactant mineral assemblage. The rates at which the other minerals in the assemblage react can be expressed in terms of ξ by defining relative reaction rates for the minerals (see below). Similarly, all reversible mass transfer can be described as a function of the overall progress variable by taking advantage of

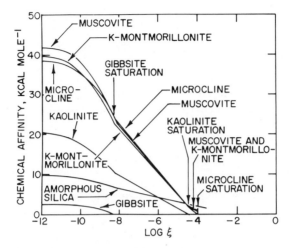

Fig. 11.4 Chemical affinities of hydrolysis reactions for the minerals shown above as a function of progress in the reaction of potassium feldspar with an aqueous phase at 25°C and 1 bar. The curves were generated using thermodynamic data taken from Helgeson (1969).

thermodynamic constraints imposed by partial and local equilibrium states established in the process.

The change in the Gibbs free energy of a system caused by a set of simultaneous reactions can be expressed in terms of the overall progress variable (ξ) by writing

$$\mathbf{A} = -\left(\frac{\partial G}{\partial \xi}\right)_{P,T} = -\sum_i \mu_i \tilde{n}_i \tag{11.31}$$

where $\mathbf{A}$ stands for the total chemical affinity of the process and $\tilde{n}_i$ represents the net change in the number of moles of the ith species caused by all of the reactions taking place in the system; that is,

$$\tilde{n}_i \equiv \frac{dn_i}{d\xi} \tag{11.32}$$

Note that in contrast to $\hat{n}_{i,j}$, $\tilde{n}_i$ is not necessarily a constant.

Equation (11.31) can also be expressed in terms of the chemical affinities of the reactions involved in a geochemical process by taking account of

$$\tilde{n}_i = \sum_j \hat{n}_{i,j} \frac{d\xi_j}{d\xi} \tag{11.33}$$

which can be combined with equations (11.20) and (11.31) to give

$$\mathbf{A} = - \sum_i \sum_j \mu_i \hat{n}_{i,j} \frac{d\xi_j}{d\xi} = \sum_j \mathbf{A}_j \frac{d\xi_j}{d\xi} \qquad (11.34)$$

Similarly, the change in Gibbs free energy of the system caused by the jth reaction can be expressed as a function of the overall progress variable by first defining

$$\mathbf{A}_j^* \equiv \mathbf{A}_j \frac{d\xi_j}{d\xi} \qquad (11.35)$$

which permits equation (11.34) to be written as

$$\mathbf{A} = \sum_j \mathbf{A}_j^* \qquad (11.36)$$

We can now write a combined statement of equations (11.23), (11.31), and (11.36) as

$$\mathbf{A}_j^* = - \sum_i \mu_i^0 \tilde{n}_{i,j} - RT \sum_i \tilde{n}_{i,j} \ln a_i \qquad (11.37)$$

where $\tilde{n}_{i,j}$ refers to the change in the number of moles of the ith species caused by the jth reaction, relative to the overall progress variable, ξ; that is,

$$\tilde{n}_{i,j} \equiv \frac{dn_{i,j}}{d\xi} = \hat{n}_{i,j} \frac{d\xi_j}{d\xi} \qquad (11.38)$$

It then follows from equations (11.25)–(11.27), (11.31), and (11.37) that

$$\mathbf{A}_j^* = \mathbf{A}_j - \sum_i (\tilde{n}_{i,j} - \hat{n}_{i,j}) \mu_i$$

$$= RT \ln (K_j/Q_j) - RT \sum_i (\tilde{n}_{i,j} - \hat{n}_{i,j}) \mu_i^0 \qquad (11.39)$$

$$- RT \sum_i (\tilde{n}_{i,j} - \hat{n}_{i,j}) \ln a_i$$

Although equation (11.39) cannot be evaluated *a priori* for a given reaction in a geochemical process, it provides a measure of the extent to which the Gibbs free energy of the system is affected by the jth reaction, relative to the progress variable for the rate-limiting step in the process.

PARTIAL AND LOCAL EQUILIBRIUM

Systems involved in geochemical processes are almost invariably in a state of partial equilibrium; that is, the system is in equilibrium with respect to at least one process (which can be described by a reversible chemical reaction) but out of equilibrium with respect to others. For example, as indicated above, the rates of homogeneous reactions in an aqueous phase generally exceed by many orders of magnitude those of heterogeneous reactions. Consequently, most dissociational reactions in a hydrothermal solution reacting with its mineralogic environment can be regarded as reversible reactions, all of which contribute to the partial equilibrium state of the system.

In contrast to partial equilibrium, local equilibrium (Thompson, 1959; Fitts, 1962) carries a spatial connotation. Local equilibrium occurs in a system when phases in contact with one another react reversibly in response to changes in chemical potential, temperature, or pressure. Local equilibrium may be established along grain boundaries or manifested by a zonal distribution of reaction products in a hydrothermal system. If, as a consequence of the reaction of a mineral assemblage with an aqueous solution, the solution becomes saturated with one or another potential reaction product, further progress in the overall reaction may or may not lead to precipitation of the mineral. For the product mineral to precipitate, the solution must be supersaturated with respect to the mineral. However, the degree of supersaturation required for precipitation depends on kinetic constraints, such as the nucleation energy required to crystallize the phase. In certain cases (such as quartz at 25°C), the kinetic barriers are so great that the potential product mineral fails to precipitate, even from highly supersaturated solutions. In others, the degree of supersaturation required to cause precipitation is negligible compared to the total mass transfer among the phases in the system. Under these circumstances, precipitation of a reaction product can be viewed as a reversible process in the context of geologic time. Ample evidence suggests that such reactions predominate in geochemical processes, which permits us to consider a mineral produced by the incongruent reaction of a hydrothermal solution with its mineralogic environment to be in local equilibrium with the solution from which it forms. The validity of this assumption is unaffected by the fact that the mass of the mineral in the system is changing as a function of progress in the overall irreversible reaction. If homogeneous equilibrium is maintained, and supersaturation with respect to all reaction products is negligible, then the rate-limiting step in the process corresponds to the irreversible reaction of the solution with the slowest mineral to react in the reactant mineral assemblage.

Reversible mass transfer from an aqueous solution to a solid reaction product in response to changes in solution composition caused by reaction of the solution with its mineralogic environment restricts relative changes in the activities of aqueous species to those compatible with $\mathbf{A}_j = 0$ for the reversible reaction. Constraints imposed on the distribution of species in the aqueous phase by a local equilibrium state of this kind are depicted schematically in Figure 11.5. The curves in the figure represent changes in the activities of the species involved in a reversible reaction with a given mineral for which $Q_j = K_j$ at all stages of progress in an irreversible reaction of an aqueous phase with other minerals. If in the actual process, the aqueous phase is supersaturated with respect to the mineral so that $Q > K$, the chemical affinity for the reaction is negative. However, if the absolute value of the affinity of the jth reaction is small compared to the total chemical affinity of the process (which is generally the case), the departure from local equilibrium has a negligible effect on the total mass transfer among the phases in the system. Accordingly, in a geologic context, comparative discrepancies in the results of mass-transfer calculations and field observations caused by departures in nature from local equilibrium constraints should be slight compared to other uncertainties involved in such calculations. A constant state of supersaturation with respect to a given product mineral constitutes a metastable state of local equilibrium, which can also be taken into account in mass transfer calculations.

All partial and local equilibrium states in a geochemical process can be represented by differential equations corresponding to the derivatives of the law of mass action. For the jth reversible reaction involving only stoichio-

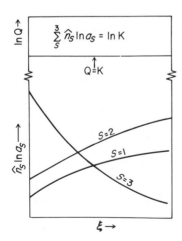

Fig. 11.5 Schematic illustrations of the relative change in the activities of three aqueous species (designated by the subscript s) permitted by constraints imposed by local equilibrium among the species and a reaction product during irreversible reaction of a hydrothermal solution with its mineralogic environment. The upper (horizontal) curve represents the activity product for the reaction (see text).

PARTIAL AND LOCAL EQUILIBRIUM

metric minerals and/or aqueous species (designated by the subscripts ψ and s, respectively), the law of mass action can be written as

$$\sum_s \hat{n}_{s,j} \ln a_s = \sum_s \hat{n}_{s,j} \ln m_s + \sum_s \hat{n}_{s,j} \ln \gamma_s = \ln K_j \qquad (11.40)$$

where $\hat{n}_{s,j}$ corresponds to the stoichiometric coefficient of the sth aqueous species in the jth reversible reaction and a_s, m_s, and γ_s represent the activity, molality, and activity coefficient of the subscripted aqueous species. Before differentiating equation (11.40) let us relate explicitly, $dn_s/d\xi$ to $dm_s/d\xi$, which can be accomplished by dividing an appropriate statement of equation (11.32) by the number of kilograms of H_2O in the system (W_{H_2O}), which leads to

$$\bar{n}_s \equiv \frac{\tilde{n}_s}{W_{H_2O}} = \frac{dm_s}{d\xi} = \frac{1}{W_{H_2O}} \frac{dn_s}{d\xi} \qquad (11.41)$$

and (by analogy)

$$\bar{n}_\psi \equiv \frac{\tilde{n}_\psi}{W_{H_2O}} = \frac{d\bar{x}_\psi}{d\xi} = \frac{1}{W_{H_2O}} \frac{dn_\psi}{d\xi} \qquad (11.42)$$

where $\bar{n}_s$ and $\bar{n}_\psi$ correspond to the change in the number of moles of the subscripted species relative to a kilogram of H_2O, and $\bar{x}_\psi$ designates the number of moles of the ψth mineral (kg H_2O)$^{-1}$ in the system. Note that equations (11.41) and (11.42) specify the same concentration scale for both minerals and aqueous species.

The equilibrium state represented in Figure 11.5 can now be described by differentiating equation (11.40) with respect to ξ, the progress variable for the overall irreversible process responsible for changing the solution composition, to give

$$\sum_s \hat{n}_{s,j} \frac{d \ln a_s}{d\xi} = \sum_s \hat{n}_{s,j} \left(\frac{\partial \ln m_s}{d\xi} + \frac{\partial \ln \gamma_s}{\partial \xi} \right)$$

$$= \sum_s \left(\frac{\hat{n}_{s,j}}{m_s} \frac{dm_s}{d\xi} + \hat{n}_{s,j} \frac{d \ln \gamma_s}{d\xi} \right)$$

$$= \sum_s \left(\frac{\hat{n}_{s,j} \bar{n}_s}{m_s} + \hat{n}_{s,j} \frac{d \ln \gamma_s}{d\xi} \right) = \frac{d \ln K_j}{d\xi} \qquad (11.43)$$

Differentiating equation (11.43) permits us to write

$$\sum_s \left(\frac{\hat{n}_{s,j} \bar{n}_s'}{m_s} + \hat{n}_{s,j} \frac{d^2 \ln \gamma_s}{d\xi^2} \right) = \frac{d^2 \ln K_j}{d\xi^2} + \sum_s \frac{\hat{n}_{s,j} \bar{n}_s^2}{m_s^2} \qquad (11.44)$$

where $\bar{n}_s'$ stands for the first derivative of $\bar{n}_s$; that is,

$$\bar{n}_s' = \frac{d\bar{n}_s}{d\xi} = \frac{d^2 m_s}{d\xi^2} \qquad (11.45)$$

Equations (11.43) and (11.44) represent explicitly constraints on reversible mass transfer imposed by partial and/or local equilibrium states in geochemical processes. They refer to any reversible reaction involving stoichiometric species such as

$$Al_2Si_2O_5(OH)_4 \text{ (kaolinite)} + 6H^+ \rightleftharpoons 2Al^{3+} + 2SiO_2(aq) + 5H_2O$$

$$(11.46)$$

or

$$ZnCl_3^- \rightleftharpoons Zn^{2+} + 3Cl^- \qquad (11.47)$$

Including provision in equations (11.43) and (11.44) for solid solution among reaction products is discussed elsewhere (Helgeson et al., 1970).

The logarithmic derivatives of the activity coefficients in equations (11.43) and (11.44) can be expressed as functions of ξ, pressure P, and temperature T. It thus follows that

$$\frac{d \ln \gamma_s}{d\xi} = \left(\frac{\partial \ln \gamma_s}{\partial \xi} \right)_{P,T} + \left(\frac{\partial \ln \gamma_s}{\partial P} \right)_{T,\xi} \frac{dP}{d\xi} + \left(\frac{\partial \ln \gamma_s}{\partial T} \right)_{P,\xi} \frac{dT}{d\xi}$$

$$= \left(\frac{\partial \ln \gamma_s}{\partial \xi} \right)_{P,T} + \frac{(\bar{V}_s - \bar{V}_s^0)}{RT} \frac{dP}{d\xi} - \frac{(\bar{H}_s - \bar{H}_s^0)}{RT^2} \frac{dT}{d\xi} \qquad (11.48)$$

where $(\bar{V}_s - \bar{V}_s^0)$ and $(\bar{H}_s - \bar{H}_s^0)$ stand for the relative partial molal volume and enthalpy of the sth aqueous species. In contrast, the equilibrium constant is independent of ξ. Hence, for an unrestricted standard state with respect to temperature and pressure we can write,

CONSERVATION OF MASS AND CHARGE

$$\frac{d \ln K_j}{d\xi} = \left(\frac{\partial \ln K_j}{\partial P}\right)_T \frac{dP}{d\xi} + \left(\frac{\partial \ln K_j}{\partial T}\right)_P \frac{dT}{d\xi}$$

$$= -\frac{\Delta V^0_{r,j}}{RT} \frac{dP}{d\xi} + \frac{\Delta H^0_{r,j}}{RT^2} \frac{dT}{d\xi} \qquad (11.49)$$

where $\Delta V^0_{r,j}$ and $\Delta H^0_{r,j}$ refer to the standard molal volume and enthalpy of the jth reaction. Equations similar to, but more complicated than equations (11.48) and (11.49) can be derived for $d^2 \ln \gamma_s/d\xi^2$ and $d^2 \ln K_j/d\xi^2$. Note that for isobaric/isothermal processes, equations (11.48) and (11.49) reduce to

$$\frac{d \ln \gamma_s}{d\xi} = \left(\frac{\partial \ln \gamma_s}{\partial \xi}\right)_{P,T} \qquad (11.50)$$

and

$$\frac{d \ln K_j}{d\xi} = 0 \qquad (11.51)$$

CONSERVATION OF MASS AND CHARGE

A geochemical process in a closed system takes place with no exchange of matter between the system and its surroundings. Consequently, the extent to which components are redistributed among the phases is constrained by

$$\sum_\psi \nu_{\epsilon,\psi} \bar{n}_\psi + \sum_s \nu_{\epsilon,s} \bar{n}_s = 0 \qquad (11.52)$$

and

$$\sum_s Z_s \bar{n}_s = 0 \qquad (11.53)$$

where $\bar{n}_\psi$ and $\bar{n}_s$ are defined by equations (11.41) and (11.42), $\nu_{\epsilon,\psi}$ and $\nu_{\epsilon,s}$ refer to the number of moles of the ϵth element in 1 mole of the subscripted species, and Z_s designates the charge on the sth aqueous species. Note that differentiation of these equations leads to

$$\sum_\psi \nu_{\epsilon,\psi} \bar{n}_\psi' + \sum_s \nu_{\epsilon,s} \bar{n}_s' = 0 \qquad (11.54)$$

and

$$\sum_s Z_s \bar{n}_s' = 0 \qquad (11.55)$$

The rate at which a given mineral reacts with an aqueous phase compared to the corresponding reaction rate for some other mineral is referred to as the relative reaction rate for the two minerals. If we designate this quantity for the rth reactant mineral ($r = 1, 2, \ldots \hat{r}$) relative to that of the $\hat{r}$th as $\theta_{r/\hat{r}}$ we can write,

$$\theta_{r/\hat{r}} = \frac{\bar{n}_r}{\bar{n}_{\hat{r}}} \qquad (11.56)$$

Relative reaction rates depend on differences in the total surface area of the minerals as well as differences in their reaction rates (cm^2 of surface area)$^{-1}$. The magnitude of $\theta_{r/\hat{r}}$ is thus a function of both the relative abundance of the minerals in the rock and the relative extent to which each reaction is controlled by chemical kinetics at the reactant surface or diffusional transfer of material through the aqueous phase.

The rate-limiting step in the hydrolysis of feldspar has been the subject of considerable discussion and controversy (Lagache, 1965, 1976; Wollast, 1967; Helgeson, 1971, 1972; Paces, 1973; Busenberg and Clemency, 1976; Petrovic, 1976), but mounting evidence suggests that surface control of reaction rates probably predominates in geochemical processes (Lagache, 1965, 1976; Petrovich, 1976; Berner, 1978; Grandstaff, 1978; Aagaard and Helgeson, 1977, 1979). Regardless of whether the rate-limiting step is controlled by diffusion or surface kinetics, the total number of moles of a given element (kg H$_2$O)$^{-1}$ transferred by reaction of a rock with a solution

$$\bar{n}_{t,\epsilon} = \sum_r \nu_{\epsilon,r} \theta_{r/\hat{r}} \bar{n}_{\hat{r}} \qquad (11.57)$$

and (for constant relative reaction rates)

$$\bar{n}_{t,\epsilon}' = \sum_i \nu_{\epsilon,r} \theta_{r/\hat{r}} \bar{n}_{\hat{r}}' \qquad (11.58)$$

If we now designate product minerals with the subscript ϕ ($\phi = 1, 2 \cdots \Phi$) and reactant minerals with the index r (as above), we can write equations (11.52) and (11.54) as

$$\sum_\phi \nu_{\epsilon,\phi} \bar{n}_\phi + \sum_s \nu_{\epsilon,s} \bar{n}_s = - \sum_r \nu_{\epsilon,r} \theta_{r/\hat{r}} \bar{n}_{\hat{r}} \qquad (11.59)$$

and

$$\sum_\phi \nu_{\epsilon,\phi}\bar{n}_\phi' + \sum_s \nu_{\epsilon,s}\bar{n}_s' = -\sum_r \nu_{\epsilon,r}\theta_{r/\hat{r}}\bar{n}_{\hat{r}}' \qquad (11.60)$$

If both the relative rate of reaction and $\bar{n}_r$ are constant, equation (11.60) reduces to zero.

Although H_2O is almost invariably consumed or produced during reaction of an aqueous solution with a silicate mineral assemblage, in a closed system the change is usually negligible compared to the total number of moles of H_2O in the solution. However, conservation of mass in a system open to H_2O requires modification of the equations given above to include provision for the changing mass of H_2O as well as the effect of the change on the molalities of the aqueous species in solution. This can be accomplished by combining the first identities in equations (11.41) and (11.42) with the mass transfer equations summarized above. In addition, if a given component other than H_2O is abstracted from or added to the system (such as CO_2 boiling off from a hydrothermal solution), equation (11.59) must be amended to

$$\sum_\phi \nu_{\epsilon,\phi}\bar{n}_\phi + \sum_s \nu_{\epsilon,s}\bar{n}_s = -\sum_r \nu_{\epsilon,r}\theta_{r/\hat{r}}\bar{n}_{\hat{r}} - \sum_f \nu_{\epsilon,f}\bar{n}_f \qquad (11.61)$$

where $\nu_{\epsilon,f}$ stands for the number of moles of the ϵth element in one mole of the fth fugitive component and $\bar{n}_f$ represents the change in the molality of the component in the aqueous phase defined by

$$\bar{n}_f \equiv \frac{dm_f}{d\xi} \qquad (11.62)$$

CALCULATION OF MASS TRANSFER

For the sake of clarity, a large number of variables were omitted in deriving the equations summarized above to promote conceptual appreciation for the thermodynamic constraints imposed by partial and local equilibrium on mass transfer among minerals and aqueous solutions in geochemical processes. These omissions include explicit provision for changes in the activity of H_2O, solid solution, oxidation/reduction reactions, fluid flow, fixed chemical potentials of components, departures from local equilibrium, calculation of changes in activity coefficients, and evaluation of the pressure and temperature dependence of equilibrium constants. The modifications required to provide for a number of these have been summarized in detail elsewhere, together with the grand matrix equation for the general case of an aqueous solution reacting with its mineralogic environment (Helgeson et al. 1970). The grand matrix equation is composed of linear differential equations representing reversible mass transfer and conservation of mass and charge like those for the simple case derived above. However, it can be condensed by substituting the differential equations representing reversible mass transfer into those describing conservation of mass, which simplifies the computational procedure by eliminating $\bar{n}_s$ for aqueous complexes from the unknown coefficient column vector. In either case, the matrix equation can be represented by

$$\bar{\beta} = \overline{\Gamma^{-1}} \bar{\lambda} \qquad (11.63)$$

where $\overline{\Gamma}$ denotes the matrix, $\bar{\beta}$ refers to the column vector of unknown reaction coefficients, and $\bar{\lambda}$ stands for the reactant mineral (or known coefficient) column vector. Evaluation of this equation for a solution of a given composition can be carried out quickly with the aid of a computer by first calculating the distribution of species in the solution at the outset of the reaction. Evaluation of the matrix equation then yields values of $\bar{n}_s$ and $\bar{n}_s{}'$ at the beginning of the process. The numerical values of these derivatives depend on the composition of, and the distribution of species in the aqueous phase. They are thus path-dependent variables that cannot be described by exact differentials. However, their integrals can be represented by Taylor's expansions of the form

$$m_s = m_s{}^0 + \bar{n}_s \Delta\xi + \frac{\bar{n}_s{}'(\Delta\xi)^2}{2!} + \cdots \qquad (11.64)$$

and (after the appearance of the ϕth product mineral),

$$\bar{x}_\phi = \bar{x}_\phi{}^0 + \bar{n}_\phi \Delta\xi + \frac{\bar{n}_\phi{}'(\Delta\xi)^2}{2!} + \cdots \qquad (11.65)$$

where $m_s{}^0$ and $\bar{x}_\phi{}^0$ stand for the molality of the sth aqueous species and the number of moles of the ϕth mineral (kg H_2O)$^{-1}$ in the system at the current stage of reaction progress, and m_s and $\bar{x}_\phi$ refer to these variables after a small increment of reaction progress represented by $\Delta\xi$. The values of m_s computed from equation (11.64) after the first increment of reaction progress permit a new matrix equation to be evaluated, which yields new values of $\bar{n}_s$ and $\bar{n}_s{}'$. These values can then be used to compute from equation (11.64) the molalities of aqueous species and the number of moles of the

CALCULATION OF MASS TRANSFER 591

ϕth mineral in the system after the second increment of reaction progress. Repetition of this procedure allows calculation of the changes in solution composition and distribution of species in the aqueous phase for the entire reaction process. If no minerals are in equilibrium with the aqueous solution at the outset of the process, $\bar{n}_\phi = \bar{n}_\phi' = \bar{x}_\phi = 0$ at $\xi = 0$. However, the changes in solution composition caused by irreversible reaction of the aqueous phase with the mineral assemblage may cause the solution to become saturated with one or another mineral in the system. Provision for reversible precipitation of the product mineral is then incorporated in the calculations. Detection of the stage of reaction progress where the chemical affinity of the hydrolysis reaction for a given mineral equals zero can be accomplished by calculating chemical affinities of hydrolysis reactions for all potential reaction products at each stage of reaction progress. After finding the value of ξ at which the solution becomes saturated with one of them, the matrix equation is enlarged to accommodate $\bar{n}_\phi$ and the reversible mass transfer equation for the mineral. This procedure provides automatically for the possibility that the mineral will dissolve at some later stage of reaction progress.

A Numerical Example—The Hydrolysis of Feldspar

To illustrate construction of the matrix equation for an actual process, let us consider the reaction of 1 mole of pure microcline with a solution containing 1000 g of H_2O and an initial concentration of 10^{-3} mole HCl (kg $H_2O)^{-1}$ at 25°C and 1 bar. The system is closed and composed of five elements (potassium, aluminum, silicon, oxygen, and hydrogen) in addition to chlorine, which does not participate in the reaction. We can thus write five statements of equation (11.59) to describe conservation of mass among the species involved in the reaction process. For purposes of illustration, let us assume that only five aqueous complexes form to significant degrees during the reaction, that is, $Al(OH)^{2+}$, $Al(OH)_4^-$, H_4SiO_4, $H_3SiO_4^-$, and H_2O. We can now write four statements of equation (11.40) to describe homogeneous equilibrium in the aqueous phase. The dissociational reactions appear as

$$Al(OH)^{2+} \rightleftharpoons Al^{3+} + OH^- \qquad (11.66)$$

$$Al(OH)_4^- \rightleftharpoons Al^{3+} + 4OH^- \qquad (11.67)$$

$$H_4SiO_4 \rightleftharpoons H_3SiO_4^- + H^+ \qquad (11.68)$$

$$H_2O \rightleftharpoons H^+ + OH^- \qquad (11.69)$$

for which

$$\frac{a_{Al^{3+}} a_{OH^-}}{a_{Al(OH)^{2+}}} = \frac{m_{Al^{3+}} \gamma_{Al^{3+}} m_{OH^-} \gamma_{OH^-}}{m_{Al(OH)^{2+}} \gamma_{Al(OH)^{2+}}} = K_{Al(OH)^{2+}}, \quad (11.70)$$

$$\frac{a_{Al^{3+}} a_{OH^-}^{-4}}{a_{Al(OH)_4^-}} = \frac{m_{Al^{3+}} \gamma_{Al^{3+}} m_{OH^-}^{-4} \gamma_{OH^-}^{-4}}{m_{Al(OH)_4^-} \gamma_{Al(OH)_4^-}} = \beta_{Al(OH)_4^-}, \quad (11.71)$$

$$\frac{a_{H^+} a_{H_3SiO_4^-}}{a_{H_4SiO_4}} = \frac{m_{H^+} \gamma_{H^+} m_{H_3SiO_4^-} \gamma_{H_3SiO_4^-}}{m_{H_4SiO_4} \gamma_{H_4SiO_4}} = K_{H_4SiO_4} \quad (11.72)$$

and

$$a_{H^+} a_{OH^-} = m_{H^+} \gamma_{H^+} m_{OH^-} \gamma_{OH^-} = K_{H_2O} \quad (11.73)$$

where $K_{Al(OH)^{2+}}$ and $K_{H_4SiO_4}$ stand for the stepwise dissociation constants for the subscripted species, $\beta_{Al(OH)_4^-}$ represents the overall dissociation constant for $Al(OH)_4^-$, and K_{H_2O} refers to the activity product constant for H_2O. The total number of differential equations we can write to describe conservation of mass and reversible mass transfer in the reaction process is thus nine, which is equal to the total number of unknown values of $\bar{n}_s$ for the species in the aqueous phase (Al^{3+}, $Al(OH)^{2+}$, $Al(OH)_4^-$, H_4SiO_4, $H_3SiO_4^-$, H_2O, K^+, H^+, and OH^-). Of the species considered in the present calculation, only K^+ and Cl^- fail to form complexes to a significant degree at 25°C and 1 bar.

Calculation of the distribution of species in the solution at the outset of the reaction requires evaluation of mass balance equations for aluminum and silicon in the aqueous phase, which can be written as

$$m_{t,Al} = m_{Al^{3+}} + m_{Al(OH)^{2+}} + m_{Al(OH)_4^-} \quad (11.74)$$

and

$$m_{t,Si} = m_{H_4SiO_4} + m_{H_3SiO_4^-} \quad (11.75)$$

where $m_{t,Al}$ and $m_{t,Si}$ stand for the total molalities of the subscripted elements in solution. Taking account of equations (11.70)–(11.72) we can thus write,

$$m_{t,Al} = m_{Al^{3+}} \left(1 + \frac{\gamma_{Al^{2+}} m_{OH^-} \gamma_{OH^-}}{K_{Al(OH)^{2+}} \gamma_{Al(OH)^{2+}}} + \frac{\gamma_{Al^{3+}} m_{OH^-}^{-4} \gamma_{OH^-}^{-4}}{K_{Al(OH)_4^-} \gamma_{Al(OH)_4^-}} \right) \quad (11.76)$$

and

$$m_{t,Si} = m_{H_4SiO_4}\left(1 + \frac{K_{H_4SiO_4}\gamma_{H_4SiO_4}}{m_{H^+}\gamma_{H^+}\gamma_{H_3SiO_4^-}}\right) \quad (11.77)$$

which define the total concentrations of aluminum and silicon in the solution in terms of the molalities and activity coefficients of four species: Al^{3+}, H_4SiO_4, OH^-, and H^+.

Activity coefficients and dissociation constants for aqueous species can be computed from thermodynamic data and equations. Employing data and equations summarized by Helgeson (1969) for this purpose, let us compute the distribution of species in the solution at the outset of the reaction process. Because the solution is so acid, we can write as a close approximation,

$$m_{H^+} \approx m_{t,HCL} = 10^{-3} \quad (11.78)$$

and thus

$$m_{OH^-} \approx \frac{K_{H_2O}}{10^{-3}} = \frac{10^{-14}}{10^{-3}} = 10^{-11} \quad (11.79)$$

which follows from equation (11.73) for dilute solutions in which activity coefficients can be regarded as unity without introducing unacceptable uncertainties in the calculation. Aside from H_2O and Cl^- (which does not participate in the reaction process), no other species are present in significant concentrations in the aqueous phase prior to reaction of the solution with microcline. However, after an infinitesimal amount of microcline dissolves congruently in the solution, say 10^{-10} moles, we can evaluate equations (11.76) and (11.77) to define $m_{Al^{3+}}$ and $m_{H_4SiO_4}$. The ionic strength of the solution is affected negligibly by the dissolution of 10^{-10} moles of microcline, which permits us to again regard all activity coefficients as unity. Because the initial solution is so acid, essentially all of the dissolved aluminum and silicon in solution after reaction of 10^{-10} moles of microcline is present as Al^{3+} and H_4SiO_4, respectively. This can be demonstrated by rearranging equations (11.76) and (11.77) and combining the resulting expressions with the values of m_{H^+} and m_{OH^-} computed above and the dissociation constants for $Al(OH)^{2+}$ and $Al(OH)_4^-$, which leads (for $m_{t,Al} = 10^{-10}$) to

$$m_{Al^{3+}} = \frac{10^{-10}}{1 + \dfrac{10^{-11}}{10^{-9.3}} + \dfrac{10^{-44}}{10^{-32.7}}}$$

$$= \frac{10^{-10}}{1 + 10^{-1.7} + 10^{-11.3}} = 10^{-10} \qquad (11.80)$$

and (because it follows from the stoichiometry of microcline that $m_{t,Si} = 3m_{t,Al} = 3 \times 10^{-10} = 10^{-9.5}$),

$$m_{H_4SiO_4} = \frac{10^{-9.5}}{1 + \dfrac{10^{-9.6}}{10^{-3}}} = \frac{10^{-9.5}}{1 + 10^{-6.6}} = 10^{-9.5} \qquad (11.81)$$

We can now evaluate equations (11.70)–(11.72) to obtain the molalities of the other species in solution after dissolution of 10^{-10} moles of microcline, which leads to

$$m_{Al(OH)^{2+}} = \frac{10^{-10} 10^{-11}}{10^{-9.3}} = 10^{-11.7} \qquad (11.82)$$

$$m_{Al(OH)_4^-} = \frac{10^{-10} 10^{-44}}{10^{-32.7}} = 10^{-21.3} \qquad (11.83)$$

and

$$m_{H_3SiO_4^-} = \frac{10^{-9.5} 10^{-9.6}}{10^{-3}} = 10^{-16.1} \qquad (11.84)$$

Because the potassium ion does not associate appreciably at 25°C and 1 bar in the solution under consideration, $m_{K^+} = m_{t,K} = 10^{-10}$.

We now have all the information necessary to construct the matrix equation for the "initial" state of the system; that is, after an infinitesimal amount of reaction has taken place, which is defined in this case as 10^{-10} mole microcline dissolved in a solution containing 1000 g of H_2O and 10^{-3} mole HCl. For a small increment of progress in the reaction of microcline with this solution, we can write as a close approximation,

$$\frac{d \ln \gamma_s}{d\xi} \approx 0 \qquad (11.85)$$

CALCULATION OF MASS TRANSFER

and because temperature and pressure are constant, equation (11.51) applies to the process. Hence, equation (11.43) can be written for the four dissociational reactions [reactions (11.66)-(11.69)] as

$$\frac{\bar{n}_{Al^{3+}}}{m_{Al^{3+}}} + \frac{\bar{n}_{OH^-}}{m_{OH^-}} - \frac{\bar{n}_{Al(OH)^{2+}}}{m_{Al(OH)^{2+}}} = \frac{\bar{n}_{Al^{3+}}}{10^{-10}} + \frac{\bar{n}_{OH^-}}{10^{-11}} - \frac{\bar{n}_{Al(OH)^{2+}}}{10^{-11.7}} = 0$$

(11.86)

$$\frac{\bar{n}_{Al^{3+}}}{m_{Al^{3+}}} + \frac{4\bar{n}_{OH^-}}{m_{OH^-}} - \frac{\bar{n}_{Al(OH)_4^-}}{m_{Al(OH)_4^-}} = \frac{\bar{n}_{Al^{3+}}}{10^{-10}} + \frac{4\bar{n}_{OH^-}}{10^{-11}} - \frac{\bar{n}_{Al(OH)_4^-}}{10^{-21.3}} = 0$$

(11.87)

$$\frac{\bar{n}_{H^+}}{m_{H^+}} + \frac{\bar{n}_{H_3SiO_4^-}}{m_{H_3SiO_4^-}} - \frac{\bar{n}_{H_4SiO_4}}{m_{H_4SiO_4}} = \frac{\bar{n}_{H^+}}{10^{-3}} + \frac{\bar{n}_{H_3SiO_4^-}}{10^{-16.1}} - \frac{\bar{n}_{H_4SiO_4}}{10^{-9.5}} = 0$$

(11.88)

and

$$\frac{\bar{n}_{H^+}}{m_{H^+}} + \frac{\bar{n}_{OH^-}}{m_{OH^-}} = \frac{\bar{n}_{H^+}}{10^{-3}} + \frac{\bar{n}_{OH^-}}{10^{-11}} = 0 \qquad (11.89)$$

Statements of equation (11.59) for the five elements involved in the reaction process appear (in the order, aluminum, silicon, oxygen, hydrogen, and potassium) as

$$\bar{n}_{Al^{3+}} + \bar{n}_{Al(OH)^{2+}} + \bar{n}_{Al(OH)_4^-} = -\bar{n}_{KAlSi_3O_8} \qquad (11.90)$$

$$\bar{n}_{H_4SiO_4} + \bar{n}_{H_3SiO_4^-} = -3\bar{n}_{KAlSi_3O_8} \qquad (11.91)$$

$$\bar{n}_{Al(OH)^{2+}} + 4\bar{n}_{Al(OH)_4^-} + 4\bar{n}_{H_4SiO_4} + 4\bar{n}_{H_3SiO_4^-} + \bar{n}_{H_2O} + \bar{n}_{OH^-} = -8\bar{n}_{KAlSi_3O_8}$$

(11.92)

$$\bar{n}_{Al(OH)^{2+}} + 4\bar{n}_{Al(OH)_4^-} + 4\bar{n}_{H_4SiO_4} + 3\bar{n}_{H_3SiO_4^-} + 2\bar{n}_{H_2O} + \bar{n}_{H^+} + \bar{n}_{OH^-} = 0$$

(11.93)

and

$$\bar{n}_{K^+} = -\bar{n}_{KAlSi_3O_8} \qquad (11.94)$$

which can be combined for $\bar{n}_{KAlSi_3O_8} = -1$ with equations (11.86)-(11.89) to give equation (11.95), which is shown in Table 11.1. Successive differentiation of this equation yields matrices identical to that in equation (11.95), but the upper part of the reactant column vector reduces to zero and the lower (reversible mass transfer) part takes on numerical entries equal (in the case of the first derivative) to the right side of equation (11.44). In each derivative of equation (11.95), the reactant coefficient column vector contains the appropriate values of $\bar{n}_s'$ or its derivatives.

As indicated above, equation (11.95) can be simplified by first combining equations (11.86)-(11.93) to eliminate $\bar{n}_{Al(OH)^{2+}}$, $\bar{n}_{Al(OH)_4^-}$, $\bar{n}_{H_3SiO_4^-}$, and $\bar{n}_{H_2O}$. For example, substituting equations (11.86) and (11.87) into equation (11.90) results in

$$\bar{n}_{Al^{3+}}\left(1 + \frac{m_{Al(OH)^{2+}} + m_{Al(OH)_4^-}}{m_{Al^{3+}}}\right)$$

$$+ \bar{n}_{OH^-}\left(\frac{m_{Al(OH)^{2+}} + 4m_{Al(OH)_4^-}}{m_{OH^-}}\right) = -\bar{n}_{KAlSi_3O_8}$$

$$= (1 + 10^{-1.7})\bar{n}_{Al^{3+}} + 10^{-0.7}\bar{n}_{OH^-} = 1 \qquad (11.96)$$

Similarly, equations (11.88) and (11.91) can be combined to give

$$\bar{n}_{H_4SiO_4}\left(1 + \frac{m_{H_3SiO_4^-}}{m_{H_4SiO_4}}\right) - \frac{m_{H_3SiO_4^-}\bar{n}_{H^+}}{m_{H^+}} = -3\bar{n}_{KAlSi_3O_8}$$

$$= (1 + 10^{-16.6})\bar{n}_{H_4SiO_4} - 10^{-13.1}\bar{n}_{H^+} = 3 \qquad (11.97)$$

Rather than solving equation (11.95) or evaluating simultaneously a set of equations like equations (11.96) and (11.97), it will suffice for the purpose of illustrating the procedure involved in calculating the molalities of aqueous species after a small increment of reaction progress if we compute by inspection approximate values of $\bar{n}_s$ at the outset of the reaction. Because essentially all of the aluminum and silica in solution during the early stages of the reaction is present as Al^{3+} and H_4SiO_4, it follows from the stoichiometry of microcline and equations (11.96) and (11.97) that $\bar{n}_{Al^{3+}} \approx \bar{n}_{t,Al} = 1$ and $\bar{n}_{H_4SiO_4} \approx \bar{n}_{t,Si} = 3$. Maintenance of charge balance in the process thus requires the reaction to proceed as

$$KAlSi_3O_8 \text{ (microcline)} + 4H_2O + 4H^+ \rightarrow K^+ + Al^{3+} + 3H_4SiO_4 \quad (11.98)$$

so $\bar{n}_{H^+} = \hat{n}_{H^+(11.98)} = -4$ and $\bar{n}_{OH^-}$ is given by [from equation (11.89)]

Table 11.1 Equation (11.95)

	Al^{3+}	$Al(OH)^{2+}$	$Al(OH)_4^-$	H_4SiO_4	$H_3SiO_4^-$	H_2O	K^+	H^+	OH^-		Reaction coefficient column vector		Reactant column vector
Conservation of mass													
Al	1	1	1	0	0	0	0	0	0		$\underline{n}_{Al^{3+}}$		1
Si	0	0	0	1	1	0	0	0	0		$\underline{n}_{Al(OH)^{2+}}$		3
O	0	1	4	4	4	1	0	0	1		$\underline{n}_{Al(OH)_4^-}$		8
H	0	1	4	4	3	2	0	1	1	$\times$	$\underline{n}_{H_4SiO_4}$	$=$	0
K	0	0	0	0	0	0	1	0	0		$\underline{n}_{H_3SiO_4^-}$		1
Reversible mass transfer													
$Al(OH)^{2+}$	10^{-10}	$10^{-11.7}$	0	0	0	0	0	0	10^{-11}		$\underline{n}_{H_2O}$		0
$Al(OH)_4^-$	10^{-10}	0	$10^{-21.3}$	0	0	0	0	0	$10^{-11.6}$		$\underline{n}_{K^+}$		0
H_4SiO_4	0	0	0	$10^{-9.5}$	$10^{-16.1}$	0	0	10^{-3}	0		$\underline{n}_{H^+}$		0
H_2O	0	0	0	0	0	0	0	10^{-3}	10^{-11}		n_{OH^-}		0

$$\bar{n}_{OH^-} = \frac{10^{0.60} 10^{-11}}{10^{-3}} = 10^{-7.4} \qquad (11.99)$$

Similarly (from equations 11.86 through 11.88),

$$\bar{n}_{Al(OH)^{2+}} = 10^{-11.7} \left(\frac{1}{10^{-10}} + \frac{10^{-7.4}}{10^{-11}} \right) = 10^{-1.7} \qquad (11.100)$$

$$\bar{n}_{Al(OH)_4^-} = 10^{-21.3} \left(\frac{1}{10^{-10}} + \frac{10^{0.6}\, 10^{-7.4}}{10^{-11}} \right) = 10^{-11.3} \qquad (11.101)$$

and

$$\bar{n}_{H_3SiO_4^-} = 10^{-16.1} \left(\frac{10^{0.5}}{10^{-9.5}} + \frac{10^{0.6}}{10^{-3}} \right) = 10^{-6.1} \qquad (11.102)$$

If we omit the second and higher order derivatives, these calculations permit evaluation of equation (11.64) to compute values of m_s for the various species after the first increment of reaction progress. For example, if we let $\Delta\xi$ be 10^{-9}, the molality of $Al(OH)^{2+}$ after the first increment is given by

$$m_{Al(OH)^{2+}} = 10^{-11.7} + 10^{-1.7} \times 10^{-9} = 10^{-10.8} \qquad (11.103)$$

Similar calculations for the other aqueous species provide all of the input data necessary to evaluate equation (11.95) or its condensed analog for the next increment of reaction progress.

Although at the outset of the reaction, the pH of the solution considered above is low and $Al(OH)^{2+}$, $Al(OH)_4^-$, and $H_3SiO_4^-$ contribute negligibly to the hydrolytic process, with continued reaction the solution pH increases and these species become important. The approximations represented by setting $\bar{n}_{Al^{3+}} = 1$ and $\bar{n}_{H_4SiO_4} = 3$ are then invalid and the matrix equation must be solved to compute accurate values of $\bar{n}_s$. Similarly, in most cases the assumption that $\gamma_s = 1$ and the approximation represented by equation (11.85) are not warranted. Matrix equations incorporating provision for changing ionic strength, activity coefficients, solid solution, oxidation/reduction reactions, and changing activity of H_2O can be solved quickly for a given increment of reaction progress with the aid of a computer and any one of several standard algorithms (Helgeson et al., 1970; Wolery, 1978).

Undesirable consequences of ill-conditioned matrices in mass transfer equations can be avoided by invoking predictor/corrector methods and curve crawling techniques in the evaluation algorithms, subdividing the matrix, or solving integrals of the differential equations (Crerar, 1975; Wolery, 1978). For example, Karpov et al. (1973), Wood (1974), Crerar (1975), and others avoid differential equations by employing finite difference methods and optimal programming techniques to determine the distribution of components corresponding to the minimum Gibbs free energy of the system at each stage of reaction progress. Although this approach can be used to evaluate simultaneously a set of material balance and law of mass action equations representing conservation of mass and local and partial equilibrium states in the system, the set of equations is nonlinear. In contrast, the corresponding set of differential equations summarized above is linear, which simplifies considerably the computation of mass transfer in multicomponent systems.

Calculation of $\bar{n}_s$, $\bar{n}_\phi$, and their derivatives permit evaluation of the Taylor's expansion represented by equation (11.64) for all of the species in the system. The distribution of species in the aqueous phase is calculated after each increment of reaction progress represented by a small but finite change in ξ from ξ to $\xi + \Delta\xi$. The new molalities of Al^{3+}, $Al(OH)^{2+}$, $Al(OH)_4^-$, H_4SiO_4, $H_3SiO_4^-$, K^+, H^+, and OH^- in solution can then be used to compute the solution composition, ionic strength, and activity coefficients of the species after the increment of progress represented by $\Delta\xi$. This permits evaluation of new matrix equations to compute values of $\bar{n}_s$ and its derivatives for $\xi + \Delta\xi$, which in turn permits calculation of the distribution of species and solution composition after another increment of reaction progress ($\xi + 2\Delta\xi$).

Candidates for reaction products resulting from the hydrolysis of microcline in the example considered above include

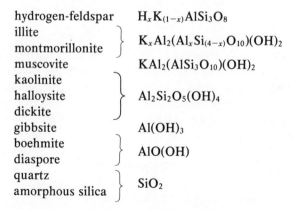

and others, some of which are metastable at 25°C and 1 bar. By computing the chemical affinities of hydrolysis reactions for all of these minerals after each increment of reaction progress, we can determine the exact stage at which the solution becomes saturated with respect to one or more of the phases. If the reaction product precipitates reversibly as microcline continues to dissolve (i.e., negligible supersaturation occurs so that the solution essentially maintains equilibrium with the reaction product), the matrix must be enlarged by one row and one column. For example, if the reaction remains congruent until the solution becomes saturated with respect to gibbsite, a column would be added to the matrix in equation (11.95) for $\bar{n}_{gibbsite}$, and a row would be inserted for the differential equation derived from the law of mass action for the reaction describing equilibrium between gibbsite and the aqueous phase; that is,

$$Al(OH)_3 \text{ (gibbsite)} \rightleftharpoons Al^{3+} + 3OH^- \qquad (11.104)$$

for which

$$a_{Al^{3+}} a_{OH^-}{}^3 = K \qquad (11.105)$$

Differentiating equation (11.105) with respect to ξ at constant pressure and temperature leads to

$$\frac{\bar{n}_{Al^{3+}}}{m_{Al^{3+}}} + \frac{3\bar{n}_{OH^-}}{m_{OH^-}} + \frac{d \ln \gamma_{Al^{3+}}}{d\xi} + 3\frac{d \ln \gamma_{OH^-}}{d\xi} = 0 \qquad (11.106)$$

which imposes the constraint of reversible mass transfer between the aqueous solution and gibbsite on further reaction progress.

DISCUSSION

Provision for precipitation of metastable phases can be made in mass transfer calculations by suppressing the appearance of stable polymorphs in the computer routine used to detect local equilibrium states. Similarly, supersaturation states and specific kinetic constraints can be incorporated by allowing selected reactions to have negative chemical affinities, which may or may not be expressed as functions of ξ. The consequences of fluid flow and open-space filling can be taken into account by precluding destruction of reaction products in the calculations. Otherwise, the simultaneous dissolution and precipitation of different reaction products corresponds to replacement of one mineral by another.

Observations of the relative extent to which minerals in a rock have been destroyed by interaction of the minerals with an aqueous solution are not

DISCUSSION

necessarily indicative of the relative rate at which the minerals reacted with the aqueous phase. Even though a given mineral may react more slowly, it may also equilibrate with the aqueous phase before the other minerals in the rock. Mass transfer calculations indicate that this is the primary reason why potassium feldspar commonly survives the weathering process at the expense of plagioclase. Results of such calculations for the reaction of potassium feldspar and coexisting albite with H_2O at 25°C and 1 bar are shown in Figure 11.6. The calculations indicate that reaction of the two minerals with $\theta_{r/\dot{r}} = 1$ leads to the sequential appearance of gibbsite, kaolinite, muscovite, sodium montmorillonite, and potassium feldspar. It can be seen that the potassium feldspar in the reactant mineral assemblage equilibrates long before albite, despite the fact that $\theta_{r/\dot{r}}$ was set to unity in the calculations. Note that the overall process leads to dissolution and precipitation of >0.3 g/kg H_2O of reaction products during the process.

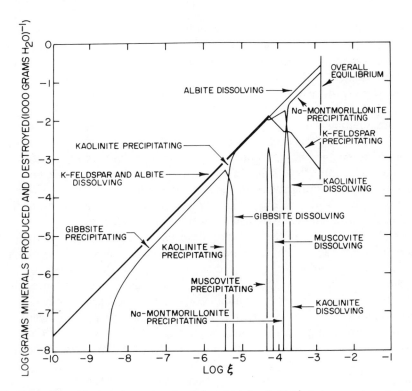

Fig. 11.6 Mass of minerals produced and destroyed (kg $H_2O)^{-1}$ during reaction of albite and coexisting potassium feldspar with an aqueous phase at 25°C and 1 bar. The curves were computed for a relative reaction rate ($\theta_{r/\dot{r}}$) equal to 1.0 using thermodynamic data taken from Helgeson (1969).

Little is known of the difference in the rates/cm² of surface area at which various silicates react with aqueous solutions. However, it seems likely that the rates are controlled by surface reactions and are comparable for minerals in a given structural class. This certainly seems to be true in the case of feldspars. Experimental data reported by Busenberg and Clemency (1976) leave little doubt that the dissolution of the reactant mineral is the primary rate-limiting step in the incongruent hydrolysis of feldspar. Their data also indicate that albite reacts only 1.5 times faster/cm² of surface area than potassium feldspar at 25 °C and 1 bar, and that the relative reaction rate varies insignificantly with reaction progress.

The choice of relative reaction rates to be used in mass-transfer calculations is usually based on trial and error evaluation of the consequences of selecting different rates, or notions of relative rates based on relative abundances of minerals in geologic systems. For example, a trial calculation can be carried out with $\theta_{r/f}$ set equal to the relative surface area of the reactant minerals in the rock, which means that all reaction rates/cm² of surface area are taken to be equal. If the results of the calculation fail to agree quantitatively with field or laboratory data, the value of $\theta_{r/f}$ can be adjusted and the calculation repeated, which permits another comparison with observational data. Repetition of this procedure leads to convergence of the values of $\theta_{r/f}$ with field or laboratory observations. The same approach can be used to refine ambiguities in choosing the initial composition of an aqueous solution corresponding to the composition of the actual aqueous phase responsible for a given geochemical process. The compositions of fluid inclusions, geothermal waters, and interstitial fluids in rocks afford a basis for selecting a trial solution composition. The results of calculations for this solution composition can then be compared quantitatively with observations, which affords a basis for refining the initial composition of the aqueous phase used in subsequent mass-transfer calculations.

Results of isobaric/isothermal mass transfer calculations for two hydrothermal systems are shown in Figures 11.7 and 11.8, which depict the computed mass of minerals produced and destroyed during reaction of "typical" ore-forming solutions with granitic rocks. The sequence and relative abundance of the alteration products and sulfides resulting from these calculations closely approximate those in hydrothermal ore deposits. It can be seen in Figure 11.8 that the reaction leads to two generations of pyrite, which result from intricate changes in the oxidation state of the system as the reaction proceeds.

In principle, mass transfer calculations can be carried out for a wide variety of geochemical processes, including those in which temperature and pressure are variables. As in the isothermal/isobaric case, the first partial derivative on the right side of equation (11.48) can be represented in the

DISCUSSION

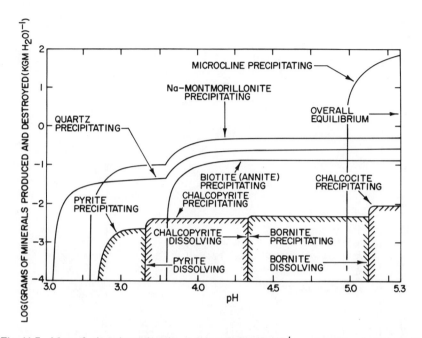

Fig. 11.7 Mass of minerals produced and destroyed (kg $H_2O)^{-1}$ as a function of solution pH during reaction of a granitic rock with a potential ore-forming solution at 200°C and 15 bars (modified after Helgeson, 1970 and reproduced with permission from the Mineralogical Society of America).

calculations by equations derived from extended Debye-Hückel theory (Helgeson et al., 1970; Helgeson and Kirkham, 1974b; Helgeson, Kirkham, and Flowers, 1979). The remaining terms in equation (11.48) and those in equation (11.49) can be expressed in terms of equations of state and thermodynamic data summarized elsewhere (Helgeson and Kirkham, 1974a, 1976; Walther and Helgeson, 1977; Helgeson et al., 1978; Helgeson, Kirkham, and Flowers, 1979). However, the extent to which changes in pressure and temperature are coupled to reaction progress must be specified in the calculations. If pressure and temperature are constant, the calculations can be carried out in the manner described above. If not, mass transfer among minerals and aqueous solutions caused by changes in pressure and temperature can be expressed by specifying a geothermal gradient characteristic of a given cooling process, such as reversible or irreversible adiabatic expansion of hydrothermal fluids. In the simplest case, cooling occurs along a geothermal profile represented by

$$P = a + bT \qquad (11.107)$$

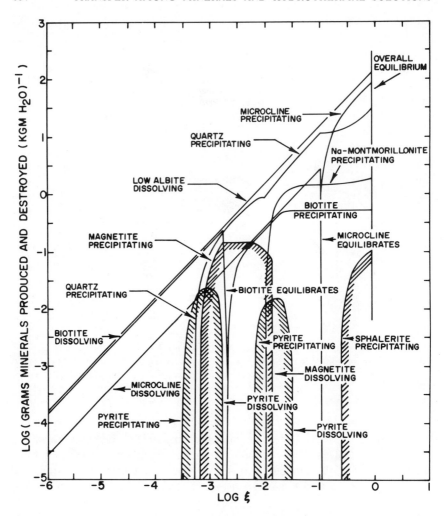

Fig. 11.8 Mass of minerals produced and destroyed (kg H$_2$O)$^{-1}$ as a function of progress in the reaction of a granitic rock with a potential ore-forming solution at 300°C and 85 bars (modified after Helgeson, 1970 and reproduced with permission from the Mineralogical Society of America).

where a and b are constants. Hence

$$\frac{dP}{d\xi} = b\frac{dT}{d\xi} \tag{11.108}$$

After carrying out an isothermal/isobaric mass-transfer calculation to characterize the overall equilibrium state resulting from the irreversible reaction of a hydrothermal solution with its mineralogic environment at a given pressure and temperature, we can then let $dT/d\xi = 1$ and examine the consequences of cooling along the geothermal profile represented by equation (11.107).

The consequences of various thermodynamic constraints such as constant enthalpy, entropy, or volume on reactions among minerals and hydrothermal solutions can also be assessed quantitatively with the aid of mass-transfer calculations. For example, if a given hydrothermal process is constrained to take place at constant enthalpy, $dT/d\xi$ becomes an unknown. However, it can be calculated by enlarging the matrix by one row and column to accommodate $dT/d\xi$ and a heat conservation equation for the process. The extent to which thermodynamic components are redistributed among phases in irreversible adiabatic processes can then be evaluated.

CONCLUDING REMARKS

Reaction paths on activity or chemical potential diagrams are vector quantities that can be described in terms of linear algebra. Work is currently in progress to determine the vector properties of various reaction paths and their implications with respect to chemical affinity and its change as a function of reaction progress. Nonequilibrium phase relations in hydrothermal systems can be represented quantitatively on chemical affinity diagrams, which afford a convenient frame of reference for evaluating the thermodynamic consequences of different reaction paths. Other efforts are being directed toward a better understanding of the relation between reaction kinetics and the chemical affinities of hydrolysis reactions. The rates of many of these reactions in nature are probably controlled to a large extent by the thermodynamic properties of the reactant minerals and the aqueous species involved in the reaction process. In some instances rate control may be exercised by the chemical affinity of the reaction. For example, in the case of a reaction near equilibrium, the rate ($\hat{r}$) can be expressed as (Prigogine, 1967)

$$\hat{r} = \frac{d\xi}{dt} = \frac{k\mathbf{A}}{RT} \qquad (11.109)$$

where $\mathbf{A}$ refers to the chemical affinity of the reaction, t denotes time, and k stands for the rate constant given by

$$k = \hat{A}e^{-\Delta E^{\ddagger}/RT} \tag{11.110}$$

where $\hat{A}$ designates the pre-exponential factor and $\Delta E^{\ddagger}$ corresponds to the activation energy for the reaction.

Mass-transfer calculations combined with field and laboratory observations afford a better understanding of the chemistry of processes responsible for hydrothermal phase relations. Each mass-transfer calculation is a computer experiment, which produces quantitative results. Comparison of these results with fluid inclusion analyses and the relative abundances of minerals in hydrothermal systems offers the means to integrate theoretical geochemistry with geologic reality.

ACKNOWLEDGMENTS

The research reviewed in this chapter has been supported over the past 10 years by NSF grants GP-4140, GE-9758, GU-1700, GU-2190, GA-828, GA-11285, GA 21509, GA-25314, GA-35888, GA-36023, DES 74-14280, and EAR 77-14492, the donors of the Petroleum Research Fund administered by the American Chemical Society under the auspices of PRF grants 5356-AC2 and 8927 AC2C, and funds received from Northwestern University, the Miller Institute and the Committee on Research at the University of California, Berkeley, the Lawrence Livermore and Berkeley Laboratories (LLL Grant 3701203 and ERDA Contracts W-7405-ENG-48 and UCB-ENG-4288), the Kennecott Copper Corporation, and the Anaconda Company. I am also indebted to many colleagues and friends who contributed substantially to the development of the theoretical approach and ideas summarized above, notably R. M. Garrels, F. T. Mackenzie, T. H. Brown, D. H. Kirkham, A. Nigrini, R. H. Leeper, D. Thorstenson, T. A. Jones, R. Beane, M. Lafon, D. Norton, P. Aagaard, and D. Bird. In addition I would like to express my appreciation to D. Crerar, H. L. Barnes, D. Norton, T. H. Brown, P. Aagaard, D. Bird, K. Jackson, and G. Flowers for their constructive reviews of the manuscript and many helpful suggestions for improvement. Thanks are also due D. Aoki for her enviable patience and skill in typing the manuscript.

REFERENCES

Aagaard, P. and H. C. Helgeson (1977) Thermodynamic and kinetic constraints on the dissolution of feldspars: *Geol. Soc. Am. Abstracts with Programs,* **9,** 873.

―――― and H. C. Helgeson (1979) Thermodynamic and kinetic constraints on reaction rates among minerals and aqueous solutions. I. Theoretical considerations: *Am. J. Sci.,* **279** (in press).

REFERENCES

Berner, R. A. (1978) Rate control of mineral dissolution under earth surface conditions: *Am. J. Sci.*, **278**, 1235-1252.

Brady, J. B. (1977) Metasomatic zones in metamorphic rocks: *Geochim. Cosmochim. Acta*, **41**, 113-126.

Brownell, D. H., Jr., S. K. Garg, and J. W. Pritchett (1975) Computer simulation of geothermal reservoirs: *Paper SPE 5381, 45th Annual Calif. Regional Meeting SPE-AIME*, Ventura, Calif.

Busenberg, E. and C. V. Clemency (1976) The dissolution kinetics of feldspars at 25°C and 1 atm CO_2 partial pressure: *Geochim. Cosmochim. Acta*, **40**, 41-50.

Cathles, L. M. (1976) A physical model for pluton driven ground water convection: *Geol. Soc. Am., Abst. Prog.*, **8**, 805-806.

Cheng, P. and K. H. Lau (1974) Steady state free convection in an unconfined geothermal reservoir: *J. Geophys. Res.*, **79**, 4425-4431.

Crerar, D. A. (1975) A method for computing multicomponent chemical equilibria based on equilibrium constants: *Geochim. Cosmochim. Acta*, **39**, 1375-1384.

De Donder, T. (1920), *Lecons de Thermodynamique et de Chimie-Physique*: Paris: Gauthier-Villars.

_____ and P. Van Rysselberghe (1936) *Thermodynamic Theory of Affinity: A Book of Principles*, Stanford: Stanford University Press.

Donaldson, I. G. (1970) The simulation of geothermal systems with a simple convective model: *Geothermics*, Special Issue 2, part 1, 649-654.

Droubi, A., C. Cheverry, B. Fritz, and Y. Tardy (1976a) Geochimie des eaux et des sels dans les sols des polders du lac Tchad: Application d'un modele thermodynamic de simulation de l'évaporation: *Chem. Geol.*, **17**, 165-177.

_____, P. Vieillard, G. Bourrie, B. Fritz, and Y. Tardy (1976b) Étude theorique de l'alteration des plagioclases bilans et conditions de stabilite des mineraux secondaires en function de la pression partielle de CO_2 et de la temperature (0°C á 100°C): *Sci. Geol. Bull.* (France), **29**, 45-62.

Elliott, D. (1973) Diffusion flow laws in metamorphic rocks: *Geol. Soc. Am. Bull.*, **84**, 2645-2664.

Faust, C. F. and J. W. Mercer (1975) Mathematical modeling of geothermal systems: *Proceedings, Second United Nations Symposium on the Development and Use of Geothermal Resources, San Francisco*, **3**, 1635-1642.

Fisher, G. W. (1970) The application of ionic equilibria to metamorphic differentiation: an example: *Am. J. Sci.*, **29**, 91-103.

_____ (1973) Nonequilibrium thermodynamics as a model for diffusion-controlled metamorphic processes: *Am. J. Sci.*, **273**, 897-924.

_____ (1975) The thermodynamics of diffusion-controlled metamorphic processes: in *Mass Transport Phenomena in Ceramics*, A. R. Cooper and A. H. Heuer, eds., New York: Plenum Press, p. 111-122.

_____ and D. Elliott (1974) Criteria for quasi-steady diffusion and local equilibrium in metamorphism: in *Geochemical Transport and Kinetics*, A. W. Hofmann, B. J. Giletti, H. S. Yoder, and R. A. Yund, eds., Washington, D.C.: Carnegie Institution Pub. 634, 231-241.

Fitts, D. D. (1962) *Nonequilibrium Thermodynamics: A Phenomenological Theory of Irreversible Processes in Fluid Systems*, New York: McGraw-Hill.

Fletcher, R. C. and A. W. Hofmann (1974) Simple models of diffusion and combined diffusion-infiltration metasomatism: in *Geochemical Transport and Kinetics*, A. W. Hofmann, B. J. Giletti, H. S. Yoder, and R. A. Yund, eds., Washington, D.C.: Carnegie Institution Pub. 634, 231-241.

Fouillac, C. G., Michard, and G. Bocquier (1977) Une methode de simulation de l'évolution des profils d'alteration: *Geochim. Cosmochim. Acta,* **41,** 207-213.

Frantz, J. D. and H. K. Mao (1976) Bimetasomatism resulting from intergranular diffusion: I. A. Theoretical model for monomineralic reaction zone sequences: *Am. J. Sci.,* **276,** 817-840.

─── and A. Weisbrod (1973) Infiltration metasomatism in the system K_2O-Al_2O_3-SiO_2-H_2O-HCl: *Carnegie Institution Year Book,* **72,** 507-515.

Fritz, B. and Y. Tardy (1976a) Séquence des minéraux secondaires dans l'altération des granites et roches basiques; modéles thermodynamiques: *Bull. Soc. Geol. France,* **7,** 7-12.

─── and ─── (1976b) Predictions of mineralogical sequences in tropical soils by a theoretical dissolution model: in *Proceedings Int. Symp. on Water-Rock Interaction,* J. Cadek and T. Paces, eds., Prague: Geol. Survey, p. 409-416.

Grandstaff, D. E. (1978) Changes in surface area and morphology and the mechanism of forsterite dissolution: *Geochim. Cosmochim. Acta.* **42,** 1899-1902.

Helgeson, H. C. (1967) Solution chemistry and metamorphism: in *Researches in Geochemistry,* Vol. II, P. H. Abelson, ed., New York: Wiley, p. 362-404.

─── (1968) Evaluation of irreversible reactions in geochemical processes involving minerals and aqueous solutions—I. Thermodynamic relations: *Geochim. Cosmochim. Acta,* **32,** 853-877.

─── (1969) Thermodynamics of hydrothermal systems at elevated temperatures and pressures: *Am. J. Sci.,* **267,** 729-804.

─── (1970) A chemical and thermodynamic model of ore deposition in hydrothermal systems: in *Fiftieth Anniversary Symposia,* B. A. Morgan, ed., *Min. Soc. Am. Spec. Paper 3,* 155-186.

─── (1971) Kinetics of mass transfer among silicates and aqueous solutions: *Geochim. Cosmochim. Acta,* **35,** 421-469.

─── (1972) Kinetics of mass transfer among silicates and aqueous solutions: Correction and clarification: *Geochim. Cosmochim. Acta,* **36,** 1067-1070.

─── (1974) Chemical interaction of feldspars and aqueous solutions: in *The Feldspars,* W. S. MacKenzie and J. Zussman, eds., Manchester University Press, p. 184-217.

───, T. H. Brown, A. Nigrini, and T. A. Jones (1970) Calculation of mass transfer in geochemical processes involving aqueous solutions: *Geochim. Cosmochim. Acta,* **34,** 569-592.

───, J. M. Delany, H. W. Nesbitt, and D. K. Bird (1978), Summary and critique of the thermodynamic properties of rock-forming minerals: *Am. J. Sci.,* **278A.**

───, R. M. Garrels, and F. T. McKenzie (1969) Evaluation of irreversible reactions in geochemical processes involving minerals and aqueous solutions—II. Applications: *Geochim. Cosmochim. Acta,* **33,** 455-481.

─── and D. H. Kirkham (1974a) Theoretical prediction of the thermodynamic behavior of aqueous electrolytes at high pressures and temperatures. I. Summary of the thermodynamic/electrostatic properties of the solvent: *Am. J. Sci.,* **274,** 1089-1198.

─── and ─── (1974b) Theoretical prediction of the thermodynamic behavior of aqueous electrolytes at high pressures and temperatures. II. Debye-Hückel parameters for activity coefficients and relative partial molal properties: *Am. J. Sci.,* **274,** 1199-1261.

─── and ─── (1976) Theoretical prediction of the thermodynamic behavior of aqueous electrolytes at high pressures and temperatures. III. Equation of state for aqueous species at infinite dilution: *Am. J. Sci.,* **276,** 97-240.

———, D. H. Kirkham, and G. C. Flowers (1979) Theoretical prediction of the thermodynamic behavior of aqueous electrolytes at high pressures and temperatures. IV. Calculation of activity coefficients, osmotic coefficients, and apparent molal and standard and relative partial molal properties to 5 kb and 600°C: *Am. J. Sci.*, **279** (in press).

Hofmann, A. (1972) Chromatographic theory of infiltration metasomatism and its application to feldspars: *Am. J. Sci.*, **272**, 69-90.

——— (1973) Theory of metasomatic zoning, a reply to Dr. D. S. Korzhinskii: *Am. J. Sci.*, **273**, 960-964.

Joesten, R. (1977) Evolution of mineral assemblage zoning in diffusion metasomatism: *Geochim Cosmochim Acta*, **41**, 649-670.

Karpov, I. K., L. A. Kaz'min, and S. A. Kashik (1973) Optimal programming for computer calculation of irreversible evolution in geochemical systems: *Geokhimiya*, 603-611 (*Geochem. Int.*, **1974**, 464-470).

Korzhinskii, D. S. (1959) *Physiochemical Basis of the Analysis of the Paragenesis of Minerals*, New York: Consultants Bureau.

——— (1970) *Theory of Metasomatic Zoning*. J. Agrell, trans., London: Oxford University Press.

——— (1973) Theory of metasomatic zoning, a reply to Dr. Albrecht Hofmann: *Am. J. Sci.*, **272**, 69-90.

Lafon, G. M. and F. T. Mackenzie (1971) Early evolution of oceans—A weathering model: *Bull. Am. Assoc. Petrol. Geol.*, **55**, 348-360.

Lagache, M. (1965) Contribution a l'etude de l'alteration des feldspaths dans l'eau, entre 100 et 200°C, sous diverses pressions de CO_2, et application à la synthese des mineraux argileux: *Bull. Soc. Fr. Min. Crist.*, **88**, 223-253.

——— (1976) New data on the kinetics of the dissolution of alkali feldspars at 200°C in CO_2 charged water: *Geochim. Cosmochim. acta*, **40**, 157-161.

Lasseter, T. J. and P. A. Witherspoon (1975) Multiphase multi-dimensional simulation of geothermal reservoirs: *Proceeding. United Nations Symposium on the Development and Use of Geothermal Resources*, San Francisco, May 20-29, **3**, 1715-1724.

Mercer, J. W. and G. F. Pinder (1974) Finite element analysis of hydrothermal systems: in *Finite Elements in Fluid Flow*. J. T. Oden et al., eds., Huntsville: University of Alabama Press, 401-414.

———, ——— and I. G. Donaldson (1975) A Galerkin-finite element analysis of the hydrothermal system at Wairakei, New Zealand: *J. Geophys. Res.*, **80**, 2608-2621.

Norton, D. (1972) Concepts relating anhydrite deposition to solution flow in hydrothermal systems: *Proc. 24th Int. Geol. Cong.*, Sec. 10, 237-244.

——— (1977) Fluid circulation in the earth's crust: *Am. Geophys. Union Mono.*, **20**, 693-700.

——— (1978) Sourcelines, sourceregions, and pathlines for fluids in hydrothermal systems related to cooling plutons: *Econ. Geol.*, **73**, 21-28.

——— (1979) Transport processes related to copper-bearing porphyritic plutons: Fluid and heat transport in pluton environments typical of the southeastern Arizona porphyry copper province: in *Geology of the Porphyry Copper Deposits*, S. R. Titley, ed., Tucson: Univ. of Arizona Press (in press).

——— and R. Knapp (1977) Transport phenomena in hydrothermal systems: The nature of porosity: *Am. J. Sci.*, **277**, 913-936.

——— and J. Knight (1977) Transport phenomena in hydrothermal systems: Cooling plutons: *Am. J. Sci.*, **277**, 937-981.

_____ and H. P. Taylor, Jr. (1979) Quantitative simulation of the thermal history of crystallizing magmas on the basis of oxygen isotope data and transport theory: An analysis of the hydrothermal system associated with the Skaergaard Intrusion: *J. Petrol.*, **20** (in press).

Paces, T. (1973) Steady-state kinetics and equilibrium between ground water and granitic rocks: *Geochim. Cosmochim. Acta*, **37**, 2641-2663.

Petrovich, R. (1976) Rate control in feldspar dissolution. II. The protective effect of precipitates: *Geochim. Cosmochim. Acta*, **40**, 1509-1521.

Prigogine, I. (1955) *Introduction to Thermodynamics of Irreversible Processes*, New York: Wiley.

_____ and R. Defay (1954) *Chemical Thermodynamics*, D. H. Everett, trans., London: Jarrold and Sons.

Sarazan, G., C. Fouillac, and G. Michard (1976) Etude de l'altération des rockes granitiques sous climat tempéré: *Geochim. Cosmochim. Acta*, **40**, 1481-1486.

Straus, J. M. and G. Schubert (1977) Thermal convection of water in a porous medium: Effects of temperature- and pressure-dependent thermodynamic and transport properties: *J. Geophys. Res.*, **82**, 325-333.

Thompson, J. B., Jr. (1959) Local equilibrium in metasomatic processes: in *Researches in Geochemistry*, ed. P. H. Abelson, Wiley and Sons, New York, 427-457.

Villas, R. N. and D. Norton (1977) Irreversible mass transfer between circulating hydrothermal fluids and the Mayflower stock: *Econ. Geol.*, **72**, 1471-1504.

Walther, J. V., and H. C. Helgeson (1977) Calculation of the thermodynamic properties of aqueous silica and the solubility of quartz and its polymorphs at high pressures and temperatures: *Am. J. Sci.*, **277**, 1315-1351.

Weare, J. H., and J. R. Stephens, and H. P. Eugster (1976) Diffusion metasomatism and mineral reaction zones: General principles and application to feldspar alteration: *Am. J. Sci.*, **276**, 767-816.

Wolery, T. J. (1978) Some chemical aspects of hydrothermal processes at mid-oceanic ridges—a theoretical study: I. Basalt-sea water reaction and chemical cycling between the oceanic crust and the oceans. II. Calculation of chemical equilibrium between aqueous solutions and minerals: *Ph.D. thesis*, Northwestern University, Evanston, Ill.

Wolery, T. J., and N. H. Sleep (1976) Hydrothermal circulation and geochemical flux at mid-ocean ridges: *J. Geol.*, **84**, 249-275.

Wollast, R. (1967) Kinetics of the lateration of K-feldspar in buffeted solutions at low temperature: *Geochim. Cosmochim. Acta*, **31**, 635-648.

Wood, J. (1974) A numerical analysis of diffusional mass transfer during the hydrolysis of K-feldspar: *Trans. Am. Geophys. Union*, **44**, 699.

12

Thermal Aspects of Ore Deposition

DENIS NORTON
University of Arizona

LAWRENCE M. CATHLES
Pennsylvania State University

The principal thermal aspect of hydrothermal ore deposition is the generation of a hydrodynamic system by thermal anomalies in the crust and dispersion of this anomaly by heat conduction and fluid convection. Consequently, thermal aspects of ore deposition are inseparably coupled to the manner in which fluids flow through fractured host rocks and to heat transfer away from a central thermal source. Environments of hydrothermal ore deposits are characterized by complex tectonics and plutonism—the former producing continuous fractures capable of transmitting fluids through otherwise impermeable rocks and the latter generating a fluid potential field. Recognition of these factors in this class of deposits began with geologic observations of Kemp (1905), Van Hise (1901), and Lindgren (1907). Interest now centers on attempts to quantitatively understand the transfer of material and heat during the process of ore formation, with emphasis on the relationships between the thermal history of an ore-forming system and the associated process of fluid flow.

HEAT TRANSFER PROCESSES

Fluid flow through fractured, permeable rocks can be adequately described by the empirical relationships between fluid flow potential, rock and fluid properties, and the mass flux of fluid given by Darcy's Law (Hubbert, 1940):

$$q = -\frac{k}{\nu}(\rho g \hat{z} + \nabla P), \qquad (12.1)$$

in which q is the fluid mass flux in g/cm$^2 \cdot$sec and ∇ is the gradient operator which gives the magnitude and direction of the most rapid change in hydrostatic pressure, P. The coordinate system is such that $\hat{z}$ is a unit vector directed vertically; $z = 0$ at the surface, and z is negative downward from the surface into the earth.

If there is no fluid flow, the mass flux, q, is zero, which requires the hydrostatic condition

$$\rho g \hat{z} + \nabla P = 0. \qquad (12.2)$$

Equation (12.2) has an interesting consequence that becomes more apparent if written in vector form; each line of the following represents a separate equation:

$$\begin{bmatrix} 0 \\ 0 \\ \rho g \end{bmatrix} + \begin{bmatrix} \partial P/\partial x \\ \partial P/\partial y \\ \partial P/\partial z \end{bmatrix} = \begin{matrix} 0 \\ 0 \\ 0 \end{matrix} \qquad (12.3)$$

The last line defines the pressure at any depth, when there is no fluid flow, as being equal to the hydrostatic pressure due to the weight of fluid above the location in question. This becomes clearer when the equation is integrated

$$P(z) = -g \int_0^z \rho(z)dz + P(z = 0) \qquad (12.4)$$

Since z is negative with depth, (12.4) indicates pressure increases with depth, and, if the fluid density $\rho(z) =$ constant, (12.4) states that $P(z) = P(z = 0) - \rho g z$, which is the normal hydrostatic increase in pressure with depth.

HEAT TRANSFER PROCESSES

A further consequence of (12.3) is that for no fluid flow $q = 0$, $\partial P/\partial x = \partial P/\partial y = 0$, indicating that pressure can only vary with depth and not with horizontal location. If P can vary only with depth, equation (12.4) indicates that fluid density must vary only with z also. Conditions of no fluid flow require fluid density to be constant on horizontal planes. Then fluid flow is an inevitable consequence of fluid density variations on a horizontal plane at a given elevation. The magnitude of this flow is the only remaining question, and from equation (12.1) we see that the fluid mass flux depends on the permeability of the rocks and viscosity of the fluid.

Consider the condition where an igneous body is emplaced into water-saturated rocks, and fluids at the contact, or within the crystallizing body, are several hundred degrees hotter than fluids removed from the influence of the intrusion but lie on the same horizontal plane. The fluid in the vicinity of the pluton will be less dense than the fluid away from the influence of the hot intrusion. From equation (12.1) and the fact that $\partial P/\partial x$ and $\partial P/\partial y$ in equation 12.3 are nonzero, we see that fluid flow is a required condition.

Fluid flow in the crust is a natural, unavoidable consequence of thermal anomalies that generate fluid density variations in a horizontal plane. Thermally forced convective flow will be associated with virtually all igneous intrusives that intrude a water-saturated environment, although some situations may be characterized by low fluid flow due to low permeabilities of the pluton environment.

Variation in fluid density, as a function of temperature at constant pressure, and solute concentrations are defined by the fluid coefficient of thermal expansion, α:

$$\alpha = -\frac{1}{\rho}\left(\frac{\partial \rho}{\partial T}\right)_{P,m_i} \tag{12.5}$$

Assuming α is a constant, density may be calculated throughout the thermal anomaly and surrounding regions with the approximate integrated form of equation (12.5)

$$\rho = \rho_0(1 - \alpha T) \tag{12.6}$$

Substituting (12.6) into (12.1) yields

$$q = -\frac{k}{\nu}(\rho_0(1 - \alpha T)g\hat{z} + \nabla P) \tag{12.7}$$

Fluid flow occurs in response to a perturbation in the normal hydrostatic pressure gradients in a manner that disperses the abnormal pressure variations. Therefore, the hydrostatic pressure gradients tend to be preserved in spite of temperature-related density anomalies. An approximation of vertical flow rates that are produced by an igneous intrusion in the crust can be made from equation (12.7) if ∇P is set equal to $(-\rho_0 g)$, the normal hydrostatic pressure gradient for cool (20°C) water. With this substitution (12.7) becomes

$$q_z \approx \frac{k}{\nu} g \rho_0 \alpha T \qquad (12.8)$$

If we take the temperature anomaly caused by the pluton, T, to be 750°C and the top of the pluton to be 2 km below the surface, e.g., $P \approx 200$ bars, the estimated mass fluid flux can be calculated from (12.8) for $\nu = 9 \times 10^{-3}$ cm^2/sec, $\rho_0 = 1$ g/cm^3, k = 1 mdarcy = 10^{-11} cm^2, $\alpha = 1.2 \times 10^{-3}$ °C^{-1}, and $g = 10^3$ cm/sec^2; then

$$q_z \approx 10^{-6} \text{ g/cm}^2 \cdot \text{sec} \qquad (12.9)$$

Fluid density perturbations are also related to variations in the solute concentrations in aqueous solutions. Compositions of both fluid inclusions in quartz crystals and mineral assemblages associated with ore deposits indicate that aqueous solutions range from about 10^{-2} m to > 3 m in solute concentration (see Table 8-2), principally as NaCl and other alkali chloride salts. An analysis of the effects of salinity on low-temperature groundwater flow problems is given by Nield (1968), Veronis (1968), and Rubin (1973). These studies and data on apparent molal volume of alkali-metal salts (Ellis and Golding, 1963) indicate that the solute effect on solution density is sufficient to partially compensate for the density anomaly due to anomalous temperature gradients. For example, a 3 m (17.5 wt. %) NaCl solution at 350°C and 200 bars will have the same density as pure H_2O at a 75° cooler temperature. The 35 wt. % NaCl solutions observed in the porphyry copper plutons would have the same density as pure H_2O at about 150° cooler temperature. Additional factors include the viscosity and density of fluids in the NaCl-H_2O system, the distribution and source of high salinity fluids in the pluton, and the variation in concentration of these fluids with time. A quantitative evaluation of salinity-related density perturbations will not be presented. However, even a brine-like hydrothermal fluid column will probably be less dense than solutions occurring away from the pluton environment along the same horizontal plane, provided the brine is hot. Just as with thermal density anomalies, a solute anomaly should generate a fluid potential and disperse itself by flow, and, since the solute presents a concentration gradient, diffusion of salt through the aqueous phase tends to disperse the solute anomaly.

HEAT TRANSFER PROCESSES

Fluid flow resulting from purely thermal effects of fluid density contributes to cooling. As fluid convects through a thermal anomaly, energy is dispersed upward and away from the source, lowering the temperature and decreasing the rate of fluid circulation. Temperature variations as a function of time in these environments may be described by the following two-dimensional conservation of energy equation:

$$C\rho_r \frac{\partial T}{\partial t} = K\nabla^2 T - q_z \frac{\partial H}{\partial z} - q_y \frac{\partial H}{\partial y} \qquad (12.10)$$

$$\quad\quad\quad\quad \text{conduction} \quad\quad\quad \text{convection}$$

where $C\rho_r$ is the heat capacity of the rock in cal/cm³·°C. This equation provides for heat transfer both by conduction and by convection of aqueous fluids during the postmagmatic cooling period but does not account for heat sources or sinks due to chemical reactions. The temperature distribution in and around a thermal anomaly is determined by the relative magnitude of the conductive heat flux,

$$J_c = -K\nabla T \qquad (12.11)$$

and the convective heat flux,

$$J_f = qH \qquad (12.12)$$

which corresponds respectively to the temperature variation terms due to conduction and convection in equation (12.10). For the case calculated in equation (12.9), and assuming a fluid heat content $H = 940$ cal/g,

$$J_{f(z)} = 9.4 \times 10^{-4} \text{ cal/cm}^2 \cdot \text{sec}. \qquad (12.13)$$

And if we assume the thermal conductivity, $K = 6 \times 10^{-3}$ cal/cm·sec·°C, and that the temperature decreases, at the most, 750° over a distance of 0.5 km,

$$J_{c(z)} = 9 \times 10^{-5} \text{ cal/cm}^2 \cdot \text{sec} \qquad (12.14)$$

Therefore, where rock permeabilities are sufficient, the transport of heat by convection is at least as important as dispersion of heat by conduction. Naturally, as the heat is dispersed, the magnitude of the fluid potential decreases because the density perturbation is decreased and is distributed over larger distances.

Since the convective fluid flux is directly proportional to the permeability, a system with permeability as low as 10^{-12} cm^2 would generate a convective flux equivalent to the conductive thermal flux. In situations where rock permeabilities are considerably less than 10^{-12} cm^2, conduction will become more important. Such low permeabilities are probably realized in very deep (under 15 km) regions below hydrothermal ore-forming environments. The transfer of heat by pure conduction also becomes more important in blocks of rock between continuous fractures where insignificant fluid flow is realized (Norton and Knapp, 1977).

The case where only conductive heat flux occurs has been thoroughly analyzed for numerous rock geometries (Lovering, 1935; Jaeger, 1968). In conductive cooling the isotherms are distributed approximately parallel to the surface of the thermal anomaly during the entire cooling history. Wall-rock temperatures are a function of the width of the thermal source, and temperatures equal to 20% of the initial temperature difference extend outward from the intrusive midplane a distance 2.5 times the body half-width. The temperatures outside the intrusive boundary never exceed 50% of the initial anomalous temperature during conductive cooling.

In view of the permeable nature of hydrothermal systems (Norton and Knapp, 1977; Villas, 1975; Villas and Norton, 1977), fluid convection may be expected whenever a hot pluton intrudes fluid-saturated rocks. In this situation, the fluid flux through fractures defines the temperature distributions and isotherms controlled by fluid fluxes, permeability, and fluid potential variations. The principles of conservation of momentum, equation (12.1), conservation of energy, equation (12.10), and conservation of mass

$$\nabla \cdot q = 0 \qquad (12.15)$$

together with an equation of state such as (12.6), initial conditions, and geologically reasonable boundary conditions describe the thermal and fluid evolution of an intrusive region. Numerical solutions for a similar system of equations are presented by Holst and Aziz (1972), Elder (1967), and Wooding (1957). Studies that account for the presence of an igneous body in the domain and evaluate the effect of temperature and pressure-dependent fluid properties are described by Cathles (1977) and Norton and Knight (1977).

The nature of circulatory fluid flow and temperature variations is presented for a simplistic pluton environment. The geologic situation is approximated by an igneous body emplaced 2.75 km below the surface at an initial temperature 750°C above ambient, Figure 12.1. This body is 2.75 km high and 2 km across and extends to infinity in the third dimension. Initially, temperature is assumed not to differ from the normal thermal gradient in the region intruded, for example, 30°C/km, and fluid fluxes are taken to be zero. The permeability of the pluton is approximately 0.07 mdarcy (7×10^{-13} cm^2) and that of the host rocks is 0.13 mdarcy (1.3×10^{-12} cm^2). Initially, the pluton is considered to be an elastic solid with continuous fractures that are equivalent in abundance and width of opening to the above bulk permeabilities (Norton and Knapp, 1977). The variations in fluid streamlines and anomalous temperatures are then computed for a series of intervals—20,000, 40,000, and 100,000 years after intrusion and the results shown in Figure 12.1. Isotherms in the lower portion of the system tend to be close against the sides of the pluton, due to the influx of ambient fluids toward its side contacts. For larger fluid fluxes, this effect becomes even more pronounced, as does the extensions of isotherms upward into the overlying host rocks. The temperature distribution around a particular streamline[1] is characterized by large positive thermal gradients in zones where fluid flows into the side of the body and by steep negative gradients where fluid flows out the top of the pluton.

A significant feature of this cooling process is the systematic variation of fluid flux with temperature—a natural consequence of the physics involved. A theoretical basis for the enduring concept of mineralization "pulses" and complex paragenesis of alteration and ore deposition may, in part, be attributed to this feature in that a large thermal anomaly associated with the emplacement of igneous masses and concomitant rock fracturing leads to relatively high temperatures and large fluid flux through the fractures (see also Knapp, 1978; Knapp and Knight, 1977). Decreases in both temperature and fluid flux follow as the thermal anomaly is dispersed, but any new igneous event proportionately increases the fluid flux. In situations where subsequent intrusives are spatially related to the same major vein system, additional masses of hydrothermal fluid are sent along the vein.

In the case of anisotropic permeability, in particular those geologic situations of a few large open fractures around the intrusive, the isotherms would be distorted outward along the fractures. A feeling for the thermal gradients and effects along a single vein can be obtained from Toulmin and Clark (1967). A few fractures with large openings tend to produce permeabilities much greater than a few millidarcies and, hence, larger fluid fluxes. Detailed consideration of the thermal aspects of ore deposition requires a notion of the magnitude of thermal energy sources, of their dis-

---

[1] Gradients between the streamlines define the fluid flux since $q = \nabla \times \Psi$ and $q_z = -\partial \Psi / \partial y$. The stream function, Ψ, is tangential to the mass flux vector. The closer together the streamlines, the greater the mass flux, q.

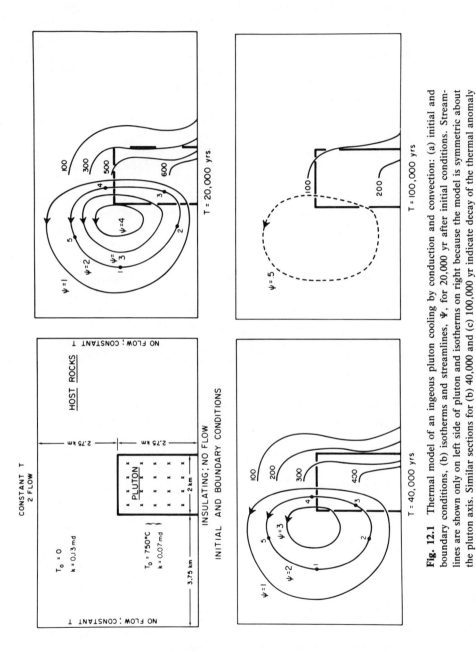

Fig. 12.1 Thermal model of an ingeous pluton cooling by conduction and convection: (a) initial and boundary conditions, (b) isotherms and streamlines, Ψ, for 20,000 yr after initial conditions. Streamlines are shown only on left side of pluton and isotherms on right because the model is symmetric about the pluton axis. Similar sections for (b) 40,000 and (c) 100,000 yr indicate decay of the thermal anomaly and decrease in fluid fluxes.

THERMAL ENERGY SOURCES

Heat Content

Thermal energy in the crust is generated by conduction of heat and emplacement of magmas from subcrustal regions, radioactive decay of ^{40}K, ^{238}U, and ^{232}Th, stress release along fault zones, exothermal mineral-solution reactions, and tidal energy. The most important of these sources is the subcrust through both conduction and transfer of energy by magmatic and related processes; other sources contribute less than a few calories of energy per gram of rock, in a fixed time increment, whereas several hundred calories per gram of rock are released by the magmatic processes in the same time interval.

The total thermal energy related to an igneous body is determined by the mass and composition of the magma. The distribution of energy is determined by the body shape. Total thermal energy released by a magma cooling to ambient temperature and interacting with aqueous solutions is due to the heat of crystallization of the magma, the heat content of the melt and solids, the heat of hydrolysis reactions, and the heat of associated reactions in the aqueous phase:

$$Q = M \, \Delta H_x + M \, C(T_i - T_0) + M_r \, \Delta H_r, \qquad (12.16)$$

where the heat capacity of the melt and solid, C, is in cal/g·°C, the heat of crystallization of solid phases from the melt, ΔH_x, is in cal/g, heat due to hydrolysis reactions, ΔH_r, is in cal/g, and the masses of the total rock and altered rock, M and M_r, respectively, are in grams. Aqueous reactions among ions and complexes and oxidation-reduction reactions are not considered. Thermal properties of systems analogous to basaltic, granitic, and albite-H_2O type systems are summarized in Table 12-1. Average heat capacities of the solids used in the calculation are suitable because, over the entire temperature range, the mineral heat capacities vary by only approximately 10%. Assuming the heat capacities of solid and melt are equal will not introduce significant errors in the calculation since geologic evidence suggests that most plutonic bodies were emplaced at temperatures below their liquidus and that, during 90% of their cooling history in the upper crust, were dominantly crystalline bodies.

Heat of crystallization is a function of magma composition, especially

Table 12-1 Thermal properties of igneous rocks

	Basalt[c]	Granite[c]	Albite-H_2O
Heat capacity, cal/g·°C	0.22	0.26	0.27
Density, g/cm^3	2.86	2.78	2.62
ΔH_x, cal/g	100	65	50[a]
$T_{solidus}$,[b] °C	900	700	1000
$T_{liquidus}$,[b] °C	1100	950	—
$C(T_{liq} - T_0)$, cal/g	242	247	270
ΔH_r^d, cal/g		44	—
Total heat cal/gr,			
Unaltered	342	312	310
50% altered		356	—

[a] 1 wt. % H_2O.
[b] at 2 kb pressure.
[c] H_2O saturated.
[d] Table 12-3.

with respect to the volatile components (Kesler and Heath, 1968). Data on the albite-H_2O systems (Burnham and Davis, 1974) permit the effect of dissolved water on ΔH_x to be calculated, Table 12-1. At depths in the crust less than 5 km, approximately 3 wt. % H_2O may be dissolved in granitic melts; therefore, $\Delta H_x = 50$ cal/g is reasonable for these magma types.

The heat of hydrolysis reactions that occur during the postmagmatic stages of cooling are due to (1) irreversible dissolution of reactant igneous minerals, (2) the sum of the heats of reversible precipitation or dissolution of hydrothermal product minerals, (3) the heats of association or dissociation of aqueous complexes, and (4) the heats of oxidation-reduction reactions between aqueous components. The enthalpy of reactions for equilibrium hydrolysis of rock forming silicates for reactions such as

$$\text{mineral} + H^+ \rightleftharpoons \text{aqueous uncomplexed ions}$$

may be expressed by

$$\Delta H_r = \Delta H_f(\text{reactants}) - \Delta H_f(\text{products}). \tag{12.17}$$

The enthalpy is the heat absorbed by the reaction, e.g., $-\Delta H_r$ is exothermic. In general, these reactions for silicate and oxide minerals, except for quartz, have large negative enthalpies of reaction, $\sim -10^4$ cal/mole, which decrease with increasing temperature (Helgeson, 1969), whereas the sulfide minerals and quartz have positive enthalpies of reaction of similar magnitude. The enthalpies of dissociation for aqueous ionic complexes typically are an order of magnitude less than those of the solid phases (Helgeson, 1969); however, oxidation reactions, particularly those involving aqueous species of sulfur and oxygen, have large negative enthalpies of reaction, $\sim -10^4$ cal/mole. Heats from oxidation reactions might be significant in hydrothermal systems where oxidation of reduced sulfur to sulfate is quantitatively important, for example, thermal spring environments.

The enthalpy of hydrolysis reactions has been calculated for the alteration of quartz diorite to quartz + sericite and to potassium feldspar + biotite at approximately 300°C, with the results shown in Tables 12-2 and 12-3, respectively. In the calculations the assumption is made that the heat of reaction due to the irreversible dissolution of minerals is equal to the heat of reaction due to the reversible dissolution and that the heat sources or sinks in the aqueous phase were minor. Data on abundances of minerals used in the calculation were selected from natural alteration assemblages (Villas, 1975). The data indicate that those overall reactions that produce these alteration phases are exothermic and that the energy released ranges from 44 to 68 cal/g.[2] The reader should also note that as the percentage of rock altered decreases, the enthalpy due to hydrolysis decreases. Also, the larger hydrolysis energy will tend to be released at the low temperatures, where the mass ratio, original minerals destroyed/alteration minerals produced, is the greatest. This latter effect is related directly to the large exothermic heats of reaction associated with mineral dissolution. Thus, the effect of the hydrolysis would be to extend the duration of the thermal anomaly, particularly for the lower temperature stages.

The total heat content of a pluton body that crystallizes at liquidus temperature, cools to ambient temperatures, and is partially altered by hydrothermal fluids, is estimated to be on the order of 360 cal/g for granite and, for its unaltered equivalents, 310 cal/g (Table 12-1). The total thermal energy associated with emplacement of felsic plutons that were subsequently altered by hydrothermal solutions is on the order of 10^{15} kcal/km^3 pluton; this estimate would be approximately 15% less for unaltered rock masses.

Shape

The shape and size of igneous bodies determine the distribution and quantity of thermal energy in the hydrothermal environment. The subsurface shapes of plutonic igneous masses are indirectly known. Basaltic bodies

---

[2] Note sign change such that heat sources are positive.

Table 12-2 Quartz diorite altered to quartz-sericite

Minerals in original rock	(wt. %)	Mass minerals, M_i, destroyed (g/100 g rock)	Heat of hydrolysis $M_i \Delta H_i$ (cal/g rock)
Albite	38	34	−18
Anorthite	17	17	−40
Quartz	16	−30	−25
Annite	11	10	−12
Potassium feldspar	18	−11	5
Muscovite	—	−20	29
Chalcopyrite	—		
Pyrite	—	−1	−7
Totals	100	−49	$\Delta H_r = \Sigma M_i \Delta H_i = -68$

Table 12-3 Quartz diorite altered to biotite + potassium-feldspar

Minerals in original rock	(wt. %)	Mass minerals, M_i, destroyed (g/100 g rock)	Heat of hydrolysis $M_i \Delta H_i$ (cal/g rock)
Albite	42	22	−14
Anorthite	28	17	−37
Quartz	12	−5	−4
Annite	10	−12	11
Potassium feldspar	8	−18	7
Chalcopyrite	—	−2	3
Pyrite	—	−2	−4
Totals	100	−61	$\Delta H_r = \Sigma M_i \Delta H_i = -44$

appear to have simple dikelike forms extending to great depths, whereas the granitic types are elongate to equant in plan view and extend only to a fraction of their diameter in depth. A combination of seismic, gravity, and heat flow data have led to the concept that batholithic masses in particular may be only a few kilometers thick (Hamilton and Myers, 1967). Estimates of the thickness of the Sierra Nevada Batholith range 5-15 km (Thompson and Talwani, 1964; Everden, 1965; Mikumo, 1965; Lachenbruch et al., 1966;

Wollenberg and Smith, 1964). The combined arguments of their independent geophysical observations, together with data on the distribution of contact metamorphic effects around intrusive bodies in deeply eroded terrains (Hamilton and Myers, 1967), suggest that the maximum diameter to thickness ratio of intrusive bodies is equal to or greater than unity. Extrapolating from these conclusions only requires the observation that the batholiths are masses of coalesced stocks. Igneous plutons in ore districts are often inferred in cross section to be the upper parts of huge subsurface masses extending to great depths (Sillitoe, 1972)—a hypothesis that is apparently in conflict with the batholith studies reported by Hamilton and Myers (1967).

Time and Space Distribution

Total thermal energy in a hydrothermal system is also determined by the temporal and spatial distribution of igneous events. Multiple igneous events generally occur in hydrothermal ore districts. Therefore, an overlap in the hydrothermal events seems inevitable since the radiometric ages of associated intrusives are generally separated by less than 1 m.y. (Moore and Lanphere, 1971; Ohmoto et al., 1966; and Page and McDougall, 1972). Such intrusions are often adjoining so that convective heat transfer from one source may be along the same continuous fractures as an earlier thermal source. Although the interaction of convective fluid systems has not been analyzed in detail, the intrusion of a second heat source into the model domain will certainly prolong the duration of circulating fluid flow, and, to the extent that continuous fractures retain their identity during the second igneous event, fluid/rock ratios will be proportionately increased even though the flow would undoubtedly change and become aligned with respect to the new heat source.

Convective fluids from intrusive bodies spaced at distances greater than the widths of their convective fluid systems will not be related. The typical spacing of active volcanic centers at modern convergent lithospheric plate boundaries is estimated to be on the order of 70 km (Marsh and Carmichael, 1974). This spacing is well beyond the limit of convective fluid circulation. Widths of convective systems are controlled by the depth to which flow occurs, e.g., the hydrothermal system in the model case is equal in width to the height of the confining domain, 5.50 km (Figure 12.1).

The time distribution of igneous events is virtually unknown on the scale of interest to the thermal model studies since the resolution of age dating methods is inadequate to detect intrusive events on the scale of $\pm 50,000$ years.

THERMAL ASPECTS OF ALTERATION AND SULFIDE DEPOSITION

The amount of hydrothermal fluid circulated through a rock volume can be approximated simply. Consider that the heat in a volume of an intrusive is lost entirely to circulating fluid, for example, none of the heat is conducted away. Then

$$Q(\text{gain fluid}) = Q(\text{loss pluton})$$

or

$$C_f M_f (T - T_0) = CM(T - T_0) + M \Delta H_x + M_r \Delta H_r \qquad (12.18)$$

where C_f is the heat capacity of the circulating fluid, see Table 12-4, M_f is the mass of fluid that circulates through the pluton as it cools from some temperature T to ambient T_0, and ΔH_x and ΔH_r are the enthalpy of crystallization and hydrolysis reactions, respectively. The mass ratio of altered rock to original rock, M_r/M, is taken to be 0.5; then

$$\frac{M_f}{M} = \frac{C}{C_f} + \left(\frac{\Delta H_x + 0.5 \Delta H_r}{T - T_0} \right) \qquad (12.19)$$

Substituting values for $C = 0.26$ cal/g·°C, $C_f = 1.3$ cal/g·°C, $T_0 = 80$°C, $T = 950$°C, $\Delta H_x = 50$ cal/g, and $\Delta H_r = 45$ cal/g, gives the fluid/rock ratio, $M_f/M = 0.28$. This approximation indicates that each gram of pluton requires, on the average, 0.3 g hydrothermal fluid to disperse the anomaly by convective heat transfer. In a pluton with uniformly distributed permeability, each gram of igneous rock would react with 0.3 g fluid. This is equivalent to an atomic oxygen ratio of 0.4, which is comparable in magnitude to predicted ratios based on $\delta^{18}O$ values (Taylor, 1971). Inhomogeneous permeability only affects the calculated local fluid/rock ratio; equation (12.19) does not depend on permeability, k.

Inhomogeneous permeability is expected in ore deposits and, indeed, is demonstrated where the distribution of alteration and ore occur along a few continuous structures. Locally, along flow channels fluid/rock ratios may exceed 100/1 due to this effect. However, from the overall viewpoint, the variations in permeability merely increase or decrease the fluid flux and hence the heat flux. A corresponding shorter or longer cooling period results, and no substantial variation in fluid/rock ratios is expected—the exception being at some lower limit of permeability where conductive heat transfer predominates. Concepts regarding the fluid flow paths and sources are discussed in Norton (1978). The following discussion describes conditions only at fixed locations in the system.

The extreme temperature variations encountered by convecting fluid along streamlines that pass into or along the margins of heat sources are fundamental to the ore deposition processes. The processes of aqueous ionic diffusion, irreversible reaction between aqueous fluids and minerals, equilibrium precipitation of alteration products, association and dissociation of metal-chloride complexes, exchange of oxygen isotopes, shifts in Rb/Sr ratios, and sulfide deposition all depend directly or indirectly on the pressure-temperature conditions along a fluid flow path. Those parameters most directly related to the overall alteration and sulfide deposition processes may be evaluated for locations along an average streamline where the pressure at any point in the pluton environment is defined by the integral of equation (12.1):

$$P = P_0 - g \int_{z=0}^{z} \rho(T, P)dz - \int_{z=0}^{z} \frac{\nu(T, P)}{k(z)} \cdot q_z dz \qquad (12.20)$$

where fluid density and viscosity are allowed to be functions of temperature and pressure, and the fluid flux, q, and permeability, k, are functions of position in the system. Calculation of pressure in this manner for the model pluton, Figure 12.2, at $t = 20,000$ years for the streamline $\Psi = 2$, defines the conditions along a flow path. A fluid packet circulates around the convection system in approximately 150 years, during which time a negligible change occurs in the temperature or pressure along the circuit. Therefore, the system, for the time duration of one fluid circuit, can be considered quasi-steady state for purposes of discussion. The principal variation in conditions at a fixed point on a flow path, with respect to time, is decreasing temperature. The calculation of pressure in the model system using equation (12.20) and parameter values used in the model showed that the increase in pressure with depth is very near hydrostatic in a convecting system (see also Cathles, 1977). Variations of parameters with respect to pressure along a streamline are only considered qualitatively, with the exception of fluid properties tabulated in Table 12-4. The pressure becomes important if the fluid passes into the supercritical region of the H_2O system because of its effect on the properties of water that directly affect the reactions between solutions and rock and fluid flow rates; hence, conclusions regarding chemical mass transfer along the segment 3-4 (Figure 12.2) on the streamlines must be regarded as qualitative estimates.

As fluid circulates downward and toward the pluton-host contact zone, 1-2 (Figure 12.2), temperature and pressure increase until the fluid flows beneath the relatively high-temperature fluid column related to the pluton's top surface. This high-temperature zone with lower fluid density contributes to a slightly lower pressure zone at depth. The fluid moving

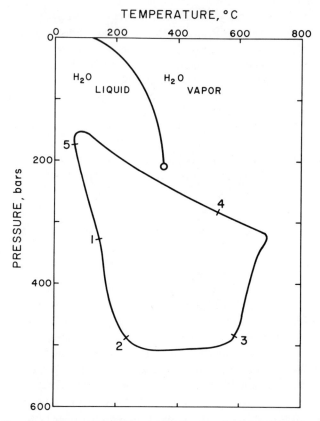

Fig. 12.2 Temperature-pressure conditions encountered by circulatory fluids along streamline, $\Psi = 2$, Figure 12.1, 20,000 yr after initial conditions. Fluid paths plotted with respect to pure H_2O system. Note slightly lower pressures along 3-4-5, or a small concentration of solute would extend the two-phase surface downward across the fluid path 3-4 and generate a boiling zone. A geothermal gradient of 30°C/km is assumed and added appropriately to the "perturbation" temperatures contained in Figure 12.1.

from 1 to 2 will tend to precipitate rock-forming silicates and sulfate alteration phases but will tend to become undersaturated with respect to metal sulfides. However, the irreversible consumption of hydrogen ions by fluid-rock interactions, together with the buffering effect of hydrothermal biotite on the fluid phase, results in local deposition of pyrite and chalcopyrite (Helgeson, 1970a). Therefore, sulfide ores in this region should be closely associated in time and space with the alteration silicate phases. Significant increases in the fluid mass flux, due to a factor of 10 decrease in fluid viscosity, are realized in the high-temperature–low-pressure regions.

Table 12-4 Values of temperature-pressure dependent parameters around a streamline

Temperatures calculated, normal hydrostatic pressure gradient (100 bar/km) assumed.

Parameter	Time = 20,000 yrs., $\Psi = 2$ Points Along Streamline					Reference
	1	2	3	4	5	
Temperature (°C)	120	225	550	500	60	Burnham et al. (1969), Keenan and Keyes (1969)
Pressure (bars)	325	490	480	280	175	
Fluid properties[a]						
density (g/cm^3)ρ	0.97	0.87	0.19	.10	1.0	Schmidt (1969)
viscosity (cm^2/sec) $\nu \times 10^3$	2.5	1.4	2.1	3.2	5	
alpha (°C^{-1}) $\alpha \times 10^3$	0.8	1.2	4.1	2.8	0.6	Helgeson and Kirkham (1974b)
dielectric constant	55	32	~4	~2	65	Helgeson and Kirkham (1974b)
enthalpy (cal/g)H	125	230	730	744	64	Schmidt (1969)

[a] H$_2$O system.

Table 12-5 Relative variation in some reaction parameters along streamline of Figure 12.2.

	Time = 20,000 urs., Ψ = 2 Points Along Streamline				
	Influx		Efflux		
Parameter	1-2	3	4	5-1	Reference
Reaction rate constant					
Feldspars	1	10	5	1	Helgeson (1970b)
Equilibrium hydrolysis constants					Helgeson (1969)
Feldspars ⎫ Biotite ⎪ Magnetite ⎬ Clays ⎪ Anhydrite ⎭			Decreasing	Increasing	
Pyrite ⎫ Chalcopyrite ⎭			Increasing	Decreasing	
Metal chloride ⎫ complexes ⎬ (dissociation) ⎭			Increasing	Decreasing	

The fluid path, across the pluton-host rock contact and into the pluton, is through a zone of nearly constant pressure and increasing temperatures (stages 2-3). In traversing stages 1-3 a decrease in the dielectric constant of H_2O contributes to an increase in stability of aqueous ionic complexes and increased solubility of mineral phases. This increase is partially counteracted by increases in ion activity coefficients in response to change in the Debye-Hückel parameters at constant ionic strength along the path from 500 to 640°C, whereas from 375 to 500°C decreases in activity coefficient occur due to the opposite affects (see Helgeson and Kirkham, 1974a,b).

Flow upward through the pluton and then into overlying host rocks (stages 3-4-5) appears to be associated with decreasing stability of aqueous metal-chloride complexes and decreasing solubility of metal sulfide minerals. These zones should contain large masses of sulfides deposited in response to decreasing temperatures as opposed to deposition by alteration mineral reactions, such as along path 1-2-3. At the same time, the solubility of alteration phases increase (Table 12-5), due to increasing equilibrium constants, but decrease in response to irreversible reactions that tend to consume hydrogen in this region (Knight, 1977).

These variations along the streamline discussed indicate that a general feature of convective flow around igneous plutons is the localization of alteration and sulfide phases in response to changing pressure and temperature conditions, irreversible and reversible reactions between solution and rocks, and the association or dissociation of aqueous ions. These coupled processes suggest the distribution of ore minerals, at or near the side contacts of plutons and near the top contacts, and reflect the significance of convective circulation of hydrothermal fluids. In particular, the chemical reactions and conditions in the inflow region (side contacts) of an igneous body tend to favor the deposition of silicate alteration phases, e.g., kaolinite, sericite, potassium feldspar, biotite. The equilibration of these phases with the convecting solution in turn favors the deposition of pyrite and chalcopyrite. The interdependence of the sulfide and silicate reactions in the inflow region suggests a close spatial and temporal association of the respective phases should be realized in the analogous zones in nature. Conditions near the outflow zone (top contact) tend to favor the dissolution of silicate phases but cause precipitation of sulfide phases. A proportion of the sulfide minerals in the outflow region are predicted to occur in continuous fracture systems and not to closely associate with alteration silicate phases.

The evolution of the hydrothermal system with time is indeed more complex than the simple flow model suggests. A temporal variation in permeability is clearly suggested by both the chemical mass transfer theory and observations in nature of partially filled flow channels. The intrusive mass prior to crystallization is apparently impermeable, although one of the biggest uncertainties in the study of hydrothermal systems is the variation and magnitude of permeability.

ACKNOWLEDGMENTS

The authors wish to thank Jerry Knight, Lynn McLean, and H. L. Barnes for critically reviewing this manuscript. Support for the research was derived from National Science Foundation Grant GA41136 (to Denis Norton) and Kennecott Copper Corporation (to Lawrence Cathles).

REFERENCES

Burnham, C. W. and N. F. Davis (1974) The role of H_2O in silicate melts: II. Thermodynamic and phase relations in the system $NaAlSi_3O_8$ - H_2O to 10 kilobars, 700° to 1100°C: *Am. J. Sci.*, **274**, 902-940.

———, J. R. Holloway, and N. F. Davis (1969) Thermodynamic properties of water to 1000°C and 10,000 bars: *Geol. Soc. Am., Special Paper 132.*

Cathles, L. (1977) An analysis of the cooling of intrusives by ground-water convection which includes boiling: *Econ. Geol.,* **72**, 804-826.

Elder, J. W. (1965) Physical processes in geothermal areas: in, *Terrestrial Heat Flow: Geophysical Monograph Series 8,* W. H. K. Lee, ed. Baltimore: Am. Geophys. Union.

——— (1967) Steady free convection in a porous media: *J. Fluid Mech.,* **27**, 29-48.

Ellis, A. J. and R. M. Golding (1963) The solubility of carbon dioxide above 100°C in water and in sodium chloride solutions: *Am. J. Sci.,* **261**, 47-60.

Everden, J. F. (1965) Depth of emplacement of granitic plutons of the Sierra Nevada (abst.): *Trans. Am. Geophys. Union,* **69**, 3471-3478.

Hamilton, W. and W. Myers (1967) The nature of batholiths: *U.S. Geol. Surv., Prof. Paper 554 C.*

Helgeson, H. C. (1969) Thermodynamics of hydrothermal systems at elevated temperatures and pressures: *Am. J. Sci.,* **267**, 729-804.

——— (1970a) A chemical and thermodynamic model of ore deposition in hydrothermal systems: *Min. Soc. Am. Special Paper 3,* 155-186.

——— (1970b) Kinetics of mass transfer among silicates and aqueous solutions: *Geochim. Cosmochim. Acta,* **35**, 421-469.

——— and D. Kirkham (1974a) Theoretical predictions of the thermodynamic behavior of aqueous electrolytes at high pressures and temperatures; I: Summary of the thermodynamic/electrostatic properties of the solvent: *Am. J. Sci.,* **274**, 1089-1198.

——— and ——— (1974b) Theoretical predictions of the thermodynamic behavior of aqueous electrolytes at high pressures and temperatures; II: Debye-Hückel parameters for activity coefficient and relative partial molal properties of the source: *Am. J. Sci.,* **274**, 1199-1261.

Holst, P. H. and K. Aziz (1972) Transient three-dimensional natural convection in confined porous media: *Int. J. Heat Mass Transfer,* **15**, 73-90.

Hubbert, M. King (1940) The theory of ground-water motion: *J. Geol.,* **48**, Part 1, 785-943.

Jaeger, J. C. (1968) Cooling and solidification of igneous rocks: in *Basalts,* **11**, 503-536.

Keenan, J. H. and F. G. Keyes (1969) *Steam Tables,* New York: Wiley.

Kemp, J. F. (1905) The problem of metalliferous veins: *Presidential Address to New York Academy of Sciences.*

Kesler, S. E. and S. A. Heath (1968) The effect of dissolved volatiles on magmatic heat sources at intrusive contacts: *Am. J. Sci.,* **266**, 824-839.

Knapp, R. B. (1978) Consequences of heat dispersal from hot plutons: *Unpublished Ph.D. dissertation,* University of Arizona.

——— and J. E. Knight (1977) Differential thermal expansion of pore fluids: Fracture propagation and micro earthquake production in hot pluton environments: *J. Geophys. Res.,* **82**, 2515-2522.

Knight, J. E. (1977) A thermochemical study of alunite, enargite, luzonite, and tennantite deposits: *Econ. Geol.,* **72**, 1321-1336.

Lachenbruch, A. H., H. A. Wollenberg, G. W. Green, and A. R. Smith (1966) Heat flow and heat production in the central Sierra Nevada, preliminary results (abs): *Trans. Am. Geophys. Union,* **47**, 179.

Lindgren, W. (1907) The relation of ore deposition to physical conditions: *Econ. Geol.,* **2**, 105-127.

REFERENCES

Lovering, T. S. (1935) Theory of heat conduction applied to geological problems: *Bull. Geol. Soc. Am.,* **46,** 69-94.

Marsh, B. and I. Carmichael (1974) Benioff Zone magnetism: *J. Geophys. Res.,* **79,** 1196-1206.

Mikumo, T. (1965) Crustal structure in central California in relation to the Sierra Nevada: *Seismol. Soc. Am. Bull.,* **55,** 65-83.

Moore, W. J. and M. A. Lanphere (1971) The age of porphyry-type copper mineralization in the Bingham mining district, Utah—a refined estimate: *Econ. Geol.,* **66,** 331-334.

Nield, D. A. (1968) Onset of thermohaline convection in a porous medium: *Water Resource Res.,* **4,** 553-560.

Norton, D. (1978) Sourcelines, sourceregions and pathline, for fluids in hydrothermal systems related to cooling plutons: *Econ. Geol.,* **73,** 21-28.

_____ and R. Knapp (1970) Transport phenomena in hydrothermal systems: Nature of porosity: *Am. J. Sci.,* **277,** 913-936.

_____ and J. Knight (1977) Transport phenomena in hydrothermal systems: Cooling plutons: *Am. J. Sci.,* **277,** 937-981.

Ohmoto, H., S. R. Hart, and H. C. Holland (1966) Studies in the Providencia Area, Mexico; II: K-Ar and Rb-Sr ages of intrusive rocks and hydrothermal minerals: *Econ. Geol.,* **61,** 1205-1213.

Page, R. W. and I. McDougall (1972) Ages of mineralization of gold and porphyry copper deposits in the New Guinea Highlands: *Econ. Geol.,* **67,** 1034-1048.

Rubin, H. (1973) Effect of nonlinear stabilizing salinity profiles on thermal convection in a porous medium layer: *Water Resources Res.,* **9,** 211-221.

Schmidt, E. (1969) *Properties of Water and Steam in SI-units,* New York: Springer-Verlag.

Sillitoe, R. H. (1972) The tops and bottoms of porphyry copper deposits: *Econ. Geol.,* **68,** 799-815.

Taylor, H. P., Jr. (1971) Oxygen isotope evidence for large-scale interaction between meteoric ground waters and tertiary granodiorite intrusions, Western Cascade Range. Oregon: *J. Geophys. Res.,* **76,** 7855-7873.

Thompson, G. A. and M. Talwani (1964) Crustal structure from Pacific Basin to central Nevada: *J. Geophys. Res.,* **69,** 4813-4837.

Toulmin, P. and S. P. Clark (1967) Thermal aspects of ore formation: in *Geochemistry of Hydrothermal Ore Deposits:* H. L. Barnes, ed., New York: Holt-Rhinehart, and Winston, 437-464.

Veronis, G. (1968) Effect of stabilizing gradient of solute on thermal convective: *J. Fluid Mech.,* **34,** 315-336.

Villas, N. R. (1975) Fracture analysis, hydrodynamic properties, and mineral abundance in altered igneous rocks at the Mayflower Mine, Park City District, Utah: Unpublished Ph.D. dissertation, University of Utah.

_____, and Norton, D. (1977) Irreversible mass transfer between circulating hydrothermal fluids and the Mayflower stock: *Econ. Geol.,* **72,** 1471-1504.

Van Hise, C. R. (1901) Some principles controlling the deposition of ores: *Trans. Am. Inst. Min. Eng.,* **30,** 27.

Wollenberg, H. A. and Smith, A. R. (1964) Radioactivity and radiogenic heat in Sierra Nevada plutons: *J. Geophys. Res.,* **69,** 3471-3478.

Wooding, R. A. (1957) Steady state free thermal convection of liquid in a saturated permeable medium: *J. Fluid Mech.,* **2,** 273-385.

13

Explored Geothermal Systems

A. JAMES ELLIS
New Zealand Department of Scientific and Industrial Research

Over the last 20 years geothermal systems in many parts of the world have been explored by wells drilled to considerable depths (over 2.5 km). Analyses have been made of water and steam samples collected from various depths or discharged from wells, revealing a wide range of solute concentrations and compositions. Extensive petrologic examinations have been carried out on rock cores or cuttings obtained during drilling.

Although there are few situations where metal ores are depositing in economic quantities from present-day hot spring or fumarole discharges, direct observations of the physical, chemical, and mineralogic conditions occurring within geothermal areas enable a quantitative check to be made of laboratory-derived models of natural hydrothermal solutions. Precise information can be derived on the nature of the deep fluid, including pH, ionic concentrations, gas partial pressures, and redox conditions. The degree of saturation of waters with various minerals can be obtained and relationships established between the deep hot-water compositions and particular secondary mineral assemblages. The natural partition of elements between rock and water phases after long periods of reaction can be observed. Geothermal wells also provide an excellent means of undertaking controlled hydrothermal experiments under natural reaction conditions (Ellis, 1960), but as yet little use has been made of this facility.

Natural thermal waters cover a wide temperature range but this chapter concentrates mainly on the chemistry of hot-water systems with measured subsurface temperatures in excess of about 100°C. Lower temperature waters occur in many geologic situations where there is circulation of meteoric water to deep levels or where water is stored at deep levels in rock aquifers (see Chapter 4). The most extensive subsurface exploration of high-temperature geothermal systems has been undertaken in Chile, El Salvador, Iceland, Indonesia, Italy, Japan, Mexico, New Zealand, the Philipines, Taiwan, Turkey, the United States, and the U.S.S.R. The different chemical and mineralogic conditions revealed by drilling in various geothermal areas are reviewed. For greater detail, the reader is referred to Ellis and Mahon (1977).

OCCURRENCE OF GEOTHERMAL SYSTEMS

Descriptions of many geothermal areas were included in a series of papers presented at United Nations Symposia on the Development and Use of Geothermal Resources at Pisa, Italy in 1970 (Geothermics, 1970) and at San Francisco in 1975 (U.N., 1975)

Many of the areas of the world in which geothermal systems with temperatures of at least 100°C have been observed in surface activity or in wells are shown in Figure 13.1. High-temperature geothermal systems exist in regions with a crustal heat flow several times the world average and where temperatures of several hundred degrees occur within a few kilometers of the surface. There is a concentration of geothermal areas around the margins of the Pacific in areas of active andesitic volcanism where crustal material becomes melted as it is thrust down to deep levels beneath a continental mass (subduction zones). They are also associated with the recent volcanic areas in zones of crustal spreading as in the mid-Atlantic Iceland situation, and on the southwest coast of California and Mexico. Rift valley systems stretching from Uganda through the Red Sea, contain several extensive geothermal systems in Uganda, Kenya, and Ethiopia. High-temperature water systems are also found in the orogenic belt of recent mountain building and volcanism stretching from northern Italy through Greece, Turkey, the Caucasus, and to S.W. China.

Most geothermal fields are associated with structures determined by tectonic activity such as block faulting, graben formation, rift valleys, or calderas. Particularly favorable locations for geothermal fields are at the intersections of faults bordering major structural blocks. Extensive local faulting is common.

Geothermal systems occur in a wide variety of rock types. In recent volcanic areas there is an association mainly with andesitic, dacitic, and

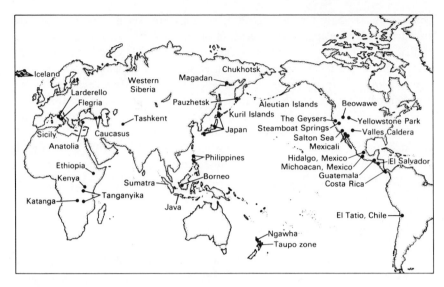

Fig. 13.1 World map showing some of the major high-temperature geothermal areas.

rhyolitic rocks rather than with basaltic eruption centers. Intrusion of rhyolitic rocks is common (even in Iceland, rhyolitic rocks occur near hydrothermal areas). Geothermal fields associated with Cenozoic tectonic activity occur in many types of metamorphic and sedimentary rocks.

Extensive aquifers of water at temperatures in excess of 100°C have been encountered in several sedimentary basins. Jones (1970) reported water temperatures of 120°C at depths between 3 and 4 km over much of the Texas and Louisiana coastal areas of the Gulf of Mexico basin, and Makarenko et al. (1970) noted similar occurrences in vast stratified artesian basins in the West Siberian, Turanian, and Skiphian plateaus of the Soviet Union.

Underground hot-water storage can be on a vast scale. The Western Siberian thermal water areas cover 3×10^6 km², and Ilinskaya (1962) reported a 2×10^4 km² area in Central Asia near Tashkent. Drilling investigations in the volcanic area hot-water system at Wairakei, New Zealand showed that it existed to depths of at least 2-3 km and was of much greater areal extent (at least 50 km²) than was predictable from surface evidence. During the removal of approximately 1 km³ of 250°C water, there were changes of only 1-2% in the ionic concentrations in the discharge waters.

In several countries, hot-spring systems have changed little during recorded history over several hundred years. The age of the Great Geysir,

Iceland, is considered to be at least 10,000 years (Barth, 1950). Geologic evidence produced by Grindley (1965) shows that hydrothermal activity at Wairakei, New Zealand existed approximately 500,000 years ago. An age of at least 10^6 years is suggested for the Steamboat Springs, Nevada, hydrothermal area (White, 1974), and Craig (1962) showed evidence from carbon isotope ratios that in this area descending meteoric water spent between 30,000 and 300,000 years underground. Both in terms of areal extent and time of persistence, hydrothermal areas are important geological features.

CHARACTERISTICS OF SYSTEMS

The hot water stored within geothermal systems has several possible origins (see below). It may be meteoric water that has circulated to great depths and has become heated. If the local high geothermal gradient is caused by a magmatic intrusion, water may be liberated from the magma as it cools and crystallizes, and this water could become part of the thermal water system. In areas of active metamorphism, waters expelled during the recrystallization and dehydration of hydrous rock minerals under elevated temperatures and pressures could be encountered as surface springs. Thermal water in sedimentary areas could consist largely of altered connate waters that were trapped in the sediments during their formation, but these waters (e.g., oil field brines) usually do not have sufficiently high temperatures to bring them within the present review.

In an area of high heat flow, the normal passage of meteoric water to considerable depths in faulted or porous rock is affected by the density differences developed as the water becomes heated conductively. A convective water flow cycle may be set up through the rock system, and in favorable situations water that has been heated to high temperatures at deep levels may return to the surface. The favorable situations include (a) a geologic structure that allows water to circulate to deep levels; (b) a sufficiently slow passage of water, and a large exposed rock surface area to allow sufficient heat exchange to occur; and (c) a return path to the surface through faults, fissures, or permeable strata.

For a system to be of practical significance as a geothermal energy source, storage of hot fluid in near-surface horizons is necessary. Many geothermal systems have an impermeable "cap" rock preventing a rapid outflow from the convection system, and causing a lateral spread of hot water in the formations beneath. In the Imperial Valley geothermal area of California, an impermeable clay sequence presented a barrier both to hot-water outflow from porous sandstones and to conductive heat flow (Helgeson, 1968). In the Wairakei and the Broadlands fields, near-surface impermeable mudstones, and rhyolite flows, respectively, have caused lateral spread and storage of hot waters in porous pumice breccias. Facca and Tonani (1967) pointed out the natural tendency for geothermal systems to become self-sealing through the near-surface deposition of silica and calcite. This, in conjunction with evidence from isotopic (Craig, 1962) and chemical balance (Ellis, 1970a) considerations, leads to a model in which brief periods (order of 10^3 years) of major water flow from a system alternate with long periods (10^4–10^5 years) of conductive water heating and minor outflow. Periods of major flow may be triggered by tectonic activity or by hydrothermal explosion. There is evidence in the form of major silica terraces that in certain thermal areas surface activity was much greater in the past then at present. Drilling in some of these areas of low water flow has revealed a present-day major underground hot-water system.

Steam or Hot Water—A Model

In most geothermal systems, deep wells (over 500 m) reach hot liquid water, but in three situations—Larderello, Italy, The Geysers, California, and Matsukawa, Japan—wells over a wide area consistently produce steam at temperatures of approximately 230–250°C. In these fields the steam was initially saturated, and in some cases wells produced minor amounts of water, but there was a rapid trend to dry steam production. These areas were described as vapor-dominated systems by White et al. (1971).

In hot-water geothermal systems with rocks of low permeability, individual well discharges may gradually increase in enthalpy, until in some cases only steam, or steam with very minor amounts of water, is produced. The high rate of discharge apparently creates in the vicinity of the well a steam or vapor-dominated pocket.

Models explaining the behavior of vapor-dominated systems such as Larderello or The Geysers have been proposed by James (1968) and by White et al. (1971). Figure 13.2 explains the balance existing between steam and water phases in geothermal systems.

Meteoric water penetrating to depths of several kilometers absorbs heat and rises again through natural fissures, along the route ABCD. It enters one, or a succession of permeable zones DE, in which either water or steam may dominate at a particular depth, depending on the relative permeabilities of the upflow path CD, and that of the natural outflow route EF. Outflow along the path EF depends on its permeability and the pressure at the top of the zone DE. In an all-liquid system there is a "thermo-artesian" pressure (Studt, 1958) in DE controlled by the different water densities in the hot- and cold-water columns, and at any point the total pressure ex-

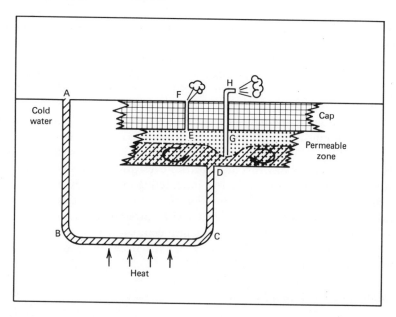

Fig. 13.2 A simple model showing some of the features of a geothermal area (see text).

ceeds that of saturated steam at the water temperature. Mass outflow through EF can not exceed the inflow through CD if an all-liquid system is to persist. If it is exceeded, a steam phase forms in DE, and the steam pressure will decrease until the mass outflow through EF is reduced to equal the mass inflow. A dynamic balance is created with a lowered hot-water level in the permeable horizon; the water is boiling, and convection cycles in the water are set up. Steam condenses on the cooler top surface of the "cap" rock, creating saturated steam conditions in the vapor-dominated portion. Temperatures in the upper levels of the steam zone should not greatly exceed 236 °C, independent of the water temperature, as saturated steam has a maximum enthalpy at this temperature (James, 1968).

A well, such as GH, drilled into the permeable zone disturbs the natural balance. If it taps a natural vapor-dominated zone, saturated steam is initially discharged, but the steam pressure is gradually reduced. As most of the heat in the reservoir is stored in the rock, steam temperatures are affected only slightly, with a result that the steam flow becomes "dry" or superheated.

A well that intersects liquid water may continue to discharge a high output of water if the upflow and reservoir permeabilities are high. Poor permeability may result in a general, or local draw-down (as shown) of water levels in the reservoir zone, and the well will discharge an increasing proportion of steam.

Figure 15.3, a cross-section of the western part of the Broadlands field (Grindley, 1970), provides a comparison with the model. The basement graywacke and overlying ignimbrite rocks are of low permeability except along fracture zones. The sequence above consists of pumiceous pyroclastic rocks and old lake sediments interstratified with impermeable rhyolite flows and ignimbrite flows. Two aquifers are present, the Waiora formation of pumice tuff breccias and sediments, and the Rautawiri pumiceous breccia, while the Ohaki rhyolite and the overlying Huka lake sediments form a cap rock structure. An active intersecting fault pattern creates fracture zones of high permeability and wells intercepting fault zones have high discharge rates (approximately 250 tons/hr). Several wells with open casings in hot but low-permeability rocks give comparatively lower discharges that change over periods of weeks or months from being predominantly water to predominantly steam.

Examples of Geothermal Systems

Table 13-1 gives representative examples of geothermal systems explored by wells in various parts of the world. Information is given on the rock types, maximum depth of drilling, maximum temperatures, total dissolved solutes, major ionic solutes, water pH (surface), and the gases associated with the discharges. The examples demonstrate variable nature of geothermal systems and geothermal fluids.

Hot waters in Iceland fall into two general types (Bodvarsson, 1961); with high-temperature waters (over 200°C) of appreciable salinity and dissolved CO_2 and H_2S content in areas of Quaternary basalts (e.g., Hveragerdi), and in the older Tertiary basalt areas lower temperature waters (less than 150°C) of low salinity and with N_2 as the main dissolved gas (e.g., Reykjavik). The Reykjanes geothermal waters are derived from seawater, and they provide the opportunity of observing how high temperatures in a basaltic aquifer modify the water composition.

The Wairakei, Broadlands, Pauzhetsk, and El Tatio areas are typical of many high-temperature, dilute, near-neutral pH, chloride water systems occurring around the Pacific Basin in Quaternary volcanic sequences which include permeable pumice, breccia, or volcanic ash horizons. The Steamboat Springs area in contrast is predominantly within compact igneous rocks. Yellowstone Park is associated with Quaternary rhyolite flows. The Matsao, Taiwan area occurs in recent andesitic rocks, but sulfur-containing strata have created a highly acidic water system.

Very high-temperature, high-salinity brines occur near the Salton Sea,

California, within a thick series of deltaic sediments, but rhyolite intrusions occur nearby (Helgeson, 1968). In Baja California the Mexicali geothermal system in a similar geologic setting is of similar temperature but with a lower water salinity (Alonso, 1966).

The Ngawha geothermal system in the far north of New Zealand is in a sequence of impermeable organic-rich sediments and the waters have very high carbon dioxide, boron, and ammonia concentrations. Nearby Quaternary basalts are a likely heat source. Low rock permeability caused the discharges from a test well to trend rapidly from predominantly water to almost dry steam. A very similar type of system exists at Sulphur Bank, California (White and Roberson, 1962).

The Kizildere field of Turkey occurs in Pliocene-Miocene sandstones and limestones overlying a Paleozoic marble basement. It is a water system of only moderate temperature, of low salinity, and of high carbonate concentration. There is no obvious volcanic association, but the area is one of intense tectonic activity (Ten Dam and Erentöz, 1970).

Matsukawa is an example of a vapor-dominated system occurring in a recent volcanic area. In contrast, Larderello and The Geysers steam fields have no clear association with recent volcanism, but for example, a major Quaternary rhyolite body exists approximately 5-7 km northeast of The Geysers.

The areas of Larderello, The Geysers, Ngawha, Sulphur Bank, and Kizildere, with sedimentary and metamorphic rock associations all occur in zones of mercury mineralization, reflecting the solubility and steam volatility of this element and the fact that it is found at higher abundance levels in organic-rich sediments than in volcanic rocks. A review of metal mineralization associated with hot-spring water areas is found in Chapter 15.

FLUID COMPOSITIONS

Steam/Liquid Distribution

Before examining the compositions of particular geothermal fluids, the relative solubilities of various chemical species in water and in saturated steam may be noted. Gas solubilities change markedly with temperature, and for most gases, at a constant partial pressure, the equilibrium concentrations in liquid water decrease as temperatures are raised from ambient conditions. The concentrations pass through a minimum at temperatures frequently in the range 50-150°C, and at higher temperatures concentrations increase steadily. If, however, a mass distribution coefficient is used,

Table 13-1 Examples of geothermal areas explored by drilling

Area	Rock types	Max. drilled depth (m)	Max. temp. (°C)
Hveragerdi, Iceland	Quaternary basalt	1200	232
Reykjavik, Iceland	Tertiary basalt	2200	146
Reykjanes, Iceland	Quaternary and Recent basalt	1750	290
Wairakei, New Zealand	Quaternary rhyolite, andesite	2300	265
Broadlands, New Zealand	Quaternary rhyolite; Mesozoic graywacke	2420	300
Pauzhetsk, Kamchatka, U.S.S.R.	Quaternary dacite, andesite	800	190
El Tatio, Chile	Quaternary and Tertiary rhyolite, andesite; Mesozoic sediments	600	260
Yellowstone Park, U.S.A.	Quaternary rhyolite	330	240
Steamboat Springs, Nevada	Granodiorite; Tertiary and Quaternary andesite, rhyolite	175	172
Matsao, Taiwan	Quaternary and Recent andesite; Miocene sandstone	1500	293
Salton Sea, California	Tertiary sediments (Quaternary rhyolite)	2470	360
Mexicali, B.C., Mexico	Sediments; Quaternary basalts, granite	2600	370
Ngawha, New Zealand	Cretaceous sediments (basalt)	600	235
Kizildere, Turkey	Tertiary sandstones, limestones; schist, marble	1000	220
Matsukawa, Japan	Quaternary and Tertiary andesite, dacite	1200	280
Larderello, Italy	Mesozoic and Tertiary clays, limestone, anhydrite; Permian schist	1600	260
The Geysers, California	Mesozoic graywacke; serpentine, basalt	2800	285

gas solubility shows a simple relationship with the relative densities (ρ) of the vapor and liquid phases (Ellis and Fyfe, 1957). A plot of the logarithm of the distribution coefficient, A versus log (ρ_L/ρ_V) is a convenient way of comparing the solubilities of different gases at low partial pressures over a wide temperature range.

$$A = \frac{n_g^L/n_{H_2O}^L}{n_g^V/n_{H_2O}^V} \qquad (13.1)$$

Main solutes (total dissolved solids, g/kg, surface sample)	Water pH (surface, cold)	Main gases present	References
Na^+,Cl^- (0.9)	9.5	CO_2,N_2,H_2S,H_2	Sigvaldason (1963)
Na^+,Cl^-,HCO_3^- (0.4)	8.6	N_2	Sigvaldason (1963)
Na^+,Cl^-,Ca^{2+} (40)	5.8	CO_2,N_2	Björnsson et al. (1970)
Na^+,Cl^-, (4.5)	8.2	CO_2,H_2S	Ellis and Mahon (1964)
Na^+,Cl^-, (4.1)	8.3	CO_2,CH_4,H_2S	Browne and Ellis (1970)
$Na^+,Ca^{2+},Cl^-,SO_4^{2-}$ (3.0)	8.9	CO_2,H_2S	Averyev, Naboko and Piip (1961)
Na^+,Cl^- (15)	7.0	CO_2,N_2,H_2S	W.A.J. Mahon (pers. comm.)
Na^+,Cl^-,HCO_3^- (1.7)	8.8	CO_2,H_2S	Honda and Muffler (1970)
Na^+,Cl^-,HCO_3^- (2.0)	8.8	CO_2,H_2,H_2S	Sigvaldason and White (1962)
Na^+,Cl^-,Ca^{2+} (30)	2.5	CO_2,H_2S	Chen (1970)
Na^+,Ca^{2+},K^+,Cl^- (350)	5.5	CO_2,CH_4,H_2S	Helgeson (1968)
Na^+,Cl^-,K^+ (17)	5.5	CO_2,H_2S	Mercado (1966)
B,Cl^-,Na^+,HCO_3^- (8)	7.5	CO_2,H_2S,NH_3	Ellis and Mahon (1966)
$Na^+,HCO_3^-,SO_4^{2-},Cl^-$ (4.8)	9.0	CO_2,H_2S	R. James (pers.comm.)
Na^+,SO_4^{2-},Fe^{2+} (1-4)	3-6	CO_2,H_2S	Sumi (1970)
Steam only		CO_2,H_2S,H_3BO_3,NH_3	ENEL (1970)
Steam only		CO_2,N_2,CH_4,H_2S	Kruger and Otte (1973)

n is the number of moles of gas or water in either the liquid (L) or the vapor (V) phase. Figure 13.3 summarizes the solubility of many gases of geologic interest (CO_2, H_2S, CH_4, NH_3, H_2, N_2, O_2). The differences between gas distribution coefficients become less at high temperatures, but the relative solubilities do not change above 100°C. The results for carbon dioxide are from Ellis and Golding (1963); for hydrogen sulfide from Kozintseva (1965); for methane from Culberson and McKetta (1951); for

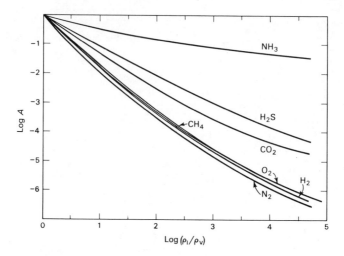

Fig. 13.3 The solubility of various gases expressed as a function of the mass distribution coefficient A with respect to the relative densities ρ of liquid and steam phases.

ammonia from Jones (1963); and for the last three gases from Pray et al. (1952). For the reactive gases NH_3, H_2S, and CO_2, only the nonionized fraction in solution is relevant to the gas/liquid equilibrium.

Even at 350 °C all of the gases show a strong preferential weight distribution toward the steam phase (NH_3 to a lesser extent). For example, in separating 5% of steam from 250 °C water containing dissolved gases, over 75% of CO_2 and H_2S is lost from the water at equilibrium, but it retains approximately 85% of the NH_3.

Solutes not normally considered volatile from dilute, low-temperature aqueous solution become measurably volatile at high temperatures. The volatility of many species is expressed to a good approximation by a distribution coefficient K_d.

$$K_d = c_V/c_L = (\rho_V/\rho_L)^n \qquad (13.2)$$

c is the weight concentration and n is a constant for each solute. The values of n were for H_3BO_3, 0.885; SiO_2, 1.9; $LiCl$, 3.4; $NaOH$, 4.1; $NaCl$, 4.4; $CaCl_2$, 5.5; Na_2SO_4, 8.4; $CaSO_4$, 8.4 (Styrikovich et al., 1960). In addition K_d for HF is 0.28 (unpublished Chemistry Division work). More detailed information is available for the steam/water distribution for SiO_2 (Martynova, 1972), for NaCl (Sourirajan and Kennedy, 1962), and for H_3BO_3 (Wilson, 1974).

Apart from "normal" gases, at temperatures below 300 °C, SiO_2, H_3BO_3, and HF are the principal constituents volatile in steam. At high temperatures, HF becomes a weaker acid and also the pH of geothermal waters tends to be lower (see below). HF becomes a more important steam constituent in very high temperature and/or acidic systems. In very acid geothermal systems, HCl may also be volatile from hot solutions.

Water Compositions

White et al. (1963) gave a detailed survey of the compositions of subsurface waters of all types. The compositions of thermal spring waters were reviewed by Waring et al. (1965) and their classification was discussed by White (1957), and by Ellis and Mahon (1964).

Table 13-2 gives the concentrations of major constituents in waters discharged from wells in several areas. Solute concentrations are in terms of milligrams per kilogram (ppm) in the waters separated from the discharges at atmospheric pressure and cooled to ambient temperatures for analysis. As high temperature waters boil off steam as they rise to the surface, the waters are concentrated to an extent that depends on the original water temperature and the enthalpy of the discharge. For example, a well tapping 250°C water discharges at normal atmospheric pressure a mixture of approximately 29.5% steam and 70.5% water at 100°C. In addition to concentration, there are changes in the water pH and in the relative concentration of acid and base constituents, due to the loss of gases such as CO_2 and H_2S into the steam phase. In Table 13-2, for many constituents with pH-dependent equilibria the concentrations are given as totals of ionic and molecular forms (e.g., total CO_2 equals CO_2 + HCO_3^- + CO_3^{2-} expressed as CO_2).

In areas of boiling hot springs with major water outflows, the general chemical nature of the spring and well water has been found to be similar, except for elements that are controlled in concentration by rapidly reversible temperature-dependent equilibria. In a given area the ratio of chloride to lithium, cesium, fluoride, bromide, iodide, arsenic, or boron in waters from deep wells often differs little from that in major surface springs (Ellis and Mahon, 1964). In contrast, concentrations of most of the major rock-forming elements change between the deep conditions and the natural outflow points.

Many spring waters characterized by high-acidity, high-sulfate, and low-chloride concentrations are of a superficial nature and represent surface waters or perched aquifers that are heated by steam flows. Steam-heated meteoric waters under stagnant subsurface conditions also may produce low-salinity, high-bicarbonate waters.

Beneath the levels of boiling and atmospheric oxidation, waters in high-temperature geothermal systems are often alkali chloride solutions at a pH within 1–2 units of the neutral water pH for the temperature. The salinity

Table 13-2 Compositions of waters from geothermal wells

Source	Depth (m)	Source temp. (°C)	pH (20°C)	Li
Well G-3 Hveragerdi, Iceland	650	216	9.6	0.3
Well Reykjavik, Iceland	600	100	8.6	<0.1
Well 8 Reykjanes, Iceland	1754	253	5.75	—
Well 44 Wairakei, New Zealand	695	255	8.4	14.2
Well 13 Broadlands, New Zealand	1080	260	8.6	12.6
Well 4 Pauzhetsk, Kamchatka, U.S.S.R.	276	195	8.2	—
Well 7[a] El Tatio, Chile	867	255	7.31	47.5
Well Y6[b] Yellowstone Park, U.S.A.	152	195	7.08	1.9
Well GS-2 Steamboat Springs, Nev., U.S.A.	175	160	8.8	5.7
Well E-205 Matsao, Taiwan	1500	245	2.4	26.0
Well IID-1 Salton Sea, Cal., U.S.A.	1600	~340	~5.5	320
Well 8 Mexicali, B.C., Mexico	1310	355	—	31
Well 1 Ngawha, New Zealand	590	230	7.4	12.2
Well 1A Kizildere, Turkey	415	200	9.0	4.5
Well MR1[c] Matsukawa, Japan	945	250	3.5	—

Ppm in waters separated from discharge at atmospheric pressure								
Na	K	Rb	Cs	Mg	Ca	F	Cl	Br
212	27	0.04	<0.02	0.0	1.5	1.9	197	0.45
95	1.5	<0.02	<0.02	0.0	0.5	—	31	—
14920	2105	—	—	1.75	2320	0.15	28700	—
1320	225	2.8	2.5	0.03	17	8.3	2260	6.3
980	200	2.2	1.3	0.02	2.4	4.5	1668	5.3
986	105	—	—	3.5	52	—	1633	3
5000	840	8.6	17.9	0.09	203	2.5	9100	—
356	22	—	—	0.04	6.6	12	347	—
655	73	1.2	1.9	0.0	4.4	1.0	871	1.5
5490	900	12.0	9.6	131	1470	7.0	13400	—
54000	23800	100	20	100	40000	—	184000	700
11200	3420	—	—	11	720	—	21600	26
950	77	0.8	0.4	—	28	0.85	1625	—
1280	135	0.0	0.33	0.11	3.0	23.7	117	1.2
80	50	—	—	10	28.8	—	5.4	—

Table 13-2 (*Continued*)

	Ppm in waters separated from discharge at atmospheric pressure						
				Total			
Source	SO$_4$	As	SiO$_2$	B	NH$_3$	CO$_2$	H$_2$S
Well G-3 Hveragerdi, Iceland	61	—	480	0.62	0.1	55	7.3
Well Reykjavik, Iceland	16	—	155	0.02	0.1	58	—
Well 8 Reykjanes, Iceland	43.5	—	898	12.2	—	2	1.0
Well 44 Wairakei, New Zealand	36	4.8	690	28.9	0.15	19	1.0
Well 13 Broadlands, New Zealand	6.5	3.2	750	48.0	1.9	117	—
Well 4 Pauzhetsk, Kamchatka, U.S.S.R.	78	—	300	—	0.6	19	—
Well 7[a] El Tatio, Chile	29	—	810	210	3.1	32	—
Well Y6[b] Yellowstone Park, U.S.A.	—	—	252	4.2	0.15	300	—
Well GS-2 Steamboat Springs, Nev., U.S.A.	132	2.7	—	36	0.0	125	1.1
Well E-205 Matsao, Taiwan	350	3.6	639	106	36	2	—
Well IID-1 Salton Sea, Cal., U.S.A.	10	—	—	498	500	—	—
Well 8 Mexicali, B.C., Mexico	0	2	1420	37.0	—	1180	—
Well 1 Ngawha, New Zealand	17	—	460	1200	46	62	—
Well 1A Kizildere, Turkey	770	38	263	26.2	2.5	1350	—
Well MR1[c] Matsukawa, Japan	535	—	387	21	—	0	0

Atomic Ratios						
Cl/B	Cl/F	Cl/Br	Na/K	Na/Li	Na/Ca	Ref.
97	55	800	13	200	250	Chemistry Division, D.S.I.R., and G. E. Sigvaldason (personal communication)
400	—	—	110	>300	330	Chemistry Division, D.S.I.R., and G. E. Sigvaldason (personal communication)
719	10^5	—	12.0	—	11.2	Björnsson et al., (1970)
23.9	145	850	10.0	28	135	Chemistry Division, D.S.I.R.
10.6	199	715	8.3	23.5	710	Chemistry Division, D.S.I.R.
—	—	1200	16.0	—	33	Vakin et al., (1970)
13.2	1540	—	10.1	31.8	43	Chemistry Division, D.S.I.R.
25.2	15.5	—	27.5	57	94	Truesdell and Singers (1971)
7.4	470	1300	15.3	35	260	Sigvaldason and White (1961) with Br, and As added from spring analyses (White, personal communication)
38.5	1030	—	10.4	64	6.5	Chemistry Division, D.S.I.R.
108	—	590	3.9	51	2.35	D. E. White (personal communication) and Chemistry Division, D.S.I.R.
178	—	1860	5.6	108	27.2	Mercado (1966)
0.41	1000	—	21	23.5	59	Ellis and Mahon (1966)
1.36	2.65	220	16.1	86	740	Chemistry Division, D.S.I.R.
0.08	—	—	2.7	—	4.8	Sumi and Maeda (1970)

[a] Atmospheric pressure = 0.6 bars.
[b] Concentrations corrected to original down-hole conditions.
[c] Also contains 508 ppm Fe and 29 ppm Al.

of the waters varies widely from field to field, as do the relative ionic concentrations.

Boron concentrations may be exceptionally high in proportion to chloride in geothermal areas associated with sedimentary rocks, for example, Ngawha, and Kizildere. The Broadlands geothermal area shows increasing boron concentrations from west to east, which correlate with an increasing association of the deep waters with greywacke and argillite rocks (Mahon & Finlayson, 1972).

The rare alkalis, lithium, rubidium and cesium, are concentrated mainly in the waters of rhyolitic or andesitic environments or of sedimentary areas of comparable rock compositions. For example, in the Ngawha area the local sediments have a lithium content similar to that of average rhyolites. The high concentrations of rubidium and cesium in the Matsao water shows how effectively these elements can be concentrated from andesite rocks into a high-temperature acidic solution.

Matsao waters have a high sulfate acidity, which appears to be due to their reaction with sulfur-containing strata (see below). A similar situation occurs at shallow levels in the Rotokaua, New Zealand, field, but deep drilling penetrated to neutral-pH, dilute alkali chloride water beneath the sulfur-bearing horizon (Ellis, 1970a). At Matsukawa the small quantities of high-sulfate, low-chloride water discharged at an early stage from some wells was apparently oxidized condensate percolating through a steam-dominated zone intersected by the wells. With the exception of these types of situations, sulfate concentrations in deep high-temperature waters are usually low (even for the Reykjanes system, which is based on seawater).

Information on minor metal concentrations in geothermal well waters is given in Chapter 15 (Table 15-1) and further data on low-temperature waters may be obtained in Chapter 4. Except in very high salinity or acidic geothermal waters, the concentrations of ore-forming elements are very low (frequently less than 0.01 ppm). The major rock constituents magnesium, aluminum, iron, and manganese are also present at very low levels in dilute high-temperature chloride waters. Marked differences in heavy metal concentrations occur between the Salton Sea and Cerro Prieto areas (similar geologic and temperature situations, but markedly different salinities), with exceptionally high concentrations of heavy metals being found in the deep, hot brines of the Salton Sea field. The acidic waters at Matsao contain appreciable zinc and lead concentrations.

Steam Compositions

From the hot-water areas in Table 13-1 (the first 15 areas), wells discharged a mixture of steam and liquid water in a proportion that depended on the original underground temperature. However, in some areas of low rock permeability, there was gradually a trend to increasing steam/water ratios. At Larderello, The Geysers, and Matsukawa (largely), after a brief period of discharge, the wells produced only steam. For steam flows compositions do not vary with the collection pressure, but for mixed steam/water discharges there are variations with the separation pressure due to the distribution of solutes between phases.

The steam phase contains most of the gases originally dissolved in the deep hot water. Table 13-3 gives analytic data for steam separated from well discharges at atmospheric pressure. The steam fraction in the discharge is given and also the calculated total concentration of gases in the deep underground water. This calculation has significance only where the inflow to the well consists of a single liquid phase. (Where the well inflow includes steam separated in the country rock the discharge enthalpy will be in excess of that for water at the down-hole temperature as measured or estimated by the silica geothermometer (see below).

The gas content of steam from wells in different areas varies widely, from a minimum of approximately 0.01 mole % to many tens of percent, with carbon dioxide frequently making up over 80% of the total gases. In most volcanic hydrothermal areas, hydrogen sulfide is usually the next most abundant gas, but in the areas of sedimentary or metamorphic rocks it may be exceeded by methane. The gases associated with the lower temperature (approximately 100°C) thermal waters of the Tertiary Iceland basalt areas are mainly nitrogen. The steam composition at The Geysers is quite distinct from that in the Larderello area, with a relatively high proportion of methane and hydrogen and a low boric acid concentration. From the concentration of boric acid in the Larderello steam, for a separation temperature of 220°C, the concentration in a deep-liquid phase would be about 1%.

For some geothermal areas analyses are available for the inert gases helium, neon, argon, krypton, and xenon (Table 13-4). For Yellowstone and Lassen hot springs, the ratios of concentrations of inert gases (except helium) suggested that they originated from a solution of air in cold water (Mazor and Wasserburg, 1965). However, the ratio of Ar/He of about 60 for these areas (in comparison with approximately 7000 expected for cold, air-saturated water) showed the presence of radiogenic helium. Even lower Ar/He ratios of about 20, and 0.65, were found for Wairakei and Larderello wells, respectively (Wasserburg et al., 1963). The presence of radiogenic argon was also demonstrated.

The ratio N_2/Ar in the geothermal gases is higher than would be expected from the outgassing of a saturated solution of air in water formed at ambient temperatures. The ratio rises to as high as 800 at Larderello,

Table 13-3 Composition of steam in geothermal discharges

Source	Depth (m)	Steam[a] fraction in discharge	Total[a] gases in steam (mole %)	Total gases in deep hot water (mole %)
Well G-3 Hveragerdi	400	0.20	0.015	0.003
Reykjavik wells	500	[c]	—	—
Average well Wairakei	650	0.32	0.063	0.020
Well 11 Broadlands	760	0.35	0.61	0.22
Well 2 El Tatio	650	0.24	0.11	0.026
Well 1 Ahuachapan	1195	0.23	0.007	0.0016
Well E205 Matsao	1500	~0.3	0.15	0.045
Wells Salton Sea	1500-1800	~0.4	0.1-1.0	0.04-0.4
Well 1 Ngawha	585	0.33	~20 (variable)	—
Well MR-2 Matsukawa	1080	high	0.22	—
Wells Larderello	500	1.00	2.0	—
Wells The Geysers	200-2000	1.00	1.0 (average)	—

[a] At atmospheric pressure.
[b] Total hydrocarbons, mainly methane.
[c] Most Reykjavik wells produce only water.

indicating a source of nitrogen other than atmospheric. Hulston and McCabe (1962a) correlated the excess nitrogen in the Wairakei samples with the hydrogen concentration of the gases, suggesting that both may be derived from organic material contained in the rocks. However for Yellowstone, from a complete integration of mass outflows of constituents in steam and water, Gunther (1973) found that the outputs of nitrogen and argon could be correlated with the proportions expected from a solution of air in cold water.

Where wells have a single liquid inflow, it is possible to relate the concentration of gases in the deep solution to partial pressures $P_g = XK_g$, where X is the mole fraction of gas in solution, and at low gas pressures $K_g = P_W/Az_W$ where P_W is the water vapor pressure and z_W is the compressibility factor for steam ($z_W = P_W V/RT$).

In Table 13-5 approximate partial pressures of individual gases are shown for several geothermal systems. The maximum partial pressures of gases in the steam discharges of Larderello and The Geysers were calculated by assuming perfect gas laws and equating the water vapor pressure with saturated steam pressure at the discharge temperature. The partial pressures of most gases vary widely between fields, but the partial pressure of hydrogen P_{H_2} is usually of the magnitude of 0.1 bars. An exception is El Tatio, but for this area the partial pressures of all of the nonacidic gases

FLUID COMPOSITIONS

			Gas composition (mole %)				
CO_2	H_2S	HC^b	H_2	$(N_2 + Ar)$	NH_3	H_3BO_3	Ref.
73.7	7.3	0.4	5.7	12.9	—	—	G. E. Sigvaldason (pers. comm.)
0.0	0.0	0.04	0.01	99.2	—	—	Sigvaldason (1973)
91.7	4.4	0.9	0.8	1.5	0.6	0.05	Chemistry Division, D.S.I.R.
94.8	2.1	1.2	0.2	1.5	0.2	—	Chemistry Division, D.S.I.R.
99	0.7	0.01	0.03	0.2	—	—	Chemistry Division, D.S.I.R.
50-80	4	—	10-40	2-10	—	—	R. B. Glover and G. E. Sigvaldason (personal communication)
92	5	0.7	0.8	1.5	—	—	Chemistry Division, D.S.I.R.
~90	Remainder H_2S with minor HC, N_2 and H_2						Helgeson (1968)
94	0.7	4.0	0.5	0.8	0.04	—	Chemistry Division, D.S.I.R.
81.8	14.1		Remainder			—	Nakamura and Sumi (1967)
94.1	1.6	1.2	2.3	0.8	0.8	0.33	ENEL (1970)
74-87	0.4-7	5-6	1-1.5	1.3-3	5	0.005	Kruger and Otte (1973); Koenig (1970)

are particularly low, and appreciable gas loss from the hot water is suspected as it flows laterally across the field in a shallow aquifer.

Deep Fluid Compositions

It is usual to collect separated steam and water phases from a well discharge because of the difficulty in taking a composite sample of a non-homogeneous fluid traveling at velocities approaching the speed of sound. The water phase is collected at a temperature lower than the original deep conditions and steam formation causes concentration of salts and depletion of gases. The loss of acidic gases and the consequent rise in the water pH causes a complex interplay of acid-base ionization equilibria and of complex ion species.

For a well drawing on a single liquid phase, from the separate analyses of vapor and liquid phases separated from the discharge at a known pressure, the composition of the deep hot solution can be calculated in terms of ionic and molecular species. No oxidation occurs under good collection conditions, which is in contrast to the oxidized nature of some natural surface flows.

A computer program was written by Truesdell and Singers (1971) to calculate deep fluid compositions, ionic ratios, temperatures, and some

Table 13-4 Inert gas and nitrogen concentrations in gases associated with geothermal discharges (cm^3/m^3 total gas)

	He	Ne	Ar	Kr	Xe	N_2	Ref.
Average Wairakei wells	~1.5	—	250	—	—	15000	Hulston and McCabe (1962a)
Wairakei Well 18	12	—	242	—	—	15000	Wasserburg et al., (1963)
Average Yellowstone and Lassen Hot Spring gases	12.9	1.0	820	0.15	0.033	43000	Mazor and Wasserburg (1965)
Yellowstone wells, gas phase	127[a]	0.3	300	0.07	0.01	—	Mazor and Fournier (1973)
Larderello	12.6	—	8.3	—	—	5500	Wasserburg et al., (1963)

[a] one result only

Table 13-5 Calculated partial pressures of gases associated with either the hot water or steam tapped by geothermal wells

Area	Temp. (°C)	Pressure (bar)					
		P_{CO_2}	P_{H_2S}	P_{CH_4}	P_{H_2}	P_{N_2}	P_{NH_3}
Hveragerdi	216	0.14	0.005	0.004	0.07	0.1	—
Wairakei	260	1.0	0.02	0.04	0.04	0.1	0.0003
Broadlands	260	9.0	0.1	0.5	0.11	1.0	0.002
El Tatio	220	1.5	0.002	7×10^{-4}	0.002	0.02	—
Ahuachapan	230	0.06	0.002	—	0.06	0.02	—
					−0.2	−0.07	
Mexicali	340	7	0.2	1.3	0.11	1.7	—
Larderello[a]	220	0.4	0.007	0.006	0.01	0.004	0.004
The Geysers[a]	230	0.2	0.01	0.01	0.003	0.005	0.01

[a] Approximate partial pressures in steam phase

solubility relationships from input data such as that contained in Tables 13-2 and 13-3. The program used known variations in equilibrium constants for acid/base, complex ion, and gas and solid solubility reactions; ionic activity coefficients calculated by the extended Debye-Hückel equation; and standard thermodynamic properties of steam and water.

Table 13-6 gives the calculated deep fluid compositions for some geothermal areas at an early stage of production. The pH of the deep hot water is usually lower than that measured in the cooled water sample; it also varies considerably between geothermal fields (see below). The bicarbonate ion in the deep waters may be partly replaced by borate and silicate as CO_2 and H_2S are lost from the water into the steam phase. An example is

$$H_4SiO_4 + HCO_3^- \rightleftharpoons CO_2 + H_3SiO_4^- + H_2O \qquad (13.3)$$

At high temperatures there is an increasing tendency for ions to become complexed to lower charged species. HF and HSO_4^- become weak acids at high temperatures, which is important for many deep waters with appreciable concentrations of these species. In the Matsao field, the deep water contains appreciable concentrations of the acids HF, HCl, and HSO_4^-, and the pH is controlled mainly by the following equilibria:

$$4S + 4H_2O \rightleftharpoons H^+ + HSO_4^- + 3H_2S \qquad (13.4)$$

$$Mg^{2+} + HSO_4^- \rightleftharpoons MgSO_4(aq) + H^+ \qquad (13.5)$$

The high stability of $MgSO_4$ and similar complexes at high temperatures has a tendency to increase the acidity of the solution.

Origin of Water and Solutes

The origin of chemicals in natural thermal waters has been argued widely. The view that volcanic spring waters are entirely magmatic fluids is not in agreement with evidence from hydrogen and oxygen isotope ratio determinations (Craig, 1963; White, 1974; and also Chapter 6), or from the compositions of dissolved inert gases. There is quantitative evidence that high-temperature thermal waters have predominantly a local meteoric water origin, but the methods of examination do not exclude the possibility of 5-10% of other water being present.

A magmatic origin for elements such as boron, fluorine, nitrogen, arsenic, lithium, potassium, rubidium, and cesium, which are found in unusually high concentrations in thermal waters, has been strongly sup-

Table 13-6 Concentrations of solution species in well waters as analyzed (top figure) and in the original deep high-temperature water (lower figure in each set); concentrations in 10^{-3} $m\text{kg}^{-1}$

Source	Temp. (°C)	pH	Na^+	K^+	Ca^{2+}
Well G-7 Hveragerdi	20	9.45	9.9	0.6	0.02
	195	8.3	8.1	0.5	0.01
Well 8 Reykjanes	20	5.75	675	56	61
	270	5.0	354	33	39
Well 24 Wairakei	20	8.3	54	5.4	0.3
	250	6.6	37	3.7	0.2
Well 13 Broadlands	20	8.6	42.7	5.1	0.06
	255	6.3	28.6	3.5	0.04
Well 7 El Tatio	20	6.75	212	21	5.3
	254	5.9	133	14	3.6
Well E205 Matsao	20	2.4	243	23	37
	245	2.9	159	16	27

Source	Temp. (°C)	Mg^{2+}	Cl^-	HCl	F^-
Well G-7 Hveragerdi	20	0.02	5.7	0.00	0.08
	195	0.00	4.7	0.00	0.06
Well 8 Reykjanes	20	0.08	845	0.0	0.01
	270	0.04	460	0.02	0.00
Well 24 Wairakei	20	0.00	62	0.00	0.44
	250	0.00	43	0.00	0.29
Well 13 Broadlands	20	—	47.2	0.00	0.24
	255	—	31.6	0.00	0.13
Well 7 El Tatio	20	0.007	244	0.00	—
	254	0.004	155	0.00	—
Well E205 Matsao	20	5.4	386	0.00	0.09
	245	3.1	259	0.58	0.00

Source	Temp. (°C)	HF	CO_3^{2-}	HCO_3^-	CO_2
Well G-7 Hveragerdi	20	0.00	0.18	1.1	0.00
	195	0.00	0.01	2.0	0.11
Well 8 Reykjanes	20	0.00	0.00	0.003	0.01
	270	0.01	0.00	0.05	32.6
Well 24 Wairakei	20	0.00	0.01	0.38	0.00
	250	0.03	0.00	0.73	8.8
Well 13 Broadlands	20	0.00	0.08	2.59	0.01
	255	0.04	0.00	3.07	102.6
Well 7 El Tatio	20	—	0.00	0.68	0.2
	254	—	0.00	0.31	18.1
Well E205 Matsao	20	0.29	0.00	0.00	0.00
	245	0.27	0.00	0.00	41.1

Table 13-6 (*Continued*)

Source	Temp. (°C)	$H_2BO_3^-$	H_3BO_3	NH_3	NH_4^+
Well G-7 Hveragerdi	20	0.04	0.02	0.001	0.001
	195	0.01	0.04	0.002	0.000
Well 8 Reykjanes	20	0.00	1.18	—	—
	270	0.00	0.77	—	—
Well 24 Wairakei	20	0.31	2.4	0.01	0.09
	250	0.00	1.9	0.07	0.01
Well 13 Broadlands	20	0.90	3.55	0.01	0.10
	255	0.00	3.07	0.14	0.02
Well 7 El Tatio	20	0.07	18.5	0.00	0.12
	254	0.01	12.7	0.06	0.03
Well E205 Matsao	20	0.00	10.0	0.00	2.2
	245	0.00	7.2	0.00	1.5

Source	Temp. (°C)	$H_3SiO_4^-$	H_4SiO_4	HS^-	H_2S
Well G-7 Hveragerdi	20	1.7	2.9	0.00	0.00
	195	0.4	3.3	0.08	0.00
Well 8 Reykjanes	20	0.00	15.7	0.00	0.03
	270	0.00	10.2	0.00	0.92
Well 24 Wairakei	20	0.50	10.7	0.03	0.00
	250	0.01	7.9	0.07	0.40
Well 13 Broadlands	20	1.06	11.5	0.00	0.00
	255	0.01	8.7	0.11	1.81
Well 7 El Tatio	20	0.01	12.5	0.00	0.00
	254	0.00	8.6	0.01	0.21
Well E205 Matsao	20	0.00	10.9	0.00	0.00
	245	0.00	7.8	0.00	1.85

Source	Temp. (°C)	SO_4^{2-}	HSO_4^-	$NaSO_4^-$	KSO_4^-
Well G-7 Hveragerdi	20	0.81	0.00	0.03	0.00
	195	0.51	0.00	0.16	0.01
Well 8 Reykjanes	20	0.24	0.00	0.14	0.01
	270	0.04	0.02	0.15	0.01
Well 24 Wairakei	20	0.26	0.000	0.03	0.00
	250	0.07	0.001	0.12	0.01
Well 13 Broadlands	20	0.06	0.000	0.01	0.00
	255	0.01	0.001	0.03	0.00
Well 7 El Tatio	20	0.23	0.000	0.07	0.01
	254	0.04	0.002	0.13	0.01
Well E205 Matsao	20	2.04	0.20	0.59	0.05
	245	0.06	1.56	0.12	0.01

FLUID COMPOSITIONS

Table 13-6 (*Continued*)

Source	Temp. (°C)	CaSO$_4$	MgSO$_4$	NaCl	KCl
Well G-7 Hveragerdi	20	0.00	0.00	0.00	0.00
	195	0.00	0.01	0.02	0.00
Well 8 Reykjanes	20	0.09	0.00	4.8	0.4
	270	0.08	0.01	88	4.0
Well 24 Wairakei	20	0.00	0.00	0.3	0.0
	250	0.01	0.00	1.2	0.1
Well 13 Broadlands	20	0.00	0.00	0.03	0.0
	255	0.00	0.00	0.94	0.1
Well 7 El Tatio	20	0.02	0.00	0.6	0.1
	254	0.03	0.00	12	0.8
Well E205 Matsao	20	0.78	0.11	0.9	0.1
	245	0.13	0.82	16	1.0

ported as they are known to be concentrated in the residual fluids from crystallizing magma. However, experimental evidence now suggests that the concentration of these and other elements into a high-temperature water phase can also occur during hydrothermal alteration and recrystallization of rocks at moderate temperatures.

Facca and Tonani (1964) suggested that the carbon dioxide, hydrogen sulfide, hydrocarbons, and ammonia present in Larderello steam could come from simple decomposition reactions in the hot sedimentary rocks. Ellis and Mahon (1964, 1967) showed by laboratory experiments that the typical elements in high-temperature natural waters could be derived from the simple reaction of volcanic rocks and water within a geothermal system. They demonstrated that the concentrations of major rock-forming elements such as silica, sodium, potassium, calcium, magnesium, aluminum, iron, and manganese were controlled in solution by pressure- and temperature-dependent mineral/solution equilibria and that other constituents such as fluoride, carbonate, and sulfate were limited by the formation of sparingly soluble salts. Transfer of appreciable proportions of chloride, boron, ammonia, fluoride, and arsenic from the rocks occurred even before there was appreciable hydrothermal alteration, an important fact suggesting that these elements (and possibly others) were not held entirely within the crystal structures of the major rock minerals but concentrated in intercrystalline material. Appreciable proportions of the lithium, rubidium, and cesium in the original rocks were released only after extensive alteration had occurred, which at the experimental times of 1–2 weeks required temperatures of 500–600 °C. However, Chelishev (1967) showed that the distribution of rubidium and cesium between either mica or feldspar and water was increasingly in favor of the solution phase as temperatures were lowered from 600 to 250 °C. An extended reaction period (2 years) at 210 °C released 10 ppm lithium into solution from a Wairakei rhyolite breccia

(R. B. Glover, personal communication), and parts per million concentrations of lithium resulted from the reaction of Iceland dolerite with 0.5 m NaCl solution at 300°C (Andrusenko and Moskalyuk, 1966).

Further work by Mahon (1967) on the reaction of water with graywacke, shale, and mudstones at temperatures of 200–300°C produced solutions with chemical characteristics (e.g., Cl/B and Cl/NH$_3$ ratios and fluoride concentrations), which were in line with those of hot waters found in these rock types. Similar experiments were reported by Kissen and Pakhomov (1967).

Following the reaction of an andesite and a shale at 400°C with pure water, 2 m NaCl, and 4 m NaCl (rock/water = 1), several parts per million of lead and copper and several tens of parts per million of manganese were found in the salt solutions, but only traces with water (Ellis, 1968). From the andesite, over 75% of lead and copper were removed during the salt solution reactions but much lesser amounts with water. The great importance of salinity in determining the concentration of many species in high-temperature natural solutions is examined further below.

The isotopic composition of an element may show its origin. For example, in the Salton Sea geothermal brines much of the lead and strontium was identified with an origin in the aquifer sediments rather than from local volcanic rocks (Doe et al., 1966). Steiner and Rafter (1966) suggested from ^{34}S/^{32}S ratios that the sulfate in Wairakei well discharges had a seawater origin, whereas the sulfide with isotopic ratios similar to that for meteoritic sulfur had an upper crust, magmatic origin. However, more recently sulfide and sulfate sulfur of a similar isotopic composition to the latter was found to be extractable from graywacke rocks that border the volcanic zone (Giggenbach, 1977). Studies of the ^{13}C/^{12}C ratios of carbon gases in geothermal areas showed that it was possible to explain their origin from mixed carbonate, organic carbon material within the rocks. The results however, did not exclude the addition of juvenile carbon to the system (Craig, 1963 and Hulston and McCabe, 1962b).

If the chemicals in geothermal waters are derived by rock leaching, there is the problem of element availability within the finite rock volume of the system, if continuous water flows persist over lifetimes of the order of 10^6 years (Ellis, 1970a). However, it is by no means certain that either water flows or water compositions have been constant throughout the lifetime of the system.

There are several alternative origins for solutes. For example, in the Reykjanes area the salt content of the waters is almost certainly derived from seawater. In sedimentary areas the incorporation of connate waters or the solution of evaporite beds may influence the composition of a geothermal water system (e.g., in the Red Sea Deep geothermal system; Miller et al., 1966). The derivation of chemicals in a geothermal water by the cycling of meteoric water through the top of a magma chamber beneath the area is not ruled out by existing evidence, and convection within the magma body would give an almost unlimited supply of chemicals.

White (1965) suggested that fine-grained clay beds may act as semipermeable membranes, allowing the preferential transport of water and certain ions across the membrane. This process would effect a concentration and influence the composition of the solution retained beneath a clay membrane sedimentary sequence. Experimental work has confirmed the operation of this process, and its particular effectiveness at low salt concentrations (Hanshaw and Coplen, 1973).

Specific Solution and Mineral Reactions

There are few elements in geothermal waters whose concentrations are not controlled by temperature- and pressure-dependent equilibria. The principal "soluble" elements with their chemistry dominated mainly by extraction and dilution processes are chlorine, bromine, iodine, boron, cesium, and arsenic. Some specific reactions are now reviewed. Unless noted, equilibrium data discussed below are for saturated water vapor pressures, since at the depths of observations in geothermal systems pressures do not greatly exceed these values.

SODIUM AND POTASSIUM

Sodium is usually the dominant cation in high-temperature geothermal waters. A systematic variation in the ratio sodium/potassium with temperature occurs in all but the acidic geothermal systems, and in many areas within a wide variety of rock types it has been possible to make a precise correlation between Na/K and water temperature (White, 1965; Ellis and Mahon, 1967; Ellis, 1970A; and Fournier and Truesdell, 1973).

$$K^+ + \text{albite} \rightleftharpoons Na^+ + \text{K-feldspar} \tag{13.6}$$

$$K^+ + Na_xCa_y\text{-feldspar} \rightleftharpoons xNa^+ + yCa^{2+} + \text{K-feldspar} \tag{13.7}$$

$$2K^+ + 6SiO_2 + Al_2O_3 + H_2O \rightleftharpoons 2KAlSi_3O_8 + 2H^+ \tag{13.8}$$

Simple ion exchange may occur: potassium may react with a plagioclase feldspar to form potassium feldspar, and there may be direct precipitation of potassium feldspar. An empirical correlation of available data from geothermal waters, oil-well waters, laboratory experiments, and fluid inclusion measurements led Fournier and Truesdell (1973) to conclude that reactions of the second type dominate, and that for geothermal waters above 100°C, a good correlation between log K^* and reciprocal absolute temperature was obtained.

$$\log K^* = \log (Na/K) + \tfrac{1}{3} \log (Ca^{1/2}/Na) \qquad (13.9)$$

Figure 13.4 from Fournier and Truesdell (1973) summarizes the good correlation obtained for a wide variety of waters. In individual areas, an even better correlation between temperature and alkali ratios is obtained (e.g., for Broadlands; Mahon and Finlayson, 1972). Only for near-neutral pH waters of high temperature and with log ($Ca^{1/2}$/Na) values less than 0.5 can the simple ratio Na/K be used as an accurate geothermometer.

Hydrothermal recrystallization of volcanic or other quartz-feldspathic rocks has a tendency to produce potassium feldspar, potassium mica, and albite. This is observed both in hydrothermally altered rocks from deep wells and in laboratory experiments (Ellis, 1969) above 200°C.

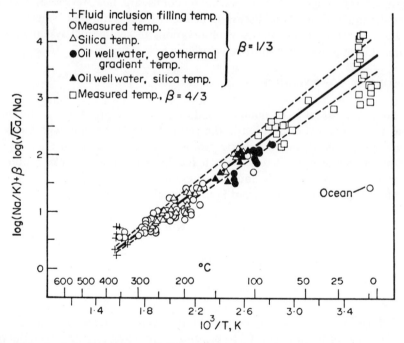

Fig. 13.4 A relationship between alkali ion concentration ratios and temperature. The two dashed lines show +15° and −15°C with respect to the mean curve. (From Fournier and Truesdell, 1973.)

FLUID COMPOSITIONS

$$3KAlSi_3O_8 + 2H^+ \rightleftharpoons KAl_3Si_3O_{10}(OH)_2 + 2K^+ + 6SiO_2 \qquad (13.10)$$

From the equilibrium equations (13.6), (13.7), and (13.10), for the coexistence of albite, potassium mica, and potassium feldspar, at each temperature unique values of Na/K, K/H, and Na/H occur in high-temperature, low-calcium solutions. The hydrogen ion activity is related to the concentrations of the sodium and potassium ions but Na/K ratios are essentially independent of pH (Ellis, 1970a). Figure 13.5 relates the pH of some deep geothermal waters to (Na + K) concentrations and shows good agreement between field measurements and the above model.

LITHIUM, RUBIDIUM, AND CESIUM

The concentration of rare alkalis in geothermal waters reflects their abundance in the surrounding rocks, with waters in basaltic areas having low concentrations compared with those in the rhyolitic and andesitic areas. Lithium and rubidium tend to decrease in concentration in waters migrating to the surface due to incorporation of the ions in low-grade hydrothermal alteration products such as clays and zeolites. Hydrother-

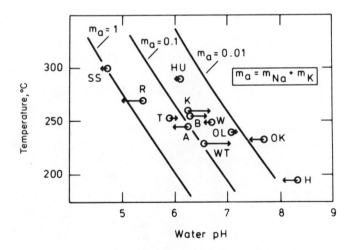

Fig. 13.5 The calculated trends with temperature in the pH of waters of three different salinities for equilibrium with potassium feldspar, albite, and potassium mica. Circled points show the pH of waters in various geothermal systems, with a connecting arrowhead to indicate the theoretical pH for the salinity of the water ($m_{Na} + m_K = m_a$). The areas are: SS, Salton Sea; R, Reykjanes; HU, Hatchobaru; T, El Tatio; K, Kawerau; B, Broadlands; A, Ahuachapan; W, Wairakei; OL, Olkaria, Kenya; WT, Waiotapu; OK, Orakei Korako; H, Hvergerdi. (From Ellis and Mahon, 1977.)

mally formed lepidolite has been noted in shallow, altered rocks (Bargar et al., 1973). In individual areas the Na/Rb ratios follow trends in the Na/K ratios, but the differences in Na/Li ratios between deep and surface waters are less marked. The isotopic ratio ^{6}Li/^{7}Li tends to increase in hot waters migrating to the surface, presumably through the preferential inclusion of the isotope ^{7}Li in hydrothermal alteration minerals (H. J. Svec and Ellis, unpublished).

CALCIUM

Geothermal waters contain the ions of several sparingly soluble calcium salts, for example, $CaCO_3$, $CaSO_4$, CaF_2. Most geothermal waters at deep levels are near saturation with calcite (Ellis, 1970a) and this mineral is frequently precipitated from the waters when they boil and lose carbon dioxide. The tendency to precipitate calcite in natural channelways and in drill pipes is particularly marked for waters containing high carbon dioxide concentrations.

The following equations express the relationship between carbon dioxide and calcium and hydrogen ion activities in the water, where K_{a_1} is the first ionization constant of carbonic acid, and K_c the equilibrium constant for the solution of calcite:

$$CaCO_3 + H_2CO_3 \rightleftharpoons Ca^{2+} + 2HCO_3^- \quad (13.11)$$

$$K_C = a_{Ca} \cdot a_{HCO_3}^2 / K_{CO_2} \cdot a_{CO_2} \quad (13.12)$$

$$= a_{Ca} \cdot a_{CO_2} \cdot K_{a_1}^2 \cdot K_{CO_2} / a_{H^2} \quad (13.13)$$

In addition, for the coexistence of potassium and sodium feldspar and potassium mica [equations (13.6) and (13.10)] $a_{Na}/a_H = K_{fm}$. If geothermal waters are in equilibrium with calcite, potassium mica and sodium and potassium feldspars, then at a temperature the product $m_{Ca} \cdot m_{CO_2}$ should increase in proportion to m_{Na}^2 (or to m_K^2).

$$m_{Ca} \cdot m_{CO_2} = \frac{K_C \cdot m_{Na}^2 \cdot \gamma_{Na}^2}{K_{a_1}^2 \cdot K_{fm}^2 \cdot \gamma_{Ca}} \quad (13.14)$$

At constant carbon dioxide concentration, for a given temperature, calcium concentrations should vary approximately with the square of the sodium ion (or potassium ion) concentrations. Low salinity waters should have high Na/Ca ratios, and the converse for high salinity waters. This is in general agreement with analytic data in Table 13-2. For a given sodium ion concentration and temperature, waters with high carbon dioxide concentrations will tend to have low calcium concentrations.

MAGNESIUM

In high-temperature geothermal waters of low salinity, magnesium concentrations are extremely low and frequently in the range 0.01–0.1 ppm. They appear to be controlled by the following type of reaction (Ellis, 1971):

$$xCa^{2+} + \text{Mg-silicate} + (2 - 2x)H^+$$
$$\rightleftharpoons Mg^{2+} + x\text{Ca-silicate} + y \cdot SiO_2 \qquad (13.15)$$

The solutions are usually close to saturation with calcite, and common calcium silicates are montmorillonite, wairakite, or epidote. Chlorite is a common secondary magnesium-containing mineral.

For New Zealand geothermal fields with water temperatures in the range 235–255°C, with sodium concentrations of 0.02–0.04 m, and with P_{CO_2} ranging 1–30 bars, Ellis (1971) showed that the value of log a_{Mg}/a_H^2 for the deep solutions averaged 6.2 and showed no major trend with P_{CO_2}. As a_{Ca}/a_H^2 for calcite saturation is inversely related to m_{CO_2}, the atomic ratio a_{Ca}/a_{Mg} should be highest in waters with low m_{CO_2}. For a constant log a_{Mg}/a_H^2 of approximately 6.0 at 250°C, m_{Mg} will be approximately 0.001, 0.10, and 10.0, at sodium concentrations of 0.01, 0.1, and 1.0 m, respectively (see Figure 13.5).

SILICA

Silica solubilities in high-temperature water were reviewed in Chapter 9. Fournier and Rowe (1966) and Mahon (1966) showed conclusively that high-temperature geothermal waters (above about 180°C) were saturated with silica in equilibrium with quartz. The concentration of silica in a sample of geothermal well water can be used to give an accurate estimate of the underground temperature (e.g., Fournier, 1970; Mahon and Finlayson, 1972). The high-temperature equilibrium solutions contain only monomeric silica species.

Metastable crystalline and amorphous forms of silica persist in contact with water to temperatures of at least 300°C and exhibit definite solubilities which are much higher than for quartz. As geothermal waters are concentrated by steam loss as they approach the surface in channels or drillpipes, they pass from a stage of saturation with quartz and undersaturation with amorphous silica, to a condition of supersaturation with quartz, and eventually with amorphous silica. For example, the temperatures at which the solubility of amorphous silica is exceeded are 95, 150, and 200°C, for waters that were originally at underground temperatures of 200, 250, and 300°C, respectively. For these waters, growth of silica polymers in solution, with the associated tendency toward amorphous silica deposition, does not begin until there is appreciable supersaturation with amorphous silica. It is only under near-surface conditions where the temperatures are too low for the quartz equilibrium to be rapidly reversible, and where solutions become appreciably supersaturated with amorphous silica that extensive deposition of opal and silica occurs in the rocks or about outflows. The reactions involving silica polymerization and deposition are complex and detailed kinetic studies have been made observing the effects of pH, supersaturation, solution composition, presence of nuclei, and flow rates (Kitahara, 1960; Yanagase et al., 1970; and Rothbaum and Wilson, 1977).

Ellis and Mahon (1964) showed that at temperatures between 250 and 350°C the solubility of silica from volcanic igneous rocks was initially approximately equal to that for amorphous silica solubility. Arnorsson (1975) showed that Iceland thermal waters at temperatures up to approximately 150°C in basalt rocks dissolved silica and deposited chalcedony on cooling or dilution to retain equilibrium with this mineral. Hot waters moving into fresh volcanic rocks could cause extensive solution of silica and create concentration levels that would later require quartz or chalcedony to be deposited to attain equilibrium. The lateral migration of hot water from a reservoir system may therefore be hindered by the tendency of processes of solution and deposition to seal flow channels.

AMMONIA

In sedimentary rock environments such as Ngawha, or Salton Sea, ammonia concentrations are higher than in geothermal systems contained within volcanic rocks. The ammonia distribution (solution/rock) is highest at high temperatures, and at low pHs (e.g., at Matsao). Whether this is an equilibrium distribution or simply more effective extraction is not clear, although under cooler near-surface conditions, ammonia is concentrated into low-grade hydrothermal alteration products (Erd et al., 1964).

FLUORIDE

The concentrations of fluoride in geothermal waters appear to be limited by the solubility of fluorite, which in the presence of silica is of the order of 10 ppm fluoride at temperatures of 200-300°C. High fluoride concentrations in general correlate with low calcium concentrations, in New Zealand geothermal waters (Mahon, 1964). Low calcium and high fluoride concentrations in geothermal waters are favored by low salinity, high carbon dioxide concentration, and high temperature.

Total fluoride concentrations may be increased in low pH waters due to the increased proportion of un-ionized HF, and the possible formation of SiF_6^{2-}, or AlF_6^{3-}.

CALCIUM SULFATE

Most geothermal waters of temperature 200–300°C and of near-neutral pH and moderate salinity (0.02–0.1 m) have sulfate concentrations in the range 10–100 ppm. As the solubility of calcium sulfate (anhydrite) in high-temperature water is very low (Chapter 9), an inverse correlation between calcium and sulfate concentrations is to be expected, except at low pHs where there is an appreciable proportion of total sulfate present as HSO_4^-. Very-high-temperature or very-high-salinity systems may have extremely low sulfate concentrations (e.g., Mexicali and Salton Sea).

IRON, MANGANESE, ETC.

The general salinity/pH relationship affects the maximum concentration of metals such as iron, manganese, copper, and lead, that can be carried in geothermal solutions. The equilibrium solution of metal ions from oxides or sulfides involves the hydrogen ion, e.g., $MO + 2H^+ \rightleftharpoons M^{2+} + H_2O$; or $MS + 2H^+ \rightleftharpoons M^{2+} + H_2S$.

Within the limits of element availability from the surrounding rocks a general relationship is to be expected between the concentration of iron, manganese, copper, lead, etc., and $(a_H)^2$, or in near-neutral pH waters $(m_{Na})^2$.

REDOX CONDITIONS

The waters of many high-temperature geothermal areas have hydrogen present at a partial pressure, P_{H_2}, of about 0.1 bars (Table 13-5). For the gaseous reaction $2H_2O \rightleftharpoons 2H_2 + O_2$ the logarithm of the equilibrium constant, $\log K_{H_2O} = -45.9, -43.5, -41.3,$ and -39.2 at 225, 250, 275, and 300°C, respectively. If $P_{H_2} = 0.1$, P_{O_2} values are $10^{-42.5}$, 10^{-40}, $10^{-37.5}$, and 10^{-35} bars, respectively. Seward (1974), and Kusakabe (1974) confirmed that P_{O_2} values of this order occurred in the Broadlands and Wairakei fields at approximately 250–260°C, and were close to the conditions for pyrite and pyrrhotite coexistence under the prevailing solution compositions.

SULFUR SPECIES

From the ratio of sulfide to sulfate and the pH of the deep hot waters, the partial pressure of oxygen necessary for equilibrium can be calculated from standard thermodynamic data (Ellis, 1967; Kusakabe, 1974). For example, at Wairakei underground P_{O_2} values of 10^{-36}–10^{-35} bars are required. This suggests a more oxidizing situation than is found by direct measurement in the system, and the sulfide and sulfate may not have attained equilibrium. A similar situation apparently exists in the deep Broadlands waters (calculations of Seward, 1974).

Ellis and Giggenbach (1971) gave results for the sulfur hydrolysis equilibrium

$$4S + 4H_2O \rightleftharpoons 3H_2S + H_2SO_4 \tag{13.16}$$

The equilibrium constant K_s at saturated water vapor pressures was approximately $10^{-11.6}$ at 200°C, $10^{-8.7}$ at 250°C, and $10^{-6.4}$ at 300°C, where $K_s = m_{H_2S}^3 \cdot m_{H^+} \cdot m_{HSO_4^-}$. For example, at 250°C the solution contains approximately 0.03 m H_2S and 0.01 m H_2SO_4. These values are in approximate agreement with concentrations of sulfide and sulfate in the Matsao well waters (Table 13-2).

At temperatures of 250–300°C, dilute solutions of ions such as $S_2O_3^{2-}$, $S_4O_6^{2-}$, and $S_5O_6^{2-}$ disproportionate into sulfide and sulfate (Pryor, 1960; Ellis and Golding, unpublished).

$$2S_5O_6^{2-} + 8H_2O \rightleftharpoons 5SO_4^{2-} + 6H^+ + 5H_2S \tag{13.17}$$

However, Giggenbach (1974) demonstrated by high-temperature spectrophotometry that at temperatures less than 200°C, at near-neutral pHs, and with the total sulfur species concentrations at a level that brings the system close to the point of coexistence with sulfur, thiosulfate and several polysulfide species can exist, such as the S_4^{2-} ion and its dissociation products S_2^- or S_3^-. These intense blue radicals may account for the color of the cooling and partly oxidized waters of hot pools in many geothermal areas. With equal and equilibrium concentrations of sulfide and sulfate of, for example, 0.01 m, the point of coexistence with sulfur occurs at approximately pH 7 at 150°C, pH 3.6 at 200°C, and pH 0.7 at 250°C. Except in very acid waters or in solutions with very high total dissolved sulfur, species other than sulfide and sulfate are unlikely to be present in geothermal waters at temperatures in excess of approximately 200°C.

ISOTOPIC EQUILIBRIA

The isotopic compositions of the elements hydrogen, oxygen, sulfur, and carbon in geothermal fluids and minerals have received considerable attention. Isotopic exchange reactions have been used to estimate underground temperatures, and to examine the extent of equilibration between minerals and solutions. For a review, see Panichi and Gonfiantini (1977).

Kusakabe (1974) used laboratory-calibrated isotope fractionation factors for oxygen in the HSO_4^--H_2O system (Mizutani and Rafter, 1969) and for sulfur in the HSO_4^--H_2S system (Robinson, 1973) to estimate underground temperatures in the Wairakei area. Similar estimates were made by Mizutani (1972) for the Otake area.

In both studies the estimated oxygen isotopic temperatures were several tens of degrees higher than temperatures measured in the wells but the sulfur isotopic temperatures were very much higher (100–200°C) than measured. Kusakabe suggested that the latter results, and the average $\delta^{34}S = 15$ ‰ in Wairakei fluids arose from sulfate from basement graywacke rocks being reduced in part to sulfide under the prevailing redox conditions, but equilibrium not being achieved. Equilibrium between oxygen in water and in sulfate was more rapid (a reaction half-life of approximately 1 year at 250°C and pH 7).

The $^{13}C/^{12}C$ ratios of CO_2 and CH_4 in geothermal fluids have been used as a geothermometer (Hulston and McCabe, 1962b), by assuming that isotopic equilibrium exists in the reaction

$$CO_2 + 4H_2 \rightleftharpoons CH_4 + 2H_2O$$

The temperatures were in reasonable agreement with those measured. However, as experiments show that this reaction is extremely slow at temperatures of 250°C, further work is required to confirm that equilibrium is achieved in geothermal systems.

Hydrogen isotope exchange between the molecules H_2O, H_2, CH_4, and H_2S offers possibilities for geothermometry in geothermal systems. For example, J. R. Hulston (1976) estimated a temperature of 250°C for the Wairakei system from the measured hydrogen isotopic fractionation between H_2 and H_2O.

Extensive oxygen isotope exchange occurs between rocks and high-temperature geothermal waters. In the Salton Sea field Clayton et al. (1968) showed that carbonates were in isotopic equilibrium with the waters at temperatures as low as 100°C, and most fine-grained country rock silicates were close to equilibrium with the waters at temperatures above

150°C. Detrital quartz however showed little tendency to exchange oxygen even at 340°C.

In the Broadlands geothermal field oxygen isotopic equilibrium exists between fissure-grown quartz, adularia and calcite at temperatures of 250-290°C (Blattner, 1975); also for secondary quartz and illite at temperatures of 160-270°C (Eslinger and Savin, 1973). Clayton and Steiner (1975) found a general approach toward oxygen isotopic equilibrium for whole rock samples from the Wairakei system and equilibration between calcite and water at temperatures above 200°C (see Chapter 15 for sulfur isotope equilibria).

HYDROTHERMAL ALTERATION

Field Observations

The mineral assemblages associated with hydrothermal alteration in many geothermal fields have been investigated by the use of drill cores or cuttings; also the chemical environment in which the alteration minerals have been produced can be deduced from well discharges.

HVERAGERDI, ICELAND

The area occurs in basaltic tuffs with minor intrusive rhyolites (Sigvaldason, 1963). Below an acid leached zone, montmorillonite was a major mineral to nearly 100 m; but at deeper levels chlorite was the most abundant mineral and calcite and zeolites were common. Beneath 250 m laumontite was the major zeolite, persisting to the highest temperature (230°C) together with epidote, pumpellyite, and clinozoisite. This deep alteration occurred at a water pH of about 8.3 and with $m_{CO_2} \simeq 2 \times 10^{-4}$.

REYKJAVIK, ICELAND

Sigvaldason (1963), described the alteration of Tertiary basalts in a 2200 m well. Alteration commenced with montmorillonite, calcite, and zeolites, with mordenite occurring at temperatures of 120°C. Iron-rich chlorite was abundant from approximately 200 m until the base of the well, and was accompanied by laumontite, calcite, and albitization of plagioclase. Epidote as a vein and replacement mineral was found throughout the hottest zone (145°C) from 1000-2000 m, but was considered by Seki (1972) to be a relic from previous higher temperatures.

REYKJANES, ICELAND

The rocks consist of an alternating sequence of recent basalt, basaltic tuffs, and hyaloclastic breccias and tuffs (Björnsson et al., 1970; Tomasson and Kristmannsdottir, 1972). Hydrothermal alteration was found to attack the basaltic glass first, then olivine and pyroxene, and finally plagioclase. The main water flows were within fractures in the basalt layers, and alkali feldspars were formed in the upflow zones.

Beneath a near-surface oxidized zone of opal and hematite, down to 400–600 m, zeolites occurred at temperatures up to 230°C (principally mordenite, stilbite, analcime; minor wairakite) together with montmorillonite (up to 200°C). Prehnite was found beneath the zeolite zone, whereas epidote first appeared at deeper levels and at temperatures over 200°C. With increasing temperature, there was a trend from montmorillonite to mixed-layer chlorite-montmorillonite (minor illite), to swelling chlorite. Beneath the oxidized zone, quartz, pyrite, calcite, and anhydrite occurred throughout. The deep 250–280°C waters were approximately of pH 5.5 and $m_{CO_2} \simeq 0.02$.

WAIRAKEI, NEW ZEALAND

The sequence of hydrothermal alteration minerals at Wairakei was summarized by Steiner (1968, 1974). The original rocks were glassy rhyolitic volcanics consisting of pumiceous tuffs and breccias, ignimbrites, and rhyolite flows. Andesite occurs locally at depth in the western part of the field. Fluid rises mainly through fissure zones and migrates laterally in permeable breccia formations. Highest temperatures (up to 265°C) and degree of rock alteration occurred above flow fissures, and alteration patterns showed a gradation in mineralogy from the fissure toward the cooler surface, and in some cases also below the fissure where temperatures decreased again. Flow fissures tended to contain quartz, potassium feldspar, and wairakite at points where boiling occurred. A shallow acid-leached zone of kaolin, alunite, and opaline silica occurred at the surface.

The susceptibility of minerals to alteration decreased in the series: glass, magnetite, hypersthene, hornblende, andesine, biotite, (quartz was unaffected). In general, potassium and silica were added to the altered rocks, and sodium and calcium removed. Figure 13.6 from Steiner (1968) summarizes typical mineralogic trends with temperature and depth in the field.

Silicic volcanic glass first converts to calcium montmorillonite which reacts with potassium from solution to form interstratified illite-montmorillonite in the temperature range 130–230°C, and illite at temperatures

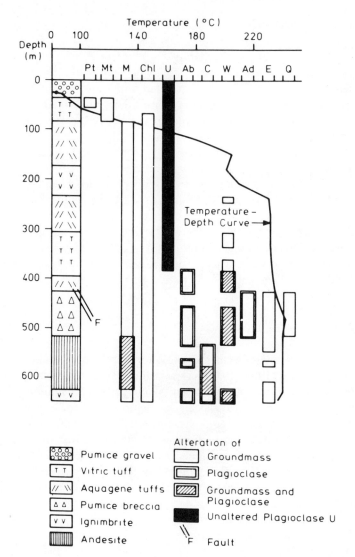

Fig. 13.6 Diagram showing data for Well 225 Wairakei, including the stratigraphic sequence, faults, fresh rock, hydrothermal alteration minerals, temperatures, and water vapour pressures. (From Steiner, 1968.)

Abbreviations for hydrothermal minerals are: Ab, albite; Ad, adularia; C, calcite; Chl, chlorite; E, epidote; M, micaceous clay; Mt, montmorillonite; Pt, ptilolite; Q, quartz; and W, wairakite.

greater than 230°C. Below the montmorillonite zone (120-130°C) iron-rich chlorite and micaceous clays were ubiquitous, and pyrite or pyrrhotite were also present throughout.

Calcium zeolites trended from mordenite at temperatures near 100°C, through laumontite (150-200°C), to wairakite at higher temperatures, an increasing thermal stability demonstrated by Coombs et al. (1959), and Seki (1972).

At the maximum temperatures of 250-260°C, alteration in permeable rhyolitic volcanics consisted principally of potassium mica, chlorite, wairakite, albite, potassium feldspar, quartz, and epidote. Calcite was not common, except in the andesitic rock. pHs were approximately 6.5, and $m_{CO_2} \simeq 0.01$ in the original 250-260°C water.

BROADLANDS, NEW ZEALAND

The hydrothermal alteration in this area is described in detail in Chapter 15. A common mineral assemblage in the deep aquifer conditions (approximately 260°C) was potassium mica, potassium feldspar, albite, chlorite, calcite, pyrite, and quartz. Wairakite, iron-rich epidote, and zoisite were rare. The water was of approximately pH 6.0-6.4, and with $m_{CO_2} \simeq 0.12$ (Browne and Ellis, 1970).

PAUZHETSK, U.S.S.R.

Averyev et al. (1961) and Naboko (1968) distinguished successive depth zones of alteration—formation of kaolin, montmorillonite, zeolites, feldspars, and of propylitization. The most intense alteration occurred in the porous members of the local andesitic and dacitic rocks. Under a silicified surface zone containing kaolinite and alunite there was widespread development of laumontite and adularia, with minor calcite and chlorite (30-250 m). From 240 m downward, laumontite and adularia diminished, and development of chlorite and calcite was widespread, together with zeolites, hydromicas, pyrite and gypsum. Albite formed from the decomposition of plagioclase beneath the zone of boiling and CO_2 loss where the temperatures approached 200°C, and $m_{CO_2} \simeq 0.002$. Adularia was formed principally in the zone of boiling and CO_2 loss.

YELLOWSTONE PARK, U.S.A.

Honda and Muffler (1970) described the rock alteration observed in an experimental 64 m well in Yellowstone Park. At the highest temperature of 170°C, the alteration mineral assemblage formed from original rhyolite obsidian sands and flow rhyolite was: quartz, analcime, mordenite, calcite, montmorillonite, celedonite, chlorite, and iron oxides.

STEAMBOAT SPRINGS, U.S.A.

The mineralogy of this area was described by Schoen, White and Hemley (1974). Granodiorite and andesite rock was altered by waters of up to 170°C to form at the maximum depth investigated (120 m) an assemblage of illite (or sericite), quartz, iron-rich chlorite, potassium feldspar, albite, and minor calcite. In cooler zones chlorite plus illite trended into mixed layer illite-montmorillonite. Albite and calcite was formed from calcic plagioclase below approximately 60 m. Water up-flow in the area was controlled by fissures, in which quartz was depositing (and calcite at depths below about 50 m): potassium-feldspar appeared in zones of boiling and CO_2 loss above about 100 m. A near-surface oxidized zone contained opaline silica, kaolinite, and alunite. The pH of the 170°C underground water was approximately 6.1 and $m_{CO_2} \simeq 0.15$ (Schoen and White, 1965).

MATSAO, TAIWAN

The patterns of hydrothermal alteration in this area of andesitic rocks were related to flow fissures. Waters at temperatures of up to 290°C were intensely acidic (pH 3) at some levels, probably due to the occurrence of bands of sulfur in the rocks. Water of near-neutral pH occurred at other levels.

Acidic alteration consisted of kaolin, quartz (or opal), pyrophyllite, or alunite (near surface only). Transitional pH zones contained the minerals gypsum or anhydrite, pyrite, illite, and calcite. Higher pH conditions (probably pH 5-6) at highest temperatures (250-290°C) produced the minerals, epidote, potassium feldspar, illite, magnesium chlorite, quartz, and pyrite ($m_{CO_2} \simeq 0.02$). Vein minerals were quartz, calcite, prehnite, pyrrhotite, and chalcopyrite (Chen, 1970; P.R.L. Browne, personal communication).

OTAKE—HATCHOBARU, JAPAN

In the Hatchobaru part of this field individual wells to depths of 1000-1200 m tapped waters at 250-280°C, which were either of near-neutral pH or of high acidity. The Otake sector of the field produced only near-neutral pH waters of 200-250°C (Hayashi, 1973).

In the zone of acid waters (pH 3-5) pyroxene andesites were altered to a leached alunite-cristobalite assemblage near the surface; at greater depths and at approximately 100–150°C to a kaolinite-quartz assemblage; and at highest temperatures to kaolinite, dickite, pyrophyllite, and quartz. With sulfate concentrations of approximately 0.001 m, at the acidity of the waters alunite would be stable only at low temperatures (Hemley et al., 1969).

The near-neutral pH waters at Otake, and in some parts of Hatchobaru produced at increasing temperatures of rock alteration (up to 260°C), mineral assemblages containing successively heulandite, laumontite or wairakite; as temperatures increased montmorillonite reacted to form first random interstratified montmorillonite-sericite, then sericite; while the coexisting chlorite had a decreasing iron content as temperatures increased. Quartz and adularia were found principally in fissure flow zones, while pyrite occurred throughout. At the highest water temperatures the mineral assemblage was albite, adularia, montmorillonite-sericite, epidote, anhydrite, chlorite, calcite, rutile, and pyrite (pH $\simeq$ 6.0, and $m_{CO_2} \simeq 0.01$).

An intermediate zone of lower permeability and often of partly altered rock was found to exist between the acid and alkaline parts of the Hatchobaru area, with minerals such as montmorillonite, chlorite, quartz, and anhydrite occurring at temperatures of 100–200°C.

KATAYAMA, ONIKOBE, JAPAN

Pliocene-Pleistocene dacite, andesite lavas and tuff breccias were altered at the maximum depths (700 m) and water temperatures (210°C) to chlorite, calcite, leucoxene, wairakite, calcite, quartz, and pyrite. At shallower depths, the transition of laumontite to wairakite occurred at temperatures of 105–175°C (Seki et al., 1969). $m_{CO_2} \simeq 0.01$ in the deep solution, and the pH of the well discharge waters ranged 5.1–6.9.

SALTON SEA, U.S.A.

The geothermal system exists in a series of Pliocene and Quaternary river sands, silts and clays of rather uniform mineralogic composition (quartz, calcite, potassium feldspar, plagioclase, illite, dolomite, and kaolinite). At temperatures rather below 100°C, the conversion of montmorillonite to illite commenced through a mixed-layer mineral, the conversion being completed at approximately 210°C. Chlorite, calcite, and CO_2 were formed from the reaction of dolomite, ankerite, and kaolinite with hot brine at temperatures below 180°C, whereas over approximately 290–310°C, iron-rich epidote was abundant and secondary potassium feldspar occurred in veins and by replacement of plagioclase. At temperatures rather over 300°C in the argillaceous sand aquifer at approximately

1200 m, the common hydrothermal alteration mineral assemblage was quartz, epidote, chlorite, potassium feldspar, albite, potassium mica, and pyrite (Muffler and White, 1969). The water pH was about 4.7, and $m_{CO_2} \simeq 0.01$ (Helgeson, 1967).

The transformation of low-grade hydrated and carbonate sedimentary minerals to secondary feldspars, epidote, chlorite, carbon dioxide, and water may be considered a metamorphic alteration process (Muffler and White, 1969). Yet the equilibrium mineral assemblage is similar to that produced at comparable temperatures by hydrothermal alteration of crystalline or glassy igneous rocks. Approach to hydrothermal equilibrium in quartz-feldspathic rock systems is therefore observable in geothermal systems from both low-grade and high-grade mineralogic starting materials.

MATSUKAWA, JAPAN

Wells drilled to depths of 1200 m intercepted andesite and dacite welded tuff and pyroxene andesite (Sumi, 1969; Nakamura et al., 1970). The waters produced were with one exception strongly acidic (surface pHs of 3-5), and contained high sulfate (usually 500-1500 ppm). In the top 150 m, rock alteration produced saponite, chlorite, and sericite but approaching the active upflow zones at depth there was a progressive sequence of more intense alteration passing through chlorite, montmorillonite, kaolinite, and alunite zones. Intersecting the current alteration was an older metastable pattern of pyrophyllite and dickite alteration presumably remaining from a higher temperature period in the field's history.

Associated with the outer chlorite zone were laumontite, hydromica, calcite, quartz, pyrite, anhydrite, and gypsum. In the montmorillonite zone, hydromica, calcite, and anhydrite occurred. Quartz and pyrite were associated both with the kaolin and with the alunite alteration zones. Sumi considered that the water entering the base of the field was of approximately pH 3-4 and at approximately 250-280°C.

According to the results of Hemley et al. (1969), for the Matsukawa solution sulfate concentrations, alunite would be stable at pHs less than about 3, or 4, at 300 or 200°C, respectively, in agreement with field measurements. In the hottest zones $m_K = 0.001$-0.01, so that $m_K/m_H = 0.1$-10; i.e., conditions in which kaolinite is stable with respect to either muscovite or montmorillonite (see Figure 5.8).

LARDERELLO, ITALY

Although the wells produce only steam, Marinelli (1969) suggested that originally there had been hot water in the area and convective circulation.

In the basement rocks of quartzites and slates, typical hydrothermal alteration minerals were adularia, zeolites, chlorite, quartz, and anhydrite. Wairakite was found in siltstone in contact with a production horizon.

SUMMARY

The formation and persistence of secondary minerals depends on many factors, including the reactivity of primary rock minerals, reaction time, temperature, pressure, solution compositions, rock permeability and porosity, and water flow.

The temperature, mineral reactivity, and rock permeability factors are demonstrated by the field examples, such as, the persistence of plagioclase in lower temperature or low-permeability hydrothermal systems. The nucleation and growth of particular phases is also dependent on the starting materials (glass, feldspar, or clays); for example, albite usually arises from decomposition of a more calcic plagioclase, whereas potassium feldspar can form directly on rock surfaces from solution. The relative thermal stability of various zeolite minerals, and of clay-micaceous minerals shown by the field examples can be reconciled with experimental and thermodynamic information (e.g., Coombs et al., 1959; Seki, 1969).

Observations in geothermal areas emphasize the significance of the water pH in characterizing alteration mineral assemblages. High-temperature waters that contain sulfate acidity through oxidation or contact with sulfur produce a series of minerals (kaolinite, alunite, dickite, pyrophyllite) quite distinct from waters which have their pH at deep levels controlled by interaction with aluminosilicate minerals. In the latter case even at the highest temperatures and salinities observed, water pHs probably do not fall much below 5.

The relative activity of the hydrogen ion to metal ions in solution defines many aluminosilicate, sulfide, and carbonate mineral stabilities at constant temperature, pressure, and concentrations of other solutes. Hemley and Jones (1964) and Helgeson (1967) presented examples of mineral stability diagrams in terms of a_{Na}/a_H, a_K/a_H, a_{Ca}/a_H^2, a_{Mg}/a_H^2, using both experimental and field information.

An apparent problem arose from noting the different mineral alteration patterns in the Wairakei and Broadlands fields (similar temperatures, 250–260°C, water salinities and pH, and comparable original rock materials). In the former area the high-temperature mineral assemblage included abundant epidote and wairakite and rare calcite, whereas at Broadlands wairakite and epidote were rare minerals (absent in many holes) yet calcite was abundant. Figure 13.7 from Ellis (1970b) shows how the solute con-

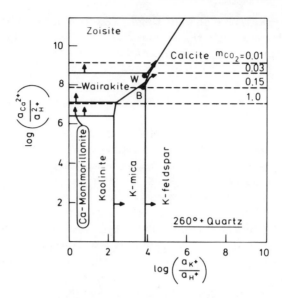

Fig. 13.7 Stability diagram for potassium and calcium minerals at 260 °C in terms of the solution ion ratios a_K/a_H, and a_{Ca}/a_H^2, and various m_{CO_2} values (Ellis 1970b). W = Wairakei; B = Broadlands.

centrations are related to, for example, the equilibrium calcium and potassium minerals. Mineral stability areas are outlined as functions of ionic activities and the values of a_{Ca}/a_{H^2} for precipitation of calcite at various m_{CO_2} concentrations are shown as dashed lines. For a particular m_{CO_2}, values of log (a_{Ca}/a_{H^2}) above this line do not occur. The figures of m_{CO_2} = 0.01, and 0.15, refer to the deep-water conditions at Wairakei, and Broadlands, respectively, and the water composition points are close to the conditions for stability of zoisite, wairakite, and potassium feldspar (Wairakei); and potassium mica, potassium feldspar, and calcite (Broadlands).

With increasingly higher m_{CO_2}, the stability fields of zoisite and wairakite disappear. Calcium zeolites and zoisite (or epidote) are likely to be found only in association with waters with low m_{CO_2}. This effect of CO_2 activity was recognized earlier by Zen (1961) for low-grade regional metamorphism.

A feature of geothermal fields in the top 1000 m is the loss of carbon dioxide from waters due to steam formation as they rise through the country rock. Referring to Figure 13.7, the consequent rise in water pH would cause Broadlands water to become saturated with calcite with the first loss of steam, and while calcite is precipitating it is unlikely that sufficient calcium supersaturation would occur to enable wairakite or epidote to

SUMMARY

form. For Wairakei, carbon dioxide loss from the water and a pH rise may result first in the growth of wairakite and subsequently of epidote before the water composition point reaches the calcite precipitation line (this line rises proportionally to the pH change while the water composition point rises proportionally to twice the pH change). Wairakite is a common fissure mineral at Wairakei, accompanying quartz and potassium feldspar.

On Figure 13.8, a stability diagram for sodium and potassium minerals at 260°C (Browne and Ellis, 1970), the Broadlands water composition point moves into the field of potassium feldspar after appreciable steam ($+\ CO_2$) loss. This is due to the increasing water pH and the shift in the mineral phase boundaries with temperature. Potassium feldspar is a common alteration mineral at the depths of boiling in geothermal fields. In the reverse situation, the heating of a flowing water aquifer by steam ($+CO_2$) condensation would encourage albite and/or potassium mica formation by lowering the pH and raising the water temperature. Waters cooled by conduction or by slight dilution would tend to form potassium mica. More extensive examples of the relationships between water chemistry and hydrothermal alteration mineralogy were given by Helgeson (1967), and by Browne and Ellis (1970).

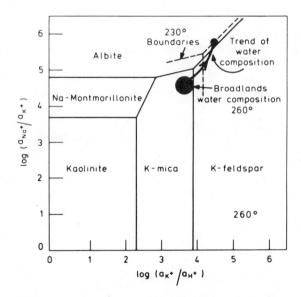

Fig. 13.8 Stability diagram for sodium and potassium minerals at 260°C full lines, and 230°C (broken lines), in terms of the solution ion ratios a_{Na}/a_H, and a_K/a_H (Browne and Ellis, 1970).

REFERENCES

Alonso, H. (1966) La Zona Geotermica de Cerro Prieto, Baja California: *Bol. Soc. Geol. Mex.,* **29,** 17-47.

Andrusenko, N. I. and A. A. Moskalyuk (1966) Experiments on the hydrothermal treatment of dolerites with a bearing on the genesis of Iceland spar deposits: *Geokhimiya,* **9,** 1119-1123.

Arnorsson, S. (1975) Application of the silica geothermometer in low temperature hydrothermal areas in Iceland: *Am. J. Sci.,* **275,** 763-784.

Averyev, V. V., S. I. Naboko, and B. I. Piip (1961) Contemporary hydrothermal metamorphism in regions of active volcanism: *Dokl. Akad. Nauk S.S.S.R.,* **137,** 407-410.

Bargar, K. E., M. H. Beeson, R. O. Fournier, and L. J. P. Muffler (1973) Present-day deposition of lepidolite from thermal waters in Yellowstone National Park: *Am. Mineral.,* **58,** 901-904.

Barth, T. F. W. (1950) *Volcanic Geology of Hot Springs and Geysers of Iceland,* Publication No. 587, Washington, D.C.: Carnegie Institution.

Björnsson, S., S. Arnorsson, and J. Tomasson (1970) Exploration of the Reykjanes thermal brine area: *Geothermics (Special Issue 2),* **2(Pt. 2),** 1640-1650.

Blattner, P. (1975) Oxygen isotope composition of fissure-grown quartz, adularia, and calcite from Broadlands geothermal field, New Zealand: *Am. J. Sci.,* **275,** 785-800.

Bodvarsson, G. (1961) Physical characteristics of natural heat resources in Iceland: *Proc. U.N. Conference on New Sources of Energy, Rome 1961,* **3,** 449-455.

Browne, P. R. L. and A. J. Ellis (1970) The Ohaki-Broadlands hydrothermal area, New Zealand: Mineralogy and related geochemistry: *Am. J. Sci.,* **269,** 97-131.

Chelishev, N. F. (1967) Laboratory studies in the distribution of rare alkalis between potassium minerals and aqueous solutions at high temperatures and pressures: *Dokl. Akad. Nauk S.S.S.R.,* **175,** 205-207.

Chen, C. H. (1970) Geology and geothermal power potential of the Tatun volcanic region: *Geothermics (Special Issue 2),* **2(Pt. 2),** 1134-1143.

Clayton, R. N., L. J. P. Muffler, and D. E. White (1968) Oxygen isotope study of calcite and silicates of the River Ranch well, Salton Sea geothermal area, California: *Am. J. Sci.,* **266,** 968-979.

——— and A. Steiner (1975), in *Wairakei Thermal Area New Zealand; its Sub-Surface Geology and Hydrothermal Rock Alteration,* N.Z. Geological Survey Bull. 90.

Coombs, D. S., A. J. Ellis, W. S. Fyfe, and A. M. Taylor (1959) The zeolite facies with comments on the interpretation of hydrothermal synthesis: *Geochim. Cosmochim. Acta,* **17,** 53-107.

Craig, H. (1962) C^{12}, C^{13}, and C^{14} concentrations in volcanic gases: *J. Geophys. Res.,* **67,** 1633.

——— (1963) The isotopic geochemistry of water and carbon in geothermal areas, in *Nuclear Geology on Geothermal Areas,* E. Tongiorgi, ed., Pisa: Consiglio Nazionale della Richerche, Laboratoria di Geologia Nucleare, p. 17-23.

Culberson, O. L. and J. J. McKetta (1951) Phase equilibria in hydrocarbon-water systems. III The solubility of methane in water at pressures to 10000 psi: *J. Petroleum Technol.,* **3,** 223-226.

Doe, B. R., C. E. Hedge, and D. E. White (1966) Preliminary investigation of the source of lead and strontium in deep geothermal brines underlying the Salton Sea geothermal area: *Econ. Geol.,* **61,** 462-483.

REFERENCES

Ellis, A. J. (1960) Mordenite synthesis in a natural hydrothermal solution: *Geochim. Cosmochim. Acta,* **19,** 145-146.

―― (1967) The chemistry of some explored geothermal systems, in *Geochemistry of Hydrothermal Ore Deposits,* H. L. Barnes, ed., New York: Holt, Rinehart and Winston, p. 465-514.

―― (1968) Natural hydrothermal systems and experimental hot-water/rock interaction: reactions with NaCl solutions and trace metal extraction: *Geochim. Cosmochim. Acta,* **32,** 1356-63.

―― (1969) Present-day hydrothermal systems and mineral deposition: *Proc. Ninth Commonwealth Mining and Met. Congress (Mining and Petroleum Section)* 1-30, London: Inst. Mining and Metallurgy.

―― (1970a) Quantitative interpretation of chemical characteristics of hydrothermal systems: *Geothermics (Special Issue 2),* **2(Pt. 1),** 516-528.

―― (1970b) Chemical processes in hydrothermal systems—A review, in *Proc. Symp. Hydrogeochemistry and Biogeochemistry,* Vol. 1, Washington, D.C.: The Clarke Co. (published 1973), p. 1-26.

―― (1971) Magnesium concentrations in the presence of magnesium chlorite, calcite, carbon dioxide, quartz: *Am. J. Sci.,* **271,** 481-489.

―― and W. S. Fyfe (1957) Hydrothermal chemistry: *Rev. Pure Appl. Chem. (Aust.),* 7, 261-316.

―― and W. F. Giggenbach (1971) Hydrogen sulphide ionization and sulphur hydrolysis in high temperature solution: *Geochim. Cosmochim. Acta,* **35,** 247-260.

―― and R. M. Golding (1963) The solubility of carbon dioxide above 100°C in water and in sodium chloride solutions: *Am. J. Sci.,* **261,** 47-60.

―― and W. A. J. Mahon (1964) Natural hydrothermal systems and experimental hot water/rock interactions: *Geochim. Cosmochim. Acta.,* **28,** 1323-1357.

―― and ―― (1966) Geochemistry of the Ngawha hydrothermal area: *New Zealand J. Sci.,* **9,** 440-456.

―― and ―― (1967) Natural hydrothermal systems and experimental hot water/rock interactions, Pt. 2: *Geochim. Cosmochim. Acta,* **31,** 519-538.

―― and ―― (1977) *Geochemistry and Geothermal systems:* New York: Academic Press.

ENEL (1970) *Larderello and Monte Amiata: Electric power by endogenous steam,* 1-42, Rome: Ente Nazionale per l'energia Elettrica.

Erd, R. C., D. E. White, J. J. Fahey, and D. E. Lee (1964) Buddingtonite, an ammonium feldspar with zeolitic water: *Am. Mineral.,* **49,** 831-850.

Eslinger, E. V. and S. M. Savin (1973) Mineralogy and oxygen isotope geochemistry of the hydrothermally altered rocks of the Ohaki-Broadlands, New Zealand geothermal area: *Am. J. Sci.,* **273,** 240-267.

Facca, G. and F. Tonani (1964) Theory and technology of a geothermal field: *Bull. volcanol.,* **27,** 1-47.

―― and ―― (1967) The self-sealing geothermal field: *Bull. Volcanol.,* **30,** 271-273.

Fournier, R. O. (1970) Silica in thermal waters: laboratory and field investigations, in *Proceedings Sympos. Hydrogeochemistry and Biogeochemistry,* Vol. 1, Wash., D.C.: The Clark Co., 122-140.

―― and J. J. Rowe (1966) Estimation of underground temperatures from the silica content of water from hot springs and wet-steam wells: *Am. J. Sci.,* **264,** 685-697.

―― and A. H. Truesdell (1973) An empirical Na-K-Ca geothermometer for natural waters: *Geochim. Cosmochim. Acta,* **37,** 1255-1275.

Geothermics (1970) Special Issue 2, Proceedings of the United Nations Symposium on the Development and Utilization of Geothermal Resources. Pisa: CNR-Instituto Internazionale per le Richerche Geotermiche.

Giggenbach, W. F. (1974) Equilibria involving polysulfide ions in aqueous solutions up to 240°C: *Inorg. Chem.*, **13**, 1724-1730.

―― (1977) The isotopic composition of sulphur in sedimentary rocks bordering the Taupo Volcanic Zone: in *Geochemistry 1977*, A. J. Ellis and A. P. W. Hodder, eds., Wellington, New Zealand Dept. Sci. and Ind. Res. Bull. **218**, 57-64.

Grindley, G. W. (1965) The geology, structure and exploitation of the Wairakei geothermal field, Taupo, New Zealand: *Bull. N.Z. Geol. Surv., n.s. 75,* 1-131.

―― (1970) Subsurface structures and relation to steam production in the Broadlands geothermal field: *Geothermics (Special Issue 2)*, **2(Pt. 1),** 248-261.

Gunther, B. D. (1973) Aqueous phase—gaseous phase material balance studies of argon and nitrogen in hydrothermal features at Yellowstone National Park: *Geochim. Cosmochim. Acta,* **37,** 495-513.

Hanshaw, B. B. and T. B. Coplen (1973) Ultra filtration by a compacted clay membrane—II. Sodium ion exclusion at various ionic strengths: *Geochim. Cosmochim. Acta,* **37,** 2311-2328.

Hayashi, M. (1973) Hydrothermal alteration in the Otake geothermal area, Kyushu: *J. Japan Geotherm. Energy Assoc.,* **10,** 9-46.

Helgeson, H. C. (1967) Solution chemistry and metamorphism, in, *Researches in Geochemistry Vol. 2,* P. H. Abelson, ed., New York: Wiley, p. 362-402.

―― (1968) Geologic and thermodynamic characteristics of the Salton Sea geothermal system: *Am. J. Sci.,* **266,** 129-166.

Hemley, J. J., P. B. Hostetler, A. J. Gude, and W. T. Mountjoy (1969) Some stability relations of alunite: *Econ. Geol.,* **64,** 599-612.

―― and W. R. Jones (1964) Chemical aspects of hydrothermal alteration with emphasis on hydrogen metasomatism: *Econ. Geol.,* **59,** 538-567.

Honda, S. and L. J. P. Muffler (1970) Hydrothermal alteration in core from research drill hole Y-1, Upper Geyser Basin, Yellowstone National Park, Wyoming: *Am. Mineral.,* **55,** 1714-1737.

Hulston, J. R. (1976) Isotopic work applied to geothermal systems at the Institute of Nuclear Sciences, New Zealand: *Geothermics,* **5,** 89-96.

―― and W. J. McCabe (1962a) Mass spectrometer measurements in the thermal areas of New Zealand, Pt. 1. Carbon dioxide and residual gas analyses: *Geochim. Cosmochim. Acta,* **26,** 383-398.

―― and ―― (1962b) Mass spectrometer measurements in the thermal areas of New Zealand Pt 2. Carbon isotopic ratios: *Geochim. Cosmochim. Acta,* **26,** 399-410.

Ilinskaya, N. (1962) Reported from Ekonomicheskaya gazeta by *New Scientist* Jan. 3, 1963, p. 28.

James, R. (1968) Wairakei and Larderello; geothermal power systems compared: *New Zealand J. Sci.,* **11,** 706-719.

Jones, M. E. (1963) Ammonia equilibria between vapor and liquid aqueous phases at elevated temperatures: *J. Phys. Chem.,* **67,** 1113-1115.

Jones, P. H. (1970) Geothermal resources of the Northern Gulf of Mexico Basin: *Geothermics (Special Issue 2)*, **2(Pt. 1),** 14-26.

Kissen, I. G. and S. I. Pakhomov (1967) Effect of high temperatures on the chemical composition of ground waters: *Geochem. Int.,* **4,** 295-308.

Kitahara, S. (1960) The polymerization of silicic acid obtained by hydrothermal treatment of quartz and the solubility of amorphous silica: *Rev. Phys. Chem. Japan,* **30,** 131-137.

Koenig, J. B. (1970) Geothermal exploration in the Western United States: *Geothermics (Special Issue 2),* **2(Pt. 1),** 1-13.

Kozintseva, T. N. (1965) The solubility of hydrogen sulfide in water and solutions under high temperatures, in, *Geochemical Investigation in the Field of Higher Pressures and Temperatures,* N. I. Kitarov, ed., Moscow: Nauka Publishing Office, p. 121-134.

Kruger, P. and C. Otte (1973) *Geothermal Energy,* Stanford: Stanford University Press.

Kusakabe, M. (1974) Sulphur isotopic variations in nature. 10. Oxygen and sulphur isotope study of Wairakei geothermal well discharges: *New Zealand J. Sci.,* **17,** 183-192.

Mahon, W. A. J. (1964) Fluorine in the natural thermal waters of New Zealand: *New Zealand J. Sci.,* **7,** 3-28.

_____ (1966) Silica in hot water discharged from drill holes at Wairakei, New Zealand: *New Zealand J. Sci.,* **9,** 135-144.

_____ (1967) Natural hydrothermal systems and the reaction of hot water with sedimentary rocks: *New Zealand J. Sci.,* **10,** 206-221.

_____ and J. B. Finlayson (1972) The chemistry of the Broadlands geothermal area New Zealand: *Am. J. Sci.,* **272,** 48-68.

Makarenko, F. A., B. F. Mavritsky, B. A. Lokshin, and V. I. Kononov (1970) Geothermal resources of the USSR and prospects for their practical use: *Geothermics (Special Issue 2)* **2(Pt. 2),** 1086-1091.

Marinelli, G. (1969) Some geological data on the geothermal areas of Tuscany: *Bull. Volcanol.,* **33,** 1-15.

Martynova, O. I. (1972) Distribution of dissolved silicon dioxide between water and steam: *Teploenergetika,* **12,** 51-53.

Mazor, E. and R. O. Fournier (1973) More on noble gases in Yellowstone National Park hot waters: *Geochim. Cosmochim. Acta,* **37,** 515-525.

Mazor, E., and G. J. Wasserburg (1965) Helium, neon, argon, krypton, and xenon in gas emanations from Yellowstone and Lassen National Parks: *Geochim. Cosmochim. Acta,* **29,** 443-454.

Mercado, S. (1966) Aspectos quimicos del approvechamiento de la energia geotermica; Campo geotermico Cerro Prieto, B.C., Mexico, D.F., Comision Federal de Electricidad, p. 27.

Miller, A. R., C. D. Densmore, E. T. Degens, J. C. Hathaway, F. T. Manheim, P. F. McFarlin, R. Pocklington, and A. Jokela (1966) Hot brines and recent iron deposits in deeps of the Red Sea: *Geochim. Cosmochim. Acta,* **30,** 341-359.

Mizutani, Y. (1972) Isotopic composition and underground temperature of the Otake geothermal water, Kyushu, Japan: *Geochem. J. (Japan),* **6,** 67-73.

_____ and T. A. Rafter (1969) Oxygen isotopic composition of sulphates. 3. Oxygen isotopic fractionation in the bisulphate ion-water system: *New Zealand J. Sci.,* **12,** 54-59.

Muffler, L. P. J. and D. E. White (1969) Active metamorphism of Upper Cenozoic sediments in the Salton Sea geothermal field and the Salton Trough, Southeastern California: *Geol. Soc. Am. Bull.,* **80,** 157-182.

Naboko, S. I. (1968) Chemical composition of the real mineral forming solutions: *Akad. Nauk S.S.S.R. (Siberian Branch), Geol. Geophys. Sect., No. 3* (99), 3-13.

Nakamura, H. and K. Sumi (1967) Geological study of Matsukawa geothermal area, Northeast Japan: *Bull. Geol. Survey Japan,* **18,** 58-72.

———, ———, K. Katagiri, and T. Iwata (1970) The geological environment of Matsukawa geothermal area, Japan: *Geothermics (Special Issue 2)* **2(Pt. 1),** 221-231.

Panichi, C. and R. Gonfiantini (1977) Environmental isotopics in geothermal studies: *Geothermics,* **6,** 143-162.

Pray, H. A., C. E. Schweickert, and B. H. Minnich (1952) Solubility of hydrogen, oxygen, nitrogen and helium in water: *Ind. Eng. Chem.,* **44,** 1146-1151.

Pryor, W. A. (1960) The kinetics of disproportionation of sodium thiosulfate to sodium sulfide and sulfate: *J. Am. Chem. Soc.,* **82,** 4794-4797.

Robinson, B. W. (1973) Sulphur isotope equilibrium during sulphur hydrolysis at high temperatures: *Earth and Planetary Sci. Letters,* **18,** 443-450.

Rothbaum, H. P. and R. D. Wilson (1977) Effect of temperature and concentration on the polymerization rate of silica in geothermal waters, in, *Geochemistry 1977,* A. J. Ellis and A. P. W. Hodder, eds., Wellington, New Zealand Dept. Sci. and Ind. Res. Bulletin **218,** 37-43.

Schoen, R. and D. E. White (1965) Hydrothermal alterations in GS-3 and GS-4 drillholes, main terrace, Steamboat Springs, Nevada: *Econ. Geol.* **60,** 1411-1421.

Schoen, R., D. E. White, and J. J. Hemley (1974) Argillization by descending acid at Steamboat Springs, Nevada: *Clays and Clay Minerals,* **22,** 1-22.

Seki, Y. (1969) Facies series in low-grade metamorphism: *J. Geol. Soc. Japan,* **75,** 255-266.

——— (1972) Lower-grade stability limit of epidote in the light of natural occurrences: *J. Geol. Soc., Japan,* **78,** 405-413.

———, H. Onuki, K. Okumura, and I. Takashima (1969) Zeolite distribution in the Katayama geothermal area, Onikobe, Japan: *Jap. J. Geol. and Geography,* **40,** 63-79.

Seward, T. M. (1974) Equilibrium and oxidation potential in geothermal waters at Broadlands, New Zealand: *Am. J. Sci.,* **274,** 190-192.

Sigvaldason, G. E. (1963) Epidote and related minerals in two deep geothermal drillholes, Reykjavik and Hveragerdi, Iceland: *U.S. Geol. Survey Prof. Paper,* **450-E,** 77-79.

——— (1973) Geochemical methods in geothermal exploration: in, *Geothermal Energy,* 12 (Earth Sciences) H. C. H. Armstead, ed., Paris: UNESCO. p. 49-60.

——— and D. E. White (1961) Hydrothermal alteration of rocks in two drillholes at Steamboat Springs, Washoe County, Nevada: *U.S. Geol. Survey Prof. Paper,* **424-D,** 116-122.

——— and ——— (1962) Hydrothermal alteration in drillholes GS-5 and GS-7 Steamboat Springs, Nevada: *U.S. Geol. Survey Prof. Paper,* **450-D,** 113-117.

Sourirajan, S. and G. C. Kennedy (1962) The system H_2O-NaCl at elevated temperatures and pressures: *Am. J. Sci.,* **260,** 115-141.

Steiner, A. (1968) Clay minerals in hydrothermally altered rocks at Wairakei, New Zealand: *Clays and Clay Minerals,* **16,** 193-213.

——— (1974) Interaction between hot water and silicic rocks in the active geothermal area at Wairakei, New Zealand: *I.A.G.C. Symposium Water-Rock Interaction,* Prague: Czech. Geological Survey.

——— and T. A. Rafter (1966) Sulfur isotopes in pyrite, pyrrhotite, alunite and anhydrite from steam wells in the Taupo Volcanic Zone, New Zealand: *Econ. Geol.,* **61,** 1115-1129.

Studt, F. E. (1958) The Wairakei hydrothermal field under exploitation: *N.Z. J. Geol. Geophys.,* **1,** 703-723.

Styrikovich, M. A., D. G. Tskhvirashvili, and D. P. Nebieridze (1960) Solubility of boric acid in saturated steam: *Dokl. Akad. Nauk S.S.S.R.,* **134,** 615-617.

Sumi, K. (1969) Zonal distribution of clay minerals in the Matsukawa geothermal area, Japan: *Proc. Int. Clay Conf. Tokyo,* **1,** Jerusalem: Israel Universities Press.

―――― and K. Maeda (1970) Hydrothermal alteration of main productive formation of the steam for power at Matsukawa, Japan. in, *Proceedings Sympos. Hydrogeochemistry and Biogeochemistry,* Vol. 1, Wash., D.C.: The Clarke Co., p. 221-228.

Ten Dam, A., and C. Erentöz (1970) Kizildere geothermal field—Western Anatolia: *Geothermics (Special Issue 2),* **2(Pt. 1),** 124-129.

Tomasson, J., and H. Kristmannsdóttir (1972) High temperature alteration minerals and thermal brines, Reykjanes, Iceland: *Contrib. Mineral. Petrol.,* **36,** 123-134.

Truesdell, A. H. and W. Singers (1971) Computer calculation of downhole chemistry in geothermal areas: *Chemistry Division, Rept 2136,* Petone, New Zealand: Dept Scientific and Industrial Research.

U.N. (1975) *Proc. U.N. Symp. Devel. Use Geothermal Resources, 2nd, San Francisco, California, May 1975.*

Vakin, E. A., B. G. Polak, V. M. Sugrobov, E. N. Erlikh, V. I. Belousov, and G. F. Pilipenko (1970) *Geothermics (Special Issue 2),* **2(Pt. 2),** 1116-1133.

Waring, G. A.; revised by R. A. Blankenship and R. Bentall (1965) Thermal springs of the United States and other countries of the world—A summary: *U.S. Geological Survey Prof. Paper, 492.*

Wasserburg, G. J., E. Mazor, and R. E. Zartman (1963) Isotopic and chemical composition of some terrestrial natural gases: in, *Earth Sciences and Meteoritics,* J. Geiss and E. D. Goldberg, eds., Amsterdam: North Holland, p. 219-240.

White, D. E. (1957) Thermal waters of volcanic origin: *Bull. Geol. Soc. Am.,* **68,** 1637-1658.

―――― (1965) Saline waters of sedimentary rocks, in, *Fluids in subsurface Environments—a Symposium, Memoir Amer. Assn. Petrol. Geologists,* **4,** 342-366.

―――― (1974) Diverse origins of hydrothermal ore fluids: *Econ. Geol.,* **69,** 954-973.

――――, J. D. Hem, and G. A. Waring (1963) Chemical composition of sub-surface waters: *U.S. Geol. Survey Prof. Paper,* **440-F,** 1-67.

――――, L. J. P. Muffler, and A. H. Truesdell (1971) Vapor-dominated hydrothermal systems compared with hot-water systems: *Econ. Geol.,* **66,** 75-97.

―――― and Roberson, C. E. (1962) Sulphur Bank, California, a major hot-spring quicksilver deposit: in, *Petrologic Studies, Geol. Soc. Am.,* Buddington Volume, 397-428.

Wilson, J. S. (1974) Environmental aspects of the multi-purpose development of geothermal resources : in, *Water 1975, Am. Instit. Chem. Eng. Symp. Series,* No. 136. **70,** 782-787.

Yanagase, K., K. Yamaguchi, T. Yanagase, Y. Suginohara, S. Kozawa, and H. Yamazaki (1970) Colloidal silica in hot-spring water. *Nippon Kagaku Zasshi,* **91,** 1141-1148.

Zen, E-an (1961) The zeolite facies: an interpretation. *Am. J. Sci.,* **259,** 401-409.

14

Fluid Inclusions as Samples of Ore Fluids

EDWIN ROEDDER
U. S. Geological Survey

Far too commonly the ore-forming fluid for a deposit is simply assumed to have been rich only in those constituents now present, even in proportion to their abundance in the deposit! Most ore fluids contained, in addition to the ore elements deposited, large amounts of volatile constituents and soluble salts that passed through the deposit leaving almost no trace. Fortunately, however, a trace can usually be found in the form of fluid inclusions, providing a record, albeit complex, fragmentary, and minuscule. Regardless of their origin and history, inclusions do represent actual samples—with rare exceptions the only samples we have—of fluids existing at some time in the geologic history of the ore deposit. As such they are important clues in understanding the geologic modus operandi.

When crystals grow or recrystallize in a fluid medium of any kind, growth irregularities of many sorts trap small portions of the fluid in the solid crystal. Although they are frequently overlooked, they can be found in the minerals of most rocks, including ores, and hence give evidence of the almost universal presence of a fluid phase during the geologic events yielding these materials. The sealing off of such irregularities may occur during the growth of the surrounding crystal, yielding *primary* fluid inclusions, or by recrystallization along fractures at some later time, yielding *secondary* inclusions.

Fluid inclusions larger than 0.1 mm are comparatively uncommon, and those larger than 1 mm are rare, although museum specimens with volumes of tens or even hundreds of cubic centimeters are known. Most fluid inclusions are less than 0.01 mm, and electron microscopy has revealed large numbers down to a minimum of about 2×10^{-5} mm. Presumably there is a size continuum down to single water molecules trapped along grain boundary dislocations, at $\sim 2 \times 10^{-7}$ mm.

In total volume, fluid inclusions seldom comprise more than a few tenths of one percent of the sample, even though they may be exceedingly numerous. A few examples show as much as five volume percent inclusions in individual crystals or zones of crystals, and one percent is not rare. Many samples of ordinary white minerals such as quartz and calcite contain $\sim 10^9$ inclusions/cm^3, and owe their white color to the inclusions, but as the average size is only 10^{-3} mm, the actual volume of inclusions is only $\sim 0.1\%$. As water is the major component of most fluid inclusions (plus smaller quantities of CO_2), the amount of inclusion fluid present may cause significant errors in some water determinations, and actually makes completely unambiguous water determinations impossible, regardless of the methods used. Any solutes present in the inclusion fluid are normally reported as part of the total analysis of the sample; not uncommonly fluid inclusions contribute as much as 100 ppm of each of several nonvolatile constituents to the analysis of separated "pure" minerals. As an extreme example, fluid inclusions containing crystals of NaCl are so common in some granitic feldspar that single-crystal X-ray photographs of the feldspar show diffraction lines of NaCl as well (Roedder and Coombs, 1967).

The importance of the nature of the samples used cannot be overemphasized, and the difficulty in obtaining usable material is a major problem in most inclusion studies. The nature of the samples used is the prime consideration in the selection of suitable methods of study, and in evaluating the precision, the accuracy, and most important, the significance of any measurements obtained. All of the many chemical analyses of fluid inclusions that have been made—the author's included—are subject to serious limitations. The chemical manipulations may be reasonably straightforward, but the small sample size, and wide variation in composition of both the inclusions and the host mineral make the analyses far from routine, and impose large analytic uncertainties.

Much more serious, however, are such problems as the possible multiplicity of origin of the inclusions extracted, and drastic contamination and/or loss during extraction. Such factors have not been adequately evaluated in many studies. As the volume varies with the cube of the radius, a few relatively large inclusions may carry far more fluid than thousands of smaller ones (Figure 14.1); this simple fact alone makes it almost impossible to obtain truly duplicate samples for any analysis test.

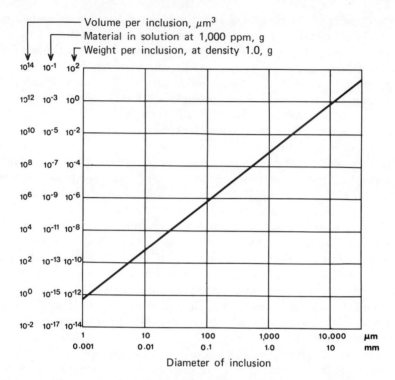

Fig. 14.1. Volume and weight of spherical inclusions. (Adapted from Roedder, 1958, *Econ. Geol.*, **53**, 238.)

As a result of the irregularities inherent in the processes of trapping of fluid inclusions detailed below, the abundance of inclusions in adjacent samples from any given deposit may vary over many orders of magnitude. In some deposits, fluid-inclusion studies are made possible only by the fortuitous discovery of suitable sample material. Furthermore, the sample requirements for individual methods of fluid inclusion study vary widely, from as little as 10^{-13} g for a number of the optical methods (Roedder, 1972), to 10^{-3} or even 1.0 g for some wet-chemical procedures. Cross-checking one method against another is occasionally possible, but the variations in the nature of the samples, and in the sample requirements for the various methods, may make such comparisons difficult to interpret.

A large and rather scattered bibliography reveals that the study of fluid inclusions has a long history. As a result of the inherently qualitative nature of many aspects of fluid-inclusion study, the older literature, by authors such as Zirkel (1876) and Sorby (1858), still provides a surprisingly large amount of useful data. Fluid inclusions have provided focal points for many heated arguments, because both their merits and their shortcomings as recorders of geologic history are frequently overstated or even stated erroneously in the literature. Even today, some believe that proof of the secondary origin of many inclusions, and of the possibility of later addition or loss (leakage) of some material, effectively or totally negate the significance of all inclusions. One can draw an analogy between the development of radioactive-age determination and fluid-inclusion research. In both, more careful and quantitative study has led to more and more valuable results, but has also revealed the existence of numerous complexities, which necessitate reinterpretation of earlier work. When resolved, these very complexities may actually make the methods more powerful.

Most of the important papers prior to 1953 (over 400) are summarized by Smith (1953). One exception is a Russian book (Ermakov, 1950; also transliterated Yermakov) now available in translation, that summarizes the extensive Russian work in this field.[1] Other books on the subject are in French (Deicha, 1955), and in Ukrainian (Kalyuzhnyi, 1960). A series of large symposia on fluid inclusions have been held in the Soviet Union (1963, 1965, 1968, 1973, 1976, 1978) and the proceedings of the first four have been published (in Russian). The world literature up to approximately 1970, particularly that dealing with the methods of study used and the data obtained on the composition of fluid inclusions, has been summarized in English (Roedder, 1972). The present chapter contains much material from this summary, and the reader is referred there for additional details, supporting data, and references. The world literature on inclusion research of all types (including silicate melt inclusions) is summarized in English in a yearly abstract and translation publication started in 1968 (Roedder, 1968).

In spite of all of the problems inherent in obtaining accurate data on the composition, and the homogenization and freezing temperatures of fluid inclusions, capable of unambiguous interpretation, these little droplets of fluid present too tantalizing a lead toward a real understanding of the ore-forming process to be ignored. Admittedly, the data at hand are still seriously limited, but with the application of the great advances in techniques of study and analysis now available, considerable progress can be anticipated.

This chapter is aimed specifically at the problems of interpretation of the significance of fluid-inclusion data. Examples will be chosen from work on hydrothermal mineral deposits wherever possible, but as good data are scarce, other types of samples will be discussed where applicable.

---

[1] As a result of recent awakening of interest in the field, particularly in the U.S.S.R., the total world literature on fluid inclusions has more than quadrupled since 1960. More than half of these recent papers are in Russian sources.

MECHANISM OF TRAPPING

The inclusions in any given sample are seldom of only one generation. Commonly the primary inclusions are sparsely distributed and large, but the secondary inclusions in the same sample are small and very numerous. As the trapping of these two fluids may have occurred millions of years apart, the fluids may have grossly different compositions, making the distinction between primary and secondary origin of critical importance. The most commonly available criteria for distinguishing these origins are listed in Table 14-1. As none of the criteria are absolute, and many are only suggestive or are applicable only to material from certain deposits, they *must be applied with care* and with awareness of the considerable ambiguity that exists. It is unfortunate but true that many inclusions simply do not permit application of any of the criteria.

Primary Inclusions

Any process that interferes with the growth of a perfect crystal may cause the trapping of primary inclusions. For example, a period of rapid crystal growth can result in a porous, dendritic layer that succeeding slower growth covers with solid, impervious layers, thereby trapping many inclusions. This type of growth is common in fluorite. Crystal growth studies have shown that a change in the concentration of even minor constituents in the fluid can affect crystal perfection grossly, and in either direction. If a hollow, shell-like, skeletal, or cavernous crystal later becomes complete, large primary inclusions may be trapped. If material is supplied to the growing faces by mass flow of the fluid (as opposed to diffusion), large inclusions may be trapped as a result of temporary starvation of the centers of faces, relative to the faster-growing edges that have easier access to the moving nutrient solution. This phenomenon has been observed in the growth of synthetic crystals and may be the mechanism whereby natural quartz and halite crystals sometimes contain a series of very large inclusions, flattened parallel to a crystal face, with only thin septa of crystal separating each chamber.

Most crystals grow as a series of almost parallel blocks (lineage structure). If some of these grow faster than others, the surfaces of the crystal can become rough, with many angular reentrants; later growth covering these reentrants yields negative crystal cavities. Such negative crystal shape is frequently but erroneously stated to be proof of primary origin (see Table 14-1). However, primary inclusions trapped by such processes are generally large and isolated or randomly arrayed, with no obvious plan or arrangement; such characteristics are valid criteria for primary origin.

Table 14-1 Criteria for the origin of fluid inclusions (revised from Roedder, 1976)

Criteria for primary origin

I. Based on occurrence in a single crystal with or without evidence of direction of growth or growth zonation.
   A. Occurrence as a single (or a small three-dimensional group) in an otherwise inclusion-free crystal (Roedder, 1965b, Figure 10 and 1972, Plate 6).
   B. Large size relative to that of the enclosing crystal e.g., with a diameter $\sim \geq 0.1$ that of crystal, and particularly several such inclusions.
   C. Isolated occurrence, away from other inclusions, for a distance of $\sim \geq 5$ times the diameter of the inclusion.
   D. Occurrence as part of a random, three-dimensional distribution throughout the crystal (Roedder and Coombs, 1967, Plate 4, Figures A and B).
   E. Disturbance of otherwise regular decorated dislocations surrounding the inclusion, particularly if they appear to radiate from it (Roedder and Weiblen, 1970, Figure 9).
   F. Occurrence of daughter crystals (or accidental solid inclusions) of the same phase(s) as occur as solid inclusions in the host crystal or as contemporaneous phases.
II. Based on occurrence in a single crystal showing evidence of direction of growth:
   A. Occurrence beyond (in the direction of growth), and sometimes immediately before extraneous solids (the same or other phases) interfering with the growth, where the host crystal fails to close in completely. (Inclusion may be attached to the solid or at some distance beyond, from imperfect growth; Roedder, 1972, Plate 1.)
   B. Occurrence beyond a healed crack in an earlier growth stage, where new crystal growth has been imperfect (Roedder, 1965b, Figures 18 and 19; Roedder et al., 1968, Figure 15).
   C. Occurrence between subparallel units of a composite crystal (Roedder, 1972, Frontispiece upper right).
   D. Occurrence at the intersection of several growth spirals, or at the center of a growth spiral visible on the outer surface.
   E. Occurrence, particularly as relatively large, flat inclusions, parallel to an external crystal face, and near its center (i.e., from "starvation" of the growth at the center of the crystal face), e.g., much "hopper salt".
   F. Occurrence in the core of a tubular crystal (e.g., beryl). This may be merely an extreme case of previous item.
   G. Occurrence, particularly as a row, along the edge from the intersection of two crystal faces.
III. Based on occurrence in a single crystal showing evidence of growth zonation (as determined by color, clarity, composition, X-ray darkening, trapped solid inclusions, etch zones, exsolution phases, etc.).
   A. Occurrence in random, three-dimensional distribution, with different concentrations in adjacent zones (as from a surge of sudden, feathery or dendritic growth).

Table 14-1 (*Continued*)

Criteria for primary origin

  B. Occurrence as subparallel groups (outlining growth directions), particularly with different concentrations in adjacent growth zones, as in previous item (Roedder, 1965b, Figure 11).
  C. Multiple occurrence in planar array(s) outlining a growth zone (Roedder and Coombs, 1967, Plate 4E). (Note that if this is also a cleavage direction, there is ambiguity.)
IV. Based on growth from a heterogeneous (i.e., two-phase), or a changing fluid.
  A. Planar arrays (as in III-C), or other occurrence in growth zones, in which the compositions of inclusions in adjacent zones are different (e.g., gas inclusions in one and liquid in another, or oil and water (Roedder et al., 1968, Figure 9).
  B. Planar arrays (as in III-C) in which trapping of some of the growth medium has occurred at points where the host crystal has overgrown and surrounded adhering globules of the immiscible, dispersed phase (e.g., oil droplets or steam bubbles).
  C. Otherwise primary-appearing inclusions of a fluid phase that is unlikely to be the mineral-forming fluid, e.g., mercury in calcite, oil in fluorite (Roedder, 1972, Plate 9, Figure 2) or air in sugar (Roedder, 1972, Plate 9, Figure 4).
V. Based on occurrence in hosts other than single crystals.
  A. Occurrence on a compromise growth surface between two nonparallel crystals. (These inclusions have generally leaked, and could also be secondary.)
  B. Occurrence within polycrystalline hosts, e.g., as pores in fine grained dolomite, cavities within chalcedony-lined geodes ("enhydros"), vesicles in basalt, or as crystal-lined vugs in metal deposits or pegmatites. (These latter are among the largest "inclusions," and have almost always leaked.)
  C. Occurrence in noncrystalline hosts (e.g., gas bubbles in amber; vesicles in pumice).
VI. Based on inclusion shape or size.
  A. In a given sample, larger size and/or equant shape.
  B. Negative crystal shape—this is valid *only* in certain specific samples and is a negative criterion in others.
VII. Based on occurrence in euhedral crystals, projecting into vugs (suggestive, but far from positive—see Roedder, 1967a, p. 523).

Criteria for secondary origin

I. Occurrence as planar groups outlining healed fractures (cleavage or otherwise) that come to the surface of crystal (note that movement of inclusions with recrystallization can cause dispersion—Roedder, 1971a, Figure 11; see also III-C above).
II. Very thin and flat; in process of necking down.

Table 14-1 (*Continued*)

Criteria for secondary origin

III. Primary inclusions with filling representative of secondary conditions.
  A. Located on secondary healed fracture, hence presumably refilled with later fluids (Kalyuzhnyi, 1971).
  B. Decrepitated and rehealed following exposure to higher temperatures or lower external pressures than at time of trapping; new filling may have original composition but lower density (Roedder, 1965a, Figure 18).

Criteria for pseudosecondary origin

  I. Occurrence as with secondary inclusions, but with fracture visibly terminating within crystal (Roedder, 1965b, Figures 18 and 19; Roedder et al., 1968, Figures 12, 14, and 15. See also III-C under "Primary" above.)
 II. Generally more apt to be equant and of negative crystal shape than secondary inclusions in same sample (suggestive only).
III. Occurrence as a result of the covering of etch pits cross-cutting growth zones (Roedder, 1972, Plate 1, Figure 8).

Anything that disturbs an otherwise perfect crystal face can also cause the trapping of primary inclusions. Thus a surface crack may cause imperfect growth and trapping of inclusions, but solid particles are by far the most common source of interference. Further supply of nutrient is hindered at the point of contact, whereas growth of the free part of the face continues. The object may be pushed along, leaving a trail of inclusions behind it, but generally the crystal grows over it, thus trapping a *solid inclusion*, with or without some of the surrounding fluid medium.

If the environment from which the host crystal is growing is a heterogeneous system of two fluids (e.g., vapor bubbles or immiscible globules of oil in an aqueous liquid), inclusions of one or the other and sometimes of both fluids may be trapped. The immiscible globules may be the *cause* of trapping because they preferentially wet the surface of the host mineral (e.g., oil on fluorite from southern Illinois). A similarly misleading situation may arise with the trapping of gas or vapor bubbles in a boiling (or effervescing) solution. Bubbles of steam or other gases appear to nucleate more readily on solid surfaces. The growing crystals may become coated with tiny gas bubbles, each of which shields the surface with which it is in contact from further growth, and so becomes enclosed. This processs should be particularly effective if sudden small pressure drops cause periodic effervescence (Barabanov, 1958). It can also occur as a result of the crystallization of the host mineral enriching the surface layers of fluid in gaseous components to the point that a new gas phase develops there. Zirkel (1873, p. 86) reports a hauynite with an estimated 3.6×10^{11} gas inclusions/cm^3, presumably trapped as bubbles of gas forming in the silicate magma from which the hauynite grew. Roedder and Coombs (1967) report similar phenomena in the trapping of immiscible globules of dense, highly saline aqueous fluids from a granitic melt, and in the trapping of immiscible globules of low-density, carbon dioxide-rich vapor from the boiling of the dense saline phase. In each case small amounts of the other fluid phase are trapped in at least part of the inclusions.

In the OH vein at Creede, Colorado, as in several epithermal veins, the inclusions in all minerals are rather uniform in filling density throughout the mine, except at the top level of the mine, where large primary inclusions in quartz crystals have a wide range of ratios of gas to liquid (Roedder, 1970a, p. 52-56). Independent evidence indicates that this portion of the vein formed at pressures and temperatures just about on the boiling curve for such fluids, so it seems probable that the ore fluids actually were boiling at this level and that the growing quartz trapped some gas and some liquid inclusions. Such examples of steam bubbles attaching themselves to crystal surfaces, and being enclosed, either with or without some liquid, are not uncommon in locations where the extrapolated P-T conditions indicate boiling to have been feasible.

Pseudosecondary and Secondary Inclusions

All primary inclusions are surrounded by host mineral that grew at about the same time the fluid was trapped. Secondary inclusions are those that form by any process after the crystallization of the host is essentially complete. Thus if a crystal is fractured in the presence of a fluid in which it has a finite solubility, the fluid will enter the fracture and start to dissolve and recrystallize the host crystal until a minimum of new surface is left, trapping new secondary inclusions in the process. Although the crystal immediately surrounding the inclusion has crystallized at the same time as the fluid was trapped, the bulk of the crystal has not, and the meaning of the term "secondary" is apparent.

Ordinarily one thinks of primary inclusions as forming during the growth of the crystal and of secondary inclusions as forming at some later time, from entirely different fluids. There is, however, a zone of overlap between the two terms. If a crystal fractures *during* its growth, the fluids from which it was growing (and which may be in the process of being trapped as primary inclusions in the rim of the crystal) will enter the fracture and become trapped in the core of the crystal. Such inclusions have been termed *pseudosecondary* (sometimes "primary-secondary" in the Soviet literature). If there has been a change in the composition of the fluids between the growth of the core and the rim, adjacent inclusions in the core may have different compositions; with such data, and the obvious occurrence in a healed fracture, the apparently secondary inclusion frequently will be disregarded. Its true origin can only be recognized if the healed crack can be traced to an abrupt ending, at a former growth face, *within* the crystal. Sample material that presents positive proof of pseudosecondary origin is rare, but in the author's opinion, a rather large percentage of the healed cleavage or curving fracture planes, now outlined by inclusions, that occur in most euhedral crystals from vugs, are actually pseudosecondary rather than secondary. The fluorite crystals from southern Illinois, and from Hansonburg, New Mexico, contain many good examples (Roedder, 1965b, photos 17 and 18; Roedder et al., 1968, Figures 12, 14, and 15.)

One additional mechanism for the formation of pseudosecondary inclusions is found in *hydrothermal leaching*. This process is probably more common than generally recognized (Barton et al., 1963), and may yield large fluid inclusions, visibly cutting across growth zones. The leaching of a crystal seldom involves simple removal of a uniform layer; for a variety of reasons, certain lines, zones, or areas are more soluble and dissolve at a much more rapid rate, yielding etch pits that may be 1000 times as deep as they are wide (Nielson and Foster, 1960). Gross differences in etching rates may even be caused by X-irradiation prior to etching. Nielson and Foster (1960) propose that impurities that have diffused to dislocations cause higher solubility and yield the pits. Deep etch pits, following invisible dislocations, frequently form in the seed plates used in synthetic quartz crystal manufacture. Similar deep etch pits may form on the tops of surface bumps on the crystals (presumably the tops of growth spirals) and at the intersections of several such spirals.[2]

The most commonly used criterion for assigning a primary origin is the occurrence of an inclusion in a crystal projecting into an open vug or vein. Unfortunately, there are many ways in which such a crystal may be fractured, thus trapping presumably secondary inclusions (Roedder, 1967a, p. 523), and it is not always possible to tell whether the crystal broke during its growth (yielding pseudosecondary inclusions) or afterward (yielding secondary inclusions).

[2]Fission-fragment tracks in minerals may localize similar deep etch pits, but of a different order of magnitude, as they are limited in length to the penetration of the fission fragments, approximately 20 μm. Although it is not *known* that fluid inclusions form by natural leaching of fission fragment tracks, it is logical to assume that this has occurred.

NATURE OF MATERIAL TRAPPED

Normal Fluid Inclusions

RELATIONSHIP TO FLUID AS A WHOLE

If the fluids trapped in an inclusion in a vein mineral are not representative of the fluid moving through the vein at the time of trapping, can studies of inclusions be of any value whatsoever? It can be assumed, in the general case, that they are *not* identical, for several reasons mentioned below. But the important point is the magnitude of the differences. Statements have been made that fluid inclusions are not actually samples of the ore fluids, but rather the "last residue" or "the final spent fraction" from the crystallization of the ore (Ingerson, 1954). A series of *primary* inclusions, trapped in zoned crystals of ore or coprecipitated gangue minerals, from zones representing the earliest start of mineralization to the last fraction of a millimeter before mineralization ended with euhedral crystals protruding into the open vein, can hardly represent a "final spent fraction." The ore or gangue mineral may continue to crystallize out on the walls of the inclusions after trapping, but there is good evidence that the amounts of such precipitation are generally very small and reversible upon reheating.

Regardless of the process whereby the fluid became saturated with respect to the host mineral for the inclusions—whether by loss of gases, mixing with other fluids, loss of heat, or reaction with wall rocks or earlier vein minerals—the fluid passing the protruding crystal *is* the fluid from which that crystal is growing, and if that fluid is trapped in the crystal as a fluid inclusion, that inclusion is a sample of the ore-forming fluid. After the fluid has passed through the ore deposit and on into a barren vein and has undergone the total of chemical transactions that occur between it and the ore body or its wallrocks, it can be considered a spent residue, but only with respect to that deposit. It may still produce other mineral deposits at higher levels, for example, near-surface or hot spring manganese deposits.

BOUNDARY LAYER EFFECTS

It has long been known that at equilibrium liquids are different in composition adjacent to a surface—that is, any interface with another phase—than they are throughout their bulk. For low-molecular-weight liquids, there is evidence of some effects for several hundred angstroms (i.e., ~ 0.03 μm). The observed effects are mainly structural and as such may be detected by changes in properties (Adamson, 1960), but compositional differences are generally not even mentioned.

Under *nonequilibrium* conditions, as in the steady-state growth of a crystal from a liquid, considerable concentration gradients are set up near the growing face of the crystallizing material. Miers (1903) measured the concentrations using a total reflection technique, and found that a zone at the surface actually had a higher concentration than the fluid as a whole. Berg (1938) found the concentration gradients to be largest near the center of the face and detected effects to a distance of approximately 0.5 mm into the liquid. This early work uncovered several interesting problems that have not been adequately explored (Buckley, 1951).

One would expect, intuitively, that concentration gradients would occur near a fast-growing crystal, forming from a strong solution of the same substance. All atoms, ions, or molecular groupings that are to add to the crystal must go through the layer of fluid at the surface of the crystal. This fluid in turn, and all other materials in solution in it, must continually be pushed away by the growing crystal. But to the best of the author's knowledge, such effects have not been studied under the conditions believed to be appropriate to ore-mineral crystallization, that is, relatively slow crystallization at elevated temperatures where diffusion is fast, from fluids perhaps high in extraneous dissolved substances, but relatively dilute with respect to the crystallizing phase. In such a situation, surface effects would presumably extend over very small distances, and hence be of very minor significance to the gross composition of the fluid trapped in inclusions. The only discordant notes here are some experimental data reported by Barnes et al., (1969), indicating that the fluids trapped in some fairly large synthetic inclusions were slightly more dilute than the bulk of the fluid in the experiment. To the writer's knowledge, no one has attempted to explain or corroborate these results.

Heterogeneous Systems

LIQUIDS PLUS SOLIDS

There is abundant evidence that at least at certain times in the growth history of many crystals, solid particles were present in suspension in the fluids from which they grew. In some deposits this solid matter stems from the crushing of coarse vein filling or the dispersion of newly formed fine-grained material. The coarser part will fall through the fluid to rest on any available surface, but if there is any vertical component to the laminar fluid movement, the finest material may be winnowed out and carried upward. Additional small particles may stem from spontaneous nucleation in the fluid. Vertical settling of such particles in horizontal vugs and open veins gives evidence of a low or zero horizontal flow component, and Barton et al. (1971) were able to estimate the vertical flow rates of hydrothermal fluids in a vein by quantizing this relationship for hematite flakes.

Many crystals show multiple zones or "ghosts" of small solid inclusions. It may be important to distinguish between particles that settle out physically, and those that have actually nucleated and crystallized out of the fluid, although it is not uncommon for the settling particles to act as nuclei for further growth. Either may be accidentally trapped in fluid inclusions forming at that time, and hence may be confused with *daughter minerals* that crystallize out of the fluids trapped in inclusions after sealing.[3] The distinction is based mainly on the phase ratios in the inclusions; a mineral that occurs in only part of the fluid inclusions, in highly variable amounts (particularly if it also occurs as simple solid inclusions in the host), is probably an accidental solid inclusion, whereas daughter minerals are apt to occur in a regular ratio to other phases. Unambiguous primary fluid inclusions are commonly found in zones of solid inclusions, attached to some of the grains.

LIQUIDS PLUS LIQUIDS

The trapping of two immiscible liquids, already mentioned in the section "Mechanism of Trapping" above, is far from rare. Droplets of oils, immiscible in aqueous solutions, are common in crystals from vugs in a number of sedimentary terrains. If an emulsion of droplets of oil in water is present at the surface of a crystal when it is fractured, this mixture flows into the crack, and secondary or pseudosecondary oil-plus-water inclusions may form (see Figure 19 in Roedder, 1965b).[4] Two other types of immiscibility are frequently recorded in inclusions in igneous rocks: silicate/silicate immiscibility (Roedder and Weiblen, 1972) and silicate/sulfide immiscibility (Roedder and Weiblen, 1970).

LIQUIDS PLUS GASES

The trapping of gas or vapor bubbles with the liquid from an originally heterogeneous, two-phase system (i.e., primary gas) invalidates most attempts to use such inclusions for geologic thermometry by the homogenization method, since this method is based on trapping of a homogeneous single-fluid phase. On the other hand, the very presence of simultaneously formed inclusions of both liquid and vapor gives the best evidence available

---

[3] Thus the striking series of crystals found in inclusions in Volynian topaz (e.g., see Roedder, 1972, plate 12), and reported as daughter crystals by various Soviet workers, are claimed by Voznyak (1968) to be accidental solid inclusions.

[4] Even liquid mercury is sometimes trapped as a fluid in fluid inclusions in calcite (Roedder, 1962a, p. 46).

that the fluids were actually on the boiling curve, thus providing an excellent anchor point from which speculations on the environment of ore deposition may be extrapolated (Barton and Toulmin, 1961). Primary gas inclusions tend to be large in hydrothermal systems, but may be exceedingly minute in silicate melts. A primary gas inclusion trapped from a hydrothermal fluid seldom is free from the liquid, which may occur only as tiny fillets in the reentrants around the inclusion, or may not be visible at all until frozen.

If carbon dioxide is present in the fluids, its limited solubility, particularly at low temperatures (Takenouchi and Kennedy, 1964, 1965), commonly results in immiscibility. A number of studies of the inclusions in ore deposits have shown evidence that two immiscible fluids, one rich in CO_2 and the other in water, were present in the vein (Kalyuzhnyi, 1955; Ypma, 1963; Roedder, 1963, p. 188; Rutherford, 1968). However, an immiscible condition at room temperature does not require immiscibility under the conditions of trapping; because the mutual solubilities of CO_2 and H_2O change considerably with temperature, formerly homogeneous fluids trapped in inclusions can, and frequently do, separate on cooling to form two immiscible fluids, supercritical CO_2 and subcritical H_2O. As with solid inclusions, the constancy of the ratios of volumes of the phases in several inclusions provides the main criterion to distinguish trapping of a heterogeneous mixture from later separation of an originally homogeneous one.

Recrystallization of Gels and Poorly Organized Materials

If a crystal such as quartz forms from a gel in which the other major constituent is water, this water might be enclosed in the crystal as fluid inclusions,[5] and similarly, if the growing crystal incorporates extraneous hydroxyl ions, these might later migrate to form aqueous fluid inclusions. Although both of these mechanisms are potentially valid, neither seem very likely in most geologic materials (Roedder, 1967a, p. 527-528).

LEAKAGE INTO OR OUT OF INCLUSIONS

The possibility of movement of fluid into or out of inclusions after they are trapped is of great concern to any intelligent use of the data from fluid inclusions. Several reports of laboratory measurements of leakage of fluid inclusions under pressure gradients have been widely quoted (Kennedy,

---

[5] Suggested by Charles Meyer, personal communication, 1963.

1950, reporting unpublished work of Grunig; Skinner, 1953; McCulloch, 1959). They show that within hours to weeks fluid can be moved into or out of fluid inclusions by establishing large pressure gradients (270–1550 atm/<1 mm). Later work and a critical review of all the evidence for leakage (Roedder and Skinner, 1968) have shown that at least some of the earlier experiments indicating leakage may be explained by microfractures introduced during sample preparation, and that most inclusions have not leaked. One exception is the possible diffusion of hydrogen under some conditions. Another is the emptying of inclusions during deformation (Ypma, 1963) or by decrepitation from intrusion of dikes (Lokerman, 1962). Such exceptions can provide valuable evidence on the chronologic sequence of events at the site of an ore body.

CHANGES SINCE TRAPPING

Phase Changes from an Originally Homogeneous Fluid

SHRINKAGE

The most prominent single feature of fluid inclusions is the occurrence of a vapor or gas bubble that may be moveable under the influence of gravity or a thermal gradient, and, if small enough (less than several microns), may be in constant motion that is mistakenly called "Brownian" motion. As detailed above, a number of inclusions do give evidence of trapping of heterogeneous systems that is of considerable value in understanding the conditions of trapping, although such inclusions are relatively rare. Minerals from ordinary rocks and ores commonly show millions of inclusions per cubic centimeter, every one a simple two-phase system of liquid and gas. These two phases result from the fact that we are examining the inclusions at a lower temperature than that at which they were trapped. As the volume coefficient of thermal expansion for most minerals is one to three orders of magnitude lower than that for water, on cooling from the temperature of trapping to room temperature the container for the inclusion shrinks much less than the fluid inside. As soon as the pressure in the inclusion drops below the total vapor pressure of the multicomponent fluid at that temperature, at equilibrium (and hence the volume of the fluid should be less than that of the inclusion), a bubble should nucleate and grow (see Figure 14.2). Sorby, in a long and classic paper on fluid inclusions (1858) proposed that this process could be reversed by simply heating the inclusions to the temperature of disappearance of the bubble to ascertain the trapping temperature.

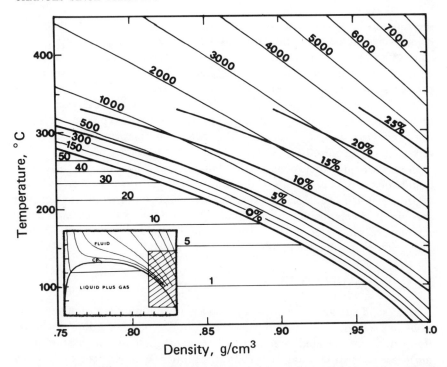

Fig. 14.2. Temperature-density diagram for the system H_2O, with isobars, from Fisher (1976). The diagram is an enlargement of the most commonly used part of the more general diagram (see inset). The heavy line at 0% is the boundary of the two-phase field for pure H_2O and hence represents the homogenization temperatures for H_2O inclusions with the stated densities. The effects of additions of 5 to 25 wt. % NaCl on this two-phase boundary are shown, from Haas (1971). These boundaries are plotted in terms of density, not degree of fill as in the original data from Lemmlein and Klevtsov (1961). (There is an heretofore apparently unreported error in the high-pressure part of the tabular data of Lemmlein and Klevtsov; graphical intercomparisons reveal that all data entries for pressures ≥ 1500 atm and degrees of fill $<85\%$ should have a degree of fill 2.5% larger than as stated in the original Russian.)

IMMISCIBILITY

There is no real dividing line between the separation of a vapor bubble by shrinkage of the fluid and separation of an immiscible fluid. If the inclusion is pure water and the temperature is 100°C, the bubble will be steam with a low density (0.0006 g/cm³), but if the temperature is around 370°C, the steam density will be approximately 0.32 g/cm³, over 500 times greater. Most gases that may be present in the fluid, such as CO_2 or CH_4, will preferentially enter the vapor phase. The density of this vapor may then exceed that of the liquid and the "gas bubble" can be seen to sink in the "liquid." Takenouchi and Kennedy (1964) have determined this density inversion point for various isobars in the system H_2O-CO_2, and state quite appropriately: "The pointlessness of continuing the normal 1-atmosphere distinction between the gas and the liquid on to higher pressures is strikingly evident here."

If a high enough concentration of CO_2 exists in the CO_2-rich vapor phase on cooling, it also may split again into two fluids, representing liquid CO_2 and gaseous CO_2. Generally this occurs at temperatures below the critical point for pure CO_2, 31.0°C, unless other gases are present.

In certain fluid inclusions a small amount of a new dense, solute-rich liquid phase separates from the liquid of the inclusions *on heating* (Roedder, 1967a, p. 532). The significance of this immiscibility in terms of specific fluid composition is not known, but in view of the probable phase relations of salt-water systems, such immiscibility with rising temperature under constant volume conditions may be expected.

CRYSTALLIZATION ON THE WALLS

Most solid substances exhibit an increase in solubility with temperature. Thus one would expect that during the natural cooling of a fluid inclusion trapped at an elevated temperature, crystallization of the host mineral must occur, as the fluid is certainly saturated with respect to it when trapped. Generally this crystallization occurs on the walls, rather than as a separate crystal. Ermakov (1950) describes some special cases in which a definite coating forms inside the inclusion. He indicates that the dissolving of this coating on heating may be remarkably slow, often requiring hours to reach equilibrium, even though the volume of the "reaction vessel" is measured in micrometers. In most cases, however, the amount of solution is immeasurably small. In many inclusions from ore deposits, there is no evidence of dissolution of the walls on heating, or of crystallization on cooling, even though the presence of minute details on the walls of the inclusions may make this qualitative test rather sensitive.

DAUGHTER MINERALS

Not infrequently the originally homogeneous fluids trapped in inclusions become saturated with respect to phases other than just that of the walls, and nucleate new crystals in the inclusions, called daughter minerals. By far the most common of these is NaCl, but other minerals, such as carbonates and sulfides are also found (see Table 1 in Roedder, 1972). Identification of daughter minerals is possible by a variety of methods (Roedder, 1972). Frequently the same phases occur in the deposit outside of the inclusions (except for the water-soluble ones, which are generally lost to the groundwater, yielding an indeterminate part of the vugs we now see). The volume of each daughter mineral, particularly of phases other than NaCl, is frequently very small, and the resolving power of the microscope limits their detection in small inclusions. As the daughter minerals are part of the composition of the original homogeneous fluid, their identification and measurement is important.

METASTABILITY

Inclusions are very small systems to consider, even from an atomic viewpoint. Thus an inclusion of 10 μm size containing a 30 wt. % solution of a salt may have only 10^{11}–10^{12} "molecules" of that salt; at 10 ppm PbS it would contain only 25 million "molecules" of PbS, enough to form a crystal 300 unit cells on an edge. Hence it is not surprising that metastability—resulting from failure to nucleate new but stable phases—becomes a problem that can be observed in (and frequently interferes with) many types of inclusion studies (Roedder, 1971a). Significant degrees of metastability (usually stated in terms of temperature) have been recognized in the nucleation of gas bubbles and daughter minerals on cooling previously homogeneous inclusions to room temperature. There are homogeneous, single-phase fluid inclusions, generally formed by trapping at less than ~70°C, that have persisted without formation of the stable bubble at rather high negative pressure for millions of years (Roedder, 1967b). Sometimes freezing and thawing the liquid will cause the nucleation of the small bubble that should be present. Salt crystals generally nucleate, even in very tiny inclusions, but other phases present in smaller amounts may be missing in the smaller inclusions.

On freezing inclusions, similar failure to nucleate newly stable phases has been observed for ice, and even more noticeably for crystals of $NaCl \cdot 2H_2O$ or $CO_2 \cdot 5.75H_2O$, as well as for the gas bubble after its elimination by expansion of the liquid on freezing. Such metastability can yield erroneous results, but it can also be useful in obtaining information on the composition of inclusions (Roedder, 1963, 1967b, 1971a).

Changes Due to Recrystallization of the Host Mineral: Necking Down and Coalescence

As originally trapped, many fluid inclusions have relatively large surface areas, for example, narrow healed cracks, and the long tubes characteristic of the primary inclusions in some minerals. If the host mineral is at all soluble in this fluid, processes of recrystallization immediately start to reduce the high surface energy of the system. Because diffusion has to move the host material only across part of the inclusion, it is not a major rate-limiting step, but apparently the low solubility of the host mineral is a major limitation. Lemmlein pointed out in a series of papers (1929, and particularly Lemmlein and Kliya, 1952) that inclusions formed by cracking crystals of water-soluble salts permit these changes to be observed and photographed with time-lapse photography. Reentrants in the walls become filled with material dissolved from the main part of the walls, so that the whole inclusion may coalesce into a single, more equant cavity. Some elongated inclusions gradually develop bulges separated by thin necks, which eventually become sealed off (Figure 14.3). The net result of the necking down is the formation of several smaller inclusions with a smaller total amount of surface energy. Presumably a similar process, on a much smaller scale, occurs with the reorganization of solvent and solute molecules along dislocations, to yield a string of minute inclusions.

Recrystallization of the host makes no difference to the total composition of the system, but if any phase change has occurred before the necking down or coalescence, the different new inclusions trapped may have grossly different compositions and densities. This possibility is of considerable significance to geologic thermometry based on inclusion filling temperatures because the single gas bubble may be trapped in one of the new inclusions and the other new inclusions may nucleate their own gas bubbles with further drop in temperature (Figure 14.4). If a daughter mineral has

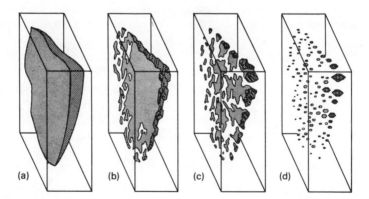

Fig. 14.3. Healing of a crack in a quartz crystal, resulting in secondary inclusions. Solution of some of the curved surfaces having non-rational indices, and redeposition as dendritic crystal growth on others, eventually results in the formation of sharply faceted, negative crystal inclusions. If this process occurs with falling temperature, the individual inclusions will have a variety of gas/liquid ratios (see Figure 14.4).

CHANGES SINCE TRAPPING

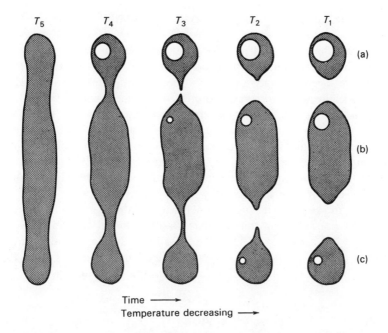

Fig. 14.4. Necking down of a long tubular inclusion. The original inclusion, trapped at temperature T_5, breaks up during slow cooling to form three separate inclusions, (a), (b), and (c). Upon reheating in the laboratory, inclusion a would homogenize above the true trapping temperature T_5; inclusion b would homogenize above T_3; inclusion c would homogenize between T_2 and T_3.

formed before an inclusion starts to coalesce, it is frequently left behind, completely surrounded by the host mineral. Quartz crystals from the Colombian emerald mines frequently show sharp cubes of NaCl, completely enclosed in the host quartz. A few such cubes show very thin tubes, sometimes almost at the limit of resolution of the microscope, connecting them with a liquid inclusion many micrometers away.

Regardless of the original shape, if the host mineral has any finite solubility in the fluid of the inclusion under the conditions at which it is held after trapping, recrystallization will generally occur. Such recrystallization is probably of very general occurrence, and most inclusions now have different shapes than when originally trapped, although the differences can be relatively small if cooling has been fast, if the original shape was close to the stable shape, as in the case of some negative crystals, or if the solubility is very low (Roedder, 1971a). Thus oil inclusions in fluorite frequently have round shapes that appear to be that of the original round drops, but inclusions of the immiscible water phase with them have undergone considerable change in shape by recrystallization (Roedder, 1972, Plate 9, Figure 3).

The final shape assumed by those fluid inclusions in which the composition permits recrystallization may be either smoothly rounded and globular, or faceted negative crystals. The balance between two competing but exceedingly weak forces determines which form will have the lowest ΔG, i.e., the smallest amount of surface (spherical), or the lowest-energy surfaces (negative crystal facets). Different composition inclusions in the same crystal, from two generations, may have the two types of shapes. Although negative crystal shape is frequently assumed to be a valid criterion of primary origin for the inclusions, many obviously secondary inclusions recrystallize to form very sharply faceted negative crystals, and etch pits and tubes are frequently lined with bright crystal faces. Inclusion shape may give at least a rough indication of inclusion age, and hence, indirectly, of inclusion origin. Thus the steps in the healing of microfractures in quartz to form planes of inclusions has been described by Tuttle (1949), and used by Wise (1964) to help in understanding microjointing.

COMPOSITION OF FLUID INCLUSIONS[6]

Methods Used and Their Validity

Many different techniques have been used but there is no panacea for the problems of determining the composition of fluid inclusions. Except for the very large inclusions, whose analyses are relatively easy, no known method or combination of methods will give an accurate, unambiguous analysis of any inclusion in any given mineral. Due to the extremely wide range in sample material and in the accuracy of the methods used, the available published analyses of fluid inclusions must be examined with considerable care.

As the size of the inclusion to be analyzed decreases, the error and ambiguity in the individual results increase rather rapidly, simultaneously with an equally rapid decrease in the number of determinable constituents. This uncertainty stems from the fact that the sample volume (and weight) of an inclusion is a function of the cube of its radius (see Figure 14.1). From this it might appear that only large inclusions should be analyzed. Unfortunately, large fluid inclusions (greater than 1 cm), permitting standard analytic techniques, occur in few localities. If a given type of

---

[6] This section is a very brief summary of another paper (Roedder, 1972). Additional details on the techniques used and their limitations, and the 2400 analyses published to that date, as well as the full bibliography, will be found there.

geologic process is to be studied in several occurrences, or if the complexities in a given geologic occurrence are to be unraveled by a study of the fluid inclusions present, it is necessary to find them in many samples and portions of samples. Even inclusions a single millimeter in diameter are generally too rare for such a study. In addition, a number of studies indicate that the probability of leakage increases rapidly in larger-sized inclusions. A resolution of this problem must be based on a compromise between the tremendous abundance of the smaller ones, and the ease and analytic reliability of work on the larger ones. The simple expedient of taking a sample containing a large number of small inclusions, to get the same amount of fluid, can be used in those cases where most of the inclusions are of one generation (Roedder, 1958), but unless done with great care, this procedure introduces many additional problems such as contamination (discussed below) that have apparently caused gross errors in some published analyses.

Methods of analysis applicable to very small inclusions are obviously needed, and the petrographic microscope, with various accessories, provides a number of them. The microscope methods are based on the optical identification of the phases present from the rather surprising variety of properties exhibited, and although the methods are only qualitative or semiquantitative at best, they are generally very easy and quick to use. As is the case with most optical identifications, there is some ambiguity, particularly if only one parameter is used, but additional independent optical methods are generally available. Most significantly, these methods are exceedingly sensitive, and are suitable for use on single inclusions weighing $< 10^{-9}$ g ($< 10 \mu m$ diam).

By the very nature of the origin and occurrence of inclusions, samples that are truly duplicates, in either quantity or composition of the fluids contained, are practically impossible to obtain. A single large inclusion of one generation in a sample can hold as much fluid as a million smaller ones of another generation in the same sample, and there is much evidence that the composition of the fluids changed even *during* the formation of many crystals. Thus each of a series of 20 recognizable individual zones in a single-zoned sphalerite crystal from Creede, Colorado, was found to contain fluids trapped at a specific and recognizably different combination of temperature and salinity (Roedder, 1976).

In addition, only a very small percentage of the inclusions in a given sample will possess the optimum combination of large size, clear display of phases present, and lack of ambiguity in assignment of origin. As a result of these considerations, it is very important to obtain as much information as possible about a given inclusion by nondestructive tests, before a destructive test is used.

Nondestructive methods yielding data on the composition of the liquid phase include estimations or determinations of its viscosity, color, surface energies against other phases (i.e., wetting characteristics), absorption of IR light, index of refraction, thermal expansion, various phase changes on heating or freezing, and critical phenomena. Solid phases in inclusions are generally so small that only optical methods are adequately sensitive for their identification, but a surprising number of parameters can be obtained, including the behavior on heating or freezing. Gross salinity, usually expressed as weight percent NaCl equivalent, is commonly obtained from the depression of the freezing point of the fluid (Roedder, 1962b). Nondestructive laser-excited Raman spectroscopy of individual solid, liquid, or gaseous phases in inclusions has been shown to be a powerful method for a limited number of constituents (Rosasco and Roedder, 1976).

Destructive methods of many types have been used on inclusions, with varying degrees of success. The ultimate goal would be a complete and accurate quantitative analysis of the total contents of a single inclusion—gas, liquid, and solid—for major constituents, for minor constituents such as H^+(pH), e^-(Eh), and trace elements, for gases, and for the isotopic composition of such elements as hydrogen, sulfur, oxygen, and carbon. As this goal is hardly feasible now, all present analyses are only partial at best. Comparatively few of them have quantitative statements of both water and salts, which are needed to calculate concentrations.[7]

Four separate steps are generally involved, any one of which may be a major source for error (e.g., from contamination or loss): *opening* the inclusions (generally by decrepitation, crushingm, or ball-milling in air or vacuum); *separation* of the inclusion constituents from the host (generally by evaporation of volatiles into a gas stream or vacuum and leaching of the residue of soluble salts with water or acids); *collection and concentration* (generally by freezing or absorption of volatiles and electrodialysis or filtration and evaporation for soluble salts); and *analysis*. The most common analytic methods used recently for gases are volumetry, barometry, gas chromatography, and mass spectrometry, and many inclusion analyses in the literature involve only determination of the gases. For soluble salts, flame photometry, atomic absorption, colorimetry, emission spectrography, and neutron activation are probably the most suitable methods at this time. The procedures used for the actual analyses present fewer problems than the other three steps, although a considerable number of recent papers still report analyses made by rather crude, semiquantitative methods.

[7] A method that has been used extensively in the U.S.S.R. involves determination of the amount of water by counting inclusions and estimating their average size in what are believed to be representative portions of polished plates, and then extrapolating to the whole sample used for crushing. The method is subject to serious errors.

The liquid is generally found to be an aqueous solution with total salt concentrations ranging from practically zero to over 40 wt. %. The salts consist of major amounts of Na^+, K^+, Ca^{2+}, Mg^{2+}, Cl^-, and SO_4^{2-}, with lesser amounts of Li^+, Al^{3+}, BO_3^{3-}, PO_4^{3-}, SiO_3^{2-}, HCO_3^-, CO_3^{2-}, and many others. Many individual ions in this list may predominate in specific occurrences, although Na^+ and Cl^- are generally the most abundant.

Solid daughter minerals can be identified by various normal petrographic techniques, as well as by single crystal X-ray and electron-microprobe procedures if they can be removed from the inclusion by careful crushing. However, these minute grains are exceedingly difficult to find in the crushed debris. Metzger et al. (1976) have shown that some daughter minerals can be found and identified on broken surfaces by scanning electron microscopy, using a nondispersive solid-state X-ray detector.

The pressure of the gases in inclusions can be estimated by the simple expedient of crushing while immersed in a fluid. The vapor bubbles in many inclusions are thus shown to be almost solely water vapor, as they collapse instantaneously and completely when exposed to atmospheric pressure (Roedder, 1970a). On the other hand, highly compressed gases, such as liquid CO_2 (at ~70 atm), will practically explode into the liquid. Although obviously only qualitative or semiquantitative, the crushing test is very useful, as it is quick and exceedingly sensitive. As little as 10^{-14} g of a relatively nonsoluble gas (less than 10^9 molecules) may be detected in this manner, as it will form an easily visible gas bubble, several micrometers in diameter at atmospheric pressure.

Examples of Compositional Data Obtained and Geologic Interpretations of Them

Individual reports of analyses have been presented in many different forms, making intercomparisons difficult, except by some artifice such as normalizing to 10,000 ppm Cl (Roedder, 1967a, p. 548). There are three general types of analyses: *ratio analyses,* in which only the ratios of several constituent ions are stated; *leach analyses,* in which the inclusions are analyzed, but the amounts of water (and hence the concentrations of the solutions), are unknown or can only be crudely estimated; and *quantitative analyses,* in which both the amounts of water, and of ions, are determined, permitting a statement of analysis in the form of the actual composition of the inclusion fluid.

PEGMATITES AND QUARTZ VEINS

Although these may differ genetically from normal hydrothermal veins, a large proportion of the chemical analyses of fluid inclusions reported in the literature have been from such samples, covering the whole spectrum from crystals of quartz and other minerals in vuggy, zoned pegmatites through to the massive white quartz veins, lenses, and pods so characteristic of low-grade metamorphic rocks.[8] There may well be a continuum between such barren quartz veins and the gold-quartz veins associated with low-grade metamorphic rocks. Total salt concentrations range over nearly two orders of magnitude (~ 0.4–40 wt. %), and the ratios of many constituents to each other vary just as widely. Sodium is the major cation, and Na/K (atomic) ranges ~ 2–70. This ratio probably reflects mainly the composition of codeposited (or coexisting) feldspars in the host rock (Orville, 1963), but there are few data relating the two compositions. The best examples are those of Poty et al., (1974) on Alpine fissure minerals, and of Weisbrod and Poty (1974) on the Mayres pegmatite.

The Ca/Mg and Cl/SO$_4$ ratios vary even more than Na/K. In analysis of leachates from crushed minerals, even very small amounts of sulfide minerals may make major contributions to the SO_4^{2-} found, by oxidation during leach. It is thus best to consider the SO_4^{2-} values as limiting maxima in each case. HCO_3^- is considerably higher than Cl^- in some of the analyses, but is not always determined, and the distinction between it and free CO_2 released during crushing under a liquid is not always clear. Free CO_2 is a common constituent, depending upon the gross composition of the host rocks.

As a result of the extensive optical study of the phases present in inclusions in pegmatitic quartz, it is apparent that the total composition can vary even more than indicated above, and individual anions or cations may have considerably higher concentrations in specific examples. Thus there have been many observations and even detailed studies of rather large crystals of NaCl (and KCl) in the inclusions in pegmatite minerals. (In spite of this, statements have been quoted and requoted that pegmatite minerals do not show these crystals.) It is hoped that eventually it may be possible to relate such fluid inclusions in pegmatites to similar inclusions in associated metamorphic rocks.

Surprising amounts of quartz and even silicate minerals such as topaz form as daughter minerals or crystallize on the walls of some inclusions and can be redissolved on heating (see, however, footnote 3). Some inclusions in granites approach hydrous, saline-silicate melts at homogenization temperatures (Roedder and Coombs, 1967). A number of

---

[8] The reason for this emphasis in the literature stems mainly from the numerous studies of fluid inclusions in quartz in the U.S.S.R. in the last few decades, as part of their extensive exploration for radiograde quartz crystal deposits (Ermakov, 1950).

fluoride, chloride, and borate daughter minerals have been found in inclusions in pegmatites in the U.S.S.R., including not only halite and sylvite, but also elpasolite (K_2NaAlF_6), cryolite (Na_3AlF_6), fluorite, fluoborates of the avogadrite-ferruccite group [$(K,Cs)BF_4$-$NaBF_4$], teepleite ($Na_2B_2O_4 \cdot 2NaCl \cdot 4H_2O$), caracolite (approximately $PbOHCl \cdot Na_2SO_4$), and numerous doubtful or unidentified phases, indicating high concentrations of several ions. The more concentrated inclusions are generally high in chlorides, and several unnamed new chloride phases have been found, including $FeCl_2 \cdot 2H_2O$, and a chloride of zinc and aluminum (Ermakov, 1965; Kalyuzhnyi, 1958, 1960; Lyakhov, 1966; Slivko, 1958; Kalyuzhnyi and Voznyak, 1967).

NORMAL HYDROTHERMAL DEPOSITS

The total concentration of salts in solution in the inclusion fluids from normal (i.e., "magmatic") hydrothermal deposits ranges from <1 to >50 wt. %. Relatively few of the chemical analyses include a valid determination of H_2O content, along with the salts, but freezing stage studies on many deposits show that most such ore fluids fall in the range of 0-5 wt. % NaCl equivalent. Sodium chloride daughter crystals, indicating concentrations of salts of 30, 40, or even over 50 wt. %, are absent in most hydrothermal deposits. They are common, however, in many porphyry copper deposits (but not all; e.g., see Nash and Cunningham, 1974) and also occur in some epithermal deposits. These are frequently stated as coming in part from near-surface boiling of brines. However since the amount of water that can be lost by boiling on simple pressure release is very limited, large increases in salt concentration by boiling can only occur under special conditions in which heat is added to a more or less static fluid (White et al., 1971). Even a very small amount of boiling can cause gross changes in the chemistry of the fluids, and precipitation of ore, so the inclusion evidence for it is important.

The fluid compositions—and the temperatures of homogenization—may also vary in space and time within a given deposit. Although many deposits show a general decrease in both variables with stage of deposition, a few show reversals (see Ermakov, 1950, and references in Roedder, 1968). Variations in space are most evident in large districts, and particularly in porphyry copper deposits (Nash and Theodore, 1971; Roedder 1971b), where the central cores generally contain high-temperature, high-salinity inclusions, and the inclusions in the peripheral deposits have much lower values.

Although a large number of thermometric studies have been made of inclusions in normal hydrothermal deposits, there are comparatively few moderately accurate and complete chemical analyses of well-documented samples. (Both the methods of cleaning and extraction, and the analytic procedures used, should be appropriate to the samples, but frequently neither are even stated.) As a result of the variation in salinity, the concentrations of individual constituents will differ greatly. For this reason, it is best to compare analyses by using ratios of constituents. Table 14-2 shows the three most important atomic ratios, Na/K, Ca/Mg, and Cl/SO$_4$ for inclusions from a series of districts, calculated from some of the more complete analyses in the literature. Similar data are available on a much wider range of deposits, including those of antimony-arsenic, Iceland spar, mercury, barite, etc. (see Roedder, 1972, Tables 3-6). The Na/K ratio is probably controlled in most of these deposits by wallrock alteration reactions, but it is not easy to distinguish between variations due to changes in the fluids at their source and those due to differing mineralogy or degrees of armoring of the vein walls. In addition, the contemporaneity of a given fluid and wallrock assemblage introduces a major ambiguity.

The Ca/Mg ratio is pertinent to various wallrock alteration problems such as sericitization of plagioclase, chloritization of pyriboles, and dolomitization, but the same limitations apply. Calcium analyses on leachates from fluorite are maxima, of course, due to leaching of the host mineral.

The major anion is usually chloride, by a factor of 2-10, although HCO_3^- is seldom determined and may exceed chloride, particularly in mercury deposits. (Free CO_2 was present as a separate phase in some mercury-ore fluids; Roedder, 1963, p. 188.) The SO_4^{2-} values (actually total S as SO_4) in analyses of inclusions in sulfide hosts are only valid as maxima, due to oxidation of sulfide even during rapid leaching. Free hydrogen sulfide is found in inclusions only rarely, and then usually in low concentrations. Information on the distribution between the various sulfur species would be valuable, but most analyses do not (and cannot) separate these. A new method, involving nondestructive laser Raman spectrocopy (Rosasco and Roedder, 1976) shows some promise of providing data on the SO_4^{2-} concentration in selected single, unopened inclusions.

The gases in inclusions in hydrothermal minerals have been examined by a wide variety of techniques, including micromanipulation of the released bubble in a sequence of liquid solvents, particularly in the U.S.S.R. (Dolgov, 1968), and in the West by gas chromatography (Cuney et al., 1976) or by mass spectrometry (Ohmoto, 1971; Barker, 1974) or a combination of the latter (Kvenvolden and Roedder, 1971). There are still many serious problems in obtaining a valid sample of gas for analysis due to a variety of causes (see Roedder, 1972, p. 32-36), and although the gas composition possibly has great importance in that it controls (or is controlled by) the equilibria between the various ionic species in the liquid, the interpretation of these data can be ambiguous. Thus even if the difficult problems of extraction and analysis are completely overcome, there is still a problem in deciding at what point or points between the temperature (and time) of trapping and that of analysis each of the many possible intergas reactions was quenched (Barker, 1965). Even the complete lack of noncondensible gases in some hydrothermal fluids (e.g., in the boiling ore fluids at Creede, Colorado; Roedder, 1970a, p. 52-56) provides useful information in that it suggests sufficient boiling has occurred to sweep out all noncondensible gases.

MISSISSIPPI VALLEY-ALPINE-TYPE DEPOSITS[9]

There have been as many complete chemical analyses made of inclusions from this type of deposit as from all normal hydrothermal-type deposits together. There are significant variations among these analyses, but perhaps the most significant feature is their amazing similarity, on a worldwide basis (Roedder, 1967c, p. 354). They seem to be quite different from those in the normal hydrothermal deposits. A considerable body of freezing data (Roedder, 1976) show that most Mississippi Valley inclusions have concentrations of 15-20% or more of salts, and the chemical analyses always yield high Na/K, Na/Ca, and Ca/Mg ratios (Table 14-2). Hall and Friedman (1963) show further that there have been systematic changes in concentration and in several of the more significant atomic ratios during the formation of ores in the southern Illinois and upper Mississippi Valley districts. In particular, these deposits show a rise in the Mg/Na ratio and a decline in both the total concentration and the D/H ratio toward the end of the deposition. Changes in the ratio Na/Cl are not consistent in the two districts.

Regardless of the locality, the relative abundance of ions is Cl > Na > Ca > K > Mg > B in almost all Mississippi-Valley-type deposits. The exceptions are mainly explicable as very-late-stage minerals with much more dilute fluids in the inclusions, or as the result of analytical problems, such as leaching of the host mineral. The sulfate analyses are maximum values only, due to possible oxidation of minute amounts of sulfides, even in some apparently clean fluorite samples. Although the sulfate analyses vary widely, one important conclusion can be reached from them: total sulfur as sulfate in these fluids is far less than chloride. Even though these

---

[9] Several authors have pointed out the close similarity between the Mississippi-Valley-type deposits and the Alpine-type lead-zinc deposits such as Mezica, Bleiberg, and Cave del Predil, and have used the two type names synonymously. A more detailed review of the inclusion data on these deposits is available (Roedder, 1976).

Table 14-2 Atomic ratios (or their ranges) for inclusion fluids from various ore deposits

The "less than" and "greater than" symbols arise from one element being below detection limits, or being present in the host, thus yielding analyses that are maxima.

Mine or District	Mineral host	Na/K	Ca/Mg	Cl/SO$_4$	Number of analyses	References	Sample notes
Normal hydrothermal deposits							
Rico, Colorado	Sphalerite	1.9–3.5	1.8–2.1	>2.6–>4.7	2	Roedder et al. (1963)	Green gemmy crystals; primary inclusions.
Ani Ugo, Japan	Sphalerite	9.1	4.4	>27	1	Roedder et al. (1963)	Orange, 5-cm crystal; primary inclusions.
Creede, Colorado	Sphalerite	9.1	8	>76	10	Roedder (1965b)	Average of outer red-brown and yellow-white zones; primary inclusions.
Providencia, Mexico	Sphalerite and calcite	2.33–11.1	5.5–50	>2.7–>40	15	Rye and Haffty (1969)	Primary inclusion.
Takatori, Japan	Topaz and quartz	2.50–5.26	0.36–3.26	—	6	Enjoji (1972)	Probably mainly primary inclusions (p. 91).
Tin ores, U.S.S.R.	Cassiterite and quartz	0.4–21	1.1–10	—	12–45	Sushchevskaya (1971)	Mg analyses available on 12 samples only.
Bluebell, Canada	Sulfides, calcite and quartz	1.5–33	—	—	20	Ohmoto (1971)	Origin of inclusions unspecified, but probably primary.
Climax, Colorado	Quartz (main stage)	3.33–8.37	3.85–8.07	—	5	Hall et al. (1974)	Probably both primary and secondary inclusions.
Mississippi-Valley-type deposits							
Southern Illinois	Fluorite	16–33	<4.9–<23	5–>1300	5	Roedder et al. (1963)	Various mines; large primary inclusions in vug crystals.

Location	Mineral					Reference	Notes
Southern Illinois	Fluorite	21–22	<15–<18	153–154	2	Roedder et al. (1963)	Secondary inclusions in thoroughly sheared vertical vein.
Southern Illinois	Fluorite, galena, sphalerite and quartz	18–37	3.1–4.7	36–260	12	Hall and Friedman (1963)	Quartz and late barite and witherite data excluded; Ca/Mg on sulfides only; Cl/SO_4 excludes data from sulfides; both primary and pseudosecondary inclusions.
Tri-State, Oklahoma	Sphalerite	36	4.5	>117	1	Roedder et al. (1963)	Subparallel yellow-brown 8-cm crystal group; primary inclusions.
Santander, Spain	Sphalerite	34	>6	>13	1	Roedder et al. (1963)	Clear yellow cleavage fragments; primary inclusions.
Cartagena, Spain	Sphalerite	33–40	3.4–3.6	>55–>106	2	Roedder et al. (1963)	Large yellow single crystal; primary inclusions (see Roedder, 1967a, p. 552).
Upper Mississippi Valley	Sphalerite and galena	21–36	1.0–5.6	>365	2	Hall and Friedman (1963)	Data on more dilute inclusions in late calcite excluded; both primary and pseudosecondary inclusions(?) in sphalerite, and of unknown but probably primary origin in galena.
Upper Mississippi Valley	Galena	32–36	2.8–4.5	>7.9–>32	4	Roedder et al. (1963)	Four zones from single 10-cm cube; inclusions of unknown but probably primary origin.

data for SO_4 may be high from oxidation of host minerals during leach, they are still one to two orders of magnitude less than chloride.

White (1958), on the basis of Newhouse's early qualitative analyses (1932), and later Hall and Friedman (1963, p. 903-904), point out the similarity of the solutions in the inclusions to oil-field brines of the sodium-calcium-chloride type. The quantitative inclusion analyses show that there is a strong similarity, but several important differences are apparent. The sodium-calcium-chloride oil-field brines normally have magnesium considerably greater than potassium (weight ratios), but the inclusions show magnesium equal to or slightly less than potassium. The atomic ratio Na/K is greater than 43 in nine out of twelve cited oil-field brine analyses (White et al., 1963), up to a maximum of 1,134, but in the 29 inclusion analyses noted in Table 14-2, the *maximum* Na/K ratio is 40. Most of these analyses are very close to their average Na/K of 26. White (1965) has shown that the Na/K ratio for natural hot waters is a function of the temperature. At 100°C the ratio is generally about 34. The analytic data on inclusions in general, and the Mississippi Valley deposits in particular, agree with this relationship, although leaching of evaporite deposits may also be involved (Sawkins, 1968).

In addition to these inclusions of brine, many deposits also have primary inclusions of oil or mixed oil and brine (Roedder, 1972, Plate 9, Figure 3). Theoretically it should be possible to obtain *both* pressure and temperature of formation from such coeval brine and oil inclusions (see beyond).

The origin of this type of deposit has been in debate for a long time, and many mechanisms have been proposed. Any proposed mechanism is worthy of further consideration only insofar as it explains the available facts. When all of the surprisingly uniform inclusion data are considered—the high total concentrations of salts, the composition of these salts, the presence of oil and methane, and the low temperatures of formation—along with the numerous other distinctive structural and mineralogic features of this type of deposit (Ohle, 1959), the immense areas over which they are found, and the isotopic data, it seems necessary to invoke processes grossly different from those processes responsible for most normal hydrothermal deposits. It is hoped that further analytic data, both chemical and thermometric, on inclusions from samples closely tied to the available structural data revealing possible flow directions, will be even more useful in delineating the environment—and possibly the cause—of ore deposition.

pH, Eh, AND HEAVY METAL CONTENT OF INCLUSION FLUIDS

Although each of these three items is of great interest, very few usable data are available. Many quantitative measurements of inclusion pH have been reported in the literature, but most are probably invalid, as they are usually made on the water solution obtained by leaching the crushed mineral. Such solutions do not contain gases under high pressure, as originally present; the inclusion fluid has been diluted as much as 5000-fold; and the solution has been in contact with the host mineral and possible impurity minerals that may affect the pH. A few measurements made directly on inclusion fluids, in part with simple litmus, are probably more significant (see Roedder, 1972, p. JJ38). Most of these appear to fall within one unit of neutral, although Kalyuzhnyi (1957) obtained values as low as 4.3 for large multiphase inclusions in pegmatitic topaz that evolved much gas upon opening. Extrapolations from such measured pH values at room temperature to the lower-density, high-pressure, high-temperature conditions of origin are possible but hazardous.

The only recorded measurements of Eh on inclusion fluids were by Petrichenko and Shaydetskaya (1968; also Petrichenko and Slivko, 1971), who used miniature electrodes inserted into a pair of 1-mm holes drilled into very large brine inclusions in rock salt from the Donbass, U.S.S.R.

Any large inclusion could be measured with existing Eh equipment, as miniature and subminiature Eh electrodes are available, but the problems of extraction of the fluid precludes most such measurements; in addition, the problems mentioned for pH measurement generally are applicable here as well.

It should be possible to calculate the Eh from detailed studies of the composition of the inclusions. Thus any multivalence state component or group theoretically could be used—Fe^{2+}/Fe^{3+}, CO/CO_2 HS^-/SO_4^{2-} or other similar pairs—but even if accurate analyses of such pairs can be made (together with pH measurements) there is little assurance that the original high-temperature equilibrium will be quenched in, or that true room-temperature equilibrium will be obtained. Some indications of the state of oxidation of the fluids may be obtained from daughter minerals, such as pyrite, hematite, and gypsum (or anhydrite). Each of these is known to occur individually in inclusions from hydrothermal deposits.

Unfortunately, the Eh of an inclusion fluid would be strongly affected by any loss of H_2 (or H_2S) from the system. Although the evidence against gross leakage of major constituents from inclusions is commanding (Roedder and Skinner, 1968), there is no proof that small amounts of hydrogen cannot leak out and increase the oxidation state of the remaining fluids. One of the major difficulties in calling on this mechanism to explain the presence of hematite flakes in inclusions is that of maintaining an adequate "hydrogen sink" outside the inclusion to drive the diffusion process. As many ore-forming processes apparently take place at low oxidation states, i.e., high partial pressures of hydrogen, the most expectable change in the state of the fluid bathing the exterior of the crystal would be toward more oxidizing conditions, particularly as the erosion surface approaches the deposit and oxygenated surface waters are involved. Even this should have only a small effect, however, because diffusion rates are greatly reduced at surface temperatures, and because groundwaters penetrating ore deposits are rather effectively buffered with respect to oxygen by reaction with sulfides.

The heavy metal content of fluid inclusions is of considerable importance in understanding the mechanism of ore-metal transport, and in making semiquantitative guesses as to the plumbing system needed to bring the ore in. Other than the Salton Sea brines and some hot springs discussed by Weissberg et al. (Chapter 15) the most direct natural evidence we have on the ore fluids comes from inclusions. The occurrence of solid opaque daughter minerals (presumably ore minerals) in such inclusions has been used to estimate the amount of precipitation of ore metals to be expected by *temperature decrease alone*. The generally rare occurrence and small size of these opaque daughter minerals indicate that the amount of precipitation of ore metals from the ore fluids must be rather low, usually 1-10 ppm (Roedder, 1960b), but there are some high-temperature inclusions that have precipitated over 1% of sulfides (Roedder, 1963, p. 181). The amounts of heavy metals *still in solution* in the inclusion fluid appear to be high, 100-10,000 ppm each of copper, manganese, and zinc, on the basis of neutron activation analyses of two carefully selected inclusion samples, a fluorite from southern Illinois and a quartz from Creede, Colorado (Czamanske et al., 1963). These authors examined the various possible sources of contamination and found them to be relatively negligible, but the unexpectedly high values obtained are difficult to explain and should be verified. A series of more recent papers (see Roedder, 1968) list analyses for one or more heavy metals, but usually without adequate statements of the cleaning techniques used before extraction, and the magnitude of the blanks obtained. The most notable exceptions are Rye and Haffty (1969) who found maximum base-metal concentrations of 890 ppm zinc and 530 ppm copper in inclusions in quartz from Providencia, Mexico; Pinckney and Haffty (1970) who found 10-1040 ppm zinc and (0)-350 ppm copper in inclusions from fluorite, quartz, and barite from southern Illinois; Ohmoto (1971) who found 20-2200 ppm zinc in stage II and III minerals from the Bluebell mine, British Columbia; and Hall et al. (1974) who found 200-4600 ppm zinc and 80-4000 ppm copper in quartz from Climax, Colorado.

Although chloride complexes are known for many metals at low to moderate temperatures, as yet unknown interactions with the high concentrations of salts in inclusion fluids may be responsible for many unexpectedly high solubilities for heavy metals at high temperatures. For example, Borina (1963) measured the apparent solubility of $CaMoO_4$ in solutions of

KCl and NaCl; she found as much as 0.015% dissolved at 400°C. Several other studies have indicated high apparent solubilities of heavy metals in strong chloride solutions. Thus Bryatov and Kuz'mina (1961) dissolved 2350 ppm lead (from galena crystals) at 400°C, and Anderson and Burnham (1964) report on the order of 1000-2000 ppm of metallic gold (or platinum) dissolved. High-temperature and supercritical aqueous sulfate fluids can hold large amounts of uranium, copper, and nickel sulfates in solution (see Marshall and Jones, 1963, and other papers in that series). Until valid theoretical procedures are developed for calculating solubilities in the hot, concentrated, multicomponent systems characteristic of nature, it appears that conclusions concerning the mechanisms of transport and deposition of ore minerals must be based at least in part upon the evidence found in the fluid inclusions. In any case, there is considerable need for further laboratory experimentation in such systems, using both the natural "visual autoclaves" provided by the inclusions themselves and the newly developed techniques for experimental measurement of hydrothermal solubility.

ISOTOPIC COMPOSITION OF INCLUSION FLUIDS

Within the last few years there has been a great increase in studies of the isotopic ratios of hydrogen, oxygen, sulfur, and carbon in either the inclusion fluids or their host minerals, or both. Many of these have been made on specific ore bodies, with the goal of an understanding of the origin of the ores and the ore fluids, and the processes involved in ore deposition. In addition, several studies have been made of the isotopic ratios of various phases in synthetic mixtures in the laboratory, to determine the isotopic fractionation factors and the phenomena affecting them (for references, see Roedder, 1968).

These data can be used in a variety of ways, and provide a very powerful tool, but certainly are not a panacea for ore deposition problems. They are most effective in limiting the processes involved, rather than identifying them. Each isotopic ratio is a single parameter that must be explicable in terms of any valid model of ore formation, but no one such parameter is definitive. A series of such isotopic parameters, however, or even better such a series along with independent data of other types, may be much more effective in limiting the possible models. Probably the most important single result of the recent extensive isotopic work on inclusion fluids is the verification of mixing of fluids from different sources, as a surprisingly common phenomenon in the formation of ore deposits, for example, the Bluebell lead-zinc deposit (Ohmoto, 1971); the Pasto Bueno tungsten deposit (Landis and Rye, 1974); the Climax molybdenum deposit (Hall et al., 1974); the Casapalca silver-lead-zinc-copper deposit (Rye and Sawkins,

1974); two Mississippi Valley districts (Hall and Friedman, 1963); and various porphyry coppers (Sheppard et al., 1971). One of the fluids is generally groundwater, and some deposits seem to have been formed solely by heated groundwater.

In other ore deposits, the inclusion fluids show no evidence of mixing. For example, the author found that the D/H ratios in the inclusions from Creede, Colorado only varied over a small range, yet the freezing data showed a gross (threefold) change in salinity of the fluids during the same stages of mineralization, and Rye noted similar results at Providencia, but in a different D/H range. If a simple mixing model is postulated, with brines and fresh waters to obtain the variable salinity, the data require that in each case the two fluids are almost identical with each other in D/H ratio—a possible but rather unlikely situation.

Isotopic data may be used in many other ways. Thus gross differences have been found between the D/H ratios of water from fluid inclusions in sphalerite crystals lining open vugs at Creede, Colorado, and the groundwaters that presumably have been bathing these crystals at 30-40 atm external pressure for many millions of years (Roedder, 1960a). Although there are more convenient and sensitive tests (Roedder, 1963, p. 207), this in itself is fairly strong evidence against gross leakage of these inclusions.

Future studies should examine inclusion fluids for the isotopic ratios of other elements as well, particularly sulfur, oxygen, and carbon, using host minerals free of the element studied, to avoid interference from exchange reactions with the host mineral upon cooling. Such isotopic exchange between fluid and ore or country rock during mineralization is useful, however, in establishing some limits on the total mass of the ore-forming fluid.

TEMPERATURE, PRESSURE, AND DENSITY OF ORE FLUIDS

Principles, Assumptions, and Limitations of Inclusion Thermometry

As mentioned above in the section titled "Changes Since Trapping," Sorby (1858) showed that the gas bubbles present in the fluid of most inclusions were the result of differential shrinkage from the higher temperature of trapping to the temperature of observation, and that the temperature of trapping can be estimated by heating the sample to the point that the bubbles disappear (see Figure 4 in Roedder, 1972, and Figure 14.2). It is readily apparent that *any* reversible phase change that occurs on cooling an originally homogeneous fluid phase—formation of a gas phase, crystallization on the walls or as daughter minerals, splitting into immiscible fluids, etc., provides a thermometer. The limitations on the use of shrinkage bubbles were detailed first by Sorby. The following discussion will center about such bubbles, but the remarks are generally applicable to any phase change.

Since Sorby's time, there have been many restatements of the assumptions and limitations of the homogenization method. The method and the data obtained have been hotly debated at times, and as a result of ignorance and partisanship, an unfortunately large number of erroneous statements have been made in the process. The limitations and assumptions are relatively simple in nature; the problems arise in attempts to estimate the validity of given samples, and to estimate the magnitude of the errors that may occur. The major assumptions are as follows:

1. The fluid trapped upon sealing of the inclusion was a single, homogeneous phase. A single inclusion is inadequate to fulfill this requirement, but when many inclusions in a sample all show apparently the same phase ratio, it is presumed that a homogeneous fluid was trapped. Unless the heterogeneity was very uniform and of small dimensions relative to the inclusion dimensions (e.g., a colloidal dispersion), individual inclusions formed from heterogeneous systems can be expected to trap different ratios of the phases present, and all further phase changes on cooling will be superimposed on these original differences. As discussed above in the section "Heterogeneous Systems," such evidence of original two-phase systems gives us valuable information on the environment, but such inclusions will generally give erroneous temperature data (see also discussion of pressure and density below). The trapping of primary gas (either vapor of the fluid itself, i.e., boiling, or a separate, more gaseous constituent such as CO_2 in H_2O) may cause particularly large positive errors in the more common case of homogenization in the liquid phase (see Figure 14.2). In the author's experience, trapping of primary gas is a relatively rare phenomenon, but in those situations where it does occur, there may be thousands of inclusions showing evidence of it.

It is particularly important to be able to verify whether or not boiling has actually occurred in those ore deposits were inclusions with widely variable gas/liquid ratios are found, since boiling can cause ore deposition. In addition to boiling, variable gas/liquid ratios can be caused by trapping at different times from fluids under different *P-T* conditions, by leakage of part of the inclusions, or by necking down, but such processes can usually be recognized by microscopy. In the simplest case, boiling will result in two types of inclusions, representing the trapping of either the liquid phase or the vapor phase. The latter (also called "steam" inclusions) contain a very little liquid and a very large bubble at room temperature, and on heating they will homogenize in the vapor phase by evaporation of the liquid, at the *same* temperature as the liquid inclusions. (This determination will be accurate only when the steam inclusion has a narrow reentrant into which the last bit of fluid phase is concentrated by capillarity.) Small amounts of liquid are sometimes trapped with the steam, just as small amounts of vapor (i.e., a small bubble) can be trapped with the liquid. Both such types of inclusions will obviously yield homogenization temperatures that are higher than the trapping temperature. It is possible, although hazardous, to use the *minimum* temperatures determined on a large number of such inclusions as the true trapping temperature, as these are the ones that are most likely to have trapped pure liquid or pure vapor.

2. The cavity in which the fluid is trapped does not change in volume after sealing. Change can and does occur by several mechanisms—crystallization on the walls or in the fluid itself, thermal contraction of the host mineral (and precipitated minerals) on cooling, and dilational change from internal or external pressure. Ermakov (1950), and others place much emphasis on crystallization on the walls or of daughter minerals reducing the volume of the cavity accordingly and presume that this can cause rather gross inaccuracies in the homogenization temperatures. Two factors tend to minimize error from this cause. First, precipitated daughter minerals (and precipitated matter on the walls) will have no effect on the volume if they redissolve on reheating. Second, the volume change upon crystallization or dissolution of a solute is not simply equal to the volume of the solid solute. For example, Benson et al. (1953) and Copeland et al. (1953) show that the behavior of NaCl in supercritical water is remarkably anomalous, in that it has a very large and negative molal volume, of the order of magnitude of 2 l/mole. Crystallization of NaCl under such conditions would increase the apparent volume. Ermakov also lays great stress on the thermal-expansion characteristics of the mineral host. This would have a very small effect on the size of the bubble at room temperature (minerals with larger thermal coefficients having somewhat smaller bubbles), but as the contraction on cooling is exactly reversed on heating, it should have no effect on the homogenization temperature. The last mechanism, dilational change, theoretically has an effect on the homogenization temperature, but can be neglected in practice. The host mineral for the inclusion is formed and the inclusion is sealed at an elevated pressure. Thus the inclusion volume should expand somewhat upon release of this external pressure upon the host as the sample is brought to the surface of the earth. Upon reheating, however, at surface pressures, the host mineral is placed under considerable strain by the pressure inside the inclusion not being balanced from outside, so the difference in volume of the cavity between that at the time of sealing and that at the time of homogenization in the laboratory would be at least double the volume compressibility of the host for the pressures involved. As these compressibilities are very small, however, the total effect is insignificant at the current stage in geologic thermometry.

If the laboratory technique is not correct, the walls of low-temperature inclusions in some soft minerals may be permanently stretched, yielding *reproducible* but erroneously high homogenization temperatures (Larson et al., 1973).

3. Nothing is added or lost from the inclusion after sealing. This has been discussed under the heading "Leakage into or out of Inclusions." The author believes that leakage occurs, but is relatively rare except in those rocks where crushing or other deformation has occurred, or where very high-pressure gradients have been set up (as in the vicinity of a near-surface dike).

A special case, equivalent to leakage, is the necking down of large inclusions into several smaller ones, mentioned above in the section "Changes Since Trapping." Where this necking down and resealing has occurred after the formation of a gas bubble (or any other new phase), the two resulting inclusions will have different homogenization temperatures (Figure 14.4). The one that traps the bubble will have a homogenization temperature above the original trapping temperature—just as though it had trapped a primary gas bubble—and the other one will have too low a homogenization temperature (actually it will show the temperature at which the necking down sealed it). The writer believes that necking down (and other changes in inclusion shape) has occurred very commonly in nature, and accounts for a good part of the scatter in results that is so apparent in practically all careful thermometric studies. Necking down would be expected to occur much faster at elevated temperatures. Thus the amount of necking down that occurs in the first 50°C of cooling after trapping could be far greater than in the next 50°C, even though the time involved for the second interval might be much greater. If such a process occurs, it might not result in sufficiently large variations in the gas/liquid ratios to cause these particular inclusions to be disregarded in advance.[10]

---

[10] Avoiding abnormal inclusions, as recommended by Ermakov and Kalyuzhnyi (1957), and others, represents an attempt to eliminate those inclusions whose homogenization temperatures would not be valid, due to necking down, primary gas, or leakage. It is perfectly possible, however, to have abrupt changes in conditions yield valid inclusions, some of which would appear abnormal. The line is sometimes difficult to draw, and the subjectivity involved has caused some doubts about the validity of the method ever since Sorby was accused of introducing bias into his samples by this selection procedure.

Those groups of inclusions that give very consistent temperatures—and many such have been reported—would merely be inclusions that, by virtue of an originally more or less stable shape, did not neck down. The balance of the scatter found in the homogenization temperatures of apparently contemporary inclusions is believed to be a result of the lack of true contemporaneity, occasional leakage, experimental errors, etc.

4. The effects of pressure are insignificant, or are known. There is much confusion concerning this subject, and the problem is further complicated by the fact that the magnitude and nature of the pressure corrections that must be applied differ with the method used for determining the filling temperature (i.e., homogenization vs decrepitation). Basically, pressure is of concern only in that it controls the density of the fluids above the boiling curve (Figure 14.2). If a fluid inclusion is trapped along the boiling curve (i.e., under *P-T* conditions such that it was either in equilibrium with a vapor or gas phase, or almost so), the homogenization temperature equals the trapping temperature and no pressure correction is needed. Such an inclusion would start to form a gas bubble as soon as it is cooled below the trapping temperature. In this case, it makes no difference whether the gas bubble that forms is of the same composition as the liquid (e.g., pure water), or is a completely different composition (e.g., CO_2).

If an inclusion is trapped at a *P-T* combination above the boiling curve, a bubble does not form in it on cooling until the pressure and temperature have dropped to the boiling curve. As the density stays essentially constant under these conditions, the change in *P* and *T* inside the inclusion must be represented by a vertical line on Figure 14.2. This difference in temperature is the pressure correction; it must be added to the homogenization temperature to obtain the true formation temperature.

If the pressure can be calculated from geologic field data on the depth of cover, the measured homogenization temperature can be corrected for pressure to yield the trapping or formation temperature.

Too frequently this pressure correction is made under the assumption that the fluid in the inclusion either is pure water, or has the same vapor pressure, thermal expansion, and compressibility as pure water. This is generally not true in nature, as most of the fluids in inclusions are NaCl brines. For these the two-phase field, bounded by the boiling curve, is raised above that of water roughly in proportion to the concentration of salts. As the system $NaCl$-H_2O is reasonably close to the natural systems, and as this is the only one for which extensive data are available, a plot of a portion of it is shown in Figure 14.5 taken from the work of Keevil (1942), and Sourirajan and Kennedy (1962). If the salinity is unknown, the possible errors in pressure correction from this factor are minimized if a

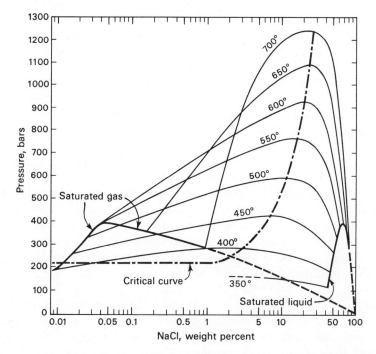

Fig. 14.5. P-X diagram of coexisting liquid and gas phases in the system NaCl-H$_2$O. (Data on the saturated liquid phase from Keevil, 1942; all other data from Sourirajan and Kennedy, 1962.)

10% salinity is assumed (see Roedder, 1971b, Figure 31). Smith (1953, p. 91) has pointed out that the difference between the temperature of homogenization of gas + liquid phases and a higher temperature at which a daughter mineral may disappear represents a minimum pressure correction.

There are comparatively few data on the density of phases in this system (Copeland and others, 1953; Benson et al., 1953; Lemmlein and Klevtsov, 1961). The addition of NaCl causes a marked rise in the temperature of the two-phase field in Figure 14.2. Thus an inclusion containing 30% gas at room temperature would homogenize at 302 °C if filled with pure water, or at 460 °C if filled with a 30% NaCl solution. The addition also causes a drastic lowering of the vapor pressures (e.g., at 400 °C, 20% NaCl drops the vapor pressure of a fluid of density of 0.7 from about 1000 to 250 bars). The compressibility of the fluid is also changed radically, so that the pressure correction for inclusions of saline fluids can be either greater or less than that for pure water inclusions. The difference is large for inclusions with high filling temperatures. Thus the pressure correction for an inclusion homogenizing in the liquid phase at 350°C, and formed at 2000 bars, would be 265°C for pure water and 180°C for a 20% NaCl solution (i.e., the formation or trapping temperature would be 615 or 530°C). Roedder and Kopp (1975) have shown that inclusions in synthetic quartz can give relatively accurate formation temperatures, even when the pressure corrections are large, as long as the pressure and composition are known. If the composition is not known, errors from this cause can be minimized by assuming 10% NaCl (Roedder, 1971b, p. 111).

5. The origin of the inclusion is known. Inclusions can only give information on the environment under which they were formed. If they are primary, they indicate the conditions under which the enclosing mineral formed; if secondary, they indicate some later conditions. Conclusive evidence of secondary origin is easy to find for many inclusions; as shown in the first part of this chapter, equally conclusive evidence for primary origin is relatively rare. There are many cases, however, where data on known secondary inclusions have proven useful.

Although the numerous limitations and assumptions detailed above may seem prohibitive, there are many samples, from many deposits, that are apparently adequate to fulfill the requirements, and it is the author's opinion that of all of the geologic thermometers now known, fluid inclusions provide perhaps the most accurate, and certainly the most generally applicable method. Of the five items discussed above, the next-to-the-last one, concerning pressure corrections, is probably the greatest single source of numerical error, and the last one, concerning origin of the inclusions, is probably the greatest single source of ambiguity in interpretation.

Methods and Apparatus Used for Thermometry

The *homogenization method* (first proposed by Sorby, 1858), in which the inclusions are examined with the microscope while being heated, is the most generally accepted method, and probably the most accurate. Its use is most seriously limited by the difficulty of achieving an adequately high, known, and controllable temperature in a sample under such conditions that high magnifications and adequate lighting may be used. There are many designs of microscope hot stages in the older literature (see Roedder, 1972, p. JJ27–28), and several more designs have been published since then (see references in Roedder, 1968 and onward). Some of these have certain definite advantages, but the instrumental problems are still far from solved. A number of the descriptions of hot stages each indicate that relatively high precision may be obtained, but cross-checking may reveal discrepancies of 50°C or more in accuracy, with variations in sample size or nature. *Meaningful* calibration procedures are not easy to develop. Fortunately the errors are largest in the high-temperature range, where other considerations such as inclusion composition and pressure corrections result in larger uncertainties anyway. A review of problems of sample selection and preparation for thermometry, inclusion microscopy, and heating stage calibration and operation is in print (Roedder, 1976) and need not be repeated here.

Far more determinations have been made by the decrepitation method than by any other procedure. The method was first proposed by Scott (1948), and developed by Smith, Peach, and coworkers (see Smith, 1953). It is used rather extensively in the U.S.S.R. (see Roedder, 1968, 1972). It is based on the rate of explosion or decrepitation of inclusions in a coarsely crushed sample with continuous increase in temperature; this rate increases rapidly at the homogenization temperature of the inclusions. Decrepitation may be detected visually, with a stethoscope, or with various electronic amplifying, integrating and counting circuits.

There has been considerable discussion in the literature concerning the theoretical significance and practical usefulness of the method. Inclusions with a high degree of filling (and hence low homogenization temperatures) have a very abrupt increase in $\Delta P/\Delta T$ at homogenization, and this is the theoretical basis for the method. The change in $\Delta P/\Delta T$ is much less abrupt or even negligible upon homogenization of lower density, higher temperature inclusions, and $\Delta P/\Delta T$ actually *decreases* upon homogenization in those inclusions with less than the critical density of filling (that is, homogenization in the gas phase). Most of the objections to the method are founded on this lack of a theoretical basis for at least part of the range, the commonly observed lack of agreement with actual homogenization data, and the occurrence of decrepitation in samples apparently free of visible fluid inclusions. Also there is some degree of subjectivity involved in eliminating "anomalous" decrepitation, and in selecting the appropriate spot on the decrepigram for the "start of decrepitation" of a given generation of inclusions. Additional causes of scatter in the results stem from the irregularity in the size of the inclusions (large ones decrepitate at lower temperatures); variations in composition (e.g., $NaCl:H_2O:CO_2$) causing gross differences in internal pressure; arrangement of inclusions in planes, making decrepitation easier; and variations in the toughness or brittleness of the mineral, resulting in large variations in the amount of overshoot (heating above the homogenization temperature) before decrepitation. There are numerous records in the literature of overshooting of 50 or 100°C without decrepitation or leakage, and even 600°C (Roedder, 1970b).

Certain liquid CO_2 inclusions in olivine may be heated 1200°C above their filling temperature without decrepitation (Roedder, 1965a). Khetchikov and Samoilovich (1970) have examined the differences between homogenization and decrepitation temperatures of synthetic quartz grown at known temperatures and pressures, and find that the decrepitation temperatures can be from 100°C too low to 160°C too high, depending on pressure of formation.

The decrepitation method is rapid, however, and integrates the results of many hundreds or thousands of inclusions. It is probably most useful as a screening method to recognize the presence of several different generations of inclusions, particularly in the opaque minerals and in the low-temperature range. Since it can reveal differences between samples, and can readily be adapted to field use, it has been used most effectively in exploration, where the "steam halo" of higher temperature inclusions surrounding a deposit provides a larger target and reveals blind deposits (see Demin, 1970; Ermakov and Kuznetsov, 1971; and numerous other references in Roedder, 1968).

Visual estimates of the degree of fill of inclusions may be made, but the volume relations of spherical inclusions can be very misleading (Figure 14.6). Far more accurate gas/liquid ratios may be obtained by measurements of inclusions with some degree of regularity, for example, on flat inclusions, where area percent equals volume percent. Although highly

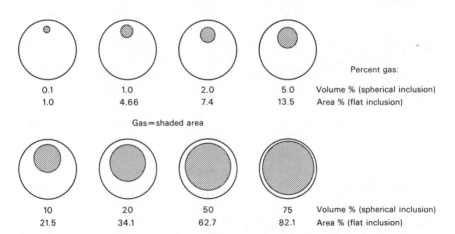

Fig. 14.6. Appearance of spherical inclusions with various percentages of gas phase. In the volume percentage figures, both the bubble and the inclusions are assumed to be spherical. The area percentages refer to thin, flat inclusions, where both bubble and inclusion are assumed to be circular disks of negligible thickness, and area percent equals volume percent.

precise calculations of the homogenization behavior of various types of inclusions have been made, variation in the composition of the fluids precludes accurate prediction of the filling temperature from room-temperature phase ratios. Little (1955) found that the predicted filling temperatures based on measured phase ratios of inclusions in cassiterites, and P-V-T-data for pure water, were as much as 160°C lower than the actual homogenization temperatures. It is much more probable that the experimentally determined "homogenization curves" (phase ratio versus temperature) for actual inclusions can be used in reverse, to determine, or at least to place limits on their compositions.

General Ranges of Temperature Data Obtained

Approximately 1300 individual reports of the use of inclusions as geologic thermometers have been published (see Roedder, 1968; and Table 1 in Roedder, 1972), and only a very brief summary can be given here. With very rare exceptions, the Mississippi-Valley-Alpine-type ore deposits show homogenization temperatures of less than 200°C, and most are in the range 100–150°C, except for some very obviously late-stage, postore gangue minerals, which frequently show temperatures below 100°C (summarized in Roedder, 1976). As the geologic evidence usually points to comparatively shallow cover during mineralization, and as the degree of filling is high, the pressure corrections to be added to these values are small, usually 10–15°C. The normal hydrothermal vein deposits show higher temperatures, which are generally in agreement with existing geologic concepts as to relative temperatures, if not the absolute temperatures. Deposits generally classed as epithermal show inclusion homogenization temperatures in the range 150–300°C, and may on occasion show evidence of boiling. Those ore deposits believed to have been formed under higher pressures and temperatures normally show homogenization temperatures up to 350 or 400°C, and, of course, have a much larger pressure correction. Pegmatites, contact metamorphic deposits, apatite deposits in alkalic rocks, and porphyry copper deposits frequently show homogenization temperatures over 500°C (Roedder, 1972, Table 7), indicating not only high temperatures, and hence pressures of formation, but also indicating very high concentrations of salts in the inclusions to permit the existence of a two-phase (liquid plus gas) system far above the critical temperature for pure water. Some of these very high-temperature inclusions show evidence also of considerable solution of the host mineral during the heating. A generally decreasing temperature is evident in many deposits consisting of multiple stages of mineralization, but sudden reversals in temperature with time apparently are not rare, and frequently are marked by abrupt changes in mineralogy. The late and last stages of mineralization in many deposits indicate much lower temperatures than the earlier ones, and not infrequently go below 100°C, even for the high-temperature deposits.

Pressure and Density Estimates

It is apparent from the discussion above, and from Figure 14.2, that the homogenization temperature of an inclusion, even if the composition is known, cannot give both temperature and pressure of formation. If either is known, the other can be obtained. Usually the pressure is merely estimated from geologic evidence of the depth of cover, at the time of formation, plus an assumption that formation pressure equals the lithostatic or the hydrostatic load. This estimated pressure is then used, along with compositional data, to calculate a pressure correction for the homogenization temperature.

Actually, fluid inclusions can give us at least some direct data on the pressures of formation. Obviously, the vapor pressure of the solution involved, at the homogenization temperature, establishes a lower limit. This can be estimated from data on the system $NaCl-H_2O$, and independent data on the approximate composition, for example, from freezing temperatures. Such pressure data can then be used to establish minimum depths, from hydrostatic pressure data for boiling columns (Haas, 1971) or representative geothermal gradients. If carbon dioxide or other gases are present, the pressures would be still higher. If the inclusions contain daughter minerals, particularly NaCl, the solution of these may be used to determine a minimum pressure, using data from Keevil (1942) shown on Figure 14.5.

If the host crystal formed from a heterogeneous system of two known, *essentially immiscible* fluid phases, each of which was trapped simultaneously, but in different inclusions, both pressure and temperature may be determined from the point of intersection of the appropriate isochores on a *P-T* plot of the systems involved. This method has been applied most frequently to inclusion pairs containing CO_2 and H_2O. Its accuracy is limited by the degree to which the two isochores approach parallelism. Quantitatively much more important limitations are imposed by the validity of the necessary assumptions of contemporaneity of trapping and of known composition. In particular, the mutual solubilities of CO_2 and H_2O may cause very major errors (see Figure 14.7). Although not yet tried in practice, the oil and brine inclusions so common in some Mississippi-Valley-type deposits have considerable potential, as the compressibilities differ greatly and the mutual solubility is probably low (Roedder, 1963, p. 176–177).

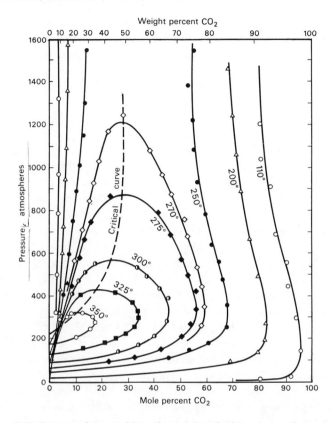

Fig. 14.7. *P-X* diagram of composition of coexisting liquid and gas phases (and, consequently, the regions of immiscibility) in the system $H_2O\text{-}CO_2$. (After Takenouchi and Kennedy, 1964, *Am. J. Sci.*, **262**, 1055–1074.)

If the two fluid phases in a *partially miscible* heterogeneous system are trapped together in various ratios in individual inclusions, both temperature and pressure can be obtained. Thus, Smith and Little (1959) proposed the use of inclusions from heterogeneous systems having a solvus, such as $H_2O\text{-}CO_2$, for a geothermometer requiring no pressure corrections, but observational difficulties may preclude its use (Roedder, 1963, p. 190).

If the inclusions formed from a homogeneous fluid mixture of CO_2 and H_2O, all simultaneously trapped inclusions will have a uniform phase ratio.

This has been observed rather frequently, and, from data on the miscibility gap in this system, plotted in Figure 14.7 (Takenouchi and Kennedy, 1964), permits at least some limits to be set on the possible range of pressure and temperature. It is apparent from this figure that the extensive miscibility of CO_2 and H_2O at temperatures over 200°C places few restrictions on the conditions of formation of a given CO_2/H_2O ratio inclusion; as the solubilities in the water phase are considerably decreased by the addition of 20 or even 6 wt. % NaCl (Takenouchi and Kennedy, 1965), the possibilities of usefulness of such inclusions are increased accordingly.

Unlike the estimates of pressure and temperature, fluid inclusions permit reasonably accurate and unambiguous estimates of the *density* of the ore-forming fluids, and there does not appear to be any other source for such information. If the relative volumes of crystal, liquid, and gas phases are determined at room temperature, by using tubular, flat, or geometrically regular inclusions, the density of the originally homogeneous fluid can be calculated. Volume changes from external or internal pressure, and from thermal expansion of the host, are negligible compared to the effects of composition and inaccuracies in phase volume measurement. Some deposits have formed from very low-density fluids, under 0.5 g/cm^3, as evidenced by large numbers of inclusions with uniformly low liquid/gas ratios. Even at these low densities, there apparently can be considerable solubility and dissociation of dissolved salts. The numerous reports of "dry gas" inclusions in some samples do not require, however, that the minerals formed from such low-density media, as only the gas phase may have been trapped during boiling or effervescence.

On the other hand, some fluid inclusions with large daughter crystals show densities as high as 1.5 g/cm^3 (e.g., 50% liquid, 40% NaCl crystal, and 10% gas, by volume at room temperature in inclusions from Bingham, Utah; Roedder, 1971b, p. 113). Most hydrothermal ore deposits have formed from fluids in the range of 0.5 to slightly over 1.0 g/cm^3. The significance of these density measurements lies in the very great effect the density has in controlling or preventing circulation and mixing of such hot brines with overlying cold fresh waters at density 1.0 (see Chapters 4 and 13, this volume).

ACKNOWLEDGMENTS

The author is indebted to many of his colleagues for stimulating discussions, and to R. W. Potter II and T. Casadevall for careful reviews of the manuscript.

REFERENCES [11]

Adamson, A. W. (1960) *Physical Chemistry of Surfaces*, New York: Interscience.

Anderson, G. M. and C. W. Burnham (1964) Solubilities of quartz, corundum, and gold in aqueous chloride and hydroxide solutions (abst.): *Geol. Soc. Am., Special Paper 82, Abstracts for 1964*, 4. (see also *Am. J. Sci.*, **263**, 494-511.)

Barabanov, V. F. (1958) The role of pressure during mineral growth in quartz-wolframite veins: *Akad. Nauk SSSR Doklady, 120* 400-403 (in Russian) transl. in *Proc. Acad. Sci. U.S.S.R. Geol. Sci. Sect.*, **120**, 565-569.

Barker, C. (1965) Mass spectrometric analysis of the gas evolved from some heated natural minerals: *Nature, 205,* 1001-1002.

―――― (1974) Composition of the gases associated with the magmas that produced rocks 15016 and 15065: *Proc Fifth Lunar Sci. Conf., Geochim. Cosmoch. Acta, Suppl. 5, vol. 2,* 1737-1746.

Barnes, H. L., J. Lusk, and R. W. Potter (1969) Composition of fluid inclusions (abst.). Abstracts of papers presented at Third International COFFI Symposium on Fluid Inclusions: *Fluid Inclusion Research―Proceedings of COFFI*, 2 (1969) 13.

Barton, P. B., Jr., P. M. Bethke, and M. S. Toulmin (1971) An attempt to determine the vertical component of flow rate of ore-forming solutions in the OH vein, Creede, Colorado: *Soc. Mining Geol. Japan, Spec. Issue 2,* 132-136 [Proc. IMA-IAGOD Meetings '70, Joint Symp. Vol.].

――――, ――――, and P. Toulmin, III (1963) Equilibrium in ore deposits: *Mineral. Soc. Am., Spec. Paper 1*, 171-185.

―――― and P. Toulmin, III (1961) Some mechanisms for cooling hydrothermal fluids: *U.S. Geol. Surv., Prof. Paper 424-D.* 348-352.

Benson, S. W., C. S. Copeland, and D. Pearson (1953) Molal volumes and compressibilities of the system NaCl-H_2O above the critical temperature of water: *J. Chem. Physics* **21**, 2208-2212.

Berg, W. F. (1938) Crystal growth from solutions: *Proc. Roy. Soc. London,* **A164,** 79-95.

Borina, A. F. (1963) Aqueous salt solutions at high pressures and temperatures as possible media of transport of ore-forming elements in hydrothermal processes: *Geokhimiya, no. 7,* 658-665 (in Russian). transl. in *Geochemistry, no. 7,* 681-690, 1963.

Bryatov, L. V. and I. P. Kuz'mina (1961) Crystallization of the sulfides of lead and zinc from aqueous solutions of chlorides: in *Growth of Crystals, 3*―Reports of the Second Conference on Crystal Growth (in English translation, 1962) New York: Consultants Bureau, p. 294-296.

Buckley, H. E. (1951) *Crystal Growth,* New York: Wiley.

Copeland, C. S., J. Silverman, and S. W. Benson (1953) The system NaCl-H_2O at supercritical temperatures and pressures: *J. Chem. Physics,* **21,** 12-16.

Cuney, M., M. Pagel, and J. Touret (1976) Analysis of gases in fluid inclusions by gas chromatography: *Bull. Soc. Fr. Minéral. Crist.,* **99** 169-177 (in French).

Czamanske, G. K., E. Roedder, and F. C. Burns (1963) Neutron activation analysis of fluid inclusions for copper, manganese and zinc: *Science,* **140,** 401-403.

―――――――
[11] English abstracts of many of the Russian references will be found in *Fluid Inclusion Research―Proceedings of COFFI*; see Roedder, 1968.

Deicha, G. (1955) *Les lacunes des cristaux et leurs inclusions fluides; signification dans la genèse des gîtes minéraux et des roches,* Paris: Masson et Cie.

Demin, Y. (1970) Structure of the aureoles of evaporation around the ore bodies of some polymetallic ore deposits of Rudni Altai (abst.), in *Collected Abstracts, IMA-IAGOD Meetings,* **1970,** Tokyo: Science Council of Japan, p. 256.

Dolgov, Yu. A. (1968) The composition of gases in processes of endogenic mineral formation: *Int. Geol. Congr. 23rd Session, 1968 (Prague) Reports of Soviet Geologists, Problem 7,* 101-111: Moscow, Izdatel, "Nauka" (in Russian). Transl. in *Fluid Inclusion Research—Proceedings of COFFI,* **4** (1971) 114-121.

Enjoji, M. (1972) Studies on fluid inclusions as the media of the ore formation: *Sci. Repts. Tokyo Kyoiku Daigaku, Sect. C,* **11,** no. 106 79-126 (in English).

Ermakov, N. P. (1950) *Research on the Nature of Mineral-Forming Solutions,* Kharkov, U.S.S.R.: University of Kharkov Press (in Russian); transl. (along with other material dated 1957-1958) in Yermakov, N. P. et al., 1965, *Research on the Nature of Mineral-Forming Solutions, with Special Reference to Data from Fluid Inclusions,* Vol. 22 of International Series of Monographs in Earth Sciences, New York: Pergamon Press.

_____ (1965) The state and activity of the fluids in granitic pegmatites of the chambered type: *Int. Geol. Congress, 22nd Session, 1964 (New Delhi), Reports of Soviet Geologists, Problem 6,* 140-160 (in Russian; English abst.).

_____ and V. A. Kalyuzhnyi (1957) The possibility of determination of real temperatures of mineral-forming solutions: *Trudy Vses. Nauch.-Issledovatel. Inst. P'ezoopticheskJournalMin. eral. Syr'ya, 1,* no. 2. 41-51 (in Russian). Transl. in *Int. Geol. Review,* **3,** 706-711, 1961, and in Ermakov, 1950.

_____ and A. G. Kuznetsov (1971) The use of thermo-barogeochemistry methods in the search for hidden ore deposits: *Fluid Inclusion Research—Proceedings of COFFI,* **4,** (1971) 122-125.

Fisher, J. R. (1976) The volumetric properties of H_2O—a graphical portrayal: *U.S. Geol. Survey J. Res.,* **4,** 189-193.

Haas, J. L., Jr. (1971) The effect of salinity on the maximum thermal gradient of a hydrothermal system at hydrostatic pressure: *Econ. Geol.* **66,** 940-946.

Hall, W. E. and I. Friedman (1963) Composition of fluid inclusions, Cave-in-Rock fluorite district, Illinois, and Upper Mississippi Valley zinc-lead district: *Econ. Geol.,* **58,** 886-911.

_____, _____, and J. T. Nash (1974) Fluid inclusion and light stable isotope study of the Climax molybdenum deposits, Colorado: *Econ. Geol.,* **69,** 884-901.

Ingerson, E. (1954) Nature of the ore-forming fluids at various stages—A suggested approach: *Econ. Geol.,* **49,** 727-733.

Kalyuzhnyi, V. A. (1955) Liquid inclusions in minerals as a geologic barometer: *Mineral. Sbornik L'vov. Geol. Obshchestva, no. 9,* 64-84 (in Russian); transl. in *Int. Geol. Rev.,* **2,** 181-195, 1960.

_____ (1956) New observations on phase transformations in liquid inclusions: *Mineral. Sbornik L'vov. Geol. Obshchestva no. 10,* 77-80 (in Russian).

_____ (1957) Results on pH measurements in solutions from liquid inclusions: *Geokhimiya, no. 1,* 77-79 (in Russian); transl. in *Geochemistry, no. 1,* 93-96.

_____ (1958) The study of the composition of captive minerals in polyphase inclusions: *Mineral. Sbornik L'vov. Geol. Obshchestva, no. 12,* 116-128 (in Russian); transl. in *Int. Geol. Rev.,* **4,** 127-138 (1962).

_____ (1960) *Methods of Study of Multiphase Inclusions in Minerals,* Kiev: Izdatel. Akad. Nauk Ukr. SSR (in Ukrainian).

REFERENCES

_____ (1971) The refilling of liquid inclusions in minerals and its genetic significance: *L'vov. Gos. Univ. Mineral. Sbornik*, **25**, 124-131 (in Russian).

_____ and D. K. Voznyak (1967) Thermodynamic and geochemical characteristics of mineralforming solutions of pegmatites of the "Zanorysh" type (from liquid inclusions in minerals): *L'vov Geol. Obshch. Mineralog. Sbornik*, **21**, no. 1, 49-61 (in Russian).

Keevil, N. B. (1942) Vapor pressures of aqueous solutions at high temperatures: *J. Am. Chem. Soc.*, **64**, 841-850.

Kennedy, G. C. (1950) "Pneumatolysis" and the liquid inclusion method of geologic thermometry: *Econ. Geol.*, **45**, 533-547.

Khetchikov, L. N. and L. A. Samoilovich (1970) The possibilities of the decrepitation method in mineral thermometry: *Akad. Nauk S.S.S.R., Izvest., Ser. Geol., 1970*, no. 7, p. 92-98 (in Russian); transl. in *Fluid Inclusion Research—Proceedings of COFFI*, **3** (1970), 94-100.

Kvenvolden, K. A., and E. Roedder (1971) Fluid inclusions in quartz crystals from South-West Africa: *Geochim. Cosmoch. Acta*, **35**, 1209-1229.

Landis, G. P. and R. O. Rye (1974) Geologic, fluid inclusion, and stable isotope studies of the Pasto Buena tungsten-base metal ore deposit, northern Peru: *Econ. Geol.*, **69**, 1025-1059.

Larson, L. T., J. D. Miller, J. E. Nadeau, and E. Roedder (1973) Two sources of error in low temperature inclusion homogenization determination, and corrections on published temperatures for the East Tennessee and laisvall deposits: *Econ. Geol.*, **68**

Larson, L. T., J. D. Miller, J. E. Nadeau, and E. Roedder (1973) Two sources of error in low temperature inclusion homogenization determination, and corrections on published temperature for the East Tennessee and Laisvall deposits: *Econ. Geol.*, **68**, 113-116.

Lemmlein, G. G. (1929) Sekundäre Flüssigkeitseinschlüsse in Mineralien: *Zeit. Krist.*, **71**, Part III, 237-256.

_____ and P. V. Klevtsov (1961) Relations among the principal thermodynamic parameters in a part of the system H_2O-NaCl: *Geokhimiya*, no. 2, 133-142 (in Russian); transl. in *Geochemistry*, no. 2, 148-158.

_____ and M. O. Kliya (1952) Distinctive features of the healing of a crack in a crystal under conditions of declining temperature: *Akad. Nauk., SSSR, Doklady, New Series, 87*, 957-960 (in Russian); transl. in *Int. Geol. Rev.*, **2**, 125-128, 1960.

Little, W. M. (1955) A study of inclusions in cassiterite and associated minerals, Ph.D. thesis, University of Toronto, (summarized in *Econ. Geol.* **55**, 485-509, 1960).

Lokerman, A. A. (1962) The possibility of study of the inter-relations of dikes and mineralization from inclusions in minerals: *Mineralog. Sbornik, L'vov Geol. Obshch.*, no. 16, 312-317 (in Russian).

Lyakhov, Yu. V. (1966) Mineral composition of multiphase inclusions in morions from Volynian pegmatites: p. 92-100 in *Research on Mineral-Forming Solutions (Materials of the First Symposium on Gas-Liquid Inclusions in Minerals, Moscow, May 17-24, 1963)*, N. P. Ermakov, ed., Moscow: Nedra Press (in Russian); also listed as *All-Union Research Inst. of Synthetic Mineral Raw Materials, Ministry of Geology U.S.S.R. Trans.*, 9.

Marshall, W. L. and E. V. Jones (1963) Aqueous systems at high temperature—X. Investigations on the systems UO_3-CuO-SO_3-D_2O, UO_3-NiO-SO_3-D_2O, and UO_3-CuO-NiO-SO_3-D_2O, 260-410°C; liquid-liquid immiscibility and critical phenomena: *J. Inorg. Nucl. Chem.*, **25**, 1021-1031.

McCulloch, D. S. (1959) Vacuole disappearance temperatures of laboratory-grown hopper halite crystals: *J. Geophys. Res.*, **64**, 849-854.

Metzger, F. W., W. C. Kelly, B. E. Nesbitt, and F. J. Essene (1977) Scanning electron microscopy of daughter minerals in fluid inclusions: *Ecol. Geol.*, **72**, 141-152.

Miers, H. A. (1903) An enquiry into the variation of angles observed in crystals; especially of potassium-alum and ammonium-alum: *Trans. Roy. Soc. London,* **202,** 459-523.

Nash, J. T. and C. G. Cunningham (1974) Fluid-inclusion studies of the porphyry copper deposit at Bagdad, Arizona: *J. Res. U.S. Geol. Survey,* **2,** 31-34.

———, and T. G. Theodore (1971) Ore fluids in the porphyry copper deposit at Copper Canyon, Nevada: *Econ. Geol.,* **66,** 385-399.

Newhouse, W. H. (1932) The composition of vein solutions as shown by liquid inclusions in minerals: *Econ. Geol.,* **27,** 419-436.

Nielsen, J. W. and F. G. Foster (1960) Unusual etch pits in quartz crystals: *Am. Mineral.,* **45,** 299-310.

Ohle, E. L. (1959) Some considerations in determining the origin of ore deposits of the Mississippi Valley type: *Econ. Geol.,* **54,** 769-789.

Ohmoto, H. (1971) Fluid inclusions and isotope study of the lead-zinc deposits at the Bluebell mine, British Columbia, Canada: *Soc. Mining Geol. Japan, Spec. Issue 2,* 93-99 [Proc. IMA-IAGOD Meetings '70, Joint Symp. Vol.].

Orville, P. M. (1963) Alkali ion exchange between vapor and feldspar phases: *Am. J. Sci.,* **261,** 201-237.

Petrichenko, O. I. and V. S. Shaydetskaya (1968) Physicochemical conditions of recrystallization of halite in rock salts: in *Mineralogical Thermometry and Barometry,* **1,** Moscow: Nauka Press (in Russian), p. 348-351.

——— and E. P. Slivko (1971) Physicochemical conditions of formation of halogene deposits of Ukraine according to the studies of inclusions in halite (abst.): in *Abstracts of Reports,* **2,** International Geochemical Congress, Moscow, 1971 (in English), p. 847-848.

Pinckney, D. M. and J. Haffty (1970) Content of zinc and copper in some fluid inclusions from the Cave-in-Rock district, southern Illinois: *Econ. Geol.,* **65,** 451-458.

Poty, B., H. A. Stalder, and A. M. Weisbrod (1974) Fluid inclusions studies in quartz from fissures of western and central Alps: *Schweiz. Min. Petr. Mitt.* **54,** 717-752.

Roedder, E. (1958) Technique for the extraction and partial chemical analysis of fluid-filled inclusions from minerals: *Econ. Geol.,* **53,** 235-269.

——— (1960a) Primary fluid inclusions in sphalerite crystals from the OH vein, Creede, Colorado (Abstract): *Bull. Geol. Soc. Am.,* **71,** *Part II,* 1958.

——— (1960b) Fluid inclusions as samples of the ore-forming fluids: *XXI Int. Geol. Congr., Proc. Sec. 16,* 218-229.

——— (1962a) Ancient fluids in crystals: *Sci. Am.,* **207,** 38-47.

——— (1962b) Studies of fluid inclusions I: Low temperature application of a dual-purpose freezing and heating stage: *Econ. Geol.* **57,** 1045-1061.

——— (1963) Studies of fluid inclusions II: Freezing data and their interpretation: *Econ. Geol.,* **58,** 167-211.

——— (1965a) Liquid CO_2 inclusions in olivine-bearing nodules and phenocrysts from basalts: *Am. Mineral.,* **50,** 1746-1782.

——— (1965b) Evidence from fluid inclusions as to the nature of ore-forming fluids: Symposium, *Problems of Postmagmatic Ore Deposition,* Prague, **2,** 375-384.

——— (1967a) Fluid inclusions as samples of ore fluids: in *Geochemistry of Hydrothermal Ore Deposits,* H. L. Barnes, ed., New York: Holt, Rinehart and Winston, p. 515-574.

REFERENCES

_____ (1967b) Metastable superheated ice in liquid-water inclusions under high negative pressure: *Science,* **155,** 1413-1417.

_____ (1967c) Environment of deposition of stratiform (Mississippi Valley-type) ore deposits, from studies of fluid inclusions: *Econ. Geol. Monograph,* **3,** 349-362.

_____, ed. (1968-onward) *Fluid Inclusion Research—Proceedings of COFFI* (an annual summary of world literature; volumes 1-5, 1968-1972 privately printed and available from the editor; volume 6, 1973, onward printed and available from the Univ. of Michigan Press.)

_____ (1970a) Application of an improved crushing microscope stage to studies of the gases in fluid inclusions: *Schweiz. Min. Petr. Mitt.,* **50,** Pt. 1, 41-58.

_____ (1970b) Laboratory studies on inclusions in the minerals of Ascension Island granitic blocks, and their petrologic significance: p. 247-258 in *Problems of Petrology and Genetic Mineralogy, V. S. Sobolev Memorial Volume II,* Yu. A. Kuznetsov, ed., Moscow, "Nauka" Press, 302 pp. (in Russian); transl. in *Fluid Inclusion Research—Proceedings of COFFI,* **5,** (1972) 129-138.

_____ (1971a) Metastability in fluid inclusions: *Soc. Mining Geol. Japan, Spec. Issue,* **3,** 327-334 [Proc. IMA-IAGOD Meetings '70, IAGOD Vol.].

_____ (1971b) Fluid inclusion studies on the porphyry-type ore deposits at Bingham, Utah; Butte, Montana; and Climax Colorado: *Econ. Geol.,* **66,** 98-120.

_____ (1972) Composition of fluid inclusions: *U.S. Geol. Surv., Prof. Paper 440 JJ.*

_____ (1976) Fluid-inclusion evidence on the genesis of ores in sedimentary and volcanic rocks: in *Handbook of Strata-bound and Stratiform Ore Deposits,* K. H. Wolf, ed., Amsterdam, Elsevier, **2,** 67-110.

_____ (1977) Changes in ore fluid with time, from fluid-inclusion studies at Creede, Colorado: in *Problems of Ore Deposition,* Fourth I.A.G.O.D. Sympos., Varna, 1974, **2,** 179-185.

_____ and D. S. Coombs (1967) Immiscibility in granitic melts, indicated by fluid inclusions in ejected granitic blocks from Ascension Island: *J. Petrology* **8,** Pt. 3, 417-451.

_____, A. V. Heyl, and J. P. Creel (1968) Environment of ore deposition at the Mex-Tex deposits, Hansonburg district, New Mexico, from studies of fluid inclusions: *Econ. Geol.,* **63,** 336-348.

_____, B. Ingram, and W. E. Hall (1963) Studies of fluid inclusions III: Extraction and quantitative analysis of inclusions in the milligram range: *Econ. Geol.,* **58,** 353-374.

_____ and O. C. Kopp (1975) A check on the validity of the pressure correction in inclusion geothermometry, using hydrothermally grown quartz: *Fortschr. Miner.,* **52,** Special Issue: I.M.A.-Papers 9th Meeting, Berlin-Regensburg, 1974, 431-446.

_____ and B. J. Skinner (1968) Experimental evidence that fluid inclusions do not leak: *Econ. Geol.,* **63,** 715-730.

_____ and P. W. Weiblen (1970) Lunar petrology of silicate melt inclusions, Apollo-11 rocks: *Geochim. Cosmochim. Acta Suppl. 1, Proc. Apollo-11 Lunar Sci. Conf.,* **1,** 801-837.

_____ and P. W. Weiblen (1972) Petrographic features and petrologic significance of melt inclusions in Apollo 14 and 15 rocks: *Geochim. Cosmochim. Acta Suppl. 3, Proc. Third Lunar Science Conf.,* **1,** 251-279.

Rosasco, G. J. and E. Roedder (1976) Application of a new laser-excited Raman spectrometer to nondestructive analysis of sulfate in individual phases in fluid inclusions in minerals (abst.): *25th Internat. Geol. Cong. Abstracts,* **3,** 812-813.

Rutherford, M. J. (1968) Geothermometry of liquid inclusions in quartz, Coronation mine, Flin Flon area, Saskatchewan: *Geol. Survey Canada Paper 68-5.*

Rye, R. O. and J. Haffty (1969) Chemical composition of the hydrothermal fluids responsible for the lead-zinc deposits at Providencia, Zacatecas, Mexico: *Econ. Geol.,* **64,** 629-643.

_____ and F. J. Sawkins (1974) Fluid inclusion and stable isotope studies on the Casapalca Ag-Pb-Zn-Cu deposit, central Andes, Peru: *Econ. Geol.,* **69,** 181-205.

Sawkins, F. J. (1968) The significance of Na/K and Cl/SO_4 ratios in fluid inclusions and subsurface waters, with respect to the genesis of Mississippi Valley-type ore deposits: *Econ. Geol.,* **63,** 935-942.

Scott, H. S. (1948) The decrepitation method applied to minerals with fluid inclusions: *Econ. Geol.,* **43,** 637-654.

Sheppard, S. M. F., R. L. Nielsen, H. P. Taylor, Jr. (1971) Hydrogen and oxygen isotope ratios in minerals from porphyry copper deposits: *Econ. Geol.,* **66,** 515-542.

Skinner, B. J. (1953) Some considerations regarding liquid inclusions as geologic thermometers: *Econ. Geol.,* **48,** 541-550.

Slivko, M. M. (1958) Inclusions of solutions in tourmaline crystals: *Trudy Vses. Nauch-Issled. Inst. P'ezooptichesk. Mineral. Syr'ya,* 2, no. 2, 63-68 (in Russian); transl. in Ermakov, 1950.

Smith, F. G. (1953) *Historical Development of Inclusion Thermometry,* Toronto: Univ. of Toronto Press.

_____ and W. M. Little (1959) Filling temperatures of H_2O-CO_2 fluid inclusions and their significance in geothermometry: *Can. Mineral.,* **6,** Part III, 380-388.

Sorby, H. C. (1858) On the microscopic structure of crystals, indicating the origin of minerals and rocks: *Geol. Soc. London Quart. J.,* **14,** Part I, 453-500.

Sourirajan, S. and G. C. Kennedy (1962) The system H_2O-NaCl at elevated temperatures and pressures: *Am. J. Sci.,* **260,** 115-141.

Sushchevskaya, T. M. (1971) Comparative characteristics of the chemical composition of tin-bearing hydrothermal solutions (from analyses of gas-liquid inclusions): in *Geochemistry of Hydrothermal Ore Formation,* V. L. Barsukov, ed., Moscow, Nauka Press, (in Russian), p. 35-60.

Takenouchi, S. and G. C. Kennedy (1964) The binary system H_2O-CO_2 at high temperatures and pressures: *Am. J. Sci.,* **262,** 1055-1074.

_____ and _____ (1965) The solubility of carbon dioxide in NaCl solutions at high temperatures and pressures: *Am. J. Sci.,* **263,** 445-454.

Tuttle, O. F. (1949) Structural petrology of planes of liquid inclusions: *J. Geol.,* **57,** 331-356.

Voznyak, D. K. (1968) Conditions of formation of topaz in syngenetic minerals in Volyn' pegmatites (abst.) in *Abstracts of reports of Third All-Union Conference on Mineralogical Thermobarometry and Geochemistry of Deep-Seated Mineral-Forming Solutions,* Moscow, Sept. 9-15, 1968: Moscow, Acad. of Sci. U.S.S.R., Ministry of Geology (in Russian), p. 48-49; Transl. in *Fluid Inclusion Research—Proceedings of COFFI,* **1,** (1968) 41.

Weisbrod, A. and B. Poty (1975) Thermodynamics and geochemistry of the hydrothermal evolution of the Mayres pegmatite (south-eastern Massif Central, France) (Part I): *Pétrologie,* **1,** 1-16 (in English).

White, D. E. (1958) Liquid of inclusions in sulfides from Tri-State (Missouri-Kansas-Oklahoma) is probably connate in origin (abst.): *Bull. Geol. Soc. Am.,* **69,** 1660.

_____ (1965) Saline waters of sedimentary rocks: *Fluids in Subsurface Environments—a Symposium,* 342-366. Mem. no. 4, Am. Assoc. Petroleum Geologists.

_____, J. D. Hem, and G. A. Waring (1963) Chemical composition of sub-surface waters; *U.S. Geol. Survey, Prof. Paper 440-F.*

REFERENCES

———, L. J. P. Muffler, and A. H. Truesdell (1971) Vapor-dominated hydrothermal systems compared with hot-water systems: *Econ. Geol.,* **66,** 75–97.

Wise, D. U. (1964) Microjointing in basement, Middle Rocky Mountains of Montana and Wyoming: *Bull. Geol. Soc. Am.,* **75,** 287–306.

Ypma, P. J. M. (1963) Rejuvenation of ore deposits as exemplified by the Belledonne metalliferous province: Thesis, University of Leiden, (in English).

Zirkel, F. (1873) *Die Mikroskopische Beschaffenheit der Mineralien und Gesteine,* Leipzig: Wilhelm Englemann.

——— (1876) Microscopical petrography: *Prof. Papers of the Engineer Dept. U.S. Army, no. 18*–Report of U.S. Geol. Exploration of the Fortieth Parallel, VI.

15

Ore Metals in Active Geothermal Systems

Byron G. Weissberg, Patrick R. L. Browne, and Terry M. Seward

New Zealand Department of Scientific and Industrial Research

The study of hot spring systems, their chemistry, precipitates, and alteration products provides insight into the processes and mechanisms involved in the formation of hydrothermal ore deposits. Hot springs at the Earth's surface have maximum temperatures of approximately 100°C and consequently seem far removed from the higher temperature processes associated with the formation of most hydrothermal ore deposits. However, drilling in hot spring areas, primarily for geothermal energy, has shown that at depths of only 1-2 km, temperatures can attain greenschist metamorphic conditions (e.g., above 350°C at Salton Sea and Mexicali) although temperatures between 150 to 300°C are more common. Furthermore, in some hot springs, surface precipitates with ore-grade concentrations of some heavy metals have been found: mercury, arsenic, and antimony are the most common although a few sinters also contain ore-grade concentrations of gold and silver. Drill cores and cuttings recovered from some geothermal areas contain small amounts of sphalerite, galena, and chalcopyrite, and other base metal minerals and at the Salton Sea and Cheleken, highly saline fluids discharged from the drillholes deposited native lead, copper, silver and mixtures of sulfur-deficient ore minerals and precipitates in the drillpipes. The temperatures and pressures in some drillholes have been measured during or soon after drilling, the compositions of fluids discharged have been analyzed, and the hydrothermal alteration assemblages in the rocks have been studied and compared with unaltered rocks in the surrounding areas (see Chapter 13). Although only a few hot spring systems have deposited mineable ores (mainly mercury or sulfur), there is increasing evidence linking the transport and deposition of ore metals and minerals in some hot spring systems with economic hydrothermal ore deposits. As exploration for geothermal resources continues, more evidence will undoubtedly become available relating the chemistry of geothermal systems to environments of hydrothermal ore formation.

Hot springs occur mainly in the Earth's mobile belts in areas of recent volcanism, high heat flow and tectonic activity (Figure 15.1). They are concentrated in the Circumpacific belt and in the zone from Spain and Italy through Yugoslavia, Turkey, the Southern Himalayas to the Indonesia arc and join with the Circumpacific belt in New Guinea. Several recent studies correlate hot spring distribution with plate tectonics and some (Tamrazyan, 1970; Tarling, 1973; Wright and McCurry, 1973; and Sillitoe, 1973) relate hydrothermal ore formation with the current plate tectonic theories.

We do not attempt a thorough review of the relationships between hot springs and ore deposits in this Chapter but will emphasise new or additional information not covered by the excellent reviews on this subject by White (1955, 1965, 1967, 1968, 1974), Dickson and Tunell (1968), Ellis (1969a), Tooms (1970), Dunham (1970), Skinner and Barton (1973), and Barnes and Hem (1973). After a brief summary of several recently described or very well known metal-rich geothermal systems, illustrating the diversity of the chemical types involved, and a discussion of the data pertaining to the transport of mercury in surface hot springs, we focus on Broadlands, New Zealand, as a case study of a metal-rich geothermal system because of the abundance of data now available on the system (one that is depth zoned—depositing a base metal suite of minerals in the altered rocks at depth and a precious metal suite of elements in surface precipitates) and because of our personal familiarity with it.

METAL-RICH GEOTHERMAL SYSTEMS

Bolivia

Ahlfeld (1974) briefly described 10 examples of late-stage antimony mineralization associated with recent or active hot springs in Bolivia.

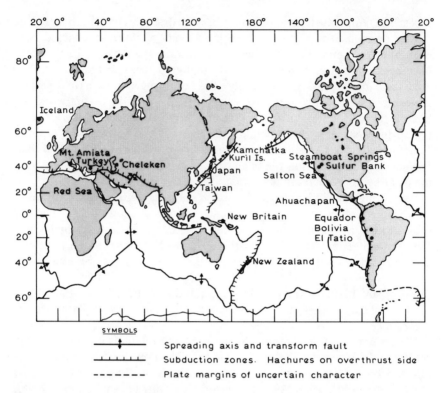

Fig. 15.1 Relationship of some metal-rich geothermal systems to plate tectonic boundaries. (Modified after Bailey et al., 1973.)

Stibnite occurs both in silica sinter and in travertine, associated, in places, with pyrite, barite, and traces of cinnabar. Unfortunately, the compositions of the Bolivian hot springs are not available.

El Tatio, Chile

At El Tatio, hot spring precipitates are rich in arsenic, antimony, barium, and strontium with one containing 3 ppm silver (Table 15-1), and the waters themselves contain up to 46 ppm arsenic. The springs are near neutral, boiling, with a chloride concentration of about 8000 ppm (A. J. Ellis, written communication 1969b) and discharge at 0.5 1/sec. The water composition of a nearby drillhole is given in Table 13-2.

Table 15-1. Metal concentrations in precipitates from geothermal wells and springs (ppm)

	As	Sb	Au	Ag	Hg	Tl	Cu	Pb	Zn	Mn	Fe		References
Broadlands, New Zealand													
Ohaki Pool	400	10%	85	500	2000	630		25	70				Weissberg (1969)
Hole 2, inside silencer	50	1000	50	2000	600?	150	2½%	400	50		1000	Ga 700, Be 400, Sn 40, V 25	Chemistry Division, D.S.I.R.
Hole 2, outside silencer	250	8%	55	200	200	1000							Weissberg (1969)
Hole 7	500	500	n.d.	100	250	250	500	50	500	200		Ga 150, Be 100, Sn 5, V 10	Chemistry Division, D.S.I.R.
El Tatio, Chile													
Spring 227	12%	1.5%	3	1	50	10		100	100	250			Chemistry Division, D.S.I.R.
Matsao, Taiwan													
Hole E-205	500	500	n.d.	25	n.d.	2	500	major	250	3000			Ellis (1972)
Red Sea, Atlantis II Deep													
126P-3M	80		0.51	33		6	2000	380	1.25%	1000	12.5%	Ni 500, Co 150, Mo 200 Ge 100, V 100, Sn 100 Cd 63, Ni 40, Co 65, Mo 125	Hendricks et al., in Degens & Ross (1969)
Average Assay			0.5	54			1.3%	0.1%	3.4%	880	29%	Ge 0.6, V 80, Ga 10	Bischoff and Manheim in Degens & Ross (1969)
Rotokawa, New Zealand													Weissberg (1969)
Hole 2	0.4%	30%	70	30	15	0.5%		50	100				
Salton Sea, California													
W-768 (130 ± 20°C)	0.1%	0.25%	n.d.	2.8%	n.d.	n.d.	major	70	n.d.	3400	6%	Ga 120, Be 370, Bi 90, Co 6	Skinner et al. (1967)
Steamboat Springs, Nevada													
Siliceous mud		4%	10	400	45		2000	400	200				White (1967)
Meta stibnite	600	2000	60	400		2000							Chemistry Division, D.S.I.R.
Uzon, eastern thermal area	11.6%			30	6700		100		500		0.72%	Mo 30, Ge 40, Ba 7000, Sr 100	Naboko and Glavatskikh (1970)
Caldera, central thermal area Kamchatka	30.15%				4700		400				5.3%	Mo 200, Ge 100, Ba 2000, Sr 300	Naboko and Glavatskikh (1970)
Waiotapu, New Zealand													
Champagne Pool	2%	2%	80	175	170	320		15	50				Weissberg (1969)

Ahuachapan, El Salvador

In the Ahuachapan thermal area, high concentrations of mercury in soils around some fumaroles (up to 30 ppm) have been noted by Koenig (1969). Incrustations of black manganese oxides (?), up to 0.25 mm thick, occur in channels draining springs at Los Toles and were also observed at Los Salitres by Sigvaldason and Cuellar (1970). Concentrations of cadmium, copper, niobium, cobalt, bismuth, and chromium were less than 2 µg/kg but molybdenum, zinc, iron, and vanadium were 6 (20), 150 (61), 27 (100) and 11 (20) µg/kg in waters from the Playon on Salitre (and one drillhole) respectively.

Japan

ARIMA SPRINGS

In Japan, several hot-spring systems transport and deposit significant quantities of metals. The Arima Springs are characterized by high sodium, calcium chloride waters (Table 15-2), moderate surface temperatures (40°C), and carry high concentrations of iron, manganese, with lesser amounts of copper, lead, and zinc (Table 15-3) and low concentrations of sulfide and sulfate. Drillholes encountered altered rhyolites containing sphalerite, galena, pyrite, calcite, and siderite: temperatures reach 133°C at 168 m. Metal-rich surface precipitates were not reported. White (1967) commented on the similarity of these waters, although dilute and of low metal content, to those of the Salton Sea.

OSOREYAMA SPRINGS

Noguchi and Nakagawa (1970) describe the high concentrations of arsenic in waters (up to 40 ppm arsenic) and surface precipitates at the Osoreyama Hot Springs, Aomori Prefecture (Table 15-2). Hottest surface waters have the highest chloride concentrations and both arsenic and HBO_2 show strong positive correlations with chloride. The acidity of the low pH waters is directly related to sulfate concentrations. They propose that the original thermal water is near neutral with high concentrations of chloride, boric acid, and arsenic, but is poor in sulfate. Arsenic in 12 hot spring precipitates ranges from 17 to 60%, excluding free sulfur. Orpiment and free sulfur occur in yellow precipitates, but realgar was not present in red-orange colored precipitates as previously reported. Up to 2.1% Pb is present in the precipitates.

Table 15-2 Compositions of waters from geothermal wells and springs (ppm)

	T(°C)	pH	Li
Apapel' Springs, Kamchatka	95	8.1	
Broadlands, New Zealand			
drillhole 2 Surface	100	8.3	11.7
Deep aquifer	261	6.2	7.93
Cheleken, USSR			
(i)	105		1.70
(ii)	70	6.0	7.90
(iii)			7.6
Dvukhyurtochnye Springs, Kamchatka	74.5	7.1	
Mendeleyev Volcano Kurile Islands			
(i) (spring)	43	2.7	n.d.
(ii) (spring)	82	1.8	0.74
(iii) (well)	150	8.5	
Osoreyama (a)	100	7.3	
Springs, Japan (b)	71.5	2.2	
Red Sea			
Atlantis II Deep	56	5.3?	
Rotokawa, New Zealand Drillhole 2	285	7.8	11
Tamagawa Springs, Japan	98	1.2	
Uzon (i)	85	5.7	
Caldera Springs, Kamchatka (ii)	98	6.9	
Waiotapu, New Zealand			
Champagne Pool	75	5.7	9.0

	Na	K	Rb
Apapel' Springs, Kamchatka	426.3	61.3	
Broadlands, New Zealand			
drillhole 2 Surface	1050	210.0	2.20
Deep aquifer	711.3	142.3	1.49
Cheleken, USSR			
(i)	11,060	56.0	0.35
(ii)	76,140	490	1.00
(iii)	70,910	510	0.4
Dvukhyurtochnye Springs, Kamchatka	638.7	17.10	

Table 15-2 (*Continued*)

		Na	K	Rb
Mendeleyev Volcano Kurile Islands				
(i) (spring)		8.0	2.5	
(ii) (spring)		675.0	73.5	0.25
(iii) (well)		2938	34	
Osoreyama (a)		3750	470	
Springs, Japan (b)		930	78	
Red Sea				
Atlantis II Deep		87,800	3040	
Rotokawa, New Zealand Drillhole 2		1785	186	
Tamagawa Springs, Japan		114	65	
Uzon	(i)	527.0	60.5	
Caldera Springs, Kamchatka	(ii)	764.0	88.8	
Waiotapu, New Zealand Champagne Pool		1220	160	

		Cs	Mg	Ca
Apapel' Springs, Kamchatka			7.0	16.0
Broadlands, New Zealand				
drillhole 2 Surface		1.70	0.10	2.20
Deep aquifer		1.15	0.07	1.49
Cheleken, USSR				
(i)		0.15	415.0	715.0
(ii)		n.d.	3,080	19,708
(iii)		0.12	3,040	20,200
Dvukhyurtochnye Springs, Kamchatka			0.83	123.0
Mendeleyev Volcano Kurile Islands				
(i) (spring)			n.d.	n.d.
(ii) (spring)		0.07	68.2	126.0
(iii) (well)			94.5	307
Osoreyama (a)			8.3	553
Springs, Japan (b)			30.0	140
Red Sea				
Atlantis II Deep			794	5390
Rotokawa, New Zealand Drillhole 2				50
Tamagawa Springs, Japan			83	210
Uzon				
Caldera Springs,	(i)		3.8	25.5

Table 15-2 (*Continued*)

		Cs	Mg	Ca
Caldera Springs, Kamchatka	(ii)		4.7	45.5
Waiotapu, New Zealand				
Champagne Pool				35

		F	Cl	Br
Apapel' Springs, Kamchatka			126.5	
Broadlands, New Zealand				
drillhole 2 Surface		7.30	1743	5.7
Deep aquifer		4.94	1180	3.9
Cheleken, USSR				
(i)			18,900.0	54.6
(ii)			157,000	526
(iii)			158,000	672
Dvukhyurtochnye Springs,				
Kamchatka		1.00	589.0	
Mendeleyev Volcano				
Kurile Islands				
(i) (spring)		n.d.	17.4	0.24
(ii) (spring)		2.2	1635	6.0
(iii) (well)			5213	
Osoreyama (a)			7025	
Springs, Japan (b)			1596	
Red Sea				
Atlantis II Deep			163,200	69
Rotokawa, New Zealand		6.6	2915	
Drillhole 2				
Tamagawa Springs, Japan		120	3240	
Uzon	(i)		904	
Caldera Springs, Kamchatka	(ii)	0.6	1216	
Waiotapu, New Zealand				
Champagne Pool		5.5	2000	7.2

	I	SO_4	As
Apapel' Springs, Kamchatka	0.03	587	2.5–3.0
Broadlands, New Zealand			
drillhole 2 Surface	0.8	8.0	8.1
Deep aquifer	0.5	5.4	5.5
Cheleken, USSR			
(i)	21.2	343	0.1
(ii)	31.7	309	0.03
(iii)	25	150	0.5

Table 15-2 (*Continued*)

	I	SO_4	As
Dvukhyurtochnye Springs, Kamchatka		812	2.8
Mendeleyev Volcano Kurile Islands			
(i) (spring)		687	
(ii) (spring)	0.7	1501	2.2
(iii) (well)		214	
Osoreyama (a)		63	
Springs, Japan (b)		885	
Red Sea			
Atlantis II Deep		954	
Rotokawa, New Zealand Drillhole 2	0.7	125	
Tamagawa Springs, Japan		1330	
Uzon (i)		229	
Caldera Springs, Kamchatka (ii)		65.8	25
Waiotapu, New Zealand			
Champagne Pool	0.4	145	4.9

	SiO_2	B	NH_3
Apapel' Springs, Kamchatka	160		
Broadlands, New Zealand			
drillhole 2 Surface	805	196	2.1
Deep aquifer	545	133	1.4
Cheleken, USSR			
(i)		108	
(ii)		84	
(iii)		109	
Dvukhyurtochnye Springs, Kamchatka	116	65	
Mendeleyev Volcano Kurile Islands			
(i) (spring)	172	1.2	1.8
(ii) (spring)	311	55	4.3
(iii) (well)	192	43	2.0
Osoreyama (a)	101	343	
Springs, Japan (b)	134	96	
Red Sea			
Atlantis II Deep		11	
Rotokawa, New Zealand, Drillhole 2	400	110	3.2
Tamagawa Springs, Japan	231		

Table 15-2 (*Continued*)

		SiO_2	B	NH_3
Uzon	(i)	350		20.0
Caldera Springs, Kamchatka	(ii)	126		20.0
Waiotapu, New Zealand				
Champagne Pool		490	29	11.5

		CO_2	H_2S	References
Apapel' Springs, Kamchatka		183		Ozerova & Lebedev, (1970)
Broadlands, New Zealand				
drillhole 2 Surface		178		Browne and Ellis, (1970)
Deep aquifer		121	107	Calc. from Browne &
Cheleken, USSR				Ellis, (1970)
(i)		522		Lebedev (1972)
(ii)		103		Lebedev (1972)
(iii)		142	7.2	Lebedev (1972)
Dvukhyurtochnye Springs,				
Kamchatka		60		Ozerova et al., (1970)
Mendeleyev Volcano				
Kurile Islands				
(i) (spring)				Brezgunov et al., (1968)
(ii) (spring)				Brezgunov et al., (1968)
(iii) (well)		24.4		Brezgunov et al., (1968)
Osoreyama (a)			14	Noguchi & Nakagawa (1970)
Springs, Japan (b)			151	Noguchi & Nakagawa (1970)
Red Sea				
Atlantis II Deep				White (1968)
Rotokawa, New Zealand		2000[a]	885[a]	Chemistry Division,
Drillhole 2				DSIR, N.Z.
Tamagawa Springs, Japan			20	Ozawa et al., (1973)
Uzon	(i)	14.6		Naboko & Glavatskikh (1970)
Caldera Springs, Kamchatka	(ii)	56.1	13.9	Naboko & Glavatskikh (1970)
Waiotapu, New Zealand				
Champagne Pool		170	6	Chemistry Division, DSIR, N.Z.

[a] Total discharge.

Table 15-3 Concentrations of minor elements in waters from geothermal wells and springs (ppm)

	Mn	Fe	Ni	Cu	Pb	Zn	Cd	Ag
Apapel' Springs, Kamchatka		0.7	0.015	0.002	0.025	0.005		
Arima Springs, Japan	61	187		0.1	0.4	0.2		
Broadlands, New Zealand								
Drillhole 2	0.0133	0.36	0.0002	0.0009	0.0013	0.001	0.00001	0.0007
Drillhole 25		0.36	0.00005	0.0013	0.0055	0.0006	0.00002	0.00025
Cerro Prieto								
Mexicali, Mexico	0.64	0.2	0.002	0.005	0.0046	0.006		0.004
Cheleken, USSR								
Wells (ii)	46.5	14.0	0.33	1.41	9.20	3.06	1.66	
(iii)		4.2		0.90	3.60	0.19	0.00	
Geysir, Iceland		0.0125	0.001			0.002		
Matsao, Taiwan								
Drillhole E205	42	220		0.05	0.6-0.9	13		
Osoreyama, Japan								
Spring (a)	1.2	0.02			0.01			
(b)	5.4	21.1			0.01			
Red Sea, Atlantis II Deep								
56° Brine	82	81		0.26	0.63	0.54		
Interstitial Brine 12op-3m	134	83		1.5	0.3	7.1		0.04
Salton Sea, California								
Drillhole No. 1, IID	1400	2290		8	102	540	2.0	1.4
Tamagawa, Japan	4.2	105		0.01	1.0	2.8		
Uzon, Kamchatka								
Spring in Central Thermal Area				0.1		0.1		0.015
Wairakei, New Zealand (average of several drillholes)	0.0007	0.012	0.001	0.0019	0.0045	0.0022	0.00055	

	Au	As	Sb		References
Apapel' Springs, Kamchatka		3	0.15–0.45		Ozerova and Lebedev (1970)
Arima Springs, Japan					Nakamura and Maeda (1961)
Broadlands, New Zealand					
Drillhole 2	0.00004	5.7	0.2	Tl 0.007, W. 0.087	Richie (1973) and
Drillhole 25				V 0.0023, Ge 0.004	Weissberg (1969)
				Sn 0.0021, Be 0.00023	
				Bi (Br 7) 0.0003	
Cerro Prieto					
Mexicali, Mexico	0.004	2	0.4		Mercado (1967)
Cheleken, USSR					Lebedev (1972)
Wells (ii)					
(iii)					
Geysir, Iceland				V 0.0151, Mo 0.0475	Arnórsson (1970)
				Ge 0.0236, Co 0.001	
Matsao, Taiwan					
Drillhole E205		3.6		Al 2.3	Ellis (1972)
Osoreyama, Japan					Noguchi and Nakagawa
Spring (a)		39.5		Al 0.7	(1970)
(b)		0.04		Al 24.2	
Red Sea, Atlantis II Deep					Brewer and Spencer, in
56° Brine				Co 0.16	Degens and Ross (1969)
Interstitial Brine 126p-3m				Mo 0.03, Al 15	
Salton Sea, California					White (1968)
Drillhole No. 1, IID	0.0	12		Tl 1.5, Sn 0.5	
				Hg 0.006, Al 4.2	
				Sr 400, Ba 235	
Tamagawa, Japan			0.4	Al 158, Hg 0.01	Ozawa et al. (1973)
Uzon, Kamchatka					Naboko and Glavatskikh
Spring in Central Thermal Area				Hg 0.012, Mo 0.014	(1970)
				Ge 0.008	
Wairakei, New Zealand (average		4.7	0.1	Be 0.00005	Ritchie (1973) and
of several drillholes)				Al 0.00035	Goguel (1975)

TAMAGAWA SPRINGS

The Tamagawa hot springs at Akita are unusual for their very high flow rates (approximately 155 l/sec. and constant for over 300 years) of near boiling acid sulfate-chloride waters (pH = 1.2) which carry high concentrations of iron (105 ppm), barium (0.2-1.0 ppm) and lead (0.2-1.6 ppm). These deposit lead-rich barite (hokutolite) with up to 21% PbO (Takano, 1969a,b, and Takano and Watanuki, 1972, 1974a,b). Arsenic in the waters ranges up to 3.4 ppm and orpiment has formed in some of the spring deposits. Nakagawa (1971) studied the solubility of orpiment in H_2S acid NaCl solutions from 15 to 80°C and concluded that it dissolves in these waters by hydrolysis according to the reaction:

$$As_2S_3 + 4H_2O \rightleftharpoons 2HAsO_2 + 3H_2S$$

deposition of As_2S_3 was due to reaction between arsenite and hydrogen sulfide. This is consistent with deposition at Tamagawa, where near boiling waters carrying 2-3.5 ppm of arsenic are low in H_2S (1-2 ppm). Slightly cooler waters (67-78°C) containing 92-240 ppm H_2S have low concentrations of arsenic (0.002-0.1 ppm).

Ozawa et al. (1973) favor a magmatic origin in which fumarolic gases containing HCl and SO_2 condense underground and mix with groundwater to explain the high sulfate and chloride, the high temperatures, and the high constant flow rates.

New Britain, Papua, New Guinea

In New Britain, Papua, and New Guinea, pyrite and marcasite deposit from hot acid to near neutral saline springs and thermal pools (Ferguson and Lambert, 1972; Ferguson, et al. 1974). At Matupi Harbor, spring waters carry up to 100 ppm iron and manganese, 2.5 ppm zinc, and slightly less than 0.1 ppm copper and lead, and form sediments enriched in iron, manganese, and zinc.

Red Sea

The Red Sea brines are important in the study of metal depositing geothermal systems; however, discussions by Degens and Ross (1969), Ross (1972), White (1968 and 1974), and Tooms (1970) make detailed review here unnecessary.

Concentrations of manganese, iron, zinc, copper, lead, and barium in the brines are much higher than in seawater, but nowhere as high as in the Salton Sea brines. Sediments forming in the deeps contain up to 20% zinc, 3% copper, 0.2% lead, 5.6 ppm gold, and 250 ppm silver (Table 15-1) providing a convincing example of present day marine syngenetic ore formation. Brine temperatures are relatively low (40-60°C) with active inflow temperatures up to 104°C, however, the brines appear to be gravitationally stable because of their high density.

The origins of the constituents in the brine appear diverse (White, 1968, 1974). Water is considered, from isotopic evidence, to be from the Red Sea and the high salinity is probably due to the dissolution of underlying evaporite beds. Metals may be derived from interaction of the hot brines with basalts or sediments at depth, with the sulfidic sulfur resulting from the organic reduction of sulfate. The elevated temperatures may be due to magmatic heating at depth or to the anomalously high geothermal gradient (10-20 times world average) and high heat flow (2-4 times normal) related to the unusually shallow crust.

Matsao, Taiwan

Geothermal exploration in the Tatun Volcanic Zone, Taiwan (Chen, 1970), includes several drillholes, one of which has formed lead-rich precipitates in the discharge pipe. A black precipitate was scraped from inside the 10-cm-diameter discharge pipe of hole E205 after a test period during which hot, acidic, highly saline water (Table 13-2) was discharged (Ellis, 1972). Crystalline components in the precipitate are galena and elemental lead; emission spectrographic analysis of the precipitate indicates additional high concentrations of arsenic, antimony, germanium, silver, copper, and barium (Table 15-1). Deposition of lead may result from the reaction $Fe + Pb^{2+} \rightleftharpoons Pb + Fe^{2+}$. The discharge water contains about 3% total dissolved solids, intermediate between Broadlands (0.3%) and the Salton Sea (35%), and lead in the water ranges 0.6-0.9 ppm with high concentrations of zinc and manganese (Table 15-3). Downhole temperatures are high, up to 293°C in hole 208, and the sodium, calcium, potassium chloride waters are similar to the Salton Sea except for lower salinity and the lower pH(2.4 at 25°C). This high acidity may derive from the hydrolysis of sulfur, which occurs in some of the rocks, through the reaction $4S + 4H_2O \rightleftharpoons 3H_2S + H_2SO_4$. Near-surface hydrothermal acid alteration consists of alunite, opal, cristobalite, kaolinite, and halloysite. A 1500-m drillhole in andesite shows chlorite, pyrite, sericite, and abundant epidote; anhydrite is a characteristic hydrothermal mineral. Calcite commonly occurs at depth suggesting that deep thermal fluids are not acid. Veinlets from a nearby drillhole (E208, 1385 m depth) consist of quartz, calcite, rare sphene, pyrite, pyrrhotite and chalcopyrite; the host andesite contains abundant adularia, quartz, chlorite and epidote, with unaltered plagioclase persisting.

Salton Sea, California

The Salton Sea geothermal area in Imperial Valley is one of the most spectacular metal-rich geothermal systems and has been described by White et al. (1963); Muffler and White (1968); Doe et al., (1966); Skinner et al. (1967); Helgeson (1967, 1968); Craig (1966, 1969); and White (1968, 1974). Only a brief review is given here.

Highly saline sodium, calcium, potassium chloride brines containing up to 35% total dissolved solids (Table 13-2) were tapped by holes drilled to 2470 m. Maximum downhole temperatures are above 360°C and the brines contain high concentrations of dissolved heavy metals (Table 15-3), particularly iron, manganese, zinc, lead, copper, and silver. Fluids discharged from the drillholes at roughly 130–170°C during a 3-month period in 1962, deposited, at the surface, several tons of an unusual and complex suite of metal sulfide minerals. The deposits consisted mostly of opaline silica (Table 15-1) that included fine-grained metal sulfide phases with a bulk composition of up to 20% copper, 7% silver, and 7% iron. However, zinc, lead, and manganese concentrations were very low despite their unusually high values in the discharge waters. Arsenic and antimony were low both in the precipitates and the discharge waters. The sulfide minerals identified in the opaline scale were digenite, bornite, chalcocite, stromeyerite, arsenopyrite, tetrahedrite, chalcopyrite, and pyrite, and also some native silver—all minerals with high metal/sulfur ratios. The fluids are sulfur deficient with respect to the dissolved metals, having an atomic ratio of total metals in solution (excluding iron) to sulfide sulfur of about 10 or 15:1; hence, only a small fraction of the metals present could deposit as sulfide minerals from the indicated bulk composition of the brines. Because the deep fluids are apparently in equilibrium with pyrite, which is abundant and ubiquitious, the low sulfide concentration of 15–30 ppm in the brine respresents equilibrium saturation values; addition of sulfide sulfur into the system would not increase the sulfide concentration in the brine but would result in precipitation of metal sulfide minerals.

Pliocene deltaic shales and siltstones at depth are now hydrothermally altered to low-grade greenschist facies rocks with secondary chlorite, potassium feldspar, potassium mica, albite, quartz, epidote, and pyrite. Pyrite, hematite, sphalerite, galena, chalcopyrite, and minor pyrrhotite, occur in cross-cutting veinlets indicating with other textures, a post-sedimentary origin and their likely deposition from the brines.

White's (1968, 1974) model for the constituents of the brines is of predominantly meteoric water dissolving evaporites to become rich in sodium chloride and then leaching local deltaic shales and siltstones to provide heavy metals, calcium, potassium, and most of the strontium and lead.

Magma probably provides the thermal energy of the system, and is possibly also the source of the sulfur and some heavy metals. CO_2 may be derived from metamorphic reactions in the sediments.

U.S.S.R.

CHELEKEN

The metal-bearing thermal brines on the Cheleken Peninsula of the eastern Caspian Sea (Turkmenia) deposit considerable quantities of native lead, sphalerite, and pyrite in drillholes and at the surface (Lebedev, 1972). The thermal waters are confined to some 20 arenaceous aquifers in the upper 1000 m of a large anticlinal fold in a late Tertiary red-bed sequence up to 2800 m thick. (Lebedev, 1968). The anticlinal crest (Chokrak Ridge) is considerably disturbed by two main fault systems, one associated with the Trans-Caspian depression. Massive pyrite and lesser amounts of galena and sphalerite occur in fracture zones along with calcite, aragonite and barite (Lebedev, 1972).

There is some natural spring discharge but most data on the hot waters are from holes drilled to obtain bromine and iodine from the waters. Lebedev (1972) delineates three main water types (Table 15-2).

1. Sodium chloride-bicarbonate-sulfate waters (up to 19,000 ppm chloride occur in the lower red beds at temperatures of 105°C and have deposited much calcite and tens of tons of barite at depth and in pipes and surface installations.
2. Sodium-calcium chloride waters in the upper red bed strata are characterized by high total dissolved salts (160,000 ppm Cl), (Table 15-2), remarkably high bromine concentrations (up to 650 ppm) and temperatures up to 80°C. Native lead with rare inclusions of native silver is being deposited in a number of wells from these waters, which have an average lead concentration of 10 mg/1 (Lebedev, 1967a). An estimated 1 ton of elemental lead precipitated in a 1000 m section of drillhole casing during a 2-year period and another 2.5 tons of lead were collected at the surface when the drillhole was flushed. Lebedev and Nikitina (1971) suggested that the lead is transported as anionic complexes such as $PbCl_3^-$, $PbCl_4^{2-}$, and perhaps as the species, $Pb(CO_3)_2^{2-}$ and $Pb(CO_3)_2Cl^{3-}$ (see Chapter 8). Presumably, the precipitation of elemental lead is due to simple electrochemical reduction by reaction with the drillhole casings.
3. Waters discharged from the near surface aquifer at the top (311 m depth in well U-1) of the upper red beds differ only from the deeper

sodium-calcium chloride waters in that they are richer in H_2S, and discharge naturally from many of the associated Chokraka springs. The waters from drillhole U-1 (Table 15-2), T $\approx$ 58°C, are depositing iron sulfide gel on wellhead installations (Lebedev et al., 1971) which initially crystallizes to greigite and mackinawite and transforms to pyrite after a number of years.

Botryoidal and reniform crusts of sphalerite, containing disseminated pyrite and galena, form in a reservoir tank where H_2S containing waters mix with metalliferous brines from the deep aquifers of the upper red beds (Lebedev, 1967b). A bulk analysis of the sphalerite-rich crust indicated iron (8.82%), lead (1.48%), cadmium (2.86%), thallium (0.40%), arsenic (1.07%), plus approximately 15% of alkalis, sulfate, carbonate, and silica (6.36%) (Lebedev, 1972).

Lebedev (1972) calculated that in 1 year the Cheleken thermal brines brought to the surface about 350 tons of lead, 50 tons of zinc, 34 tons of copper, 24 tons of cadmium, and 8 tons of arsenic.

It would be of interest to examine drill core (if available?) from aquifer horizons in the red bed formation for evidence of base metal sulfide deposition in rocks from the hot waters at depth.

KAMCHATKA REGION

Apapel' Springs. The hot springs of Apapel' occur in the median Kamchatka Range in a tributary valley of the Anavgay River. There are 14 springs in a 1-km^2 marshy area; these are related to the main Kamchatka fault system in a region of Tertiary and Quaternary volcanism. Many volcanic rocks are highly silicified and altered to clays near the fault (and springs) as is frequently observed on the margins of large grabens on Kamchatka (Vakin et al., 1970).

Shcheglov (1962) found cinnabar in one hot spring and further studies by Ozerova and Lebedev (1970) and Ozerova et al. (1971) report the presence of metacinnabar, realgar, and orpiment. Fine-grained cinnabar coats the opalite-clay rock and hydrothermal quartzite on the walls of several springs and deposits of silica gel and pyrolusite are observed where hot waters drain from some springs.

In a number of hot spring "cauldrons," rhythmic clay-sulfide skin up to 3-mm thick covers the rocks. The initial coating may be metacinnabar succeeded by cinnabar, realgar and clay; this "sandwich" may be repeated several times. Occasionally, blebs of orpiment occur in the realgar layers.

Clay-carbonate-sulfide crusts also occur and consist of thin (0.1 mm) layers of metacinnabar and cinnabar covered by amorphous silica, calcite (up to 1 cm) and, in places, pyrolusite. Small pyrite cyrstals and minor fine-grained cinnabar occur in the calcite layers.

The mercury sulfides crystallize from dilute, weakly alkaline (pH = 8.1) waters (Table 15-2) near 95°C, and discharging about 4 l/sec.

Dvukhyurtochnye Springs. These springs also occur in the mid-Kamchatka Range associated with andesitic basalts, pyroclastics and tuffaceous sediments. The dilute, near neutral springs (Table 15-2) have an estimated total discharge of 3 1/sec and a maximum temperature of approximately 74°C (Ozerova et al., 1970).

Concentrically zoned pyrite oolites (up to 3 mm diam.) occur in the bottom of a cauldron where hot water bubbles up. These oolites have a remarkably high content of mercury (2%) and arsenic (up to 2.7%) (Ozerova et al., 1971), although the mercury concentration in the hot water is low (7×10^{-4} ppm) (Ozerova et al., 1970). The mercury is apparently coprecipitated with iron sulfide gel, which subsequently recrystallizes to pyrite enclosing the mercury within its structure.

Mendeleyev Volcano. Mendeleyev is an active volcano (last eruption in 1880) on Kunashir Island at the southern extremity of the Kurile Islands chain and is composed of older andesites and andesitic basalts and younger dacitic lavas. Mercury mineralization is associated with considerable solfataric activity (Ozerova et al., 1972; Lebedev, 1970) on the northeastern flank of the volcano.

Cinnabar (containing up to 0.5% selenium) occurs as pockets up to 20 × 30 cm, as druses and as crystals up to 0.2 mm, in small vugs in a pyritic body, and may partially or entirely be replaced by metacinnabar. This pyritic ore exhibits colloform aggregates of opal-sphalerite-pyrite-marcasite and opal-pyrite-rutile (Ozerova and Dobrovol'skaya, 1969) with local replacement of sphalerite by smithsonite. Cinnabar, with contemporaneous stibnite, coats both marcasite and galena in cavities containing reniform, water-clear nodules of cristobalite (Lebedev, 1970) and covellite, chalcocite, rutile, anatase, rare native gold, native copper (Ozerova et al., 1972), and minor orpiment and realgar (Lebedev, 1970) have been reported.

Springs discharging higher up the slope on the northeastern flank of Mendeleyev volcano, are of the acid sulfate, low chloride type (Table 15-2, i) with low flow rates and temperatures from 20 to 40°C. At lower elevations (40–100 m) hot ($T \approx 90°C$), acid, sulfate-chloride waters (Table 15-2, water ii) discharge vigorously at an estimated overall rate of 60 l/sec. Along several hundred meters of coast at Goriachiy Pliazh, there are seepages (low flow rates) of near neutral chloride waters (Table 15-2, water iii) and during drilling, dilute sodium chloride waters of near-neutral pH with temperatures up to 150°C were encountered. The three types of water have deuterium concentrations similar to local meteoric water although the slightly higher deuterium concentrations and the Cl/Br ratio of the deep hot water at Goriachiy Pliazh indicate seawater entry below 400 m (Brezgunov et al., 1968).

The isotopic composition of sulfate sulfur from a surface spring at Goriachiy Pliazh (+17‰) is similar to oceanic sulfate (+18‰) near the coast of Kunashir Island (Ozerova et al., 1971) whereas heavier sulfate from another spring (+28.6‰) was attributed to isotopic enrichment of modern and relict marine sulfate by high temperature reduction of sulfate.

Uzon Caldera. An area of intense hydrothermal activity occurs in the Uzon caldera of East Kamchatka with surface waters up to 98°C and pH varying from 1 to 8 (Vakin et al., 1970). The deep waters are apparently near-neutral sodium chloride type (Naboko and Glavatskikh, 1970), but some of the hot spring discharges are enriched in sulfate due to near-surface oxidation of reduced sulfur and perhaps also to hydrolysis of elemental sulfur by hot waters.

In the eastern thermal area, amorphous, brick-red precipitates are forming as coatings and crusts on submerged boulders and on the banks of basins (Naboko and Glavatskikh, 1970). The precipitates deposited from NaCl waters (Table 15-2, i) of pH = 5.68 and $T = 85°C$, are enriched in silica and arsenic (15.34%) (Table 15-1) similar to the yellow-orange arsenic-rich precipitates of Waiotapu, N.Z. (Weissberg, 1969).

In the central (Tsentral'nyy) hot spring field, hydrothermally altered pyroclastic debris is cemented with crystalline sulfides. The ore-bearing layer lies at a 25–30-cm depth but, in places, may be traced to 7.5 m (Ozerova, Naboko and Vinogradov, 1971). The "cement" is composed predominantly of realgar and cryptocrystalline, powdery, orpiment with minor stibnite. Cinnabar forms fine-grained crusts and rare small prismatic crystals and is more common in the lower part of the ore horizon where it is also associated with black, earthy metacinnabar, and native mercury. These minerals are associated with native sulfur, pyrite, and marcasite with minor amounts of chalcopyrite (crystals) and pulverulent bornite (Naboko and Glavatskikh, 1970). Ozerova et al. (1971) also report sphalerite, covellite, and chalcocite with pyrite and marcasite at depth. Gangue minerals are opal, sulfur, gypsum, barite, clay minerals, bitumen, and calcite. Precipitation is from NaCl waters (Table 15-2, water ii) with a near-neutral pH = 6.89 and $T = 98°C$. Pavlov and Karpov (1972) also report the precipitation of native mercury, native sulfur and scorodite from the very acid surface waters with high oxidation potential and sulfate sulfur.

MERCURY IN HOT SPRINGS

Ores of mercury, and commonly associated arsenic and antimony minerals, provide the most obvious link between hot springs and ore deposits. Although most economic mercury deposits are not located near present-day hot springs or even in obvious fossil hot spring deposits, a few hot spring areas have been mined for mercury (some unsuccessfully or unprofitably) such as Ngawha, New Zealand; Skaggs Springs, Sulfur Bank, Wilbur Springs, and Coso Springs, California; and Steamboat Springs, Nevada. Cinnabar and metacinnabar in small quantities are presently being deposited in the orifices of a few hot springs, notably at Sulfur Bank, Steamboat Springs, Cedarville Hot Springs, Amedee Springs, and Boiling Springs, Idaho, and may also be actively forming at Skaggs Springs and Coso Springs, California; Ngawaha, New Zealand; the Alasehir geothermal area in Turkey (Ten Dam and Khrebtov, 1970); and at several places in the Kamchatka-Kurile region, U.S.S.R. The major element chemistry of some of these springs is listed in Table 15-2 and the chemical controls on transport and deposition mechanisms have been thoroughly discussed by Dickson (1964), White (1967), Dickson and Tunell (1968), Barnes et al. (1967), Tunell (1970), and Learned et al. (1974).

A recent contribution to the problem of mercury transport and deposition in hot springs is the analysis of mercury concentrations in the waters. Results indicate surprisingly low concentrations, for example, 2.0 $\mu g/l$ at Amedee Springs (White et al., 1970) and 1.5 $\mu g/l$ in unfiltered water at both Sulfur Bank and Wilbur Springs and 0.5 $\mu g/l$ in filtered water at Sulfur Bank (Barnes et al., 1973). Similar concentrations occur at Champagne Pool, Waiotapu, New Zealand [2.6 $\mu g/l$ unfiltered and 0.44 $\mu g/l$ filtered (Weissberg, 1975)] in waters actively depositing red-orange, metal-rich, amorphous precipitates (Table 15-1 and 15-2) (Weissberg, 1969).

Barnes et al. (1973) calculated that Wilbur Springs waters were apparently in equilibrium with cinnabar and hence the measured mercury concentration may be due to leaching of previously deposited cinnabar; but other spring waters appeared to be greatly supersaturated with respect to cinnabar and hence could not result from leaching.

Hot spring waters most closely associated with active hypogene transport and deposition of cinnabar, that is, those with good flows of neutral to slightly alkaline hot water of moderate to low salinity, appear to carry only small amounts of mercury—on the order of 2 $\mu g/l$. These low mercury concentrations represent the mercury remaining in the waters after precipitation of cinnabar or metacinnabar has taken place. The overall transport capabilities of spring waters, particularly at depth, may be greater than the analyses indicate because an unknown amount of mercury may have precipitated before sampling. Assuming that only the filterable fraction of mercury in the analyzed waters represents all of the mercury available for precipitation, for example, 1 or 1.5 µg/l, then it appears that some 60-80% of the mercury transported to the surface is precipitated in the hot spring area.

Higher mercury concentrations occur in some acid springs with little discharge and high chloride and sulfate concentrations. Davy (1974) and Davy and van Moort (1974) report mercury concentrations of 26 to 28 µg/l in three springs at Ngawha, New Zealand, associated with elemental mercury in nearby soils formed by vapor transport, and also the site of previous mining of cinnabar. Nakagawa (1974) gives a range from less than 0.01 to 26 µg/l in 55 Japanese hot springs not obviously associated with mercury deposits, where the highest mercury contents occur in acid springs of high chloride and sulfate contents. Ohta and Terai (1971) also report a range from less than 0.1 up to 30 µg/l in 10 acid hot springs around the Shirane volcano, Gunma Prefecture. These springs probably represent waters modified by surface oxidation and perhaps some concentration by evaporation rather than deeper seated water chemistry, and hence, should not be considered as primary transporting solutions.

Flow rates of some systems are: 10 to 20 l/sec at Champagne Pool, Waiotapu, New Zealand, (Weissberg, 1969); 28 l/sec at Wilbur Springs, California (Moiseyev, 1968); 3 l/sec at Sulfur Bank (White, 1967); Amedee Springs, California 70 l/sec total (White, 1967); 12 l/sec total at Boiling Springs, Idaho (White, 1967) and at Steamboat Springs, Nevada 70 l/sec total (White, 1967) of which less than 10% reaches the surface as visible hot springs. A flow rate of 30 to 35 l/sec totals about 10^9 l/yr and, if 1 µg of Hg is deposited per liter, then approximately 1 kg of mercury is deposited per year. This could easily account for the observed amounts of mercury at some hot springs. For example, at Amedee, Tunell (1970) estimated at least 181 kg of HgS occurs in the surface aprons of the springs; at a flow rate of 70 l/sec and precipitation of 1 µg/l, about 90 years of continuous deposition would be required. Sulfur Bank, the most productive mercury ore deposit directly related to hot springs (approximately 4.7×10^6 kg of mercury was mined) would require at least 10 times the present flow rate (3 l/sec) or 10 times the deposition rate of 1 µg/l, in order to account for its formation in less than 5×10^6 years. Obviously higher flow and deposition rates in the past were needed to form the Sulfur Bank deposit in a reasonable length of time, say 0.5 to 1×10^6.

BROADLANDS GEOTHERMAL SYSTEM—A CASE STUDY

In New Zealand, several metal-depositing geothermal systems occur along the east side of the Taupo Volcanic Zone at Rotokaua, Broadlands, Waiotapu, and Waimangu (Weissberg, 1969). These are characterized by amorphous arsenic- or antimony-rich sulfide precipitates high in gold, silver, mercury, and thallium forming around the orifices of natural hot springs or from steam-water mixtures discharging from geothermal drill-holes. Base-metal sulfide minerals occur in drill cores and cuttings recovered from depth at three of these areas, Broadlands, Kawerau, and Waiotapu (Browne, 1969, 1971), indicating a zoned deposition pattern, with zinc, lead, and copper sulfides at depth and a near-surface element suite of arsenic, antimony, mercury, thallium, silver, and gold. Exploration of the Broadlands geothermal system, preparatory to the installation of a geothermal power station, has produced a wealth of material for study and has stimulated research into the metal transport mechanisms operating in the system. In the remainder of this chapter we present a description of the Broadlands geothermal field as a detailed case study of a depth-zoned, potential ore-forming geothermal system.

Geology

The geothermal field, 25 km NE of Lake Taupo, is one of several hydrothermal systems in a graben ~50 km wide by 240 km long filled with more than 2000 m of Pliocene to Holocene calc-alkaline volcanic rocks consisting of silicic ash flows and air fall pyroclastics, rhyolites, and lesser quantities of dacite, basalt, andesite, and lacustrine sediments. Graben subsidence was more than 2 km in the last million years and at Broadlands has averaged 3.5 mm/yr over the last 4000 years (Browne, 1974). Surface thermal activity at Broadlands is minor, mainly the Ohaki Pool described below, and natural surface heat flow of 17.5 Mcal/sec is small compared with Wairakei (101 Mcal/sec) (Mahon and Finlayson, 1972). Thirty-six holes have now (1979) been drilled ranging between 776 and 1404 m deep, but with one 2420 m (BR15) (Figure 15.2); their combined mass output is in excess of 2.25×10^6 kg/hr at wellhead pressures between 9 and 14 bars.

Fluid channels are probably produced by natural hydraulic fracturing initiated by normal faults with only small displacement (Phillips, 1972); production wells are sited to intersect them at depth. Measured downhole temperatures generally increase with depth (maximum 307°) but local inversions occur in some holes due to inflow of colder water. However, the hot water entering the wells in the permeable production zones is generally between 250 and 270°C. Below 600 m, measured temperatures are mostly

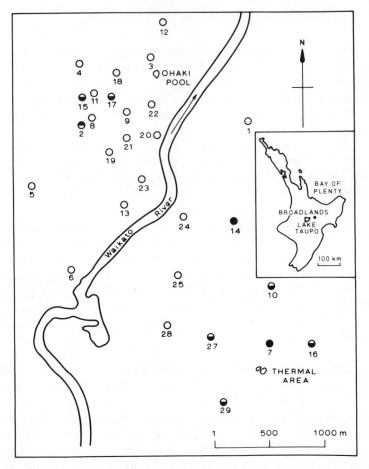

Fig. 15.2 Location of geothermal wells and hot springs at Broadlands, New Zealand. Wells marked, ◐, intersected base-metal sulfide minerals at depth; wells marked, ◑, deposited precious-metal-bearing, antimony-rich precipitates at the surface; and wells marked, ●, have both surface precipitates and base-metal sulfides at depth.

in good agreement with those calculated from silica concentrations, Na/K ratios (Mahon and Finlayson, 1972), oxygen isotopes (Eslinger and Savin, 1973, and Blattner, 1975), and fluid inclusion studies (Browne, Roedder and Wodzicki, 1974).

The stratigraphy of the field, revealed by drilling, (Figure 15.3), consists of near-horizontal Quaternary to Recent rhyolites, dacites, andesites, lacustrine tuffaceous sediments, ash flow, air fall, and water deposited pyroclastics. These overlie a very irregular but generally west-dipping

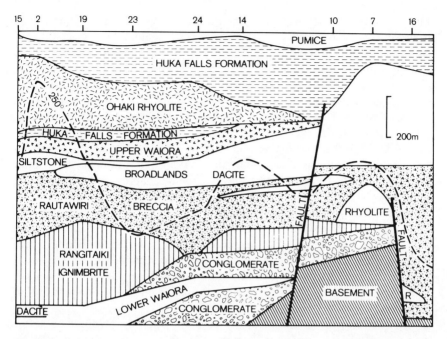

Fig. 15.3 Geologic cross-section of the Broadlands, New Zealand, geothermal field.

surface of the regional Mesozoic basement of greywacke and argillite. The basement, of great but unknown thickness, was reached by six drillholes (BR 7, 10, 14, 15, 16, and 29); similar basement rocks are widespread in the North Island and underlie silver-gold- and lead-zinc-copper bearing mineralized Miocene calc-alkaline volcanic rocks in the Coromandel Peninsula.

Sulfide Mineralogy

METAL-RICH SURFACE PRECIPITATES

Prior to drilling, the major surface thermal feature at Broadlands was Ohaki Pool, with a surface of approximately 800 m^2, at least 10 m deep, filled with clear, pale turquoise-blue water (95°C) and lined with grey-white silica sinter. The pool is on a large slightly elevated silica terrace (10,000 m^2) and discharges neutral to slightly alkaline sodium chloride-bicarbonate water (Table 15-2 and 15-3) at rates variously estimated from 4 to 12 l/sec. During 1957, a flocculent red-orange precipitate appeared in the pool waters and slowly became incorporated into the marginal sinter and overflow channel. Apart from included silica, the precipitate is an amorphous antimony sulfide (Table 15-1) nearly identical to "metastibnite" first found at Steamboat Springs, Nevada (Becker, 1888; White, 1967). The precipitate also contains sufficiently high concentrations of gold, silver, mercury, thallium, and arsenic to qualify as gold-silver ore. This element suite also occurs in several gold deposits of the Western United States, characterized by disseminated, very fine-grained, or "invisible" gold, that occur in the deposits at the Getchell, White Caps, Carlin, Cortez, Gold Acres, and the Mercur District. The formation of this precipitate at Ohaki Pool was short lived, and by 1966 a thin layer of silica sinter coated the "metastibnite" indicating that the discharge event ended before drilling began.

Similar precipitates have formed from the discharges of three drillholes (BR2, 7, and 14) and minor amounts of cinnabar in old siliceous sinter crop out approximately 250 m south of BR7. Drillhole BR3 is connected hydrologically to Ohaki Pool since its discharge caused the water level in the pool to fall and finally disappear. Since mid-1971, when the wells were shut, the water level in the pool has risen slowly to about half of its full level.

The waters forming the metal-rich antimony sulfide precipitates at the surface are undersaturated with respect to these materials at depth. No stibnite, realgar, orpiment or cinnabar has been found in any of the drillcores, although they have been carefully looked for. Experiments show that amorphous "metastibnite" precipitate quickly crystallizes when subjected to downhole conditions (Weissberg, 1969) and could not persist as an amorphous precipitate at depth. Therefore, it must form either at the surface or in the drillpipe during its ascent to the surface due to the rapid physical and chemical changes taking place, for example decrease in temperature from above 250°C to less than 100°C; concentration due to the separation of steam (approximately 35% of the mass flashes to the steam phase); loss of CO_2 and H_2S to the steam phase; or partial oxidation by atmospheric oxygen of H_2S remaining in the liquid phase.

Some elements present in the surface precipitates are partially removed from the geothermal fluid at depth, as indicated by the trace element studies of pyrite, sphalerite, and galena by Ewers (1975) as discussed below.

BASE METAL SULFIDES

The distribution of sphalerite, galena, and chalcopyrite in some Broadlands drillholes is summarized in Table 15-4. More detailed descriptions of the mineralization and associated alteration in BR7 and 16 are given elsewhere (Browne, 1969, 1971).

Table 15-4 Distribution of base-metal sulfides in Broadlands drillholes

Well no.	Depth of well (m)	Base-metal sulfide zone				Rock types in base-metal sulfide zone	Associated hydrothermal minerals
		Depth range (m)		Temperature range (°C)			
		Minimum	Maximum	Minimum	Maximum		
7	1125	802	925	272	276	rhyolite tuff, greywacke	quartz, adularia, albite, calcite, pyrite, illite, chlorite
10	1092	954	1045	265	275	conglomerate, greywacke	quartz, adularia, chlorite, illite
14	1289	1234	1235	294	294	argillite, conglomerate	quartz, adularia, chlorite, illite, pyrite
15	2421	1620	2314	286	298	ignimbrite, tuff	calcite, quartz, chlorite, adularia, albite, pyrite
16	1406	280	1393	120	276	Broadlands Dacite Rautawiri Breccia greywacke	quartz, illite, chlorite, calcite pyrite, pyrrhotite
17	1084	607	609	273	273	Upper Waiora pumiceous tuff breccia	calcite, illite, wairakite, quartz pyrite
27	1158	1031	1033	280	280	tuff	quartz, pyrite, sphene, zoisite
29	1027	914	994	265	265	tuff, tuffaceous sandstone	illite, pyrite, chlorite, sphene

The base-metal sulfides are most abundant closest to the surface and have their greatest vertical distribution in BR16 (Table 15-4). Westward from BR16, base-metal sulfides are at progressively greater depths (higher temperatures) and the thickness of the base-metal bearing zones also generally decreases. Cuttings were not recovered below 475 m in BR27 and base-metal sulfides may be more widespread here than indicated. Except for BR16, the base-metal sulfides occur at depths where the measured temperatures vary between 265 and 298°C with pressures between 58.6 and 158.8 bars. Their distribution is not obviously controlled by stratigraphy, as they occur in several lithologic units, but they are most abundant in cores that show fracturing. The converse does not apply and most fracture zones do not contain base-metal sulfides. Furthermore, the base-metal sulfide distribution is not related to either permeability or bore output; for example, BR16 is a negligible producer but BR27 a very good one.

Sphalerite is the most common base-metal sulfide but galena, or chalcopyrite, predominate locally. In relatively impermeable, nonporous rocks such as the Broadlands Dacite, sphalerite, galena, pyrite, calcite, and hydrothermal silicate minerals form veinlets. In more porous rocks, such as the Rautawiri Breccia, euhedral, unetched sulfide crystals up to 2 mm diameter are disseminated in small vugs throughout the rock (Figure 15.4). Textures indicate equilibrium between sulfides, silicates, and deep fluids. It is difficult to determine an unambiguous sulfide paragenesis from veinlet textures; sphalerite commonly encloses galena but the reverse occurs in places and it seems likely that their deposition must have been near contemporaneous (Figure 15.4).

Chalcopyrite is mainly present as inclusions in sphalerite but occasionally also forms discrete anhedral grains. Rare arsenopyrite, nickel glaucodot, cobaltite, and silver telluride occur as minute inclusions in a sphalerite grain from BR7 examined by electron microprobe (Browne, 1969). Inclusions of pentlandite in pyrrhotite from BR16 have been reported by Ewers (1975).

None of the cores contain sufficient lead, copper, or zinc to be regarded as ore. Analyzed materials were selected to avoid veins, hence, some reported values may underestimate the amount present. Zinc contents of cores from BR16 are mostly below 1%, copper averages 15 ppm with a maximum of 30 ppm and samples without visible galena usually contain less than 10 ppm lead. One sample (Br16, 302 m) contains 20% lead.

Sphalerite Compositions. Sphalerite from BR7 at 844 m contains 20 ppm silver, 7 ppm gallium with more than 100 ppm indium and strontium; analyzed sphalerites from BR7, 10, and 14 mostly contain less than 1 wt% $CuFeS_2$. Electron microprobe analysis of sphalerites from Br7, 10, 14, 15, and 16 show FeS contents between 6.7 and 22.9 mole % and MnS varies

Fig. 15.4 Stereo (SEM) photomicrographs showing textures of hydrothermal sphalerite, galena, and silicate minerals lining vugs in core samples recovered from a depth of 990 m in Broadlands drillhole 16. (a) Galena (ga) and sphalerite (sp) on a background of quartz and clays (field of view is 840 μm wide). (b) Galena on sphalerite; quartz and clays in background (field of view is 525 μm wide). (c) Galena molded around sphalerite; quartz and clays in background (field of view is 465 μm wide).

between 0.3 and 1.57 mole % (Browne and Lovering, 1973); even samples from the same depth vary (7.4-13.9 mole % FeS and 0.67-1.57 mole % MnS). Only sphalerite from Br16, 787 m, has a uniform FeS composition; it appears to be in equilibrium with, and is presumably buffered by, pyrite and pyrrhotite. Sphalerite compositions cannot be directly related to depth, host rock, temperature, or pressure, although deepest sphalerites have the highest MnS contents. Zoning (and hence crystal nonequilibrium) is shown in several grains by color and FeS variations. Since Broadlands seems to have been at a stable temperature for a long time, sphalerite zoning is thought to be due to sulfur fugacity fluctuations as pyrrhotite is mostly absent and the system is therefore not buffered with respect to sulfur (Browne and Lovering, 1973).

Fluid Inclusions in Sphalerite and Quartz. Freezing temperatures of fluid inclusions were measured in 33 primary and 60 secondary (pseudosecondary) inclusions in eight hydrothermal quartz and one sphalerite crystal (Browne, Roedder and Wodzicki, 1974). The observed freezing point depressions (0.1-0.8°C) are in agreement with the present-day Broadlands water compositions when the range of downhole CO_2 compositions (0.1-0.3 m) (Table 13-6; Table 3 of Mahon and Finlayson, 1972) are included with dissolved salts in calculating freezing point depressions, although they appear too large if only the dissolved salts in the downhole waters (3,000 ppm) are considered. This corroborates the suggestion, based on the interpretation of textural evidence, that the sphalerite crystals are in equilibrium with, and have grown from, waters of present-day compositions.

The homogenization temperatures of 177 primary fluid inclusions of hydrothermal quartz and sphalerite crystals from seven drillholes vary between 201 and 293°C and average 8°C above the measured bore temperatures; they range from 13°C below to 37°C above bore temperatures (Browne et al., 1974). Filling temperatures averaging 6°C below the corresponding measured bore temperatures are given by 198 secondary or pseudosecondary inclusions, which are more abundant and have a wider temperature spread than primary inclusions. The close agreement between filling temperatures of fluid inclusions and measured well temperatures indicates either stable temperatures in the Broadlands geothermal system through time, or the inclusions in the crystals studied are of modern origin.

Isotopic Composition of Sulfides. $\delta^{34}S$ values of sulfides from Br7 average +3.9‰ for pyrite, +3.7‰ for sphalerite and +1.4‰ for galena compared with +5.2‰, +4.6‰, and +2.7‰ for Br16 (Browne, et. al., 1975). Attainment of isotopic equilibrium, particularly below 260°C, appears slower than for chemical equilibrium. Only two galena-sphalerite pairs from more permeable zones show both isotopic and chemical equilibrium giving isotope temperatures (280°C) in good agreement with measured well temperatures (267° and 274°). This contrasts with δ^{34}S values of sulfides from Br16, 302-323 m that are markedly different from theoretical δ^{34}S equilibrium values giving temperatures well above those measured (Browne, Rafter and Robinson, 1975), (Kajiwara and Krouse, 1971); this is matched by observed chemical disequilibrium between phases. Galena from Br7 analyzed by M. H. Delevaux, U.S. Geological Survey, Denver has the following lead isotope ratios Pb^{206}/Pb^{204} of 18.966; Pb^{207}/Pb^{204} of 15.640; Pb^{208}/Pb^{204} of 28.836. These values are similar to galena and lead in greywacke and rhyolites from the Coromandel Peninsula (Cooper and Richards, 1969) where a deep-seated rather than local source is suggested.

Pyrite-Pyrrhotite. Pyrite is the most abundant and widespread sulfide mineral at Broadlands; pyrrhotite is known from 11 drillholes (Figure 15.5) and has also been deposited on well casing, but marcasite has been identified in only one core (from Br27).

Pyrite is present in concentrations up to 10% and generally forms disseminated euhedral crystals less than 2 mm in size; less commonly it occurs in veinlets, small vugs, or as dispersed irregular shaped grains. Pyrrhotite occurs as brown-black or golden crystals, up to 4 mm diameter, often in clusters. X-ray diffraction of pyrrhotite from Br16 (787 m) (Browne and Lovering, 1973) indicates anomalous pyrrhotite possibly formed from the oxidation of monoclinic pyrrhotite (Taylor, 1971). Elsewhere both monoclinic and hexagonal types occur.

Pyrrhotite and pyrite may form by replacement of primary magnetite and ferromagnesian minerals, but the abundance of pyrite and its presence in veins shows that much iron and sulfur has also been added to the systems through hydrothermal fluids.

Trace Elements in Pyrite. Ewers (1975) determined the distribution of gold, arsenic, antimony, thallium, silver, bismuth, lead, zinc, tellurium, selenium, and cobalt in the sulfide fraction, mainly pyrite, from Br16 core as a function of depth (Figure 15.5). Two main zones of sulfide deposition occur, one at approximately 300-400 m in fractured Broadlands Dacite, and the other at about 800 m in the Rautawiri Breccia (Figure 15.3).

Pyrite from the 800 m and 300-400 m zones contains up to 1.00 ppm and 1.30 ppm of gold, respectively, and near surface pyrite is highly enriched with up to 16.6 ppm gold. Arsenic, antimony, thallium, and selenium are also enriched in surface (depth 200 m) pyrite. Galena and sphalerite in the sulfide fractions account for the observed lead, zinc, silver, tellurium, selenium, and bismuth.

Rocks from the 300-400 m sulfide zone contain both adularia and

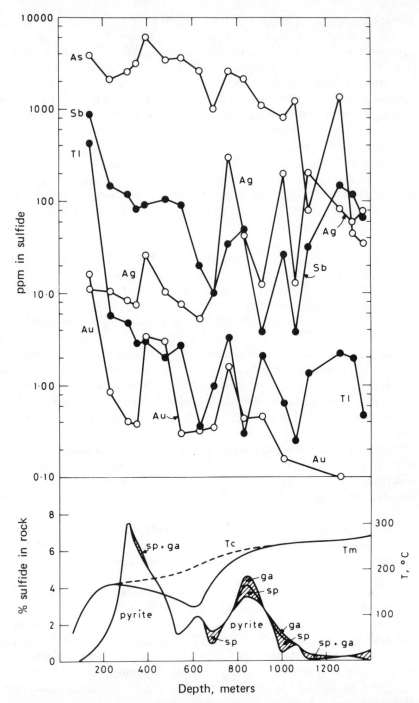

Fig. 15.5 Distribution of trace metals in pyrite with depth in Broadlands drillhole 16 (modified after Ewers, 1975). Symbols: Tm, measured temperature; Tc, calculated temperature (after Eslinger and Savin, 1973); sp, sphalerite; and ga, galena.

calcite and the occurrence of abundant sulfide minerals may be caused by boiling. The exact mechanism of pyrite precipitation is poorly known because of the lack of data on iron transport and complexing in hydrothermal solutions. Boiling causes a loss of CO_2 (and H_2S) and the consequent pH increase may decrease the equilibrium concentration of iron in solution, by precipitating pyrite or pyrrhotite. If gold, arsenic, and antimony are present as thio complexes, their coprecipitation would be further enhanced by removal of reduced sulfur (as H_2S) through boiling and to a minor extent, by precipitation of iron sulfides. Browne et al. (1975) suggested that dilution by influx of surface water in this zone may account for the etching of sphalerite and galena. There is no evidence of boiling to explain the occurrence of the sulfide zone at 800 m although this may have occurred locally.

Hydrothermal Alteration

Rocks in the field are extensively altered to secondary silicate and carbonate assemblages (Browne and Ellis, 1970; Browne, 1973; Eslinger and Savin, 1973). Originally they contained quartz and andesine with a glassy or fine-grained groundmass; primary potassium feldspar was absent but hornblende, biotite, hypersthene, magnetite, ilmenite, apatite, and zircon occurred in minor, accessory, or rare amounts. Quartz and apatite are least affected during hydrothermal alteration but the other minerals alter at differing temperatures and permeabilities (Browne and Ellis, 1970).

The stable hydrothermal assemblage at 260°C is quartz-albite-potassium mica-calcite-iron chlorite-potassium feldspar; iron epidote and wairakite are rare and zoisite was noted in only two cores. Siderite, cristobalite, kaolin (including dickite), calcium montmorillonite, leucoxene, and mordenite occur in cooler parts of the field. The minerals replace primary constituents but are also deposited directly from solution into vugs and veinlets. Secondary minerals and textures occurring at Broadlands are common to many hydrothermal ore deposits, particularly the "epithermal" gold-silver ores, suggesting that similar processes were involved in their origin.

Factors affecting the formation of hydrothermal minerals at Broadlands are (1) temperature, (2) permeability, (3) fluid composition, (4) rock composition and texture. Pressure is important in determining the depth of boiling.

Permeability is an important control since most mineralogic changes are not isochemical and the rocks must be open for exchange of constituents. More impermeable rocks, such as rhyolite or the Rangitaiki Ignimbrite, are commonly little altered, despite high temperatures, whereas in porous and permeable zones, for example the Upper Waiora and Rautawiri

Breccia formations, reaction is invariably complete and equilibrium between minerals and fluids attained.

The composition of the underground fluids in terms of sodium, potassium, and calcium activities, pH, and CO_2 concentration can be used to construct mineral stability diagrams for the system at 260°C (Browne and Ellis, 1970). Chapter 13 explains (Figure 13.7) reasons for the abundance of calcite and the paucity of epidote, zoisite, and wairakite in many wallrock alteration assemblages.

Chemistry of the Discharge Waters

MAJOR COMPONENTS

Dilute sodium chloride-bicarbonate waters discharged at Broadlands are described in detail by Mahon and Finlayson (1972). Chloride concentrations range from 1200 to 2000 ppm after wellhead separation from steam (Table 13-2). The pH (at 20°C) after loss of CO_2 and steam, varies from 7.5 to 8.5 (Table 15-2) and deep water pH is approximately 6.3 (Table 13-6). Average calculated partial pressures of gases in the deep waters are $P_{CO_2} = 12$, $P_{H_2S} = 0.1$, $P_{H_2} = 0.1$, $P_{CH_4} = 0.5$ and $P_{N_2} = 1.0$ bar. The high P_{H_2} (0.1 bar) is comparable to other geothermal systems (see Chapter 13).

The concentration of species in the deep waters at 260°C may be calculated from the analysis of the total discharge (water and steam) using equilibrium constants for all species present (for details, see Chapter 13). The deep waters of Br2 are slightly alkaline (Table 15-2) with pH = 6.2 and have a chloride ion concentration of 0.0323 m (= 1134 ppm). The total reduced sulfur ($H_2S + HS^-$) in Br2 deep water is $3.19 \times 10^{-3}\ m$ (= 102 ppm).

During initial production at Broadlands, drillholes Br2, 9, 11, and 13 tapped a single liquid phase, but with time this changed to two-phase flow with an increasing proportion of steam. A discussion of the interrelation of rock permeability, enthalpy, and water-steam discharge and draw-down in geothermal systems is given in Chapter 13.

Steam-heated, chloride-deficient groundwater discharged from Br6 on the south margin of the field and waters in the east (Br7, 10, 14, 16) have lower Cl/B ratios, probably because of addition of boron through hot water interaction with greywacke which is here closer to the surface.

OXIDATION POTENTIAL AND EQUILIBRIUM

Rock-water interactions in Broadlands are generally in equilibrium (Browne and Ellis, 1970), and equilibria between solute and gaseous species can be also related to the oxidation potential of the system and the iron sulfide mineralogy. A potential $-$pH diagram approximating Broadlands conditions (ΣS reduced = 0.004; ΣFe = 10^{-5} m, 250°C) constructed by Seward (1974) indicates an oxidation potential between -0.50 and -0.60 V based on the observed iron sulfide mineralogy and a deep water pH of 6.1. The oxidation potentials calculated from analytical data for species in Broadlands water participating in the following reactions

$$CH_4 \text{ (g)} + 2H_2O \text{ (l)} \rightleftharpoons CO_2 \text{ (g)} + 8H^+ + 8e^-$$
$$H_2 \text{ (g)} \rightleftharpoons 2H^+ + 2e^-$$

$$2NH_3 \text{ (g)} \rightleftharpoons N_2 \text{ (g)} + 6H^+ + 6e^-$$

$$H_2S \text{ (aq)} + 4H_2O \text{ (l)} \rightleftharpoons SO_4^{2-} + 10H^+ + 8e^-$$

are -0.57, -0.58, -0.57, and -0.50 V, respectively. Agreement is good, suggesting attainment of equilibrium among species and between iron sulfides and the waters. The value of Eh = -0.50 V for the sulfide-sulfate reaction is lower because sodium sulphate ion pairing has not been considered.

ORE-FORMING ELEMENTS IN THE WATERS

Metal concentrations in the waters are very low (Table 15-3), and their great enrichment in precipitates indicates that high metal contents in waters are not required to form ore-grade materials. Critical factors are deposition rate and time. A trial discharge of Br2 provided sufficient information to calculate that approximately 1-5% of the precious metals in the waters was incorporated into the surface precipitate (Weissberg, 1969).

About 800 km^3 of Br2 water (0.04 µg gold/1) would contain one million ounces of gold (32 × 10^6 g) and would require approximately 57,000 years to discharge through a geothermal system, such as Wairakei, with a natural discharge rate of 1.6 × 10^6 kg/hr (or 0.014 km^3/yr).

Similar concentrations of gold are reported in thermal waters (0.004-0.4 µg gold/l) and precipitates from thermal waters (0.005-5 ppm gold) at Yellowstone National Park, Wyoming, by Tilling et al. (1973), and Gottfried et al. (1972). They comment that sinters with high gold contents also contain correspondingly high arsenic, antimony, and mercury, but the host rhyolites contain little gold (0.1-1.3 µg gold/kg).

Seward (1973) showed that gold was probably transported in Broadlands waters as $Au(HS)_2^-$. With Cl = 1200 ppm, pH = 6.1, and total sulfide concentration = 120 mg/kg for 260°C Broadlands waters, the calculated gold solubility as $AuCl_2^-$ would be 5.6×10^{-9} ppm but measured gold concentrations (Table 15-3) are 4×10^{-5} ppm. Gold solubility as the $Au(HS)_2^-$ complex is 1.5×10^{-2} ppm. Broadlands waters appear to be undersaturated with respect to the gold thio complex at depth.

Similar calculations for lead using experimental data from Seward and Hutchison (1975) indicate that $Pb(HS)_3^-$ is the stable thio lead complex in high temperature, near neutral solutions. The solubility of galena is described by the reaction,

$$PbS + H_2S + HS^- \rightleftharpoons Pb(HS)_3^-$$

for which

$$K_p(270°C) = \frac{[Pb(HS)_3^-]}{[H_2S][HS^-]} = 4 \times 10^{-4}$$

where brackets indicate molal concentrations and unity activity coefficients are assumed; hence, in Broadlands waters

$$\log Pb = \log K_p - pK_{a1} + pH + 2 \log [H_2S] = -10.5$$

or

$$Pb = 3 \times 10^{-11} \, m = 0.006 \, \mu g/l$$

Data on Ka_1, the first dissociation constant of H_2S, are from Ellis and Giggenbach (1971). The measured concentration (averaged of six drill holes) of lead is 11 $\mu g/l$ hence, thio complexes of lead are not significant in Broadlands waters. The lead is presumably present as chloro, carbonato, or hydroxo complexes.

The silver chloride species, AgCl and $AgCl_2^-$, predominate in 0.033 m chloride solutions at 260°C (i.e. Broadlands waters). (Seward, 1976). Assuming equilibrium with Ag_2S, total silver concentrations (as chloro complexes) are approximately 0.3 $\mu g/kg$ (versus approximately 0.7 $\mu g/kg$ silver measured) as calculated from

$$Ag_2S(c) + 2H^+ + 2Cl^- \rightleftharpoons 2AgCl(aq) + H_2S(aq) \, (K^1 = 10^{-5.5})$$

and

$$Ag_2S(c) + 2H^+ + 4Cl^- \rightleftharpoons 2AgCl_2^- + H_2S(aq) \, (K^{11} = 10^{-2.1})$$

The data to calculate K^I, K^{II} and γ_{Cl^-}, used to obtain [AgCl] and [AgCl$_2^-$], are from Seward (1976), Helgeson (1969), Ellis and Giggenbach (1971), and Liu and Lindsay (1972). Hence, chloro complexing can account for the transport of silver in these waters.

ORIGIN OF WATER AND SOLUTES

Deuterium and oxygen-18 contents of water and steam at Broadlands $\delta D = -40‰$ and $\delta^{18}O = -4.5‰$ indicate a local meteoric origin of the waters (Giggenbach, 1971b). The isotopic data gave no evidence of magmatic water additions, but in any case, up to 10% magmatic contribution would not be isotopically detectable.

Experimental leaching of volcanic and sedimentary rocks by solutions at 100–600°C indicates that elements in the waters at Broadlands could derive from such interaction (Ellis and Mahon, 1964, 1967; Ellis, 1968). The isotopic composition of sulfur in the basement greywackes (Giggenbach, unpublished data) suggests the greywackes are a possible source of sulfur in addition to the possible magmatic upper crust origin suggested by Steiner and Rafter (1966).

DURATION OF THERMAL ACTIVITY

The age of the Broadlands hydrothermal system is not known but indirect evidence suggests it is possibly between about 150,000 and 500,000 years old. The suspected buried sinter (BR10, 284 m) and detrital hydrothermal minerals in a few cores of the Huka Falls Formation indicate some thermal activity was taking place at the time of its deposition. Unfortunately, the age of the Huka Falls Formation at Broadlands is not precisely known although pollen evidence suggests that at Wairakei, the formation (correlation based on similar rock lithologies) is about 500,000 years old (Grindley, 1965). The duration of hydrothermal activity at Steamboat Springs, Nevada is at least 1 million and probably as much as 3 million years (White 1974), but continuous activity cannot be proved. These periods are not greatly different from the duration of some hydrothermal ore-depositing events (Silberman et al., 1972).

It is probable that surface activity at Broadlands has been variable with the volume of natural discharge at a particular time dependent on the "plumbing" or the timing of fissure-forming earth movements. However, the fairly close agreement between fluid inclusion homogenization temperatures in sphalerite and quartz crystals and measured well temperatures (Browne et al. 1974) suggests the temperature of the Broadlands system has remained remarkably constant for a long period or that all of the crystals studied are of recent origin.

CONCLUSIONS

Hot springs represent minor ephemeral surface manifestations of considerable subsurface activity, commonly at higher temperatures. This is particularly true of the very saline hot brines that rarely reach the surface because of their high densities. Both the Salton Sea and the Red Sea brines were discovered by accident, although the lower temperature brines at Cheleken occur at the surface and have been known for a long time. Studies of these systems demonstrate the ability of hot brines to transport base metals in quantities adequate for ore formation but waters appear deficient in sulfur for efficient metal deposition. This problem may be more apparent than real provided sulfur is slowly added to the system over sufficiently long periods. Possibly, there are many unknown areas of intense hydrothermal activity within 1 km of the surface but without surface evidence of their presence. Only through geophysical work, drilling, and perhaps luck, will we learn whether hot brines are sufficiently common to account for most recognized hydrothermal base-metal ore deposits, as suggested by high salinities of fluid inclusions, or whether they are as rare as now appears.

Dilute geothermal waters, that are commonplace, may transport and deposit base metal sulfides but at rates that appear too low to account for the formation of economic base-metal ore deposits. They also may transport and deposit gold, silver, mercury, arsenic, and antimony in amounts adequate to account for many so called "epithermal" ore deposits in spite of the very low concentrations of these metals in the waters, provided sufficiently high flow rates persist for long enough times. Lifetimes and flow rates of some systems, e.g., Steamboat Springs, Nevada, and Wairakei and Broadlands, New Zealand, appear adequate. The formation of hydrothermal ore deposits may require favorable conditions persisting over millions of years instead of tens of thousands, thus accounting for the rarity of ore deposits.

ACKNOWLEDGEMENT

We would like to thank Dr. G. R. Ewers, University of Melbourne for permitting us to use his data on trace metals in Broadlands sulfides prior to publication.

REFERENCES

Ahlfeld, F. (1974) Neue Beobachtungen über die Tekonik und die Antimonlagerstätten Boliviens: *Mineral. Deposita (Berlin)*, **8**, 125-131.

REFERENCES

Arnórsson, S. (1970) The distribution of some trace elements in thermal waters in Iceland: *Geothermics, Special Issue 2, Vol. 2, Part 1*, 542-546.

Bailey, E. H., A. L. Clark, and R. M. Smith (1973) Mercury: *U.S. Geol. Survey Prof. Paper 820*, 401-414.

Barnes, H. L., S. B. Romberger, and M. Stemprok (1967) Ore Solution chemistry II. Solubility of HgS in sulfide solutions: *Econ. Geol.* **62**, 957-982.

Barnes, I. and J. D. Hem (1973) Chemistry of Subsurface waters: *Annual Review of Earth and Planetary Sciences Vol. I*, F. A. Donath, ed., Palo Alto, California: Annual Reviews, p. 157-181.

―――, M. E. Hinkle, J. B. Rapp, C. Heropoulos, and W. W. Vaughn (1973) Chemical composition of naturally occurring fluids in relation to mercury deposits in part of North-central California: *U.S. Geol. Survey Bull.* **1382-A**, 1-19.

Becker, G. F. (1888) Geology of the quicksilver deposits of the Pacific Slopes: *U.S. Geol. Survey, Monogram* **13**, 1-486.

Bischoff, J. L. and F. T. Manheim (1969) Economic potential of the Red Sea heavy metal deposits: in *Hot Brines and Recent Heavy Metal Deposits in the Red Sea*, E. T. Degens and D. A. Ross, ed., New York: Springer-Verlag, p. 535-549.

Blattner, P. (1975) Oxygen isotopic composition of fissure-grown quartz, adularia and calcite from Broadlands geothermal field, New Zealand: *Am. J. Sci.* **275**, 785-800.

Brewer, P. G. and Spencer D. W. (1969) A note on the chemical composition of the Red Sea brines in: *Hot Brines and Recent Heavy Metal Deposits in the Red Sea*, E. T. Degens and D. A. Ross, ed., New York: Springer-Verlag, p. 174-179.

Brezgunov, V. S., Dunichev, V. M., Zotov, A. V., Riznich I. I., Soyfer, V. N. and Tkachenko, R. I. (1968) Origin of the thermal water of Mendeleyev voicano, Kunashir Island. *Doklady Akad. Nauk SSSR* (in English), **179**, 112-115.

Browne, P. R. L. (1969) Sulfide mineralisation in a Broadlands geothermal drillhole, Taupo Volcanic Zone, New Zealand: *Econ. Geol.*, **64**, 156-159.

―――, (1971) Mineralisation in the Broadlands geothermal field, Taupo Volcanic Zone, New Zealand: *Society Mining Geology Japan*, special issue 2 (Proc. IMA-IAGOD Meetings '70 Joint Symp. Vol.), 64-75.

――― (1973) The Geology, mineralogy and geothermometry of the Broadlands geothermal field, Taupo Volcanic Zone, New Zealand: Unpublished Ph.D. thesis, Victoria University of Wellington.

――― (1974) Subsidence rate at Broadlands from Radiocarbon dates: *N. Z. J. Geol. Geophys.*, **17**, 494-495.

――― and A. J. Ellis (1970) The Ohaki-Broadlands hydrothermal area, New Zealand: Mineralogy and related geochemistry: *Am. J. Sci.*, **296**, 97-131.

――― and J. F. Lovering (1973) Composition of sphalerites from the Broadlands geothermal field and their significance to sphalerite geothermometry and geobarometry: *Econ. Geol.*, **68**, 381-387.

―――, T. A. Rafter, and B. W. Robinson (1975) Sulphur isotopic variations in nature, II: Sulphur isotope ratios of sulphides from the Broadlands geothermal field, New Zealand: *N. Z. J. Sci.*, **18**, 35-40.

―――, E. Roedder, and A. Wodzicki (1974) Comparison of past and present geothermal waters, from a study of fluid inclusions, Broadlands Field, New Zealand: *International Symposium on Water-Rock Interaction*, Prague. Proc., 140-149.

Chen, C. H. (1970) Geology and geothermal power potential of the Tatun Volcanic Region: *Geothermics, Special issue 2, Vol. 2, Part 2*, 1134-1143.

Cooper, J. A. and J. R. Richards (1969) Lead isotope measurements on volcanics and associated galenas from the Coromandel-Te Aroha region, New Zealand: *Geochem. J.*, **3**, 1-14.

Craig, H. (1966) Isotopic composition and origin of The Red Sea and Salton Sea geothermal brines: *Science*, **154**, 1544-1548.

_____ (1969) Discussion: Source fluids for the Salton Sea geothermal system: *Am. J. Sci.*, **267**, 249-255.

Davey, H. A. (1974) Mechanism for mercury deposition at Ngawha Springs, New Zealand: *Papers Proc. Royal Society of Tasmania*, **108**, 157-158.

_____ and J. C. Van Moort (1974) Current Mercury deposition at Ngawha Springs, New Zealand: *Search*, **5**, 154-156.

Degens, E. T. and D. A. Ross, eds. (1969) *Hot Brines and Recent Heavy Metal Deposits in the Red Sea*, New York: Springer-Verlag.

Dickson, F. W. (1964) Solubility of cinnabar in Na_2S solutions at 50-250°C and 1-1,800 bars, with geologic applications: *Econ. Geol.*, **59**, 625-635.

_____ and G. Tunell (1968) Mercury and antimony deposits associated with active hot springs in the western United States: in *Ore Deposits of the United States, 1933-1967, The Graton—Sales Volume*, J. D. Ridge, ed., New York: Amer. Inst. of Mining, Metallurg. and Petroleum Eng., Vol. II, 1673-1701.

Doe, B. R., C. E. Hedge, and D. E. White (1966) Preliminary investigation of the source of lead and strontium in deep geothermal brines underlying the Salton Sea geothermal area: *Econ. Geol.*, **61**, 462-483.

Dunham, K. C. (1970) Mineralization by deep formation waters: A review: *Trans. Inst. Mining and Metallurgy* **79**, Sect. B, B 127-B 136.

Ellis, A. J. (1968) Natural hydrothermal systems and experimental hot-water/rock interaction: Reaction with NaCl solutions and trace metal extraction: *Geochim. Cosmochim. Acta*, **32**, 1356-1363.

_____ (1969a) Present-day hydrothermal systems and mineral deposition: *Proc. Ninth Commonwealth Mining and Met. Congress*. (*Mining and Petroleum Section*) London: Inst. Mining and Metallurgy, 1-30.

_____ (1969b) Survey for geothermal development in Northern Chile. Preliminary geochemistry report, El Tatio geothermal field, Antofagasta Province: Unpublished report.

_____ (1972) Report on geothermal project Tatun volcanic zone, Taiwan: *D.S.I.R. New Zealand Report No. C.D. 5561.*

_____ and W. F. Giggenbach (1971) Hydrogen sulphide ionisation and sulphur hydrolysis in high-temperature solution: *Geochim. Cosmoschim. Acta*, **35**, 247-260.

_____ and W. A. J. Mahon (1964) Natural hydrothermal systems and experimental hot-water/rock interactions: *Geochim. Cosmochim. Acta*, **28**, 1323-1357.

_____ and _____ (1967) Natural hydrothermal systems and experimental hot-water/rock interactions (Part II): *Geochim. Cosmochim. Acta*, **31**, 519-538.

Eslinger, E. V. and S. M. Savin (1973) Mineralogy and oxygen isotope geochemistry of the hydrothermally altered rocks of the Ohaki—Broadlands, New Zealand, geothermal area: *Am. J. Sci.* **273**, 240-267.

Ewart, A., (1966): Review of mineralogy and chemistry of acidic volcanic rocks of Taupo Volcanic Zone, New Zealand: *Bull. Volcanol.*, **29**, 147-172.

Ewers, G. (1975) Volatile and precious metal zoning in the Broadlands geothermal field, New Zealand: Unpublished Ph.D. thesis, University of Melbourne.

REFERENCES

Ferguson, J. and I. B. Lambert (1972) Volcanic exhalations and metal enrichments at Matupi Harbor, New Britain, T.P.N.G.: *Econ. Geol.,* **67,** 25-37.

___, ___, and H. E. Jones (1974) Iron sulphide formation in an exhalative-sedimentary environment, Talasea, New Britain, P.N.G.: *Mineral. Deposita (Berl.),* **9,** 33-47.

Giggenbach, W. F. (1971a) Optical spectra of high alkaline sulphide solutions and the second dissociation constant of hydrogen sulphide: *Inorg. Chem.,* **10,** 1333-1338.

___ (1971b) Isotopic composition of waters of the Broadlands geothermal field: *N. Z. J. Sci.,* **14,** 959-970.

Goguel, R. L. (1977) Improved analytical values for low solubility constituents of Wairakei geothermal waters: in *Geochemistry 1977,* Bulletin 218 New Zealand Dept. of Scientific and Industrial Research.

Gottfried, D., J. J. Rowe, and R. I. Tilling (1972) Distribution of gold in igneous rocks: *U.S. Geol. Surv. Prof. Paper* **727,** 1-42.

Grindley, G. W. (1965): The geology, structure and exploitation of the Wairakei geothermal field, Taupo, New Zealand. *Bull. N. Z. Geol. Survey, n.s.,* **75.**

Helgeson, H. C. (1967) Silicate metamorphism in sediments and the genesis of hydrothermal ore solutions: *Econ. Geol., Monograph* **3,** 333-342.

___ (1968) Geologic and thermodynamic characteristics of the Salton Sea Geothermal system: *Am. J. Sci.,* **266,** 129-166.

Helgeson, H. C. (1969) Thermodynamics of hydrothermal systems at elevated temperatures and pressures: *Am. J. Sci.,* **267,** 729-804.

Hendricks, R. L., F. B. Reisbick, E. J. Mahaffey, D. B. Roberts, M. N. A. Peterson (1969) Chemical composition of sediments and interstitial brines from the Atlantis II, Discovery, and Chain Deeps: in *Hot Brines and Recent Heavy Metal Deposits in the Red Sea,* E. T. Degens and D. A. Ross, ed. New York: Springer-Verlag, p. 407-440.

Kajiwara, Y. and H. R. Krouse (1971) Sulfur isotope partitioning in metallic sulfide systems: *Can. J. Earth Sci.,* **8,** 1397-1408.

Koenig, J. B. (1969) Mercury distribution at two geothermal fields: Coso Hot Springs, California, and Ahuachapan, El Salvador: *Geol. Soc. Am. Abstracts for 1969.*

Learned, R. E., G. Tunell, and F. W. Dickson (1974) Equilibria of cinnabar, stibnite and saturated solutions in the system $HgS-Sb_2S_3-Na_2S-H_2O$ from 150° to 250°C at 100 bars, with implications concerning ore genesis: *J. Res., U.S. Geol. Surv.* **2,** 457-466.

Lebedev, L. M. (1967a) Contemporary deposition of native lead from hot Cheleken brines: *Doklady Akad. Nauk SSSR* (in English), **174,** 173-176.

___ (1967b) Modern growth of sphalerite: *Doklady Akad. Nauk SSSR* (in English), **175,** 196-199.

___ (1968) Chemical properties and ore content of hydrothermal solutions at Cheleken: *Doklady Akad. Nauk SSR* (in English), **183,** 180-182.

___ (1970) New Data on the mineralogy of pyrite ore at the Mendeleyev volcano: *Doklady Akad. Nauk SSSR* (in English), **191,** 131-134.

___ (1972) Minerals of contemporary hydrotherms of Cheleken: *Geochemistry International,* **9,** 485-504.

___ and I. B. Nikitina (1971) The chemical complexes in which lead migrates in the Cheleken thermal brines at depth: *Doklady Akad. Nauk SSSR* (in English), **197,** 229-230.

___ A. P. Polushkina, and G. A. Sidorenka (1971). The current generation of iron sulphides in Cheleken: *Doklady Akad. Nauk SSSR* (in English), **197,** 231-234.

Liu, C. and W. T. Lindsay Jr (1972) Thermodynamics of sodium chloride solutions at high temperatures: *J. Solution Chem*, **1**, 45-69.

Mahon, W. A. J. and J. B. Finlayson (1972) The chemistry of the Broadlands geothermal area New Zealand: *Am. J. Sci.*, **272**, 48-68.

Mercado, S. (1967) Geoquimica hidrotermal en Cerro Prieto B.C., Mexico: Mexico, D.F., Comision Federal de Electricidad.

Moiseyev, A. N. (1968) The Wilbur Springs Quicksilver District (California). Example of a study of hydrothermal processes by combining field geology and theoretical geochemistry: *Econ. Geol.*, **63**, 169-181.

Muffler, L. J. P. and D. E. White (1968) Active metamorphism of upper Cenozoic sediments in the Salton Sea geothermal field and the Salton Trough, southeastern California: *Geol. Soc. Am. Bull.*, **80**, 157-182.

Naboko S. I. and S. F. Glavatskikh (1970) Present-day ore deposition in the Uzon Caldera, Kamchatka: *Doklady Akad. Nauk SSSR* (in English), **191**, 190-192.

Nakagawa, R. (1971) Solubility of orpiment (As_2S_3) in Tamagawa Hot Springs, Akita Prefecture: *Nippon Kagaku Zasshi*, **92**, 154-159.

_____ (1974) The mercury content in hot springs: *Nippon Kagaku Kaishi 1974*, 71-74.

Nakamura, H. and K. Maeda (1961) Thermal waters and hydrothermal activities in Arima hot spring area, Hyogo Prefecture (Japan): *Japan Geol. Surv. Bull., n.s.*, **12**, 489-497.

Noguchi, K. and R. Nakagawa (1970) Arsenic in the waters and deposits of Osoreyama Hot Springs, Aomori Prefecture: *Nippon Kagaku Zasshi* **91(2)**, 127-31.

Ohta, N. and M. Terai (1971) The mercury content of several acid hot springs in Japan: *Bull. Chem. Soc. Japan* **44**, 1153-1157.

Ozawa, T., M. Kamada, M. Yoshida, and I. Sanemasa (1973) Genesis of acid hot spring: in *Proc. Symp. Hydrogeochemistry and Biochemistry*, **1**, Washington, D.C.: The Clark Co., 105-121.

Ozerova, N. A., N. K. Aydin'yan, M. G. Dobrovol'skaya, M. A. Shpetalenko, A. F. Martynova, V. I. Zubov, and I. P. Laputina (1972) Current mercury ore formation on Mendeleyev.

_____ and M. G. Dobrovol'skaya (1969) Mercury mineralisation at the Mendeleyev volcano: *Doklady Akad. Nauk SSSR* (in English), **187**, 154-157.

_____ Y. S. Borodoev, T. P. Kirsanova, M. T. Dmetrieva and L. N. Vjalsow (1970) Mercuriferous pyrite from Dvukhyurtochnye thermal springs at Kamchatka: *Geologia Rudnykh Mestorozhdeniy* (in Russian) **1**, 73-78.

_____ and L. M. Lebedev (1970) Mercury containing springs at Apapel, Kamchatka: in *Essays on the Geochemistry of Mercury, Molybdenum and Sulphur in Hydrothermal Processes*, Moscow: Nauka (in Russian), p. 49-64.

_____ S. I. Naboko, and V. I. Vinogradov (1971) Sulphides of mercury, antimony and arsenic forming from the active thermal springs of Kamchatka and Kurile Islands: *Proceedings I.M.A.-I.A.G.O.D. meeting, 1970, Soc. Mining Geol. Japan, special issue 2*, 164-170.

_____, N. K. Aydin'yan, M. G. Dobrovol'skaya, M. A. Shpetalenko, A. F. Martynova, V. I. Zubov, and I. P. Laputina (1972). Current mercury ore formation on Mendeleyev. *Int. Geol. Rev.*, **14**, 136-149.

Pavlov, A. L. and G. A. Karpov (1972) Physical chemistry of modern ore formation in the Uzon caldera, Kamchatka: *Doklady Akad. Nauk SSSR* (in English), **206**, 227-229.

Phillips, W. J. (1972). Hydraulic fracturing and mineralization: *J. Geol. Soc.*, **128**, 337-359.

Ritchie, J. A. (1973) A determination of some base metals in Broadlands geothermal waters: *Report No. CD.2164, Chemistry Division, New Zealand Dept. of Scientific and Industrial Research*.

Ross, D. A. (1972) Red Sea hot brine area: Revisited: *Science* **175,** 1455-1457.

Seward, T. M. (1973) Thiocomplexes of gold and the transport of gold in hydrothermal ore solutions: *Geochim. Cosmochim. Acta,* **37,** 379-399.

―――― (1974) Equilibrium and oxidation potential in geothermal waters at Broadlands, New Zealand: *Am. J. Sci.,* **274,** 190-192.

―――― (1976) The stability of chloride complexes of silver in hydrothermal solutions up to 350°C: *Geochim. Cosmochim. Acta,* **40,** 1329-1341.

―――― and M. N. Hutchison (1975) The solubility of galena in sulphide solutions up to 300°C and 1500 bar and the stability of thio lead complexes: Unpublished.

Shcheglov, I. I. (1962) Recent cinnabar deposits at Apapel Springs: *Doklady Akad. Nauk. SSSR* (in English). **145,** 141-142.

Sigvaldason, G. E. and G. Cuellar (1970) Geochemistry of the Ahuachapan thermal area, El Salvador, Central America: *Geothermics Special Issue 2, Vol. 2, Part 2,* 1392-1399.

Silberman, M. L., C. W. Chesterman, F. J. Kleinhampl and C. H. Gray Jr. (1972): K-Ar ages of volcanic rocks and gold-bearing quartz-adularia veins in the Bodie Mining District, Mono County, California: *Econ. Geol.,* **67,** 597-604.

Sillitoe, R. H. (1973) Formation of certain massive sulphide deposits at sites of sea-floor spreading: *Inst. Min. Metall., Trans., Sect. B.* **82,** B40-B45.

Skinner, B. J. and P. B. Barton (1973) Genesis of mineral deposits: *Annual Review of Earth and Planetary Sciences Vol. I,* F. A. Donath, ed., Palo Alto: Annual Reviews, 183-211.

――――, D. E. White, H. J. Rose, and R. E. Mays (1967) Sulfides associated with the Salton Sea geothermal brine: *Econ. Geol.,* **62,** 316-330.

Steiner, A. and T. A. Rafter (1966) Sulfur isotopes in pyrite, pyrrhotite, alunite and anhydrite from steam wells in the Taupo Volcanic Zone, New Zealand: *Econ. Geol.,* **61,** 1115-1129.

Takano, B. (1969a) Barium and lead content of Tamagawa hot spring waters, Akita: *Scientific Papers of the College of General Education, University of Tokyo,* **19,** 81-86.

―――― (1969b) Effect of chlorocomplex of lead on the deposition of lead-bearing barite from hot spring water: *Geochem. J. (Japan),* **3,** 117-126.

―――― and K. Watanuki (1972) Strontium and calcium coprecipitation with lead-bearing barite from hot spring water: *Geochemical Jour. (Japan)* **6,** 1-9.

―――― (1974a) Geochemical implications of the lead content of barite from various origins: *Geochem. J. (Japan),* **8,** 87-95.

―――― (1974b) Lead-bearing barite from Kawarage Mine, Akita Prefecture: *Scientific Papers of the College of General Education, University of Tokyo* **24(1),** 25-31.

Tamrazyan, G. P. (1970) Continental drift and thermal fields: *Geothermics, Special Issue 2,* **2,** *Part 2,* 1212-1225.

Tarling, D. H. (1973) Metallic ore deposits and continental drift: *Nature,* **273,** 193-196.

Taylor, L. A. (1971) Oxidation of pyrrhotites and the formation of anomalous pyrrhotite: *Carnegie Institution Wash., Yearbook* **70,** 287-289.

Ten Dam, A. and A. I. Khrebtov (1970) The Menderes Massif geothermal province: *Geothermics, Special Issue 2,* **2,** *Part I,* 117-123.

Tilling, R. I., D. Gottfried, and J. J. Rowe (1973) Gold abundance in igneous rocks: Bearing on gold mineralization: *Econ. Geol.,* **68,** 168-186.

Tooms, J. S. (1970) Review of knowledge of metalliferous brines and related deposits: *Trans. Inst. Mining and Metallurgy,* 79, Sect B, B116-B126.

Tunell, G. (1970) Mercury: in *Handbook of Geochemistry,* K. H. Wedepohl, ed., Berlin: Springer-Verlag, Vol. II/2, Chapt 80 F-H.

Vakin, E. A., B. G. Polak, V. M. Sugrobov, E. N. Erlikh, V. I. Belousov, and G. F. Pilipenko (1970) Recent hydrothermal systems of Kamchatka: *Geothermics, Special Issue* 2, 1116-1133.

Weissberg, B. G. (1969) Gold-silver ore-grade precipitates from New Zealand thermal waters: *Econ. Geol.*, **64**, 95-108.

_____ (1975) Mercury in some New Zealand waters: *N. Z. J. Sci.*, **18**, 195-203.

White, D. E. (1955) Thermal springs and epithermal ore deposits: *Econ. Geol. 50th Anniv. Vol. 1*, 99-154.

_____ (1965) Metal contents of some geothermal fluids: *Symposium on Problems of Post-magmatic Ore Deposition*, **V.II**, 432-443.

_____ (1967) Mercury and base-metal deposits with associated thermal and mineral waters: in *Geochemistry of Hydrothermal Ore Deposits*, H. L. Barnes, ed., 1st Ed., New York: Holt, Rhinehart and Winston, p. 575-631.

_____ (1968) Environments of generation of some base-metal ore deposits: *Econ. Geol.*, **63**, 301-335.

_____ (1974) Diverse origins of hydrothermal ore fluids: *Econ. Geol.*, **69**, 954-973.

_____, E. T. Anderson, and D. K. Grubbs (1963) Geothermal brine well: Mile-deep drill hole may tap ore-forming magmatic water rocks undergoing metamorphism: *Science*, **139**, 919-922.

_____, M. E. Hinkle, and I. Barnes (1970) Mercury contents of natural thermal and mineral fluids: in *Mercury in the Environment, U.S. Geol. Survey Prof. Paper 713*, 25-28.

Wright, J. B. and P. McCurry (1973) Magmas, mineralization and sea floor spreading: *Geol. Rundshau*, **62**, 116-125.

INDEX

Abbott Mercury Mine (California), 451
Acid-base reactions, 184-86
Acidity:
   activity of O_2-pH diagram, 353
   Eh-pH diagram, 352-53
   of exchange operators, 185
   of fluid inclusions, 714-15
   of geothermal waters, 643, 648, 654, 660
   of hydrothermal solutions, 129-30, 412, 531-32
      calcite solubility and, 483-85
      sedimentary waters, 149
Activity-activity diagram, 350, 352
Activity coefficients, 586, 593
Activity-composition diagram, 346
Activity of sulfur, see Sulfur, activity of
Activity of S_2-temperature diagram, 349-50
Adsorption, sulfur isotopic fractionation and, 544
Advanced argillic alteration, 204-10, 269
Affinity, chemical, 605-6
   reaction progress and, 575-82
Agricola (Georg Bauer), 3
Ahuachapan (El Salvador, Chile), 742
Ajo (Arizona), 216
Alberta Basin, 166, 245
   density of waters in, 146
   temperature in, 142
   zinc in formation waters of, 157
Alder Creek (Idaho), 218
Alkaline earth carbonate minerals, dolomite, 485-87
   solubility of, 474-90
   strontianite, 487-89
   system CO_2-H_2O, 474-77

   system CO_2-H_2O salts, 477-78
   witherite, 489-90
Alteration, 173-227
   age relative to ore, 176
   argillic, see Argillization
   Ca/H^2 ratio in, 196
   characteristics of, 173-74, 569
   chemistry of processes of, 183-92
      calculations of alteration equilibria, 186-88, 208-9
      exchange processes and exchange operators, 183-86
      presentation of mineral equilibria, 188-91
      reaction paths and mass transfer during alteration, 192
   chloritic, see Propylitic alteration
   classification of, 202-5
   in geothermal systems, 668-75
      Broadlands, New Zealand, 671, 769-70
   in greisen deposits, 226-27
   heat of, 621
   HF/H_2O ratio in, 199
   in hot springs and epithermal ore deposits, 205-7
   K/H ratio in, 188, 190, 196, 214
   K/Na ratio in, 198
   by magmatic aqueous phase, 125-32
   mass transport in, 178-82
   metamorphism contrasted to, 176-77
   metasomatic processes in, 176-82
   Na/H ratio in, 188, 190, 196
   Na/K ratio in, 710
   phase rule and, 177
   at porphyry copper deposits, 210-18, 259-60
      potassic zone: mineralogy and conditions

in, 213-15
propylitic and other types of alteration, 215-16
relative ages of, 210, 216
sequence of veins, 216
sericitic zone, 215
supergene alteration, 261-62
vertical zoning, 216-18
reaction paths of, 192
sericitic, see Sericitization
in skarn deposits, 218-26
genesis of deposits, 222-23
magnesian and calcic skarns, 219-20
relation to ore, 224-25
relation to other types of ore deposits, 225-26
structures, 218-19
volume changes, 221-22
zoning, 220-21
solfataric, 207
stability of minerals in, 193-202
decarbonation reactions, 194-95
dehydration reactions, 193-94
hydrolysis reactions, 195-97
oxidation and sulfidation reactions, 199-202
supergene, 211-12, 214-16
at porphyry copper deposits, 261-62
thermal aspects of, 624-29
around veins in granitic rocks, 207-10
zeolitic, see Zeolitic alteration
zoning of, 173, 178-82
Aluminosilicate components in magmas, thermodynamic properties of, 79-81
Aluminosilicate melts, reaction of water with, 73-75
solution of volatiles other than water in, 75-78
thermodynamics of melting and, 82-86
Aluminosilicate rocks, alteration types in, 204-5
Amedee Springs (California), 757, 758
Ammonia complexes, 410
Ammonia in geothermal waters, 664
Amphibolite facies of metamorphism, 93
Amphibolites, generation of hydrous magmas by, 98
melting relations of, 93-95, 97
Analytic techniques (or methods):
for fluid inclusions, 704-7
for ore minerals, 290-91

for sulfur and carbon isotope ratios, 511-13
Andesite, melting of, 87
Anhydrite, in geothermal waters, 665
solubility of, 490-94
Antimony complexes, 433
Apapel' Springs (USSR), 754-55
Appalachian provinces, lead isotopes of massive sulfides of, 63
Aqueous phase, magmatic, see Magmatic aqueous phase
Argillization (argillic alteration), 174-75, 202-5, 208, 261
at porphyry copper deposits, 211, 215
see also Advanced argillic alteration; Intermediate argillic alteration
Arima Springs (Japan), 742
Arsenic complexes, 433, 750
Arsenopyrite, 381
Assimilation of magmas, see Magmas, assimilation of

Barite: solubility of, 495-500
stratiform deposits of, 163, 165
Barium, in sedimentary waters, 139, 158-59, 165
Basins, epicontinental, lead isotopes of, 45-52
sedimentary, see Sedimentary basins
Bassanite, stability of, 490
Bathurst (New Brunswick, Canada), lead isotopes of, 59, 63
Battle Mountain (Nevada), 215
Bauer, Georg (Agricola), 3
Beaumont, Elie de, 5
Belitung (Indonesia), lead isotopes of, 61
Big Bug district (Arizona), 216
Bingham (Utah), 115, 118, 221-22, 257, 261
alteration in, 213, 216, 217, 225
Bingham Canyon (Utah), lead isotopes of, 52
Biotite, generation of hydrous magmas by, 99
Bisbee (Arizona), lead isotopes of, 52
Bischof, Gustav, 5
Bishop (California), 218
Blackhawk (Maine), lead isotopes of, 59, 63
Bluebell (British Columbia), 550, 716
isotopic compositions of, 266-67
Bodie (Nevada), 254, 255
Boiling (of hydrothermal solutions), 206, 481, 643, 769
fluid inclusions and, 691, 696-97, 709
NaCl and, 462
in prophyry copper deposits, 217-18

INDEX

precipitation caused by, 17-18
second (resurgent boiling), solidification of hydrous magmas and, 110-13
Boiling Springs (Idaho), 757, 758
Bolivia, hot springs in, 739
Bonneterre Dolomite, lead isotopes of, 50
Bornite, 376-78
Boulder Batholith, 249, 485
Boulder County (Colorado), 207
Brecciation, during emplacement of hydrous magmas, 113-15
Brines, 11
  sedimentary, see Sedimentary waters
Broadlands (New Zealand) geothermal area, 253, 412, 415, 453, 636, 638, 648, 759-73
  alteration, 671, 769-70
  base metal sulfides, 762-69
  chemistry of discharge waters, 770-73
  fluid inclusions in sphalerite and quartz, 766
  geology, 759-61
  isotopic composition of sulfides, 766-67
  metal-rich surface precipitates, 761-62
  Ohaki Pool, 761-62
  origin of water and solutes in, 773
  pyrite-pyrrhotite, 767
  sphalerite compositions, 764-66
  trace elements in pyrite, 767-69
Brunswick Tin Mines (New Brunswick), 226
Buchans (Canada), lead isotopes of, 57-59
Buffers, fixed-point, 284-85
  sliding-scale, 284, 285
Burnt Hill (New Brunswick), 226
Butte (Montana), alteration in, 174-75, 207-10, 214, 217
  isotopic compositions of, 257-59, 267-69
  lead isotopes of, 52

Cadmium complexes, 423-25
Calc-alkaline magmas, generation of, see Magmas, generation of calc-alkaline
Calcic skarns, 219-220
Calcite, solubility of, 478-85, 662
  solubility product of, 485, 486, 488, 489
Calcium in geothermal waters, 662-63
Calcium sulfate (anhydrite) in geothermal waters, 665
Carbonate complexes, 483-84
Carbonate host rocks, lead isotopes of ores in, 50-52
Carbonate minerals, solubility of, see Alkaline earth carbonate minerals, solubility of
Carbonates, carbon isotope ratios of sedimentary, 559
  isotopic analyses using, 512-13
Carbonatites, 484
Carbon dioxide, isotopic analyses of, 512-13
  in magmas, 77, 106-7
  solubility of, 474-77, 662, 697, 700
Carbon isotope ratios:
  analytic techniques, 511-13
  fractionation factors, 550-57
  geothermometry, 550-53, 667
  of hydrothermal ore deposits, 559-61
  of magmatic systems, 553-58
  natural range of, 510
  or organic compounds, 558
  of sedimentary carbonates, 559
Carlin (Nevada), lead isotopes of, 57
Casapalca (Peru), 266
  lead isotopes of, 61
Cascade Mountains (Oregon), 64, 249, 253
Celestite, solubility of, 488, 494-95
Central Mississippi, 156
  brines, 150
Central Tennessee District, lead isotopes of, 50
Cerro de Pasco (Peru), 207
Cesium in geothermal waters, 661-62
Chalcopyrite, 362, 363, 378
Cheleken (USSR), brines, 11
  lead and zinc in, 156, 158
  thermal area, 753-54
Chemical reactions, see Reactions
Chloride complexes, 16, 118-19, 122, 410
  deposition from, 435-36
Chlorine in magmatic aqueous phase, 116-23
Chloritic alteration, see Propylitic alteration
Christmas Copper Mine (Arizona), 220
Clausius-Clapeyron equation, 347
Climas (Colorado), 114, 226, 257-59, 716
  lead isotopes of, 55
Coast Range Batholith (British Columbia), 249
Cobar (New South Wales, Australia), lead isotopes of, 59
Cochiti (New Mexico), 207
Compaction of sediments, 140-41, 151-52
  flow of water during, 165
Complexes (complex ions), 15-16, 406-11
  ammonia, 410
  carbonate, 483-84

chloride, 16, 118-19, 122
  deposition from, 435-36
  fluoride, 473
  ligand concentrations in, 411-12
  organic, 157, 410-11
  solubilities of, 433
  stability of, 417-33
    cadmium complexes, 423-25
    copper complexes, 420-22
    ferrous chloride complexes, 418-20
    gold complexes, 430, 432, 433
    iron complexes, 417-20
    lead complexes, 425-27
    mercury complexes, 427-29
    silver complexes, 430, 431
    zinc complexes, 423-25
  sulfide, 16, 408-9
    deposition from, 437-38
  sulfur-containing, 408-10
  telluride, 452
  see also specific elements
Components (mobile components), 283
  activities of, 284-85
Comstock Lode (Nevada), 254, 255
Concentrations in ore solutions, 411-15
  anydrite, 494
  barite, 498
  calcite, 484
  dolomite, 487
  hydrogen sulfide, 123-24, 130-32, 438
  ligands, 411-12
  magmatic, 117-18
  metals, 412-15, 665
  NaCl, 462-63
  representative total (analytical), by location (table), 413-15
  sedimentary waters, 154
  silica, 470, 498
  sulfur dioxide, 124, 130-132, 438
Condensed temperature-composition diagram, 344, 345
Conduction, 615-16, 619
Connate waters, 7
  isotopic composition of, 7-8
  see also Sedimentary waters
Conservation of mass and charge, mass transfer and, 587-89
Continental arcs, lead isotopes of, 42, 43
Continental crust, epicontinental basins and platforms, 45-52
  lead isotope ratios of, 32-33, 41-42, 45-57

noncratonized upper crust, 56-57
  rejuvenated craton, 52-56
Continents, generation of magmas under, 96-101
Convecting hydrothermal systems, 9
Convection, 248-49, 269, 611, 615-16, 624-25, 635
  porphyry copper deposits associated with, 259
Convective circulation, 269
Cooling of intrusives, 616, 619, 628-29
  shape and, 621-23
Copper complexes, 123
  stabilities of, 420-22
Copper Creek, 257
Copper deposits:
  Michigan, 9
  red-bed, 421
  White Pine (Michigan), 9, 440
Copper Harbor Conglomerate, 9
Copper-iron-nickel-sulfur system, 379-80
Copper-iron-sulfur system, 374-79
Copper-iron-zinc-antimony-sulfur system, 383
Copper ores in Utah, 13
Copper-sulfur system, 372-74
Cornwall (Pennsylvania), 222
Coso Springs (California), 757
Cotta, Heinrich von, 5
Covellite, 362, 363
Craton, rejuvenated, lead isotopes of, 52-56
Creede (Colorado), 256, 415, 442, 534, 692, 705, 711, 716
  lead isotopes of, 56
Critical temperatures of hydrothermal solutions, 463-64, 478
Cubanite, 378
Cyprus-type ore deposits, 263
  lead isotopes of, 43, 57
  see also Ophiolites

Darcy's Law, 164
Darwin (California), 265-66, 520, 533, 550
Dating techniques, 10
Daubrée, Gabriel Auguste, 5
Debye-Hückel theory, 482, 498
Decarbonation, 183
  alteration and, 194-95
Dehydration (dehydration reactions), 183
  alteration and, 193-94

INDEX

of montmorillonite, 139, 141, 145, 147-48, 166
of sediments, 139-40, 145
Density of fluids, 612-15, 625, 730
sedimentary waters, 146, 164
Deposition, *see* Ore deposition
Descartes, Réné, 4
Deuterium, *see* Hydrogen isotope ratios
Dew-point techniques, 293
Diagenesis of sediments, 137-38
Diffusion, 178-82, 210, 569-70
in skarn deposits, 218
sulfur isotopic fractionation and, 544
Disequilibrium, 287-88
in chalcopyrite, 362, 363, 378
in covellite, 362, 363
in idaite, 363
in pyrite, 361-65
in sphalerite, 364-65
in sulfur isotope ratios, 533-35, 539-44
Distribution coefficient of gases, 640-42. *See also* Henry's Law; Setschenow equation
Dolomite, solubility of, 485-87
Dolomitization, 486-87
Ducktown (Tennessee), 264
Dvukhyurtochnye Springs (USSR), 755

Eagle Mountain (California), 219
East Tennessee District, lead isotopes of, 50
Echo Bay, 415, 453, 454, 533, 550
Eh-pH diagram, 352-53
Electrochemical cells, 293-94
Electron-probe microanalyzer, 290-91
Electrum-tarnish method, 293
El Salvador (Chile), 115, 211, 214, 216, 217
El Tatio (Chile), geothermal area, 638, 740
El Teniente (Chile), 130
Ely (Nevada), 215, 257
Emmons, S. F., 5
Epithermal ore deposits, 774
alteration in, 205-7
anhydrite in, 493
barite in, 498
isotopes of, 253-56
temperatures of, 727
Equilibria:
alteration, calculations of, 186-88, 208-9
fluxing agents and, 283, 294
isotopic fractionation factors, 238-41
carbon isotopic, 550-57

in geothermal systems, 667-68
of sulfur compounds, 514-21, 530-33
local (or mosaic), 177-79
mass transfer and, 579, 583-87
mass transfer, 574
metastable, 177
mineral, graphic presentation of, 188-91
partial, mass transfer and, 579, 583-87
preservation of a former state of, 288
requirements of, 286-88
subsolidus hydrolysis, 125-30
univariant equilibria, 335-40, 385-86
impurities' effect on, 386
Evaporites, 147
sedimentary waters and, 150, 158
Exchange operators, 183-86
Exchange reaction, isotopic, 238

Faulting, flow of sedimentary waters during, 165-66
Feldspar, hydrolysis of, 588, 591-600, 659
Ferrous chloride complexes, stabilities of, 418-20
Fluid inclusions, 684-730
changes since trapping, 698-704, 721
composition of, 704-18
isotopic composition, 717-18
methods of analysis, 704-7
Mississippi Valley-alpine-type deposits, 711, 714
normal (magmatic) hydrothermal deposits, 709-11
pegmatites and quartz veins, 707-9
ph, Eh, and heavy metal content, 714-17
daughter minerals in, 696, 700-1, 707
decrepitation of, 725-26
density in, 730
geothermometry of, 718-28
general ranges of temperature data obtained, 727-28
methods and apparatus, 724-27
immiscibility in, 691-92, 696, 697, 699-700, 728
interpretation of, 685-87
leakage into or out of, 697-98, 721
mechanism of trapping of, 688-93, 724
nature of material trapped, 694-97
pressure in, 720-24, 728-30
primary, 688-92
pseudosecondary and secondary, 692-93
size of, 685

Fluids:
  density of, 612-15, 625, 730
  flow of, 612-17
    alteration and sulfide deposition, 624-29
  thermal expansion of, 613
  thermal conductivity of, 615
  viscosity of, 614, 625, 626
  see also Convection; Hydrothermal solutions
Fluoride, in geothermal waters, 664-65
  in sedimentary waters, 159
Fluoride complexes, 473
Fluorine in magmatic aqueous phase, 116
Fluorite, solubility of, 471-74, 664-65
Formation waters, see Sedimentary waters
Fractionation factors, 238
  carbon isotopic, 550-57
  in geothermal systems, 667-68
  of sulfur compounds, 514-21
Fractionation processes of sulfur isotope ratios, 521-44
Franciscan Formation, 65
Free energy-composition diagram, 346-47, 350
Fukuchilite, 376, 378

Gabbs (Nevada), 219
Galapagos Rift, 493
Galapagos Rise, 45
Gaspé (Quebec), 225
Gas solubilities, 707, 710-11
  in geothermal waters, 639-43, 649
Geobarometers, 381
  sphalerite, 371
Geopressured zones, 144-45
Geothermal gradients, 470
  in sedimentary basins, 143
  at thermal contacts, 616
Geothermal systems, 632-77
  age of, 634-35, 773
  alteration in, 205-7, 668-75
  cap rocks of, 635-37
  deep fluid compositions in, 651, 654
  examples of, 638-41
  isotopic equilibria in, 667-68
  mercury in, 639, 757-58
  meteoric waters in, 635-37, 643
  models of, 636-38
  ore metals in, 738-74
  permeability of, 635-36
  rock types associated with, 633-34
  in sedimentary basins, 634
  steam compositions in, 648-51
  steam/liquid distribution in, 639-43
  structures associated with, 633
  vapor-dominated, 636-38
  world occurrence of, 633-35, 739
  see also specific geothermal areas
Geothermal waters:
  acidity of, 643, 648, 654, 660
  alkalis in, 659-63
  compositions of, 643-48, 743-49
  isotopic variations in, 244-45, 255, 667-68
  origin of, 635, 654, 657-59, 752-53
    at Broadlands (New Zealand), 773
  origin of chemicals in, 654, 657-59
  redox conditions in, 665
  specific solution and mineral reactions in, 659-68
  sulfur species in solution in, 666
Geothermometers, 289-90, 368, 381
  alkali metal ion, 659-60
  carbon isotope, 550-53, 667
  feldspar-H_2O-oxygen isotope, 251
  fluid inclusions, 718-28
    general ranges of temperature data obtained, 727-28
    methods and apparatus, 724-27
  hydrogen isotopes, 667
  invariant points, table of, 297-334
  oxygen isotopes, 667
  pyrite-pyrrhotite solvus, 361
  quartz, 469, 663
  sulfur isotope ratios, 513, 517-21, 667
Geysers, The (California), 206, 636, 639, 649
Gilman (Colorado), 257
  lead isotopes of, 55
Gneisses, hydrous magmas generated by, 98-99
Gold complexes, 771-72
  stabilities of, 430, 432, 433
Goldfield (Nevada), 207, 255
Gold-silver deposits, deposition of, 452-54, 762, 771
  isotopes of, 253-56
Gouy layer, 151
Granitic rocks, alteration around veins in, 207-10
Great Geysir (Iceland), geothermal area, 634-45
Greenschist facies, 92-93
Greigite, 359, 360
Greisen, 270

alteration in deposits of, 226-27
skarns related to, 225-26
Growth zoning, 286-87
Gulf Coast (Gulf Basin), 168, 245
  composition of waters in, 149
  flow of sedimentary waters in, 165, 166
  fluid pressures in, 144-45
  isotopic compositions of waters from, 160
  membrane-filtration in, 153
  salinity of waters in, 147
  sediments of, 139
  temperature in, 142-44
Gulf of Mexico Basin, viscosities of waters in, 146
Gypsum, solubility of, 490-500

Hall's Peak (Australia), lead isotopes of, 60
Hanover District (New Mexico), 220, 438
  lead isotopes of, 55
Hansonburg District (New Mexico), 693
  lead isotopes of, 52
  sulfur isotopes of, 548, 550
Haycockite, 378
Heat sources, 619-23
Heat transfer processes, 612-19
Hebrides (Scotland), 249, 253
Henderson (Colorado), 114, 217
  lead isotopes of, 55
Henry's Law, 475, 478. *See also* Distribution coefficient of gases; Setschenow equation
Hixbar (Philippines), 43
Homestake (South Dakota), 264
Hornblende:
  generation of hydrous magmas by, 99-100
  melting relations of, 88-91, 93-94
  water content of magmas and, 90-91
Hot springs:
  alteration in, 205-7
  lifetimes of, 10
  see also Geothermal systems
Hot-water systems, *see* Geothermal systems
Hunt, Thomas Sterry, 5
Hutton, James, 4
Hveragerdi (Iceland), 668-69
Hydration, *see* Dehydration
Hydrochloric acid in magmas, 76
Hydrofluoric acid in magmas, 76-77
Hydrogen, in geothermal systems, 770
  in magmas, 78, 125, 130
Hydrogen isotope ratios:
  determination of, 237
  in fluid inclusions, 718
  fractionation curves, 238
  in geothermometers, 667
  of massive sulfide-type deposits, 263
  of metamorphic ore deposits, 263-65
  of Mississippi Valley-type deposits, 262
  notation and standards, 237-38
  of porphyry copper deposits, 256-61
  redox reactions and, 241
  see also Isotope ratios
Hydrogen metasomatism, 129, 131, 484-85, 629, 715-16
Hydrogen sulfide, in magmas, 76, 102-3
  in magmatic aqueous phase, 123-24, 130-32
Hydrolysis:
  alteration and, 195-97
  of feldspar, 588, 591-600, 659
  heat of, 619-21
  in magmatic aqueous phase, 125-30
  of sulfur, 751
Hydrothermal alteration, *see* Alteration
Hydrothermal fluids, *see* Hydrothermal solutions
Hydrothermal mineral deposits:
  definition of, 2
  depth of, 10-11
  early speculations on origins of, 3-6
  lifetimes of, 10
  see also Ore deposition
Hydrothermal solutions (or fluids):
  acidity of, 129-30, 412, 531-32
    calcite solubility and, 483-85
    sedimentary waters, 149
  amounts of, 250-53, 624
  changes with time of, 692-93
  channelways of, 8-11
  compositions of, 11-13, 255, 663, 665, 706, 707, 709
  concentrations in, *see* Concentrations in ore solutions
  critical temperatures of, 463-64, 478
  definition of, 2
  flow rates of, 695, 758
  forces that drive, 9-10
  geochemically scarce metals in, 14-15
  isotopic composition of, 7-8, 250-53. *See also* Isotopic ratios
  leaching by, 693
  magmatic, *see* Magmatic hydrothermal solutions

788   INDEX

metal-rich, examples of, 716
sedimentary genesis of, *see* Sedimentary waters
source of dissolved constituents in, 11-15, 252-53
summary of current views on, 6-8
transport by and precipitation from, 15-19
types of, 7
see also Fluids; Magmatic aqueous phase
Hydrothermal systems:
  channelways of, 8-11
  vapor-dominated, 636
  see also Geothermal systems; *and specific topics*
Hydrous magmas, emplacement and solidification of, 104-15
  brecciation, 113-15
  crystallization, 107-9, 114-15
  porphyry copper deposits, 112-13
  quartzo-feldspathic rocks, 104-5
  second (resurgent) boiling, 110-13
  from subduction zones, 104
  telluric pressure, 112-13
  thermal effects associated with, 106
  volatile separation, 106-7

Iceland, alteration in, 206, 668-69
  geothermal area, 649, 658, 664
Idaho Batholith, 249
Idaitè, 363, 372, 378
Igneous intrusions, interactions between surface-derived waters and, 248-50
Illinois Basin, 165, 499
  composition of waters in, 149
  density of waters in, 146
  flow of water in, 164
  heat transfer in, 144
  isotopic compositions of waters from, 160
  salinity of waters in, 147
Illinois-Kentucky district, 165, 262, 499
  geothermal area, 693, 716
  lead isotopes of, 50, 51
Immiscibility of fluid inclusions, 691-92, 696, 697, 699-700, 728
Imperial Valley (California), geothermal area, 635-36
Inclusions, *see* Fluid inclusions
Indicators, sliding-scale, 281
Infiltration, 178-82
Intermediate argillic alteration, 204
Intrusions, igneous, interactions between surface-derived waters and, 248-50
Invariant points, 296
  impurities' effect on, 386
  table of, 297-334
Ion exchange reactions, 184-86
Ion probe, 291
Iron-arsenic-sulfur system, 381-82
Iron complexes, 122
  stability of, 417-20
Iron deposits, skarn-type, 225
Iron Mountain (New Mexico), 226
Iron Springs District (Utah), 14
Iron-sulfur system, 358-61
Iron-zinc-sulfur system, 365-72
Island arcs, lead isotopes of, 42, 43
  nonsubmarine, 61-63
  primitive island arc-continental margin transition, 63-65
  submarine volcanic exhalative massive sulfide deposits, 57-61
  see also Subduction zones
Island Mountain (California), 43
Isotope ratios:
  of carbon, *see* Carbon isotope ratios
  of epithermal ores, 253-56
  of fluid inclusions, 717-18
  fractionation factors, *see* Fractionation factors
  in geothermal areas, 658
  in geothermal waters, 244-45, 255, 667-68
  of hydrogen, *see* Hydrogen isotope ratios
  of hydrothermal solutions, 7-8, 250-53
  of lead, *see* Lead isotope ratios
  of magmatic waters, 247-49, 265-66
  of matamorphic waters, 7-8, 246, 248
  of meteoric waters, 241-42, 257
  notation and standards, 237-38, 509, 513
  of ocean waters, 243-44, 558
  of oxygen, *see* Oxygen isotope ratios
  of porphyry copper deposits, 256-61
  of sedimentary waters, 153-54, 245, 262
  of sulfur, *see* Sulfur isotope ratios
  of thorium, 38, 40
  of uranium, 38, 40
Isotopic exchange reaction, 238
  in geothermal systems, 667-68
  sulfur, 521-22, 525-26

Japan, hot-springs systems in, 742, 750

Kamchatka Region (USSR), geothermal

INDEX

areas, 754-56
Katayama, Onikobe (Japan), geothermal area, 673
Keeweenawan lavas, 9
Kemp, James Furman, 6
Kentucky-Illinois district, see Illinois-Kentucky district
Kizildere (Turkey), geothermal area, 639, 648
Klamath Mountains, lead isotopes of, 63-65
Kupferschiefer (Germany), lead isotopes of, 49
  sulfur isotopes of, 540-41
Kuroko-type deposits, 263, 533
  anhydrite in, 493
  lead isotopes of, 57-62
  sulfur isotopes of, 545-48

Lake George (Colorado), 226
La Motte Sandstone (Missouri), 13, 14
  lead isotopes of, 50, 51
Larderello (Italy), geothermal area, 206, 636, 639, 649-50, 657, 674-75
Laser beam microprobe, 291
Lateral secretion, 3
Launay, L. de, 5
Leaching, hydrothermal, 693
Lead:
  geochemical cycle of, 33-34
  in sedimentary waters, 139, 156
  source of, 13
Lead complexes, 753, 772
  stabilities of, 425-27
Lead isotopes (lead isotope ratios), 13, 22-66
  of carbonate rock ores, 50-52
  of continental arcs, 42, 43, 57-65
  of continental crust, 32-33, 41-42, 45-57
    epicontinental basins and platforms, 45-52
    noncratonized upper crust, 56-57
    rejuvenated craton, 52-56
  evolutionary model of, 34-42
  of geothermal systems, 767
  of island arcs, 42, 43
    nonsubmarine, 61-63
    primitive island-continental margin transition, 63-65
    submarine volcanic exhalative massive sulfide deposits, 57-61
  of island and continental arc environments, 57-65
J-type, 46
of mantle, 32

789

measurement of, in Phanerozoic rocks, 23
of Mississippi Valley-type deposits, 33, 50-52, 158
of oceanic crust, 33, 41, 43, 45
of oceanic environments, 43-45
orogene and, 42
parameters and equations of, 23, 24
of rejuvenated cratonal deposits, 52-56
of sandstone ores, 46-49
of shale ores, 49
source rocks of, 25, 32
Leadville (Colorado), lead isotopes of, 55
Ligands, see Complexes
Linchburg (New Mexico), 218-19
Lindgren, Waldemar, 6
Lithium, in geothermal waters, 661-62
Loellingite, 381

Mackinawite, 359
Maden (Turkey), 43
Magdalena (New Mexico), alteration in, 175
Magmas, 71-132
  aqueous phase of, see Magmatic aqueous phase
  assimilation of, 100-1
    in oceanic crust, 96
    quartzo-feldspathic rocks, 104-5
  generation of calc-alkaline (hydrous), 91-103
    by assimilation, 100-1
    by biotitic rocks, 99
    under continents, 96-101
    by gneisses, 98-99
    by hornblendic rocks, 99-100
    by mafic amphibolites, 98
    by mafic magmas, 97
    metal sulfides, 101-3
    modes of generation, 91-92
    subduction zones, 92-96
    sulfur dioxide from, 527
  hydrous, emplacement and solidification of, 104-15
    brecciation, 113-15
    crystallization, 107-9, 114-15
    porphyry copper deposits, 112-13
    quartzo-feldspathic rocks, 104-5
    second (resurgent) boiling, 110-13
    from subduction zones, 104
    telluric pressure, 112-13
    thermal effects associated with, 106
    volatile separation, 106-7

porphyry copper deposits generated by, 101-2, 107-8, 112-13
thermodynamic relations in, 78-91
  hornblende, 89-91
  melting, see Melting, thermodynamics of silicate components, 79-81
  water, 78-79
volatiles in, 72-78
  carbon dioxide, 77, 106-7
  carbon isotopes, 553-58
  hydrochloric acid, 76
  hydrofluoric acid, 76-77
  hydrogen, 78, 125, 130
  hydrogen sulfide, 76, 102-3
  sulfur dioxide, 77
  sulfur isotopes, 524-27
see also Magmatic aqueous phase; Magmas, water in
water in, 72-78
  abundance of, 72-73
  aluminosilicate melts, reaction with, 73-75
  during crystallization, 105-6
  differentiation, 106
  diffusion and turbulent flow of the magma, 105
  hornblende, melting relations of, 90-91
  melting, effects on, 73, 86-87
  solubilities of, in silicate melts, 74-75
  thermodynamic properties of, 78-79
  see also Magmas, hydrous
Magmatic aqueous phase, 116-25
  alkali partitioning in, 120-23
  chlorine in, 116-23
  copper in, 123
  in equilibrium with melt, 116, 119-20
  fluorine in, 116
  generation of, 116
  hydrogen sulfide in, 123-24, 130-32
  hydrolysis in, 125-30
  iron in, 122
  sulfur in, 123-25
  sulfur dioxide in, 124, 125, 130-32
Magmatic hydrothermal solutions, 4, 10, 654
  carbon isotopes of, 553-58
  isotopic variations in, 247-49, 265-66
  sulfur isotope ratios of, 523-27
Magnesian skarns, 219-20
Magnesite, solubility of, 486-87
Magnesium, in geothermal waters, 663
Manometer, 293

Mantle, lead isotope ratios of, 32
Marcasite, 359, 362, 365
Marysvale (Utah), 207
Mass action, law of, 584-85
Massive sulfide-type deposits:
  hydrogen and oxygen isotope ratios of, 263
  lead isotopes of, 57-59
  sulfur isotopes of, 545-48
  see also Kuroko-type deposits
Mass transfer, 568-606
  calculation of, 589-605
  chemical reactions and, 570-82
    chemical affinity and reaction progress, 575-82
    conservation of mass and charge and, 587-89
    partial and local equilibrium and, 579, 583-87
  supersaturation and, 583-87
Mass transport in alteration, 178-82
Matsao (Taiwan), geothermal area, 638, 648, 672, 751
Matsukawa (Japan), geothermal area, 636, 639, 648, 649, 674
Melting:
  of amphibolites, 93-95, 97
  fractionation of sulfur isotopes during, 523-27
  of plagioclase, 84-88
  thermodynamics of, 81-91
    aluminum coordination, effects of, 82-86
    andesite, 87
    micas, 89-91
    plagioclase, 84-88
    pressure, 86-88
    water, 86-87
Melts, see Aluminosilicate melts; Silicate melts
Membrane filtration, in sediments, 150-54, 245, 544
Mendeleyev Volcano (USSR), 755-56
Mercury complexes, stabilities of, 427-29
Mercury ore deposits, deposition of, 450-51
  in geothermal areas, 639, 754, 756-58
Metals, concentration in ore solutions, 412-15
Metal sulfides, magma generation and, 101-3
Metamorphic ore deposits, 520-21
  oxygen and hydrogen isotope ratios of, 263-65
Metamorphic waters, isotopic composition of, 7-8, 246, 248
Metamorphism:

INDEX 791

alteration contrasted to, 176-77
contact, 623
in subduction zones, 92-93
Metastability:
  of bornite, 376-78
  of chalcopyrite, 362, 363
  of covellite, 362, 363
  in fluid inclusions, 701
  of fukuchilite, 376
  of idaite, 363, 372
  of iron sulfides, 358-60
  of marcasite, 365
  of monoclinic pyrrhotite, 360
  of sulfate species, 533-35
  see also Disequilibrium; Mineral stabilities; and System
Meteoric waters:
  in geothermal areas, 635-37, 643
  in hydrothermal systems, 250-53
  influx into intrusions, 250
  isotopic equation for, 242, 243
  isotopic variations in, 241-42, 257
  North American, 242, 243, 257
  see also Sedimentary waters
Mexicali (Mexico), geothermal area, 639
Micas, melting relations of, 89-91
Michigan basin:
  density of waters in, 146
  isotopic compositions of waters from, 160
  salinity of waters in, 147
  viscosities of waters in, 146
Michigan copper deposits, 9
Middle Fork (Washington), 215
Mineral deposits:
  definition of, 1
  hydrothermal:
    definition of, 2
    depth of, 10-11
    early speculations on origins of, 3-6
    lifetimes of, 10
Mineral equilibria, calculation of, 651, 654
  in geothermal systems, 651, 654, 657
  graphic presentation of, 188-91
  see also Mineral stabilities
Mineralizers, 15-16
Mineral Park (Virginia), 257
Mineral (Virginia), lead isotopes of, 59, 63
Mineral stabilities, 278-390
  alteration and, 193-202
    decarbonation reactions, 194-95
    dehydration reactions, 193-94

hydrolysis reactions, 195-97
oxidation and sulfidation reactions, 199-202
arsenopyrite, 381
chalcocite, 372, 374
chalcopyrite, 378
covellite, 372, 374
cubanite, 378
digenite, 372, 374
djurleite, 374
equilibrium requirements, 286-88
Fe_2S_3, 359
Fe_2S, 358
fukuchilite, 378
greigite, 359, 360
haycockite, 378
idaite, 378
lellingite, 381
mackinawite, 359
marcasite, 359
mooihoekite, 378
pentlandite, 359
preservation of former equilibrium state, quenching, 288
pyrite, 358-65
pyrrhotite, 358-61
sphalerite, 359, 365-72
smythite, 359, 360
talnakhite, 378
tennantite, 384
tetrahedrite, 384
wurtzite, 359, 365-68
see also Metastability; System
Mississippi Valley-type deposits (or ores), 8, 9, 167, 420
  carbon isotopes of, 561
  chemical composition of ore-forming fluids in, 160-61
  deposition of, 163, 446-50, 727
  fluid inclusions in, 711, 714, 727, 728
    atomic ratios for (table), 711-13
  hydrogen and oxygen isotope ratios of, 262
  isotopic composition of ore-forming fluids in, 160
  lead isotopes of, 33, 50-52, 158
  organic complexes in, 410
  regional zoning of fluorite and strontium-bearing minerals in, 160-61
  sedimentary ore solutions in, 160-61
  sulfur isotopes of, 548-50
  temperature-salinity relations in, 161-63

Mobile components, 283
  activities of, 284-85
Molybdenum deposits, porphyry, 259-61
Molybdenum mineralization, 260
Montmorillonite, dehydration of, 139, 141, 145, 147-48, 166
Mooihoekite, 378
Morenci (Arizona), 261
Morococha (Peru), 545

Nacimiento (Peru), lead isotopes of, 48, 49
Nevada, gold-silver deposits in western, 253-54
New Britain (Papua, New Guinea), geothermal area, 750
Ngawha (New Zealand), geothermal area, 639, 757, 758
Nickel complexes, 157
Northern Pennines (England), 499

Oceanic crust, assimilation of magmas in, 96
  lead isotope ratios of, 33, 41, 43, 45
Oceanic intraplate deposits, lead isotopes of, 45
Oceanic rises, lead isotopes of, 43
Ocean waters:
  calcite solubility in, 483
  isotopic composition of, 243-44, 558
  in Kuroko deposits, 493, 495
  in Mississippi Valley-type deposits, 548
  sulfate reduction and, 539-44
Old Lead Belt (Missouri), 13
Open system, concept of, 283
Ophiolites, 250, 263
Optical examination, 288-89
Ore deposit:
  definition of, 1
  epithermal, 774
    alteration in, 205-7
    anhydrite in, 493
    barite in, 498
    isotopes of, 253-56
    temperatures of, 727
  metamorphic, 520-21
Ore deposition, 434-54
  causes of, 434
  from chloride complexes, 435-36
  duration of process of, 163, 441-42, 774
  gold-silver ores, 452-54
  mercury ores, 450-51
  of Mississippi Valley-type deposits, 163, 446-50, 727

  of porphyry copper ores, 130-31, 442-46
  replacement and, 438-41
  from sulfide complexes, 437-38
  thermal aspects of, 611-29
    alteration and sulfide deposition, 624-29
    heat transfer processes, 612-19
    sources of thermal energy, 619-23
  zoning and, 441
Organic complexes, 157, 410-11
Organic compounds, carbon isotope ratios of, 558
Orogene (Orogenic belt), lead isotope ratios and, 42
Osmosis in sediments, 145
  membrane filtration, 150-54, 245, 544
Osmotic pumping, 9
Osoreyama Springs (Japan), 742
Otake-Hatchobaru (Japan), geothermal area, 672-73
Oxidation, 183, 526, 527
  alteration and, 199-202
  of sulfides, kinetic isotopic effects during, 535-38
Oxidation-reduction reactions (redox), 183-84, 241, 665
Oxidation state, in sedimentary waters, 149-50
Oxygen isotope ratios:
  determination of, 237
  feldspar geothermometer, 251
  fractionation curves, 238
  in geothermometers, 667
  of massive sulfide-type deposits, 263
  of metamorphic ore deposits, 263-65
  of Mississippi Valley deposits, 262
  notation and standards, 237-38
  of porphyry copper deposits, 256-61
  of sedimentary waters, 159-60
  see also isotope ratios

Paleohydrology, 9
Panguna (Bougainville), 121
Park City (Utah), 534
  lead isotopes of, 55
Pasto Bueno (Peru), 266, 550
  isotopic compositions of, 269-71
Pauzhetsk (USSR), geothermal area, 638, 671
Pegmatites, fluid inclusions in, 707-9, 727
Pelagic sediments, lead isotopic compositions of, 43, 45, 60, 63-65
Pentlandite, 359
Permeability:

alteration, sulfide deposition and, 624-25
of geothermal systems, 635-36
heat transfer processes and, 613, 615-19
pH, *see* Acidity
Phanerozoic deposits, 22
lead isotopic ratios for, 23
Phase diagrams, 344-53
activity–activity, 350, 352
activity–composition, 346
activity of O_2–pH, 353
activity of S_2–temperature, 349-50
complex sulfide systems, 383
condensed temperature–composition, 344, 345
Eh–pH, 352-53
free energy–composition, 346-47, 350
from geologic observations, 187
impurities' effect on, 386
invariant points, tables of, 297-334
multicomponent systems, 350
relative importance of variables in controlling sulfide phase assemblages, 353
temperature–composition, 344
total-pressure–activity of S, 349
total-pressure–composition, 346
total-pressure–temperature, 347-49
univariant equilibria, 335-40
Phase relations:
direct determination of, 291-94
indirect determination of, 294-96
invariant points, 385
table of, 297-334
multicomponent systems, 350
of specific sulfide systems, 354-90
of sulfur, 354-58
univariant equilibria, 335-40, 385-86
see also Phase diagrams
Phase rule, 222
application to mineral deposits, 280-86
Gibbs, 177-78
mineralogical, 177-78, 190, 281, 283
Phillips, John A., 5
Pine Creek (California), 220
Pine Point deposits (Northwest Territories), 548
carbon isotopes of, 561
lead isotopes of, 50-52
sulfur isotopes of, 520
Plagioclase, geothermometer, 87-88
melting of, 84-88

Plate tectonics, 23
Porosity of sediments, 140-41
Porphyry copper deposits, 265
alteration at, 210-18, 259-60
potassic zone, mineralogy and conditions in, 213-15
propylitic and other types of alteration, 215-16
relative ages of, 210, 216
sequence of veins, 216
sericitic zone, 215
supergene alteration, 261-62
vertical zoning, 216-18
Andean, 61, 63
anhydrite in, 494
breccia pipe formation in, 114-115
consanguinity of molybdenum deposits and, 71-72
convection associated with, 259
deposition of, 130-31, 442-46
fluid inclusions in, 709, 727
in geothermal waters, 659-61
H_2S/SO_2 in, 527-29
hydrogen and oxygen isotope ratios of, 256-61
isotope ratios of, 256-61
lead isotopes of, 52, 61
magmatic generation of, 101-2, 107-8, 112-13
skarns associated with, 225
sulfur isotopes of, 545
Porphyry molybdenum deposits, 259-61
Pošepný, F., 5
Potassic alteration, 204, 259
at porphyry copper deposits, 210-11, 215
Precipitation from hydrothermal solutions, 15-19
Pressure, 470
in fluid inclusions, 707, 720-24, 728-30
of fluids in sedimentary basins, 144-46
hydrostatic, 261, 613-14
lithostatic, 260
melting in magmas and, 86-88
precipitation caused by changes in, 17-18
Progress variable, 580-82, 585
Propylitic alteration, 204, 208
at porphyry copper deposits, 211, 215-16
Providencia (Mexico), 559, 716
carbon and sulfur isotopic ratios of, 520, 550
isotopic composition of, 265-66

Pyrite, 358-65
  at Broadlands, New Zealand, 767
  composition of, 361, 767
  disequilibrium in, 361-65
Pyrrhotite, 358-61
  at Broadlands, New Zealand, 767
  indicator method, 293

Quartz, fluid inclusions in, 766
  solubility of, 456-71, 663-64
Quartz veins, fluid inclusions in veins of, 707-9
Quenching, 288

Rammelsberg (Germany), lead isotopes of, 49
Raul (Peru), 546, 547
Ray (Arizona), 214
Reactions:
  acid-base, 184-86
  decarbonation, 183
    alteration and, 194-95
  in geothermal waters, 659-68
  heat of, 619-21, 625
  hydrolysis, see Hydrolysis
  ion exchange, 184-86
  mass transfer and, 570-82
    chemical affinity and reaction progress, 575-82
  oxidation, see Oxidation
  oxidation-reduction (redox), 183-84, 241, 665
  relative rates, 288, 485
    mass transfer and, 588-89, 602
  sulfidation, and alteration, 199-202
  see also Isotopic exchange reaction
Red-bed copper deposits, 421
  lead isotopes of, 46, 48
Red Mountain (Arizona), 213, 217
Redox reactions, see Oxidation-reduction reactions
Red Sea:
  brines, 19
  deposits, 43, 45, 263
    deposition of, 546, 547
    lead and zinc in, 156, 158
  geothermal area, 750-51
Reduction, 183
  hydrogen isotope ratios and, 241
  of sulfates, kinetic isotopic effects during, 539-44
Replacement, 438-41, 485
  in skarns, 218

Replacement-type deposits, sulfur isotopes of, 550
Reykjanes (Iceland), thermal area, 493, 638, 658, 668, 669
Rosebury (Australia), lead isotopes of, 60
Rotokaua (New Zealand), 648
Rubidium, in geothermal waters, 661-62

Safford (Arizona), 211, 257
St. Francois Mountains (Missouri), 249
Salinity of sedimentary waters, 147-48
  temperature and, 161-63
Salton Sea (California) geothermal area, 453, 470, 638-39, 648, 658, 752-53
  alteration in, 673-74
  lead and zinc in, 156, 158
  geothermal brines, 11, 14
  lead isotopes of, 46
Sandberger, F., 5
Sandstones, lead isotopes of, 46-49
San Juan Mountains (Colorado), 207, 249, 253
  isotope variations in, 256
  see also Creede; Sunnyside
San Manuel-Kalamazoo (Arizona), 115
  alteration in, 211, 212, 216
Santa Rita (New Mexico), 211, 214-16, 220, 225, 257-58
Sardinia deposits, sulfur isotopes of, 548, 549
Sassak (Indonesia), lead isotopes of, 61, 62
Scanning electron microscope, 291
Scheerer, T., 5
Scrope, George J. P., 5
Seawater, see Ocean waters
Sedimentary basins, 138
  fluid pressure in, 144-46
  fluid release from, 141, 165-66
  geothermal gradient in, 143
  heat transfer in, 144
  temperatures in, 142-44
  see also Sedimentary waters
Sedimentary waters:
  acidity of, 149
  barium in, 139
  density of, 146, 164
  dissolved species in, 148-50
  evaporites and, 150, 158
  flow of, 163-67
    during compaction, 165
    during evolution in thermal regime, 166-67
    during tectonic deformation, 165-66

INDEX 795

in geothermal systems, 634, 636
isotope composition of, 153-54, 245, 262
lead in, 139
migration of, 146
as ore-forming fluids, 154-63, 714
  barium and strontium, 158-59
  chemical compostion, 160-61
  fluorine, 159
  isotopic composition of water, 159-60
  lead and zinc, 156-58
  sites of ore deposition, 163
  sulfur, 155-56
  temperature-salinity relations, 161-63
oxidation state of, 149-50
salinity of, 147-48, 161-63
sources of, 139-40
temperature of, 142-44, 634
  salinity and, 161-63
viscosity of, 146-47, 164
Sediments:
  carbonate isotopes of, 559
  compaction of, 140-41, 151-52
    flow of water during, 165
  dehydration reactions of, 139-40, 145
  diagenesis of, 137-38
  membrane properties of, 150-54, 245, 544
  porosity of, 140-41
  thermal conductivity of, 143, 166
Sericitization (sericitic alteration), 174-75, 202-5, 208, 269
  at porphyry copper deposits, 210-15
Setschenow equation, 478
Seward Peninsula (Alaska), 226
Seychelles Islands, 249
Shales, lead isotopes of ores in, 49
Shasta district (California), lead isotopes of, 59, 63
Sierra Nevada batholith, 65
Silica, 498
  in geothermal waters, 663-64
  solubility of, 465-71, 663-64
Silicate components in magmas, thermodynamic properties of, 79-81
Silicate melts, reaction of water with, 73-75
  See also Aluminosilicate melts
Silicification, 202-5, 260
Silver complexes, 772-73
  stabilities of, 430, 431
Silver Mine (Missouri), 226
Skaggs Springs (California), geothermal area, 757

Skarns, alteration in deposits with, *see* Alteration, in skarn deposits
  calcic, 219-20
  definition of, 218
  magnesian, 219-20
  zoning of, 175
Smythite, 359, 360
Sodium, in geothermal waters, 659-61
Sodium chloride, boiling of hydrothermal solutions and, 462
  solubility of, 462
Solid solutions, thermodynamics of, 296
Solubilities:
  of gases, 707, 710-11
  in geothermal systems, 639-43
  of host minerals of inclusions, 701-4
  of non-ore minerals, 461-502
    alkaline earth carbonate minerals, *see* Alkaline earth carbonate minerals, solubility of
    barite, 495-500
    calcite, 478-85
    celestite, 494-95
    fluorite, 471-74
    gypsum and anhydrite, 490-94
    quartz and other SiO_2 polymorphs, 465-71
    sulfates, 490-500
    system $NaCl-H_2O$, 461-65
  of ore minerals, 404-54, 415
    copper sulfides, 442
    predicting, 434
    quality of data, 416-17
    stability of complexes, 417-33
Solubility product:
  of barite, 498
  of calcite, 485, 486, 488, 489
  of dolomite, 485, 486
  of ore minerals, 404
  of strontianite, 488
Sorby, Henry, 5
Soret effect, 154
Source rocks:
  of convecting solutions, 624
  of geochemically scarce metals, 14-15
  of lead isotopes, 25, 32
  uranium and thorium contents of, 33
  see also Sedimentary basins
Southeast Missouri district, 13
  lead isotopes of, 50-52
  sulfur isotopes of, 548, 549
Southern California batholith, 249

Sphalerite, 359, 365-72
  compositions at Broadlands, New Zealand, 764-66
  FeS content of, 369-71
  fluid inclusions in, 766
Spor Mountain (Utah), 226
Stabilities, of complexes, see Complexes, stability of
  mineral, see Mineral stabilities
Standard Mean Ocean Water (SMOW), 237
Steamboat Springs (Nevada) geothermal area, 206-7, 249, 253, 635, 638, 757, 758, 762, 773
  alteration in, 672
Strontianite, solubility of, 487-89
Strontium in sedimentary waters, 158-59
Subduction zones:
  emplacement of hydrous magmas from, 104
  generation of magmas in, 92-96
  metamorphism in, 92-93
  water in, 92-93
  see also Island arcs
Subsurface brines, see Sedimentary waters
Sulfates:
  isotopic analyses using, 512-13
  kinetic isotopic effects during reduction of, 539-44
  in sedimentary waters, 155
  solubilities of, 490-500
  see also specific sulfates
Sulfidation, alteration and, 199-202
Sulfide complexes, 16, 408-9
  deposition from, 437-38
Sulfide mineral stabilities, see Mineral stabilities
Sulfides (sulfide minerals):
  complex sulfide systems, 383
  isotopic analyses using, 512-13
  kinetic isotopic effects, during leaching, 538-39
    during oxidation, 535-38
  massive deposits of, see Massive sulfide-type deposits
  metal, magma generation and, 101-3
  methods of studying and examining, 288-91
  phase relations of specific sulfide systems, 354-90
  in sedimentary waters, 155-56
  thermal aspects of deposition of, 624-29
  trace-element distribution in, 384
Sulfosalts, 383-84

Sulfur:
  activity of:
    activity of S_2-temperature diagram, 349-50
    measurement of, 292-94
    total-pressure–activity of S_2 diagram, 349
  fugacity of, 292-94
    measurement of, 369
    as reference state, 354
  isotopic exchange reactions in, 521-22, 525-26
  in magmatic aqueous phase, 123-25
  phase relations of, 354-58
  in sedimentary waters, 155-56
Sulfur Bank (California), geothermal area, 757, 758
Sulfur dioxide, in magmas, 77
  in magmatic aqueous phase, 124, 125, 130-32
Sulfur isotope ratios, 509-50
  analytic techniques, 511-13
  in aqueous sulfur species, 530-33
  in fluid-mineral equilibria, 527-29
  fractionation factors, 514-21
  fractionation processes of, 521-44
  of geothermal systems, 766-67
  geothermometers, 513, 517-21, 667
  leaching of sulfides: kinetic isotopic effects during, 538-39
  of magmatic systems, 523-27
  of massive sulfide deposits, 545-48
  of Mississippi Valley-type deposits, 548-50
  natural range of, 510
  notation for, 509, 513
  oxidation of sulfides, kinetic isotopic effects during, 535-38
  of porphyry copper deposits, 545
  reduction of sulfates, kinetic isotopic effects during, 539-44
  sulfide-sulfate kinetics, 533-35
  of vein- and replacement-type deposits, 550
Sulfur species in solution, 408-10, 524, 530-33
  in geothermal waters, 666
Sunnyside (Colorado), 256, 415
  carbon and oxygen isotope ratios of, 520, 550
Supergene alteration, 211-12, 214-16
  of porphyry copper deposits, 261-62
Supersaturation, mass transfer and, 583-87
System:
  Al_2O_3-K_2O-Na_2O-SiO_2-HCl-H_2O, 188

INDEX 797

CaCO$_3$- CO$_2$-NaCl-H$_2$O, 482
Ca-Fe-Si-C-O-S-H-F-W-Cu-Zn, 220
CO$_2$-H$_2$O, 474-77
CO$_2$-H$_2$O salts, 477-78
copper-iron-nickel-sulfur, 379-80
copper-iron-sulfur, 374-79
copper-iron-zinc-antimony-sulfur, 383
copper-sulfur, 372-74
H$_2$O-CO$_2$, 729-30
iron-zinc-sulfur, 365-72
iron-arsenic-sulfur, 381-82
iron-sulfur, 358-61
NaCl-H$_2$O, 461-65, 720, 728
see also Metastability; Mineral stabilities; and Phase diagrams

Tactite, see Skarns
Taishu (Japan), lead isotopes of, 61
Talnakhite, 378
Tamagawa Springs (Japan), 750
Tectonic provinces, lead isotopes and, 35-36, 42
Telluric pressure, magma emplacement and, 112-13
Telluride complexes, 452
Temperature-composition diagram, 344
Temperatures:
  critical, of hydrothermal solutions, 463-64, 478
  of Mississippi Valley-type deposits, 161-63
  precipitation caused by changes in, 17
  of sedimentary basins and waters, 142-44, 634
  see also Geothermometers
Temperino (Italy), 181, 221
Tennantite, 384
Tennessee zinc deposits, 9
Tetrahedrite, 384
Thermal coductivity of sediments, 143, 166
Thermal energy sources, 619-23
Thermochemical data, 186
  sources of, 295, 354
Thermodynamic calculations, 651, 654. See also Mineral equilibria; Thermochemical data
Tholeiitic basalt, lead isotopic field of, 32
Thorium, isotope ratios of, 38, 40
  in source rocks, 33
Tin complexes, 433
Tintic (Utah), alteration in, 176
  lead isotopes of, 56

Tonga-Kermadec island arc, lead isotopes of, 60, 64
Tonopah (Nevada), 253-55
Total-pressure--activity of S diagram, 349
Total-pressure--composition diagram, 346
Total-pressure--temperature diagram, 347-49
Trace elements in sulfides, 384
Transfer, see Mass transfer
Transport, by hydrothermal solutions, 15-16. See also Mass transfer
Tri-State district, 262
  lead isotopes of, 50, 51
Tungsten ores, 271
Twin Buttes (Arizona), 220

Um Gheig (Egypt), lead isotopes of, 52
Univariant equilibria, 335-40, 385-86
  impurities' effect on, 386
Upper Mississippi Valley, see Wisconsin district
Uranium, isotope ratios of, 38, 40
  in source rocks, 33
Uranium complexes, 433
Utah copper ores, lead in, 13
Uzon Caldera (USSR), 756

Valley Copper (Arizona), 216, 217
Vanadium complexes, 157
Van Hise, Charles Richard, 5
Vapor-dominated hydrothermal systems, 206, 636-38
Viscosity, of fluids, 614, 625, 626
  of sedimentary waters, 146-47, 164
Volcanic terranes, epithermal ore deposits in, 253-56
Volume changes during mineralization, 221-22

Wairakei (New Zealand), 470
  alteration in, 206, 669-71
  geothermal area, 634-36, 638, 649, 650, 657-58, 669-71
  isotope ratios of, 245, 249, 255
Waimangu, 759
Waiotapu (New Zealand), 758, 759
Wallrocks, precipitation caused by chemical reactions of, 18
Waters:
  isotopic composition of, in sedimentary basins, 159-60
  in magmas, see Magmas- water in
  magmatic, see Magmatic hydrothermal

solutions
  metamorphic, *see* Metamorphic waters
  meteoric, *see* Meteoric waters
  ocean, *see* Ocean waters
  sedimentary, *see* Sedimentary waters
  in subduction zones, 92-93
  see also Geothermal systems; Magmatic aqueous phase
Werner, Abraham Gottlob, 4
White Pine copper deposits (Michigan), 9, 440
Wilbur Springs (California), geothermal area, 451, 757, 758
Wisconsin district (Upper Mississippi Valley), 415, 520
  isotopic composition of, 262
  lead isotopes of, 50
  sulfur isotopes of, 548, 550
Witherite, solubility of, 489-90
Wood River district (Idaho), lead isotopes of, 49, 52, 56
Wurtzite, 359, 365-68

X-ray diffraction techniques, 289

Yatani (Japan), 453
Yellowknife (Northwest Territories), 454
Yellowstone National Park (Wyoming), 249, 771
  alteration in, 206, 671-72
  geothermal area, 638, 649
Yerington (Nevada), 211

Zeolitic alteration, 195, 206, 671, 769
Zinc, in sedimentary waters, 156-58
Zinc complexes, 423-25
Zinc deposits in Tennessee, 9
Zoning:
  of alteration, 173, 178-82
    skarns, 175, 220-21
    vertical, 210, 216-18
  of crystals, 286-87
  of ore minerals, 441